ROUTLEDGE HANDBOOK OF WATER AND HEALTH

This comprehensive handbook provides an authoritative source of information on global water and health, suitable for interdisciplinary teaching for advanced undergraduate and postgraduate students. It covers both developing and developed country concerns.

It is organized into sections covering: hazards (including disease, chemicals and other contaminants); exposure; interventions; intervention implementation; distal influences; policies and their implementation; investigative tools; and historic cases. It offers 71 analytical and engaging chapters, each representing a session of teaching or graduate seminar.

Written by a team of expert authors from around the world, many of whom are actively teaching the subject, the book provides a thorough and balanced overview of current knowledge, issues and relevant debates, integrating information from the environmental, health and social sciences.

Jamie Bartram is Director of The Water Institute and Don and Jennifer Holzworth Distinguished Professor in the Department of Environmental Sciences and Engineering at the Gillings School of Global Public Health, University of North Carolina at Chapel Hill, USA.

Associate Editors: Rachel Baum (University of North Carolina at Chapel Hill, USA), Peter A. Coclanis (University of North Carolina at Chapel Hill, USA), David M. Gute (Tufts University, USA), David Kay (University of Wales, Aberystwyth, UK), Stéphanie McFadyen (Health Canada, Canada), Katherine Pond (University of Surrey, UK), William Robertson (Water Microbiology Consultant, Canada), and Michael J. Rouse (Independent International Consultant on Water Industry, UK).

ROUTLEDGE HANDBOOK OF WATER AND HEALTH

Edited by Jamie Bartram
with
Rachel Baum, Peter A. Coclanis, David M. Gute,
David Kay, Stéphanie McFadyen, Katherine
Pond, William Robertson and Michael J. Rouse

LONDON AND NEW YORK

First published 2015
by Routledge
2 Park Square, Milton Park, Abingdon, Oxon OX14 4RN

and by Routledge
711 Third Avenue, New York, NY 10017

First issued in paperback 2017

Routledge is an imprint of the Taylor & Francis Group, an informa business

British Library Cataloguing-in-Publication Data
A catalogue record for this book is available from the British Library

Library of Congress Cataloging in Publication Data
Routledge handbook of water and health / Jamie Bartram with
Rachel Baum, Peter Coclanis, David Gute, David Kay, Stéphanie
McFadyen, Kathy Pond, William Robertson, and Michael Rouse.
pages cm
Includes bibliographical references and index.
1. Waterborne infection. 2. Water – Purification.
3. Water supply – Pollution – Heath aspects. I. Bartram, Jamie.
RA642.W3R68 2015
613.2´87–dc23
2015010404

ISBN 13: 978-1-138-49530-2 (pbk)
ISBN 13: 978-1-138-91007-2 (hbk)

Typeset in Bembo
by HWA Text and Data Management, London

Recommended citation: Bartram, J., with Baum, R., Coclanis,
P.A., Gute, D.M., Kay, D., McFadyen, S., Pond, K., Robertson, W.
and Rouse, M.J. (eds) 2015. *Routledge Handbook of Water and Health*.
Routledge, London and New York.

CONTENTS

Contents

Contents

Contents

FIGURES

TABLES

ACKNOWLEDGEMENTS

This book distils much knowledge and experience – both the 'supply side' of what is known and the 'demand side' of what information is sought. While the editors and authors have provided the words, their colleagues and students, past and present, have unknowingly contributed much depth and insight. In addition to the editors, whose names appear on the book cover, and the authors, whose names appear on each chapter, recognition is due to the reviewers who provided feedback, guidance and input, often to more than one chapter and more than one iteration of chapters: Bill Anderson, Nick Ashbolt, Cristina Villanueva Belmonte, Robert Bos, Jim Durch, Sue Cavill, Jonathan Chenoweth, Joseph Cook, Andy Cotton, Ollie Cumming, Monica Emelko, Andreas Farnleitner, John Fawell, Roger Few, Jeffrey Foran, David Fuente, Rick Gelting, Michele Giddings, Louisa Gosling, Alexander van Geen, Yakir Hasit, Pascale Hoffman, Damian Hoy, Tapio Katko, Pawan Labhasetwar, France Lemieux, Ray Lloyd, John May, Graham McBride, Rachel McDonnell, Anthony McMichael, John Scott Meschke, Hannah Neumeyer, Susan Petterson, Richard Pollack, Kelly Reynolds, Angella Rinehold, Ulrike Rivett, Jeff Shaman, Otto D. Simmons III, Harry Swain, Terrence Thompson, Rich Thorsten and Tim Wade.

Rachel Baum deserves special note for keeping the project tracked, on time and up to standard while at the same time reviewing every chapter from the perspective of the user and serving as an author. I would also like to thank several anonymous reviewers of the book proposal for the publishers, as well as Tim Hardwick, Ashley Wright and colleagues at Earthscan/Routledge for their encouragement and assistance. As is the way of these things, there will be important acknowledgements remembered as we go to press – I apologise to those undeservingly unrecognized in consequence. My most profound thanks go to Jane for support and encouragement.

1

INTRODUCTION*

Jamie Bartram and Rachel Baum

THE WATER INSTITUTE, DEPARTMENT OF ENVIRONMENTAL SCIENCES AND
ENGINEERING, GILLINGS SCHOOL OF GLOBAL PUBLIC HEALTH, UNIVERSITY
OF NORTH CAROLINA AT CHAPEL HILL, NORTH CAROLINA, USA

There is worldwide recognition of the importance of water for health. Treatise after treatise states that human survival is impossible without it; that the human body comprises 60 per cent water; that civilization and civilizations depend on water; and that our contemporary health and quality of life are made possible and sustained through our management of this natural resource.

Archaeological and historical evidence convey the importance that water has had for the development of settlements and civilizations and for the health of their populations across the course of history. Excavations of ancient settlements suggest substantive investment in water management, dating back to at least 1000 BCE with the construction of aqueducts in Mesoamerica (Lucero and Fash, 2006). A common preference for wells and springs suggests that demand for 'clean' (clear) water is widespread.

Water figures prominently in religion and ritual, both historic and contemporary; and serves as a metaphor for cleanliness. It is used in the Christian baptism ritual, in Islamic ablutions for cleansing before prayer, in Judaic purification baths for certain rituals and in Shintoistic cleansing before prayer. In some Amazonian tribes a source of water used for drinking is abandoned if an animal is seen to drink from it – even if that requires moving the household; and in some parts of northern Pakistan, where fresh water comes from glacier melt, coldness of water is considered an indicator of drinkability.

Ancient written works indicate a belief in some relationship between water (cause) and health (effect). Hippocrates described at length the sources, qualities and health effects of water in *Airs, Waters, Places* (circa 400 BCE); and in the first century AD Pliny the Elder wrote extensively on the kinds of water and is attributed with the admirably succinct *in aqua sanitas* (i.e. in water there is health). This association of water with health extended to interventions intended to protect and improve health through water, with evidence for: bulk transport of cleaner water to human settlements such as the aqueducts of ancient Rome; bulk storage of water in underground tanks in the Mediterranean area; bathing facilities whether public or private; treatment of water to improve its quality through means such as settling, filtering

---

* Recommended citation: Bartram, J. and Baum, R. 2015. 'Introduction', in Bartram, J., with Baum, R., Coclanis, P.A., Gute, D.M., Kay, D., McFadyen, S., Pond, K., Robertson, W. and Rouse, M.J. (eds) *Routledge Handbook of Water and Health*. London and New York: Routledge.

and boiling; and drainage of both urban areas and marshlands – all known of through written sources and excavations (IWA, 2013).

There was debate in ancient Rome on the appropriateness of lead plumbing due to health concerns. While the credibility of there having been health effects has been questioned (Hodge, 1981), debate on the desirability of ceramic over lead pipes to resolve this health concern provides one of the first examples of response to a health problem introduced through water management itself.

Other water-related diseases were also important around the ancient Mediterranean area, including malaria and schistosomiasis, the latter which is thought to have been associated with flood irrigation from the Nile in ancient Egypt. It is unclear whether their water associations were perceived at the time, although association of marshes with ill health was, and the name *mal aria* (literally 'bad air') suggests appreciation of an external cause.

Construction of large water storage 'tanks' for water storage for irrigation in Sri Lanka, where rainfall is insufficient for rain-fed agriculture in the dry region (north and east), dates from around 400 BC. Parakrama Bahu the Great (AD 1153–1186) went as far as to say, 'Let not even a drop of rain water go to the sea without benefiting man.' He is credited with building or restoring 163 major and 2,617 minor tanks, including a 30 km² tank at Polonnaruwa, known as the Parakrama Samudra (Sea of Parakrama) which irrigated nearly 100 km² (Murphey, 1957). Over time these tanks were connected to form a network supporting sequential ('cascading') use. While the reasons for the decline of these systems are not known, one suggested factor has been malaria.

Dale's Laws, the first laws in English-speaking North America, written in 1611 for the governing of 'Virginea', include provisions that explicitly link interventions to disease reduction. They also indicate knowledge of causes of disease, and include proportionate measures depending on perceived likelihood or severity of risk – several hundreds of years before the germ theory of disease.

> There shall no man or woman, launderer or launderess, dare to wash any unclean linen, drive bucks or throw out the water or suds of foul clothes in the open street within the Pallizadoes or within forty feet of the same, nor rench and make clean any kettle, pot, or pan or such like vessel within twenty feet of the old well or new pump; Nor shall any one aforesaid, within less than a quarter of one mile from the Pallizadoes, dare to do the necessities of nature, since by these unmanly, slothful and loathsome immodesties, the whole Fort may be choked and poisoned with ill airs.

While an appreciation of the value of water for health can be recognized in these early accounts, contemporary understanding of water and health largely derives from the advances achieved since the mid-nineteenth century. This marked a period of major scientific advances, the industrial revolution, increasing urbanization in Western Europe, and the 'sanitary revolution'.

Much of the health gains achieved during the nineteenth century sanitary revolution are credited to the progressive implementation of interventions in centralized drinking water treatment and supply, and urban drainage and sanitation (management of human excreta through latrine or water-borne sewerage). These health improvements were initially achieved before the era of immunization, antibiotics, and effective treatment of the associated diseases (McKinlay and McKinlay, 1977). Indeed growth in understanding of the mechanisms linking water and health provided key contributions to the development of the science of public health and the discipline of epidemiology (Chapters 66 and 67). The associated benefits accrued across society widely. However the motivation, as exemplified by the Public Health Act of 1848 in England, related largely to prevention of spread of disease from poor to wealthy populations and to the need for a healthy workforce to sustain industrial productivity, rather than any concern for equality and shared progress (Fee and Brown, 2005). Mirroring this situation, in many countries today, access to safe drinking water (and sanitation), as well as the associated health burden, is extremely inequitable (WHO/UNICEF JMP, 2014). Similarly, intervention in water today mirrors the same potential for socially progressive measures that would extend benefits to poor and otherwise disadvantaged as well as privileged populations. The importance of equity is deeply rooted in human rights (Chapter 51) and in public health practice.

Rapidly evolving understanding of water and health

This rich history may create the illusion of a domain that is 'mature'. Indeed in the 1960s–1980s there was an often implicit, and sometimes explicit, assumption that we 'knew' all that was necessary to know about water and health, and that the remaining problems were problems of 'development'. The solution to these problems would be to implement the well-known interventions for the residual, poverty-associated, problems of water to quietly go away under the influence of general development. Subsequent events have highlighted the over-optimism of those assumptions. This parallels the overly optimistic views regarding the suppression of infectious disease that prevailed soon after the eradication of smallpox.

In the early 1960s, the emerging environmental movement brought to light new health threats – from chemical agents of concern to both health and the environment – and made the case that these two were themselves intimately connected. Rachel Carson's *Silent Spring*, published in 1962, and her research on the adverse effects of the chemical pesticide DDT, spearheaded this movement, demonstrating the connectedness of human and environmental systems, questioning the inherent goodness of technological advances, and led to a shift in the burden of proof of 'safety'. However, research and interventions on water and health focused on infectious disease, so chemical agents causing chronic disease transmitted through water remained poorly understood; and a decade after *Silent Spring*, chemical hazards were still peripheral to, and poorly understood by, many of those working on drinking water and health. The then-prevailing view was that,

> Conditions usually considered noninfective, such as cancer and artherosclerotic heart disease, also are believed to vary with water quality, but the relations are subtle and are poorly understood as yet. It is rarely possible by available technical means to control non-infective disease (apart from dental caries) by changing the quality of domestic water …
>
> *White et al., 1972, pp 161–162*

Notwithstanding the importance of the impact that attention to industrial chemicals had on risk management, it subsequently became clear that, globally, disease burdens associated with chemical contamination of water were primarily associated with the naturally occurring geogenic elements fluorine and arsenic where they occur in excess concentrations (Chapter 10). Indeed in Bangladesh, what has been described as the largest mass poisoning in history emerged, where insistence on pursuing the traditional (infectious disease prevention focused) agenda and well-established approaches (extending access to 'improved' water sources such as boreholes with hand pumps) delayed recognition of and response to such a chemical hazard (Chapter 68). While concern for industry-derived chemical contamination of water triggered concerns, other chemical hazards, ironically including some from water treatment and distribution, were progressively recognized. In an echo of the ancient Roman debate over lead in water distribution, disinfection by-products (DBPs) have attracted attention as unintended companions of water disinfection. Their regulation is complex, in part because of the implicit trade-off (chemical safety versus microbial safety), and also because the most studied compounds tend to be those regulated, encouraging the use of less-studied (but not necessarily more safe) alternatives. The difficulty in balancing these chemical risks with their infectious counterparts contributed to the cholera outbreak that began in Peru, following avoidance of chlorination, based on fears of DBPs, and swept through Latin America in the 2000s (Hanekamp, 2006). In the examples of arsenic in Bangladesh and DBP regulation, inadequate understanding of the relative risks arising from microbes versus that from chemicals complicated appropriate and effective decision making.

The discovery of *Legionella* bacteria as a cause of a life-threatening pneumonia (Legionnaires' disease) in 1976 (Chapter 8) further expanded our understanding of the scope of disease exposure routes associated with water. The role of water as a vector of disease, through droplet inhalation, again reconfirmed that engineered water systems can introduce new health hazards as well as solve them – in this case through multiplication of *Legionella* bacteria in water systems.

Notwithstanding the breadth of water-related disease and the rapid evolution in our understanding of it, the global burden of water-related disease is still dominated by faecally transmitted infection: the combination of diarrhea (including the under-nutrition which diarrhea causes and the adverse health effects of that under-nutrition), soil-transmitted helminthes and environmental (tropical) enteropathy (Chambers and Medeazza, 2014). This is true despite the great reductions reported in diarrheal disease in recent decades.

The second half of the twentieth century saw growing concern about human population growth, urbanization and the carrying capacity of planet earth, including worries about the ability to feed the burgeoning population. In large part, this concern was addressed through the 'Green Revolution' and substantive increases in irrigated agriculture. By 2000, 70 per cent of water withdrawals globally were directed to agriculture with far higher rates in some areas (UN Water, 2008). Expansion of irrigation where there was inadequate design of schemes to account for health risks – notably in sub-Saharan Africa – brought in its wake a dramatic increase in some water-related diseases, especially schistosomiasis. More recently the impacts of demands on water for agriculture have been exacerbated by a combination of declining dependability of rainfall for rain-fed agriculture; increasing water scarcity; over-exploitation of (especially groundwater) resources; and escalating demands for agricultural foodstuffs (both for direct consumption and to satisfy the meat demand of an increasingly wealthy population). One response has been greater indirect and direct use of waste water in agriculture with associated health concerns for both infectious and chemical hazards (Chapter 47).

Unsurprisingly water in its various manifestations has appeared prominently in a series of innovations and developments, fads and fashions:

- *International policy and development discourse* brought water (as part of water, sanitation and hygiene (WaSH)) to the fore with the International Drinking Water Supply and Sanitation Decade (1981–1990) and the Millennium Development Goals (MDGs) and it seems likely to remain there in light of proposals for future Sustainable Development Goals (SDGs) (Chapter 43).

- *New technologies* emerged such as membranes for water treatment, desalination to introduce new sources of water, 'smart' metering to enhance management of scarce water and pinpoint irrigation.
- *New approaches* became fashionable such as 'privatization' of urban drinking water services *à la* World Bank, which foundered on public opinion that saw water as more than simply a commodity or business, while sequential changes in approach highlighted weaknesses in community-managed water schemes.
- *Market-based solutions* became the dominant policy response and sit today in tension with *philosophies* that place greater importance on individual and community empowerment (Chapters 49 and 36).
- *New perspectives* provided different outlooks, such as the recognition of water and sanitation as a human right, which changed development discourse from charity and humanitarianism towards right bearers and duty holders, and reconfirmed the role of the state in ensuring 'progressive realization' (Chapter 51).
- *Renewed appreciation of, and focus on, behaviours* emerged, whether at the level of risk/risk-reducing behaviours of individuals (hand washing, household water treatment and safe storage (HWTS), community-led total sanitation (CLTS)).

Just as the emergence of the 'germ theory' of disease and its relationship to the attributes of water played a substantive role in the establishment of the science of public health and the discipline of epidemiology, so today water continues to provide a development space for emerging ideas and concepts. One example is the demand for evidence from randomized control trials (RCTs) to inform evidence-based policy and practice. As RCTs became acknowledged as the 'gold standard' source of evidence for clinical practice, such trials were subsequently demanded for other health interventions. However, they are quite costly and some contemporary water-related evidence calls into question the value for money of some of the evidence they provide. This is especially true for interventions which themselves have complex 'upstream' determinants, in policy, behaviour and programming (Rehfuess et al., 2009). A further complexity is that water interventions do not sit neatly and tidily within any one sector or discipline and they require inter-sectoral activity and interdisciplinary communication to be conducted effectively.

Learning from history

History teaches us that it is easy to overestimate the completeness of our understanding of water and health. Such over estimation leads to slower and less effective responses, so we need to interpret available evidence with an eye toward future knowledge and recognize that the water and health relationship is itself evolving as more information about it becomes available.

Examples of areas of weakness in our contemporary understanding include the impacts of climate and demographic changes on the management of water resources for population health.

It is widely assumed that slowing population growth worldwide will facilitate progress in drinking water supply and health. However, the number of households worldwide is increasing while population growth, in general, is slowing. Today, community-level access to drinking water is approaching saturation and coverage of household-level water access is increasing, such that household number will have a greater influence on drinking-water provision than population size in the future (Bartram et al., 2012). Nevertheless, in areas such as Sub-Saharan Africa and parts of Asia, predicted population growth still poses a problem, especially as it is coupled with rapid urbanization.

With extreme weather events increasing in number and severity in some regions, adaptation to climate change should now be a substantive concern. Nevertheless real programmatic and management change to increase resilience in water management is limited, despite the long time-horizons in water infrastructure planning, which often span many decades.

Today's responses to challenges such as those from demographic and climate change are weak, as organizations continue to underinvest in infrastructures and fail to take the opportunity of renewal to improve future resilience.

Contemporary developments also highlight the limitations inherent in addressing established and emerging water-health challenges in isolation of one another. In the later decades of the twentieth century, a long-established tendency to compartmentalize water issues isolated interdependent problems. For example, information on the relative health benefits of drinking water *quality* 'versus' *access* interventions led to competition between professionals advocating for one or other of these, rather than holistic (synergistic) consideration of the two together. A larger scale compartmentalization can be seen between 'big water' (largely water resources management) and 'WaSH'; integrated water resource management (IWRM; Chapter 42) emerged in response to the need to relate different water 'uses' (domestic, agriculture, industry, ecosystem maintenance) to one another. However, despite substantive investment in IWRM, it has had limited impact on legislation and management practices (UN Water, 2014). Debate around the development of post-2015 SDGs showed the 'WaSH' and 'big water' advocates making little effort towards integration of their perspectives, although they were then joined in the political process. The need to pursue integrated thinking and to understand the roles of water within wider contexts is also evident from the emerging 'nexus' perspective. This builds on the interdependencies among water, energy, food and arguably climate, recognizing that actions on any one of these impacts on the others and seeks to identify win–win opportunities rather than trade-off compromises in order to secure greater overall benefit (Chapter 43).

Both the least and most developed countries share some contemporary challenges. Rural populations receive inferior drinking-water services of lesser quality than their urban counterparts in countries worldwide (Bain et al., 2014), and emerging evidence suggests that there are achievable improvements to urban water safety management (Gunnarsdottir et al., 2012) that offer the potential to benefit health globally. Additionally, countries worldwide struggle to address their underinvestment in an increasing stock of critical health infrastructure, as exemplified by the fact that the American Society of Civil Engineers routinely gives the USA's water and waste water infrastructure a 'D' grade (ASCE, 2015). Describing factors such as these may unintentionally exacerbate a widespread assumption that water-health problems are largely technical in nature and that the application of optimization approaches can determine which technical solutions and policies to implement. However, advancing water and health is rarely a purely technical exercise. Indeed recent work suggests that the pace of progress, in advancing drinking water access for example, is not determined by gross national income, government effectiveness, official development assistance, renewable internal freshwater resources, female population with secondary education, population living below $1.25 a day, the Gini coefficient, under-5 child mortality rates, or the Human Development Index – leaving the tantalizing possibility that deliberate policy may have a determining impact on progress (Luh and Bartram, 2015).

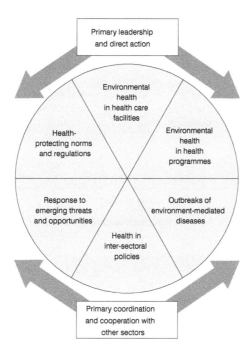

Figure 1.1 Health sector roles in water and health

Source: Rehfuess et al., 2009

Looking ahead

These contemporary challenges and opportunities point to the need for a fresh cycle of water and health initiatives, which in turn require a new generation of professionals with both the depth of knowledge and the interdisciplinary skills necessary to respond to existing and future challenges. They will have to work across compartmentalized 'silo' boundaries of professional disciplines, economic sectors and government organizations. Such a shift will create demands on health professionals to identify and respond to the health sector roles in water and health. It will also make demands on professionals, policy makers and practitioners in other sectors that in fact manage the determinants of water and health interactions (Figure 1.1).

In developing the content of this volume we have placed health at the centre and explored the factors that link it to the many facets of water – whether through the risk of disease from unsafe exposures or the achievement of health through effective interventions. Adopting a wide interdisciplinary and inter-sectoral perspective has enabled us to highlight and explore the diverse relationships grouped under 'water and health' – whether the immediate and direct (such as access to sufficient safe water for domestic use), the indirect (water's role in food production), the beneficial (health protection) or the detrimental (disease propagation), all of which are affected by complex distal factors (such as climate, poverty and demographic changes).

In adopting a water-health focus, some conventional components are less prominent – sanitation, for example, merits an entire volume of its own, but features in many different chapters. Hygiene is dealt with in as much as water is a prerequisite for many hygiene behaviours. Similarly, while industrial pollution is of concern, it receives limited

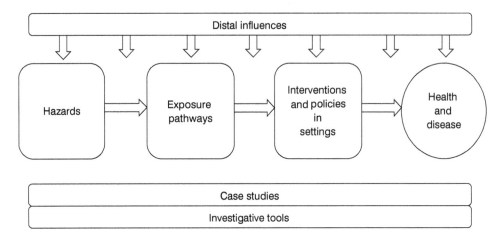

Figure 1.2 Themes addressed in the *Handbook of Water and Health*

attention, and while ecosystem services play a key role in protecting health and stewarding biodiversity, which itself contributes to health and well-being, they also receive limited attention. We judge that a water-health focus is sufficiently tight that we can adequately cover its breadth in the space available to us, sufficiently integrated and cross-cutting to add value, sufficiently complex to stimulate learning and sufficiently forward looking to merit renewed attention.

We organize the contributions that comprise this volume into eight themes (Figure 1.2). Each theme features an introductory chapter followed by a series of chapters that explore important dimensions of the topic and illustrates them with issues of contemporary and future relevance. Each theme is therefore larger and each chapter shorter than a conventional book chapter.

Theme: water-related hazards

In this part we look at the specific agents of harm to human health. Infectious agents are associated with the majority of water-related disease at a global scale and we provide one example (chapter) from each of the four categories of water-related disease proposed by White et al. in 1972. We supplement this categorization with health impacts of water carriage, and with hazards, specifically from *Legionella*, toxic cyanobacteria, chemical hazards and radionuclides in water. The emphasis is on the agents themselves, the harm they cause and how they enter the body.

Theme: exposure pathways

We begin this part by reviewing approaches to classification of water-related disease (causal agent, outcome, transmission route) and then explore four routes of concern for water-related disease: drinking water, recreational water, exposure arising through food contamination and exposure to water-related zoonoses. The emphasis in this part is on how the agents move in the environment and transmission from source to human exposure.

Theme: interventions

In this part, we explore the direct actions that are taken in order to control human exposure to water-related hazards. Individual chapters address examples of engineered interventions (drinking-water supply, drinking-water treatment, waste water treatment); behavioural interventions (household water treatment and safe storage, personal hygiene); and management interventions (Water Safety Plans, drinking-water system maintenance and sustainability, and management of chemical hazards).

Theme: implementing interventions

Much implementation centres on specific settings of concern that represent points of leverage for protecting population health. In this part we turn to settings-based approaches, specifically looking at households, schools, workplaces, health care settings, small communities/rural populations and urban areas.

Theme: distal influences

A number of remote or distal influences modify the risk of exposure to disease-causing agents and of illness as well as influence the scope, scale, impact and sustainability of interventions. Understanding these influences is important for both the adaptation of existing systems and services and for long-term policy and programming. The introductory chapter reviews the current policy focus on the idea of a 'nexus' connecting many of these as a critical challenge for the twenty-first century. We then look at the effects, projected impacts and potential adaptations required, as a result of the impact of seven critical distal influences: water scarcity, climate change, poverty, emergencies and disasters, demography (population increase, ageing and urbanization), water reuse, and war and conflict.

Theme: policies and their implementation

In this part we look at the approaches taken towards development and implementation of policy to reduce the burden of water-related disease. The first chapters look at policy at different levels – catchment based and international. Attention then turns to the tools used to support implementation of policy. These include regulation (drinking-water quality, recreational outdoor water, recreational swimming pool waters and waste water use), economic principles (subsidies and demand-driven approaches), the role of an informed public, and the application of rights-based approaches and specific challenges for women. Finally we look at the roles of health in policy making and decision taking about water management, reviewing the roles the health sector can play and the application of Health Impact Assessment as a policy and planning tool.

Theme: investigative tools

A number of analytical tools are used extensively in advancing the understanding of water health relationships and of human intervention to manage them. In this part we provide introductory-level texts on nine of these, focusing on their application to water and health. Individual chapters address: epidemiology, quantitative microbial risk assessment, burden of disease assessment, water monitoring and testing, microbial quality indicators, pollutant transport modelling, Geographic Information Systems (GIS) and spatial analysis, demand assessment, and cost–benefit analysis.

Theme: learning from history

The use of case studies and the case method have proven valuable over time in advancing our understanding of water and health and in facilitating the adoption and application of good practices. Accordingly, this part provides seven instructive case studies spanning the period from the mid-nineteenth century to the present day. The authors of the cases trace the histories of these events, evaluate the effectiveness of the responses to these events, and offer broader assessments of the roles of the events and the actions taken in enhancing our knowledge regarding water and health and in improving our approaches to risk management. The introductory chapter discusses some of the ways in which knowledge about history – and basic historical methodology – can allow public health professionals to make better policy decisions.

References

ASCE (American Society of Civil Engineers) (2015) '2013 Report Card for America's Infrastructure'. ASCE. http://www.infrastructurereportcard.org/executive-summary/

Bain, R., Wright, J., Christenson, E. and Bartram, J. (2014) 'Rural:Urban Inequalities in Post 2015 Targets and Indicators for Drinking Water', *Science of the Total Environment*, 509–513 DOI: 10.1016/j.scitotenv.2014.05.007

Bartram, J., Elliott, M. and Chuang, P. (2012) 'Getting Wet, Clean and Healthy: why households matter', *The Lancet*, 380(9837), 85–86.

Carson, R. (1962) *Silent Spring*. New York: Houghton Mifflin.

Chambers, R. and Medeazza, G. (2014) 'Undernutrition's Blind Spot. a review of fecally transmitted infections in India', *Journal of Water, Sanitation and Hygiene for Development*, 4(4), 576–589.

Dale's Laws (1611) Accessed through Duhamie.org at http://www.duhaime.org/LawMuseum/LawArticle-1416/1611-Dales-Laws-the-Colony-of-Virginia.aspx

Fee, E. and Brown, T. (2005) 'The Public Health Act of 1848,' *Bulletin of the World Health Organization*, 83(11), 866–867.

Gunnarsdottir, M.J., Gardarsson, S.M., Elliott, M., Sigmundsdottir, G. and Bartram, J. (2012) 'Benefits of Water Safety Plans: microbiology, compliance and public health', *Environmental Science and Technology*, 46, 7782–7789.

Hanekamp, J.C. (2006) 'Precaution and Cholera: a response to Tickner and Gouveia-Vigeant', *Risk Analysis*, 26(4), 1013–1019.

Hipocrates (400 BC) *Airs, Waters, Places*, trans. Francis Adams. http://classics.mit.edu/Hippocrates/airwatpl.mb.txt

IWA (2013) 'A Brief History of Water and Health from Ancient Civilizations to Modern Times'. IWA Water Wiki, http://www.iwawaterwiki.org

Lucero, L.J. and Fash, B.W. (2006) *Precolombian Water Management: Ideology, Ritual, and Power*, University of Arizona Press.

Luh, J. and Bartram, J. (2015) 'Progress Towards Universal Access to Improved Water and Sanitation and its Correlation to Country Characteristics.' *Bulletin of the World Health Organisation*.

McKinlay, J.B. and McKinlay, S.M. (1977) 'The Questionable Contribution of Medical Measures to the Decline of Mortality in the United States in the Twentieth Century', *Health and Society* 55(3), 405–428.

Murphey, R. (1957) 'The Ruin of Ancient Ceylon', *The Journal of Asian Studies*, 16(2), 181–200.

Rehfuess, E.A., Bruce, N. and Bartram, J. (2009) 'More Health for Your Buck: health sector functions to secure environmental health', *Bulletin of the World Health Organization*, 87, 880–882.

UN Water. (2008) *Water in a Changing World*, UN World Water Development Report 3. http://www.unesco.org/new/fileadmin/MULTIMEDIA/HQ/SC/pdf/WWDR3_Facts_and_Figures.pdf

UN Water. (2014) 'Sustainable Water Management is Achieving Economic, Social and Environmental Benefits, say Countries', UN Water Survey. http://www.un.org/en/sustainablefuture/pdf/un_water_report_rio03052012_clean.pdf

White, G.F., D.J. Bradley and A.U. White (1972) *Drawers of Water: Domestic Water Use in East Africa.* Chicago, IL: University of Chicago Press.

WHO/UNICEF JMP (Joint Monitoring Programme) (2014) *Progress on Drinking Water and Sanitation, Report*, Geneva: World Health Organization and UNICEF.

PART I

Water-related hazards

2

INTRODUCTION TO WATER-RELATED HAZARDS[*]

Stéphanie McFadyen

HEAD OF MICROBIOLOGICAL ASSESSMENT SECTION, WATER AND AIR
QUALITY BUREAU, HEALTH CANADA, OTTAWA, ON, CANADA

William Robertson

CONSULTANT IN WATER MICROBIOLOGY, MARTINTOWN, ON, CANADA

Learning objectives

1 Be aware of the profound links between water-related hazards and human health.
2 Appreciate the breadth of water-related hazards.
3 Distinguish between water-related hazards and risks.

Environmental factors are at the root of a significant burden of illness globally, and particularly in developing nations (Prüss-Üstün et al., 2011, 2014); water-related hazards are a key contributor to environmental factors. In general terms hazards can be defined as *potential* sources of adverse health effects or harm to people. Water-related hazards can take many forms. These hazards can arise from the obvious physical damage caused by storms, floods, water shortages or mudslides; the presence of indigenous or introduced hazardous microorganisms, chemicals or radionuclides in water; or from the more subtle yet far-reaching physical and mental harm or injury associated with issues such personal water carriage in some regions of the world. The impact of any one water-related hazard on human health can be influenced by many factors as seen in the simplified illustration shown in Figure 2.1.

It is important here to make the distinction between hazard and risk. As noted above, hazard refers to potential sources of adverse health effects or harm. Risk, on the other hand, is the likelihood or probability that a person may suffer adverse health effects or be harmed if exposed

---

[*] Recommended citation: McFadyen, S. and Robertson, W. 2015. 'Introduction to water-related hazards', in Bartram, J., with Baum, R., Coclanis, P.A., Gute, D.M., Kay, D., McFadyen, S., Pond, K., Robertson, W. and Rouse, M.J. (eds) *Routledge Handbook of Water and Health*. London and New York: Routledge.

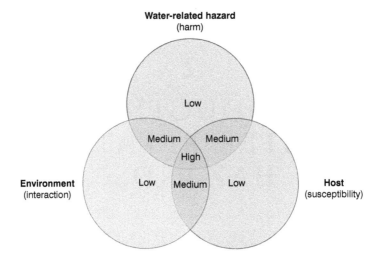

Figure 2.1 Risk triangle – factors influencing exposure, disease and physical harm from water-related hazards

Source: Adapted from WHO, 2012

to a hazard. As a simple example, consider lightning strikes: lightning is a massive electrostatic discharge and is clearly a deadly hazard, yet the chances of a person being killed by lightning depend on the amount of time spent outdoors and the presence of lightning-proof buildings. In a developed country such as Canada for example, the risk is generally less than one in a million but it can be higher in developing countries (Mills et al., 2008). In a drinking water context, the path from hazard to risk can be direct in a low resource setting where there is little to no treatment or management of a water source, or it can involve many steps to understand the impacts of risk mitigation measures. Consider an enteric virus such as norovirus, known to be a hazard in drinking water. Noroviruses cause acute gastrointestinal upset (nausea, vomiting, diarrhea) and their routes of transmission include person-to-person contact as well as contaminated water amongst others. To quantify the water-related risk from norovirus, information on concentrations in source water, the impact of water treatment barriers and a pathogen specific dose–response model is applied in a process known as microbiological risk assessment (WHO, 2011; Health Canada, 2011). More detail on this approach can be found in Chapter 56. While working to quantify the water-related risk from norovirus, it must be clearly understood that interventions to control water-borne exposures to this virus may not affect the overall burden of disease if other significant exposures routes such as food or person-to-person transmission are not controlled. In the case of pathogenic microorganisms the immune status of the exposed population or individual determines their vulnerability to infection and illness. Some recognized groups in this category include children, pregnant women, the elderly and persons with compromised or suppressed immune systems. In the case of chemicals, exposure during susceptible times of development (e.g. the developing foetus or in young and children) may result in long-lasting harm. In all cases, the most significant effects are felt when the hazard, a vulnerable population or individual and environmental considerations overlap (Figure 2.1); not surprisingly, this will also be the zone where risk is at its highest. Exposures and ensuing risks are addressed in other parts of the handbook.

An important proportion of the global disease burden is estimated to be due to inadequate water, sanitation and hygiene conditions (Prüss-Üstün et al., 2014). Access to safe water and sanitation is well recognized as an efficient way to improve human health outcomes. In this part, microbiological and chemical sources of hazards are loosely framed within a system widely known as the Bradley Classification (White et al., 1972) and are referred to as water borne, water washed, water based or water-related insect vectors (Chapters 3, 4, 5 and 6) . The categories of water-related hazards as used in this part are introduced in Chapter 3, Table 3.1. Other hazards such as such as flooding and water scarcity are addressed in the 'Distal influences' part (see Chapters 34 and 35).

The Bradley Classification of water-related infective disease has proven its usefulness over time, both in enhancing understanding of disease transmission routes and in informing appropriate preventative action. However, in a changing world, the classification is not without its limitations. Thus, in addition to describing the classic categories of water-related disease, Chapter 3 explores our current understanding of water-related hazards and proposes new concepts to enhance the simple and robust Bradley Classification system. The goal of this proposal is to ensure a more comprehensive and generally applicable system that remains relevant now and into the future.

Many waterborne and water-washed diseases are caused by faecal–oral pathogens. These microorganisms are among the most widespread contributors to the global burden of disease. In Chapter 4, the transmission routes of faecal–oral infections are explored, together with the characteristics of microorganisms and hosts that contribute to infection and immunity. Important groups of faecal–oral pathogens and their contributions to diarrheal disease are described. This information is key to understanding how the diseases caused by these faecal–oral pathogens can be reduced by improved water, sanitation and hygiene.

Chapter 5 describes water-based diseases and the responsible biological agents. The chapter introduces the term 'water-based diseases'. This not only includes the classical definition of water-based diseases, that is, those involving parasitic helminthes, for example Guinea worms and schistosomes whose life cycle is tied to aquatic invertebrates, but also a variety of other helminthes transmitted through water or raw and undercooked seafood. This definition is extended further to a variety of free-living infectious protozoa (e.g. *Naegleria fowleri*) and bacteria (e.g. *Legionella* spp.) and toxin-producing dinoflagellates, diatoms and cyanobacteria. The chapter also touches upon various methods to control the spread of microorganisms responsible for water-based diseases.

Water-related vector-borne diseases are described in Chapter 6. In this chapter the major vector-borne diseases (malaria, river blindness, Rift Valley fever, arboviral encephalitides, dengue fever and yellow fever) and their insect vectors are summarized. The chapter also reviews the variation in linkages between specific vector-borne diseases and water and offers a brief discussion of water management strategies to control vector-borne diseases.

The physical and mental hardships associated with water carriage are discussed in Chapter 7. Using examples, the role of water carriage in musculoskeletal disorders and physical injury are presented together with the various risk factors which shape this association. These risk factors include the individual (age, gender, physical and mental health), the task (load, equipment and favoured method of carriage) and the physical environment (distance, terrain and climate). Potential solutions and strategies are summarized at the individual, task and environmental levels needed to improve access to water and reduce the health risks related to water carriage.

Chapters 8 and 9 of this section provide more detail on selected examples of water-related hazards to provide a deeper understanding of the complexities and cross-cutting nature of these hazards.

Chapter 8 – Hazards from *Legionella* – expands upon the information provided in Chapter 5. It explains the role bio-aerosols play in water-related disease, primarily legionellosis; identifies specific engineered water systems (e.g. industrial, residential and health care) that have been discovered to be sources of legionellosis and other infectious respiratory diseases; and describes the role demographics play in transmission of the diseases.

Cyanobacteria and their toxins (Chapter 9) are discussed as a complex and challenging hazard that can be considered both water borne and water based. The cyanobacteria discussed in this chapter are microorganisms whose life cycle occurs in the water environment, but the health hazard they present is their production of dangerous toxins which they may passively release into the water. Thus their management is linked to ecology (controlling nutrient inputs and understanding the environmental conditions that contribute to excessive cyanobacterial growth); technology (appropriate water treatment to address the diverse array of toxic compounds produced by various species of cyanobacteria); and human behaviour (limiting human recreational contact with cyanobacterial blooms).

In Chapter 10, several high priority chemical contaminants of global relevance are discussed. Some of these contaminants are due to natural geologic occurrence, while others have industrial sources. Some key contaminants are in fact by-products of the drinking-water treatment process used to reduce the risk from microbiological pathogens, demonstrating the importance of understanding and balancing risks. This chapter also explores the spectre of chemical mixtures and endocrine disrupting chemicals. Of particular concern with chemical contaminants is the potential exposure to vulnerable populations such as young children, where the toxicities and the many unknown interactions of these chemicals may have unforeseen results.

A discussion of water-related hazards is not complete without addressing the implications of radionuclides in water. Radiation in drinking water may originate from ionizing radiation that is emitted by a number of natural or human made radioactive substances. Although the risk from radiological hazards in drinking water are expected to be small in comparison to those from microbiological or chemical hazards, there may be water sources where it is necessary to mitigate radionuclide concentrations to protect human health (WHO, 2011). For this reason the recurring theme of assessing exposure to a hazard from all potential sources is emphasized in Chapter 11.

The breadth of water-related hazards is clear from the wide-ranging topics in this part. These chapters provide insight into how physical considerations, infectious diseases, toxic chemicals and radionuclides fit in the context of water-related hazards; indicate how these hazards may interact to have a profound impact on human health; and inform the potentially broad and far-reaching impact of changes that are made to a water supply to address any one hazard.

References

Health Canada. (2011). *Guidelines for Canadian Drinking Water Quality*: Guideline Technical Document – Enteric Viruses. Health Canada, Ottawa, Canada (Catalogue No H129-6/2011E).

Mills, B., Unrau, D., Parkinson, C., Jones, B., Yessis, J., Spring, K. and Pentelow, L. 2008. Assessment of lightning-related fatality and injury risk in Canada. *Natural Hazards*, 47(2), 157–183.

Prüss-Üstün, A., Vickers, C., Haefliger, P. and Bertollini, R. (2011) Knowns and unknowns on burden of disease due to chemicals: a systematic review. *Environmental Health*, 10(9). http://www.ehjournal. net/content/10/1/9

Prüss-Üstün, A., Bartram, J., Clasen, T., Colford, J., Cumming, O., Curtis, V., Bonjour, S., Dangour, A.D., De France, J., Fewtrell, L., Freeman, M.C., Gordon, B., Hunter, P., Johnston, R.B., Mathers, C., Mausezahl, D., Medlicott, K., Neira, M., Stocks, M., Wolf, J. and Cairncross, S. (2014) Burden of disease from inadequate water, sanitation and hygiene in low- and middle-income settings: a retrospective analysis of data from 145 countries. *Tropical Medicine & International Health*, 19(8), 894–905.

White, G., Bradley, D. and White, A. (1972) *Drawers of Water*. University of Chicago Press, Chicago, IL, USA.

WHO. (2011) *Guidelines for Drinking-water Quality*, Fourth Edition. World Health Organization, Geneva, Switzerland.

WHO. (2012) *Animal Waste, Water Quality and Human Health*. Edited by: Dufour, A., Bartram, J., Bos, R. and Gannon, V. pp 257–282. IWA Publishing, London, UK.

3

BRADLEY CLASSIFICATION OF DISEASE TRANSMISSION ROUTES FOR WATER-RELATED HAZARDS[*]

Jamie Bartram

DIRECTOR, THE WATER INSTITUTE, DEPARTMENT OF ENVIRONMENTAL
SCIENCES AND ENGINEERING, GILLINGS SCHOOL OF GLOBAL PUBLIC HEALTH,
UNIVERSITY OF NORTH CAROLINA AT CHAPEL HILL, NORTH CAROLINA, USA

Paul Hunter

UNIVERSITY OF EAST ANGLIA, NORWICH, UK

Learning objectives

1 Understand the purposes and value of a disease classification system.
2 Recognize and explain the 'Bradley Classification' of water-related disease.
3 Understand an updated interpretation of the 'Bradley Classification', since its first description, and appreciate its comprehensive validity, wide applicability and future relevance worldwide.

Disease classification

A classification is literally 'The action or process of classifying something according to shared qualities or characteristics' (Oxford Dictionaries, 2014), the definition of which goes on to provide as an example 'the classification of disease according to symptoms'. However, disease classification can be based on diverse perspectives and many group

[*] Recommended citation: Bartram, J. and Hunter, P. 2015. 'Bradley Classification of disease transmission routes for water-related hazards', in Bartram, J., with Baum, R., Coclanis, P.A., Gute, D.M., Kay, D., McFadyen, S., Pond, K., Robertson, W. and Rouse, M.J. (eds) *Routledge Handbook of Water and Health*. London and New York: Routledge.

diseases according to their underlying processes (e.g. infective, allergic, metabolic, neoplastic); the primary organs affected (genitourinary, respiratory, gastrointestinal); immediate cause (toxic, microbial, genetic); and, for externally caused diseases, the type of exposure or source of the hazard (airborne, water borne, endemic, zoonotic). Infectious diseases may be classified according to the broad classification of the infecting agents (viral, bacterial, protozoal) or even by the genus of infective agents (mycobacterial, treponemal). Other classification systems may also be used for various reasons especially when intending to stress the importance of a group of diseases such as the 'neglected tropical diseases' or means of transmission such as faecal–oral diseases.

A classification is useful only in as much as it enhances understanding, communication and effective action, in this case towards disease prevention. The process and outcome of classification are useful in clarifying and making subject to criticism our 'view of the world'. In doing so and when well conducted they tend to improve comprehension, enhance the real and perceived 'balance' among perspectives and encourage critical reflection. Classifications serve to increase order, to make complex matters tractable and to encourage reflection on the principal dimensions of a theme. The process of classification often forces explicit recognition of influences that may otherwise be unrecognized or implicit.

'Water-related disease' is a diverse assemblage. The hazards or agents that are the direct cause of damage include bacteria, viruses, protozoa, helminthes, chemicals and personal physical factors. They may originate from human or animal excreta, industrial operations or be parts of natural or disturbed ecosystems. The sites of entry include ingestion but also inhalation/aspiration, wounds and perforation of mucous membranes and intact skin. The sites of damage – whether by pathogens or chemical toxins – include every organ of the body. The symptoms are diverse ranging from the acute (self-limiting diarrhoea) to the chronic (cancers, blindness and life-long infections) and recurring (malaria); and are both direct (caused by a pathogen or chemical) and indirect (such as an existing condition aggravated by the physiological effects of the water-related disease or through an intermediate state such as such as malnutrition caused or aggravated by water-related disease). Finally, epidemiological investigation identifies numerous underlying associated factors such as poverty, education, climate, demography, housing and use of basic services. Some of these may be judged to be 'causal' whereas others may be associated through a series of co-factors, and distinguishing among these is complex.

In 1972 White et al. proposed '...a new classification intended to be both more comprehensive and more precise in predicting the likely effects of changes in water supply on infective diseases' (White et al., 1972, p162), noting that 'All infections related to water supplies are included – that is, all those likely to change in incidence or severity as a result of changing water supplies.' The classification has become widely known as the 'Bradley Classification'. It comprises four broad and non-exclusive classes of water-related disease: water borne; water washed; water based; and diseases with a water-related insect vector (Figure 3.1).

A review of Figure 3.1 and the associated text of the 'Drawers of Water' study show that the classification reflects a careful handling of multiple perspectives, accounting for aspects of transmission, control and disease consequences. Evidently the sense of 'supply' – and hence the scope of the definition – was broad, including water resource-related concerns and was not restricted to a narrow interpretation of drinking water or household water.

A CLASSIFICATION OF INFECTIVE DISEASES
RELATED TO WATER

Category	Example
I. Waterborne	
a) Classical	Typhoid
b) Nonclassical	Infectious hepatitis
II. Water-washed	
a) Superficial	Trachoma, Scabies
b) Intestinal	*Shigella* dysentery
III. Water-based	
a) Water-multiplied	
percutaneous	Bilharziasis
b) Ingested	Guinea worm
IV. Water-related	
insect vectors	
a) Water-biting	Gambian sleeping sickness
b) Water-breeding	Onchocerciasis

Figure 3.1 Facsimile of the table presenting a 'Classification of Infective Diseases Related to Water'

Source: White et al., 1972

Water-borne disease

White et al. (1972) describe this class as 'where water acts as a passive vehicle for the infecting agent' (p162).

Definition/explanation

The water-borne class is, for many, the most intuitive and comprises those diseases caused by the ingestion of pathogens in water. It is fundamentally concerned with water quality and safety. Chapter 4 of this volume provides an overview of the diverse viral, bacterial, protozoal and helminthic causes of water-borne disease. While water-borne disease can be manifest as, sometimes large, outbreaks, the ability of public health surveillance systems to detect such events, even in countries with advanced health information systems, is very low and in fact the majority of water-borne disease is not associated with identified outbreaks.

Recognized links and complexities

The great majority of the agents of water-borne disease can also be transmitted by other means, all leading to ingestion. As such it overlaps with the concept of faecal–oral disease transmission (Figure 3.2). Hygiene, and especially hand washing at critical times, is an important preventive measure and is also reflected in the water-washed class of water-related diseases; and in the facts that some agents and diseases may be both water-borne and water-washed.

Changes over time requiring interpretation or implying modification of the class

Several important considerations have emerged or been clarified since 1972 that impact our interpretation of water-borne disease. These include knowledge of: a greater number of

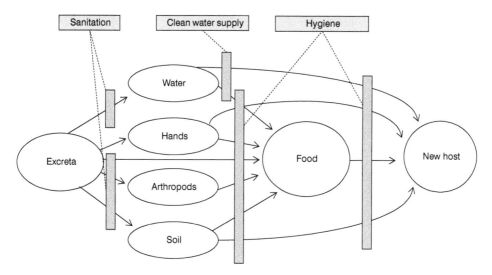

Figure 3.2 Faecal–oral disease transmission

This representation reflects the original terminology of Wagner and Lanoix (1958). It is popularly referred to as the 'F diagram' and depicted with water as Fluids, arthropods as 'Flies', hands as 'Fingers' and sometimes with the addition of Fomites.

causal agents; deterioration of water quality between source and use; the risks of animal and human faecal pollution; and understanding of chemical hazards

Knowledge of, and the perceived importance of, some individual causal agents has developed. Two agents which were historically important causes of mortality, cholera (caused primarily by toxigenic strains of the bacterium *Vibrio cholerae*) and typhoid (caused by the bacterium *Salmonella* Typhi), have declined in global importance but remain locally important where outbreaks occur. They were described by White et al. (1972) as having become established as the 'types' for discussion of water-borne diseases and criticized as not being representative in that role, which remains true today. Other agents have been recognized, including several viruses, such as rotavirus and norovirus (caliciviruses). For some viruses, while water-borne transmission may contribute to disease spread, water quality management is unlikely to provide an effective control measure as other routes of transmission dominate. Other recognized agents include the protozoans *Giardia* spp and *Cryptosporidium* spp, both of which have an environmentally resistant cyst form, the latter of which is small and has challenged some of the water treatment processes that are widespread in industrialized nations. The range of now-recognized water-borne disease agents is summarized in Table 3.1. This list is considerably longer than that of White et al. (1972).

Since the original development of the Bradley Classification, much attention has been paid to deterioration in water quality between 'source' and use. This includes aspects of primary relevance to low and middle income countries (LMICs) such as faecal pollution in water transport and household storage (Wright et al., 2004; Shields et al., 2014). Debate has included concern about the relative importance of intra-household versus community excreta exposure (VanDerslice and Briscoe, 1993; Baum et al., 2013) and the value of household water treatment and safe storage as a health intervention (Clasen et al., 2009; Schmidt and Cairncross, 2009; Hunter, 2009). While this attention does not fundamentally

Table 3.1 Pathogens transmitted through drinking water

Pathogen	Health significance	Persistence in water supplies	Resistance to chlorine	Relative infectivity	Important animal source
Bacteria					
Aeromonas hydrophila	Moderate	May multiply	Low	High	No
Burkholderia pseudomallei	High	May multiply	Low	Low	No
Campylobacter jejuni, C.coli	High	Moderate	Low	Moderate	Yes
Escherichia coli – pathogenic	High	Moderate	Low	Low	Yes
E. coli – Enterohaemorrhagic, e.g. E. coli O157	High	Moderate	Low	High	Yes
Francisella tularensis	High	Long	Moderate	High	Yes
Helicobacter pylori	Pending	Short	Pending	Pending	Yes
Legionella pneumophila	High	May multiply	Low	Moderate	No
Leptospira interrogans	High	Long	Low	High	Yes
Mycobacteria (non-tuberculous, *M. avium* complex, MAC)	Low	May multiply	High	Low	No
Salmonella Typhi	High	Moderate	Low	Low	No
Other salmonellae	High	May multiply	Low	Low	Yes
Shigella dysenteriae, S. flexneri, S. mansoni	High	Short	Low	High	No
Vibrio cholerae	High	Short to long	Low	Low	No
Viruses					
Adenoviridae (adenoviruses)	Moderate	Long	Moderate	High	No
Astroviridae (astroviruses)	Moderate	Long	Moderate	High	No

Caliciviridae (noroviruses, sapoviruses)	High	Long	Moderate	High	Potentially
Hepeviridae (hepatitis E viruses)	High	Long	Moderate	High	Potentially
Picornaviridae (enterovirus, parechovirus, hepatitis A viruses)	High	Long	Moderate	High	No
Reoviridae (rotaviruses)	High	Long	Moderate	High	No
Protozoa					
Acanthamoeba spp.	High	May multiply	Low	High	No
Cryptosporidium hominis, C. parvum	High	Long	High	High	Yes
Cyclospora cayetanensis	High	Long	High	High	No
Entamoeba histolytica	High	Moderate	High	High	No
Giardia intestinalis	High	Moderate	High	High	Yes
Naegleria fowleri	High	May multiply	Low	Moderate	No
Toxoplasma gondii	High	Moderate	Long	High	Yes
Helminths					
Dracunculus medinensis	High	Moderate	Moderate	High	No
Fasciola hepatica, F. Gigantica	Pending	Pending	Pending	Pending	Yes
Schistosoma mansoni, S.japonicum, S.mekongi, S.intercalatum, S. haematobium	High	Short	Moderate	High	Yes

Source: Adapted from WHO, 2011, p119

alter our understanding of water-borne disease, it does modify the frame of reference with which we analyse and interpret associated evidence.

White et al. (1972) note that '… animal pollution with fecal coliforms has different implications from human pollution'. Since that time the increase in numbers of livestock has continued. Large numbers may be kept close to human populations, the management of the excreta is often poor and water provides a connection between animal and human populations. Disease agents of concern because of animal to human disease transmission (e.g. *Crysptosporidium*) have been of increasing concern (reviewed in Cotruvo et al., 2004 and Dufour et al., 2012). These issues increase the scope for concern about water-related zoonoses and the range of agents of concern may increase further as understanding grows in the future. The importance of this for transmission relates to control measures in that close-to-source measures will vary between human faecal sources (sanitation) and animal sources (livestock including feedlot management and animal access to water resources).

Chemical hazards were not addressed in the original classification. This was a conscious decision:

> Conditions usually considered noninfective, such as cancer and artherosclerotic heart disease, also are believed to vary with water quality, but the relations are subtle and are poorly understood as yet. It is rarely possible by available technical means to control non-infective disease (apart from dental caries) by changing the quality of domestic water …
>
> *White et al., 1972, pp 161–162*

Since that time greatly increased attention has been paid to chemical exposure and human health including exposure through drinking water, reflected in the inclusion of an increasing number of individual chemical hazards in consecutive editions of WHO's *Guidelines for Drinking-water Quality* (WHO, 1984, 1993, 2008, 2011). Nevertheless a limited number of agents are credible contributors to substantive disease burden at a global scale, potentially only fluoride and arsenic and possibly lead. But many others may be of local or national concern depending on industrial processes, agricultural practices, geology, water treatment and distribution methods and other factors (Thompson et al., 2007 and Chapter 10). This is reflected in extensive regulations in many LMIC and industrialized countries. A major change has taken place since 1972 in that the claim that 'It is rarely possible by technical means to control [chemical hazards in water]' does not hold true. In the same way that there are multiple routes of transmission of pathogens through the faecal–oral route, so chemical exposure may be through multiple means including foodstuffs and inhalation as well as ingestion with water; and in the same way that multiple transmission routes of pathogens must be understood for efficient and effective intervention, so must diverse contributions to overall chemical exposure be understood. These typically depend on major contaminant-source type categories (Chapter 25).

Exposure and disease burden

Until recently, inhabiting a dwelling for which water was collected from an 'improved source' was used as an indicator of access to, or use of, safe water. On this basis, according to WHO and UNICEF, some 11 per cent or 780 million of the global population did not use an improved source in 2010, with the greatest number of the unserved in Asia and the highest proportion of the unserved in sub-Saharan Africa. More recent literature correcting for the proportion of dwellings in which water tests show no faecal indicator bacteria in a one-off test increase these figures to 28 per cent or 1.9 billion (in 2012, Bain et al., 2014). Further correction to account for the sanitary status of the water source itself increases them further to 47 per cent or 3 billion (Onda et al., 2012). These figures do not fully account for water quality variability. Data on water safety in other settings such as schools, workplaces, markets and health care settings (Cronk et al., 2015) are very limited but would increase these figures further still.

A recent estimate of the burden of diarrhoeal disease from exposure to inadequate drinking water sanitation and hygiene in LMICs concluded that 842 000 million diarrhoea deaths (1.5 per cent of the total disease burden and 58 per cent of diarrhoeal diseases) were caused by this cluster of risk factors. It is difficult to disentangle the various components of faecal–oral disease transmission, especially water-borne and water-washed disease (Prüss-Ustün et al., 2014).

The disease burden, including that associated with some faecal–oral diseases, includes a number of sequelae (i.e. delayed and often chronic effects consequent to the original infection) that should be taken into consideration. An example is Guillain-Barré syndrome. This is an adverse neurological outcome that affects a small proportion of people following a viral or bacterial infection such as campylobacteriosis. The long-term debilitating effects, while experienced by only a proportion of those infected, weigh heavily.

Water-washed disease

White et al. (1972) describe this class as comprising 'infections that decrease as a result of increasing the volume of available water' and also provide a formal operational definition: 'water washed infections are those whose incidence or severity can be reduced by augmenting the availability of water without improving its quality' (p162).

Definition/explanation

The water-washed disease class is fundamentally concerned with access to and use of water for personal, food and domestic hygiene. In contrast to water-borne disease, the role of water is in prevention of disease transmission rather than as a vehicle for carriage of pathogens. Water-washed diseases may be conveniently divided into two types. The first comprises faecal–oral diseases, which may also be water- and food-borne, because water serves both as a vehicle for their transmission and is necessary for adequate personal and food hygiene whereby both sufficient water and safe water are necessary for effective prevention. The second includes diseases such as trachoma (*Chlamydia trachomatis*), skin sepsis (diverse bacterial causes) and yaws (*Treponema pertenue*) where transmission is person to person and hygiene also plays a role in prevention.

Recognized links and complexities

White et al. (1972) wisely side-stepped the issue of behavior by defining the class in terms of water availability, rather than extend the definition to encompass the fact of its use. Monitoring activities over the subsequent period have provided information on water access and use (Bartram et al., 2014) but study and monitoring of actual behaviors has proven problematic: suitable indicators are hard to identify and data collection is fraught with bias (Freeman et al., 2014). Indeed the relationships among access, reliability, use and behavior are complex and poorly understood.

White et al. (1972) note that 'water has many uses, several affecting health and it seems clear that the volume employed for each activity will affect health. We do not know the quantitative relationships ...' That statement remains substantively true today, despite the studies that have taken place in the interim, although broad bands of relative adequacy for different purposes from personal hydration to domestic hygiene have been described (Howard and Bartram, 2003).

Also originating from the 'Drawers of Water' study was a depiction of the relationship of domestic water use according to the time/distance for water collection. The 'plateau curve', refined by Cairncross (1990), suggested that water use was plentiful when the source was very close to the dwelling; plateaued at around 20 litres per capita per day where a water collection trip took around 5 to 30 minutes and declined further thereafter. Recent work has confirmed much greater domestic water use when water is accessed 'on plot', but evidence for the decline when the collection burden is excessive is elusive and this now affects a small and declining population (Evans et al., 2013). Similarly there is some evidence for greater health benefits associated with dwelling-level water access (idem) and it seems likely that future policy, such as the Sustainable Development Goals (Chapter 43), will give greater emphasis to this level of access.

The notion of availability (or access) – the fundamental basis on which the class is identified – has proven complex (Bradley and Bartram, 2013). In addition to varying degrees of access (above), complexity includes the facts that access is normally defined at a household level and applied to all associated persons equally, but the experience of access varies between individuals within households; and access is also experienced and may be restricted in extra-household settings (Cronk et al., 2015).

Changes over time requiring interpretation or implying modification of the class

On trachoma, White et al. (1972) said that while '... most workers think that making water more readily available reduces the amount of clinical trachoma, little precise data is available' and that 'Most ophthalmologists would hold that prevalence [of trachoma] is also reduced by a readily available water supply, but as yet there is limited evidence to support this view.' Certainty about the role of water, sanitation and hygiene (WaSH) has increased: Stocks et al. (2014) confirm its importance in trachoma elimination; and it is reflected in the SAFE strategy for eradication of blinding trachoma (Surgery, Antibiotics, Facial cleanliness and Environment) but it has received limited attention in large-scale eradication efforts.

There are three areas in which new information or an evolving understanding of the issue affects the class of water-washed disease: hydration (reviewed in Howard and Bartram, 2003); the impact of personal hygiene; and water carriage.

There is now strong evidence that good personal hygiene is associated with decreased transmission of non-faecal–oral, as well as faecal–oral water-washed diseases. These include some respiratory diseases with large global disease burdens such as pneumonia (Aiello et al., 2008; Freeman et al., 2014).

The process involved in having to fetch water each day can also have adverse impacts on health. There is substantive evidence for skeletal injury associated with water carriage (Chapter 7), and suggestive evidence for violence on persons collecting water from shared community sources. In addition some work has shown that infectious diseases such as meningococcal disease has spread in refugee camps along the routes that people walk to fetch water (Santaniello-Newton and Hunter, 2004). While different in cause and exposure, these are associated with water access, and remedial measures are primarily related to enhancing access levels to that of the household. Other evidence suggests that this level of water access is associated with health benefits (Kayser et al., 2013; Cumming et al., 2014). As such they fit well in this class despite the word 'washed' in the titular description being inappropriate.

Exposure and disease burden

The water-washed disease class is fundamentally concerned with water access, which is monitored at a household level by WHO and UNICEF, suggesting that 56 per cent of the global population live in households with water 'on plot' (normally a piped supply) and a further 33 per cent use an 'improved source' such as a protected community source. Although these data do not account for relative accessibility, the great majority of these sources are within 30 minutes' collection time. These two categories of access serve as indicators of relative household water use (Evans et al., 2013).

Freeman et al. (2014) estimate that approximately 19 per cent of the world population washes their hands with soap after contact with excreta and that hand washing reduces the risk of diarrhoeal disease by 23 per cent (after adjustment for unblinded studies).

Water-based disease

White et al. (1972) describe this class as 'where a necessary part of the life cycle of the infecting agent takes place in an aquatic animal' (p162).

Definition/explanation

The water-based disease class is concerned with those agents of disease that pass an obligatory part of their life cycle in water. Infection may be through ingestion – as is the case with dracunculiasis – which therefore overlaps with the water-borne class; or may be across the intact or damaged (abraded, wounded) skin (percutaneous) – as is the case with schistosomiasis and leptospirosis. These are normally disease agents with more-or-less complex life cycles. Chapter 5 describes the cause, transmission and prevention of schistosomiasis.

Recognized links and complexities

White et al. (1972) divide the water-based class into two: the first comprising the helminths (schistosomes), which they consider 'water multiplying' because of proliferation in the intermediate snail host; and the second comprising guinea worm (*Dracunculus medinensis*), whose larvae infect aquatic crustaceans but do not multiply in them.

Unlike the agents associated with water-borne and water-washed disease, the helminth causes of water-based disease do not elicit protective immunity and may be associated with cumulative auto-infection. For example, someone excreting schistosomes who contaminates their own water source may (if there are suitable snail hosts present) be further infected through water contact and thereby increase their own wormload. This is important because the severity of disease is related to the intensity of infection.

Changes over time requiring interpretation or implying modification of the class

There are three important new considerations that affect the water-based disease class. These are the potential inclusion of leptospires, toxic cyanobacteria and recreational water contact.

White et al. (1972) state that the agents of water-based disease 'are all helminths, parasitic worms' and place leptospires only in the water-borne class. Leptospires enter water bodies from the urine of some rodent species and may infect humans percutaneously as well as following ingestion. Disease is sometimes associated with flooding which causes distribution of rodent urine (and associated leptospires), human exposure to water and physical injury including skin cuts and abrasions. On the basis that transmission can be associated with water contact; that the alternative/intermediate hosts are water-associated (albeit not necessarily so); and that prevention is through management of water bodies and human contact with them, they are a natural component of this class

Interest in toxic cyanobacteria emerged in the 1990s with the recognition that these organisms – that are prokaryotes (like bacteria) but photosynthetic (like eukaryotic algae) – can produce a range of potent toxins of human health concern and recognition of substantive health impacts in unusual cases where they have overgrown in drinking-water sources (Chorus and Bartram, 1999; see also Chapter 9). Disease outbreaks have been associated with ingestion, and cyanobacterial toxins are therefore included among water-borne chemical hazards (see above). However, while contact-related injury has largely been limited to animals, there is cause for human health concern and contact is discouraged where blooms occur. Since risk management relates to management of water resources, and of human contact with these agents, they are best classified with other water-based diseases.

Another newly identified agent in this class is *Naeglaria fowleri*, a rare and almost always fatal cause of primary amoebic meningitis. Interest in these organisms has been fueled by concerns about the effects of global climate change.

A further exposure that may be argued to fit in the water-based class is associated with recreational or bathing use of natural water bodies contaminated with human or animal excreta. The agents of concern do not conform to the 'complex life cycle' criterion and transmission is by ingestion. However, disease prevention relates to water resource management and control of human contact with water resources. While an argument may be made for a water-based classification, on balance we believe they fit more properly as water borne.

Here we conclude that while a complex life cycle is a frequent characteristic of water-based disease agents it is not a necessary characteristic in their definition. In arguing this we note that this relates to the aspect of disease *transmission* which is common among these agents.

The relevance of the water-based class may increase as water storage at large and small scales increases in response to population growth, increased wealth and climate change (Bradley and Bos, nd).

Exposure and disease burden

For the agents reflected in the original classification, exposure leading to disease is associated with collection of water from unprotected open sources such as rivers, lakes, ponds and streams – either through ingestion or physical contact (wading into infested waters and percutaneous infection). Around 2 per cent of the global population collects drinking water from such sources (WHO and UNICEF, 2014). To this number must be added those for whom water contact is associated with productive activity, especially irrigated agriculture and its association with schistosomiasis.

The global burden of dracunculiasis has declined substantively in response to global eradication efforts and it may become the first parasitic disease to be eradicated (WHO, 2014).

Diseases with a water-related insect vector

White et al. (1972) describe this class as 'those infections spread by insects that breed in water or bite near it' (p162) and note that 'Many tropical infections have insect vectors, and the larvae of many insects are aquatic. There thus grows up a relationship between water and many vectorborne diseases.'

Definition/explanation

The title of this class is largely self-explanatory. The disease agent itself may have no relationship itself with water; rather its insect vector either breeds in or bites near water bodies. The relationship to water is determined by the insect vector, with some preferring stagnant polluted waters and others preferring clean, fast moving waters. Globally malaria is the most important of this class, which also includes Bancroftian filariasis, onchocerciasis, dengue, yellow fever and other arboviral infections all of which are transmitted by insects with aquatic larvae. Tsetse flies, which transmit Gambian sleeping sickness, are found near rivers but do not breed in water. Chapter 6 of this volume describes the cause, transmission and prevention of some vector-borne diseases.

Recognized links and complexities

In many cases there are multiple vector insects that can transmit each of these diseases and disease epidemiology is strongly related to the locally prevalent vectors. In addition and contrary to the general assumption that it is degraded or polluted environments that favor vector proliferation, the *Simulium* flies (blackflies) that transmit onchocerciasis breed in clear mountain streams.

Some of the agents in this class have increased their range and/or disease burden (e.g. dengue).

Both general and water-related disease prevention may relate to reducing insect vector load or to reducing human exposure to insects. In some settings large-scale drainage has been very effective in disease prevention and this is true both in parts of Europe and in the south of the United States of America against malaria. Exhaustive elimination of small water containers (such as abandoned car tyres) has value in settings where they provide a suitable breeding site for the prevalent vector and where exhaustive intervention is feasible. Where biting occurs near water bodies (e.g. *Simulium*, tsetse flies), access to improved or household water may reduce exposure. Where vectors breed in household water storage containers, application of insecticides, both chemical and biological (*Bacillus thuringiensis*), to such containers is sometimes practiced (WHO, 2011). Since the development of the Bradley Classification, use of large-scale insecticide (including DDT) applications has grown and waned in popularity. General preventive measures also include case treatment, prophylaxis (including immunization) and the use of insecticide-treated bed nets.

Changes over time requiring interpretation or implying modification of the class

There is widespread interest in this class in anticipation of the impacts of climate change on vector distributions (Caminade et al., 2014).

Future value of a water-related disease classification

As noted above, a classification is useful only in as much as it enhances understanding, communication and effective action, in this case towards prevention of water-related disease.

The value of the Bradley Classification for educational purposes is self-evident from the frequency with which it is cited in teaching materials, and from its use in communication and fostering action, for example in framing and implementing health impact assessments (Chapter 53).

Experience using the classification in teaching and in policy work suggest that its perceived weaknesses include the non-exclusive nature of its classes (the fact that many diseases can be both water borne and water washed for instance is sometimes a cause of consternation); the fact that some diseases that are not referred to in the original text have emerged, re-emerged or increased in importance or visibility; and the fact that some disease causes and health outcomes that were recognized at the time (such as chemical toxins and water carriage effects) were not explicitly incorporated despite its declared comprehensive intentions.

It is noteworthy that an attempt to describe mutually exclusive classes (Feachem et al., 1977) was not well-received. Similarly an attempt to classify sanitation/excreta-related disease (Feachem et al., 1983) stumbled on technical complexity and ultimately failed to secure the widespread adoption and use of its water-related counterpart.

The success of the classification, founded on its simplicity and its successful marriage of aspects of transmission and of prevention, leads to the question of its comprehensive applicability and future value.

The Bradley Classification emerged from work in rural settings in developing regions on one continent and this has contributed to a perception that it has limited applicability in an increasingly industrialized and urbanized world. Thus despite having been demonstrated to have enduring value and utility, it is sometimes perceived as of limited global relevance rather than as providing a unifying framework. Indeed its wider applicability was suggested but not assumed, 'Although it was made necessary by the particular characteristics of the small, heavily polluted sources of the African tropics, it may have wider relevance, at any rate within tropical areas' (White et al., 1972, p163).

While it is evident that the original classification was intended to be comprehensive – 'All infections related to water supplies are included' (White et al., 1972, pp162–163) – the fact that we are readily able to incorporate both infectious diseases that have emerged or re-emerged since development of the classification, and importantly also to incorporate other water-related health outcomes of diverse non-infectious nature relating to chemical hazards, behaviors and physical injury, suggests potential comprehensiveness.

In light of actual changes in our world – in its ecosystems and in the status of access to water supplies – and changes in our understanding of water-related disease, alongside demonstration of the universal value of such a classification, suggests that the further development and application of the classification has value.

Need for new classes

It has been previously suggested (Bradley, 2009) that an additional class to accommodate aerosol-transmitted disease would be desirable. At the time of description of the original classification, inhalation-related causes were not considered. The principal hazard of concern is *Legionella* bacteria, the causal agents of legionellosis, that multiple in biofilms, especially in engineered water systems (plumbing, evaporative cooling) where nutrient and temperature conditions support their growth. They survive and multiply within a number of free-living protozoa in a way analogous to the 'complex life cycles' of many water-based diseases. *Legionella* infection occurs after inhalation of droplets or aerosols containing the bacteria that are small enough to enter the lungs, such that water serves as a medium for disease transmission analogous to the water-borne class. *Legionella* therefore already overlaps the water-borne and water-based classes from the point of view of transmission. However, management of water for disease prevention has two dimensions – minimization of proliferation (reinforcing the water-borne characterization, although concern is especially related to complex plumbed systems) and minimization of the creation and dispersal of droplets/aerosols (which is not reflected in any current class). Thus from a *transmission* perspective *Legionella* can be accommodated in the current classification, but from the perspective of *prevention* this is insufficient.

Other bacteria of health significance that multiply in water include some that were already recognized at the time of development of the Bradley Classification (such as *Pseudomonas aeruginosa*) and some that have been subsequently identified (such as *Mycobacterium avium* complex). These may be argued to meet in part the definition of water-based disease in that they proliferate in water, although in several of these cases it is unclear whether the complex life cycle generalization is observed. All of these may also be seen as water borne, as infection occurs through ingestion (except *Pseudomonas* which causes a number of infections of other sites). However, these classifications also seem insufficient in that they reflect transmission but not prevention perspectives.

As noted above, the Bradley Classification reflects a careful handling of multiple perspectives, accounting for aspects of transmission, control and disease consequences. The agents described above may be accommodated in the classification either in its original form or through the addition of a new 'inhalation' class but we consider that such a modification would allow the transmission perspective to dominate to the detriment of prevention. Here we therefore propose the addition of an alternative class to reflect the risks arising from *engineered water systems*. Such a class would capture risks that arise within those systems and for which action on the system itself is the, or one of the, principal preventive measures. Such a class would accommodate agents that multiply in plumbing systems (the bacteria noted here). It would also better capture the chemical hazards that arise from plumbing materials (such as lead and vinyl chloride monomer) which are also water borne but for which conventional water quality interventions, such as central water treatment, are inappropriate.

A further argument in favor of such a new class derives from the health hazards arising from radon. Radon is a radio-active gas emitted especially from granitic geologies and which tends to accumulate in the basements of dwellings. It is also found in water from associated aquifers. Piped drinking water supply into dwellings may contribute to accumulation in dwellings as the gas is released in water use in showers, dishwashers and washing machines. The global disease burden remains unclear, as does the contribution of water to household radon accumulation. It may be argued to be water borne since the water medium leads, albeit

Table 3.2 Proposal for a modified classification

	Original Bradley Classification			Proposed clarifications and modifications	
Class	Sub-classes		Examples	Revised sub-classes	Additional examples*
Water borne	Classical		Typhoid	Infectious	Cryptosporidium, Giardia and several viruses are important additions (see Table 3.1 for an extensive list); faecal–oral diseases associated with bathing/recreational water use
	Nonclassical		Infectious hepatitis		
				Toxic chemicals	Arsenic, fluoride (toxic effects at high exposures), lead
				Nutrient minerals	Fluoride (beneficial effects at moderate exposures)
Water access-related disease (formerly water washed)	Superficial		Trachoma, scabies	Superficial	
	Intestinal		Shigella dysentery	Intestinal	
				Respiratory	Pneumonia
				Hydration	
				Injury and violence associated with water collection	
Water based	Water multiplied percutaneous		Bilharziasis	Contact	Leptospires, Naegleria fowleri
	Ingested		Guinea worm	Ingested	Toxic algae and their toxins
Water-related insect vectors	Water biting		Gambian sleeping sickness	No change	Malaria
	Water breeding		Onchocerciasis	No change	
Engineered water system associated (new proposed class)				Inhaled	Legionella, radon
				Ingested	MAC
				Contact	Pseudomonas

* Sequelae of infectious disease apply to all infectious agents

indirectly, to human exposure. It would also logically fall into an 'inhalation' class, but once again while suitable from a transmission perspective this has little relevance to prevention.

Finally, White et al. (1972) briefly note the potential association of water hardness with arterial disease (footnote to p162) and make mention of fluorine (fluoride) in the context of its adverse effects when exposure is excessive. There is strong evidence for a protective role for fluoride against dental caries, and addition of fluoride to water supplies in areas where natural concentrations are very low is recommended by the World Health Organization and practiced in some countries. There is suggestive evidence for a protective effect of higher water hardness or components associated with water hardness. While these may all be characterized as water quality issues they reflect nutritional inadequacy rather than toxicity and there may be an argument that they be accommodated in a new sub-class of water-borne disease or an entirely new sub-class for essential minerals.

Summary

We propose here a series of explanations, refinements and modifications to the Bradley Classification that would consolidate and secure its comprehensive nature, general applicability and future relevance. Our proposals are summarized in Table 3.2, which we believe is faithful to White et al.'s (1972) original intent: to propose a '… classification intended to be both more comprehensive and more precise in predicting the likely effects of changes in water supply …'

References

Aiello, AE, Coulborn, RM, Perez, V and Larson, EL 2008. Effect of hand hygiene on infectious disease risk in the community setting: a meta-analysis. *American Journal of Public Health*, 98: 1372–1381.

Bain, R, Cronk, R, Hossain, R, Bonjour, S, Onda, K, Wright, J, Yang, H, Slaymaker, T, Hunter, P, Pruess Ustun, A, and Bartram, J 2014. Global assessment of exposure to fecal-contamination through drinking-water based on a systematic review. *Tropical Medicine and International Health*, 19(8): 917–927.

Bartram, J, Brocklehurst, C, Fisher, M, Luyendijk, R, Hossain, R, Wardlaw, T and Gordon, B 2014. Global monitoring of water supply and sanitation: a critical review of history, methods, and future challenges. *Int. J. Environ. Res. Public Health*, 11: 8137–8165, doi:10.3390/ijerph110808137.

Baum, R, Luh, J, and Bartram, J 2013. Sanitation: a global estimate of sewerage connections without treatment and the resulting impact on MDG progress. *Environ Sci Techn*, 47: 1994–2000.

Bradley, D 2009. The spectrum of water related disease transmission processes. In Institute of Medicine (ed.) *Global Issues in Water Sanitation and Health*, pp 60–73. The National Academies Press, Washington DC, USA.

Bradley, D and Bartram, J 2013. Domestic water and sanitation as water security: monitoring, concepts and strategy. *Phil Trans R Soc A*, 371. http://dx.doi.org/10.1098/rsta.2012.0420.

Bradley, D and Bos, R, nd. Water storage: health risks at different scales. http://www.worldwaterweek.org/documents/Resources/Best/Bradley-Bos.pdf, accessed 24 Jan 2015.

Cairncross, S 1990. Health aspects of water and sanitation, In Kerr C (ed.) *Community Health and Sanitation*. Intermediate Technology Publications Ltd, London, UK.

Caminade, C, Kovats, S, Rocklov, J, Tompkins, AM, Morse, AP, Colón-González, FJ, Stenlund, H, Martens, P and Lloyd, SJ 2014. Impact of climate change on global malaria distribution. *Proc Natl Acad Sci USA*, 111(9): 3286–3291. doi: 10.1073/pnas.1302089111.

Chorus, I and Bartram, J (eds) 1999. *Toxic Cyanobacteria in Water: a guide to their public health consequences, monitoring and management*. E&FN Spon, London, UK.

Clasen, T, Bartram, J, Colford, J, Luby, S, Quick, R and Sobsey, M (2009). Comment on 'household water treatment in poor populations: is there enough evidence for scaling up now?' *Environ. Sci. Technol*, 43(4): 986–992.

Cotruvo, JA, Dufour, A, Rees, G, Bartram, J, Carr, R, Cliver, DO, Craun, GF, Feyer, R and Gannon, VPJ (eds) 2004. *Waterborne Zoonoses: identification, causes and control*. IWA Publishing, London and WHO, Geneva.

Cronk, R, Slaymaker, T and Bartram, J 2015. Monitoring drinking water, sanitation, and hygiene in non-household settings: priorities for policy and practice. *International Journal of Hygiene and Environmental Health*. In press.

Cumming, O, Elliott, M, Overbo, A and Bartram, J 2014. Does global progress on sanitation really lag behind water? An analysis of global progress on community- and household-level safe water and sanitation. *PLoS ONE* 9(12): e114699, doi10.1371/journal.pone.0114699.

Dufour, A, Bartram, J, Bos, R and Gannon, V (eds) 2012. *Animal Waste, Water Quality and Human Health*. IWA Publishing, London, on behalf of the World Health Organization.

Evans, B, Bartram, J, Hunter, P, Williams, AR, Geere, JA, Majuru, B, Bates, L, Fisher, M, Overbo, A and Schmidt, WP 2013. *Public health and social benefits of at-house water supplies. Final Report*. University of Leeds, Leeds, UK.

Feachem, R, McGarry, M and Mara, D 1977. *Water Wastes and Health in Hot Climates*. Wiley and Sons, London.

Feachem, R, Bradley, D, Garelick, H and Mara, D 1983. *Sanitation and Disease: health aspects of excreta and waste management*. John Wiley and Sons, London.

Freeman, MC, Stocks, ME, Cumming, O, Jeandron, A, Higgins, JP, Wolf, J, Pruss-Üstünl A, Bonjour, S, Hunter, PR, Fewtrell, L and Curtis, V 2014. Hygiene and health: systematic review of handwashing practices worldwide and update of health effects. *Tropical Medicine and International Health* 19(8): 906–916.

Howard, G and Bartram, J 2003. *Domestic Water Quantity, Service Level and Health*. WHO, Geneva.

Hunter, P. 2009. Household water treatment in developing countries: comparing different intervention types using meta-regression. *Environ. Sci. Technol*, 43(23): 8991–8997, doi: 10.1021/es9028217.

Kayser, GL, Moriarty, P, Fonseca, C and Bartram, J 2013. Domestic water service delivery indicators and frameworks for monitoring, evaluation, policy and planning: a review. *International Journal of Environmental Research and Public Health,* 10(10): 4812–4835.

Onda, K, LoBuglio, J and Bartram, J 2012. Global access to safe water: accounting for water quality and the resulting impact on MDG progress. *Int. J. Environ. Res. Public Health*, 9(3): 880–894, doi:10.3390/ijerph9030880

Oxford Dictionaries 2014. 'Classification' http://www.oxforddictionaries.com/us/definition/american_english/classification, accessed 25 July 2014.

Prüss-Üstün, A, Bartram, J, Clasen, T, Colford Jr, JM, Cumming, O, Curtis, V, Bonjour, S, Dangour, AD, De France, J, Fewtrell, L, Freeman, MC, Gordon, B, Hunter, PR, Johnston, RB, Mathers, C, Mäusezahl, D, Medlicott, K, Neira, M, Stocks, M, Wolf, J, Cairncross, S 2014. Burden of diarrhoeal disease from inadequate water, sanitation and hygiene in low- and middle-income settings: a retrospective analysis of data from 145 countries. *Trop Med Internat Hyg*,19(8): 894–905.

Santaniello-Newton, A and Hunter, PR 2004. Management of an outbreak of meningococcal meningitis in a Sudanese refugee camp in Northern Uganda. *Epidemiology and Infection*, 124: 75–81.

Schmidt, W-P and Cairncross, S 2009. Household water treatment in poor populations: is there enough evidence for scaling up now? *Environ. Sci. Technol*, 43(4): 986–992, doi:10.1021/es802232w.

Shields, K, Bain, R, Cronk, R, Wright, J and Bartram, J 2014. Influence of supply type on relative quality of source and stored drinking water in developing countries: a bivariate meta-analysis. Submitted to *Environmental Health Perspectives* July 2014.http://ehp.niehs.nih.gov/1409002/.

Stocks, ME, Ogden, S, Haddad, D, Addiss, DG, McGuire, C and Freeman, MC 2014. Effect of water, sanitation, and hygiene on the prevention of trachoma: a systematic review and meta-analysis. *PLOS Medicine*, 11(2): e1001605, doi:10.1371/journal.pmed.1001605.

Thompson, T, Fawell, J, Kunikane, S, Jackson, D, Appleyard, S, Callan, P, Bartram, J and Kingston, P 2007. *Chemical Safety of Drinking-water: assessing priorities for risk management*. World Health Organization, Geneva.

VanDerslice, J and Briscoe, J 1993. All coliforms are not created equal: a comparison of the effects of water source and in-house water contamination on infantile diarrheal disease. *Water Resources Research*, 29(7): 1983–1995.

Wagner, EG and Lanoix, JN 1958. *Excreta Disposal for Rural Areas and Small Communities*. World Health Organization, Geneva.

White, G, Bradley, D and White, A 1972. *Drawers of Water*. University of Chicago Press, Chicago, IL, USA.

WHO 1984. *Guidelines for Drinking-water Quality. Recommendations*, first edition. World Health Organization, Geneva.

WHO 1993. *Guidelines for Drinking-water Quality. Recommendations*, second edition. World Health Organization, Geneva.

WHO 2008. *Guidelines for Drinking-water Quality*, third edition. World Health Organization, Geneva.

WHO 2011. *Guidelines for Drinking-water Quality*, fourth edition. World Health Organization, Geneva.

WHO 2014. Dracunculiasis (guinea-worm disease) Fact sheet N°359 (Revised). World Health Organization. http://www.who.int/mediacentre/factsheets/fs359/en/, retrieved 24 Jan 2015.

WHO and UNICEF 2014. *Progress on Drinking Water and Sanitation: 2014 update*. World Health Organization, Geneva.

Wright, J, Gundry, S, and Conroy, R 2004. Household drinking water in developing countries: a systematic review of microbiological contamination between source and point-of-use. *Tropical Medicine and International Hygiene*, 9(1): 106–117.

4

WATERBORNE AND WATER-WASHED DISEASE*

Mark D. Sobsey

KENAN DISTINGUISHED PROFESSOR, UNIVERSITY OF NORTH CAROLINA, GILLINGS
SCHOOL OF GLOBAL PUBLIC HEALTH, DEPARTMENT OF ENVIRONMENTAL
SCIENCES AND ENGINEERING, CHAPEL HILL, NORTH CAROLINA, USA

Learning objectives

1 Describe transmission routes of faecal–oral infections.
2 Understand the characteristics of microorganisms and hosts that contribute to infection, disease and immunity.
3 Describe important groups of faecal–oral pathogens and their contributions to diarrheal disease.

Faecal–oral infections and transmission routes: the F-diagram

Infections are caused by the proliferation or multiplication of disease-causing microorganisms in a host (Baron, 1996; Mims et al., 2001; National Institutes of Health). Faecal–oral infections result typically from the ingestion of human or animal faecal matter that contains pathogens (disease-causing microorganisms). Upon such exposure, the ingested faecal pathogens may reach susceptible sites within the host gastrointestinal tract where they initiate infection and multiply to high concentrations, often in contact with susceptible host cells. The infecting microorganisms typically use the nutrients of the host to multiply and are then shed from the body at concentrations of millions to many billions per grams of faeces. This shedding by infected hosts is part of a perpetuating cycle of faecal excretion into the environment that later may be ingested again by other humans and animals (Feachem et al., 1983; Carr, 2001). The transmission routes of faecal contamination and the ways in which people may ingest pathogens are diverse. They include water (fluids), food, fomites (surfaces), vectors such as flies, hands (fingers)

---

* Recommended citation: Sobsey, M.D. 2015. 'Waterborne and water-washed disease', in Bartram, J., with Baum, R., Coclanis, P.A., Gute, D.M., Kay, D., McFadyen, S., Pond, K., Robertson, W. and Rouse, M.J. (eds) *Routledge Handbook of Water and Health*. London and New York: Routledge.

38

and other environmental media as shown in the F-diagram (see Chapter 3, Figure 3.2). Waterborne faecal pathogens are acquired typically by ingestion of contaminated water and water-washed pathogens are those acquired by other routes of contact with faecal matter in the environment and with exposures resulting from lack of sufficient water for hand washing, bathing and other basic personal hygiene activities. This cycle of exposure and resulting infection is further exacerbated by conditions of poor sanitation and hygiene as a result of inadequate management of human and animal excreta.

Most faecal–oral microbes, including all viruses, bacteria and protozoan parasites, are potentially infectious for a host as soon as they are excreted in faeces and can reach a susceptible host who can ingest them (Feachem et al., 1983; Baron, 1996; Food and Drug Administration, 2012). However, many faecal–oral helminths are not immediately infectious when excreted and require some period of time in the environment to mature or to otherwise undergo changes over time facilitated by environmental conditions or an intermediate host, before they reach a state that is infectious for a human host. This phenomenon is referred to as latency.

Faecal–oral infections

Sites, processes and events of infection in the host: an overview

Microbes entering the body must reach a target site and come in contact with target cells to initiate infection (Mims et al., 2001). This process is facilitated by unique physical-chemical entities that are typically on the surface of the infecting microbes and act as an attachment or binding agent to cells and tissues of the host. Correspondingly, there are unique physico-chemical receptors on host cells and tissues that facilitate the attachment and multiplication of the infecting microbe and its possible entry into host cells and tissues (Baron, 1996). The success and outcomes of the process of infection and the potential for subsequent disease is mediated by both the properties of the microbial pathogen and of the host.

The typical course of events in infection and disease involves: (1) initial exposure and microbe contact with susceptible host cells and tissues at a target site in the body, (2) microbe multiplication at the initial site of infection, (3) possible development of symptoms of illness or disease as a result of pathophysiological effects of the proliferating microbes on host cell and tissues at the initial site of infection, (4) possible spread of the microbes to other parts of the body and further multiplication of the microbes at these other sites, and (5) either successful response of the host to the multiplying microbes by destroying them, eliminating them completely from the body and restoring damaged tissue with new, healthy cells (a "cure") or, alternatively, continued and uncontrolled microbial proliferation and pathophysiological effects 'that result in more severe disease and eventually death' (Baron, 1996; Casadevall and Pirofski, 1999; 2000; Mims, 2001; Kolling et al., 2012; National Institutes of Health). The time course of these events in the process of infection and possible disease outcomes vary among different microbial pathogens, from as short as a day or two to as long as weeks, or even months and years. Other outcomes of infection and disease are possible, including persistence of the microbes in the host for an extended period of time or for the life of the host. Such persistence may be with or without host shedding of the microbes from the body and with or without continued or recurrent illness or other adverse health effects.

Virulence factors and properties of microbes

Pathogenic microbes, including those infecting the gastrointestinal tract, have unique surface properties to facilitate infection. They may also possess additional surface properties and other attributes and functions that contribute to their ability to cause disease. Collectively, these are referred to as virulence factors or virulence properties (Baron, 1996; Casadevall and Pirofski, 1999; 2000; Mims, 2001). Virulence properties can be placed into distinct categories, such as microbial ligands that bind to host cells, adherence and effacement factors that facilitate close attachment and tight binding to host cell cells, invasins that facilitate entry into host cells and penetration through tissues, toxins that are either part of the surface structure of the microbe (for example, endotoxins) or are released externally by the microbes (exotoxins), and enzymes that carry out functions contributing to the processes of infection and pathophysiology (Baron, 1996; Casadevall and Pirofski, 1999; 2000; Mims, 2001).

Categories of gastrointestinal or enteric infection

A wide variety of microbial pathogens cause a range of human infections and diseases via the faecal–oral route of transmission. This is particularly true of infections caused by pathogens that multiply in the gastrointestinal tract (intestines), either extracellularly or within cells of the intestinal tissues. Some pathogens remain localized only in the intestinal tract, where they cause infection and may produce illness or disease. The most typical of these diseases are collectively referred to as diarrhea or acute gastrointestinal illness marked by multiple episodes of defaecation (bowel movements) per day, often as abnormal stools. In many cases such illness also includes nausea and multiple episodes of vomiting (Baron, 1996).

Other gastrointestinal or enteric pathogens have the ability to penetrate the lining of the intestinal tract and cause dysentery (bloody diarrhea) yet remain confined there (Baron, 1996; Mims, 2001). Yet other pathogens can spread from the intestinal tract to other tissues or organs. Some can enter the bloodstream and lymphatic system to become generalized or systemic infections causing other kinds of pathophysiological effects and disease (Baron, 1996; Mims, 2001). These pathophysiological effects include: (1) hepatitis in the liver as caused by hepatitis A and E viruses, (2) neurological disease such as encephalitis caused by enteroviruses and epileptic seizures and other neurological disorders caused by the protozoan parasite *Toxoplasma gondii*, (3) neuromuscular disease such as neuromuscular paralysis caused by polioviruses and Guillain–Barré syndrome caused by campylobacter bacteria, (4) heart disease such as myocarditis caused by certain enteroviruses, (5) kidney disease such as hemolytic uremic syndrome caused by enterohemorrhagic *E. coli*, and (5) cancers such as cancer of the stomach caused by *Helicobacter pylori* (Baron, 1996; Mims, 2001; National Institutes of Health). These systemic or generalized infections or infections of other target organs and tissues and their resulting diseases tend to be of greater severity and health impact due to more extensive tissue damage and the spread and further multiplication of the pathogen with pathophysiological effects in other parts of the host's body, often with high fever and high mortality rates.

Categories of infecting enteric microorganisms based on disease potential, severity, sequelae and disability-adjusted life years

Whether or not a faecal–oral pathogen that causes an intestinal infection in a host also causes frank illness or disease and possibly death is dependent on the properties of the pathogen, the physiological state of the human or animal host and the interactions between them

(Casadevall and Pirofski, 1999; 2000; Mims, 2001; National Institutes of Health). Some potential pathogens are more likely to cause infection without illness while others are more likely to produce infection with illness, and sometimes a high risk of death. Pathogens that are more likely to cause not only gastrointestinal infection but also obvious disease or illness in most infected hosts are sometimes called "frank" pathogens. Other pathogens that often cause infection without illness but sometimes cause illness in more susceptible hosts, especially those at high risk due to other underlying conditions or vulnerabilities, are sometimes called "opportunistic" or conditional pathogens. The ability of a microbe causing infection to also cause disease is therefore dependent on its specific properties as well as host properties. In particular, the distinction of an infecting microorganism being either non-pathogenic, opportunistic or conditional, or a frank pathogen is dependent to some extent on its virulence properties: does it possess any, and if so, which ones, how many of them, and are they expressed?

The severity of disease caused by faecal–oral pathogens differs greatly among the different specific pathogens and it is also influenced by the state of the host (Baron, 1996; Mims et al., 2001: National Institutes of Health). For some pathogens, initial infection and illness may result in later adverse health effects that are different from the initial illness and are referred to as sequelae. These sequelae can range from neurological disease (encephalitis), to neuromuscular disease (paralysis), reactive arthritis, heart disease (myocarditis), diabetes, cancer (e.g. of the stomach) and immunosuppression. The severity of disease is an important consideration not only in terms of the consequences for individual human hosts but also for estimating the disease burdens produced by faecal–oral and other pathogens in populations. Such disease burden estimates are based on disability-adjusted life years (DALYs) that take into consideration both the years of life lost (YLL) and years lived with disability or disease (YLD). DALY = YLL + YLD, which can be quantified for any infectious disease or other disease-causing agent, such as a toxic chemical (World Health Organization, 2011).

Host factors in infection and disease

Host factors and properties contribute to the extent to which an infecting microbe is either non-pathogenic, opportunistic or a frank pathogen (Baron, 1996; Mims, 2001; Casadevall and Pirofski, 1999; 2000; Kolling et al., 2012; National Institutes of Health). These host properties are genetic (such as possessing specific host cell receptors for the attachment of the pathogen to host cells), age (as some pathogens are more likely to cause disease in one age group than another, such as the very young, adults or the elderly), the presence of underlying cell and tissue damage due to external trauma or chronic conditions such as inflammation, and the presence of abnormal conditions of physiological function such as acid-base and electrolyte balance (Baron, 1996; Mims, 2001; Kolling et al., 2012). In the gastrointestinal tract these physical, chemical and biological barriers include the acid pH of the stomach, the antibacterial activity of pancreatic enzymes, bile and intestinal secretions, the peristalsis of the intestines and normal sloughing of intestinal epithelial cells, and the activities of the normal flora of the bowel that can inhibit pathogen infection. Certain hosts are more likely to become ill and suffer from disease caused by an infection because they have conditions or are in states that make them more susceptible. In particular, the very young, the very old, those with immunodeficiency and pregnant women are hosts at greater risk of developing disease as a result of an infection. Such population groups are often referred to as "sensitive", "susceptible", "vulnerable" or high risk populations (Mims et al., 2001; Kolling et al., 2012; National Institutes of Health).

Host immunity and response to infection

The immune system of the host also influences susceptibility to and outcomes of infection (Baron, 1996; Casadevall and Pirofski, 1999; 2000; Mims et al., 2001; National Institutes of Health). The extent of susceptibility to infection differs in hosts who are immunocompetent versus immunocompromised or immunosuppressed, particularly in relation to the specific conditions of the host's immune status for a specific pathogen or group of related pathogens. The immune status of the host for a specific microorganism can influence whether infection occurs at all and, if it does, what the consequences are likely to be (Baron, 1996; Mims et al., 2001; National Institutes of Health). If the host has never been infected by a specific microbe, there is no prior immunity and the host will be fully susceptible to infection. If the host has prior immunity from a previous infection with the same or similar microorganisms, such pre-existing immunity may be fully protective against reinfection with the same or similar microbes (Baron, 1996; Mims et al., 2001). Infection by certain microbes can result in lifelong immunity against reinfection causing disease, such as for hepatitis A and E viruses and polioviruses that cause poliomyelitis. In the case of infections caused by other microorganisms, immunity may be protective against reinfection for a shorter period of time (months or a few years). Such temporary immunity may decline over time. Partial immunity may result, meaning that an infection may occur again but may be of shorter duration, there may be less likelihood of illness or illness may be milder. Reinfection is possible for enteric viruses such as noroviruses and rotaviruses, bacteria such as *Salmonella* species and *Campylobacter* species, protozoan parasites such as *Cryptosporidium* and *Giardia* and helminths such as *Ascaris lumbricoides*.

Risks of infection and disease and dose–response relationships

Pathogens differ in the extent to which they can cause infection and illness when ingested as a single dose, such as in a glass of water (Teunis et al., 1996; Haas et al., 1999; WHO, 2011). However, the ingestion of even one microorganism has the potential to cause infection and possibly disease, even if the probability of infection or disease from a single exposure is low (see Chapter 56, 'Quantitative microbial risk assessment'). Microorganisms also differ in the extent to which an infection will result in illness or disease. For some pathogens, infections have a relatively high probability (50 per cent or more) of also causing illness, while for other pathogens, the probability of illness from an infection is relatively low (<10 per cent or even <1 per cent). An important factor in risk of illness from an infection is the age of the host. For some faecal–oral pathogens, infection of a host very early in life carries a high risk of illness (>50 per cent), such as infection by rotaviruses in infants, while for others, the risk of illness when infected very early in life is relatively low, such as for polioviruses (<10 per cent in infants). In contrast, the risk of illness from rotaviruses is low in adults while the risk of illness from polioviruses is high in older hosts who have not been previously infected.

Examples of highly infectious faecal–oral pathogens that have a high probability of causing infection and possibly illness from a very low dose (as few as one ingested microorganism) are the human noroviruses and rotaviruses. For these viruses the 50 per cent infectious doses (doses for which the probability of infection are about 50:50 or 50 per cent based on human volunteer studies) are perhaps one or a few virus particles (Ward et al., 1986; Teunis et al., 2008). Other faecal–oral microorganisms with relatively high infectivity (high probability of infection at relatively low doses) are some strains of the protozoan parasite *Cryptosporidium parvum* and some enteric bacteria that cause dysentery, such as *Shigella* species and enterohemorrhagic strains of *Escherichia coli*, for which 50 per cent infectious doses are in the range of 10–100 microorganisms. In contrast, some other faecal–oral pathogens are relatively low in infectivity, such as certain species of Salmonella bacteria, for which the 50 per cent infectious doses for different species are >1000 microorganisms based on human volunteer studies (Teunis et al., 1996; Haas et al., 1999). However, compilation of data from careful and timely epidemiological investigations of foodborne and waterborne outbreaks suggest that some species and strains of Salmonella are infectious and have the potential to cause disease at relatively low doses of between 10 and 100 microorganisms (Food and Drug Administration, 2012).

The infectivity dose–response relationships of many faecal–oral pathogens based on human volunteer studies have been compiled previously (Teunis et al., 1996; Haas et al., 1999; 2014) and more recent studies have developed additional infectivity and disease outcome dose–response relationships from epidemiological investigations of waterborne and foodborne outbreaks of disease where there was timely and careful investigation to identify cases and also successful efforts to isolate and quantify the pathogenic microorganisms in the incriminated exposure vehicles, such as food or water (Food and Drug Administration, 2012).

Infection and the carrier state

Humans or animals colonized with pathogens and experiencing no illness for extended periods of time are often considered "carriers" and they may continue to faecally excrete appreciable quantities of these colonizing microbes (Baron, 1996; Casadevall and Pirofski, 1999; 2000; Mims et al., 2001). The carrier state with continued faecal shedding without evidence of frank disease is a sanitation and public health concern because there are no visible indications that the host is harboring and shedding potential pathogens that could cause illness in another susceptible host that can become exposed to and infected by such faecally shed pathogens. The carrier state has been documented for a wide range of faecal–oral pathogens, but especially certain enteric bacteria and helminths.

Zoonotic pathogens

Of the many different faecal–oral viruses, bacteria, protozoan parasites and helminths that can infect and cause disease in humans, some can also infect animals, with or without causing disease in their animal hosts (Baron, 1996; Food and Drug Administration, 2012). These are referred to as zoonotic pathogens and their diseases are called zoonoses. Some of the most important zoonotic pathogens are Gram-negative enteric bacteria, including most of the *Salmonella* species, *Campylobacter* species and various strains of disease-causing *E. coli*. These bacteria infect and colonize many animals, including food animals for humans, such as poultry, cattle and sheep, often with no evidence of disease within the animals but extensive faecal shedding of these pathogens. Such faecal shedding from infected and colonized animals is especially a concern if the animal faecal wastes are used for agriculture as soil amendments or as irrigation water or if they are discharged without treatment and reach waterways used for drinking water supply, aquaculture and to irrigate fields used for growing produce that may be eaten raw, bathing or recreational use (swimming). The enteric protozoan parasites *Cryptosporidium parvum* and *Giardia lamblia (intestinalis)* are also widely present in faeces and shed by a wide range of mammals, including food animals such as cattle. Among the helminths, the pork tapeworm, *Taenia solium*, is a zoonotic pathogen for which exposure and infection can be by ingestion of faecally contaminated water, although they are not directly transmitted person to person by the faecal–oral route and require a swine host for their lifecycle. Transmission to humans is primarily from ingestion of contaminated pork.

Presence, transport, persistence and fate of faecal–oral pathogens in the environment

Faecal presence, pathogen shedding and other faecal–oral pathogen sources in the environment

Faecal–oral pathogens present in the environment typically have come from human or possibly animal faeces (Feachem et al., 1983). Pathogen concentrations in human or animal faeces can be very high, typically in the millions to billions of microorganisms per gram, and even higher for some pathogens such a rotaviruses (Feachem et al., 1983). Typically, concentrations of faecal pathogens in the environment are lower than those in human or animal faecal matter and faecal wastes shed into the environment. However, faecal waste in the environment, such as excreta in latrines or deposited on the ground or raw sewage discharged without treatment to water or land, can still have the high concentrations of faecal–oral pathogens that were present when excreted (Feachem et al., 1983). In the environment the concentrations of the pathogens generally decline due to dilution and dispersion in environmental media, die-off or inactivation and sequestration in environmental compartments, such as attaching to terrestrial soil particles or being deposited by sedimentation in aquatic sediments of water bodies. Although such faecal–oral microbes may accumulate at such sites in the environment, they can be remobilized and be released back into water and further transported. Regardless of where faecal–oral microbes are in the environment, they can be sources of exposure if humans come in contact with them

Factors influencing faecal–oral microbe transport, survival and fate in the environment

For microbes transmitted by the faecal–oral route, transport and persistence in the environment is related to risk of host exposure, infection and disease. In general, the relative survival of different classes or types of faecal–oral pathogens is helminth ova and microbial (bacterial and mycotic) spores > viruses $\simeq$ protozoan parasite cysts and oocysts > vegetative bacteria (Feachem et al., 1983). The extent to which microbes survive is dependent on the impacts or influences of a variety of physical, chemical and biological conditions or processes. In general, the time it takes for the concentrations of infectious microbes to be reduced by one half (50 per cent) in environmental media ranges from hours to days for the least resistant ones to many months or even years for the most persistent ones. It is important to keep in mind that the survival of microbes is a kinetic process, which means that the magnitude of loss of survival increases with time (Rzezutka and Cook, 2004; John and Rose, 2005; Pedley et al., 2006; Bertrand et al., 2012).

Major physical processes or conditions influencing pathogen survival are heat or thermal effects, desiccation or drying, relative humidity (when on surfaces and in the airborne state), exposure to ultraviolet radiation, microbial aggregation, adsorption to particles and surfaces and encapsulation or being embedded in other, larger particles (Sobsey and Meschke, 2003; Rzezutka and Cook, 2004; John and Rose, 2005; Pedley et al., 2006; Tang, 2009). In general faecal–oral microbes survive better at lower temperatures, including freezing. The impacts of drying and desiccation and of relative humidity on the survival of faecal pathogens vary depending on the specific microbe and on other factors, such as the chemical composition of the medium, with some surviving for only a short time and others surviving longer (Sobsey and Meschke, 2002; Tang, 2009). For example, the extent of rotavirus survival differs in water, on surfaces, in air and at different relative humidities (Ansari et al., 1991). Faecal–oral microbes are rendered non-infectious by sufficient doses of ultraviolet radiation such as present in sunlight, and they survive better when aggregated, adsorbed to particles and surfaces and when encapsulated or embedded in large particles. Survival of faecal–oral microbes in the environment is generally better near neutral pH (pH 7) where there is sufficient moisture, than at extreme low (acidic) and high (alkaline) pH levels. Other chemical factors influencing faecal–oral microbe survival include the presence of organic matter, the presence and forms of ammonia, and various enzymes. Most microbes are protected by natural organic matter and their survival is decreased by ammonia at higher pH where it predominates over ammonium ion. In general, they survive less well in the presence of enzymatic activity by protease, lipase, nuclease and amylase enzymes in environmental media and in the presence of high dissolved solids or solute concentrations, such as in seawater as compared to fresh water. Other chemicals in the environment that reduce the survival of faecal–oral pathogens include various oxidants, antibiotics and fatty acids and esters and other organic acids (Rzezutka and Cook, 2004; John and Rose, 2005; Pedley et al., 2006). Biological factors include the antagonistic effects of microbial activity and metabolism by the microbial community in environmental media, microbial predation by mechanisms such as engulfment and ingestion by higher organisms and association with or proliferation in biofilms. In summary it is important to recognize that faecal–oral pathogens are often able to survive long enough to pose risks of infection and possibly illness to humans that become exposed to them through environmental media and pathways (Rzezutka and Cook, 2004; John and Rose, 2005; Pedley et al., 2006).

Important faecal–oral pathogens, their characteristics and properties and their infections and diseases

A number of taxonomically different faecal–oral microorganisms can infect the gastrointestinal tracts of human and in some cases other animals and be shed in faeces in high numbers, typically at concentrations of millions per gram or more.

Viruses

The important faecal–oral viruses, bacteria and parasites are listed in Table 3.1, Chapter 3. All of these faecal–oral pathogens contribute to a large burden of documented disease and death that has been attributed in part to environmental exposures to these pathogens (Feachem et al., 1983; Leclerc et al., 2002; Fletcher et al., 2012; La Rosa et al., 2012; Rzezutka and Cook, 2004). Some virus groups consist of several or many different members, each of which can infect the same human host over time. Most faecal–oral viruses can infect and produce gastrointestinal illness. Ingestion of low doses of these viruses has a high probability of causing infection and in some cases a high probability of illness as well (Teunis et al., 1996; Haas et al., 1999). Of the faecal–oral viruses, the adenoviruses, enteroviruses and the reoviruses can also infect the respiratory tract to cause respiratory illness and shedding of the viruses in respiratory secretions and exudates. The enteroviruses can also cause skin rashes and conjunctivitis.

Bacteria

The important faecal–oral bacteria and bacteria groups are shown in Table 3.1, Chapter 3. In some cases there are many different strains, serogroups, genogroups and pathogenic types in a listed group, as with the viruses (Feachem et al., 1983; Baron, 1996; Leclercet al., 2002). This is the case for the different pathogenic strains, sub-groups and types of *E. coli*, the many different species, sub-groups, serotypes and strains of non-typhoid *Salmonella* species and the several strains, serogroups and pathogenic types of *Vibrio cholera*.

Among the faecal–oral bacteria pathogens, some, such as the *Shigella* species and enterohemorrhagic strains of *E. coli*, are infectious at low doses, others; such as *Campylobacter* species, at moderate doses and yet others such as *Salmonella*, the non-hemorrhagic *E. coli* and *Aeromonas hydrophila* pathogens only at relatively high doses.

Protozoan parasites

Some important faecal–oral protozoan parasites and their key properties are shown in Table 3.1, Chapter 3. The faecal–oral protozoan parasites comprise several distinct taxonomic groups and all except *Toxoplasma gondii* cause gastrointestinal infection and illness (Feachem et al., 1983; Fletcher et al., 2012). *Entamoeba histolytica* can cause dysentery (amoebiasis) with symptoms of bloody diarrhea, stomach pain and fever. Rarely, *E. histolytica* can invade the liver and cause abscesses and even more rarely infect the lungs or brain (Baron, 1996). *Toxoplasma gondii*, which has felines as its definitive host, can also infect humans and other mammalian hosts by oral ingestion of feline faeces. It has caused foodborne (from contaminated meat of infected animals) and occasionally waterborne disease. It produces a flu-like illness with body aches, swollen lymph nodes, headache, fever and fatigue. Occasionally it spreads to other sites in the body such as the brain to cause chronic or persistent infection with disease symptoms such as neurological disorders. *Cryptosporidium* species, *Giardia intestinalis*, *Toxoplasma gondii* and possibly *Entamoeba histolytica* are considered infectious at low to moderate doses. For *Cyclospora cayetanensis* the dose–response relationship is unknown but thought be low based on epidemiological evidence from common source outbreaks. Although acute gastrointestinal illness from faecal–oral protozoan parasites is often transient and self-limiting, persistent and repeated infection in children can lead to stunting and other developmental deficits. Both *Entamoeba histolytica* and *Toxoplasma gondii* can cause chronic infections with spread to other organs of the body and long-term adverse health effects, especially in immunocompromised humans. Immunity to most infections by faecal–oral protozoan parasite pathogens is protective temporarily against reinfection, but immunity can wane over time and reinfection and illness can occur later, often from different serotypes or strains within species, genus or sub-group (Baron, 1996).

Helminths

Important faecal–oral helminths are listed in Table 4.1.

All of the faecal–oral helminths in Table 4.1 are considered geohelminths because they can be present as helminth ova (eggs) in human faecal matter that has been deposited on the ground (Feachem et al., 1983; Baron, 1996). They can persist and remain infectious in moist soils, sometimes for years. After the faecally excreted eggs mature in the soil (in hours, days or weeks, depending on the helminth), human exposure typically occurs by the mature helminth larvae penetrating the skin (by walking barefoot or handling or otherwise

Table 4.1 Important faecal–oral helminths

Helminth group	Names	Zoonotic?	Infectivity	Localized or systemic illness
Nematodes (roundworms)	*Ascaris lumbricoides*, *Strongyloides stercoralis*	No	High	Localized enteric infestation but spreads via bloodstream, to heart, lungs and intestines
Hookworms	*Ancylostoma duodenale* and *Necator americanus*	No	High	Same as above
Whipworms	*Trichuris trichuria*	No	High	Localized; rare spread to other sites

coming in contact with faecally contaminated soil) or by ingestion of mature ova on faecally contaminated produce or in faecally contaminated water. The life cycle of the hookworms and roundworms typically involves the movement of the helminth larvae from the skin, through the bloodstream to the heart and lungs and then to the gastrointestinal tract by having been swallowed as larvae that migrated from the lungs to the throat. The hookworm *Ancylostoma duodenale* can also be transmitted through the ingestion of larvae. Heavy infestations with helminths can cause a range of adverse health effects, including abdominal pain, diarrhea, blood and protein loss, intestinal blockage, rectal prolapse, and impairment of physical and mental development in children. Helminth infections are widespread, especially in the developing world where hygiene and sanitation conditions are poor and there is extensive exposure to human faecal matter from environmental sources (Feachem et al., 1983; Baron, 1996). Although helminth infections are readily treatable, reinfection occurs commonly due to re-exposure from faecally contaminated environmental sources.

Contributions of different faecal–oral pathogens to diarrheal disease and death

There are many different faecal–oral pathogens that cause diarrheal disease and such disease is a major contributor to the global burden of disease as indicated in Table 3.1 of Chapter 3 and Table 4.1 (Prüss-Ustün et al., 2014). Therefore, risks of diarrheal disease and death are likely to be dramatically reduced by improved water and sanitation and hygiene in low- and middle-income settings. A study identified the microbial agents involved in cases of moderate to severe diarrheal disease in children <5 years of age (Kotloff et al., 2013). The diarrhea cases were studied in seven countries – four in sub-Saharan Africa and three in south Asia – and were identified through community health surveys linked to sentinel health care facilities. The pathogens were: pathogenic *E. coli*, *Shigella*, *Aeromonas*, non-typhoidal *Salmonella*, *Campylobacter jejuni* and *Vibrio cholera* bacteria; rotaviruses, noroviruses and adenoviruses; and the protozoan parasite *Cryptosporidium*. Additional faecal–oral pathogens that caused diarrheal disease deaths were *Entamoeba histolytica* and *Giardia*. These findings indicate that many faecal–oral pathogens are important contributors to diarrheal disease and death in developing countries. A component of the same study examined the role of water as a risk factor for the observed cases of diarrheal disease based on data obtained from household surveys in one country only (Baker et al., 2013). The risks of diarrheal disease in children were lower when using private or public tap water, always having access to the primary water source, fetching water daily and breastfeeding. They were higher when it took >30 minutes to fetch water. Overall, faecal–oral pathogens are important contributors to the global burden of diarrhea, other enteric diseases such as cholera and typhoid fever, as well as many other diseases and syndromes described above. Improved water, sanitation and hygiene can reduce their disease burden contribution.

Summary of faecal–oral infections caused by waterborne and water-washed pathogens

A large number and variety of faecally excreted pathogens can be ingested from water and other environmental sources (e.g. food, fomites and fingers) often leading to infection and illness. These faecal–oral pathogens include bacteria, viruses, protozoan parasites and helminths. Some of them pose high risks of infection and resulting disease from ingestion of very low doses of them. The risks of infection, resulting disease and death from ingestion of faecal–oral pathogens are also dependent on the specific properties of the pathogen as well as the host, including virulence properties of the pathogens and susceptibility properties of the host such as genetic factors and immunity. There are both qualitative and quantitative differences in the properties of the faecal–oral pathogens and their hosts that influence the risks of infection and potential for disease and death. In addition, the risks of infection and resulting disease from ingestion of faecal–oral pathogens are also influenced by their potential sources in the environment, such as human faecal contamination, animal faecal contamination (if they are zoonotic), other non-faecal sources (having natural aquatic habitat) and their ability to survive and be transported in the environment to again reach human hosts. The diseases caused by infections with faecal–oral pathogens range from short-term diarrhea and other symptoms of gastrointestinal illness (e.g. nausea and vomiting) that may be of short duration, to more generalized or systemic infections that involve further spread, infection and damage in other bodily sites within the host to cause more severe diseases of organs such as the liver (hepatitis), the brain (encephalitis and seizures), neuromuscular system (paralysis) and the heart (myocarditis), as well as developmental disorders and conditions in children (stunting and cognitive dysfunction). The duration and outcomes of infections and diseases by faecal–oral pathogens may range from as little as days, to weeks to months, with some infections becoming lifelong states with continued or recurrent disease conditions (chronic diseases and sequelae) that can lead to death. Overall, the faecal–oral pathogens are among the most important and widespread contributors to the global burden of disease. The waterborne and water-washed diseases caused by these faecal–oral pathogens can be appreciably reduced by improved water, sanitation and hygiene.

Key recommended readings (open access)

1 Baron, S. (1996) *Medical Microbiology*, 4th edition. Edited by Samuel Baron. University of Texas Medical Branch at Galveston. ISBN-10: 0-9631172-1-1. Available at: http://www.ncbi.nlm.nih.gov/books/NBK7627/. Description: This is an online medical microbiology textbook that describes many of the faecal–oral pathogens of concern and the processes of infection and pathogenesis for them and their diseases. Although somewhat dated, the information in this resource is still relevant and useful.

2 Carr, R. (2001) Excreta-related infections and the role of sanitation in the control of transmission. World Health Organization (WHO). *Water Quality: Guidelines, Standards and Health*. Edited by Lorna Fewtrell and Jamie Bartram. Published by IWA Publishing, London, UK. ISBN: 1-900222-28-0. Available at: http://www.who.int/water_sanitation_health/dwq/iwachap5.pdf. Description: A book chapter listing and

describing human faecal–oral pathogens, their infections, presence in faecal wastes, transmission pathways, and their control by sanitation processes and measures.

3   Feachem, R.G., D.J. Bradley, H. Garelick and D.D. Mara (eds) (1983) *Sanitation and Disease. Health Aspects of Excreta and Wastewater Management*. World Bank Studies in Water Supply and Sanitation. Published for the World Bank by John Wiley & Sons, New York. Available at: http://www-wds.worldbank.org/servlet/WDSContentServer/IW3P/IB/1999/12/23/000178830_98101911180473/Rendered/PDF/multi0page.pdf. Description: Comprehensive but somewhat dated treatise on faecal–oral and other pathogens in human and animal excreta, describing their sources, quantities, human health effects and risks, environmental survival and transport, reduction by waste treatment process and beneficial reuse of waste.

4   Mims, C.A., A. Nash and John Stephen (2001) *Mims Pathogenesis of Infectious Disease*, 5th edition. ISBN: 978-0-12-498264-2. Available at: http://ilmufarmasis.files.wordpress.com/2011/07/mims__pathogenesis_of_infectious_disease_-_5th_ed__2001.pdf. Description: This is a definitive textbook on the process of infection and the pathogenesis of infectious disease. It covers all routes of infection, the process of infection, pathogen entry and exit from hosts and the immune response.

5   National Institutes of Health. Understanding Emerging and Re-emerging Infectious Diseases. Online module. Available at: http://www.ncbi.nlm.nih.gov/books/NBK20370/. Description: This is a simple introduction to infectious diseases and pathogens, including those transmitted by faecally contaminated water and food and associated with poor hygiene.. A supporting document is also available at: http://science.education.nih.gov/supplements/nih1/Diseases/guide/pdfs/nih_diseases.pdf

References

Ansari, SA et al. (1991) Survival and vehicular spread of human rotaviruses: possible relation to seasonality of outbreaks. *Rev Infect Dis.* 13(3): 448–461.

Baker, KK et al. (2013) Quality of piped and stored water in households with children under five years of age enrolled in the Mali site of the Global Enteric Multi-Center Study (GEMS). *Am. J. Trop. Med. Hyg.* 89(2): 214–222.

Baron, S (1996) *Medical Microbiology*, 4th edition. Galveston, TX: University of Texas Medical Branch at Galveston. Available at: http://www.ncbi.nlm.nih.gov/books/NBK7627/.

Bertrand, I et al. (2012) The impact of temperature on the inactivation of enteric viruses in food and water: a review. *J Appl Microbiol.* 112(6): 1059–1074. doi: 10.1111/j.1365-2672.2012.05267.x. Epub 2012 Mar 20.

Carr, R (2001) Excreta-related infections and the role of sanitation in the control of transmission, in: World Health Organization (WHO). *Water Quality: Guidelines, Standards and Health*. Edited by Lorna Fewtrell and Jamie Bartram. London: IWA Publishing, London, UK.

Casadevall, A and Pirofski, L. (1999) Host-pathogen interactions: redefining the basic concepts of virulence and pathogenicity. *Infect Immun.* 67: 3703–3713.

Casadevall, A and Pirofski, L. (2000) Host-pathogen interactions: basic concepts of microbial commensalism, colonization, infection, and disease. *Infect Immun.* 68(12): 6511–6518.

Feachem, RG, Bradley, DJ, Garelick, H and Mara, DD (eds) (1983) *Sanitation and Disease. Health Aspects of Excreta and Wastewater Management*. World Bank Studies in Water Supply and Sanitation. New York: John Wiley & Sons. Available at: http://www-wds.worldbank.org/servlet/WDSContentServer/IW3P/IB/1999/12/23/000178830_98101911180473/Rendered/PDF/multi0page.pdf.

Fletcher, SM et al. (2012) Enteric protozoa in the developed world: a public health perspective. *Clin Microbiol Rev.* 25(3): 420–449. doi: 10.1128/CMR.05038-11. Review. Available at: http://cmr.asm.org/content/25/3/420.long.

Food and Drug Administration. (2012) *Bad Bug Book, Foodborne Pathogenic Microorganisms and Natural Toxins*, 2nd edition. Available at: http://www.fda.gov/Food/FoodborneIllnessContaminants/CausesOfIllnessBadBugBook/.

Haas, CN et al. (1999) *Quantitative Microbial Risk Assessment*, 1st edition. New York: John Wiley & Sons.

Haas, CN et al. (2014) *Quantitative Microbial Risk Assessment*, 2nd edition. New York: John Wiley & Sons.

John, DE and Rose, JB. (2005) Review of factors affecting microbial survival in groundwater. *Environ. Sci. Technol.* 39(19): 7345–7356.

Kolling, G. et al. (2012) Enteric pathogens through life stages. *Front Cell Infect Microbiol.* 2: Article 114. Available at: http://www.ncbi.nlm.nih.gov/pmc/articles/PMC3427492/

Kotloff, KL et al. (2013) Burden and aetiology of diarrhoeal disease in infants and young children in developing countries (the Global Enteric Multicenter Study, GEMS): a prospective, case-control study. *Lancet* 382: 209–222.

La Rosa, G et al. (2012) Emerging and potentially emerging viruses in water environments. *Ann 1st Super Sanita.* 48(4): 397–406. doi: 10.4415/ANN_12_04_07. Available at: http://www.ncbi.nlm.nih.gov/pubmed/23247136.

Leclerc, H et al. (2002) Microbial agents associated with waterborne diseases. *Crit Rev Microbiol.* 28(4): 371–409.

Mims, CA, Nash, A and Stephen, J (2001) *Mims Pathogenesis of Infectious Disease*, 5th edition. Available at: http://ilmufarmasis.files.wordpress.com/2011/07/mims__pathogenesis_of_infectious_disease_-_5th_ed__2001.pdf.

National Institutes of Health (n.d.) Understanding emerging and re-emerging infectious diseases. Online module. Available at: http://www.ncbi.nlm.nih.gov/books/NBK20370/.

Pedley, SM et al. (2006) Pathogens: health relevance, transport and attenuation. Ch. 3, pp 49–80, in: World Health Organization. *Protecting Groundwater for Health: Managing the Quality of Drinking-water Sources.* Edited by O. Schmoll, G. Howard, J. Chilton and I. Chorus. London: IWA Publishing.

Prüss-Ustün, A et al. (2014) Burden of disease from inadequate water, sanitation and hygiene in low- and middle-income settings: a retrospective analysis of data from 145 countries. *Trop Med Int Health.* 19(8): 894–905. doi: 10.1111/tmi.12329. Epub 2014 Apr 30. Available at: http://onlinelibrary.wiley.com/doi/10.1111/tmi.12329/pdf.

Rzezutka, A and Cook, N. (2004) Survival of human enteric viruses in the environment and food. *FEMS Microbiol Rev.* 28(4): 441–453.

Sobsey, MD and Meschke, JS. (2003) Virus survival in the environment with special attention to survival in sewage droplet and other environmental media of fecal or respiratory origin. Draft, August 21, 2003. University of North Carolina. See: http://www.unc.edu/courses/2008spring/envr/421/001/WHO_VirusSurvivalReport_21Aug2003.pdf.

Tang, JW. (2009) The effect of environmental parameters on the survival of airborne infectious agents. *J. R. Soc. Interface* 6(Suppl 6): S737–S746.

Teunis, PFM et al. (1996) The dose-response relation in human volunteers for gastro-intestinal pathogens. Report no. 284550002, RIVM, Bilthoven. Available at: http://rivm.openrepository.com/rivm/bitstream/10029/9966/1/284550002.pdf

Teunis, PFM et al. (2008) Norwalk virus: how infectious is it? *J Med Virol.* 80(8): 1468–1476.

Ward RL et al. (1986) Human rotavirus studies in volunteers: determination of infectious dose and serological response to infection. *J Infect Dis.* 154(5): 871–880.

World Health Organization (2011) *Guidelines for Drinking-water Quality*, 4th edition. Available at: http://whqlibdoc.who.int/publications/2011/9789241548151_eng.pdf.

5

WATER-BASED DISEASE AND MICROBIAL GROWTH*

Charles P. Gerba

PROFESSOR OF ENVIRONMENTAL MICROBIOLOGY,
UNIVERSITY OF ARIZONA, TUCSON, AZ, USA

Gordon L. Nichols

CONSULTANT EPIDEMIOLOGIST, PUBLIC HEALTH ENGLAND,
UK; UNIVERSITY OF EAST ANGLIA, NORWICH, UK; UNIVERSITY
OF THESSALY, GREECE; UNIVERSITY OF EXETER, UK

Learning objectives

1 Understand the term "water-based diseases".
2 Recognize the important groups of water-based microorganisms and their diseases.
3 Describe interventions to control transmission of water-based diseases.

What are water-based diseases?

According to Bradley (1977) water-based diseases (WBDs) are caused by helminths that have a life cycle involving an aquatic invertebrate animal. These helminths may be ingested in water or food (vegetable or animal) or enter the body through the skin during recreational or occupational exposures to water. Many of these diseases are also associated with defective sanitation (American Water Works Association, 2006). There is also a range of protozoa, bacteria, dinoflagellates, diatoms and cyanobacteria that live in water and infect humans directly or through ingestion of contaminated shellfish. These microorganisms and the diseases they cause are also considered water-based in this chapter and are summarized in Table 5.1.

---

* Recommended citation: Gerba, C.P. and Nichols, G. 2015. 'Water-based disease and microbial growth', in Bartram, J., with Baum, R., Coclanis, P.A., Gute, D.M., Kay, D., McFadyen, S., Pond, K., Robertson, W. and Rouse, M.J. (eds) *Routledge Handbook of Water and Health*. London and New York: Routledge.

Table 5.1 Water-based diseases, those growing in water or parasites with a life cycle that depends on water or poor sanitation

Organism	Illness	Ecology	Cases associated with	Treatment/control	Classification
Legionella pneumophila Gram-negative, rod shaped bacterium	Pneumonia; Pontiac fever	Thermotolerant; grows at 22–45 °C	Cooling towers, hot tubs, warm water springs, shower heads, misters	More resistant than enteric bacteria to chlorine. Controlled by water temperature	Waterborne from environment
Pseudomonas spp. Gram-negative, rod shaped bacterium	Ear, eye, wound and respiratory infections	Common in biofilms; grows in distilled and reverse osmosis water	Swimming pools, whirlpools, hot tubs and contact lens solutions, drinking water taps	Very sensitive to common water disinfectants	Waterborne from environment
Aeromonas spp. Gram-negative, rod shaped bacterium	Wounds contaminated from dirty water	Commonly found in sewage and the environment	Bathing or other activity in water	None	Waterborne from environment
Burkholderia cepacia Gram-negative, rod shaped bacterium	Respiratory infection in people with cystic fibrosis	Can be acquired from water but some transmission is possible between patients	Hospital visits	Can grow in some disinfectant solutions. Cleaning and hygiene in hospitals	Waterborne from environment
Burkholderia pseudomallei Gram-negative, rod shaped bacterium	Melioidosis	Common in water and soil; Southeast Asia and Northern Australia are the areas in which it is primarily found	Inhalation of contaminated dust or water droplets, ingestion of contaminated water, and contact with contaminated soil, especially through skin abrasions	Sensitive to common water disinfectants	Waterborne from environment

Organism	Disease	Occurrence	Transmission	Disinfection	Source
Mycobacterium avium complex Acid fast, rod shaped bacterium	Pulmonary disease	Most common bacteria identified in biofilms in chlorinated drinking water distribution systems and fixtures (e.g. shower heads); thermotolerant	Inhalation of water droplets from drinking water and hot tubs	Very resistant to chlorine, chloramines and ultraviolet (UV) light disinfection	Waterborne from environment
Mycobacterium avium subsp. *paratuberculosis* (MAP) Acid fast, rod shaped bacterium	Linked to Crohn's disease	Found in soil and herbivorous animals where it causes Johne's disease	Presumed exposure through water, food and the environment	Resistant to disinfection and pasteurisation	Waterborne
Mycobacterium xenopi Acid fast, rod shaped bacterium	Respiratory disease in people with underlying lung damage	Can be isolated from piped water systems	Exposure through bathing in contaminated domestic water	Resistant to residual disinfectant concentrations	Waterborne from environment
Mycobacterium kansasii Acid fast, rod shaped bacterium	Furunculosis of the feet	Can colonize stagnant water and is found in contaminated footbaths	Feet contaminated through footbaths	Cleaning and disinfection	Waterborne from environment
Plesiomonas shigelloides	Diarrhea	Common in some tropical waters	Contaminated water or food	Sensitive to disinfectants	Waterborne from environment
Vibrio cholerae Gram-negative, comma shaped bacterium	Acute watery diarrhea	Human pathogen found in coastal waters	Contamination of water and seafood	Very sensitive to common disinfectants	Waterborne from environment or person to person
Vibrio vulnificus Gram-negative, comma shaped bacterium	Severe and life-threatening wound infections. More common in men with underlying liver disease	Native to both marine and fresh waters	Infection through eating seafood or infected wound	Sensitive to disinfectants	Waterborne from environment
Vibrio spp. Gram-negative, comma rod shaped bacterium	Gastroenteritis, skin infections	Native to both marine and fresh waters; commonly associated with marine animals including copepods, shellfish and crustaceans	Drinking and bathing water	Very sensitive to all common disinfectants	Waterborne from environment

continued…

Table 5.1 continued

Organism	Illness	Ecology	Cases associated with	Treatment/control	Classification
Alexandrium spp. Dinoflagellate	Paralytic shellfish poisoning	Blooms in coastal waters	Eating shellfish	Monitoring of shellfish beds for toxin	Foodborne from environment
Dinophysis spp. and *Procentrum* spp. Dinoflagellates	Diarrheic shellfish poisoning	Blooms in coastal waters	Eating shellfish	Monitoring of shellfish beds for toxin	Foodborne from environment
Karenia spp. Dinoflagellate	Neurotoxic shellfish poisoning	Blooms in coastal waters	Eating shellfish	Monitoring of shellfish beds for toxin	Foodborne from environment
Pseudo-nitzschia spp. Diatom	Amnesic shellfish poisoning	Blooms in coastal waters	Eating shellfish	Monitoring of shellfish beds for toxin	Foodborne from environment
Protoperidinium crassipes Dinoflagellate	Azaspiracid shellfish poisoning	Blooms in coastal waters	Eating shellfish	Monitoring of shellfish beds for toxin	Foodborne from environment
Gambierdiscus spp. Dinoflagellate	Ciguatera food poisoning	Blooms in coastal waters	Eating reef fish	Avoiding consumption of reef fish	Foodborne from environment
Ostreopsis spp. Dinoflagellate	Clupeotoxism and palytoxin poisoning	Blooms in coastal waters	Eating contaminated crabs/ fish; breathing sea spray	Monitoring coastal waters	Foodborne or respiratory from environment
Cylindrospermopsis raciborskii Cyanobacterium	Cause of Palm Island mystery disease (malaise, anorexia, vomiting, headache, painful liver enlargement, bloody diarrhea and kidney damage)	Blooms in fresh water	Toxic through (in increasing order of severity) bathing, drinking and dialysis	Outbreak linked to use of copper sulphate to control bloom, which released toxin	Waterborne from environment
Microcystis aeruginosa Cyanobacterium	Hepatic failure, abdominal pain, diarrhea, vomiting, liver cancer	Blooms in fresh eutrophic waters	Toxic through (in increasing order of severity) bathing, drinking and dialysis	Notice explaining risks for inland waters	Waterborne from environment

Organism	Disease/symptom	Notes	Exposure	Control	Classification
Anabena spp. Cyanobacterium	Abdominal pain, diarrhea, vomiting	Blooms in fresh water	Potentially toxic through (in increasing order of severity) bathing, drinking and dialysis	Notice explaining risks for inland waters	Waterborne from environment
Oscillatoria spp. Cyanobacterium	Liver cancer link	Blooms in fresh water	Toxic through (in increasing order of severity) bathing, drinking and dialysis	Notice explaining risks for inland waters	Waterborne from environment
Lyngbya majuscule Cyanobacterium	Swimmer's itch (dermatitis)	A marine alga producing toxins	Recreational skin exposure	Local risk notices	Waterborne from environment
Naegleria fowleri protozoan	Primary amoebic meningoencepahlitis or PAM	Occurs in water above 35 °C; thermotolerant	Warm surface water, hot springs, drinking water	Fairly resistant to chlorine and UV light disinfection. Monitoring thermal waters	Waterborne from environment
Acanthamoeba protozoan	Eye infections among contact lens wearers primarily	Common in surface waters and distribution systems	Tap water, surface waters	Very resistant to chlorine and UV light disinfection. Monitoring thermal waters	Waterborne from environment
Dracunculus medinensis Guinea worm	Dracunculiasis	Infection occurs by ingestion of water fleas containing the organism	Unfiltered surface waters	Easily controlled by filtration of drinking water and algicides. WHO eradication plan	Water-based
Angiostrongylus spp.	Eosinophilic meningitis	Life cycle involves rats and snails	Eating uncooked snails, slugs, crabs or freshwater shrimp	Avoid uncooked risky foods. Improved sanitation	Water-based/defective sanitation/foodborne
Anisakis spp. and *Pseudoterranova* spp.	Painful stomach infection	Life cycle involves seals and fish	Infection through eating raw fish	Avoid uncooked risky foods. Freezing raw fish	Water-based/defective sanitation/foodborne
Ascaris lumbricoides	Worm infection with intestinal blockage	Life cycle requires environmental maturation of ova in faeces for a week	Contamination of water and food	Improved hygiene. Improved sanitation	Defective sanitation/foodborne

continued…

Table 5.1 continued

Organism	Illness	Ecology	Cases associated with	Treatment/control	Classification
Capillaria philippinensis	Chronic diarrhea, borborygmi, bipedal edema, anorexia and weight loss	Life cycle involves fish and birds/humans	Infection through raw fish or shellfish	Avoid uncooked risky foods. Improved sanitation	Water-based/defective sanitation/foodborne
Gnathostoma spp.	A creeping eruption and Quincke's edema (slowly migrating erythema with pruritus)	Life cycle involves water flea (Cyclops), fish or frogs and carnivorous animals	Infection through raw fish or frogs	Avoid uncooked risky foods. Improved sanitation	Water-based/defective sanitation/foodborne
Trichuris trichiura	Anaemia and diarrhea	Life cycle requires environmental maturation of ova in faeces for a week	Contaminated water and food	Improved hygiene. Improved sanitation	Defective sanitation/ foodborne
Diphyllobothrium spp. and *Nanophyetus salmincola*	Diarrhea, epigastric pain, nausea and vomiting	Life cycle involves water flea (Cyclops) and salmonid fish	Raw fish	Avoid uncooked risky foods. Improved sanitation	Water-based/defective sanitation/foodborne
Echinococcus spp.	Hydatid disease in humans	Life cycle involves dogs and ruminants	Contamination of water, or directly from dogs	Avoid uncooked risky foods. Improved sanitation	Water-based/defective sanitation/foodborne
Echinostoma spp. Fluke (digenetic trematode)	Abdominal pain, diarrhea, weight loss and tiredness	Life cycle involves snails and fish	Consumption of raw snails, clams or fish	Avoid uncooked risky foods. Improved sanitation	Water-based/defective sanitation/foodborne
Heterophyes heterophyes Fluke (digenetic trematode)	Diarrhea and abdominal pain	Life cycle through snail and estuarine fish	Raw fish	Avoid uncooked risky foods. Improved sanitation	Water-based/defective sanitation/foodborne
Opisthorchis sinensis Fluke (digenetic trematode)	Abdominal pain, nausea, fatigue, abdominal discomfort, anorexia, weight loss, jaundice	Life cycle through snail and estuarine fish	Raw fish	Avoid uncooked risky foods. Improved sanitation	Water-based/defective sanitation/foodborne
Paragonimous westermani Fluke (digenetic trematode)	Bad cough, bronchitis and blood in sputum (hemoptysis)	Life cycle involves snails, crabs and pigs/dogs/cats	Uncooked crabs/crayfish and occasionally boar meat	Avoid uncooked risky foods. Improved sanitation	Water-based/defective sanitation/foodborne

Organism	Clinical features	Life cycle	Transmission	Prevention/control	Category
Schistosoma spp. Fluke (digenetic trematode)	Colonic polyposis, bloody diarrhea (Schistosoma mansoni), hypertension, splenomegaly (S. mansoni, S. japonicum), cystitis, ureteritis, hematuria, glomerulonephritis, bladder cancer (S. haematobium), pulmonary hypertension (S. mansoni, S. japonicum, more rarely S. haematobium)	Life cycle involves snails and humans	Recreational/work activity in water	Controlled through chemicals affecting snails. Avoid risky water contact. Periodic mass treatment. Improved sanitation	Water-based
Spirometra spp. Tapeworm	Eye or cerebral sparganosis	Life cycle involves copepods, fish/amphibians/reptiles and cats and dogs	Humans are infected by consuming water containing infected copepods, or uncooked frogs/snakes	Avoid uncooked risky foods. Filter water. Improved sanitation	Water-based

Helminths that have a life cycle involving water

Helminths are parasitic worms that cause disease and illness in humans. Three important water-based helminths are the nematode (or roundworm) Guinea worm (*Dracunculus medinensis*) which causes dracunculiasis, the trematodes (or flukes) which includes schistosomiasis (*Schistosoma* spp.) and the cestode (or tapeworm) *Spirometra* responsible for sparganosis. In Bradley's original definition, all of these are organisms that are transmitted through water. However, a number of related parasites have an aquatic life cycle but are transmitted through infected food. They are therefore covered separately. All helminths regardless of source can be controlled through proper personal hygiene and sewage disposal practices.

Dracunculus medinensisis

Dracunculus medinensisis is a helminth that has unequivocally been shown to be transmitted by drinking water only (Percival et al., 2004). This nematode worm (adult female one meter long) is transmitted when people drink water containing larvae-infected water fleas (copepods). The copepods quickly die in the stomach but the larvae are liberated. They then penetrate the wall of the intestine and migrate through the body. The female worm migrates under the skin tissue to the lower limbs, forming a blister from which it eventually emerges. Millions of immature worms are then released into the water to infect more copepods.

The infection is rarely fatal but individuals become non-functional for months, causing crop losses, and social and financial problems. Infection confers little protective immunity and people can suffer repeated bouts of disease. Transmission can be prevented by simple filtration of the drinking water; even passage of water through common fabrics is sufficient. Because of major efforts towards its worldwide eradication through the World Health Organization (WHO) and the Carter Center the incidence of dracunculiasis has decreased from an estimated 3.5 million cases, largely in Africa in 1986, to 126 reported cases in 2014 (WHO, 2015a). The eradication programme includes: control of the copepod vectors using a chemical larvicide; provision of safe drinking water through filtration to remove copepods; protective walls around wells to prevent drinking water contamination from infected people; robust surveillance for the disease to identify all cases; treatment and containment in affected communities; and, health education and community mobilization to prevent infections. In 2014 new cases were reported from four endemic countries – Chad, Mali, South Sudan and Ethiopia. Eradication can be difficult in countries in conflict zones, and improved security can result in new cases being reported that were previously undetected. While efforts to eliminate remaining new cases can be onerous there is a strong possibility that within the next few years this parasite will be eradicated worldwide (WHO 2014a), which will make it the only infectious disease to be eradicated apart from smallpox.

Schistosoma spp.

Schistosomes are parasitic flukes. They can lead to chronic ill health and are the most important of the water-based diseases with over 200,000 deaths per year. Forty million cases were treated for schistosomiasis and 261 million people required preventive treatment in 2013 (WHO, 2015b). There are 26 *Schistosoma* species, six of which can infect humans (*S. intercalatum, S. haematobium, S. mansoni, S. guineensis, S. japonicum* and *S. mekongi*). In their life cycle the ova hatch into ciliated miracidia in the water and migrate to certain species of snails, which they infect. The parasite grows in the snails and liberates cercariae that can swim in water and attach and burrow into people's skin. These lose their tails and become schistosomulae that migrate through various tissues, ending up as adult worms in mesenteric veins. The worms mate and produce large numbers of eggs, some of which pass into the intestine or bladder. Much of the long-term disease is attributable to the body's immune response to these invading parasites. Following a few days after exposure a skin rash or itchy skin develops, followed after a month or two by fever, chills, cough and muscle aches. The chronic symptoms include abdominal pain, enlarged liver, blood in faeces or urine and problems passing urine, while chronic infection can also increase risks of bladder infections (Centers for Disease Control and Prevention, 2012).

Infection with schistosomes can be common locally in places with poor sanitation. Those at risk are children who swim and play in contaminated fresh water and adults who are commonly exposed to such water through their occupations. While the parasites cannot infect the gastrointestinal tract the cercaria can penetrate the mucous membranes of the mouth or throat. Intervention can be through mass treatment of the at risk population with Praziquantel, treatment to eliminate snail populations, through improved sanitation or through reducing exposure to at risk waters. The parasites thrive in slow flowing water and are strongly associated with irrigation.

Some avian species (e.g. *Schistosoma spindale* and *Trichobilharzia regent*) can give rise to itchy red spots (known as cercarial dermatitis or swimmer's itch) in people who have had contact with contaminated water. They are otherwise thought to be non-pathogenic to humans.

Spirometra spp.

Spirometra spp. are tapeworms that occur in dogs and cats and shed eggs which can pass into water and infect copepods. When infected copepods are consumed by a secondary host such as amphibious animals including frogs, the procercoid (first stage) larvae develop into plerocercoid (second stage) larvae. When these are consumed by a cat or dog the larvae develop into a tapeworm again. In humans, ingestion of the first stage larvae which are present in the copepod Cyclops results in human sparganosis, with "Sparganum" larvae forming in nodules. Infection is thought to derive from contaminated drinking water or consuming undercooked or uncooked fish, amphibians or reptiles. *S. mansonoides* is the organism most commonly diagnosed. The sparganum can migrate to subcutaneous tissue, the eye orbit, breast tissue, inner ear, lungs or pleural cavity, urinary tract, abdominal cavity and the brain and can live for up to twenty years. The tapeworm phase will not form in humans. Although the disease can occur worldwide, most infections occur in the Far East.

Helminths that have a life cycle involving water and food

A variety of helminthic infections can be transmitted through the water cycle, but this transmission is usually as a result of the parasite life cycle involving the consumption of raw produce or fish, amphibians or reptiles in an uncooked or undercooked state (WHO, 2001, 2014c, 2014d). These organisms are regarded as agents of WBD because they have life cycles that are dependent on water, although the predominant final transmission to humans is through food rather than water. Many are also included in "Diseases of defective sanitation" in the Bradley classification (Bradley, 1977). Transmission is more common in communities living in close proximity to water and where sanitation is rudimentary. A few of the organisms described here (*Ascaris*, *Trichuris* and *Echinococcus*) may be transmitted by drinking contaminated water or food but do not have a life cycle that involves intermediate hosts that live in water.

Nematodes

Angiostrongylus spp.

Angiostrongylus spp. are roundworms and *A. cantonensis* (the rat lungworm) is the most common cause of human eosinophilic meningitis. Adult *A. cantonensis* worms live in the pulmonary arteries of rats. First stage larvae are passed in the rat faeces and are ingested by an intermediate host (snail or slug). The larvae molt to form third stage larvae that are infective to mammalian hosts. Humans can acquire the infection by eating raw or undercooked snails or slugs infected with the parasite or by eating raw produce that contains a small snail or slug, or by eating infected crabs or freshwater shrimps that are infected with larvae. In humans, juvenile worms migrate to the brain, or rarely to the lungs, where the worms ultimately die. *A. costaricensis* (syn. *Parastrongylus costaricensis*) causes intestinal angiostrongyliasis and the life cycle is similar to that of *A. cantonensis*. However, *A. costaricensis* can reach sexual maturity in humans and can release eggs into the intestinal tissues where there is an intense local inflammatory reaction to degenerating eggs and larvae. Ova and larvae are not shed in faeces. People with *Angiostrongylus* spp. have usually acquired the infection in a developing country.

Anisakis spp. and Pseudoterranova spp.

Anisakis spp. and *Pseudoterranova* spp. are roundworms (helminths) which infect marine mammals, fish and crustaceans. Adult worms live in the stomach of marine mammals and eggs from these worms pass into the sea where they become embryonated and develop into second stage larvae. They then hatch and become free living. They are taken up by crustaceans and develop into third stage larvae which become infectious to fish and squid. When the fish die the larvae migrate to muscle tissue and can pass through predation from fish to fish. Infective third stage larvae develop to an adult stage in pinniped mammals (e.g. seals). Humans are an accidental host for third or fourth stage larvae through the consumption of raw fish, and freezing the fish is thought to prevent infection. If humans eat fish contaminated with third stage larvae these molt twice and develop into adult worms, penetrating into the oesophagus, gastric or intestinal mucosa and causing abdominal pain, nausea, vomiting, abdominal distension, blood in the stools and mild fever. Worms and eggs are not known to be excreted in human faeces and diagnosis is by gastroscopy. Treatment is by endoscopic removal of the worms.

Ascaris lumbricoides

Ascaris lumbricoides is the largest and most ubiquitous roundworm infecting humans. It is most common in the Far East. Fertilized ova in contaminated food or water are ingested, transformed into larval worms and enter the bloodstream through the duodenal wall. From there they migrate to the respiratory system, are coughed up, swallowed and returned to the small intestine where they mature into adult male and female worms. Females pass several thousand ova per day in faeces. They remain non-infectious until they have matured for two weeks in the environment (usually in "night soil" – composted human faeces). Infected people are often asymptomatic. When symptoms do occur they include bloody sputum, cough, fever and abdominal discomfort.

Capillaria philippinensis

Capillaria philippinensis cause intestinal infections with symptoms of chronic diarrhea, borborygmi (audible abdominal sounds caused by increased intestinal activity), bipedal edema (fluid accumulation and swelling in both feet), anorexia and weight loss. Infection is transmitted where people eat raw fish, shrimps, crabs and snails. The life cycle of *C. philippinensis* has developed through people defecating in fields or in the same body of water where they gather fish and shellfish for eating. The disease is restricted to particular geographic areas, but is common in some parts of the Philippines and Thailand.

Gnathostoma spp.

Gnathostoma spp. are nematode worms (helminths) that have a life cycle involving the water flea Cyclops, fish or frogs and carnivorous animals. Humans are usually infected through consuming infected fish or frogs in an uncooked state. Gnathostomiasis is rarely reported in travelers, although the disease remains a major public health problem in Southeast Asia. The disease presents as a creeping eruption and Quincke's edema (slowly migrating erythema with pruritus). Six species are thought to cause human gnathostomiasis: *G. hispidum*, *G. binucleatum*, *G. doloresi*, *G. nipponicum*, *G. malaysiae* and *G. spinigerum*, which is the species most commonly infecting humans.

Trichuris trichiura

Trichuris trichiura is an intestinal nematode commonly called whipworm. It is estimated there are some 700 million people infected with the parasite. Infective eggs are ingested, hatch and mature in the small intestine. Female *Trichuris* produce several thousand eggs per day. Immature eggs are passed from faeces to soil, become infective after a brief period of time and contaminate food or drinking water, although waterborne outbreaks have not been recorded. People with minor infections are often asymptomatic while those with severe infections can experience frequent and painful passage of stools containing mucous, water and blood. Infected children can become severely anemic.

Trematodes

Nanophyetus salmincola

Nanophyetus salmincola is a digenetic trematode. Human nanophyetiasis is a zoonotic disease that occurs in the coastal United States Pacific Northwest. The disease causes gastrointestinal complaints and eosinophilia (an abnormal increase in the number of eosinophils, a specific type of white blood cell). The eggs are difficult to distinguish from *Diphyllobothrium latum*. Nanophyetiasis is transmitted by the larval stage (metacercaria) that encysts in the flesh of freshwater fishes. The infection has been associated with the ingestion of incompletely cooked or home-smoked salmon or steelhead trout. With salmon and trout the parasite's cysts can survive the period spent at sea.

Echinostoma spp.

Echinostoma spp. are intestinal flukes (trematodes) and include over 100 species. *E. revolutum*, *E. malaynum*, *E. hortense* and *E. echinatum* are the species usually causing human infection. Infections are endemic in the Far East and Southeast Asia. They can live in the duodenum, causing abdominal pain, diarrhea, weight loss and tiredness. The life cycle includes a snail as the first intermediate host, usually followed by a fish as the second intermediate host (in some species this may be an amphibian or mollusc). Infection is thought to result from the consumption of raw snails, clams or fish.

Fasciola and Fasciolopsis spp.

Fasciola and *Fasciolopsis* are liver flukes common in herbivores. *F. hepatica* occurs worldwide in sheep and cattle that graze in wet pastures. *Fasciola gigantica* is found in cattle in the Middle East, Africa and Asia, and *Fasciolopsis buski* is common in pigs in areas of Indonesia and Northern Thailand and the Far East. These parasites require a snail as an intermediate host, and humans are infected through the consumption of aquatic plants contaminated with the metacercaria. For *F. hepatica* this has commonly been watercress, whereas with *F. buski* the aquatic plants are water chestnuts or water caltrop.

Other trematodes

Heterophyes heterophyes is a small fluke (digenetic trematode) that lives in the small intestine of humans, cats, foxes and dogs. The parasite causes irritation of the small intestine, resulting in diarrhea and abdominal pain. Two intermediate hosts are required for the life cycle, the first being a snail and the second being an estuarine fish. People are infected by eating raw or undercooked fish. *H. heterophyes* occurs in parts of Africa, Southeast Asia and Hawaii.

Opisthorchis spp. is a liver fluke (helminth) that is common in Southeast Asia and some parts of the former Soviet Union. The life cycle of all three species involves snails and fish, and people are infected following the consumption of uncooked fish.

Paragonimous spp. is a lung fluke (helminth) that is common in Southeast Asia, with infections occurring in Africa and Latin America and, rarely, in the United States. The life cycle of all species involves snails and crustaceans. People are infected following the consumption of uncooked crabs and occasionally uncooked wild boar meat.

Cestodes

Diphyllobothrium spp.

Diphyllobothrium spp. are fish tapeworms whose life cycle also involves the water flea Cyclops. Although many carriers are asymptomatic, overt clinical manifestations of diphyllobothriasis can include diarrhea, epigastric pain, nausea and vomiting. Humans acquire the tapeworm through the consumption of raw fish. Although fish tapeworm infections in Arctic and subarctic residents are often attributed to *Diphyllobothrium latum*, twelve other species infect humans. *Diphyllobothrium* spp. are implicated in human infections in Scandinavia, western Russia, the Baltic and the Pacific Northwest, throughout the circumpolar area and in northern communities bordering the Pacific. Larvae of *D. dendriticum* occur predominantly in salmonid fishes (e.g. Arctic char, salmon, trout, whitefish). The usual intermediate hosts of *D. latum* are pike and perch, but only rarely salmonids. *D. ursi* and *D. klebanovskii* occur predominantly in Pacific salmon, and *D. dalliae* in Alaskan blackfish. *Diphyllobothrium* ova are excreted in the faeces of infected people.

Echinococcus spp.

Echinococcus spp. cause hydatid disease in humans, with single (*E. granulosus*) or multiple (*E. multilocularis*) cyst formation. Humans become infected by ingesting eggs, with resulting release of an oncosphere in the intestine that penetrates the gut wall and migrates to various organs where it develops into a cyst. Infection results from consumption of *Echinococcus* ova through the consumption of ova in contaminated soil, drinking water or food contaminated with tapeworm eggs. There are four known species, including the two main species *E. granulosus* and *E. multilocularis*, and two rarer species, *E. vogeli* and *E. oligarthrus*. *E. granulosus* has a number of subspecies. The adult forms of most *E. granulosus* subspecies occur in most canids (dogs) and form hydatids in wild ruminants, sheep, pigs, cattle and humans. *Echinococcus multilocularis* has a number of subspecies in Europe and in North America. Foxes and to a lesser extent dogs and cats are the definitive hosts and small rodents are the intermediate hosts. *Echinococcus vogeli* is found in South America, with bush dogs the definitive hosts and rodents are the normal intermediate host. The cysts resemble *E. granulosus*, but often become septate, forming multi-chambered cysts (i.e. polycystic hydatids). *Echinococcus oligarthrus* has a life cycle involving wild cats as the definitive hosts and rodents as intermediate hosts, but human infection is rare.

Free-living microorganisms that cause water-based diseases

There are a number of free-living microorganisms that grow in natural waters or in built systems (i.e. plumbing). These do not follow the traditional faecal–oral transmission routes (see Chapter 4) and usually cause disease only when they grow to larger numbers than normal. These can include protozoa, bacteria, cyanobacteria, dinoflagellates and diatoms. Some of these pathogens reside within biofilms, some inside protozoa and other aquatic organisms, and some result from blooms within the body of water as a result of physical and chemical conditions.

Free-living protozoan pathogens that grow in water

Naegleria fowleri

Naegleria fowleri occurs in low numbers in surface waters; however, when the water temperature exceeds 35 °C (hot springs, lakes, groundwater) it can grow to large numbers (Percival et al., 2004). It is an amoeboflagellate, changing forms between cysts, amoeba and a free-swimming flagellate. Infection occurs by the flagellate form entering the nose (usually during head immersion while swimming) where it sheds its flagella. The amoeba form follows the nerves to the brain, where it begins to destroy the brain tissue, causing primary amoebic meningoencephalitis (PAM). Death usually follows within four to six days with 98 per cent mortality. It usually infects young children and young adults, although infant cases have been reported in India. Cases have been reported in Australia and the United States related to the use of treated tap water for nasal cleaning. The illness is rare, with only a little over 400 cases reported worldwide. The organism has similar resistance to free chlorine as *Giardia* cysts, but is much more resistant to UV light disinfection.

Acanthamoeba spp.

This genus of amoeba is ubiquitous in natural and artificial aquatic environments including tap water and swimming pool waters. In the developed world it is most commonly associated with eye infections linked to contact lens stored in unsterile water or tap water. If untreated *Acanthamoeba* eye infections can lead to permanent blindness. The cyst of the organisms are highly resistant to environmental stresses including drying, freezing, UV light and free chlorine up 50 mg/L. It is commonly found in chlorinated tap water where it grazes on the bacteria in biofilms. Free-living protozoa such as *Acanthamoeba* can support the intracellular growth of bacteria pathogenic to humans (e.g. *Legionella*), and can affect their survival, resistance to disinfection, and even increased virulence following passage through protozoa (Nwachuku and Gerba, 2004).

Free-living bacterial pathogens in natural waters or built systems

Legionella spp.

At least nineteen of the fifty-four species of *Legionella* have been shown to cause legionellosis (Lee and Nichols, 2010), but the species most commonly associated with illness is *L. pneumophila*, which was first recognized after an outbreak in 1976 in Philadelphia. It is responsible for a serious life-threatening pneumonia, especially in older individuals. The fatality rate may be as high as 25 per cent. A milder form, referred to as Pontiac fever, causes a low grade fever with recovery in two to five days. The disease is found worldwide with most cases occurring in the summer and fall. It also appears to occur more frequently after flooding and periods of high rainfall (Garcia-Vidal et al., 2013). It is transmitted by inhalation of aerosols from contaminated water or compost.

Almost any source of warm water is a likely source of *Legionella*. Outbreaks have been associated with cooling towers, warm groundwater, hot water heaters, shower heads, hot tubs, humidifiers, ornamental fountains, thermal springs, misters, etc. Legionellae are more resistant to chlorine than *Escherichia coli*, and can grow in waters at temperatures between 25 and 45 °C. It can be controlled in hot water systems by raising the temperature of hot water heaters to over 60 °C and to 50 °C at outlets combined with regular flushing of the distribution system. *Legionella* has been shown to be present in drinking water systems even when exposed to 0.75 mg/L of free chlorine, as it commonly resides within a protective microbial biofilm and can be protected by its intracellular association with protozoa such as *Acanthamoeba* spp. (Buse et al., 2012; Thomas and Ashbolt, 2011). *Legionella* is addressed specifically in Chapter 8.

Mycobacterium avium complex

Mycobacteria are rod-shaped bacteria that contain high levels of lipid (waxy) material. The role of water in mycobacterial diseases other than tuberculosis was reviewed by the World Health Organization in 2004 (Pedley et al., 2004). They are among the most resistant non-spore-forming bacteria to chlorine and other common water disinfectants. Environmental mycobacteria are usually removed by standard drinking water treatment processes, but generally show more resistance to chlorination than other bacteria (Nichols et al., 2004). Standard water microbiology methods do not normally recover many of these species because they grow slowly. Specific methods are available for the isolation and enumeration of environmental *Mycobacterium* spp., including decontamination and the use of selective media. They can cause infections through the contamination of hospital equipment such as endoscope washers, where water can stagnate.

The *Mycobacterium avium* complex (MAC) consists of at least twenty-eight serovars of two distinct species, *M. avium* and *M. intracellulare*. Pulmonary disease caused by MAC has dramatically increased in the United States over the last three decades (Kendall and Winthrop, 2013). The disease is most common among individuals over sixty years of age. Predisposing factors include age, chronic lung disease, bronchogenic carcinoma and AIDS (von Reyn et al., 2004). MAC can also cause pulmonary disease, osteomyelitis and septic arthritis in people with no known predisposing factors. Diseases caused by MAC can be life-threatening and infections are difficult to treat because of resistance to many antimycobacterial agents. It is believed that municipal drinking water systems are an important reservoir for MAC. They are among the most common bacteria identified in biofilms in chlorinated drinking water distribution systems and fixtures (e.g. shower heads) (Falkinham et al., 2004). Epidemiological investigations have associated water sources with infections by atypical mycobacteria. These bacteria can multiply in water that is essentially free of nutrients and they are relatively resistant to disinfection by chlorination, chloramines and UV light.

Mycobacteria can be commonly isolated from raw water samples and have also been found in up to 50 per cent of municipal and private drinking water samples. The greater occurrence of mycobacteria in disinfected water distribution systems is probably related to their greater resistance to disinfectants than other vegetative bacteria. Hospital water systems often harbor MAC and these may be a source of nosocomial infection. *M. avium* has primarily been associated with hot water systems, which in some cases may be persistently colonized by the same strain of *M. avium*.

Mycobacterium avium subsp. *paratuberculosis* (MAP) is a very slow growing bacterium that causes Johne's disease in agricultural animals and has been implicated in the causation of Crohn's disease in humans. MAP has been detected in water using polymerase chain reaction (PCR) but does not appear to grow in potable water.

Other mycobacterium

Mycobacterium xenopi is found in water and soil and can act as an opportunist pathogen. It can colonize piped water supplies, particularly warm water systems. People are probably quite commonly exposed to low numbers of this organism as a result of drinking, washing, showering and inhalation of natural aerosols. *M. xenopi* is a cause of pulmonary disease, particularly in people with underlying lung damage or immune deficiencies, although a substantial minority of cases occur in persons with no apparent underlying disorder. The organisms are of low virulence and although people are frequently infected, overt disease is uncommon except in the immunocompromised. *M. xenopi* infection can be falsely diagnosed because clinical specimens (e.g. sputum and urine) may become contaminated from water before or after specimen collection. *Mycobacterium kansasii* is a slow-growing respiratory pathogen that normally exists in water and may act as an opportunist pathogen. *M. kansasii* occurs most commonly in older men who have an underlying respiratory disease. The exact routes of transmission of this pathogen are unclear but appear to be from contaminated water including drinking water. It occurs in water distribution systems where it can survive within biofilms and it is relatively resistant to chlorine. *Mycobacterium fortuitum* is a rapid-growing mycobacteria that can colonize footbaths and infect the skin and nails of people using these, causing recurrent furunculosis (Pedley et al., 2004).

Pseudomonas aeruginosa

This genus of Gram-negative bacteria are important opportunistic pathogens in immunocompromised persons but are also capable of causing infections in healthy individuals. *P. aeruginosa* is the species most commonly associated with ear, eye, skin, wound and respiratory infections. Infections, particularly folliculitis (infection of the hair follicles), can be associated with swimming pools, whirlpools, hot tubs and contact lens solutions (Mena and Gerba, 2009), but ear infections can also occur in people swimming in natural waters. It is even capable of growth to large numbers in distilled or reverse-osmosis treated water. In hospitals transmission is related to their growth in the drinking water taps and generation of aerosols. It can be a particular problem in neonatal intensive care. Although *P. aeruginosa* can be found in drinking water it is not ubiquitous. It is sensitive to common water disinfectants and only occurs in those regions of a water distribution system where residual disinfectant concentrations are low or non-existent.

Aeromonas spp.

Aeromonas spp. are common within sewage, can cause wound infections, and have been thought to be a cause of diarrhea. Point source diarrhea outbreaks do not occur and evidence of their role in diarrheal disease is equivocal. They are commonly found in raw foods, sewage and source waters, and can bloom in drinking water distribution systems. A study of infectious intestinal diseases undertaken in the 1990s found *Aeromonas* spp. as commonly in people without diarrhea as in those experiencing symptoms (American Water Works Association, 2006).

Burkholderia cepacia

B. cepacia is a species with nine subspecies, all of which inhabit the rhizosphere. Some subspecies cause pulmonary infections in cystic fibrosis (CF) patients. The organisms are widespread in the natural environment in agricultural soil and water. Infection in non-CF patients is usually in the hospital setting, through exposure to contaminated water reservoirs or disinfectant solutions. It is rarely isolated from potable water and is readily killed by hot water (65 °C).

Burkholderia pseudomallei

Melioidosis is an infectious disease of humans and animals caused by *B. pseudomallei* found in soil and water. Chronic melioidosis is characterized by infection of the bone (osteomyelitis), respiratory tract and pus-filled abscesses in the skin or other organs. Acute melioidosis takes one of three forms: a localized skin infection; an infection of the lungs associated with fever and coughing; and septicemia (growth of the organisms in the bloodstream). Humans and animals are believed to acquire the infection by inhalation of contaminated dust or water droplets, ingestion of contaminated water, or contact with contaminated soil, especially through skin abrasions. Persons working in rice paddy fields are at higher risk. While melioidosis infections have been reported all over the world, Southeast Asia and Northern Australia are the areas in which it is primarily found. There is a clear association with increased rainfall: the number (and severity) of cases increased following heavy precipitation. Three outbreaks have been linked to contamination of the drinking water supply, where disease control measures, such as cleaning of the pipes and disinfection, led to a cessation of the outbreaks (McRobb et al., 2013).

Plesiomonas shigelloides

Plesiomonas shigelloides is a bacterial pathogen that has been implicated as a cause of diarrheal disease. It is common in the natural waters in some tropical countries. Because *P. shigelloides* has a similar cell wall lipopolysaccharide antigen to *Shigella sonnei* it has been suggested that some natural immunity to *Shigella* may be conferred by being exposed to *P. shigelloides*.

Vibrio spp.

Vibrio spp. are curve-shaped bacteria that can be found in marine, estuarine and river water. A variety of *Vibrio* spp. can cause human disease, including the halophilic *V. parahaemolyticus*, *V. vulnificus*, *V. fluvialis*, *V. hollisae* and the non-halophilic vibrios non-O1 *V. cholerae* and *V. mimicus*. These organisms proliferate in coastal waters during the summer months in response to warm water temperatures and lowered salinity, and *Vibrio* infection risks can be modeled based on these parameters (Nichols et al., 2014). People are infected through the consumption of raw or undercooked contaminated shellfish, other foods and faecally contaminated water. *Vibrio parahaemolyticus* inhabits estuarine and marine environments and can cause food poisoning through the contamination of seafood. *Vibrio vulnificus* can cause severe, soft tissue infections or septicaemia, or both, rather than diarrhea. While *Vibrio vulnificus* is often transmitted by eating sea foods it can be transmitted by open wounds during swimming or wading in marine or coastal waters (Oliver, 2005). It rapidly causes degradation of tissues and bacteremia. Infections are usually in people, more commonly men, who are immunocompromised or who have underlying liver disease. Skin infections have been associated with bathing in both marine and fresh waters. Mortality rates approach 50 per cent.

Vibrio cholerae

Cholera is a waterborne disease, passing through a faecal–oral route and causing large waterborne outbreaks, and the water route is still important in developing countries. Cholera is a disease that is characterized by acute and life-threatening diarrhea and dehydration usually in epidemic outbreaks. Cholera is transmitted through drinking water, shellfish and contaminated food. The disease occurs in pandemics, with particular strains of *V. cholerae* passing round the world through infected people and contaminated ballast water in ships. There have been seven pandemics, the most recent of which began in Indonesia in 1961 and caused a large outbreak in Peru. The disease is usually restricted to less developed countries where drinking water and waste disposal are poor, and to migrant populations associated with drought, famine and war.

   V. cholerae is commonly associated with marine animals including copepods, shellfish and crustaceans and can be transmitted by marine foods. *V. cholerae* can occur naturally in marine or coastal waters. They prefer warmer temperatures, reduced salinity and in combination with elevated pH and plankton blooms rapid multiplication of vibrios occur in coastal and estuarine aquatic environments. Serologic studies have resulted in the identification of over 200 different serological O groups within the species *V. cholerae*. In general, strains outside these serogroups (commonly referred to as "non-O1/non-O139 *V. cholerae*") are non-pathogenic or asymptomatic colonizers in humans, or cause mild, sporadic illness (such as gastroenteritis, mild wound or ear infections) in otherwise healthy hosts.

Dinoflagellates and diatoms that bloom in water and produce toxins

These organisms grow in marine, coastal and estuarine waters and produce a range of potent toxins, often in mixtures. They occur predominantly in saltwater and, under the right conditions, can produce harmful algal blooms (HABs) that can cause toxic effects in fish and other sea-life as well as humans. The organisms and their potent natural toxins can accumulate within shellfish, causing paralytic shellfish poisoning (PSP; caused by *Alexandrium* spp.), diarrheic shellfish poisoning (DSP; caused by *Dinophysis* spp. and *Procentrum* spp.), neurotoxic shellfish poisoning (NSP; caused by *Karenia* spp.), azaspiracid shellfish poisoning (AZP; caused by *Protoperidinium crassipes*) and amnesic shellfish poisoning (ASP; caused by *Pseudo-nitzschia* spp.). Some of the toxins (produced by *Gambierdiscus* spp. and *Gonyaulax tamarense*) can also accumulate by passing up the food chain to contaminate large reef fish (e.g. barracuda) that are rendered toxic (e.g. ciguatera toxins and brevitoxins). Some of these toxins (e.g. brevetoxins associated with the marine dinoflagellate, *Karenia brevis*; and other toxins with *Ostreopsis ovata*) have been shown to cause respiratory symptoms through exposure to aerosols of seawater, particularly in asthmatics (see Table 5.1).

Other dinoflagellates

A number of related organisms may be implicated in disease but the evidence is poor. *Pfiesteria piscicida* is a dinoflagellate that resides in estuarine waters and has been linked to mass fish kills. The initial links with skin and neurological problems in exposed humans have not been corroborated.

Cyanobacteria

Cyanobacteria, photosynthetic bacteria and commonly referred to as blue-green algae, grow as blooms or mats, mostly within freshwater bodies, and like dinoflagellates are called harmful algal blooms (HABs). There are a large variety of species. Many of these produce potent toxins that are capable of causing acute and chronic disease in mammals, including humans. The toxins include microcystins, nodularins, anatoxins, saxitoxins, aplysiatoxins, cylindrospremopsins, beta-methyl-amino-l-alanine (BMAA) and lipopolysaccharides (see Table 9.1 in Chapter 9). Algal blooms are more commonly found in eutrophic inland waters (eutrophic waters have a high concentration of nutrients). Human health risks arise if the water is consumed untreated, if people bathe or participate in water contact sports in waters with a scum or heavy bloom and if contaminated water is used in renal dialysis. There have been some notable outbreaks of disease associated with cyanobacterial toxins with a high mortality rate in dialysis patients. The risks through long-term exposure to contaminated drinking water may be greater than occasional recreational exposure to cyanobacterial blooms while bathing in natural waters. More information on cyanobacteria and their toxins are presented in Chapter 9.

Key recommended readings

1   WHO (2004) Waterborne zoonoses: identification, causes and control. Edited by J.A. Cotruvo, A. Dufour, G. Rees, J. Bartram, R. Carr, D.O. Cliver, G.F. Craun, R. Fayer and V.P.J. Gannon. World Health Organization, Geneva. Chapters 3, 18 and 19

respectively provide an extensive review of water-related diseases, the major helminth zoonoses in water and fascioliasis in particular.

2 WHO (2003) *Guidelines for Safe Recreational Water Environments, Volume 1: Coastal and Fresh Waters.* World Health Organization, Geneva. Chapters 5 and 7 respectively provide a comprehensive summary of free-living microorganisms and algae and cyanobacteria found in coastal and estuarine waters.

3 WHO (2011) *Guidelines for Drinking-water Quality*, fourth edition. World Health Organization, Geneva. Chapter 11 provides an extensive summary of free-living bacteria, protozoa and helminths found in drinking water.

References

American Water Works Association. (2006) *Waterborne Pathogens.* Denver, CO: American Water Works Association.

Bradley, D. (1977) Health aspects of water supplies in tropical countries. In Feachem, R., McGarry, M. and Mara, D. (eds) *Water, Wastes and Health in Hot Climates.* London: John Wiley.

Buse, H.Y., Schoen, M.E. and Ashbolt, N.J. (2012) Legionellae in engineered systems and use of quantitative microbial risk assessment to predict exposure. *Water Research.* 46: 921–933.

Centers for Disease Control and Prevention. (2012) Parasites – Schistosomiasis. http://www.cdc.gov/parasites/schistosomiasis/index.html (accessed December 2014).

Falkinham, J.O., Nichols, G., Bartram, J., Dufour, A. and Porteals, F. (2004) Natural ecology and survival in water of mycobacteria of public health significance. In Pedley, S., Bartram, J., Rees, G., Dufour, A. and Cotruvo, J.A. *Pathogenic Mycobacteria in Water: A Guide to Public Health Consequences, Monitoring and Management.* Geneva: WHO.

Garcia-Vidal, C. et al. (2013) Rainfall is a factor for sporadic cases of *Legionella pneumonia. PLOS ONE* 8(4): e61036. Doi:10.1371/journal.pone.0061036.

Kendall, B.A. and Winthrop, K.L. (2013) Update on the epidemiology of pulmonary nontuberculus mycobacteria infections. *Seminars in Research. Critical Care Medicine.* 34: 87–94.

Lee, J., Nichols, G.L. (2010) Legionnaires' disease. In Ayers, J., Harrison, R., Maynard, R. and Nichols, G.L. (eds) *Environmental Medicine.* London: Hodder.

McRobb, E. et al. (2013) Meliodosis from contaminated bore water and successful UV sterilization. *American Journal of Tropical Hygiene.* 89: 367–368.

Mena, K.D. and Gerba, C.P. (2009) Risk assessment of *Pseudomonas aeruginosa* in water. *Reviews of Environmental Contamination and Toxicology.* 201: 71–115.

Nichols, G.L., Andersson, Y., Lindgren, E., Devaux, I. and Semenza, J.C. (2014) European monitoring systems and data for assessing environmental and climate impacts on human infectious diseases. *Int J Environ Res Public Health.* 11(4): 3894–3936.

Nichols, G.L., Ford, T., Bartram, J., Dufour, A. and Porteals, F. (2004) Introduction. In Pedley, S., Bartram, J., Rees, G., Durfor, A. and Cotruvo, J.A. *Pathogenic Mycobacteria in Water: A Guide to Public Health Consequences, Monitoring and Management.* London: IWA.

Nwachuku, N. and Gerba, C.P. (2004) Health effects of *Acanthamoeba* spp. and its potential for waterborne transmission. *Reviews in Environmental Contamination and Toxicology.* 180: 93–131.

Oliver, J.D. (2005) Wound infections caused by *Vibrio vulnificans* and other marine bacteria. *Epidemiology and Infection.* 133: 383–391.

Pedley, S., Bartram, J., Rees, G., Durfor, A. and Cotruvo, J.A. (2004) *Pathogenic Mycobacteria in Water: A Guide to Public Health Consequences, Monitoring and Management.* London: IWA.

Percival, S.L. et al. (2004) *Microbiology of Waterborne Diseases.* San Diego, CA: Academic Press.

Thomas, J.M. and Ashbolt, N.J. (2011) Do free-living amoebae in treated drinking water systems present an emerging health risk? *Environmental Science and Technology.* 45: 860–869.

von Reyn, C.F., Pozniak, A., Haas, W. and Nichols, G. (2004) Disseminated infection, cervical adenitis and other MAC infections. In Pedley, S., Bartram, J., Rees, G., Durfor, A. and Cotruvo, J.A. *Pathogenic Mycobacteria in Water: A Guide to Public Health Consequences, Monitoring and Management.* London: IWA.

World Health Organization. (2001) Water-related diseases – Ascarisis. http://www.who.int/water_sanitation_health/diseases/ascariasis/en/ (accessed April 2015).

World Health Organization. (2014a) Dracunculiasis (guinea-worm disease). http://www.who.int/dracunculiasis/en/ (accessed July 2015).

World Health Organization. (2014b) Schistosomiasis. Fact sheet N°115. Updated February 2014. http://www.who.int/mediacentre/factsheets/fs115/en/ (accessed July 2015b).

World Health Organization. (2014c) Foodborne trematodiases. Fact sheet N°368. Updated April 2014. http://www.who.int/mediacentre/factsheets/fs368/en/ (accessed April 2015).

World Health Organization. (2014d) Echinococcus. Fact sheet N°377. Updated March 2014. http://www.who.int/mediacentre/factsheets/fs377/en/ (accessed April 2015).

6

WATER-RELATED INSECT VECTORS OF DISEASE*

Arne Bomblies

PhD, Assistant Professor of Civil and Environmental Engineering,
University of Vermont, Burlington, Vermont, USA

Learning objectives

1 Identify several prominent vector-borne diseases and their arthropod vectors.
2 Understand the fundamental reasons for varying degrees of connection between vector-borne disease and water.
3 Relate water management strategy to disease vector ecology.

A vector-borne disease results from the transmission of illness-causing agents (bacteria, viruses, protozoa, and worms) by arthropod carriers of the disease ("vectors"), such as mosquitoes, ticks, and flies. Snails that play a role in schistosomiasis transmission are sometimes also considered vectors. Common vector-borne diseases and their vectors include malaria (*Anopheles* mosquitoes), river blindness (*Simulium* black flies), Rift Valley fever (*Aedes* and *Culex* mosquitoes), dengue (*Aedes* mosquitoes), yellow fever (*Aedes* mosquitoes), arboviral encephalitides (various species of mosquitoes), Chagas disease (*Triatoma* kissing bugs), Lyme disease (*Ixodes* ticks), plague (various fleas, predominantly *Xenopsylla*), and sleeping sickness (*Glossina* tsetse flies).

Several of these vector-borne diseases have close connections to standing water due to the ecology of the vector. For example, all mosquitoes develop in water and need water for their larval and pupal stages. Therefore, where the absence of water prevents the development of mosquitoes, mosquito-borne disease cannot be stably transmitted (the occasional transient infected mosquito transported through wind or man-made conveyances notwithstanding). Mosquitoes have specific developmental habitat preferences, meaning that specific mosquitoes will not lay eggs in just any standing water. Some mosquitoes of the genus *Aedes* are adapted to develop in small collections of water in pockets of trees or rocks. Individuals of many

* Recommended citation: Bomblies, A. 2015. 'Water-related insect vectors of disease', in Bartram, J., with Baum, R., Coclanis, P.A., Gute, D.M., Kay, D., McFadyen, S., Pond, K., Robertson, W. and Rouse, M.J. (eds) *Routledge Handbook of Water and Health*. London and New York: Routledge.

epidemiologically important *Aedes* species will also oviposit (lay eggs) and develop in small anthropogenic water bodies such as metal cans or discarded tires but are generally not found in large bodies of water. Other *Aedes* mosquitoes develop in expanses of floodwater that can result in an increase in their populations following heavy rain. Some important malaria vector mosquitoes of the genus *Anopheles* develop almost exclusively in small, turbid pools of standing water left by monsoon rains. Others of this genus prefer swampy environments, while still others develop in the edges of rivers and lakes. Malaria, river blindness, and Rift Valley fever are three dominant vector-borne diseases that are closely linked with the availability of water, and thus can be sensitive to rainfall variability and water infrastructure development. Other diseases with connections to water include arboviral encephalitides, dengue, and yellow fever. Some arthropod vectors, such as the kissing bugs that transmit Chagas disease in parts of Central and South America or the deer ticks that transmit Lyme disease, have little or no connection to water because the vector's life cycle does not depend on standing water.

Malaria

Background

Malaria is caused by single-celled protozoan parasites of the genus *Plasmodium*. In 2010, there were an estimated 219 million cases worldwide resulting in 660,000 deaths (World Health Organization, 2012). Most (90 percent) of the malaria deaths are in Africa, primarily because of the anthropophilic (human-biting) nature of the dominant African vector mosquitoes *Anopheles gambiae* and *Anopheles arabiensis* (collectively termed *Anopheles gambiae sensu lato*), as well as the predominance in Africa of *Plasmodium falciparum*, the most deadly of the five human malaria species (*P falciparum*, *P vivax*, *P malariae*, *P ovale*, *P knowlesi*).

The malaria parasite cycles between human and mosquito hosts, with several stages. The Plasmodium life cycle is depicted in Figure 6.1. Mosquitoes inject saliva prior to taking a blood meal to prevent clotting. Infectious sporozoites in the mosquito's saliva are injected into the skin, from where lymph transports them into the human bloodstream. These sporozoites metamorphose into merozoites which invade the liver and then red blood cells, causing them to be destroyed. A further change leads to gametocytes, which when ingested by another mosquito taking a blood meal sexually reproduce in the vector midgut, gestate as an oocyst, and then rupture releasing sporozoites that migrate to the salivary glands and make that mosquito infectious to another human, thus completing the transmission cycle.

The life cycle of the mosquito is depicted in Figure 6.2. An egg is deposited in standing water. It hatches after 1–2 days, and progresses through four larval stages and a pupal stage before emerging as an adult mosquito. The development time depends on water temperature, and is approximately one week.

Anopheles mosquitoes exist in much of the world, including regions in which malaria is not currently prevalent such as the temperate zones of the northern hemisphere. In malaria endemic regions, however, temporal trends in malaria transmission are closely tied to the population dynamics of the vector mosquitoes.

Connection with water

Anomalously wet climatic conditions can lead to a dramatic increase in mosquito abundance, and epidemics can arise when regions of normally marginal transmission receive much rain (Kiszewski and Teklehaimanot, 2004). In much of the seasonal transmission zone of

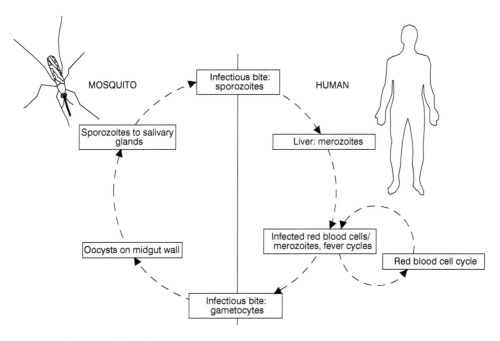

Figure 6.1 Plasmodium life cycle

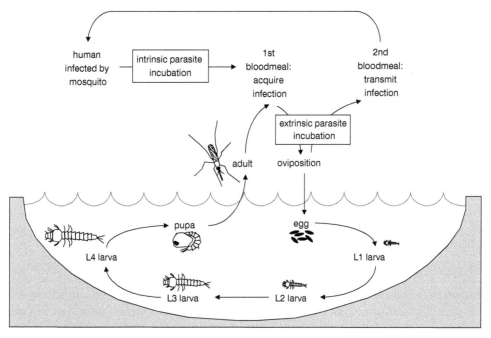

Figure 6.2 Life cycle of the mosquito

Africa, *Anopheles gambiae sl* mosquitoes develop in the shallow ephemeral pools that form as a result of monsoon rainfall (Figure 6.2). Because these pools must persist longer than the temperature-dependent development rate of the mosquito larvae in that pool, repeated rainfall or very heavy rainfall favor exponential growth of mosquito populations, and longer dry conditions kill the cohorts of developing larvae within the pools (Bomblies et al., 2008). However, it has been noted in African environments that very high intensity rainfall can wash larvae out of developmental habitat, diminishing the effect of rainfall-induced population amplification (Paaijmans et al., 2007). In addition, in wetter areas such as much of the equatorial rainforest of the world, mosquito abundance may not be limited by water availability. Instead, it may be limited by predators and parasites (Service, 1977), or potentially sunlight or nutrient availability. The connection of malaria dynamics to climate and hydrological factors does not fit one single mold. The local transmission of the disease depends greatly on the ecology of the primary vectors. For example, in contrast to the dominant environmental determinants of African mosquito populations, the Asian vector mosquito *Anopheles culicifacies* population has been responsible for a major epidemic during a dry period. *An culicifacies* develops along the banks of rivers, and tends to have low numbers during normal years (Carter, 1929; Rajendram et al., 1950; Wijesundera, 1988). However, during the period 1934–1935 two successive monsoons failed in Ceylon (now Sri Lanka), and there river flow slowed to a series of standing water pools, allowing a much greater surface area for mosquito development. Consequently, a devastating malaria epidemic struck Ceylon during a dry period (Reiter et al., 2004). Other malaria vectors throughout the world have their own specific behavioral ecology that can affect the population response to rainfall. Therefore, the relationship of mosquito abundance to water depends greatly on the local vector's specific behavioral ecology.

In addition to rainfall variability, hydraulic development has been known to influence malaria incidence. For example, the construction of the Bargi dam in India resulted in 2.4 times more cases than before (Keiser et al., 2005). *Anopheles* mosquitoes do not typically develop in the open waters of lakes or reservoirs, preferring the shallow shores of the lakes that may be protected from waves by emergent vegetation, or any small rain fed pools that may form and persist near the water body as a result of increased soil wetness. A river whose moving water is not attractive to mosquitoes for oviposition, once dammed can contain large areas of vegetated fringe that can constitute suitable habitat. Moreover, it may be the case that water impoundments attract human activity for irrigation, washing, recreation, etc. The indirect effects of these impoundments can therefore promote transmission by encouraging vector–human contact.

River blindness

Background

Blackflies of the genus *Simulium* are responsible for transmitting river blindness, also known as onchocerciasis (World Health Organization, 1995). The roundworm *Onchocerca volvulus*, the infectious agent of the disease, enters the skin in the larval stage during the vector's blood meal where it develops into an adult form. The larvae migrate to host skin tissue, form subcutaneous nodules and mature into adults. The adult female produces thousands of larval worms in the human host. These migrate into the eyes and upper layers of skin, causing severe itching and lesions. Disease symptoms, which include skin rash, also include vision changes (keratitis) that can ultimately lead to blindness. Small, filarial stages of the worm's life cycle can then enter another blackfly vector when it ingests a blood meal.

River blindness is endemic in Central America, northern South America, and much of sub-Saharan Africa. Although recent interventions to reduce transmission in the Americas have been very successful, the disease is still prevalent in Africa. Current control programs in the Americas as well as Africa rely primarily on the drug ivermectin to treat infected people, and transmission has declined. In 2013, the Pan American Health Organization announced that river blindness had been eliminated from Colombia (West et al., 2013).

Connection with water

The blackfly vectors of river blindness develop in whitewater rapids, and larval flies can be found clinging to submerged rocks and other objects in contact with fast-moving water. Because of the blackfly's larval habitat, the disease foci are centered on fast-flowing streams. Dam construction on slower-moving, non-whitewater streams has been tied to outbreaks of river blindness, because the newly-constructed spillways channel fast-flowing water from the reservoir and thus offer the vector excellent developmental habitat (Jobin, 1999, pp 307–311). A dam spillway may create whitewater conditions in otherwise slow-moving water that is unsuitable as blackfly developmental habitat. The disease foci are normally near fast-flowing streams, and primarily affect permanent residents because repeat inoculations are usually necessary to contract the disease. Travelers and other occasional visitors to disease foci are not at high risk of disease.

Rift Valley fever

Background

Phlebovirus causes Rift Valley fever (RVF), a severe zoonotic (transmissible between humans and animals) illness in humans and domestic animals (World Health Organization, 2010). Mosquitoes of the genera *Aedes* and *Culex* vector the disease, primarily amongst domestic animals such as goats, sheep, cows, and camels. Virus has also been isolated from mosquitoes of genus *Anopheles* and *Mansonia*. Human disease is often associated with direct contact of infected meat or blood, but direct inoculation from infected mosquitoes can also occur. The majority of human cases are mild, with flu-like symptoms, but a small percentage of cases lead to a reduction of vision, neurological damage, or hemorrhagic fever symptoms which can lead to death.

RVF outbreaks have occurred in several sub-Saharan African countries as well as in the Arabian Peninsula, and many other countries in Africa that have been spared a significant outbreak have reported isolated cases or serological evidence of presence of the virus. RVF remains endemic in southern and eastern Africa, largely because the vectors are capable of transovarial (parent-to-offspring) transmission of the virus. Virus is passed into the eggs of *Aedes* mosquitoes, as shown by the detection of virus in adult male and female *Aedes lineatopennis* reared in the lab from field-collected specimens in Kenya (Linthicum et al., 1985). Other *Aedes* mosquitoes have this capability as well; therefore where *Aedes* is the dominant vector, epidemics occur following rainfall-induced increases in mosquito abundance. However, some vectors of the virus, such as *Culex* mosquitoes that act as primary vectors in Egypt and Arabia, are not capable of transovarial transmission. Hence, the occurrence of an epidemic in those areas requires reintroduction of the virus through transport of infected animals or wind-borne transport of mosquitoes (Gerdes, 2004).

Connection with water

Major RVF epidemics have been associated with climatic anomalies and resultant flooding. For example, the El Niño event of 1997–1998 caused significant flooding in East Africa. As a result of the increased mosquito abundance, 89,000 people were infected with RVF and 250 people died (Gerdes, 2004). In 2006 and 2007, heavy rains in Kenya and Somalia caused flooded conditions that resulted in markedly increased mosquito abundance, which triggered a RVF epidemic that claimed 75 lives in Kenya and likely more in Somalia and Sudan.

Hydraulic development can also lead to epidemics. The construction of the Aswan High Dam in Egypt in 1971 created many acres of flooded land suitable for mosquito development, whose exploding numbers caused an epidemic of 200,000 cases and nearly 600 human deaths. Because of the regionally-varying dominant vectors, in arid portions of northern and western Africa and Arabia, the epidemics are associated with the building of dams, whereas in eastern and southern Africa, heavy rainfall is usually the cause of a RVF outbreak (Gerdes, 2004).

Other water-related vector-borne diseases

Arboviral encephalitides

Endemic malaria has been eradicated in temperate zones of North America and Europe, but some arthropod-borne diseases still affect these populations, primarily the arboviral encephalitides (Lindsey et al., 2014). This category includes West Nile virus (introduced to North America in 1999), eastern equine encephalitis, Japanese encephalitis, La Crosse encephalitis, St Louis encephalitis, and western equine encephalitis. Most encephalitis viruses are vectored by mosquitoes (but some are transmitted by ticks), and the mosquito-transmitted viruses are therefore tied to water. For example, West Nile virus exhibits slightly higher than normal transmission following above average rainfall in the eastern United States (Landesman et al., 2007) but in some cases shows marked local increases during droughts. For example, Andreadis et al. (2004) report that in Connecticut, arid conditions concentrated *Culex* vectors in the few structures that retain water during droughts such as catch basins, and they suggest that concentrated foci of viral activity ensue. West Nile virus shows slightly lower than normal transmission following above-average rainfall in the western United States (Landesman et al., 2007). Taken together, these results show a loose, heterogeneous connection to water availability. Some primary vectors of these arboviruses (e.g. *Culex* and *Aedes* mosquitoes) develop in flood and swamp water, and their abundance can be influenced by rainfall and hydrologic variables (Shaman et al., 2002).

Culiseta melanura, the primary vector mosquito responsible for maintaining enzootic transmission of eastern equine encephalitis among avian hosts, develops primarily in acidic bogs, and seeks out developmental sites among the root mats of trees, notably white cedar trees in swampy environments. These sites are typically surface expressions of groundwater and remain flooded throughout the year, resulting in little connection of mosquito abundance to rainfall variability.

The behavioral ecology of vector mosquitoes of arboviral encephalitides thus strongly influences the connection with water, and in some cases disease dynamics may not be connected with hydrologic variability.

Dengue and yellow fever

These two diseases are grouped together for the purpose of this book, because both diseases are caused by flaviviruses that are transmitted by *Aedes* mosquitoes (World Health Organization,

2009). Also, both are very significant diseases (an effective yellow fever vaccine exists and has greatly reduced disease burden). *Ae aegypti* serves as the primary vector for both diseases, although other species have also been implicated. *Ae aegypti* mosquitoes develop in collections of water in small containers which are mostly of anthropogenic origin such as flower vases, toilet tanks, or discarded tires. For this reason, the disease does not show the pronounced connection with rainfall variability that some mosquito-borne diseases exhibit, nor are there strong spatial connections of disease foci with naturally-occurring standing water.

Water management

The demand for water and hydroelectric power is increasing in many parts of the world that are affected by vector-borne disease. Knowledge of the fundamental connections of vector-borne disease with water can influence the design and operation of reservoirs, irrigation schemes, agricultural operations, and stormwater management so as to minimize the public health threats. For example, in the Tennessee Valley of the United States, reservoir water levels were fluctuated to prevent malaria-transmitting *Anopheles* mosquito developmental habitat from persisting (Derryberry and Gartrell 1952). Combined with case management and improved housing, this strategy contributed to the elimination of malaria from the southeastern United States in the early 20th century. Currently, well-designed reservoirs and irrigation schemes also limit the effect on malaria transmission. For example, intermittently irrigated rice cultivation has been shown to reduce malaria by persistent habitat reduction (Keiser et al., 2002), and new engineering design of small dams and siphons has also contributed to malaria control (Konradsen et al., 2004). Water management for public health goals also extends beyond concerns about malaria. For example, onchocerciasis (Jobin, 1999), schistosomiasis (Steinmann et al., 2006), and Rift Valley fever (Linthicum et al., 1985) also exhibit sensitivity to dams and water management and have been the subjects of design improvements. Water-related development projects and management strategies must be designed with significant consideration of the specific habitat preferences of disease vectors in order to mitigate introduced disease burden resulting from the development. These habitat preferences can be highly variable between species and connected to water in different ways. Therefore, public health risk can be significantly reduced with an understanding of the water/disease connections.

Key recommended readings (open access)

1 World Health Organization 2012 malaria report. http://www.who.int/malaria/ publications/world_malaria_report_2012/en/. This is an authoritative summary on the state of malaria in the world, and the progress made in combating the disease by intervention efforts.
2 http://www.cdc.gov/parasites/onchocerciasis/. A summary of the parasite and epidemiology of river blindness, from the US Centers for Disease Control and Prevention.

References

Andreadis, TG, Anderson, JF, Vossbrinck, CR and Main, AJ (2004) Epidemiology of West Nile virus in Connecticut: a five-year analysis of mosquito data 1999–2003. *Vector-borne and Zoonotic Diseases* 4(4): 360–378.

Bomblies, A, Duchemin, JB and Eltahir, EAB (2008) The hydrology of malaria: model development and application to a Sahelian village. *Water Resources Research* 44(12): W12445, doi:10.1029/2008WR006917.

Carter, HF (1929) Further observations on the transmission of malaria by Anopheline mosquitoes in Ceylon. *Ceylon Journal of Science* 2(4): 159–176.

Derryberry, OM and Gartrell, FE (1952) Trends in malaria control program of the Tennessee Valley Authority. *American Journal of Tropical Medicine and Hygiene* 1(3): 500–507.

Gerdes, GH (2004) Rift Valley fever. *Revue scientifique et technique de l'office international des epizooties* 23(2): 613–623.

Jobin, W (1999) *Dams and Disease. Ecological design and health impacts of large dams, canals, and irrigation systems.* E&FN Spon: London, 580 pp.

Keiser, J, Utzinger, J and Singer, BH (2002) The potential of intermittent irrigation for increasing rice yields, lowering water consumption, reducing methane emissions, and controlling malaria in African rice fields. *Journal of the American Mosquito Control Association* 18(4): 329–340.

Keiser, J, Caldas de Castro, M, Maltese, MF, Bos, R, Tanner, M, Singer, BH and Utzinger, J (2005) Effect of irrigation and large dams on the burden of malaria on a global and regional scale. *American Journal of Tropical Medicine and Hygiene* 72(4): 392–406.

Kiszewski, A and Teklehaimanot, A (2004) A review of the clinical and epidemiological burdens of epidemic malaria. *American Journal of Tropical Medicine and Hygiene* 71(suppl.2): 128–135.

Konradsen, F, van der Hoek, W, Amerasinghe, FP, Mutero, C and Boelee, E (2004) Engineering and malaria control: learning from the past 100 years. *Acta Tropica* 89(2): 99–108.

Landesman, WJ, Allan, BF, Langerhans, RB, Knight, TM and Chase, JM (2007) Inter-annual associations between precipitation and human incidence of West Nile virus in the United States. *Vector-borne Zoonotic Diseases* 7(3): 337–343.

Lindsey, NP, Lehman, JA, Staples, JE and Fischer, M (2014) West Nile virus and other arboviral diseases – United States, 2013. *MMWR* 63(24): 521–526.

Linthicum, KJ, Davies, FG and Kairo, A (1985) Rift Valley fever virus (family *Bunyaviridae*, genus *Phlebovirus*). Isolations from Diptera collected during an inter-epizootic period in Kenya. *Journal of Hygiene* 95: 197–209.

Paaijmans, KP, Wandago, MO, Githeko, AK and Takken, W (2007) Unexpected high losses of *Anopheles gambiae* larvae due to rainfall. *PLoS ONE* 2(11): e1146. doi:10.1371/journal.pone.0001146.

Rajendram, S, Abdul Cader, MIIM and Visvalingam, T (1950) Malaria eradication in Ceylon. *Nature* 166: 486.

Reiter, P, Thomas, CJ, Atkinson, PM, Hay, S, Randolph, SE, Rogers, DJ, Shanks, GD, Snow, RW and Spielman, A (2004) Global warming and malaria: a call for accuracy. *Lancet Infectious Diseases* 4(6): 323–324.

Service, MW (1977) Mortalities of the immature stages of species B of the *Anopheles gambiae* complex in Kenya: comparison between rice fields and temporary pools, identification of predators, and effects of insecticidal spraying. *Journal of Medical Entomology* 13: 535–545.

Shaman, J, Stieglitz, M, Stark, C, Le Blanck, S and Cane, M (2002) Using a dynamic hydrology model to predict mosquito abundances in flood and swamp water. *Emerging Infectious Diseases* 8(1): 8–13.

Steinmann, P, Keiser, J, Bos, R, Tanner, M and Utzinger, J (2006) Schistosomiasis and water resources development: systematic review, meta-analysis, and estimates of people at risk. *The Lancet Infectious Diseases* 6(7): 411–425.

West, S, Munoz, B and Sommer, A (2013) River blindness eliminated in Colombia. *Ophtalmic Epidemiology* 20(5): 258–259.

Wijesundera, MS (1988) Malaria outbreaks in new foci in Sri Lanka. *Parasitology Today* 4(5): 147–150.

World Health Organization (1995) *Onchocerciasis and Its Control*. Report of a WHO Expert Committee on onchocerciasis control. WHO Technical Report Series 852.

World Health Organization (2009) *Dengue: guidelines for diagnosis, treatment, prevention and control.* Geneva: World Health Organization.

World Health Organization (2010) *Rift Valley Fever.* Fact sheet N°207. World Health Organization. May. http://www.who.int/mediacentre/factsheets/fs207/en/.

World Health Organization (2012) *World Malaria Report*, Geneva, available at http://www.who.int/malaria/publications/world_malaria_report_2012/en/.

7

HEALTH IMPACTS OF WATER CARRIAGE*

Jo Geere

PHYSIOTHERAPY LECTURER, FACULTY OF MEDICINE AND HEALTH
SCIENCES, UNIVERSITY OF EAST ANGLIA, NORWICH, NORFOLK, UK

Learning objectives

1 Draw on existing evidence to explain why the relationship between musculoskeletal disorders and water carriage is important.
2 Describe factors which may influence the association between water carriage and musculoskeletal disorders, in particular to explain potential differences between households.
3 Integrate the principles of development effectiveness into strategies, which aim to improve access to water and minimize health risks associated with water carriage.

The situation

Even with recent increases in the global population able to access safe drinking water (Bartram et al., 2012), many people continue to rely on out of house sources to obtain water for household consumption and to maintain hygiene (Evans et al., 2013; WHO and UNICEF, 2014a). Transport or carriage of water filled containers from an external water source or shared water supply point to the home is therefore a necessary and regular routine for many people (Sorenson et al., 2011).

Different methods of water carriage have been observed between and within different regions of the world (Evans et al., 2013), and will be influenced by the resources available to household members, environmental factors, traditional or cultural norms and personal preferences. For example, particularly in rural areas of many African countries, water carriage is perceived as a domestic chore for women or children and is often performed by head loading (Hemson, 2007; Porter et al., 2013). Women in African countries commonly balance a 20–25 liter bucket of water on their head and transport it over distance, often on poorly maintained paths or along busy roadsides (Geere et al., 2010a). Alternative methods of

* Recommended citation: Geere, J. 2015. 'Health impacts of water carriage', in Bartram, J., with Baum, R., Coclanis, P.A., Gute, D.M., Kay, D., McFadyen, S., Pond, K., Robertson, W. and Rouse, M.J. (eds) *Routledge Handbook of Water and Health*. London and New York: Routledge.

water carriage include the use of shoulder or head straps to secure a container against the back (Hoy et al., 2003), containers directly rolled along the ground (Geere et al., 2010b), or pushed with a handle (for example the 'hippo water roller'), buckets simply carried at the side of the body or attached to a pole or yoke (Evans et al., 2013), wheelbarrows loaded with several containers (Geere et al., 2010b), or the use of manpowered or animal drawn carts and motorized vehicles. This chapter will focus on water carriage and transport methods accomplished through human work, rather than by animal driven or motorized vehicles.

As a functional task, water carriage requires coordinated movement of the body and sufficient strength to lift and balance or move water filled containers. Disorders affecting the musculoskeletal system can therefore reduce an individual's ability to carry water, as they commonly cause pain, reduce strength and impair movement (Brooks, 2006). This is important to acknowledge because in low and middle income regions, where a greater proportion of the population will access water from out of house sources, the burden of disease from musculoskeletal disorders (MSKs) is huge and growing (Hoy et al., 2014). The burden is increasing because of population growth and ageing, as well as the increasing incidence of traumatic injury from motor vehicle accidents (Nantulya and Reich, 2002), obesity linked to changing lifestyle behaviors (Nugent, 2008; Berenbaum and Sellam, 2008) and manual labor performed in poor working conditions (Messing and Östlin, 2006; Fathallah, 2010).

Water carriage itself may also contribute to this growing burden of MSK disease in low and middle income regions (Lloyd et al., 2010; Evans et al., 2013; Hoy et al., 2014). Water carriage and MSKs should be considered as having a bi-directional relationship; MSKs of any cause may limit physical capacity to carry water filled containers, whilst water carriage may expose people to greater risk of physical injury, or over a period of time may have direct impact on the musculoskeletal system through the effects of repetitive loading (Geere et al., 2010$_4$).

Proposed post-2015 sustainable global development targets include 'universal basic access to drinking water' and 'halving the proportion of the population without access at home to safely managed drinking water services by 2030' (WHO and UNICEF, 2014b). Even if they are achieved, by definition these goals indicate that millions of people will continue to rely on out of house sources and water carriage to support a basic level of water service in 2030. This means that the physical work of water carriage should not be ignored, as for many it is the critical step of the water supply chain which determines the quantity of water available at home.

Current evidence

To develop well-informed strategies for improving access to water from out of house sources, the relationships between water carriage and MSKs or physical injury and disability should be more clearly understood. However, little empirical research has been undertaken to date to investigate such relationships. A systematic search of peer reviewed literature (Evans et al., 2013) identified seven studies which focused on water carriage. Four studies reported descriptive statistics related to water carriage (Thompson et al., 2000; Hemson, 2007; Geere et al., 2010a; Sorenson et al., 2011), and two studies used qualitative research methods, one reporting children's perceptions of health in relation to water carriage (Geere et al., 2010b) and the other exploring gender issues (Sultana, 2009). Finally, Lloyd et al. (2010) reported pain and rating of perceived exertion experienced by women during head loading in a laboratory setting. A common conclusion of the studies was that water carriage can impact on general health of the water carrier and that pain commonly experienced during water carriage is most likely due to disorders or strain of the musculoskeletal system. No large scale epidemiological studies were found which had used an appropriate study design to quantify the association between water carriage and physical health outcomes such as self-report of pain, physical functioning or disability.

Since the review, one study has been reported which did measure the association between these factors and water carriage (Evans et al., 2013; Geere and Hunter, 2014). Initial analyses have been reported and found an increased risk of pain in the upper back or hands among people with a history of water carriage, as well as the association of water carriage with a particular pattern of pain reported in multiple body areas. The findings will be discussed in light of relevant scientific evidence below.

Because of the limited amount of research specifically investigating water carriage and MSKs or physical injury, it is reasonable to draw on other sources of evidence, for example laboratory based studies of physical loading, or occupational health research investigating risk factors similar to those which occur during water carriage. It is essential to recognize that there are limitations in extrapolating data derived in a particular setting with a specific population, to different settings and populations (Beaglehole et al., 1993, p76); however, some useful comparisons can be made to inform policy and practice and base decisions on the available evidence.

Risk factors for MSKs or physical injury related to water carriage

Factors which influence the physical health impact of water carriage can be categorized as those pertaining to the individual, the water carriage task or the environment. This is consistent with both the 'bio-psychosocial model' of health (Waddell, 1992) and a 'participatory ergonomic systems' approach to activity analysis (Grandjean, 1988; Buckle, 2005). The bio-psychosocial model recognizes that health is influenced by the inter-relationship between physical, psychological and social factors, which should therefore be considered as potential risk factors in epidemiological studies investigating MSKs or physical injury. The 'participatory ergonomic systems' approach recognizes that individual, task related and environmental factors will influence the relationship between work and health, and that all stakeholders can contribute to problem solving. Such factors may have a detrimental or positive impact on health outcomes and access to water. Evaluation of both risk factors and protective mechanisms relevant to water carriage will highlight potential strategies for improving access to water.

Individual

Age

Changes to the structure and functioning of the musculoskeletal system occur so commonly over time that they are associated with ageing (Hamerman, 1997). Peak strength is attained in adulthood, and common age related changes, such as cervical spondylosis or arthritis (Belachew et al., 2007) result in weaker musculoskeletal tissues in older age, such that lower loads and minor trauma may result in injury or tissue stress to produce local or referred pain. Changes in the structure and shape of musculoskeletal tissues such as the intervertebral discs may encroach on neighboring pain sensitive structures and become a further mechanism for referred pain (Twomey and Taylor, 1988). Children also have reduced capacity for physical loading, particularly before they reach skeletal maturity (Nuckley et al., 2002, 2007). There is some debate about what constitutes 'normal' age related changes and what can be ascribed to injury or abnormal physical strain (Taylor and Finch, 1993). However, death, serious spinal injury and advanced degenerative changes in the cervical spine have been reported as a consequence of head loading (Levy, 1968; Joosab et al., 1994; Jager et al., 1997). In households and communities which do not have at-house water supply and where healthy, younger adults are absent, for example due to employment migration, illness or early death (Bicego et al., 2003; Cheng and Siankam, 2009), older adults and children may have to regularly carry water and therefore are inherently at greater risk of physical injury.

A recent study conducted in South Africa, Ghana and Vietnam (Evans et al., 2013; Geere and Hunter, 2014) found that pain in the upper back or hands was associated with current or previous history of water carriage, as was a particular pattern of pain in multiple body areas, including the head, upper back, chest, hands and feet. The association was stronger for water carriage by head loading compared to other methods. In light of studies such as those cited above, the finding of specific pain sites and patterns associated with water carriage is interpreted as being due to the detrimental effects of axial spinal loading, most plausibly linked to deformation affecting soft tissues in the cervical spine or upper back. Interestingly, a second key pattern of self-reported pain was slightly negatively associated with water carriage, but not significantly associated with a particular water carriage method. The pattern of correlated pain areas included neck, arm, back and leg pain. This suggests that there can be different underlying MSKs or biomechanical triggers of pain affecting people who carry water, some of which are increased and some decreased by water carriage, and it suggests that head loading is more likely to be problematic. This is consistent with the view that individual risk factors will have a significant influence on the health impact of water carriage and result in subgroups of people who are more vulnerable to particular detrimental effects, whilst others can experience different and slightly beneficial effects (Geere and Hunter, 2014).

Gender

In many countries, water carriage, particularly by head loading, is considered a task for women or children and not for men (Sultana, 2009; Geere et al., 2010a; Sorenson et al., 2011). Yet studies have found women to have a higher prevalence of neck pain compared to men (Fejer et al., 2006) and female gender has been identified by some as a risk factor for persisting symptoms and disability as a result of neck injury (Cote et al., 2004). Women who are size matched to men have proportionally smaller, less dense skeletons and less muscle mass, particularly in the cervical spine (Stemper et al., 2008; Lang, 2011). This suggests that despite the social expectation in some regions that water carriage is a woman's role (Sultana, 2009; Geere et al., 2010a), men are anatomically more suited to the task, particularly when it is performed by head loading.

Comorbidity

MSKs, health status, physical injury or disability may reduce a person's capacity to easily or safely carry water. Globally, the burden of musculoskeletal pain is huge and the bulk of this is known to occur in low and middle income countries (Vos et al., 2012; Hoy et al., 2014). The burden of MSKs in low and middle income countries increased by 60 percent between 1990 and 2010 (Hoy et al., 2014). Low back pain causes more disability than any other condition globally, whilst neck pain and osteoarthritis are particularly large contributors to the burden of disease (Hoy et al., 2014).

Particularly in Africa, communities which typically have low levels of at-house water supply may also be those in which debilitating chronic disease, such as HIV, is highly prevalent (WHO, 2014). In low income regions generally, rates of physical injury due to road traffic accidents (Nantulya and Reich, 2002; WHO, 2004, p10) or violence are higher (WHO, 2002, p68) and support services for people with physical disability are poor (Groce et al., 2011; Bunning et al., 2014). Some communities now additionally contend with the impact of diseases related to modern 'lifestyle' choices, such as smoking, poor diet and high alcohol consumption (Doyal and Hoffman, 2009). These factors may combine to produce vulnerable households, comprised of people whose ability to physically access quantities of water sufficient for their health needs is seriously compromised.

Psychosocial factors

Common psychological conditions such as stress, anxiety or depression can influence physical health (Alford, 2007), particularly pain perception or disability related to common MSKs (Waddell, 2006). The way in which an individual, or cultural group, interpret signs and symptoms of a health condition will influence what they attribute their symptoms to and how they report and react to MSKs (Waddell, 2006). This may influence reporting of symptoms and levels of disability, and therefore MSK prevalence rates derived from survey data. Health beliefs will influence how people try to manage MSKs and cope with activities which they associate with symptoms, such as water carriage. Behaviors will also be influenced by gender (Sultana, 2009) and family roles (Hemson, 2007) expected within one's social group and broader cultural practice. Psychosocial factors may therefore influence symptom reporting and practice of water carriage, and be one of the main limitations to generalizing research findings from the original study setting and population. However, addressing health beliefs and changing behaviors may provide an effective avenue for mitigating the health impacts of water carriage.

Task

Whilst most ergonomic research has been conducted in high income countries and in occupational settings, several factors which have been found to influence physical injury risk and prevalence or incidence of MSKs (Adams et al., 2002; Buckle and Devereux, 2002; Leclerc et al., 2004; Walker-Bone and Cooper, 2005; Porter et al., 2013) also commonly occur during water carriage. For example repeated forward bending of the low back, elevation of the arms and sustained compressive tissue loading are MSK risk factors which occur whilst lifting, head loading and lowering filled water containers. Such factors are further influenced by the load characteristics, available equipment and preferred carriage method. Water is an inherently unstable substance to carry, and handling may be made more heavy and awkward by the use of worn and poor quality equipment, such as dilapidated wheelbarrows and containers without appropriate handles and lids (Geere et al., 2010a).

The results of occupational studies indicating risk of injury with specific movements may be positively biased by the 'healthy worker effect'; people in paid employment are likely to be a healthier cohort than those who are not employed (Li and Sung, 1999). Cohorts of people who perform water carriage typically include those who are not in paid employment, have a greater age range of people and include individuals whose health may be compromised by comorbidity or malnutrition (Geere et al., 2010a). In developing countries, substantial numbers of people are employed in heavy work (Hoy et al., 2014) and water carriage may be performed in addition to other manual labor. Risk of injury due to the physical work of water carriage may therefore be greater than one would expect from existing occupational health data.

Capacity for physical adaptation to repetitive or sustained loading stress and repair after injury will be influenced by individual factors, but also by how well load characteristics are matched to the individual's health and strength. In particular, load weight and stability will have a substantial impact on a person's ability to safely carry water. Subjective indicators of individual stress such as the 'modified Borg scale' or 'rating of perceived exertion' scale (Finch et al., 2002) may be useful to evaluate work effort and risk of physical injury during specific tasks.

Work patterns, such as frequency and pacing of water collection trips, will also determine whether sufficient time for recovery and tissue adaptation or repair occurs. Frequency of collection trips may be strongly influenced by household size and demand for water quantity. Other social and environmental factors may affect timing and pacing of water carriage. For example, unreliable water supplies may lead to collection of as much water as possible when it is available, within a short space of time and with insufficient rest periods.

Environmental

Environmental factors are also likely to influence the physical work and physiological stress associated with water carriage, particularly distance to the water point, path incline or quality and climate. Sudden, unexpected movements may occur during water carriage, for example as people move quickly to avoid road traffic, or as a result of slipping on poor quality paths or where there are no footpaths. This can subject regions of the spine or other musculoskeletal tissues to strain at much higher peak velocities and loading than would otherwise be expected (Adams et al., 2002). Water carriage will pose greater risks of physical injury through such mechanisms for older people, adults with advanced degenerative changes from a past history of head loading or injury and for children who have not reached physical maturity. Risk of physical injury may also be increased by exposure to social and physical risk factors during the water collection journey, such as inter-personal violence and road traffic. Communities with low levels of at-house water supply may also be stressed by crowding, poverty, violence and poor traffic safety, particularly in urban or densely populated areas (WHO, 2002; WHO, 2004).

Potential solutions

Messages about the association between water carriage and MSKs should be conveyed cautiously at present, as conclusions may change in the light of new evidence. The introduction of 'solutions' or suggestions to try alternative water carriage practices should also be sensitively handled. Particularly where water carriage is already a valued household or community role, it may have a positive effect on quality of life and sense of well-being. Water carriage may create and maintain opportunities for positive social interaction and regular physical activity, thereby increasing cardiovascular fitness, and for some individuals resilience to injury and related disability through better muscle functioning. It may also reduce the risk of becoming overweight or obese. The existing evidence does not yet provide a clear picture and is not conclusive.

However, in moving forward, key principles of development effectiveness must be employed in strategies which aim to reduce risk and impact of MSKs during water carriage (Hoy et al., 2014). Effective development principles are described in international agreements and include community ownership and inclusive development partnerships, alignment, harmonizing and mainstreaming, and delivering results and mutual accountability (Hoy et al., 2014). Particularly because bio-psychosocial, cultural and environmental factors as described above will influence the practice and health impacts of water carriage, lessons learned from international development should be applied to strategies to target support for individuals or communities, and improvements to the water carriage task and environment. Some examples of how this could be done are provided below.

Individual

Because existing data suggests that the impact of water carriage will be most detrimental for vulnerable individuals, inclusive community development partnerships should be utilized to identify vulnerable individuals or households (Gona et al., 2006) and develop contextually relevant support strategies (Hartley, 2006). Community based approaches have been successfully implemented to support people with disabilities, by mobilizing local resources to provide respite, transport or to complement government or non-governmental organization (NGO) rehabilitation services (Kuipers et al., 2008). In the context of water carriage, community based support might take the form of an affordable water cart delivery scheme to vulnerable (or to all) households, managed co-operatively or through local government and community partnerships. Whilst there may be considerable challenges to adopting community based approaches in areas where there is a lack of civil society organizations or community cohesion, it is a model which has demonstrated success in addressing inequalities and social stigma (Hartley et al., 2009).

Task

Attempts to reduce the impact of water carriage have included the development of equipment to change the method of water carriage or redistribute loading on the body. Existing data suggests that head loading in particular may be problematic for some (Geere and Hunter, 2014) and supports the value of such efforts. However, cost can be a prohibitive barrier to up-take of new devices which will particularly affect the poorest and most vulnerable community members. Again, inclusive community development should be utilized to help evaluate the feasibility and acceptability of new devices in the design and trial phase of products and could be used to explore opportunities for local production, sharing and distribution or maintenance of equipment. Depending on local contexts and the physical environment, improved design, quality or availability of wheel barrows, trolleys, carts or bicycles could make a considerable difference to the work of accessing water from out of house sources.

New devices may increase the opportunities for individuals to vary their methods of water carriage, and avoid head loading if or when necessary. The value of alternative equipment may be from increasing the options and choice of water carriage methods available, rather than in completely replacing traditional methods, so that individuals can better manage their own work in relation to episodic MSK symptoms. For example, headache can be exacerbated during the menstrual cycle, which may be a key time when an alternative water carriage method to head loading is of most value to some women. Alternative methods may also be crucial to reduce loading but facilitate normal mobility and function subsequent to injury and at a particular phase of healing. For example, loading through the neck or traction through the shoulder region may be particularly problematic in the early recovery phase of whiplash injury from motor vehicle accidents (Taylor and Twomey, 2005).

More reliable, well maintained and equitably delivered water services could increase opportunities for work pacing and increase an individual's sense of control over water carriage. This may be particularly important, as work pacing and a sense of control over organization of work tasks have been shown to reduce MSK symptoms and disability in occupational settings (Linton, 2001). In the water sector, mobile devices seem to offer great opportunities for better monitoring, evaluation and communication between service providers and users to facilitate better management of water services. More comprehensive monitoring could provide the information necessary to demonstrate the true impact of water service improvements to governments or development donors, and encourage adequate investment and resource mobilization. Evaluation of health impacts and benefits, including MSKs, could be aligned with mobile monitoring of water service delivery. Other social and economic benefits could also be monitored in this way to demonstrate the full impact of water service improvements and changes to water carriage practices. Better quality monitoring, evaluation and communication is consistent with the development principle of delivering results and mutual accountability.

Environmental

Initiatives which aim to reduce the burden of MSKs or reduce the need for water carriage, should not be delivered as an independent or 'vertical programme', but harmonized with existing national and regional systems, processes and projects (Hoy et al., 2014). In this way local resources and expertise can be strengthened and used efficiently to achieve multiple outcomes. For example, initiatives to create safer communities, roads and built environments (Briggs, 2003), will reduce the risk of traumatic accidents, violence and physical injury during water carriage. Environmental initiatives and policies which aim to improve mobility for people with disabilities, such as providing improved pathways and building access, better transport and reduced distance to shared water points or sanitation facilities, or water harvesting at home, can facilitate the use of equipment or assistive devices during water carriage to also benefit non-disabled people; an example of inclusion and mainstreaming (WHO, 2011).

Aligning access to water, sanitation, energy and housing as a 'package for human development' has been utilized in community development strategies to upgrade environmental conditions in urban slums (Chikoti, 2012). Particularly in many rural regions, collecting fuel from the local environment is a household chore in addition to water carriage (Hemson, 2007). Improved energy provision may substantially reduce load carriage work overall and therefore the cumulative impact of water carriage.

Conclusion and suggestions for future research

The achievement of the Millennium Development Goal (MDG) 7, which was to reduce by half the proportion of people without access to safe water, should be recognized and celebrated (WHO and UNICEF, 2014a). However, much work remains to extend the success further and improve access to safe drinking water for all. In particular, the limitations of the current methods of global data collection to monitor 'access to safe drinking water' should be acknowledged and the concept of 'access' should be defined in more detail (Devi and

Bostoen, 2009). Progress toward the MDG for safe drinking water is monitored by national household surveys, in which 'access to safe drinking water' is assumed if householders report an 'improved' source as their main supply of drinking water (Devi and Bostoen, 2009). An 'improved' source is one which is likely to be protected from contamination by pollutants. However, this limited definition gives no indication of how and whether sufficient quantities of water are accessed relative to the household's needs.

It is proposed here that monitoring of '*access* to safe water' should include at least some indicators of the following attributes:

- the methods of water carriage used and whether they ensure that sufficient quantities of water are accessed to meet the household's requirements
- the distance over which water is typically carried from the usual and alternative water supply points, and/or time taken for water collection trips
- the ease, physical comfort and safety with which water is transported from main and alternative supply points to the home
- the reliability and sustainability of the water source and methods of water carriage.

Future research should incorporate such information to facilitate more detailed and critical analysis of the extent to which access to water has been sustainably improved in specific regions and communities. It should also evaluate health outcomes including those which indicate the prevalence and severity of MSKs. Such detail is essential to implement, evaluate and monitor the success of interventions which aim to improve access to water, particularly in areas which continue to rely on water carriage as the final link in the water supply chain. For people affected by MSKs of any cause, greater control and choice over how, where and when they obtain water may be particularly important. Contextually appropriate ways to mitigate the risk factors for MSKs and physical injury during water carriage from out of house sources must also be explored.

Key recommended readings (open access)

1 Geere, J. A., Hunter, P. R. & Jagals, P. 2010a. Domestic water carrying and its implications for health: a review and mixed methods pilot study in Limpopo Province, South Africa. *Environmental Health: A Global Access Science Source*, 9: 52. The paper reviews the potential physical health impacts of water carriage and provides initial evidence from a mixed methods pilot study conducted in South Africa in 2008.

2 Groce, N., Bailey, N., Lang, R., Trani, J. F. & Kett, M. 2011. Water and sanitation issues for persons with disabilities in low- and middle-income countries: a literature review and discussion of implications for global health and international development. *Journal of Water & Health*, 9, 617–627. The paper reviews what is currently known about access to water and sanitation for persons with disabilities in low and middle income countries from the perspective of both international development and global health, and identifies existing gaps in research, practice and policy.

3 Porter, G., Hampshire, K., Dunn, C., Hall, R., Levesley, M., Burton, K., Robson, S., Abane, A., Blell, M. & Panther, J. 2013. Health impacts of pedestrian head-loading: a review of the evidence with particular reference to women and children in sub-Saharan Africa. *Social Science & Medicine*, 88, 90–97. The paper reviews the effects of head loading in sub-Saharan Africa.

References

Adams, M., Bogduk, N., Burton, K. & Dolan, P. 2002. *The Biomechanics of Back Pain*. Edinburgh, Churchill Livingstone.

Alford, L. 2007. Findings of interest from immunology and psychoneuroimmunology. *Manual Therapy*, 12, 176–180.

Bartram, J., Elliott, M. & Chuang, P. 2012. Getting wet, clean, and healthy: why households matter. *The Lancet*, 380, 85–86.

Beaglehole, R., Bonita, R. & Kjellstrom, T. 1993. *Basic Epidemiology*. Geneva, World Health Organization.

Belachew, D. A., Schaller, B. J. & Guta, Z. 2007. Cervical spondylosis: a literature review with attention to the African population. *Archives of Medical Science*, 4, 315–322.

Berenbaum, F. & Sellam, J. 2008. Obesity and osteoarthritis: what are the links? *Joint Bone Spine*, 75, 667–668.

Bicego, G., Rutstein, S. & Johnson, K. 2003. Dimensions of the emerging orphan crisis in sub-Saharan Africa. *Social Science & Medicine*, 56, 1235–1247.

Briggs, D. 2003. *Making a Difference: indicators to improve children's environmental health*. Geneva, World Health Organization.

Brooks, P. M. 2006. The burden of musculoskeletal disease – a global perspective. *Clinical Rheumatology*, 25, 778–781.

Buckle, P. 2005. Ergonomics and musculoskeletal disorders: overview. *Occupational Medicine*, 55, 164–167.

Buckle, P. W. & Devereux, J. J. 2002. The nature of work-related neck and upper limb musculoskeletal disorders. *Applied Ergonomics*, 33, 207–217.

Bunning, K., Gona, J. K., Mung'ala-Odera, V., Newton, C., Geere, J., Hong, C. S. & Hartley, S. 2014. Survey of rehabilitation support for children 0–15 years in a rural part of Kenya. *Disability and Rehabilitation*, 36, 1033–1041.

Cheng, S. T. & Siankam, B. 2009. The impacts of the HIV/AIDS pandemic and socioeconomic development on the living arrangements of older persons in sub-Saharan Africa: a country-level analysis. *American Journal of Community Psychology*, 44, 136–147.

Chikoti, P. U. 2012. People's experience of urban development on the ground. Informality, slum upgrading, land management, and legislation in Malawi. Urban Landmark Conference. Johannesburg.

Cote, P., Cassidy, J. D., Carroll, L. J. & Kristman, V. 2004. The annual incidence and course of neck pain in the general population: a population-based cohort study. *Pain*, 112, 267–273.

Devi, A. & Bostoen, K. 2009. Extending the critical aspects of the water access indicator using East Africa as an example. *International Journal of Environmental Health Research*, 19, 329–341.

Doyal, L. & Hoffman, M. 2009. The growing burden of chronic diseases among South African women. *CME: Your South African Journal of CPD*, 20, 456–458.

Evans, B., Bartram, J., Hunter, P., Rhoderick Williams, A., Geere, J., Majuru, B., Bates, L., Fisher, M., Overbo, A. & Schmidt, W.-P. 2013. *Public Health and Social Benefits of At-house Water Supplies Final Report*. Leeds, University of Leeds.

Fathallah, F. A. 2010. Musculoskeletal disorders in labor-intensive agriculture. *Applied Ergonomics*, 41, 738–743.

Fejer, R., Kyvik, K. & Hartvigsen, J. 2006. The prevalence of neck pain in the world population: a systematic critical review of the literature. *European Spine Journal*, 15, 834–848.

Finch, E., Brooks, D., Stratford, P. W. & Mayo, N. E. 2002. *Physical Rehabilitation Outcome Measures*. Ontario, Lippincott Williams & Wilkins.

Geere, J. & Hunter, P. R. 2014. Personal history of water carriage is associated with self-reported pain location and ratings of general health. Water and Health Conference. Where science meets policy, University of North Carolina.

Geere, J. A., Hunter, P. R. & Jagals, P. 2010a. Domestic water carrying and its implications for health: a review and mixed methods pilot study in Limpopo Province, South Africa. *Environmental Health: A Global Access Science Source*, 9: 52.

Geere, J. L., Mokoena, M. M., Jagals, P., Poland, F. & Hartley, S. 2010b. How do children perceive health to be affected by domestic water carrying? Qualitative findings from a mixed methods study in rural South Africa. *Child: Care, Health and Development*, 36, 818–826.

Gona, J. K., Hartley, S. & Newton, C. R. 2006. Using participatory rural appraisal (PRA) in the identification of children with disabilities in rural Kilifi, Kenya. *Rural and Remote Health*, 6, 553.

Grandjean, E. 1988. *Fitting the Task to the Man. A textbook of occupational ergonomics*. London, Taylor Francis.

Groce, N., Bailey, N., Lang, R., Trani, J. F. & Kett, M. 2011. Water and sanitation issues for persons with disabilities in low- and middle-income countries: a literature review and discussion of implications for global health and international development. *Journal of Water & Health*, 9, 617–627.

Hamerman, D. 1997. Aging and the musculoskeletal system. *Annals of the Rheumatic Diseases*, 56, 578–585.

Hartley, S., Finkenflugel, H., Kuipers, P. & Thomas, M. 2009. Community-based rehabilitation: opportunity and challenge. *The Lancet*, 374, 1803–1804.

Hartley, S. E. 2006. *CBR as Part of Community Development. A poverty reduction strategy*. London, University College London, Centre for Child Health.

Hemson, D. 2007. The toughest of chores: policy and practice in children collecting water in South Africa. *Policy Futures in Education*, 5, 315–326.

Hoy, D., Toole, M. J., Morgan, D. & Morgan, C. 2003. Low back pain in rural Tibet. *The Lancet*, 361, 225–226.

Hoy, D., Geere, J., Davatchi, F., Meggitt, B. & Barrero, L. H. 2014. A time for action: Opportunities for preventing the growing burden and disability from musculoskeletal conditions in low- and middle-income countries. *Best Practice & Research Clinical Rheumatology*, 28: 1–17.

Jager, H. J., Gordon-Harris, L., Mehring, U. M., Goetz, G. F. & Mathias, K. D. 1997. Degenerative change in the cervical spine and load-carrying on the head. *Skeletal Radiol*, 26, 475–481.

Joosab, M., Torode, M. & Rao, P. V. 1994. Preliminary findings on the effect of load-carrying to the structural integrity of the cervical spine. *Surgical and Radiologic Anatomy*, 16, 393–398.

Kuipers, P., Wirz, S. & Hartley, S. 2008. Systematic synthesis of community-based rehabilitation (CBR) project evaluation reports for evidence-based policy: a proof-of-concept study. *BMC International Health and Human Rights*, 8, 3.

Lang, T. F. 2011. The bone-muscle relationship in men and women. *Journal of Osteoporosis*, http://dx.doi.org/10.4061/2011/702735

Leclerc, A., Chastang, J. F., Niedhammer, I., Landre, M. F. & Roquelaure, Y. 2004. Incidence of shoulder pain in repetitive work. *Occupational and Environmental Medicine*, 61, 39–44.

Levy, L. F. 1968. Porter's neck. *British Medical Journal*, 2, 16–19.

Li, C.-Y. & Sung, F.-C. 1999. A review of the healthy worker effect in occupational epidemiology. *Occupational Medicine*, 49, 225–229.

Linton, S. J. 2001. Occupational psychological factors increase the risk for back pain: a systematic review. *Journal of Occupational Rehabilitation*, 11, 53–66.

Lloyd, R., Parr, B., Davies, S. & Cooke, C. 2010. Subjective perceptions of load carriage on the head and back in Xhosa women. *Applied Ergonomics*, 41, 522–529.

Messing, K. & Östlin, P. 2006. *Gender Equality, Work and Health: A review of the evidence*. Geneva, WHO Press.

Nantulya, V. N. & Reich, M. R. 2002. The neglected epidemic: road traffic injuries in developing countries. *British Medical Journal*, 324, 1139–1141.

Nuckley, D. J., Van Nausdle, J. A., Eck, M. P. & Ching, R. P. 2007. Neural space and biomechanical integrity of the developing cervical spine in compression. *Spine*, 32, E181–E187.

Nuckley, D. J., Hertsted, S. M., Ku, G. S., Eck, M. P. & Ching, R. P. 2002. Compressive tolerance of the maturing cervical spine. *Stapp Car Crash Journal*, 46, 431–440.

Nugent, R. 2008. Chronic diseases in developing countries health and economic burdens. *Annals of the New York Academy of Sciences*, 1136, 70–79.

Porter, G., Hampshire, K., Dunn, C., Hall, R., Levesley, M., Burton, K., Robson, S., Abane, A., Blell, M. & Panther, J. 2013. Health impacts of pedestrian head-loading: a review of the evidence with particular reference to women and children in sub-Saharan Africa. *Social Science & Medicine*, 88, 90–97.

Sorenson, S. B., Morssink, C. & Campos, P. A. 2011. Safe access to safe water in low income countries: water fetching in current times. *Social Science & Medicine*, 72, 1522–1526.

Stemper, B. D., Yoganandan, N., Pintar, F. A., Maiman, D. J., Meyer, M. A., Derosia, J., Shender, B. S. & Paskoff, G. 2008. Anatomical gender differences in cervical vertebrae of size-matched volunteers. *Spine*, 33, E44–E49.

Sultana, F. 2009. Fluid lives: subjectivities, gender and water in rural Bangladesh. *Gender, Place & Culture*, 16, 427–444.

Taylor, J. & Twomey, L. 2005. Whiplash injury and neck sprain: a review of its prevalence, mechanisms, risk factors and pathology. *Critical Reviews in Physical and Rehabilitation Medicine*, 17, 285–299.

Taylor, J. R. & Finch, P. 1993. Acute injury of the neck: anatomical and pathological basis of pain. *Annals of the Academy of Medicine, Singapore*, 22, 187–192.

Thompson, J., Porras, I. T., Wood, E., Tumwine, J. K., Mujwahuzi, M. R., Katui-Katua, M. & Johnstone, N. 2000. Waiting at the tap: changes in urban water use in East Africa over three decades. *Environment and Urbanization*, 12, 37–52.

Twomey, L. & Taylor, J. 1988. Age changes in the lumbar spinal and intervertebral canals. *Paraplegia*, 26, 238–249.

Vos, T., Flaxman, A. D., Naghavi, M., Lozano, R., Michaud, C. & Murray, C. J. L. 2012. Years lived with disability (YLDs) for 1160 sequelae of 289 diseases and injuries 1990–2010: a systematic analysis for the Global Burden of Disease Study 2010. *The Lancet*, 380, 2163–2196.

Waddell, G. 1992. Biopsychosocial analysis of low back pain. *Baillière's Clinical Rheumatology*, 6, 523.

Waddell, G. 2006. Preventing incapacity in people with musculoskeletal disorders. *British Medical Bulletin*, 77–78, 55–69.

Walker-Bone, K. & Cooper, C. 2005. Hard work never hurt anyone – or did it? A review of occupational associations with soft tissue musculoskeletal disorders of the neck and upper limb. *Annals of the Rheumatic Diseases*, 64, 1112–1117.

WHO. 2002. Chapter 1. Violence – A global public health problem. In: Krug, E. G., Dahlberg, L. L., Mercy, J. A., Zwi, A. B. & Lozano, R. (eds.) *World Report on Violence and Health*. Geneva, WHO Press.

WHO. 2004. Chapter 3. Risk Factors. In: Peden, M., Scurfield, R., Sleet, D. A., Mohan, D., Hyder, A. A., Jarawan, E. & Mathers, C. (eds.) *World Report on Road Traffic Injury Prevention*. Geneva, WHO Press.

WHO. 2011. Chapter 6. Enabling Environments. In: *World Report on Disability*. Geneva, World Health Organization and the World Bank.

WHO. 2014. Data on the size of the HIV/AIDS epidemic: Data by WHO region. Geneva, WHO.

WHO & UNICEF. 2014a. Progress on drinking water and sanitation – 2014 update. Geneva, World Health Organization and UNICEF.

WHO & UNICEF. 2014b. *WASH POST-2015: proposed targets and indicators for drinking-water, sanitation and hygiene*. Geneva, WHO Press.

8

HAZARDS FROM *LEGIONELLA**

Richard Bentham

ASSOCIATE PROFESSOR, FLINDERS UNIVERSITY,
ADELAIDE, SOUTH AUSTRALIA, AUSTRALIA

Learning objectives

1 Appreciate the role of bioaerosols in water-borne disease.
2 Identify water systems in the built environment that may be sources of disease.
3 Understand the importance of the exposure to different water sources in the demographic transmission of disease.

Legionella and disease

Environmental ecology

The *Legionella* family (*Legionellaceae*) is a group of bacteria found in a wide and diverse range of environmental niches, but in particular in fresh water. In the natural environment they are heterotrophs, organisms that feed on complex organic molecules such as amino acids. To source these organic nutrients they have developed a range of different strategies. All of these strategies depend on deriving their nutrients from other organisms in either natural or anthropogenic water systems. As a result they are always part of a complex community in the environment and not easy to isolate as pure cultures in the laboratory. Another important ecological feature is that they have a general growth range of between 20 and 45 °C, but survive at both higher and lower temperatures, though 55 °C is widely regarded as the upper limit for survival (Fields et al., 2002; WHO, 2007, Chs. 1, 2).

Legionella are colonisers of biofilm (microbial slimes) and it is suggested that they do not multiply as free living planktonic organisms in the water column. Attachment to surfaces is therefore an important part of their ecology. Within attached biofilm *Legionella* have been shown to interact with protozoa, algae and other bacteria. They may also draw nutrients

---

* Recommended citation: Bentham, R. 2015. 'Hazards from *Legionella*', in Bartram, J., with Baum, R., Coclanis, P.A., Gute, D.M., Kay, D., McFadyen, S., Pond, K., Robertson, W. and Rouse, M.J. (eds) *Routledge Handbook of Water and Health*. London and New York: Routledge.

from the biofilm matrix and multiply independently of other microorganisms (WHO, 2007, Ch. 1, 2; Taylor et al., 2009).

The interaction with protozoa is particularly import from a health and water perspective. Protozoa are a large and diverse group of simple single celled eukaryotic organisms. Some of these organisms graze on bacteria and organic matter contained in biofilms. In adverse environmental conditions some of this group of protozoa, the amoebae, may form dormant 'cysts'. These cysts are resistant to desiccation and to high concentrations of disinfectants (up to 50 mg/L free chlorine) (Newsome et al., 2001; Thomas et al., 2004; WHO, 2007, Chs. 1, 2; Taylor et al., 2009).

Usually amoebae ingest bacteria and digest them in food vacuoles. Under some environmental conditions the ingestion of *Legionella* bacteria results in a different outcome. The *Legionella* bacteria have a sophisticated pathway of response to amoebal ingestion. Effectively, *Legionella* disables the amoebal ingestion process and then recruits the amoebal biochemistry for replication of *Legionella* cells within the vacuole. This culminates in the release of cytotoxins which break open the amoebal host releasing large numbers of *Legionella* bacteria (Newsome et al., 2001; Fields et al., 2002; WHO, 2007, Ch.1).

Human disease

The primary human disease caused by *Legionella* is Legionnaires' disease a severe pneumonia that may have a fatality rate of as much as 25 per cent of those affected. The secondary cause of disease is Pontiac fever, a non-fatal flu-like illness with a short incubation period and short duration. Initial symptoms may present as a cold/flu-like malaise, including coughing, muscle aches, dizziness and nausea. These progress to a 'full blown' pneumonia with respiratory distress, spiking fever, disorientation and multi-organ involvement. *Legionella* infections are not laterally (person to person) transmitted. The introduction of the *Legionella* bacteria into the lung from contaminated water is an undisputed cause of infection. This may be via inhalation of aerosol or by aspiration of water droplets. Aerosol inhalation appears to be the route of transmission of disease for the majority of reported cases of disease, though the contribution of aspiration through ingestion of contaminated water is clearly significant. Cases of disease are usually individuals who are immune compromised by smoking, existing respiratory disease, age (>50 years unless there are other predisposing factors), high alcohol intake or deliberate immune suppression (transplant recipients). By far the highest risk factor is smoking regardless of other considerations. The incubation period before disease is evident can be anything from 1 to 21 days after exposure, though the median is between 5 and 7 days. This long incubation period makes identifying a source of exposure problematic (Fields et al., 2002; WHO, 2007, Ch. 1).

To cause disease the *Legionella* bacteria must enter the alveolus of the lung, the smallest pockets of gaseous exchange. Aerosol is able to penetrate these areas as the droplet size is usually 5 um or less. Droplet sizes of greater magnitude than 5 um are unlikely to enter the alveolus. Within the alveolus are lung phagocytes; these are amoeba-like white blood cells (lymphocytes) that act as a barrier to microbial infection (Fields et al., 2002; WHO, 2007).

In the case of Legionnaires' disease the immune system of compromised individuals has poor recognition of the *Legionella* bacteria. The *Legionella* bacteria invade the phagocyte using the same mechanism as employed for amoebic infection. This 'case of mistaken identity' results in a severe pneumonia. The release of cytotoxins to break open the host cell walls after *Legionella* multiplication also results in non-specific damage to other lung tissues. These toxins may also be transported through the vascular system and damage other organs (WHO, 2007).

Water systems

Legionnaires' disease is an infection that is exclusively associated with engineered water systems. Engineered water systems are those that supply industrial, commercial and domestic supplies. The design and operation of built environment water systems may readily fulfil all of the requirements for the organisms colonization, multiplication, aerosoliation and transmission to susceptible individuals.

By their nature engineered water systems have a large surface area, and in some areas very large surface area to volume ratios. This presents the opportunity for biofilm formation and so the colonization by *Legionella* and protozoa. The presence of biofilm in engineered water systems is well established and an ongoing concern regarding maintenance and provision of water quality (Newsome et al., 2001; Thomas et al., 2004).

Legionella colonization of water distribution systems will inevitably result in transmission of *Legionella* to building water supplies. Once colonized these systems may then elevate water temperature and distribute to outlets where aerosolization or aspiration can occur.

The combination of large surface areas for biofilm growth, the resistance of protozoan host to disinfection, elevated water temperatures and delivery of aerosol/contaminated water explains why Legionnaires' disease is so strongly associated with engineered water systems (Berk et al., 1998; Newsome et al., 2001; WHO, 2007).

The range of water systems and the configurations that contribute to risk of disease are detailed in the next section.

The built environment

Industrial water systems

In particular, cooling water systems and cooling towers have been associated with disease and were implicated as the source of the first recognized outbreak of disease in 1977 (Field et al., 2002). Cooling towers are associated with heat rejection from large buildings. Their function is usually to cool refrigerant for such applications as air-conditioning and computer suites. Essentially cooling towers cool water by evaporation to the atmosphere. The cooled water is then piped to a heat load and then returned to the cooling tower. The point of heat exchange between water and coolant is typically a large surface area at optimal temperatures for *Legionella* and protozoan growth. Warmed water from the heat exchange process is returned to the cooling tower. At the cooling tower deliberate water splashing and the use of fans maximize the cooling of the water. Unfortunately it also maximizes the production and distribution of aerosol (WHO, 2007, Chs. 2, 5).

Although aerosol release is minimized, the volumes released by cooling towers are considerable. Between 0.001 and 0.01 per cent of the circulating volume (or 1 and 10 mL) per litre passing though the tower is lost to the atmosphere. This aerosol has been demonstrated to travel large distances and multiple outbreaks have demonstrated *Legionella* infection caused at between 1 and 1.5 km from the source and in some instances much further. This clearly demonstrates the inhalation of aerosol as a cause of respiratory disease (Brown et al., 1999).

An interesting conundrum is that in the case of cooling tower outbreaks of disease only *Legionella pneumophila* Serogroup 1, a specific strain, has been associated with disease. Though other strains and organisms are commonly present in cooling water systems they have not been proven to cause disease outbreaks. So far this phenomenon is unexplained.

Other industrial systems have also been sources of disease. The predetermining factors have always been generation of aerosol and operating temperatures between 20 and 50 °C. Examples of such systems are water based metal working fluids, high pressure hoses and process cooling waters (WHO, 2007, Ch. 5).

Potable/residential water systems

Potable water systems and in particular warm water systems are a significant cause of exposure and disease. Though domestic hot water systems appear to be a cause of disease, data on the extent of the problem is still very vague. Hot water systems by design circulate large volumes of water at temperatures conducive to *Legionella* and protozoa multiplication. Systems in larger buildings are often recirculating which prevents the normal 'flushing' of the organisms from the system and aids their distribution. The outlets from these sources are clearly the point of transmission of disease (Bollin et al., 1985; Newsome et al., 2001; Fields et al., 2002).

Showers, faucets, humidifiers and other outlets provide areas for biofilm colonization and aerosol production. The outlets themselves are usually at some distance from the last area of temperature control. Recirculating hot water systems will normally operate at water temperatures above the growth and survival of *Legionella* bacteria (<60 °C). After leaving the recirculation the water temperature is usually tempered to less than 50 °C to avoid scalding (Bollin et al., 1985; WHO, 2007).

Evidence suggests that the mode of transmission from these systems may be via aerosol inhalation, but may also be via aspiration. This means that a potable supply may pose a risk of infection either by inhalation of the organisms or ingestion of contaminated water followed by aspiration (Fields et al., 2002; WHO, 2007).

Epidemiological data suggest that Legionnaires' disease from these types of systems is usually associated with direct exposure to the source. Spa pools and 'hot tubs' are common causes of disease from this category of systems. A range of different *Legionella* strains and species may cause infection (Fields et al., 2002). Clearly the contamination comes from the potable supply, but then enters a nutrient rich circulating environment with optimal growth temperatures. The aeration and circulation of heated water is conducive to the development of biofilms, growth of *Legionella* and protozoa and production of aerosol. In contrast to cooling towers, aerosol does not travel far from these devices. However, there is unavoidable direct exposure to aerosol of all individuals using or in close proximity to the spa pool. There exists the possibility of ingestion and aspiration as a source of infection from these devices (den Boer et al., 2002; WHO, 2007).

Other aerosol generating equipment has also been associated with disease, such as decorative fountains and grocery misting devices. In these instances often the location of lighting within or close to the devices raises water temperature above 20 °C facilitating the multiplication of the bacteria and subsequent dissemination of the contaminated aerosol. Infections are usually associated with close proximity to these devices; long distance travel of aerosol has not been reported (WHO, 2007).

Health care systems

Given the predisposing risk factors for disease it is no surprise that health care premises are a major source of notified disease. Infection is most commonly associated with the potable water system and contaminated showers and faucets, though other devices such as humidifiers, hydrotherapy pools and ice machines have also been identified as sources of infection.

Within the health care setting it is important to keep water temperatures low enough to minimize the risk of scalding. Scalding causes more cases of mortality and morbidity each year than *Legionella* infections. This means water is delivered at temperatures conducive to *Legionella* growth (43–45 °C). The large recirculating systems delivering water that has been deliberately tempered to outlets in these facilities is an obvious source of both aerosol inhalation and ingestion and aspiration (WHO, 2007).

Health issues

Community acquired disease

Community acquired disease occurs both as sporadic cases and as outbreaks. In sporadic cases the sources are often not identified due to the multiplicity of possible exposures of the affected individual. The source may be a cooling water system, a shower, a spa pool, a decorative fountain or contaminated potable water. The mode of transmission may be aerosol inhalation or aspiration; the relative contributions of these two exposure routes have not been determined. Combined with the long incubation period and the relatively common colonization of systems, definitively identifying the source of a single case is rare.

Outbreaks of disease are usually 'explosive' with a number of cases being identified within a short period due to a point exposure. Cooling tower associated outbreaks will commonly result in cases spread over significant areas. The clear mode of transmission is by aerosol inhalation, and climatic factors such as humidity and sunlight will affect how far aerosol may be transmitted. Outbreaks may range from a few individuals to hundreds of cases. Though hampered by the large number of cooling towers operating in the built environment and the incubation period, the identification of a source is quite common. This is because the multiple cases allow a 'triangulation' of the likely exposure area based on where individuals have been during the incubation period (Fields et al., 2002; WHO, 2007).

Outbreaks from spa pools, fountains and misting devices are usually (but not always) smaller than those from cooling towers. Instances where large numbers of people have been affected are limited to trade displays where systems were not disinfected and there was an unusually large number of people exposed to the system. Aerosol generation and travel is more limited and as a result the affected individuals have usually had either direct contact or been in close proximity to the infected source. Both aerosol inhalation and aspiration are potential routes of transmission of the disease from these sources (den Boer et al., 2002).

Nosocomial disease

Rather than sporadic or outbreak scenarios, health care facilities tend to have prolonged ongoing cases of disease. Systems may be systemically or locally contaminated, that is, the whole system has colonization by the organism or only one or a few outlets. The variability of the immune status of the occupants and the complexity of the water systems and their potential for systemic or local contamination often leads to small numbers of cases over prolonged periods. This pseudo-sporadic presentation of disease means that outbreaks may remain undetected for prolonged periods (Fields et al., 2002; WHO, 2007).

Aged care, organ transplant and oncology facilities form a subset of high risk areas within health care premises. This is due to the immune compromised status of the population. It is also evident that individuals who have undergone throat and neck surgery are prone to infection. This is probably as a result of a compromised swallowing reflex that increases the likelihood of aspiration as a route of transmission (WHO, 2007).

A range of different *Legionella* species and strains have been associated with nosocomial disease; this is most probably a function of the immune status of those exposed. Although *Legionella pneumophila* infection predominates, other *Legionella* species may also cause infection (WHO, 2007).

Summary

Legionnaires' disease is an infection associated with contamination of water supplies in the built environment, and could be considered a 'first world disease' in that identifiable disease is limited to countries where complex water systems are more prolific. Whether this 'first world' definition is a product of the jurisdictional capabilities for disease surveillance or provision of sophisticated water systems is uncertain. The factors leading to disease are the requirements of these systems to unintentionally provide a suitable niche for *Legionella* colonization. Water held between 20 and 50 °C in systems with large surface areas to support biofilm growth that release aerosol or provide contaminated water that may be aspirated are the features common to all cases of disease in the built environment.

Key recommended readings (open access)

1   World Health Organization (2007) Legionella and the Prevention of Legionellosis. http://www.who.int/water_sanitation_health/emerging/legionella.pdf. A comprehensive review of the organism, its ecology and disease presentations. Epidemiology and approaches to managing the colonization of the built environment and minimizing potential for disease.
2   Fields, B.S., Benson, R.F., Besser, R.E. (2002) Legionella and legionnaires' disease: 25 years of investigation. *Clinical Microbiology Review* 15(3): 506–526. http://cmr.asm.org/content/15/3/506.full.pdf+html. A microbiological review of the organism. Includes its ecology, disease pathogenesis and epidemiology as well as sources of disease and factors contributing towards outbreaks.

References

Berk, S.G., Ting, R.S., Turner, G.W., et al. (1998) Production of respirable vesicles containing live Legionella pneumophila cells by two *Acanthamoeba* spp. *Appl. Environ. Microbiol.* 64(1): 279–286.

Bollin, G.E., Plouffe, J.F., Para, M.F., and Hackman, B. (1985) Aerosols containing Legionella pneumophila generated by shower heads and hot-water faucets. *Appl. Environ. Microbiol.* 50: 1128–1131.

Brown, C.M., Nuorti, P.J., Breiman, R.F., Hathcock, A.L., Fields, B.S., Lipman, H.B., Llewellyn, G.C., Hofmann, J., and Cetron, M. (1999) A community outbreak of legionnaire's disease linked to hospital cooling towers: an epidemiological method to calculate dose of exposure. *Internat. J. Epidemiol.* 28: 353–359.

den Boer, J.W., Yzerman, E.P., Schellekens, J., Lettinga, K.D., Boshuizen, H.C., Van Steenbergen, J.E., Bosman, A., Van den Hof, S., Van Vliet, H.A., Peeters, M.F., Van Ketel, R.J., Speelman, P., Kool, J.L., and Conyn-Van Spaendonck, M.A. (2002) A large outbreak of legionnaires' disease at a Dutch flower show. *Emerg. Infect. Dis.* 8: 1.

Fields, B.S., Benson, R.F., and Besser R. (2002) Legionella and legionnaires' disease: 25 years of investigation. *Clin Microbiol Rev.* 15(3): 506–526.

Newsome, A.L., Farone, M.B., Berk, S.G., and Gunderson, J.H. (2001) Free living amoebae as opportunistic hosts for intracellular bacterial parasites. *J. Eukaryot. Microbiol.* 48: 13S–14S.

Taylor, M., Ross, K., and Bentham, R. (2009) *Legionella*, protozoa, and biofilms: interactions within complex microbial systems. *Microb. Ecol.* 58: 538–547.

Thomas, V., Bouchez, T., Nicolas, V., Robert, S., Loret, J.F., and Lévi, Y. (2004) Amoebae in domestic water systems: resistance to disinfection treatments and implication in Legionella persistence. *J. Appl. Microbiol.* 97: 950–963.

World Health Organization (2007) *Legionella* and the Prevention of Legionellosis. http://www.who.int/water_sanitation_health/emerging/legionella.pdf.

9

TOXIC CYANOBACTERIA*

Ron W. Zurawell

WATER QUALITY SCIENTIST, ALBERTA MONITORING EVALUATION
AND REPORTING AGENCY, EDMONTON, ALBERTA, CANADA

Learning objectives

1 Understand the environmental conditions contributing to the excessive growth and predominance of cyanobacteria in surface waters.
2 Understand the diverse array of toxic compounds produced by various species of cyanobacteria.
3 Appreciate the potential risk and symptoms of exposure to cyanotoxins.

Cyanobacteria occurrence and toxicity

Cyanobacteria, commonly called blue-green algae, are an assemblage of photosynthetic bacteria that inhabit nearly all surface waters and moist terrestrial environments. In the context of evolution, cyanobacteria are considered archaic organisms that predate algae and plants on earth. As such, they have developed unique adaptations for nutrient uptake, buoyancy regulation, and persistence during unfavorable growing conditions. A physically stable water column, warm surface water temperatures (18–30 °C), high nutrient concentrations (phosphorus [P] and nitrogen [N]), low ratios of N relative to P, high alkalinity/pH, and low carbon dioxide (CO_2) availability, are advantageous for growth and reproduction of cyanobacteria, allowing them to outcompete algae. Cyanobacteria often dominate plankton communities seasonally in eutrophic temperate lakes or perennially in eutrophic subtropical lakes and can produce large surface accumulations or 'blooms' that impede both recreation and domestic water uses (Sommer et al., 1986; Zohary and Robarts, 1989).

It is important to note that excessive cyanobacterial growth and formation of surface blooms are completely natural phenomena in surface waters situated in nutrient-rich basins (e.g. Prairie Provinces and States of Canada and the U.S.). However, anthropogenic eutrophication – increased nutrient loading or 'fertilization' of surface waters – resulting from

---

* Recommended citation: Zurawell, R.W. 2015. 'Toxic cyanobacteria', in Bartram, J., with Baum, R., Coclanis, P.A., Gute, D.M., Kay, D., McFadyen, S., Pond, K., Robertson, W. and Rouse, M.J. (eds) *Routledge Handbook of Water and Health*. London and New York: Routledge.

human development and activities in watersheds (e.g. agriculture, deforestation, urban and industrial development), is contributing to a global increase in the occurrence and severity of blooms. As a result, many oligotrophic (nutrient deficient) and mesotrophic (low nutrient) surface waters across North America are gradually becoming more nutrient enriched and are experiencing ecological shifts to cyanobacteria-dominated aquatic communities.

It is also important to consider the influence climate warming may have on the ecology of surface waters. A straightforward implication of warming climate is higher surface water temperatures leading to increasing growth of common bloom-forming species, as rate of cyanobacteria reproduction peaks at about 30°C, above which the rate declines. Less obvious is the impact warming climate has on the ice-free period for north temperate lakes. An earlier spring ice-melt can result in rapid stratification and lengthier periods of water column stability. This not only influences growth rate, but more importantly the onset, severity and duration of blooms. Evidence suggests that climate is a stronger determining factor of cyanobacteria abundance in deeper, meso- and mildly eutrophic surface waters than shallow excessively nutrient-rich, hyper-eutrophic systems (Taranu et al., 2012).

Cyanobacterial blooms can cause numerous issues that: (1) impede recreation (e.g. offensive odors, reduced water clarity); (2) limit its use for drinking water (i.e. production of non-toxic odorous substances geosmin, 2-methylisoborneol and -cyclocitral impart taste and odor to finished drinking water); and (3) threaten natural aquatic biota like sport fish (i.e. decay of cyanobacterial blooms can lead to increasing ammonia concentrations and depletion of dissolved oxygen) (Paerl, 1988; Kenefick et al., 1992). However, it is the production of potent hepato- (liver) and neuro-toxins by several species and strains of cyanobacteria that present the greatest potential risk to humans and domestic and wild animals. The community composition – abundance and growth stage of various species and strains – of cyanobacteria plays a key role in determining the type and concentrations of toxins present at a given time. Cyanobacteria communities and hence blooms are both temporally and spatially dynamic; and they can be complex (i.e. comprising many species) or quite simple (i.e. dominated by one or two species). Notably, toxins are often present at varying concentrations in water prior to the formation of surface blooms and while some blooms can be extremely toxic possessing high concentrations of several toxins simultaneously, other blooms may contain trace levels or even none at all.

Hepato- (liver) toxins

It is generally acknowledged that microcystin (MCYST) is the most prevalent toxin produced by cyanobacteria, but rather than being a single toxin of concern, MCYSTs comprise a group of more than 80 described analogues. MCYSTs are small monocyclic peptides composed of seven amino acids (Figure 9.1). Numerous toxic and non-toxic analogues result from alterations and substitutions of constituent amino acids. The cyclic structure of MCYSTs imparts stability to heat and, thus, boiling is not an effective means against the risk of exposure to these toxins in drinking water.

MCYSTs exert toxicity by binding to and inhibiting specific enzymes – protein phosphatase types 1 and 2A – that play key roles in a multitude of cell functions including cell division, cell-to-cell signaling and cell metabolism (Key recommended reading #3). The primary target organs of MCYSTs and other cyanotoxins are shown in Table 9.1. MCYSTs are water soluble and, as a result, an active uptake mechanism is required for toxicity. Toxin transport is actively mediated by certain organic anion transporter polypeptides that are expressed in mammalian liver cells – this explains the liver-specific toxicity of MCYSTs (Dietrich and Hoeger, 2005; König et al., 2006). Another polypeptide that is capable of MCYST uptake is

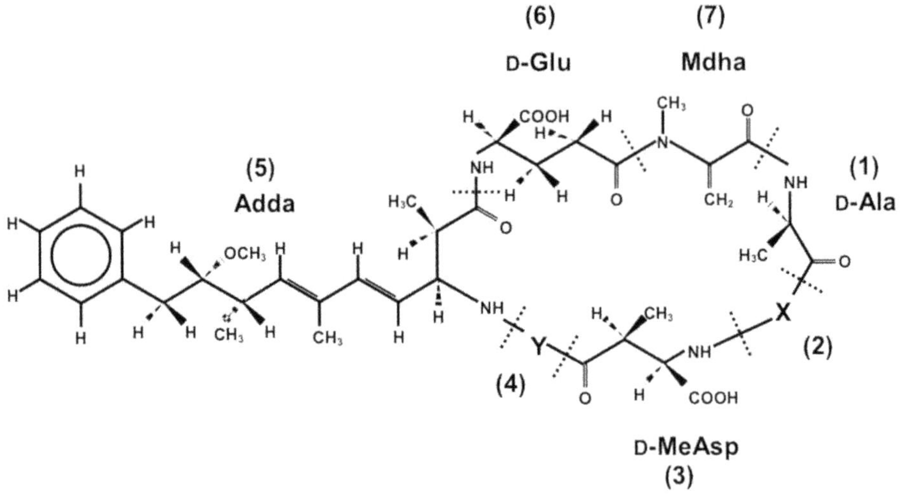

Figure 9.1 Generalized chemical structure of MCYST

Position (1) is D-Alanine; (2) X is a variable L-amino acid; (3) is D-erythro-β-methylaspartic acid; (4) Y is another variable L-amino acid; (5) is Adda, (2S, 3S, 8S, 9S)-3-amino-9-methoxy-2,6,8-trimethyl-10-phenyldeca-4,6-dienoic acid; (6) is D-Glutamic acid and (7) is N-methyldehydroalanine (from Zurawell et al., 2005)

expressed in a variety of cells including bile duct epithelial, kidney and intestinal epithelial cells, but predominantly in blood capillary endothelium of the brain (Fischer et al., 2005; Lee et al., 2005). This finding corroborates evidence indicating the toxin can cross the blood–brain and blood–cerebrospinal fluid barriers and may explain observed acute neurotoxicity in those fatally exposed to MCYST in Caruaru, Brazil in 1996 (Pouria et al., 1998; Dietrich and Hoeger, 2005). More recently, several additional transporter polypeptides have been identified in the testes and spermatogonia of animals and humans (Klaassen and Aleksunes, 2010; Svoboda et al., 2011; Zhou et al., 2012). This provides some explanation of decreased testicular and epididymis weights, decreased sperm concentration, viability and motility, increased sperm abnormalities and changes in reproductive hormone levels reported in animals (Kirpenko et al., 1981; Ding et al., 2006; Li et al., 2008, 2011; Chen et al., 2011, 2013).

A number of bloom-forming cyanobacteria (Table 9.1) produce MCYSTs and it's likely the cosmopolitan distribution of these species is responsible for the high global prevalence of these toxins. As a result, the World Health Organization (WHO) has recommended drinking and recreational water guidelines for MCYSTs (see Key recommended reading #3). MCYSTs severely damage the structure and function of liver cells, representing a significant threat to humans, domestic animals and wildlife. Exposure to elevated toxin levels may induce symptoms including gastrointestinal pain, nausea, vomiting, severe diarrhea, headaches and fever. They have been shown to act as tumor promoters, and chronic exposure to MCYSTs has been linked to an increased incidence of primary liver (hepatocellular carcinoma) and colorectal cancers in rural human populations in China; however, it is not clear whether dietary, genetic, lifestyle factors associated with colorectal cancer, as well as other potential biological and chemical contaminants (e.g. aflatoxin), were considered (Zhou et al., 2002). Other studies have also shown MCYSTs to induce oxidative DNA damage in liver cell isolates, which suggests the ability to initiate cancer (Žegura et al., 2003).

Table 9.1 Cyanotoxins and their primary target organs

Toxin	Cyanobacterial general/species	Health effects
Cyclic peptides		
Microcystins	Microcystis aeruginosa, M. flos-aquae, M. wesenbergii, *Dolichospermum (formerly Anabaena) flos-aquae, *D. circinalis, *D. lemmermannii, Planktothrix agardhii (Oscillatoria) and P. rubescens,	Liver lesions, tumor promoter, possible carcinogen
Nodularins**	Nodularia, *Dolichospermum sp., Planktothrix (Oscillatoria)	Liver lesions, tumor promoter
Alkaloids		
Anatoxin-a	*Dolichospermum flos-aquae, D. circinalis, D. lemmermannii, D. macrosporum, D. planctonicum, D. spiroides, Aphanizomenon flos-aquae, Cylindrospermum sp., Planktothrix agardhii, Raphidiopsis sp. and Oscillatoria sp.	Affects nerve synapse, interrupting normal neural impulses to muscles
Anatoxin-a(S)	*Dolichospermum flos-aquae, D. lemmermannii, D. spiroides and (possibly) P. agardhii	Affects nerve synapse, interrupting normal neural impulses to muscles
Cylindrospermopsin	**Chrysosporum (formerly Anabaena) bergii, **Chrysosporum (formerly Aphanizomenon) ovalisporum, Cylindrospermopsis raciborskii, Umezakia natans, Raphidiopsis curvata	Affects liver and possibly kidney; possibly genotoxic and carcinogenic
Lyngbyatoxin-a, aplysiatoxins	Lyngbya	Dermatitis and eye irritation, gastrointestinal symptoms
Saxitoxin, neo-saxitoxin	Aph. flos-aquae, D. circinalis, Cylindrospermopsis raciborskii and Lyngbya wollei	Affects nerve axons

Note: *Dolichospermum sp. and **Chrysosporum sp. represent recent taxonomic classification revisions

Cylindrospermopsin

Cylindrospermopsin is a tricyclic alkaloid that is highly water-soluble, heat-stable and potentially more persistent than MCYSTs in varying light and pH conditions. It is known to be produced by several bloom-forming species of cyanobacteria (Table 9.1) originally reported in tropical waters. However, it appears that one of the primary producers, *Cylindrospermopsis raciborskii*, is increasingly being reported in temperate waters. Cylindrospermopsin causes extensive damage primarily to liver and kidney and although the mode of toxicity is not completely understood, it appears the toxin inhibits protein synthesis by interfering with protein elongation (Falconer, 2005). Exposure to cylindrospermopsin has been shown to induce morphological changes in red blood cells in animals, which may be linked to the cylindrospermopsin-mediated effects on the liver and kidney. Animal toxicity studies also suggest that cylindrospermopsin may be carcinogenic. Symptoms of cylindrospermopsin exposure include nausea, vomiting, diarrhea, abdominal tenderness, pain and acute liver failure. Notably, clinical symptoms may immediately not show up following exposure, but may occur several days later. Currently, few countries have implemented drinking or recreational water quality guidelines for cylindrospermopsin.

Neurotoxins

Cyanobacterial neurotoxins represent a group of compounds that generally act to disrupt the normal propagation of neural impulses to muscles, causing paralysis and death via respiratory failure and asphyxia. Neurotoxins include: anatoxin-a, anatoxin-a(s) and the (neo-) saxitoxins (commonly known as paralytic shellfish poisons). These compounds are potentially lethal to humans and have been deemed responsible for animal deaths in North America, the United Kingdom, Scandinavia and Australia – the LD_{50} values (the dose lethal to 50 percent of experimental populations) for anatoxin-a, anatoxin-a(s) and saxitoxin based on intraperitoneal injection in mice is 200, 20, and 10 $\mu g/kg$, respectively (Carmichael, 1994; Sivonen and Jones, 1999).

Anatoxin-a is a low molecular weight secondary amine produced by several species of cyanobacteria (Table 9.1). It mimics acetylcholine – the principal neurotransmitter within the peripheral nervous system and neuro-modulator within the central nervous system. Anatoxin-a has a greater ($\approx$20 times) affinity than acetylcholine for receptors on the post-synaptic membranes of muscles. Unlike acetylcholine however, binding is irreversible and anatoxin-a is not hydrolyzed by acetylcholinesterase (James et al., 2007). Affected muscles continue to be stimulated causing muscular twitching, fatigue and, eventually, paralysis. Intoxication may induce muscle cramping, decreased movement, collapse, exaggerated abdominal breathing, cyanosis, convulsions and paralysis. Severe overstimulation of respiratory muscles can result in rapid death from respiratory arrest and asphyxia. Compared to hepatotoxic cyanotoxins, anatoxin-a is relatively unstable in the environment as it is rapidly degraded by sunlight (oxygen independent photodegradation) and bacteria belonging to genus *Pseudomonas*. The rapid degradation of anatoxin-a following the collapse of cyanobacterial blooms makes monitoring environmental levels of this toxin difficult. Due to the sporadic occurrence and difficulty monitoring this toxin, few countries have implemented water quality guidelines for anatoxin-a.

Anatoxin-a(s) is a naturally occurring organophosphate produced principally by species of *Dolichospermum* (formerly *Anabaena*) (Table 9.1). Like other organophosphates including paraoxon (the active metabolite of the parathion insecticide) and the chemical-warfare nerve agent sarin, anatoxin-a(s) irreversibly binds to and inhibits acetylcholinesterase (Mahmood and Carmichael, 1986). Although anatoxin-a(s) is chemically unrelated to anatoxin-a, both toxins result in similar over-stimulation of the post-synaptic membrane, blocking subsequent nerve impulse leading ultimately to fatigue and paralysis (Carmichael, 1994). Anatoxin-a(s) is highly toxic to mammals and causes symptoms consistent with that of synthetic organophosphates exposure such as hyper-salivation [the 's' in anatoxin-a(s) was coined for excessive salivation occurring in intoxicated animals and is consistent with symptoms in humans exposed to sarin nerve agent] lacrimation, urinary incontinence, defecation and convulsions. Death by respiratory arrest/asphyxiation can be rapid when respiratory muscles are affected. Like anatoxin-a, anatoxin-a(s) degrades relatively quickly in the environment, but the rate is influenced by pH – it would be expected to decompose more quickly in alkaline waters as opposed to neutral or acidic conditions. This characteristic, along with the lack of commercially available standards required for analytical assessment are reasons anatoxin-a(s) is infrequently monitored in surface waters and few countries have implemented water quality guidelines.

Saxitoxins comprise a group of related carbamate alkaloids including non-sulphated (saxitoxin/neosaxitoxin), single sulphated (gonyautoxins) and double sulphated (C1 and C2 toxins) variants referred to as paralytic shellfish poisons (PSPs). PSPs are part of a larger group of associated shellfish-related poisons. These toxins occur during blooms of eukaryotic (dinoflagellate) algae – so called 'red tide' events – in marine coastal regions that are responsible for seasonal closures of shellfish harvesting. Specifically, the dinoflagellate algae *Alexandrium fundyense* produces saxitoxin, but other species can produce different toxic compounds (e.g. *Karenia brevis* produces brevetoxin, as known as neurotoxic shellfish poison; the diatom *Pseudonitzschia*, produces domoic acid, as known as amnesic shellfish poison). Interestingly, within freshwaters these PSPs are produced by several cyanobacteria including *Aphanizomenon*, *Dolichospermum*, *Lyngbya* and *Cylindrospermopsis* (Table 9.1) and not eukaryotic algae. Saxitoxin inhibits nerve impulse propagation along axons by blocking sodium ion entry into nerve cells through sodium channels. This effectively suppresses stimulation of muscles. Mild PSP intoxication causes slight tingling and numbness of the lips. With higher doses, tingling and numbness moves to the extremities, leading to loss of control and flaccid paralysis, leaving the affected individuals calm and conscious through the progression of symptoms. Like other cyanobacterial neurotoxins, death from respiratory arrest can occur. Saxitoxin is water-soluble and heat-stable (boiling water and cooking won't destroy it) and like anatoxin-a(s) is more persistent in acidic environments compared to alkaline environments. Reported half-lives for saxitoxin vary from 9 to 28 days and closely related gonyautoxins may persist beyond three months in natural waters (Jones and Negri, 1997). Commercial standards of saxitoxin are available and screening tests and quantitative analyses of water and food are common.

Other compounds of concern

In addition to the above groups of toxins, cyanobacteria are known to produce other potentially toxic compounds. Dermatotoxic alkaloids – lyngbyatoxin and aplysiatoxins – can elicit severe dermatitis and eye irritation on contact, while ingestion or inhalation can cause mild to severe nose and throat irritation. Lipopolysaccharides expressed by most cyanobacteria act as general skin irritants upon contact. These compounds can produce symptoms (including red, blotchy, raised skin rash) similar to – and thus mistaken for – common swimmer's itch resulting from infection by *Schistosoma* cercaria. Moreover, ingestion of cyanobacterial dermato- and hepato-toxins can produce symptoms similar to acute schistosomiasis including: headache, fever, severe abdominal pain and diarrhea.

Given similarities in symptoms to other water-borne (and even food-borne) illnesses, determining the cause of human illness resulting from recreating in cyanobacterial infested waters can be difficult. This problem is made even more difficult by the high degree of variability in the appearance of cyanobacteria (influenced by species morphology and pigmentation and growth stage) and their blooms (influenced by degree of accumulation, wind dispersal, natural senescence, light induced photolysis and decay). Positive identification of cyanobacteria requires training and familiarity and is critical to understanding risk to human health.

Summary

Cyanobacteria and their toxins can have a severe impact on both recreational water activities and drinking water supplies. Acute as well as long-term health impacts may result from exposure to cyanotoxins. It is generally recognized that conditions leading to excessive cyanobacteria growth in fresh water include a combination of high nutrient concentrations and favorable environmental conditions (e.g. light, temperature, stratification). Yet it remains difficult to predict when a cyanobacterial bloom will occur and if it will contain toxins. The current long-term strategies to control or minimize the growth of cyanobacteria in water sources focus on limiting nutrient inputs to the water. Ongoing, systematic monitoring and assessment of water supplies for factors contributing to cyanobacterial blooms is used to manage the risks. Many countries have developed protocols for managing the risks to bathers when blooms occur in recreational water, as well as response plans for reducing the concentration of cyanotoxins in drinking water. Some of the options for securing a safe drinking water supply may include effective water treatment, switching to an alternate water supply, or changing the location (distance and depth) of the water supply intake.

Key recommended readings

1 Botana, L.M. (2007) *Phycotoxins: Chemistry and Biochemistry*, Blackwell Publishing, 368 pp. Available at: http://onlinelibrary.wiley.com/book/10.1002/9780470277874. This book presents information on 16 phycotoxins including those of cyanobacterial origin. While major emphases are given to (bio) chemistry, aspects of origin, toxicology and analytical methodology are also addressed.
2 Federal Environment Agency (Umweltbundesamt) (2012) *Current Approaches to Cyanotoxin Risk Assessment, Risk Management and Regulations in Different Countries*, Chorus, I. (ed.), 151 pp. Available at: http://www.uba.de/uba-info-medien-e/4390.html. This is a compilation of regulatory approaches to cyanotoxins in several countries around the world. Contributions do not necessarily represent authorized government positions, but rather the personal views of the authors.
3 World Health Organization (WHO) (1999) *Toxic Cyanobacteria in Water: A Guide to Their Public Health Consequences, Monitoring and Management*, Chorus, I. and Bartram, J. (eds.), E & FN Spon, London, 416 pp. Available at: http://www.who.int/iris/bitstream/10665/42827/http://apps.who.int/iris/bitstream/10665/42827/1/0419239308_eng.pdf?ua=1. Prepared by an international group of experts, this book presents a comprehensive review on the topic of cyanobacteria and their associated toxins. Chapters are dedicated to examining the need to protect drinking water, recreational waters and other water supplies from contamination by toxic cyanobacteria and to control their impact on human health.

References

Carmichael, W.W. (1994) 'The toxins of cyanobacteria', *Sci. Am.*, 270, pp. 78–86.

Chen, L. et al. (2013) 'The interactive effects of cytoskeleton disruption and mitochondria dysfunction lead to reproductive toxicity induced by microcystin-LR', *PLOS ONE*, 8 (1), pp. 1–11 (available at: www.plosone.org/article/info%3Adoi%2F10.1371%2Fjournal.pone.0053949).

Chen, Y. et al. (2011) 'Decline of sperm quality and testicular function in male mice during chronic low-dose exposure to microcystin-LR', *Reprod. Tox.*, 31, pp. 551–557.

Dietrich, D. and Hoeger, S. (2005) 'Guidance values for microcystins in water and cyanobacterial supplement products (blue-green algal supplements): A reasonable or misguided approach?', *Toxicol. Appl. Pharmacol.*, 203, pp. 273–289.

Ding, X. et al. (2006) 'Toxic effects of *Microcystis* cell extracts on the reproductive system of male mice', *Toxicon*, 48, pp. 973–979.

Falconer, I.R. (2005) *Cyanobacterial Toxins of Drinking Water Supplies: Cylindrospermopsins and Microcystins.* Boca Raton, FL: CRC Press.

Fischer, W.J. et al. (2005) 'Organic anion transporting polypeptides expressed in liver and brain mediate uptake of microcystin', *Toxicol. Appl. Pharmacol.*, 203, pp. 257–263.

James, K.J. et al. (2007) 'Anatoxin-a and analogues: Discovery, distribution, and toxicology', in Botana, L.M. (ed.) *Phycotoxins: Chemistry and Biochemistry*. New York: Blackwell Publishers, pp. 141–158.

Jones, G.J. and Negri, A.P. (1997) 'Persistence and degradation of cyanobacterial paralytic shellfish poisons (PSPs) in freshwaters', *Water Res.*, 31, pp. 525–533.

Kenefick, S.L. et al. (1992) 'Odorous substances and cyanobacterial toxins in prairie drinking water sources', *Water Sci. Technol.*, 25, pp. 147–154.

Kirpenko, Y.A. et al. (1981) 'Some aspects concerning remote after effects of blue-green algae toxin impact on animals' in Carmichael, W.W. (ed.) *The Water Environment: Algal Toxins and Health*. New York: Plenum Press, pp. 257–270.

Klaassen, C.D. and Aleksunes, L.M. (2010) 'Xenobiotic bile acid and cholesterol transporters: Function and regulation', *Pharmacol. Rev.*, 62, pp. 1–96 (as cited in Zhou et al., 2012).

König, J. et al. (2006) 'Pharmacogenomics of human OATP transporters', *J. Naunyn-Schmiedeberg's Arch. Pharmacol.*, 372, pp. 432–443.

Lee, W. et al. (2005) 'Polymorphisms in human organic anion-transporting polypeptide 1A2 (OATP1A2): Implications for altered drug disposition and central nervous system drug entry', *J. Biol. Chem.*, 280, pp. 9610–9617.

Li, D. et al. (2011) 'Toxicity of cyanobacterial bloom extracts from Taihu Lake on mouse, *Mus musculus*', *Ecotoxicology*, 20, pp. 1018–1025.

Li, Y. et al. (2008) 'The toxic effects of microcystin-LR on the reproductive system of male rats in vivo and in vitro', *Reprod. Toxicol.*, 26, pp. 239–245.

Mahmood, N.A. and Carmichael, W.W. (1986) 'The pharmacology of anatoxin-a(s), a neurotoxin produced by the fresh-water cyanobacterium *Anabaena flos-aquae* NRC 525-17', *Toxicon*, 24, pp. 425–434.

Paerl, H.W. (1988) 'Nuisance phytoplankton blooms in coastal, estuarine and inland waters', *Limnol. Oceanogr.*, 33, pp. 823–847.

Pouria, S. et al. (1998) 'Fatal microcystin intoxication in haemodialysis unit in Caruaru, Brazil', *Lancet*, 352, pp. 21–26.

Sivonen, K. and Jones, G. (1999) 'Cyanobacterial toxins' in Chorus, I. and Bartram, J. (eds.) *Toxic Cyanobacteria in Water: A Guide to Their Public Health Consequences, Monitoring and Management.* E & FN Spon, London, pp. 41–111.

Sommer, U. et al. (1986) 'The PEG-model of seasonal succession of planktonic events in fresh waters', *Arch. Hydrobiol.*, 106, pp. 433–471.

Svoboda, M. et al. (2011) 'Organic anion transporting polypeptides (OATPs): Regulation of expression and function', *Curr. Drug Metab.*, 12, pp. 139–153.

Taranu, Z. et al. (2012) 'Predicting cyanobacterial dynamics in the face of global change: The importance of scale and environmental context', *Global Change Biol.*, 18, pp. 3477–3490.

Žegura, B. et al. (2003) 'Microcystin-LR induces oxidative DNA damage in human hepatoma cell line HepG2', *Toxicon*, 41, pp. 41–48.

Zhou, L. et al. (2002) 'Relationship between microcystin in drinking water and colorectal cancer', *Biomed. Environ. Sci.*, 15, pp. 166–171.

Zhou, Y. et al. (2012) 'The toxic effects of microcystin-LR of rat spermatogonia in vitro', *Toxicol. Lett.*, 212, pp. 48–56.

Zohary, T. and Robarts, R.D. (1989) 'Diurnal mixed layers and the long-term dominance of *Microcystis aeruginosa*', *J. Plank. Res.*, 11, pp. 25–48.

Zurawell, R.W. et al. (2005) 'Hepatotoxic cyanobacteria: A review of the biological importance of microcystins in freshwater environments', *J. Toxicol. Environ. Health. Part B*, 8, pp. 1–37.

10

CHEMICAL HAZARDS*

Lisa Smeester, Andrew E. Yosim and Rebecca C. Fry

Department of Environmental Sciences and Engineering,
Gillings School of Global Public Health, University of
North Carolina, Chapel Hill, North Carolina, USA

Learning objectives

1 Describe several high priority chemical contaminants of global relevance found in drinking water.
2 Detail sources and routes of exposure of those contaminants.
3 Understand the potential health effects of such exposures in human populations.

Introduction

Contaminants in drinking water can be microbiological, radiological, physical or chemical. In this chapter, we summarize the health effects associated with exposure to high priority chemical contaminants. A thorough understanding of the detrimental health effects caused by unsafe water is imperative as it can influence prioritization for remediation, and guide decisions related to protecting public health. It is important to note that while drinking water is one route of exposure for many chemicals, it is not always the dominant route of exposure. Thus, risk management decisions consider which targets for mitigation (water, food, environment, industry, etc.) will have the most impact.

The list presented here is based on a prioritization of chemicals affecting drinking water developed by the World Health Organization (WHO) to help in prioritizing risk management decisions (WHO, 2007). The contaminants in Table 10.1 include those given highest risk priority due to their natural occurrence in the environment (specifically, fluoride, arsenic and selenium), those likely to be present in drinking-water sources worldwide, and those known to cause adverse health effects. These priority contaminants can be categorized by three broad classifications: inorganic contaminants, organic contaminants and contaminants of emerging concern.

* Recommended citation: Smeester, L., Yosim, A.E. and Fry, R.C. 2015. 'Chemical hazards', in Bartram, J., with Baum, R., Coclanis, P.A., Gute, D.M., Kay, D., McFadyen, S., Pond, K., Robertson, W. and Rouse, M.J. (eds) *Routledge Handbook of Water and Health*. London and New York: Routledge.

Table 10.1 Guideline values for priority chemical contaminants in drinking water

	WHO guideline values (mg/L)	Exposure sources	Health considerations and physiologic targets
Inorganic contaminants			
Arsenic	0.010 mg/L*	Naturally occurring; industries	Cardiovascular, dermal, gastrointestinal, hepatic, neurologic/neuro-behavioral, respiratory, and reproductive effects. Carcinogenic
Fluoride	1.5 mg/L	Naturally occurring; industries	Dermal, gastrointestinal and musculoskeletal effects
Lead	0.01 mg/L	Naturally occurring; industries; production and distribution	Cardiovascular, gastrointestinal, hematological, musculoskeletal neurologic/neuro-developmental, renal and reproductive effects. Probable carcinogen
Nitrates (NO3-)	50 mg/L**	Agriculture; human settlement; industries	Hematological effects. Probable carcinogen
Nitrites (NO2-)	3 mg/L	Agriculture; human settlement; industries	Hematological effects. Mutagenic. Probable carcinogen
Selenium	0.010 mg/L	Naturally occurring; industries	Dermal, dental and hair/nail effects
Organic contaminants			
Disinfection by-products			
Bromodichloromethane	0.060 mg/L	Industries; production and distribution; water treatment	Hepatic, renal effects. Possible carcinogen
Bromoform	0.100 mg/L	Industries; production and distribution; water treatment	Hepatic, neurologic and renal effects
Chloroform	0.300 mg/L	Industries; production and distribution; water treatment	Cardiovascular, developmental, hepatic, neurologic, renal and reproductive effects. Possible carcinogen

Dibromochloromethane	0.100 mg/L	Industries; production and distribution; water treatment	Hepatic, neurologic and renal effects. Possible carcinogen
Chloroacetic acid	0.02 mg/L	Industries; production and distribution; water treatment	Cytotoxic. Genotoxic. Possible carcinogen
Dichloroacetic acid	0.05 mg/L*	Industries; production and distribution; water treatment	Cytotoxic. Genotoxic. Possible carcinogen
Trichloroacetic acid	0.2 mg/L	Industries; production and distribution; water treatment	Cytotoxic. Genotoxic. Possible carcinogen
Organochloride pesticides			
Aldrin + Dieldrin	0.00003 mg/L	Agriculture; industries	Neurologic, reproductive and thyroid effects
Chlordane	0.0002 mg/L	Agriculture; industries	Hepatic and neurologic effects. Possible carcinogen
Dichlorodiphenyltrichloroethane (DDT) + metabolites	0.001 mg/L	Agriculture; industries; public health use as vector control (DDT)	Hepatic, immunologic, renal and reproductive effects. Possible carcinogen
Endrin	0.0006 mg/L	Agriculture; industries	Hepatic effects
Lindane	0.002 mg/L	Agriculture; industries	Hepatic and renal effects. Possible carcinogen

Note: *provisional (data are insufficient to ensure guideline value is achievable under a wide range of circumstances); ** short term exposure

Source: Adapted from WHO, 2011e; IARC, 2014

Inorganic contaminants

Arsenic

Arsenic is a water contaminant currently ranked as the highest priority contaminant of the Agency for Toxic Substances and Disease Registry (ATSDR) and is highly ranked by other agencies as well (ATSDR, 2011). Currently, many countries base their regulatory limits for arsenic in drinking water upon the WHO's provisional guideline value of 0.010 mg/L (WHO, 2011e).

It represents a critical health issue and is a significant cause of disease as it contaminates the water of more than 100 million people worldwide (Uddin and Huda, 2011). Inorganic arsenic (iAs) has been detected in drinking water sources in several areas of the world, including Bangladesh, Mexico, India, Vietnam and the United States (ATSDR, 2007a).

Arsenic can accumulate in certain groundwater drinking sources as a result of natural geologic occurence (ATSDR, 2007a). Some populations are exposed to arsenic through consumption of drinking water that is contaminated with arsenical pesticides or naturally occurring mineral deposits (NTP, 2011). Anthropogenic activities can also cause arsenic to accumulate in the environment, including metal mining and smelting, coal combustion, wood combustion and waste incineration (ATSDR, 2007a). Food can also be a source of arsenic with iAs as the predominant form found in meats, rice and cereal; organic arsenic as the predominant form in seafood (CDC, 2009; IARC, 2012). Occupational exposure to inorganic arsenic can also occur for workers involved in metal smelting, wood preservation, semiconductor manufacturing or glass production, among others (ATSDR, 2007a).

Health effects of inorganic arsenic include carcinogenesis (IARC, 2004; IARC, 2014). Cancers that have been associated with arsenic exposure include those of the digestive tract, kidney, liver, lung, skin, urinary bladder, and lymphatic and hematopoietic systems (NTP, 2011). Exposure to iAs has also been linked to non-cancer health effects, including adverse effects on memory and intellectual function, heart disease, liver hypertrophy, neurobehavioral effects, respiratory system disease, diabetes and several reproductive effects, including pregnancy complications and adverse effects on the developing fetus (Kapaj et al., 2006).

Exposure to arsenic in drinking water can be reduced through a variety of treatment processes, including chemical coagulation/filtration, adsorption, ion exchange resins, reverse osmosis and membrane filtration (EPA, 2000).

Fluoride

Fluorides are a class of naturally occurring and synthetic compounds that are derived from the highly reactive nature of fluorine (ATSDR, 2003a). The current WHO guideline for fluoride in drinking water is 1.5 mg/L (WHO, 2011e). Fluoride is considered essential by the WHO in preventing tooth decay (WHO, 2004; ATSDR, 2003a). However, at high levels, fluoride may lead to serious adverse health effects. Thus estimates of the dose of fluoride from all sources are very important.

In general, the toxicity of fluoride increases with its solubility, as it can be more readily absorbed following oral ingestion (ATSDR, 2003a). Approximately 50 million people worldwide have a drinking water source where the naturally occurring fluoride is at or around a concentration of 1 mg/L, with elevated levels occurring in Africa, China, the Middle East and some South Asian regions such as India and Sri Lanka (WHO, 2005a; Ayoob and

Gupta, 2006). Additionally, approximately 355 million people worldwide receive artificially fluoridated water (WHO, 2005a) as a public health measure to prevent tooth decay.

Fluoride is widely distributed in the environment and found naturally in drinking water at low levels, generally between 0.01 and 0.3 mg/L (WHO, 2004). Other sources that contribute to fluoride can be anthropogenic in nature, such as smelters, industrial manufacturing and textile factories, fertilizer plants and burning of fluoride-rich coal (Ayoob and Gupta, 2006). Community water supply fluoridation programs were initiated for the prevention of tooth decay (Peckham and Awofeso, 2014). Paradoxically, there is a fine line between benefit and risk as research suggests concentrations up to 1 mg/L are beneficial for optimal dental health, but concentrations above 1.5 mg/L increase the risk of dental fluorosis. At higher concentrations and over long periods of fluoride exposure, more serious adverse health effects such as skeletal fluorosis may develop (Ayoob and Gupta, 2006).

Adverse health effects are seen with both acute and long term exposure to elevated levels of fluoride. Acute toxicity is rare and usually due to accidental water contamination, fire or explosions. While initial acute toxicity may manifest in gastrointestinal symptoms such as nausea, vomiting and gastric distress (Ayoob and Gupta, 2006), the most detrimental effects are those on the skeletal system. The skeletal system is the primary site of accumulation in the body, allowing for the beneficial effects of fluoride in tooth decay prevention. Yet over-accumulation may result in irreversible dental fluorosis, or even the more serious skeletal fluorosis. Dental fluorosis manifests at concentrations in water between 1 and 3 mg/L and is indicated by hypocalcification, striations or stains on teeth. Skeletal fluorosis may occur with levels of 3–6 mg/L of fluoride (WHO, 2004) in water, leading to hardening and calcification of the bones which makes them more brittle, an increased risk of fracture, improper accumulation of bone tissue in and around joints, and in the most severe cases ossification of ligaments and cartilage (Meenakshi and Maheshwari, 2006).

It is both difficult and expensive to reduce naturally high levels of fluoride in drinking water sources. The best option in such a case is often to find an alternative source of water with lower levels of fluoride. Although challenging, exposure can be reduced through a variety of technologies, adsorption being the most common practice. Other techniques to reduce the fluoride content of drinking water include membrane filtration systems such as reverse osmosis or nanofiltration (Meenakshi and Maheshwari, 2006; Mohapatra et al., 2009).

Lead

Lead (Pb) is a heavy metal that occurs naturally in the environment and is collected from ore mined from the earth's crust (ATSDR, 2007b). The provisional WHO guideline value is 10 ug/L (0.010 mg/L) (WHO, 2011e; Villanueva et al., 2014). Most lead-related environmental exposures come from anthropogenic sources, with primary routes of exposure being food, drinking water, cigarette smoke and occupational exposures (ATSDR, 2007b). Comparatively, exposure to naturally occurring environmental levels of lead in drinking water is relatively low (WHO, 2011e); roughly 80 percent of an individual's daily intake of lead is attributable to non-water sources (WHO, 2011b).

Higher levels of contamination of potable water supplies can occur around lead mining sources, but occurs primarily from old pipes and pipe fittings containing lead. Additionally, lead compounds are found in many polyvinyl chloride (PVC) pipes and can be leached from them as well (WHO, 2011b). Lead exposure poses the greatest threat to children, the developing fetus and pregnant women. Lead has a higher relative gastrointestinal absorption in children (Ziegler et al., 1978) and can cross the blood–brain barrier, making it particularly detrimental to neurodevelopment (Sanders et al., 2009). Additionally, the mobilization of lead stored in bones is known to increase during pregnancy, and lead from the mother's bloodstream can cross the placental barrier (IARC, 2006).

Lead is a potent, acute toxicant, and the long term health effects of lead exposure are also a public health concern. Lead is a cumulative poison, a well-established neurotoxicant (Mason et al., 2014) and probable human carcinogen (IARC, 2006). Lead exposure has been associated with an increased risk of lung, stomach and urinary bladder cancers in human populations (IARC, 2006; ATSDR, 2007b).

As noted above, most lead exposure can be attributed to sources other than drinking water. Lead in drinking water is primarily due to leaching from plumbing systems and drinking water distribution pipes. In such cases, exposure can be reduced through controlling corrosion in the distribution system; replacing lead service lines as well as any lead-containing faucets or fittings in a house; installing carbon-based or other treatment devices that will remove lead at the tap; or flushing water that has been standing in pipes for several hours by allowing it to run.

Nitrate and nitrite

Nitrate (NO3-) and nitrite (NO2-) ions occur naturally as part of the environmental nitrogen cycle. Human activities such as certain agricultural practices (including inorganic potassium or ammonium nitrate fertilizer and organic nitrate livestock manures), wastewater treatment, nitrogenous waste products in human and other animal excreta, and discharges from industrial processes and motor vehicles are the most common sources of both nitrate and nitrite (WHO, 2007). The WHO guideline value is currently 50 mg/L in water (WHO, 2011a).

Nitrates occur naturally in both soil and water, with surface water concentrations ranging from 0 to 18 mg/L (WHO, 2011a). Globally, the mean concentration of nitrate in water has risen approximately 36 percent over the past two decades, with regions such as the Eastern Mediterranean and Africa experiencing contamination that has more than doubled. Nitrites may be introduced into drinking water via the distribution systems, either by oxidation of ammonia by nitrifying bacteria during stagnation in pipes or by an improper chloramination disinfection process (EPA, 1993).

The primary impact of excessive nitrate consumption on human health is its effect on blood hemoglobin during reduction to nitrite. Nitrite causes the oxidation of blood hemoglobin to methemoglobin, which is unable to bind oxygen and results in a condition called methemoglobinemia (WHO, 2011a). Since methemoglobinemia reduces the amount of oxygen that can be carried in the blood, hypoxia-related symptoms such as labored breathing, cyanosis, decreased mental acuity and fatigue are seen. Historically it was believed infants were at greatest risk of water-related nitrate health effects; however, a more recent assessment of the epidemiological evidence suggests it is more likely a combination of other sources of nitrate exposure (e.g. processed baby foods) as well as genetic deficiencies in hemoglobin structure or metabolic enzymes which leaves them particularly vulnerable (Richard et al., 2014).

Mechanistically, there is a concern that the formation of N-nitroso compounds through endogenous nitrosation of nitrates and nitrites may contribute to carcinogenesis. Epidemiological studies have not demonstrated an association between nitrate/nitrite contaminated drinking water and cancers of the gastrointestinal system (esophagus, stomach, bladder and colon), nevertheless nitrate and nitrite are classified as group 2A probable carcinogens based on animal studies (IARC, 2010). It should also be noted that while epidemiological studies have shown a positive correlation between levels of nitrate and hypertrophy of the thyroid gland (van Maanen et al., 1994), effects on thyroid hormone dysregulation have been inconclusive as nitrate was associated with lowered thyroid stimulating hormone (TSH) among women (van Maanen et al., 1994), while increased levels of TSH was reported in a children's cohort (Tajtakova et al., 2006).

Reducing exposure to nitrate from drinking water generally includes management of activities within the watershed/aquifer to control agricultural inputs, treatment to decrease nitrate levels in the water supply, and management of nitrification in the distribution system. Wells that are located in agricultural areas are susceptible to nitrate and nitrite contamination, particularly shallow wells. Water containing levels of nitrate and/or nitrite above guideline levels should not be used to prepare formula or other foods for infants. For wells that persistently have nitrate or nitrite test results above guideline or regulatory levels, installing a drinking water treatment device, using an alternative drinking water source, or relocating or drilling a deeper well to reach a safe supply should be considered (Health Canada, 2013).

Selenium

Selenium is an essential nutrient with antioxidant properties found naturally in the environment in rocks and soil, as well as surface and ground waters, and is released through both natural processes and human activities (ATSDR, 2003b). Selenium exposure occurs mainly through food. Low levels of exposure also occur through air and via drinking water (WHO, 2011c). Elevated drinking water concentrations of selenium may occur in water sources near mineral deposits and those contaminated by agricultural irrigation drainage (ATSDR, 2003b). However, even in high selenium areas, the relative contribution of selenium from drinking water is likely to be small in comparison with that from locally produced food (WHO, 2011e).

Nonetheless the WHO has established a provisional drinking water guideline value for selenium of 0.04 mg/L (40 ug/L) (WHO, 2011e).

Like fluoride, selenium offers a protective health benefit at low levels. Thus various national and international organizations have established recommended daily intakes of selenium. However, exposure to selenium at excessive levels over the long term can lead to selenosis. Symptoms of selenosis, or selenium poisoning, include brittle hair and resulting hair loss, nail deformities, skin discoloration, dental discoloration and decay (ATSDR, 2003b; WHO, 2011c).

A number of technologies can be used to ensure the selenium content of drinking water is minimized, including membrane filtration, electrodialysis, ion exchange and adsorbents such as activated alumina (Kapoor et al., 1995; WHO, 2011c). Other options include using alternative sources, and blending low-selenium sources with high-selenium sources.

Organic contaminants

Disinfection by-products

Unintended chemical contamination can arise from efforts to deliver safe drinking water through disinfection. One of the most prevalent barriers to safe drinking water is microbiological contamination (see Chapter 5). Given the global scope of waterborne pathogen-related morbidity and mortality, disinfection of drinking water supplies is a critical step in reduction of disease burden. Disinfection is frequently achieved through the addition of chemicals to water. There are many chemical disinfectants; among the most common are chlorine, chlorine dioxide and chloramines. Such chlorine-containing disinfectants are strong oxidants and can react with naturally occurring organic matter found in drinking water, leading to the formation of disinfection by-products (DBPs) such as trihalomethanes (THMs) and haloacetic acids (HAAs).

Because chlorine is the most cost-effective and widely used disinfectant, it is not surprising that its THM by-products bromodichloromethane, bromoform, chloroform and dibromochloromethane are the most commonly monitored in drinking water. While the WHO does not address an overall guideline value, regulatory limits for total THMs have been set by the US EPA (80 ug/L), the European Union and Canada (100 ug/L) (WHO, 2005b). The WHO guidelines levels for each of the THM by-products can be found in Table 10.1. HAAs are the second most common group of DBPs formed after disinfection with chlorine (Hua and Reckhow, 2007). The US EPA currently regulates five HAAs (known as the HAA_5) at a limit of 0.06 mg/L, measured as the sum of the concentrations of five monitored HAA species: chloroacetic acid, dichloroacetic acid, trichloroacetic acid, bromoacetic acid and dibromoacetic acid. Many municipal bodies are switching to chloramine as a chlorine alternative, as chloramination produces lower concentrations of THMs compared to chlorination. However, chloramine is associated with increased formation of nitrogenated and iodinated DBPs (Richardson and Postigo, 2012).

Toxicity studies using animal models indicate chronic exposure to specific THMs is associated with the formation of fatty liver cysts (WHO, 2005b). Mammalian *in vitro* assays have shown genotoxic and cytotoxic (Plewa et al., 2010), as well as clastogenic (Escobar-Hoyos et al., 2013) effects upon exposure to specific HAAs. While there is no definitive toxicologic data indicating carcinogenicity of THMs, some epidemiological studies have found an increased risk of cancers of the colon and urinary bladder associated with THMs (Villanueva et al., 2004). However, such epidemiological research should carefully consider that DBPs are, by nature, mixtures themselves, and people are likely co-exposed to DPBs and other contaminants present in drinking water supplies.

An emerging class of DBPs, iodinated DBPs (iodo-DBPs), are primarily formed as a reaction between chloraminated or chlorinated drinking water and iodide from either natural or anthropogenic sources including point-of-use water treatment (Barceló, 2012). Although iodo-DBPs are not currently regulated by the US EPA (EPA, 2013), mammalian *in vitro* studies have shown that, as a group, iodo-DBPs are more genotoxic and cyotitic then their chlorinated or brominated counterparts (Plewa et al., 2004; Richardson et al., 2008). In fact, iodacetic acid is the most genotoxic DBP studied in mammalian cells (Richardson et al., 2008).

Although much effort is made to reduce the presence of DBPs in drinking water, such actions must never compromise the effectiveness of disinfection. A number of strategies exist to reduce the formation of DBPs, the most effective method being source management to reduce the presence of DBP precursors such as natural organic matter (NOM) and synthetic compounds which react with oxidizing disinfectants to produce DBPs (EPA, 2007).

Organochlorine pesticides

All pesticides are toxic by their very definition; however, one class of pesticides, organochlorine pesticides (OCP), are of particular concern due to well documented detrimental health effects (Crinnion, 2009) and because their persistence in the environment leads to ongoing exposure risk. OCPs are highly toxic, minimally biodegradable, and as they are fat-soluble they are difficult to excrete, leading to bioaccumulation in adipose tissue (Crinnion, 2009; Seth, 2014). OCPs are biomagnified, meaning they become more concentrated as they move up the food chain (Crinnion, 2009). These properties lead to many OCPs being banned by the Stockholm Convention (UNEP, 2001); however, OCPs are still used in developing countries due to low cost of production and for public health reasons, such as the use of dichlorodiphenyltrichloroethane (DDT) to combat malaria (WHO, 2011d).

Pesticides are used globally in agriculture, as well as for public health purposes such as disease vector control in many developing nations. Pesticide runoff can directly contaminate water bodies and residue may leach through the soil into groundwater (Seth, 2014). In industrial nations, many pesticides are not routinely detected in drinking water and, of those that are, it is at such low concentrations that they do not appear to impact human health in most cases (Fawell, 2012). However, OCPs have been detected in surface- and ground drinking water sources of India (Kaushik et al., 2012), China (Bao et al., 2012; Wu et al., 2014a, 2014b), Pakistan (Azizullah et al., 2011) and Turkey (Aydin et al., 2013), as well as in the Aral Sea Drainage Basin in Central Asia (Tornqvist et al., 2011), in some cases exceeding the maximum admissible levels.

Current regulatory limits for individual OCPs vary. The WHO sets guideline values for aldrin and dieldrin combined, chlordane, DDT and its metabolites combined, endrin and lindane (see Table 10.1), whereas the European Union sets regulatory limits for pesticides, regardless of chemical class, at 0.0001 mg/L per pesticide, not to exceed a total of 0.0005 mg/L for all pesticides detected in drinking water.

Health effects of OCP exposure include impaired lipid metabolism and obesity, as well as endocrine and metabolic dysregulation (Androutsopoulos et al., 2013). Epidemiological studies also indicate a link between exposure to the DDT metabolite dichlorodiphenyldichloroethylene (DDE) and type 2 diabetes (Taylor et al., 2013). Particularly vulnerable to the effects of OCP exposure are pregnant women and their fetuses. Studies have shown detectable levels of OCPs in breast milk (Fujii et al., 2012), and that exposure to OCPs may lead to dysregulation of maternal thyroid hormones (Lopez-Espinosa et al., 2009). Moreover, studies have shown that OCPs can cross the placental barrier directly affecting the fetus (Waliszewski et al., 2000). Even at extremely low levels of *in utero* exposure there is risk of neural tube birth defects (Ren et al., 2011), as well as decreased birth weight and head circumference (Lopez-Espinosa et al., 2011).

While traditional water treatment processes are generally ineffective at reducing the levels of organochlorine pesticides in drinking water (Miltner et al., 1989), the most common methodology is the use of adsorbents such as granular activated carbon (GAC). A number of other strategies are effective including advanced oxidation processes, ultra or nanofiltration, and reverse osmosis (EPA, 2011).

Contaminants of emerging concern

Contaminants of emerging concern (CEC) are a broad category of pollutants not commonly included in monitoring programs despite having the potential to be released into the environment and result in adverse health effects from exposure. "Emerging" does not necessarily mean a new pollutant. While many CECs have been in use for decades, they may be of emerging concern because recent advances in detection technology have made it possible to detect extremely low levels of substances in drinking water or due a better understanding of their potential to cause harm to human health (Sauve and Desrosiers, 2014). One group of CECs that has gained widespread attention is endocrine disruptors (EDs), which include numerous industrial chemicals and pharmaceuticals and personal care products (PPCPs). The endocrine system is responsible for maintaining the body's homeostasis, as well as playing a role in reproduction and development. Endocrine disrupting chemicals (EDCs) interfere with the normal function of the endocrine system, and may lead to adverse health effects.

Because of their widespread use, EDCs enter the environment through a variety of sources including industrial discharge, agricultural and surface water runoff, effluents from wastewater treatment plants and even through excretion (Caliman and Gavrilescu, 2009). EDCs can be persistent and may accumulate through the recycling of wastewater effluents to be used as drinking water, as standard waste water treatments do not entirely remove many of these chemicals (Caliman and Gavrilescu, 2009). Endocrine disrupting CECs have been detected in both ground and surface waters, as well as finished water, in countries across the globe (Hemminger, 2005; Benotti and Snyder, 2009; Gavrilescu et al., 2015).

EDCs interfere with hormone signaling in various ways, including mimicking endogenous hormones to activate a response, as well as binding to a receptor thus blocking endogenous hormones from interaction and signaling. Industrial chemicals used in the manufacturing of plastics such as Bisphenol A (BPA) and phthalates, brominated flame retardants (BFR), and PPCPs such as reproductive hormones and steroids, musk fragrances, and antimicrobials, are all known EDs (Rahman et al., 2009; Witorsch and Thomas, 2010).

Potential health effects of BPA include infertility, reproductive and sexual dysfunction, adverse birth outcomes such as miscarriage, shorter gestational age, low birth weight, and metabolic disorders such as diabetes and insulin resistance (Rochester, 2013). Additionally, animal studies indicate there is concern that pre- and perinatal exposures may lead to neural alterations as well as altered development and accelerated puberty (NTP-CERHR, 2008).

Phthalates have been shown to reduce testosterone levels in animal models, leading to malformations of the external genitalia and internal reproductive organs, as well as other "demasculinizing" effects (Foster, 2006). A review of epidemiological evidence suggests exposure to certain phthalates may affect reproductive health, including shorter gestational age, premature breast development, altered semen quality and motility, and shortened male anogenital index (Hauser and Calafat, 2005).

In vitro studies of BFRs have been shown to dysregulate several endocrine signaling pathways (Ren and Guo, 2013). A meta-review of epidemiological data suggests exposure to BFRs, specifically polybrominated diphenyl ethers (PBDEs), may lead to reproductive health effects, alteration in thyroid function, developmental issues and diabetes (Kim et al., 2014).

Mechanistic studies in zebrafish have shown synthetic musks to have antiestrogenic effects (Schreurs et al., 2004); however, despite detectable levels of synthetic musk fragrances found in blood plasma and breast milk (Hutter et al., 2005, 2010), to date little is known about potential health effects in humans.

The antimicrobials triclosan (TCS) has been shown to disrupt thyroid hormone signaling in animal models, possibly hinting to a role in thyroid mediated developmental disorders. While there is a paucity of data on endocrine dysregulation in human subjects, TCS is found at detectable levels in urine, and human breast milk (Dann and Hontela, 2011).

While traditional water treatment processes do not fully remove the full range of EDCs, further treatment with ozone, granular activated carbon (GAC) adsorption, and membranes are effective mitigation strategies (Rahman et al., 2009; Caliman and Gavrilescu, 2009).

Although the levels of individual industrial use and PPCP EDCs currently found in drinking water are unlikely to result in adverse health effects, the potential for additive effects is not known. Chemicals in mixtures interact in complex ways. More research and scientific evidence is needed to identify the human and environmental effects of chronic low level exposure to such mixtures.

Conclusions

This chapter details some of the chemical contaminants found in drinking water which have potential for detrimental human health effects. The sources of these contaminants can vary from natural geologic occurrence to industrial sources and from the by-products of the drinking water treatment process used to reduce the risk from microbiological contaminants. The health effects upon exposure can include both cancer and non-cancer endpoints. Of particular concern is the potential exposure to humans during susceptible times of development, such as the *in utero* period representing a critical time of fetal development or in young children where the toxicities of these chemicals may result in long-lasting harm to human health. More research on the health impacts is required for these vulnerable groups and timing of exposure. For every chemical it is important to consider whether water is the dominant route of exposure as this information is critical to target mitigation strategies.

Key recommended readings (open access)

1 Villanueva, C. M., Kogevinas, M., Cordier, S., Templeton, M. R., Vermeulen, R., Nuckols, J. R., Nieuwenhuijsen, M. J. & Levallois, P. 2014. Assessing exposure and health consequences of chemicals in drinking water: current state of knowledge and research needs. *Environ Health Perspect*, 122, 213–221. Concise overview of known and emerging water contaminants, highlighting the toxicity and carcinogenicity of chemicals, associated health effects, and exposure assessment recommendations.
2 Fawell, J. 2012. Chemicals in the water environment. Where do the real and future threats lie? *Ann Ist Super Sanita*, 48, 347–353.
   Discussion of the complexity of water contamination issues including variation in distribution from diverse water sources, the paradox of chemicals offering both beneficial and harmful effects in a dose dependent manner, and the need for advanced analytical methods to address emerging contaminants.

References

Androutsopoulos, V. P., Hernandez, A. F., Liesivuori, J. & Tsatsakis, A. M. 2013. A mechanistic overview of health associated effects of low levels of organochlorine and organophosphorous pesticides. *Toxicology*, 307, 89–94.

ATSDR. 2003a. Toxicological Profile for Fluorides, Hydrogen Fluoride, and Fluorine. Agency for Toxic Substances and Disease Registry. http://www.atsdr.cdc.gov/toxprofiles/tp11.pdf

ATSDR. 2003b. Toxicological Profile for Selenium. Agency for Toxic Substances and Disease Registry. http://www.atsdr.cdc.gov/toxprofiles/tp92.pdf

ATSDR. 2007a. Toxicological Profile for Arsenic. CAS#: 7440-38-2, i-500. Agency for Toxic Substances and Disease Registry. http://www.atsdr.cdc.gov/toxprofiles/tp2.pdf

ATSDR. 2007b. Toxicological Profile for Lead. Agency for Toxic Substances and Disease Registry. http://www.atsdr.cdc.gov/toxprofiles/tp13.pdf

ATSDR. 2011. *The Priority List of Hazardous Substances That Will Be the Subject of Toxicological Profiles.* Available: http://www.atsdr.cdc.gov/SPL/index.html.

Aydin, M. E., Ozcan, S., Beduk, F. & Tor, A. 2013. Levels of organochlorine pesticides and heavy metals in surface waters of Konya closed basin, Turkey. *Scientific World Journal*, 2013, 849716.

Ayoob, S. & Gupta, A. K. 2006. Fluoride in drinking water: a review on the status and stress effects. *Critical Reviews in Environmental Science and Technology*, 36, 433–487.

Azizullah, A., Khattak, M. N., Richter, P. & Hader, D. P. 2011. Water pollution in Pakistan and its impact on public health – review. *Environ Int*, 37, 479–497.

Bao, L. J., Maruya, K. A., Snyder, S. A. & Zeng, E. Y. 2012. China's water pollution by persistent organic pollutants. *Environ Pollut*, 163, 100–108.

Barceló, D. 2012. *Emerging Organic Contaminants and Human Health*, New York, Springer.

Benotti, M. J. & Snyder, S. A. 2009. Pharmaceuticals and endocrine disrupting compounds: implications for ground water replenishment with recycled water. *Ground Water*, 47, 499–502.

Caliman, F. & Gavrilescu, M. 2009. Pharmaceuticals, personal care products and endocrine disrupting agents in the environment – a review. *CLEAN – Soil, Air, Water*, 37, 277–303.

CDC (Centers for Disease Control and Prevention). 2009. *Fourth national report on human exposure to environmental chemicals*. Department of Health and Human Services, Atlanta, GA.

Crinnion, W. J. 2009. Chlorinated pesticides: threats to health and importance of detection. *Altern Med Rev*, 14, 347–359.

Dann, A. B. & Hontela, A. 2011. Triclosan: environmental exposure, toxicity and mechanisms of action. *J Appl Toxicol*, 31, 285–311.

EPA, U.S. 1993. Methods for determination of inorganic substances in environmental samples. U.S. Environmental Protection Agency. http://monitoringprotocols.pbworks.com/f/EPA600-R-63-100.pdf

EPA, U.S. 2000. Technologies and Costs for Removal of Arsenic from Drinking Water. http://water.epa.gov/drink/info/arsenic/upload/2005_11_10_arsenic_treatments_and_costs.pdf

EPA, U.S. 2007. Simultaenous Compliance Guidance Manual for the Long Term 2 and Stage 2 DBP Rules. http://www.epa.gov/ogwdw/disinfection/stage2/pdfs/guide_st2_pws_simultaneous-compliance.pdf

EPA, U.S. 2011. Finalization of Guidance on Incorporation of Water Treatment Effects on Pesticide Removal and Transformations in Drinking Water Exposure Assessments. http://www.epa.gov/pesticides/science/efed/policy_guidance/team_authors/water_quality_tech_team/wqtt_dw_treatment_effects_removal_transformation.htm#tab_4

EPA, U.S. 2013. Basic Information about Disinfection Byproducts in Drinking Water: Total Trihalomethanes, Haloacetic Acids, Bromate, and Chlorite. http://water.epa.gov/drink/contaminants/basicinformation/disinfectionbyproducts.cfm

Escobar-Hoyos, L. F., Hoyos-Giraldo, L. S., Londono-Velasco, E., Reyes-Carvajal, I., Saavedra-Trujillo, D., Carvajal-Varona, S., Sanchez-Gomez, A., Wagner, E. D. & Plewa, M. J. 2013. Genotoxic and clastogenic effects of monohaloacetic acid drinking water disinfection by-products in primary human lymphocytes. *Water Res*, 47, 3282–3290.

Fawell, J. 2012. Chemicals in the water environment. Where do the real and future threats lie? *Ann Ist Super Sanita*, 48, 347–353.

Foster, P. M. 2006. Disruption of reproductive development in male rat offspring following in utero exposure to phthalate esters. *Int J Androl*, 29, 140–147; discussion 181–185.

Fujii, Y., Ito, Y., Harada, K. H., Hitomi, T., Koizumi, A. & Haraguchi, K. 2012. Comparative survey of levels of chlorinated cyclodiene pesticides in breast milk from some cities of China, Korea and Japan. *Chemosphere*, 89, 452–457.

Gavrilescu, M., Demnerova, K., Aamand, J., Agathos, S. & Fava, F. 2015. Emerging pollutants in the environment: present and future challenges in biomonitoring, ecological risks and bioremediation. *N Biotechnol*, 32, 147–156.

Hauser, R. & Calafat, A. M. 2005. Phthalates and human health. *Occup Environ Med*, 62, 806–818.

Health Canada. 2013. Guidelines for Canadian Drinking Water Quality: Guideline Technical Document — Nitrate and Nitrite. Health Canada, Ottawa, Ontario (Catalogue No H144-13/2-2013EPDF).

Hemminger, P. 2005. Damming the flow of drugs into drinking water. *Environ Health Perspect*, 113, A678–A681.

Hua, G. & Reckhow, D. A. 2007. Comparison of disinfection byproduct formation from chlorine and alternative disinfectants. *Water Res*, 41, 1667–1678.

Hutter, H. P., Wallner, P., Moshammer, H., Hartl, W., Sattelberger, R., Lorbeer, G. & Kundi, M. 2005. Blood concentrations of polycyclic musks in healthy young adults. *Chemosphere*, 59, 487–492.

Hutter, H. P., Wallner, P., Hartl, W., Uhl, M., Lorbeer, G., Gminski, R., Mersch-Sundermann, V. & Kundi, M. 2010. Higher blood concentrations of synthetic musks in women above fifty years than in younger women. *Int J Hyg Environ Health*, 213, 124–130.

IARC. 2004. IARC working group on the evaluation of carcinogenic risks to human. Some drinking-water disinfectants and contaminants, including arsenic. *IARC Monogr Eval Carcinog Risks Hum*, 84, 1–512.

IARC. 2006. Inorganic and organic lead compounds. *IARC Monogr Eval Carcinog Risks Hum*, 87.

IARC. 2010. Ingested nitrate and nitrite, and cyanobacterial peptide toxins. *IARC Monogr Eval Carcinog Risks Hum*, 94.

IARC. 2012. *IARC monographs on the evaluation of carcinogenic risks to humans – volume 100C: Arsenic, metals, fibres, and dusts.* WHO, Geneva.

IARC. 2014. IARC Monographs. 1–111. http://monographs.iarc.fr/ENG/Classification/ClassificationsAlphaOrder.pdf

Kapaj, S., Peterson, H., Liber, K. & Bhattacharya, P. 2006. Human health effects from chronic arsenic poisoning – a review. *J Environ Sci Health A Tox Hazard Subst Environ Eng*, 41, 2399–2428.

Kapoor, A., Tanjore, S. & Viraraghavan, T. 1995. Removal of selenium from water and wastewater. *International Journal of Environmental Studies*, 49, 137–147.

Kaushik, C. P., Sharma, H. R. & Kaushik, A. 2012. Organochlorine pesticide residues in drinking water in the rural areas of Haryana, India. *Environ Monit Assess*, 184, 103–112.

Kim, Y. R., Harden, F. A., Toms, L. M. & Norman, R. E. 2014. Health consequences of exposure to brominated flame retardants: a systematic review. *Chemosphere*, 106, 1–19.

Lopez-Espinosa, M. J., Murcia, M., Iniguez, C., Vizcaino, E., Llop, S., Vioque, J., Grimalt, J. O., Rebagliato, M. & Ballester, F. 2011. Prenatal exposure to organochlorine compounds and birth size. *Pediatrics*, 128, e127–e134.

Lopez-Espinosa, M. J., Vizcaino, E., Murcia, M., Llop, S., Espada, M., Seco, V., Marco, A., Rebagliato, M., Grimalt, J. O. & Ballester, F. 2009. Association between thyroid hormone levels and 4,4'-DDE concentrations in pregnant women (Valencia, Spain). *Environ Res*, 109, 479–485.

Mason, L. H., Harp, J. P. & Han, D. Y. 2014. Pb neurotoxicity: neuropsychological effects of lead toxicity. *Biomed Res Int*, 2014, 840547.

Meenakshi & Maheshwari, R. C. 2006. Fluoride in drinking water and its removal. *J Hazard Mater*, 137, 456–463.

Miltner, R. J., Baker, D. B., Speth, T. F. & Fronk, C. A. 1989. Treatment of seasonal pesticides in surface waters. *Journal (American Water Works Association)*, 81, 43–52.

Mohapatra, M., Anand, S., Mishra, B. K., Giles, D. E. & Singh, P. 2009. Review of fluoride removal from drinking water. *J Environ Manage*, 91, 67–77.

NTP. 2011. National Toxicology Project 12th report on carcinogens. Arsenic and inorganic arsenic compounds. *Rep Carcinog*, 12, 50–52.

NTP-CERHR. 2008. National Toxicology Program Center for the Evaluation of Risks to Human Reproduction. Monograph on the Potential Human Reproductive and Developmental Effects of Bisphenol A. NIH Publication No. 08-5994. http://ntp.niehs.nih.gov/ntp/ohat/bisphenol/bisphenol.pdf

Peckham, S. & Awofeso, N. 2014. Water fluoridation: a critical review of the physiological effects of ingested fluoride as a public health intervention. *Scientific World Journal*, 2014, 293019.

Plewa, M. J., Simmons, J. E., Richardson, S. D. & Wagner, E. D. 2010. Mammalian cell cytotoxicity and genotoxicity of the haloacetic acids, a major class of drinking water disinfection by-products. *Environ Mol Mutagen*, 51, 871–878.

Plewa, M. J., Wagner, E. D., Richardson, S. D., Thruston, A. D., Jr., Woo, Y. T. & McKague, A. B. 2004. Chemical and biological characterization of newly discovered iodoacid drinking water disinfection byproducts. *Environ Sci Technol*, 38, 4713–4722.

Rahman, M. F., Yanful, E. K. & Jasim, S. Y. 2009. Endocrine disrupting compounds (EDCs) and pharmaceuticals and personal care products (PPCPs) in the aquatic environment: implications for the drinking water industry and global environmental health. *J Water Health*, 7, 224–243.

Ren, A., Qiu, X., Jin, L., Ma, J., Li, Z., Zhang, L., Zhu, H., Finnell, R. H. & Zhu, T. 2011. Association of selected persistent organic pollutants in the placenta with the risk of neural tube defects. *Proc Natl Acad Sci U S A*, 108, 12770–12775.

Ren, X. M. & Guo, L. H. 2013. Molecular toxicology of polybrominated diphenyl ethers: nuclear hormone receptor mediated pathways. *Environ Sci Process Impacts*, 15, 702–708.

Richard, A. M., Diaz, J. H. & Kaye, A. D. 2014. Reexamining the risks of drinking-water nitrates on public health. *Ochsner J*, 14, 392–398.

Richardson, S. & Postigo, C. 2012. Drinking water disinfection by-products. *In:* Barceló, D. (ed.) *Emerging Organic Contaminants and Human Health.* Springer, Berlin and Heidelberg.

Richardson, S. D., Fasano, F., Ellington, J. J., Crumley, F. G., Buettner, K. M., Evans, J. J., Blount, B. C., Silva, L. K., Waite, T. J., Luther, G. W., Mckague, A. B., Miltner, R. J., Wagner, E. D. & Plewa, M. J. 2008. Occurrence and mammalian cell toxicity of iodinated disinfection byproducts in drinking water. *Environ Sci Technol*, 42, 8330–8338.

Rochester, J. R. 2013. Bisphenol A and human health: a review of the literature. *Reproductive Toxicology*, 42, 132–155.

Sanders, T., Liu, Y., Buchner, V. & Tchounwou, P. B. 2009. Neurotoxic effects and biomarkers of lead exposure: a review. *Rev Environ Health*, 24, 15–45.

Sauve, S. & Desrosiers, M. 2014. A review of what is an emerging contaminant. *Chem Cent J*, 8, 15.

Schreurs, R. H., Legler, J., Artola-Garicano, E., Sinnige, T. L., Lanser, P. H., Seinen, W. & Van Der Burg, B. 2004. In vitro and in vivo antiestrogenic effects of polycyclic musks in zebrafish. *Environ Sci Technol*, 38, 997–1002.

Seth, P. 2014. Chemical contaminants in water and associated health hazards. *In:* Singh, P. P. & Sharma, V. (eds) *Water and Health.* Springer, India.

Tajtakova, M., Semanova, Z., Tomkova, Z., Szokeova, E., Majoros, J., Radikova, Z., Sebokova, E., Klimes, I. & Langer, P. 2006. Increased thyroid volume and frequency of thyroid disorders signs in schoolchildren from nitrate polluted area. *Chemosphere*, 62, 559–564.

Taylor, K. W., Novak, R. F., Anderson, H. A., Birnbaum, L. S., Blystone, C., Devito, M., Jacobs, D., Kohrle, J., Lee, D. H., Rylander, L., Rignell-Hydbom, A., Tornero-Velez, R., Turyk, M. E., Boyles, A. L., Thayer, K. A. & Lind, L. 2013. Evaluation of the association between persistent organic pollutants (POPs) and diabetes in epidemiological studies: a national toxicology program workshop review. *Environ Health Perspect*, 121, 774–783.

Tornqvist, R., Jarsjo, J. & Karimov, B. 2011. Health risks from large-scale water pollution: trends in Central Asia. *Environ Int*, 37, 435–442.

Uddin, R. & Huda, N. H. 2011. Arsenic poisoning in Bangladesh. *Oman Med J*, 26, 207.

UNEP. 2001. Stockholm Convention on Persistent Organic Pollutants. http://chm.pops.int/Portals/0/Repository/convention_text/UNEP-POPS-COP-CONVTEXT-FULL.English.PDF

Van Maanen, J. M., Van Dijk, A., Mulder, K., De Baets, M. H., Menheere, P. C., Van Der Heide, D., Mertens, P. L. & Kleinjans, J. C. 1994. Consumption of drinking water with high nitrate levels causes hypertrophy of the thyroid. *Toxicol Lett*, 72, 365–374.

Villanueva, C. M., Cantor, K. P., Cordier, S., Jaakkola, J. J., King, W. D., Lynch, C. F., Porru, S. & Kogevinas, M. 2004. Disinfection byproducts and bladder cancer: a pooled analysis. *Epidemiology*, 15, 357–367.

Villanueva, C. M., Kogevinas, M., Cordier, S., Templeton, M. R., Vermeulen, R., Nuckols, J. R., Nieuwenhuijsen, M. J. & Levallois, P. 2014. Assessing exposure and health consequences of chemicals in drinking water: current state of knowledge and research needs. *Environ Health Perspect*, 122, 213–221.

WHO. 2004. Fluoride in drinking-water. Background document for development of WHO Guidelines For Drinking-water Quality. http://www.who.int/water_sanitation_health/dwq/chemicals/fluoride.pdf

WHO. 2005a. Nutrients in drinking water. http://www.who.int/water_sanitation_health/dwq/nutrientsindw.pdf

WHO. 2005b. Trihalomethanes in drinking-water. Background document for development of WHO Guidelines for Drinking-water Quality. http://www.who.int/water_sanitation_health/dwq/chemicals/THM200605.pdf?ua=1

WHO. 2007. Chemical safety of drinking-water: Assessing priorities for risk management. http://www.who.int/water_sanitation_health/dwq/dwchem_safety/en/

WHO. 2011a. Chemical hazards in drinking-water: nitrate and nitrite. Background document for development of WHO Guidelines for Drinking-water Quality. http://www.who.int/water_sanitation_health/dwq/chemicals/nitratenitrite_background.pdf?ua=1

WHO. 2011b. Lead in drinking-water. Background document for development of WHO Guidelines for Drinking-water Quality. http://www.who.int/water_sanitation_health/dwq/chemicals/lead.pdf

WHO. 2011c. Selenium in drinking-water. Background document for development of WHO Guidelines for Drinking-water Quality. http://www.who.int/water_sanitation_health/dwq/chemicals/selenium.pdf

WHO. 2011d. The use of DDT in malaria vector control. WHO position statement. http://www.who.int/malaria/publications/atoz/who_htm_gmp_2011/en/

WHO. 2011e. *Guidelines for Drinking-water Quality*, fourth edition. World Health Organization. http://whqlibdoc.who.int/publications/2011/9789241548151_eng.pdf?ua=1

Waliszewski, S. M., Aguirre, A. A., Infanzon, R. M. & Siliceo, J. 2000. Carry-over of persistent organochlorine pesticides through placenta to fetus. *Salud Publica Mex*, 42, 384–390.

Witorsch, R. J. & Thomas, J. A. 2010. Personal care products and endocrine disruption: a critical review of the literature. *Crit Rev Toxicol*, 40 Suppl 3, 1–30.

Wu, C., Luo, Y., Gui, T. & Huang, Y. 2014a. Concentrations and potential health hazards of organochlorine pesticides in (shallow) groundwater of Taihu Lake region, China. *Sci Total Environ*, 470–471, 1047–1055.

Wu, C., Luo, Y., Gui, T. & Yan, S. 2014b. Characteristics and potential health hazards of organochlorine pesticides in shallow groundwater of two cities in the Yangtze River Delta. *CLEAN – Soil, Air, Water*, 42, 923–931.

Ziegler, E. E., Edwards, B. B., Jensen, R. L., Mahaffey, K. R. & Fomon, S. J. 1978. Absorption and retention of lead by infants. *Pediatr Res*, 12, 29–34.

11

RADIONUCLIDES IN WATER[*]

R. William Field

PROFESSOR, DEPARTMENT OF OCCUPATIONAL AND ENVIRONMENTAL
HEALTH, DEPARTMENT OF EPIDEMIOLOGY, COLLEGE OF PUBLIC
HEALTH, UNIVERSITY OF IOWA, IOWA CITY, IA, USA

Learning objectives

1 Understand the basic types of radiation and radiation dose.
2 Develop a basic knowledge of the primary sources of radionuclides in water and their associated health risks.
3 Compare and contrast the drinking water guidelines established by the World Health Organization, the European Directive, and the U.S. Environmental Protection Agency.

Introduction

Radionuclides are unstable forms of elements (i.e., radioactive isotopes). Water quality can be adversely impacted by various sources of radionuclides including naturally occurring, technologically-enhanced, and man-made. However, the contribution of ingested radionuclides in drinking water to an individual's overall radiation exposure is typically low and generally results from naturally occurring radionuclides (WHO, 2008).

Types of radiation

The types of radiation encountered from these sources include primarily alpha (α), beta (β), and gamma (γ) radiation. All three types of radiation are considered ionizing radiation, in the form of particles or waves, because they have sufficient energy to remove the electrons from atoms. Alpha particles are composed of two neutrons and two protons and can only travel a few centimeters in air. While these particles generally cannot penetrate skin, if the radionuclides that emit α-particles are internally deposited through breathing or ingestion, the high density of ionizations along their short path distributes a much higher radiologic

---

[*] Recommended citation: Field, R.W. 2015. 'Radionuclides in water', in Bartram, J., with Baum, R., Coclanis, P.A., Gute, D.M., Kay, D., McFadyen, S., Pond, K., Robertson, W. and Rouse, M.J. (eds) *Routledge Handbook of Water and Health*. London and New York: Routledge.

dose to tissues as compared to β-particles or γ-rays. In fact, α-particles are unique among environmental carcinogens because they produce a higher rate of double-strand DNA breaks compared with other types of ionizing radiation, which makes DNA repair more difficult. Beta particles are light and fast electrons with a mass of about 1/2000th of a proton. Beta particles have greater penetrating power in comparison to α-particles, but have lower capability to ionize tissues. Gamma rays are highly penetrating electromagnetic radiation that travel at the speed of light. Gamma rays, which do not have a charge or mass, can ionize atoms in the body directly or cause secondary ionizations upon transference of their energy to atomic particles such as electrons.

Radiation units

The international system (SI) of unit of radiation activity in water is becquerel per liter (Bq/L) of water. The Bq is defined as the activity of a quantity of radioactive material in which one nucleus decays per second. In the United States, radiation in water is measured in units of picocuries per liter (pCi/L). One pCi/L is equal to 2.22 radioactive disintegrations per minute per liter of water. One Bq/L equals one decay per second per liter or approximately 27 pCi/L.

Radiation dose

The radiation dose imparted from ingesting a particular radionuclide is directly related to several factors including: the activity of radiation ingested; the degree of absorption from the gut and intestines; the organs or tissues to which the radionuclide or radioactive decay products are transported; the duration of time in that location prior to excretion or decay (e.g., biological half-life); the type of radionuclide ingested (e.g., α emitter); and the radiosensitivity of the organ or tissue.

The *absorbed dose* refers to how much energy is deposited in the tissue or organ by the radiation. The SI unit for absorbed dose is the gray (Gy). The non-SI unit used in the U.S. for radiation absorbed dose is the rad, where 100 rad = 1 Gy.

The *dose equivalent* is the product of the absorbed dose and a radiation weighting factor that reflects the potential damage caused by the type of radiation (i.e., α, β, γ). A weighting factor is used since some types of radiation (e.g., α-particles) are more biologically damaging to live tissues or organs even when the absorbed dose from other types of radiation is equal.

The sievert (Sv) is the SI unit for *effective dose* (i.e., the sum of all the dose equivalents received by all tissues or organs) weighted by the different biologic effects (i.e., sensitivity) to radiation for different organs and tissues in the body. One sievert is equivalent to the traditional unit of 100 rem (roentgen equivalent man). Since the Sv is such a large unit, it is often expressed in terms of millisieverts (mSv; where 1000 mSv = 1 Sv).

Sources of radionuclides in water supplies

Naturally-occurring radionuclides

Radionuclides found in the natural environment, such as potassium-40 and other radionuclides in the primordial series (i.e., terrestrial origin), are considered naturally-occurring radioactive materials (NORM). In a series of radioactive nuclides, each member of the series is formed by the decay of the nuclide before it. The series ends with a stable nuclide. The primary naturally occurring radionuclides of human health concern in water originate from the long lived radioactive decay chains of uranium-238 (uranium series), uranium-235 (actinium series), and thorium-232 (thorium series) (NAS 2012). From these series, some of the frequently detected radionuclides include uranium, radium-228, radium-226, and radon-222.

Low levels of naturally occurring radionuclides are found in rocks and soils. The radionuclides, or their precursor decay product, can have a very long radioactive half-life (e.g., the radioactive half-life of uranium-238 is approximately 4.5×10^9 years) that serves as a constant source of NORM. The radionuclides enter water supplies from underground deposits that are in contact with groundwater, from run-off (e.g., erosion) into surface water supplies, and from deposition in surface water supplies from the air (e.g., windblown dust). The degree of release of radionuclides from natural deposits is influenced by numerous factors including mineralogy of the host rock (e.g., distribution and concentration of the parent element within the rock matrix), the mineral grain size, the solubility of radionuclide, the release of the radionuclide by weathering (e.g., leaching), the pH and chemical composition of the water passing through the deposit, the contact time between the water and deposit, the type of radioactive decay (e.g., the α recoil of an element from substrate to water), and flow pathways. In some cases, radionuclides can increase in concentration to the point they pose health concerns. However, the presence of one decay product (e.g., radium-226) in a decay series does not allow the accurate prediction of the presence of the precursor radionuclide (e.g., uranium-238) in the decay chain, or vice versa, due to differences in their chemical properties such as solubility.

Technologically-enhanced naturally occurring radionuclides

Technologically-enhanced naturally occurring radioactive materials (TENORM) include the same NORM radionuclides, but at enhanced (i.e., elevated) concentrations or newly occurring due to human activities. TENORM contamination can result from movement of radionuclides into surface and groundwater from numerous activities including, but not limited to, uranium mining and processing (e.g., *in situ* leaching, surficial and shaft mining), phosphate mining and phosphate fertilizer production, metal mining and production (e.g., aluminum, copper, rare earths, titanium, zircon), oil and gas production (e.g., oil drilling, hydraulic fracturing), soil conditioning (e.g., use of NORM contaminated materials), coal mining and energy generation (e.g., coal ash), ash generation, water and waste water treatment (e.g., treatment residues), and geothermal energy production. Human activities often disrupt the normal processes of radionuclide release previously described and result in radionuclide (e.g., radium-226) discharges from TENORM sources that can increase waterborne radionuclide concentrations by several orders of magnitude.

Man-made radionuclides

There are numerous sources of anthropogenic (i.e., man-made) sources of radionuclides that have increased waterborne radionuclide concentrations locally, regionally, and worldwide. These sources include nuclear power plants, nuclear weapons production and reprocessing (e.g., Hanford, Washington, U.S.; Savannah River, South Carolina, U.S.; Mayak Production Association, Russia), permitted medical and industrial releases, nuclear weapons testing (e.g., Pacific Ocean and Nevada Test Site, U.S.; Semipalatinsk test site, Republic of Kazakhstan; Mururoa and Fangataufa atolls, French Polynesia), commercial fuel reprocessing (e.g.,

La Hague plant, France; Thermal Oxide Reprocessing Plant, Sellafield, England; Mayak Production Association, Russia), geological radioactive waste repositories, and nuclear accidents (e.g., Mayak, Russia; Chernobyl, Ukraine; Fukushima, Japan). Some of the man-made radionuclides of public health concern include cesium-137, iodine-131, iodine-129, plutonium-239, strontium-90, and uranium-235. For example, above ground nuclear testing increased anthropogenic radionuclides with longer half-lives in water worldwide including tritium (i.e., Hydrogen-3), plutonium-239 and 240, cesium-137, and strontium-90. Additional details about the various radionuclides are provided elsewhere (U.S. EPA, 2014; Weinhold, 2012; WHO, 2006).

Examples of waterborne radionuclides of health concern

Uranium

Natural uranium is a mixture of three separate radionuclides including uranium-238, uranium-235, and uranium-234. Approximately, 99.27 percent of uranium mass occurs as uranium-238 with 0.72 percent and 0.0055 percent occurring as uranium-235, and uranium-234, respectively. However, these relative crustal abundances of uranium differ from the uranium isotopic ratio found in groundwater. For example, since uranium-234 decays at a faster rate than uranium-238, it releases more alpha particles per unit time that enhances the preferential leaching of uranium-234. Typical ratios of uranium-234/uranium-238 activity in groundwater range from 0.9 to 1.3, but can vary frequently from 0.5 to 40.

Uranium is not considered an external radiation hazard since uranium decays by α-emission and α-particles generally do not penetrate the dead layer of human skin. While uranium found in drinking water supplies is usually from leaching of natural deposits, uranium is also released into water supplies through technological activities like uranium mining, phosphate production, combustion of coal and other energy sources, and the use of phosphate fertilizer, etc. (i.e., TENORM). If the uranium in the soil or rocks is oxidized, it can be transported great distances in groundwater.

While the majority of drinking water supplies that have been measured generally exhibit uranium concentrations below 1 μg/L (25 mBq/L) (Cothern and Lappenbusch, 1983; Smedley et al., 2006), surveys of groundwater supplies performed in numerous countries, including China, Finland, Germany, India, Japan, Portugal, India, Uzbekistan, Mongolia, and the United States, have reported occurrences of concentrations exceeding 1 μg/L (ATSDR, 2013; EFSA, 2009; Nriagu et al., 2012; Wu et al., 2014). In rare occurrences, uranium concentrations as high as 700 μg/L to 2,020 μg/L have been measured in wells drilled in uranium-rich rock (e.g., Canada) (Betcher et al., 2011; WHO, 2006, 2014).

Soluble forms of uranium have a higher probability of entering the bloodstream if ingested. The primary site of deposition of uranium in the body is in bones (~66 percent) with lower depositional percentages in the liver and kidneys. Without further ingestion, uranium has a biological half-life in the bones (i.e., time required for half of the uranium to leave the bones) of 70–200 days and half-lives of 1–2 weeks for uranium deposited in other areas of the body (ATSDR, 2013). Uranium that is absorbed in the body is eliminated in the urine. The primary documented adverse health effect of ingesting uranium is kidney damage, which is attributed to the chemical toxicity of uranium rather than by the α-particles emitted by uranium. While neither the International Agency for

Table 11.1 Drinking water guidelines[a] by the World Health Organization (WHO), European Directive, and U.S. Environmental Protection Agency (EPA)

Parameter	World Health Organization	European Directive	U.S. EPA
	Guidance level	Derived concentration	Maximum contaminant limit (MCL)
Uranium	30 μg/L[b]	–	30 μg/L (0.67 pCi/μg)[c]
Uranium-238	–	3.0 Bq/L	–
Uranium-234	–	2.8 Bq/L	–
Combined radium-228 and radium-226	–	–	5 pCi/L (0.19 Bq/L)
Radium-228	0.1 Bq/L	0.2 Bq/L	–
Radium-226	1.0 Bq/L	0.5 Bq/L	–
	Screening level		
Gross alpha[d]	0.5 Bq/L	0.1 Bq/L	15 pCi/L (0.58 Bq/L)
Gross beta	1.0 Bq/L	0.1 Bq/L	4 mrem/year (40 μSv/yr)
	Guidance level	Parametric value	Proposed maximum contaminant level
Radon-222	100 Bq/L	100 Bq/L[e]	300 pCi/L (11 Bq/L)

Notes:

[a] Descriptions of guidance terms are presented in the text
[b] WHO designated the guideline value as provisional because of the scientific uncertainties about uranium's toxicity (WHO, 2014)
[c] U.S. EPA conversion from μg/L
[d] Excludes uranium and radon
[e] Remedial action is considered to be justified on radiological protection grounds when radon concentrations exceed 1,000 Bq/L

Research on Cancer (IARC) nor the U.S. EPA have classified natural uranium as a known human carcinogen (Group 1), uranium is considered a Group 1 carcinogen under IARC's category of α-particle-emitting radionuclides that are internally deposited (IARC, 2012). However, its carcinogenicity has not been convincingly demonstrated in human studies. The WHO and U.S. EPA have set the maximum contaminant limit (MCL) and guidance level for uranium at 30 μg/L (Table 11.1) (WHO, 2011). The European Union has set a derived concentration level for uranium-238 and 234 at 3.0 Bq/L and 2.8 Bq/L, respectively (CEU, 2013). As mentioned above, a derived concentration is a radionuclide-specific concentration related to the 0.1 mSv dose criterion.

Radium-226 and radium-228

Radium has four naturally occurring isotopes including radium-228, radium-226, radium-224, and radium-223. The most common radium isotope, radium-226 (1,600 year radioactive half-life, emits α-particles and gamma rays) is formed in the uranium-238 decay series. Radium-228 (5.75 year half-life, β-emitter) and radium-224 (3.6 day half-life, α-emitter) are both decay products in the thorium-232 decay series. Radium-223 (11.4 day half-life, α-emitter) is formed in the uranium-235 series. Because radium-224 and radium-223 have relatively short radioactive half-lives and lower relative abundance as compared to radium-226 and radium-228, they are considered a lower health risk.

Like uranium, higher concentrations of radium are generally not found in groundwater unless a local TENORM activity is contaminating nearby surface water. In a 2011 survey, researchers measured the concentrations of radium in pretreated water samples from 15 major aquifer systems in the U.S. that serve 45 states. The survey found exceedences of the U.S. EPA's MCL in seven principal aquifer systems, located primarily on the East Coast and Midwest. The exceedences occurred for 39 (~4 percent) of the 971 samples for which both radium-226 and radium-228 were analyzed (Szabo et al., 2012). The researchers noted that radium occurrence was greatly influenced by the geochemical properties of the aquifer systems (e.g., sorption, desorption, and exchange). For example, slight increases in radium mobility were noted in aquifers with poor sorptive capacity as well as frequent occurrences of elevated radium in anoxic water (e.g., low dissolved-O_2). Globally, waterborne radium concentrations have been shown to vary greatly and in some locations significantly exceed the WHO's guidance level of 0.1 Bq/L (Lauria et al., 2012). Water supplies also have the potential to be contaminated by radium due to activities including coal mining, phosphate production, fertilizer use, hydraulic fracturing, and oil and gas drilling. Additional details regarding radium as TENORM are available elsewhere (UNSCEAR, 2010). Radium can also adsorb to water and pipe scales and present a TENORM hazard to workers as well as increased radon levels in drinking water.

IARC has classified radium-228, radium-226, and radium-224 as Group 1 carcinogens (i.e., known carcinogenic to humans) (IARC, 2012). Since radium is a chemical analogue of calcium, a small percentage (~20 percent) of ingested radium deposits in bones where it has the potential to cause bone cancer (i.e., osteosarcoma). Protracted ingestion of radium has also been linked with increases in aplastic anemia, leukemia, and lymphoma (U.S. EPA, 2011). The major adverse health effect related to radium-226 from a population risk perspective is from the initial decay product of radium-226, namely radon-222 (see below). The European Directive's derived concentration and WHO's guidance level for radium-226 and radium-228 vary slightly (Table 11.1), while the U.S. EPA has set the MCL for combined radium-226 and radium-228 at 0.19 Bq/L (5 pCi/L) (Table 11.1) with a maximum contaminant level goal (MCLG) of zero (U.S. EPA, 2014).

Radon-222

Radon-222 (hereafter referred to as radon) is an invisible and odorless naturally occurring radioactive noble gas. It is the only gaseous decay product in the U-238 decay series, and is therefore usually more mobile. In most geographic locations, the major source of elevated indoor radon concentrations is from entry of the gas into a home, or other building, from radium-226 containing rocks and soils that are beneath or around the substructure of a building. Radon enters a building through penetrations (e.g., cracks, sump pump, etc.) in

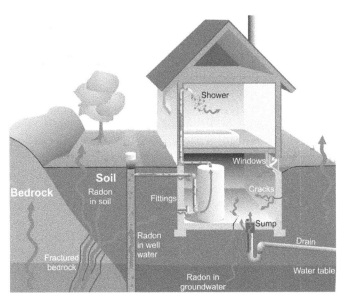

Figure 11.1 How radon enters houses

a structure's foundation (Figure 11.1). However, the second largest contributor to indoor radon, and primary source of radon gas in some cases, is from off-gassing of radon during the use of groundwater in a home. In the United States, off-gassing of waterborne radon from groundwater sources (i.e., wells) accounts for about 1 to 2 percent of the total indoor air radon concentrations (U.S. EPA, 2012) in homes.

Radon is moderately soluble in water; solubility decreases with increasing water temperature. Because radon has a relatively short radioactive half-life of 3.8 days, it is generally not transported large distances in groundwater. Private well water supplies and small community groundwater supplies, especially those located in consolidated rock, typically have the highest radon concentrations. The radon concentration in a particular well is influenced primarily by the lithology of the aquifer solids near the well. Generally, granitic-type rock aquifers have the highest radon concentrations. For example, the radon concentrations in wells from granitic rock type aquifers generally exceed 8,000 pCi/L and have been reported as high as 1,300,000 pCi/L (Milvy and Cothern, 1990). Another type of crystalline aquifer, metamorphic rock aquifers, can also have high concentrations of radon depending on the type of rock originally metamorphosed. Information on testing water for radon is available elsewhere (WHO, 2009).

Radon that is dissolved in groundwater supplies off-gasses as the water is used in the home for showering, bathing, dishwashing, etc. The U.S. National Academy of Science (NAS, 1999) estimates the transfer coefficient of radon between water and air in homes in North America is 1.0×10^{-4} (i.e., 10:000 to 1 transfer ratio). However, the transfer factor can vary significantly and does not represent short term ratios in specific rooms where the water is utilized (e.g., bathroom) (Nazaroff et al., 1987; Vinson et al., 2008). While ingestion of waterborne radon was estimated to cause approximately 20 deaths from stomach cancer in the United States in 1999 (NAS, 1999), the greatest health risk from waterborne radon is from the off-gassing of radon into the air of a building. As radon gas undergoes further decay in the air, it produces a series of solid radioactive decay products that can be inhaled (WHO, 2009). Two of the α-particle emitting decay products, polonium-218 and polonium-214, deliver the majority of the radiation dose to the lining of the lung (i.e., respiratory epithelium). The α-particles can cause changes in the cells of the lining of the lung (e.g., double strand DNA breaks, etc.) that have the potential to lead to lung cancer.

The WHO has set a reference level of 100 Bq/m³ for indoor air (WHO, 2011). An estimated 90 percent of the radiation dose attributable to radon in drinking water originates from inhalation after the off-gassing of radon. As well, the majority of radon in indoor air is due to the ingress of radon from the soil rather than degassing from the drinking water. As a result, WHO recommends that screening levels for radon in water should be based on the WHO member states national airborne reference levels. However, WHO notes that in geographic areas where high radon concentrations in drinking water occur, it is prudent to assess whether reducing the radon concentrations, such as aeration of the water prior to entering the home, is justified.

The European Union (CEU, 2013) has set a parametric value of 100 Bq/L for radon in water (Table 11.1). A parametric value indicates a level at which member states should determine if exceedences present a human health risk that requires remedial action (CEU, 2013). European member states are allowed to set a level for radon greater than 100 Bq/L, up to 1,000 Bq/L. Remedial action is considered justifiable, on radiological protection grounds, where radon in water concentrations exceed 1,000 Bq/L.

The U.S. EPA estimates that the additional indoor radon exposure caused by off-gassing of radon from water containing 300 pCi/L (11.1 Bq/L) would cause an incremental lifetime cancer risk of about two cancers for every 10,000 persons or approximately 80 avoided lung cancer deaths annually for approximately 27,000 affected U.S. public water systems (U.S. EPA, 2012). The estimated number of additional lung cancer deaths does not include radon exposures from private water supplies. In consideration of the relative risks of waterborne radon as compared to radon gas from ground sources (U.S. EPA 2003) as well as the expense that would be incurred by utilities to meet the proposed guideline, the U.S. EPA proposed in 1999 that there be both a MCL of 300 pCi/L (11.1 Bq/L) plus an alternative maximum contaminant level (AMCL) of 4,000 pCi/L (148 Bq/L).

The AMCL would contribute about 0.4 pCi/L (14.8 mBq/L) to indoor air, which is equivalent to the average outdoor radon gas concentration in the U.S. (Federal Register 1999). The AMCL would be applied only if a state or water system developed and implemented an EPA-approved multimedia mitigation (MMM) program to address radon in indoor air. The MMM program guidelines included requirements for state or water systems to follow to promote efforts for radon gas reduction including mitigation of existing homes that have indoor radon gas concentrations above the U.S. EPA's action level of 4 pCi/L (150 Bq/m³) and promotion of building new homes radon-resistant (see U.S. EPA, 2012 for more details). While the U.S. EPA's proposed waterborne radon rules have not been promulgated in the U.S., some U.S. states have adopted guidance levels for public water supplies (e.g., Massachusetts, North Carolina).

Cesium-137

Cesium-137 is an anthropogenic radionuclide, with a 30 year half-life, produced during nuclear fission (i.e., the splitting of a nucleus into at least two other nuclei) of various isotopes of uranium, plutonium, and thorium. Cesium-137 decays by β decay that is shortly followed by the emission of a γ ray from its short lived decay product, barium-137m (the "m" indicates it is a metastable nuclear isomer that decays very quickly). Two other isotopes of cesium, cesium-135 and cesium-134, are often considered less of a health concern because of their decay characteristics. For example, cesium-135, a β emitter, with a half-life of 2.3 million years has very low specific activity (i.e., number of decays per unit mass or volume). Cesium-134, a β emitter, has a half-life of 2.1 years so does not persist in the environment as long as cesium-137. However, determining cesium-137/cesium-134 ratios may help to identify the source and age of cesium in water.

If cesium-137 is ingested, it behaves in a similar manner to potassium with the majority of it absorbed into the bloodstream through the intestines. It tends to concentrate in muscles where it has a biological half-life around 70 to 100 days. Since cesium-137 decays by β emission and γ emission, via barium-137m, it has the potential to cause cancer. IARC has classified internalized radionuclides that emit β-particles as carcinogenic to humans (Group 1). Cesium-137 is released through routine processes at nuclear facilities (e.g., nuclear power plants, nuclear fuel reprocessing), from nuclear detonations (e.g., fallout from above ground nuclear testing), and from nuclear accidents (e.g., Chernobyl in the Ukraine). Direct discharges of cesium-137 and cesium-134 into the ocean and atmospheric fallout from Japan's Fukushima Nuclear Power Plant disaster represent the largest radioactive cesium releases to the ocean to date (i.e., 2015). Ingestion of bioaccumulated cesium-137 in seafood is the major exposure pathway related to the occurrence of cesium-137 in seawater. The U.S. EPA has set a MCL of 4 mrem/yr (40 μSv/yr) and a MCLG of zero for gross beta emitters like cesium-137, tritium, strontium-90, and cobalt-60.

Drinking water guidelines

The WHO (2011) has developed a four step risk strategy for screening radionuclides in water. The first step is adoption of a 0.1 mSv (10 mrem) individual dose criterion (IDC) for ingestion of drinking water for one year. The IDC is based on the recommendation from the International Commission on Radiological Protection (ICRP, 2000) that it is prudent to restrict the prolonged component of the individual dose to 0.1 mSv in any given year. An IDC of 0.1 mSv/year is a dose the WHO (2011) expects will not produce any detectable adverse health effects. WHO's screening levels are not enforceable limits, but rather guidance levels that if exceeded trigger further investigation. The second step is to determine if initial screening measurements for gross alpha activity and gross beta activity exceed 0.5 Bq/L and 1 Bq/L, respectively (Table 11.1). If they do not exceed these values, further action is not required. In step three, if either of the screening levels is exceeded, individual radionuclides need to be identified, their concentrations need to be determined, and compared with guidance levels set by the WHO. Step four determines whether additional actions are needed to reduce the dose. See WHO (2011) for additional details.

In 2013, the Council of the European Union approved a new directive that included guidance for radionuclides in water (CEU, 2013). Member states have four years to develop national legislation based on the directive including setting 'parametric values' for the monitoring of radionuclides in water intended for human consumption. A parametric value is not a limit value, but rather a value at which member states should consider if exceedences present a human health risk that require remedial action (CEU, 2013). The European directive also provides 'derived concentrations' for radioactivity in water intended for human consumption for both natural and artificial radionuclides based on a calculated dose of 0.1 mSv and annual intake of 730 liters (Table 11.1).

U.S. National Primary Drinking Water Regulations for radionuclides were promulgated in 1976 under the Safe Drinking Water Act (SDWA) and revised in 2000 to include uranium (U.S. EPA, 2014) (Table 11.1). The revised rule, which was enacted in 2003, sets enforceable

MCLs and MCLGs for four categories of radionuclides including uranium, radium, gross α-emitters (e.g., excludes uranium and radon, but includes radium, polonium, and thorium), gross β emitters (e.g., strontium-90) and photon emitters (i.e., x-rays and γ-rays) including cesium-137, cesium-134, and cobalt-60. The MCLG, which is a non-enforceable public health goal, is zero (i.e., no toxicity) for each category. U.S. EPA's water regulations apply to community water systems that have at least 15 service connections or that serve 25 or more individuals all year.

Reduction of radionuclides in drinking water

Water utilities often choose to use an alternative water supply or to blend the water in a controlled manner with another water source, with lower concentrations of radionuclides, in cases where the original water supply has unacceptably high concentrations of radionuclides. When these replacement sources are not available, water treatment options like coagulation, sedimentation, and sand filtration can remove up to 100 percent of the suspended radionuclides. Water softening, using lime-soda ash, is also effective at reducing radionuclides with removal efficiencies related to the specific radionuclide and the proportion of radionuclides associated with particulates. Additional, and often more expensive, methods to reduce radionuclide concentrations include charcoal filtration, reverse osmosis, ion exchange, and use of natural zeolites (WHO, 2011).

Radon is easily removed from water supplies by simple aeration, but care must be taken to ensure that water plant operators or members of the public are not exposed to radon gas from this potentially large source. Radon can also be removed from water using adsorption to granular activated charcoal, but the amount of granular charcoal required often limits the attractiveness of this option (WHO, 2011). A comprehensive summary of the removal of dissolved radionuclides is available elsewhere (Brown et al., 2008).

Key recommended readings

1 CEU, The Council of the European Union (2013) 'Council Directive 2013/51/ Euratom', *Official Journal of the European Union*, 296, pp. 12–22. Available at: http:// ec.europa.eu/energy/nuclear/radiation_protection/radioactivity_in_drinking_water_ en.htm (accessed: 24th November 2014). The Council of the European Union's requirements for the protection of the health of the general public with regard to radionuclides in water intended for human consumption.

2 Kiger, P.J. (2013) 'Fukushima's radioactive water leak: what you should know, *National Geographic Magazine*. Available at: http://news.nationalgeographic.com/news/ energy/2013/08/130807-fukushima-radioactive-water-leak/ (accessed: 24th November 2014). News story about release of radioactive water from Fukushima provides an example of the potential for manmade radionuclide releases into water.

3 U.S. EPA, U.S. Environmental Protection Agency (n.d.) *Basic Information about Radionuclides in Drinking Water.* Available at: http://water.epa.gov/drink/contaminants/ basicinformation/radionuclides.cfm (accessed: 24th November 2014). U.S. EPA site provides an overview of some of the radionuclides in water, their health effects, information on testing for radionuclides in water, current water regulations, radionuclide removal from drinking water, and links to additional information.

4 Weinhold, B. (2012) 'Unknown quantity: regulating radionuclides in tap water', *Environmental Health Perspectives*, 120(9), pp. a350–a356. Available at: http://ehp.niehs.

nih.gov/120-a350/ (accessed: 24th November 2014). Bob Weinhold, an environmental journalist, discusses some gaps in our knowledge about the sources, exposures, and associated health effects of radionuclides in water. He also highlights some of the radionuclides that are gaining the most research attention and provides an example of a fairly unexplored mechanism of toxicity.

5 WHO, World Health Organization. (2011) *Guidelines for Drinking-water Quality*, fourth edition, Geneva, Switzerland: WHO Press. Available at: http://whqlibdoc.who.int/publications/2011/9789241548151_eng.pdf (accessed: 24th November 2014). WHO report provides guidelines for drinking water quality to protect public health including recommendations for managing the risk from hazards that may compromise the safety of drinking water.

References

ATSDR, Agency for Toxic Substances and Disease Registry, Division of Toxicology and Human Health Sciences, Environmental Toxicology Branch (2013) *A Toxicological Profile for Uranium*. Atlanta, GA: U.S. Department of Health and Human Services, Public Health Service. Available at: http://www.atsdr.cdc.gov/toxprofiles/tp150.pdf (accessed: 17th September 2014).

Betcher, R.N., Gascoyne, M., and Brown, D. (2011) 'Uranium in groundwaters of southeastern Manitoba, Canada', *Canadian Journal of Earth Sciences*, 25(12), pp. 2089 2103.

Brown, J., Hammond, D., and Wilkins B.T. (2008) 'Handbook for assessing the impact of a radiologic incident on levels of radioactivity in drinking water and risks to operatives at water treatment works', HPA-RPD-040, Chilton, Didcot, Oxfordshire, UK: Health Protection Agency, Radiation Protection Division. Available at: http://dwi.defra.gov.uk/research/completed-research/reports/DWI70-2-192_radionuclides.pdf (accessed: 29th January 2015).

CEU, The Council of the European Union (2013) 'Council Directive 2013/51/Euratom', *Official Journal of the European Union*, 296, pp. 12–22.

Cothern, C.R., and Lappenbusch, W.L. (1983) 'Occurrence of uranium in drinking water in the U.S.', *Health Physics*, 45(1), pp. 89–99.

EFSA, The European Food Safety Authority (2009) 'Uranium in foodstuffs, in particular mineral water', *The European Food Safety Authority Journal*, 1018, pp. 1–59. Available at: http://www.efsa.europa.eu/en/efsajournal/doc/1018.pdf (accessed: 17th September 2014).

Federal Register (1999) Vol. 64, No. 211. National Primary Drinking Water Regulations: Radon-222, Proposed Rule (November 2, 1999), 59246-59378. [64 FR 59246], Available at: http://www.gpo.gov/fdsys/pkg/FR-1999-11-02/html/99-27741.htm, (accessed: 19th May 2015).

IARC, International Agency for Research on Cancer (2012) 'A review of human carcinogens: radiation; internalized α-particle emitting radionuclides', Vol. 100D, Lyon, France: International Agency for Research on Cancer. Available at: http://monographs.iarc.fr/ENG/Monographs/vol100D (accessed: 17th September 2014).

ICRP, International Commission on Radiological Protection (2000) 'Protection of the public in situations of prolonged radiation exposure. Recommendations of the International Commission on Radiological Protection', *ICRP Publication 82, Annals of the ICRP*, 29(1–2).

Lauria, D.C., Rochedo, E.R., Godoy, M.L., Santos, E.E., and Hacon, S.S. (2012) 'Naturally occurring radionuclides in food and drinking water from a thorium-rich area', *Radiation and Environmental Biophysics*, 51(4), pp. 367–374.

Milvy, P., and Cothern, C.R. (1990) 'Scientific background for the development of regulations for radionuclides in drinking water', in Cothern, C.R., and Rebers, P.A. (eds) *Radon, Radium, and Uranium in Drinking Water*. Chelsea, MI: Lewis Publishers, Inc., pp. 1–16.

NAS, National Academy of Sciences, Committee on the Risk Assessment of Exposures to Radon in Drinking Water, Board of Radiation Effects Research, Commission on Life Sciences, National Research Council (1999) *Risk Assessment of Radon in Drinking Water*. Washington, DC: National Academy Press. Available at: http://www.nap.edu/openbook.php?record_id=6287 (accessed: 18th September 2014).

Nazaroff, W.W., Doyle, S.M., Nero, A.V., and Sextro, R.G. (1987) 'Potable water as a source of 463 airborne 222Rn in U.S. dwellings: a review and assessment', *Health Physics*, 52, pp. 281–295.

Nriagu, J., Nam, D.H., Ayanwola, T.A., Dinh, H., Erdenechimeg, E., Ochir, C., and Bolormaa, T.-A. (2012) 'High levels of uranium in groundwater of Ulaanbaatar, Mongolia', *Sci Total Environ*, 414, pp. 722–726.

Smedley, P.L., Smith, B., Abesser, C., and Lapworth, D. (2006) 'Uranium occurrence and behaviour in British groundwater', (CR/06/050N) (unpublished), British Geological Survey. Available at: http://nora.nerc.ac.uk/7432/ (accessed: 17th September 2014).

Szabo, Z., DePaul, V.T., Fischer, J.M., Kraemer, T.F., and Jacobsen, E. (2012) 'Occurrence and geochemistry of radium in water from principal drinking-water aquifer systems of the United States', *Applied Geochemistry*, 27(3), pp. 729–752.

UNSCEAR, United Nations Scientific Committee on the Effects of Atomic Radiation (2010) Sources and effects of ionizing radiation. UNSCEAR 2008 report to the General Assembly with scientific annexes. Volume I. Annex B: *Exposures of the Public and Workers from Various Sources of Radiation*. New York: United Nations. Available at: http://www.unscear.org/docs/reports/2008/09-86753_Report_2008_Annex_B.pdf (accessed: 18th September 2014).

U.S. EPA, U.S. Environmental Protection Agency (2003) 'EPA assessment of risks from radon in homes', Air and Radiation Report EPA 402-R-03-003. Available at: http://www.epa.gov/radiation/docs/assessment/402-r-03-003.pdf (accessed: 19th September 2014).

U.S. EPA, U.S. Environmental Protection Agency (2011) *Radium*. Washington, DC: U.S. Environmental Protection Agency. Available at http://www.epa.gov/radiation/radionuclides/radium.html (accessed: 17th September 2014).

U.S. EPA, United States Environmental Protection Agency (2012) Office of Water (4607M), Report to Congress: 'Radon in drinking water regulations'. EPA 815-R-12-002, 2012. Available at: http://water.epa.gov/lawsregs/rulesregs/sdwa/radon/upload/epa815r12002.pdf (accessed: 11th September 2014).

U.S. EPA, United States Environmental Protection Agency (2014) U.S. Environmental Protection Agency, electronic resource, 'Water: basic information about regulated drinking water contaminants', Washington, DC. Available at: http://water.epa.gov/drink/contaminants/basicinformation/radionuclides.cfm (accessed: 19th September 2014).

Vinson, D.S., Campbell, T.R., and Vengosh, A. (2008) 'Radon transfer from groundwater used in showers to indoor air', *Applied Geochemistry*, 23, pp. 2676–2685.

Weinhold, B. (2012) 'Unknown quantity: regulating radionuclides in tap water', *Environmental Health Perspectives*, 120(9), pp. a350–a356. Available at: *http://ehp.niehs.nih.gov/120-a350/* (accessed: 2nd November 2014).

WHO, World Health Organization (2006) *Guidelines for Drinking-water Quality*, Vol. 1, Recommendations, third edition, First Addendum to Third Edition, Chapter 9, Radiological aspects, 197–209, Geneva, Switzerland: WHO Press. Available at: http://www.who.int/water_sanitation_health/dwq/gdwq0506.pdf (accessed: 2nd November 2014).

WHO, World Health Organization (2008) *Guidelines for Drinking-water Quality, Incorporating First and Second Addenda to Third Edition*. Vol. 1. Recommendations. World Health Organization, Geneva, Switzerland. Available at: http://www.who.int/water_sanitation_health/dwq/fulltext.pdf?ua=1 (accessed: 29th January 2015).

WHO, World Health Organization (2009) *WHO Handbook on Indoor Radon: A Public Health Perspective*, Zeeb, H., and Shannoun, F. (eds). Available at: http://whqlibdoc.who.int/publications/2009/9789241547673_eng.pdf (accessed: 2nd November 2014).

WHO, World Health Organization (2011) *Guidelines for Drinking-water Quality*, fourth edition, Geneva, Switzerland: WHO Press. Available at: http://whqlibdoc.who.int/publications/2011/9789241548151_eng.pdf (accessed: 2nd November 2014).

WHO, World Health Organization (2014) 'Chemical hazards in drinking-water: uranium', Water Sanitation Health, electronic resource. Available at: http://www.who.int/water_sanitation_health/dwq/chemicals/uranium/en/ (accessed: 2nd November 2014).

Wu, Y., Wang, Y., and Xie, X. (2014) 'Occurrence, behavior and distribution of high levels of uranium in shallow groundwater at Datong basin, northern China', *The Science of the Total Environment*, 472 (15), pp. 809–817.

PART II

Sources of exposure

12

INTRODUCTION TO EXPOSURE PATHWAYS[*]

Katherine Pond

Robens Centre for Public and Environmental Health, Department of Civil and Environmental Engineering, University of Surrey, Guildford, UK

Learning objectives

1 Describe the main exposure pathways of hazards/agents of disease relevant to water.
2 Understand some of the factors that influence the survival and movement of harmful agents in the environment.
3 Understand how agents of disease can be transmitted from the source to the host.

Despite enormous progress, contamination of food and water by microbial pathogens, chemical compounds or radiologic agents still affects the health of millions of people in both the developing and developed world leading to severe morbidity and mortality in vulnerable populations. The risks to human health from contact with water, for example, is not just a consequence of the *quality* of the water but also of the way the water is used and how that use shapes the way in which we come into contact with the water and thus the disease agents.

It is useful to distinguish between the terms *hazard, sources, exposure pathways* and *transmission routes*.

A *hazard* is something that does harm (agents such as pathogens and toxic chemicals for example). However, a hazard also includes the absence of protective measures and even the absence of controls over protective measures. These hazards could include drowning associated with poor management of water, a lack of available power supplies to run drinking-water treatment works, lack of trained staff to run laboratory facilities, inadequate protection of drinking water sources and harmful points of abstraction (see Bartram et al., 2009).

The *source* of the agent or hazard is the reservoir where that agent is found. This may be for example, a body of water, an animal, food or sewage.

Exposure pathways are the means by which an individual comes into contact with the hazard. Exposure may or may not lead to adverse health effects and is determined by concurrent

---

[*] Recommended citation: Pond, K. 2015. 'Introduction to exposure pathways', in Bartram, J., with Baum, R., Coclanis, P.A., Gute, D.M., Kay, D., McFadyen, S., Pond, K., Robertson, W. and Rouse, M.J. (eds) *Routledge Handbook of Water and Health*. London and New York: Routledge.

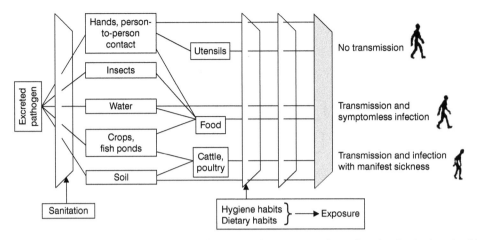

Figure 12.1 Various transmission routes of pathogens from excreta of an infected individual to a healthy individual

Source: Adapted from Carr, 2001

exposures to the same or different hazard, susceptibility and immunity of the individual. Exposure pathways may be primary (direct contact) or secondary (inhalation).

Transmission routes are the ways in which an individual or group acquires the disease-causing pathogen. The transmission route could be, for example, through drinking contaminated water (e.g. Cryptosporidium) or through the faecal–oral route (hepatitis A virus). Transmission routes may be direct (through animal to animal, or human to animal physical contact) or more commonly indirect (through, e.g., vehicles (or fomites), vectors, air currents, dust particles, water, food or faecal–oral contact). Figure 12.1 illustrates how pathogens are spread from excreta of an infected individual to a healthy individual showing the various possible transmission routes.

The exposure pathways discussed in this part are: drinking water, recreational water and food. Waterborne zoonoses are also discussed as a composite of a source with one or more transmission routes and exposure pathways. Waterborne zoonoses can originate from infected livestock, other animals (including humans) or the environment.

The diversity of infectious agents, their survival and behaviour in the environment and pathways of transmission have caused problems for public health engineers for hundreds of years in attempting to design barriers against disease. In order to help develop intervention strategies, diseases are classified into their ecological and environmental characteristics, rather than their disease characteristics. This forms four transmission categories as defined by White et al. (1972; also see Chapter 3). Similarly, chemical hazards may be categorised, from a water and health perspective, according to the principal sources associated with human exposure (WHO, 2011; see Chapter 10).

Pathogen fate, survival and transport

Many hazards that affect water and food safety are naturally occurring – soils, geology and topography of the land can all contribute hazards that can contaminate water and food products. Factors such as climate, hydrology, land use and other activities can also contribute by affecting movement and transport of the hazard through the environment. Understanding

Table 12.1 Factors affecting survival of pathogens in surface water

	Solar radiation	Temperature	Salinity	Predation
Cryptosporidium	Medium (+)	High (+)	Medium (+)	Low (+)
Giardia	Medium (+)	High (+)	Medium (+)	Low (+)
Campylobacter	High (+)	High (+)	Medium (+)	Low (+)
E. coli 0157:H7	High (+)	High (none)	Medium (+)	Low (+)
Enterovirus	High (+)	High (+)	Medium (+)	Low (+)
Norovirus	Likely high (+)	Likely high (+)	Unknown – likely medium (+)	Low (+)

Source: Adapted from Pond et al., 2004

the role of the environment in transmission of human and animal pathogens and the factors that affect their survival is central to mitigation against their spread. There are many factors that determine pathogen survival. These include: (1) their virulence and pathogenicity properties, (2) their ability to change unpredictably, (3) their interrelationships with different hosts and dynamic natural and anthropogenic environments (Sobsey and Pillai, 2009). Environmental factors such as salinity, pH, temperature, relative humidity and sunlight and the ability to bind to soil are known to influence the survival times of bacteria, viruses and fungi (see for example Sheridan and McDowell, 1998; Pond et al., 2004). Many factors may affect the potential risk from pathogens in water – climate change, urban development, flooding, abstraction of groundwater for example. Other potential sources are leaking sewers, fly-tipping and even accidental or deliberate release of pathogens or chemicals from industry or 'bio-terrorism'. Table 12.1 illustrates some of the factors influencing pathogen inactivation in surface waters.

Research has shown how pathogens and chemicals move through soils and provides information on their quantity at source and in vulnerable 'receptors' such as rivers, reservoirs and shallow wells (see, for example, Pond et al., 2004). Understanding this has proved to be imperative in protecting drinking-water sources. A well-known case is that of arsenic in Bangladesh. The main source of arsenic in drinking water is arsenic-rich rocks through which the water has filtered. It may also occur because of mining or industrial activity in some areas. Historically, surface water sources in Bangladesh have been contaminated with human and animal excreta and the associated pathogens, causing a significant burden of disease and mortality. In this particular tragedy tube-wells were installed to provide 'pure water' to prevent morbidity and mortality from gastrointestinal disease. However, the water from the millions of tube-wells that were installed was not tested for arsenic contamination, resulting in thousands of people being exposed to the chemical (Smith et al., 2000). This is further discussed in Chapter 68.

Catchment scale models have been developed to investigate changes in land use (which affect concentrations of pathogens and chemicals, survival of pathogens and movement) and concentrations of pathogens or chemicals in water (see, for example, Silgram et al., 2008; Kay et al., 2010).

As we are developing more knowledge about the survival and transport of pathogens, some traditional beliefs are being questioned. For example, it has long been believed that groundwater is free from pathogens and does not require treatment before it can be consumed. This is because it is thought that pathogen numbers are reduced by physical and biological processes in the soil, and if they do reach the aquifer then they are retained around the point of entry by the physico-chemical properties of the organism and the rock (Pond and Pedley, 2011). Some pathogenic micro-organisms, particularly viruses, have significantly greater survival times than the single time that has been used to assess the risk to groundwater (Charles et al., 2009). In addition, the geology of the aquifer plays an important part in the transport of the pathogens. For example, water can travel several kilometres in a day through a highly fissured karst aquifer, meaning that a source of contamination which is several kilometres away may still be a risk to the quality of the groundwater. This has implications for risk assessment and risk management activities.

The focus in this part is on transmission routes and exposure pathways. Once a person is exposed to a hazard, the likelihood of disease will depend on interactions between the source, the environment, the human host and the pathogen. Preventing disease requires interventions that break the transmission routes. How we do this is discussed in the part on interventions. So when we are considering the sources of disease agents or hazards, what is the scale of the problem?

Exposure pathways and transmission routes

Drinking water (see Chapter 13)

Exposure to poor quality drinking water is still worryingly common – not only in developing countries but also in the developed world. In developing countries, although poor water quality is one of the key areas of risk for human health, enormous progress has been made – between 1990 and 2012, for example, 2.3 billion people around the world gained access to an improved drinking-water source. During that same time, the number of children who died from diarrhoeal diseases – strongly associated with poor water, sanitation and hygiene – fell from approximately 1.5 million to just over 600,000 annually (WHO, 2014). However, major outbreaks of disease attributed to contaminated drinking water still occur even in countries where water quality is often taken for granted (e.g. Walkerton, Canada, Chapter 69 and Milwaukee, USA, Chapter 70). Small or 'private' water supplies are particularly vulnerable to contamination often due to their lack of protection, and communities using these supplies are often not able to apply adequate treatment methods to the water for reasons such as finance; isolation/remoteness of the supply and adequate technical knowledge. During 2009–2010, 33 drinking water-associated outbreaks were reported in the USA, resulting in 1,040 cases of illness, 85 hospitalizations and nine deaths. The majority (58 per cent) of outbreaks were of *Legionella*; *Campylobacter* accounted for 12 per cent of outbreaks and 78 per cent of illnesses. The most commonly identified outbreak deficiencies in drinking water-associated outbreaks were *Legionella* in plumbing systems (57.6 per cent), untreated groundwater (24.2 per cent) and distribution system deficiencies (12.1 per cent) (Anon, 2013).

Recreational water (see Chapter 14)

Recreational water-associated disease outbreaks result from exposure to infectious pathogens or chemical agents in treated recreational water venues (e.g. pools and hot tubs or spas) or untreated recreational water venues (e.g. lakes and coastal areas). Transmission is by ingestion, dermal contact or inhalation. For some chemicals these three exposure pathways combine (for example with disinfectant by-products; Richardson et al., 2011). In the USA, for example in 2009–2010, 81 recreational water-associated disease outbreaks were reported to Centers for Disease Control and Prevention's (CDC's) Waterborne Disease and Outbreak Surveillance System (WBDOSS) via the National Outbreak Reporting System (NORS) and resulted in over 1,326 cases of illness and 62 hospitalizations; no deaths were reported. Among the 57 outbreaks associated with pools, hot tubs or spas, 24 (42 per cent) were caused by *Cryptosporidium*. Among the 24 outbreaks associated with untreated recreational water, 11 (46 per cent) were confirmed or suspected to have been caused by cyanobacterial toxins (Hlavsa et al., 2014; Chapter 9).

Food (see Chapter 15)

Food can act as a source of exposure to bacteria, viruses, protozoa and parasitic worms – consumption of contaminated food can transmit *E. coli*, hepatitis A, norovirus, *Trichinella*, *Giardia*, *Sarocystis* and *Cryptosporidium* for example (Questier, 2011). Despite advances in food science and technology, foodborne illness is a rising cause of morbidity across the globe. Up to 30 per cent of the population in industrialized countries may contract foodborne illness every year (FAO/WHO, 2003). Factors that have been attributed to foodborne outbreaks include cross contamination with faecal matter of both domestic as well as wild animals, contact with contaminated water, as well as the use of untreated manure or sewage, lack of field sanitation, poorly or unsanitized transportation vehicles, and contamination by handlers (Brackett, 1999). Direct and indirect use of wastewater is also contributing to outbreaks and is further discussed in Chapter 47.

Zoonoses (see Chapter 16)

Zoonoses are defined by the Pan American Health Organization as 'diseases and infections which are transmitted naturally between vertebrate animals and man' (Pan American Health Organization, 2001). Transmission may occur by a number of pathways and exposure may arise from indirect contact through food or drink to direct contact through occupational exposure on farms, from pets or through leisure pursuits.

Foodborne zoonotic pathogens are transmitted through consumption of contaminated food or drinking water. Infectious agents in foodstuffs include bacteria such as *Salmonella* and *Campylobacter*, viruses such as norovirus or hepatitis A virus, and parasites such as *Trichinella*. *Campylobacter* is one of the species of greatest public health importance. *Campylobacter jejuni* and *C. coli* (thermophilic campylobacters) can be found in a wide range of livestock (especially poultry) and wildlife species. *Campylobacter* was first confirmed to cause human illness in 1972, and by 1986 it became recognized as the most commonly reported gastrointestinal pathogen in the UK, ahead of Salmonella. *C. jejuni* accounts for approximately 90 per cent of human infection. Transmission to humans is through the faecal–oral route, usually by the consumption of contaminated foods or water. In 2012, there were 72,592 laboratory reports of *Campylobacter* in the UK. This is an increase of 0.5 per cent from 2011 (Defra, 2013).

Non-foodborne zoonotic diseases such as malaria, West Nile virus and Lyme disease can be spread by vectors such as mosquitoes, ticks, flies, fleas and lice. Verotoxin-producing *Escherichia coli* (*E. coli*) can be acquired through contact with infected farm animals but it also can be transmitted through the environment, for example Verotoxin-producing *E. coli* in contaminated recreational water.

Many believe that zoonoses or other non-human reservoirs will be the source of most emerging waterborne human pathogens (Sobsey and Pillai, 2009). Indeed, about 75 per cent of the new diseases that have affected humans over the past 10 years have been caused by pathogens originating from an animal or from products of animal origin (e.g. West Nile virus) (EFSA, 2015). Many of these diseases have the potential to become global problems.

The 'One Health' (CDC, 2013) concept recognizes that the health of humans is connected both to the health of animals and the environment and strongly promotes the collaboration of the human health, veterinary health and environmental health communities to achieve successful public health interventions. There are many documented examples of this concept which are relevant to this text in terms of looking at sources of exposure to harmful agents. For example, CDC describe the value of co-operation between professionals to tackle Rift Valley fever, which is caused by a virus that is transmitted by mosquitoes. Since it was discovered in 1930, the virus has caused many devastating outbreaks in Africa and the Middle East. These outbreaks often occur in years of very heavy rainfall and flooding, providing a perfect environment for the infected mosquito eggs in the soil to hatch. Infected animals have high levels of virus in their blood, making it easy for mosquitoes that bite the animals to then become infected with the virus. This chain of transmission must be broken to prevent disease in people and animals. Predicting when heavy rains are coming would allow time to prevent Rift Valley fever illnesses in animals and humans by vaccinating the animals.

Conclusion

When reading the following chapters the reader should be mindful that the concept of co-operation and collaboration has become more important in recent years because many factors have changed the relationships between humans, animals and the environment. These changes have caused the emergence and re-emergence of many diseases. Co-operation and collaboration between various disciplines is thus more essential than ever if we are to respond effectively to health problems at the human–water–animal–food interface. The following chapters explore the sources of the hazards/agents that cause disease. Knowing how these agents move in the environment and their transmission routes are key to developing effective and sustainable interventions, discussed in subsequent parts.

References

Anon (2013). Surveillance for waterborne disease outbreaks associated with drinking water and other non-recreational water — United States, 2009–2010. *Morbidity and Mortality Weekly*, September 6, 2013 / 62(35), 714–720.

Bartram, J., Corrales, L., Davison, A., Deere, D., Drury, D., Gordon, B., Howards, G., Rinehold, A. and Stevens, M. (2009). *Water Safety Plan Manual: step by step risk management for drinking-water supplies.* Geneva: WHO.

Brackett, R.E. (1999). Incidence, contributing factors, and control of bacterial pathogens in produce. *Postharvest Biology and Technology*, 15(3), 305–311.

Carr, R. (2001). Excreta-related infections and the role of sanitation in the control of transmission, in Fewtrell, L. and Bartram, J. (eds) *Water Quality: Guidelines, standards and health.* London: WHO/IWA Publishing. http://www.who.int/water_sanitation_health/dwq/whoiwa/en/

CDC (2013). One Health in Action. http://www.cdc.gov/onehealth/in-action/index.html

Charles, K.J. et al. (2009). Assessment of the stability of human viruses and coliphage in groundwater by PCR and infectivity methods. *Journal of Applied Microbiology*, 106, 1827–1837.

Defra and Public Health England (2013). Zoonoses Report, UK 2012. https://www.gov.uk/government/uploads/system/uploads/attachment_data/file/236983/pb13987-zoonoses-report-2012.pdf (accessed 20/01/2015).

EFSA (2015). Zoonotic diseases. http://www.efsa.europa.eu/en/topics/topic/zoonoticdiseases.htm

FAO/WHO (2003). Foodborne disease monitoring and surveillance systems. Paper prepared by the Government of Malaysia. FAO/WHO Regional Conference on Food Safety for Asia and Pacific Seremban, Malaysia, 24–27 May 2004. http://www.fao.org/docrep/meeting/006/j2381e.htm (accessed 22/05/2015).

Hlavsa, M.C., Roberts, V.A., Kahler, A.M., Hilborn, E.D., Wade, T.J., Backer, L.C. and Yoder, J.S. (2014). Recreational water-associated disease outbreaks — United States, 2009–2010. *Morbidity and Mortality Weekly*, January 10, 2014 / 63(01), 6–10.

Kay, D., Anthony, S., Crowther, J., Chambers, B.J., Nicholson, F.A., Chadwick, D., Stapleton, C.M. and Wyer, M.D. (2010). Microbial water pollution: A screening tool for initial catchment-scale assessment and source apportionment. *Science of the Total Environment*, 408(23), 5649–5656.

Pan American Health Organization (2001). *Zoonoses and Communicable Diseases Common to Man and Animals*. 3rd edition. Washington, DC: PAHO.

Pond, K. and Pedley, S. (2011). Water and environmental health. In Battersby, S. (ed.) *Clay's Handbook of Environmental Health*. 20th edition. London and New York: Spon Press.

Pond, K., Rueedi, J. and Pedley, S. (2004) MicroRisk literature review – Pathogens in drinking water sources. http://www.microrisk.com/uploads/pathogens_in_drinking_water_sources.pdf

Questier, J. (2011). Food and food safety. In S. Battersby (ed.) *Clay's Handbook of Environmental Health*. Twentieth edition. London: Spon Press.

Richardson, S.D., DeMarini, D.M., Kogevinas, M., Fernandez, P., Marco, E., Lourencetti, C., Ballesté, C., Heederik, D., Meliefste, K., McKague, A.B., Marcos, R., Font-Ribera, L., Grimalt, J.O. and Villanueva, C.M. (2011). What's in the pool? A comprehensive identification of disinfection by-products and assessment of mutagenicity of chlorinated and brominated swimming pool water. *Environmental Health Perspectives*, 118(11), 1523–1530.

Sheridan, J.J. and McDowell, D.A. (1998). Factors affecting the emergence of pathogens on foods. *Meat Science*, 49(Suppl. I), S151–S167.

Silgram, M., Anthony, S.G., Fawcett, L. and Stromqvist, J. (2008). Evaluating catchment-scale models for diffuse pollution policy support: Some results from the EUROHARP project. *Environmental Science and Policy*, 11(2), 153–162.

Smith, A.H., Lingas, E.O. and Rahman, M. (2000). Contamination of drinking-water by arsenic in Bangladesh: A public health emergency. *Bulletin of the World Health Organization*, 78(9), 1093–1103.

Sobsey, M. and Pillai, S.D. (2009). Where future emerging pathogens will come from and what approaches can be used to find them, besides VFARs. *Journal of Water and Health*, 7(Suppl 1), S75–S93.

White, G.F., Bradley, D.J. and White, A.U. (1972). *Drawers of Water: domestic water use in East Africa*. Chicago, IL: University of Chicago Press.

WHO (2011). *Guidelines for Drinking Water Quality*. 4th edition. Geneva: World Health Organization.

WHO (2014). Investing in water and sanitation: increasing access, reducing inequalities. *UN-Water Global Analysis and Assessment of Sanitation and Drinking-Water GLAAS 2014 Report*. Geneva: World Health Organization.

13

DRINKING WATER CONTAMINATION[*]

Christine Stauber and Lisa Casanova

ASSISTANT PROFESSORS, SCHOOL OF PUBLIC HEALTH,
GEORGIA STATE UNIVERSITY, ATLANTA, GA, USA

Learning objectives

1 Describe important chemical, microbiological and radiological agents found in drinking water.
2 Identify key health concerns associated with exposure to drinking water.
3 Discuss challenges with measuring and estimating exposure via drinking water.

Global burden of waterborne disease

Estimates suggest that exposure to contaminated water, inadequate sanitation and lack of sufficient water for hygiene causes 1.5 percent of all deaths (Prüss-Ustün et al., 2014). Significant improvements have been made in the past two decades; however, drinking water remains a major source of exposure for chemical, microbiological and (to lesser extent) radiological agents, although food may also represent an equal or greater opportunity for exposure to chemical and microbiological agents. Globally, more than 700 million people lack access to improved sources of drinking water (World Health Organization, 2014). This lack of access to improved water results in increased opportunities for exposure to pathogens that cause a wide array of infectious diseases including diarrheal disease and helminth infections, among many other illnesses (Prüss, et al., 2002; Straif et al., 2009; World Health Organization and UNICEF, 2014). Prüss-Ustün et al. (2014) estimate more than half a million deaths from diarrheal disease were attributable to inadequate drinking water in 2012 alone. Even households that have achieved access to improved sources of drinking water are not guaranteed that the water be free of harmful microorganisms or chemicals (Onda et al., 2012). Recent global estimates suggest tens of millions of people are exposed via drinking water to harmful chemicals such as naturally occurring arsenic (Smith et al., 2000;

[*] Recommended citation: Stauber, C. and Casanova, L. 2015. 'Drinking water contamination', in Bartram, J., with Baum, R., Coclanis, P.A., Gute, D.M., Kay, D., McFadyen, S., Pond, K., Robertson, W. and Rouse, M.J. (eds) *Routledge Handbook of Water and Health*. London and New York: Routledge.

Mead, 2005). Protecting, treating and adequately storing drinking water remain essential mechanisms to reduce exposure to a variety of contaminants that pose serious health risks.

Over the last two decades, substantial gains have been made to provide access to improved drinking water worldwide (World Health Organization, 2014). However, there is a growing interest in disaggregating the details behind the statistics. There is emerging evidence that the greatest health gains are achieved by gaining access to a continuous piped supply as compared to other types of improved access (Brown et al., 2013; Wolf et al., 2014). A more nuanced approach to measuring the role of improved drinking water and its impact on health is needed (Onda et al., 2012; Kayser et al., 2013; Wolf et al., 2014).

Chemical hazards associated with drinking-water exposure

Short-term exposure to chemicals in drinking water is often not of immediate health consequence with some exceptions (including nitrate) (World Health Organization, 2011). Long-term exposure to chemicals in water has been linked to non-infectious diseases, including cancers attributed to arsenic (Straif, et al., 2009) and trihalomethanes (Villanueva et al., 2003). Although there are many potential chemicals that can and have been found in drinking water and an equal number of different health effects, there are relatively few chemicals that are likely to be present in any one source of drinking water at any one point in time. To assess the potential risk of exposure that a chemically contaminated water source poses, it is best to determine the origin of the chemical contamination in the environment (see Table 13.1). Managing exposure to different chemicals often calls for different approaches such as source water selection to avoid naturally occurring sources of contamination such as arsenic-contaminated groundwater, or pollution control for anthropogenic sources, such as minimizing creation of disinfection byproducts during the water treatment process.

In the *Guidelines for Drinking-Water Quality* (4th edition) the World Health Organization has provided tolerable daily intake (TDI) in drinking water for hundreds of potential chemical contaminants (World Health Organization, 2011). The TDI are guidelines derived as the amount of chemical that can be consumed (in drinking water) over a lifetime with minimal risk to health (World Health Organization, 2011). However, health effects data are often not available for human exposures and therefore the majority of these guidelines are based on data collected from animals. As a result, there is a significant amount of uncertainty surrounding these values, especially at long-term low dose exposures likely to be found in

Table 13.1 Sources of chemicals found in drinking water

Source of chemical	Example sources	Example chemical	Suggested tolerable daily intake in drinking water
Naturally occurring	Rocks, soil, geologic formation	Arsenic	10 μg/L
Industrial sources	Mining, manufacturing, sewage, runoff	Benzene	10 μg/L
Agricultural	Fertilizers, pesticides	Atrazine	100 μg/L
Water treatment	Coagulation, disinfection	Trihalomethanes such as chloroform	300 μg/L

Source: Adapted from Table 8.1 in World Health Organization (2011)

drinking water. Therefore, these TDIs were designed to be conservative guidelines limiting the potential risk to the consumer.

Globally, naturally occurring arsenic (As) is the chemical of most concern for exposure via drinking water. Since the early 1990s, exposure to As via drinking water (wells) was identified as a concern in Bangladesh (Smith et al., 2000; Mead, 2005). Through subsequent years, this discovery has been called the worst mass poisoning in history as estimates suggest that 20–50 million in Bangladesh alone are at risk of exposure to potentially disease causing concentrations of As in drinking water (> 50 μg/L) (Smith et al., 2000; Mead, 2005). Even in the United States, estimates suggest that millions may be exposed to drinking water that exceeds the U.S. Environmental Protection Agency's standard of 10 μg/L (Mead, 2005). Long-term As exposure has been associated with skin, bladder and lung cancer as well as other chronic diseases such as kidney disease (Smith et al., 2000; Mead, 2005; Straif et al., 2009).

Microbiological hazards associated with drinking-water exposure

Globally, drinking water that is contaminated due to inadequate source water protection, treatment and safe storage is associated with infectious diarrheal disease, a leading cause of death in children less than five years of age (Prüss-Ustün et al., 2014). Significant advances in provision of safe drinking water over the last two decades notwithstanding, more than 700 million remain at risk from consuming water from unimproved drinking-water supplies (World Health Organization, 2014). Exposure to microbiologically contaminated drinking water results in billions of cases of diarrheal disease and more than half a million deaths each year (Straif et al., 2009; Prüss-Üsten et al., 2014; World Health Organization, 2014). Major etiologic agents of concern include viruses, bacteria and parasites; a brief list of those agents commonly associated with transmission via drinking water is shown in Table 13.2. Even improved drinking-water supplies may pose microbial risk as there is growing evidence that these improved supplies are not always microbiologically safe. Bain et al. (2014) estimated that improved supplies that are susceptible to contamination may leave more than one billion people at risk for exposure to fecal contamination and pathogens via contaminated drinking water.

In the United States, substantial improvements in reducing infectious disease spread through drinking water can be attributed to laws such as the Safe Drinking Water Act and the Surface Water Treatment Rule which increased protection of source waters used for drinking water. These laws have also increased requirements for drinking-water treatment. However, outbreaks continue to highlight vulnerabilities in water supply and treatment.

Table 13.2 Etiologic agents of concern

Bacteria	Viruses	Protozoa	Helminths
Campylobacter	Enteroviruses	Cryptosporidium	Dracunculus medinensis
Escherichia coli	Hepatitis A	Cyclospora	
Legionella	Hepatitis E	Entamoeba	
Salmonella	Noroviruses	Giardia	
Shigella	Rotaviruses		
Vibrio cholerae			

Source: Adapted from Table 7.1 in World Health Organization (2011)

In their most recent surveillance efforts, the Centers for Disease Control and Prevention identified 33 outbreaks associated with exposure to contaminated drinking water (Centers for Disease Control and Prevention, 2013). Two pathogens, *Legionella* and *Campylobacter*, were responsible for the majority of outbreaks and illnesses, respectively, highlighting potential trends in drinking-water exposure in piped systems – premise plumbing and small systems. Water supplies most often associated with waterborne disease outbreaks in the United States were private wells and small community supplies where treatment deficiencies were found (Centers for Disease Control and Prevention, 2013). For infectious diarrheal disease, the main risk factor for the outbreak was identified as untreated water or treatment deficiency. For infectious respiratory illness (*Legionella*), the spread of this infection remains difficult to control as often the organism multiplies in premise plumbing systems in buildings (Centers for Disease Control and Prevention, 2013).

Radiological hazards associated with drinking-water exposure

While the risk from exposure to radiological agents is small compared to microbiological and chemical risks, exposure to these contaminants is of concern as it may result in cancer. The primary radiological hazard associated with drinking water is radon (World Health Organization, 2011). A naturally occurring gas, formed in certain geologic conditions, radon can seep into groundwater. It is important to note that drinking water may not be the main route of exposure to radon as indoor air may have a major role. In addition to radon, other radionuclides can be associated with contamination of drinking water but make up a smaller portion of exposure via drinking water globally (World Health Organization, 2011).

Estimating drinking-water exposure

Epidemiologic studies attempting to link drinking-water exposures to health outcomes pose a number of challenges for making causal inferences. Causal links to outcomes are complicated by the fact that outcomes of chronic chemical exposure, such as cancer, are multi-causal. Even in the case of infectious diarrheal disease, it can be difficult to separate drinking-water exposure from a complex causal web of fecal–oral transmission pathways that include food, hands, fomites, soil and other environmental sources. Accurate exposure assessment for contaminants in drinking water is a major challenge for epidemiologic studies and risk assessments; water is a ubiquitous exposure occurring through multiple pathways (ingestion, inhalation, dermal), and a source of both acute and chronic hazards. Assessment of chemical, microbial and radiological hazards via drinking water typically need two main pieces of information:

1　the occurrence of contaminants in the source water, including levels, frequency, and duration;
2　individual water use and consumption patterns and related behaviors.

For ideal exposure assessments, both types of information would be available. However, many times this information is difficult to gather; exposure assessments of water have multiple sources of variability and uncertainty, and often rely on a variety of assumptions. For example, for the purpose of drinking-water risk assessment, in the United States it is commonly assumed that adults consume one to two liters of water a day, live for 70 years and weigh 70 kg (United States Environmental Protection Agency, 2011) when estimating our exposure to a variety of contaminants in drinking water. Different assumptions may be used for vulnerable groups such as very young or very old people, pregnant women and immune-compromised individuals who may be at increased risk for exposure or poor health outcomes as a result of drinking-water exposures. The very young are often at increased risk of exposure due to smaller body mass and higher (per kilogram) consumption compared to adults. Susceptibility also results from increased risk of poor health outcomes associated with immature or weakened immune systems as is common in the very young or old. As a result of their vulnerability, increased protection measures should be considered to reduce exposure to chemical, microbiological and radiological agents. In addition to relying on assumptions about the individual, estimating the concentration of contaminants in water is also challenging. There is significant spatial and temporal variability in drinking-water contamination, and there may be limited data available to characterize levels and variability, particularly over large spatial and time scales. Studies of chemical contaminants pose unique challenges because the relevant timeframe of exposure can be quite long, spanning years or decades (Villanueva et al., 2007).

Epidemiologic studies of drinking-water exposure may employ a variety of approaches to assign exposure values which may involve direct and indirect measurements. Common approaches include using residential history as a proxy for exposure to specific water sources (Koivusalo et al., 1997; Villanueva et al., 2007), self-report of water sources (Doyle et al., 1997), self-report of water consumption patterns (Strauss et al., 2001), sampling of water for contaminants at the source (Moe et al., 1991) or in individual households (Clasen et al., 2008), or direct measurement of biomarkers of exposure in individuals (Backer et al., 1999). Some epidemiologic studies have used randomized controlled trial designs, assigning exposure groups by providing some households with filters designed to eliminate any contaminants in water at the household level, essentially creating a group unexposed to contaminants that may be present (Payment et al., 1991; Colford et al., 2005). All of these approaches have advantages and disadvantages and the approach selected should be optimized for the epidemiologic problem at hand.

Key recommended readings (open access)

1 Prüss, A., Kay, D., Fewtrell, L. and Bartram, J. 2002. Estimating the burden of disease from water, sanitation and hygiene at the global level. *Environmental Health Perspectives*, 110(5): 537–542. This article provides a description of the link between water, sanitation and hygiene and variety of diseases.

2 Mead, M.N. 2005. Arsenic: in search of an antidote to a global poison. *Environmental Health Perspectives*, 113(5): A378–A386. Environews. This article is a thoughtful discussion of the magnitude of the arsenic contamination problem, with global and local perspectives.

3 Onda, K., LoBulgio, J. and Bartram J. 2012. Global access to safe water: accounting for water quality and the resulting impact on MDG progress. *International Journal of Environmental Research and Public Health*, 9(3): 880–894. This article provides an update to progress on the Millennium Development Goal progress with an analysis of the difference between improved and safe drinking water.

References

Backer, L.C., Ashley, D.L., Bonin, M.A., Cardinali, F.L., Kieszak, S.M. and Wooten, J.V. (1999) Household exposures to drinking water disinfection by-products: whole blood trihalomethane levels. *Journal of Exposure Analysis and Environmental Epidemiology*, 10: 321–326.

Bain, R., Cronk, R., Hossain, R., Bonjour, S., Onda, K., Wright, J., Yang, H., Slaymaker, T., Hunter, P., Prüss-Ustün, A. and Bartram, J. (2014), Global assessment of exposure to faecal contamination through drinking water based on a systematic review. *Tropical Medicine & International Health*, 19: 917–927. doi: 10.1111/tmi.12334

Brown, J., Hien, V.T., McMahan, L., Jenkins, M.W., Thie, L., Liang, K., Printy, E. and Sobsey, M.D. (2013) Relative benefits of on-plot water supply over other 'improved' sources in rural Vietnam. *Tropical Medicine & International Health*, 18(1): 65–74.

Centers for Disease Control and Prevention (2013) Surveillance for waterborne disease outbreaks associated with drinking water and other nonrecreational water – United States, 2009–2010. *Morbidity and Mortality Weekly Report*, 62(35): 714–720.

Clasen, T.F., Thao, D.H., Boisson, S. and Shipin, O. (2008) Microbiological effectiveness and cost of boiling to disinfect drinking water in rural Vietnam. *Environmental Science & Technology*, 42: 4255–4260.

Colford, J.M., Wade, T.J., Sandhu, S.K., Wright, C.C., Lee, S., Shaw, S., Fox, K., Burns, S., Benker, A. and Brookhart, M.A. (2005) A randomized, controlled trial of in-home drinking water intervention to reduce gastrointestinal illness. *American Journal of Epidemiology*, 161: 472–482.

Doyle, T.J., Zheng, W., Cerhan, J.R., Hong, C.-P., Sellers, T., Kushi, L. and Folsom, A. (1997) The association of drinking water source and chlorination by-products with cancer incidence among postmenopausal women in Iowa: a prospective cohort study. *American Journal of Public Health*, 87: 1168–1176.

Kayser, G.L., Moriarty, P., Fonseca, C. and Bartram, J. (2013) Domestic water service delivery indicators and frameworks for monitoring, evaluation, policy and planning: a review. *International Journal of Environmental Research and Public Health*, 10(10): 4812–4835.

Koivusalo, M., Pukkala, E., Vartiainen, T., Jaakkola, J.J. and Hakulinen, T. (1997) Drinking water chlorination and cancer – a historical cohort study in Finland. *Cancer Causes & Control*, 8: 192–200.

Mead, M.N. (2005) Arsenic: in search of an antidote to a global poison. *Environmental Health Perspectives*, 113(5): A378–A386.

Moe, C., Sobsey, M., Samsa, G. and Mesolo, V. (1991) Bacterial indicators of risk of diarrhoeal disease from drinking-water in the Philippines. *Bulletin of the World Health Organization*, 69(3): 305–317.

Onda, K., LoBuglio, J. and Bartram, J. (2012) Global access to safe water: accounting for water quality and the resulting impact on MDG progress. *International Journal of Environmental Research and Public Health*, 9: 880–894.

Payment, P., Richardson, L., Siemiatycki, J., Dewar, R., Edwardes, M. and Franco, E. (1991) A randomized trial to evaluate the risk of gastrointestinal disease due to consumption of drinking water meeting current microbiological standards. *American Journal of Public Health*, 8: 703–708.

Prüss, A., Kay, D., Fewtrell, L. and Bartram, J. (2002) Estimating the burden of disease from water, sanitation and hygiene at the global level. *Environmental Health Perspectives*, 110(5): 537–542.

Prüss-Ustün, A., Bartram, J., Clasen, T., Colford, J.M., Jr., Cumming, O., Curtis, V., Bonjour, S., Dangour, A.D., De France, J., Fewtrell, L., Freeman, M.C., Gordon, B., Hunter, P.R., Johnston, R.B., Mathers, C., Mausezahl, D., Medlicott, K., Neira, M., Stocks, M., Wolf, J. and Cairncross, S. (2014) Burden of disease from inadequate water, sanitation and hygiene in low- and middle-income settings: a retrospective analysis of data from 145 countries. *Tropical Medicine & International Health*, 19(8): 894–905.

Smith, A.H., Lingas, E.O. and Rahman, M. (2000) Contamination of drinking-water by arsenic in Bangladesh: a public health emergency. *Bulletin of the World Health Organization*, 78: 1093–1103.

Straif, K., Benbrahim-Tallaa, L. and Baan, R. (2009) World Health Organization International Agency for Research on Cancer Monograph Working Group. A review of human carcinogens—part C: metals, arsenic, dusts, and fibres. *Lancet Oncology*, 10: 453–454.

Strauss, B., King, W., Ley, A. and Hoey, J.R. (2001) A prospective study of rural drinking water quality and acute gastrointestinal illness. *BMC Public Health*, 1: 8–14.

United States Environmental Protection Agency. (2011) *Exposure Factors Handbook: 2011 edition*, vol. EPA/600/R-09/052F. Washington, DC: National Center for Environmental Assessment.

Villanueva, C.M., Fernandez, F., Malats, N., Grimalt, J.O. and Kogevinas, M. (2003) Meta-analysis of studies on individual consumption of chlorinated drinking water and bladder cancer. *Journal of Epidemiology and Community Health*, 57: 166–173.

Villanueva, C.M., Cantor, K.P., Grimalt, J.O., Malats, N., Silverman, D., Tardon, A., Garcia-Closas, R., Serra, C., Carrato, A. and Castano-Vinyals, G. (2007) Bladder cancer and exposure to water disinfection by-products through ingestion, bathing, showering, and swimming in pools. *American Journal of Epidemiology*, 165: 148–156.

Wolf, J., Pruss-Ustun, A., Cumming, O., Bartram, J., Bonjour, S., Cairncross, S., Clasen, T., Colford, J.M., Jr., Curtis, V., De France, J., Fewtrell, L., Freeman, M.C., Gordon, B., Hunter, P.R., Jeandron, A., Johnston, R.B., Mausezahl, D., Mathers, C., Neira, M. and Higgins, J.P. (2014) Assessing the impact of drinking water and sanitation on diarrhoeal disease in low- and middle-income settings: systematic review and meta-regression. *Tropical Medicine & International Health*, 19(8): 928–942.

World Health Organization. (2011) *Guidelines for Drinking-Water Quality* – 4th edition. Geneva: WHO Press.

World Health Organization and UNICEF. (2014) Progress on drinking water and sanitation, *Joint Monitoring Programme for Water Supply and Sanitation Progress on Drinking Water and Sanitation: 2014 Update*. Geneva: WHO Press.

14

RECREATIONAL WATER CONTAMINATION[*]

Marc Verhougstraete

POSTDOCTORAL RESEARCH ASSOCIATE, THE UNIVERSITY OF ARIZONA,
COLLEGE OF PUBLIC HEALTH, TUCSON, ARIZONA, USA

Jonathan Sexton

RESEARCH SPECIALIST, THE UNIVERSITY OF ARIZONA, COLLEGE
OF PUBLIC HEALTH, TUCSON, ARIZONA, USA

Kelly Reynolds

ASSOCIATE PROFESSOR, THE UNIVERSITY OF ARIZONA, COLLEGE
OF PUBLIC HEALTH, TUCSON, ARIZONA, USA

Learning objectives

1 Understand the important microbiological agents found in recreational water.
2 Learn to recognize the key health concerns associated with point and nonpoint sources of recreational water pollution.
3 Examine the different types of recreational water and associated human behaviors.

Introduction

Recreational waters can be divided into two main types: surface waters (e.g. oceans, lakes, rivers, hot springs, streams and beaches) and enclosed systems (e.g. pools, spas, water parks, splash pads, fountains and hot tubs). Recreational activities include primary and secondary contact. Primary contact involves direct contact with water (e.g. swimming, surfing or diving). Secondary contact involves indirect contact with water (e.g. boating or fishing). While a plethora of contaminants and pollution types exist in recreational waters around

---

[*] Recommended citation: Verhougstraete, M., Sexton, J. and Reynolds, K. 2015. 'Recreational water contamination', in Bartram, J., with Baum, R., Coclanis, P.A., Gute, D.M., Kay, D., McFadyen, S., Pond, K., Robertson, W. and Rouse, M.J. (eds) *Routledge Handbook of Water and Health*. London and New York: Routledge.

the world, this chapter will focus on waterborne microorganisms and human pathogens in surface water systems. This chapter will address the multiple pathogens associated with recreational water, waterborne pathogen sources, and the human health effects associated with waterborne pathogens and recreational water activities.

Recreational water pathogens

Researchers have identified many microorganisms in recreational waters which have the potential to impact human health including bacteria, protozoa, helminths and viruses (Table 14.1). Additionally, surface water types (i.e. small springs and streams to lakes and oceans) have unique microbial communities depending on a multitude of factors including source water, depth, size, turbidity and flow rate (Maier et al. 2009). Microbial concentrations in the various surface water systems can have a wide range, as illustrated in Figure 14.1, with photosynthetic and heterotrophic organisms measured in various water systems. In general, higher concentrations of microorganisms are found closer to the shoreline. Likewise, greater exposures to pathogens generally occur near shore through primary contact. Bacterial concentrations decrease as water depth increases due to the availability of organic matter, nutrients, sunlight and animal influence.

In marine waters, commonly detected pathogens include bacteria (*Vibrio cholerae, V. parahaemolyticus* and *V. vulnificus, Clostridium perfringens and Mycobacterium avium* complex) and viruses (enteroviruses, hepatitis A viruses, Norwalk viruses, reoviruses, adenoviruses and rotaviruses). The most common pathogens in freshwaters around the world include *Shigella, Campylobacter, Leptospira, Giardia*, trematode worms and *E. coli* O157:H7 (Pond 2005; Craun et al. 2005). The protozoa *Cryptosporidium* and *Giardia* have been detected in marine water but are more frequently detected in freshwater, especially in rivers and streams (Furness et al. 2000; Coupe et al. 2006). These highly stable organisms can survive for periods greater than 160 days in freshwater (Medema 1997) under variable sunlight exposure, presence of other microorganisms and a wide range of temperatures. Sources of these two pathogens include wastewater effluent and wild and domestic animals.

Warm and brackish waters support the growth of *Naegleria fowleri* (*N. fowleri*), a water based pathogen independent of fecal contamination, which causes primary amoebic meningoencephalitis (PAM). *N. fowleri* has been identified throughout the world and poses the most significant risk to bathers who submerge their head in naturally occurring hot springs. While most *N. fowleri* infections occur in tropical or subtropical climates, infections have been reported in areas with a continental climate, and rare PAM infections have been reported in absence of hot spring exposure (De Jonckheere 2011). Reported *N. fowleri* infections are low throughout the world, but infections have a 95 percent mortality rate. As of 2011, only 235 PAM cases have been reported worldwide: 24 in Europe, 19 in Australia, 9 in New Zealand, 43 in Asia, 6 in Africa, 122 in North America and 12 in South America (De Jonckheere 2011).

Microbial water quality monitoring includes the collection and processing of multiple samples per site. Prone to lengthy collection, processing and result reporting stages, this process can lead to delays in public notification. Methods and practices used around the world to reduce the time from sample collection to notification including rapid testing methods, on-site testing, predictive methods and electronic signage. Quantitative polymerase chain reaction (qPCR) quantifies DNA of a specific organism and is emerging as the preferred method for routine monitoring. This method can be used to detect fecal indicator organisms or source specific markers in less than four hours, but requires extensive

Table 14.1 Key pathogens identified in recreational water

Organism	System most prolifically identified	Infectious dose (ID_{50})
Bacteria		
Campylobacter	Beaches; surface waters contaminated by animals	8.90E+02
E. coli O157	Enclosed systems; surface waters	3.20E+03
Shigella spp.	Enclosed systems; surface waters	1.50E+03
Legionella spp.	Enclosed systems – spas	1.20E+01
Leptospira	Beaches; surface waters contaminated by animals	ND
Staphylococcus aureus and Methicillin-resistant S. aureus (MRSA)	Beaches/beach sand; surface waters; enclosed systems	9.10E+06
Mycobacterium avium complex	Enclosed systems – hot tubs	1.00E+03
Vibrio vulnificus	Surface waters	ND
Protozoa and helminths		
Cryptosporidium	Enclosed systems; surface waters	1.20E+01
Giardia	Beaches; surface water; pools	3.50E+01
Microsporidia (emerging)	Enclosed systems	ND
Naegleria fowleri	Enclosed systems; hot springs; surface waters	1.30E+02
Trematode worm	Fresh surface waters	
Viruses		
Human adenovirus	Enclosed systems	1.10E+00
Echovirus	Enclosed systems	9.20E+02
Hepatitis A and E	Surface waters; enclosed systems	ND
Coxsackievirus	Enclosed systems	ND
Norovirus	Enclosed systems; beaches; surface waters	ND

Note: ND: Not defined

Sources: Pond (2005), Sinclair et al. (2009), Craun et al. (2005), and QMRA (n.d.)

training and expensive reagents (Griffith and Weisberg 2011). While qPCR has the potential to replace current monitoring methods, low pathogen concentrations limit the use of direct monitoring and many genes commonly used to identify pathogens are found in other organisms (e.g. bacteriophage), resulting in false positives. Immunomagnetic separation adenosine triphosphate (IMS-ATP) rapidly measures ATP with a bioluminescence which allows for the quantification of bacteria, but this technique has a high rate of false positives. Measuring optical brighteners, another rapid technique, uses a fluorometer to detect dyes

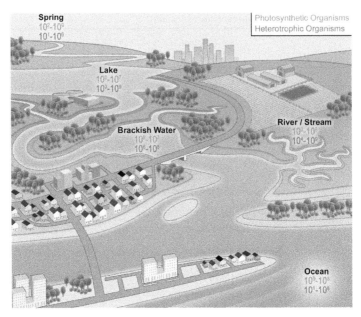

Figure 14.1 Concentrations of photosynthetic and heterotrophic organisms in different types of water in colony forming units (CFU)/100 ml

Source: Adopted from Maier et al. (2009)

from detergents in wastewater, but this technique has a lower sensitivity than microbial assessments. Empirical predictive models fit algorithms to previously analyzed water quality parameters to produce real-time estimates of water quality.

Another practice of recreational water quality monitoring to reduce cost and effort is to use waterborne pathogen surrogates, known as fecal indicator organisms (Chapter 4). Ideally, fecal indicator organisms are: 1) present only when fecal pathogens are present; 2) present in greater concentrations than pathogens; 3) more resistant to treatment and environmental surroundings than pathogens; 4) easily detected; 5) representative of a specific group of pathogens; and 6) do not grow in aquatic environments (Bonde 1966; Dufour 1984). The established fecal indicator organisms most commonly used for recreational water quality monitoring are *Escherichia coli* (*E. coli*) in freshwater and enterococci in marine water.

Recreational water and health

Epidemiological studies and waterborne disease surveillance performed during the second half of the 20th century (Prüss 1998; Pond 2005) have provided vital information for the development of recreational water quality policy and protection of human health (Chapter 45). These studies measured fecal indicator bacteria in marine and fresh water systems to demonstrate that exposure to impaired water quality had a measurable effect on human health. Results from these studies identified that adverse health effects exhibited stronger correlations with the fecal indicator bacteria *E. coli* in freshwater and enterococci in marine water.

Epidemiological studies performed at beaches throughout the world suggest that a relationship exists between gastrointestinal symptoms and fecal indicator bacteria concentrations in recreational water, as supported by consistent dose–response relationships. Epidemiological studies relevant to recreational water activities identified more frequent illness incidences in swimmers than in non-swimmers. More specifically, they identified that fecal indicator organism (*E. coli* and enterococci) densities at beaches were statistically related to increased rates of gastrointestinal illness and infections of the eyes, ears and the upper respiratory system in swimmers. Risk assessment of waters impacted by different fecal sources (e.g. animals or humans) has not been thoroughly assessed, yet it remains a significant concern for scientists and public health officials as exposure to recreational waters contaminated with any type of feces increases the risk of illness (Prüss 1998; Soller et al. 2010).

Pollution sources

The greatest potentials for pollution transport in surface recreational waters are the physical movement of water itself and bathers. Overland flows concentrate pollutants and rapidly transport them down gradient where they enter larger systems and become magnified in aquatic environments, impacting recreational surface water quality as either point source pollution or nonpoint source pollution. The largest point sources of pollution include sewage effluents, combined sewer overflow systems, industrial effluents and concentrated animal feeding operations. The major nonpoint sources of pollution include improperly functioning sanitation systems (leaking sewers, on-site septic systems or industrial waste systems) and runoff from agriculture or urban landscapes. Bather shedding of pathogens into waters during immersion is another major source of contamination to recreational waters (Gerba 2000). Recreational waters may be impacted by one or more pollution source.

Point source impacted waters have a stronger direct correlation between fecal indicator organisms and gastrointestinal illness in comparison to beaches impacted by nonpoint source pollution (Colford et al. 2007). Many pathogens have been identified in point source impacted waters including coxsackievirus, human polyomavirus, *Salmonella*, *Campylobacter* and *Staphylococcus aureus* (Bolton et al. 1999; Goodwin & Pobuda 2009; Korajkic et al. 2011). Implementation and enforcement of water protection laws such as the Clean Water Act (US 1972), the Water Framework Directive (European Union 2000) and the Protection of the Environment Operations Act (Australia 1997) have vastly improved the management of point source discharges (i.e. combined sewer overflows and industrial discharges) and have resulted in the reduction or elimination of pollution in all types of water where these regulations have been applied.

As efforts continue to reduce contamination from identified point sources, regulators and stakeholders have begun the arduous task of addressing nonpoint sources of pollution. Stormwater runoff in developed countries and a lack of infrastructure in developing countries have demonstrated the significant effects of nonpoint source pollutants on water quality throughout the world. Nonpoint source pollution leads to poor water quality primarily when precipitation inundates soil or treatment systems, then produces contaminated overland flow which enters surface waters and beaches. While there is increased risk of gastrointestinal illness in bathers at beaches impacted by nonpoint sources of pollution, the correlation between gastrointestinal illness and fecal indicator bacteria is not clearly established (Colford et al. 2007; Fleisher et al. 2010). This is likely due to intermittent pathogen loading stemming from unpredictable stormwater runoff.

Surface waters impacted by nonpoint source pollution often reflect the surrounding watershed. For instance, water systems surrounded by urban development tend to have a strong association with human pathogens including polyomavirus, adenoviruses, norovirus and human sewage specific DNA markers. Rurally located systems are usually impacted by pathogens associated with agriculture animal waste such as *Cryptosporidium*, *Salmonella*, *Giardia* and *E. coli* O157. Impaired water quality can also result from failing infrastructure systems. A significant correlation exists between failing wastewater systems and fecal indicator bacteria concentrations at beaches (Lipp et al. 2001), but far less research has been conducted on the impacts of failing infrastructure compared to contaminated runoff.

Human behavior

Human behaviors greatly influence exposure pathways in recreational waters, and thus influence health outcomes. The greatest risk for pathogen exposure and illness occurs in children and immunocompromised populations (Sanborn & Takaro 2013). Children have poorly developed immune systems, usually have longer exposure times, participate in more high-risk behaviors and are exposed to greater concentrations of contaminants in the shallow water. Recreational activity in the swash zone, the area where waves break on the sand, is a human behavior specific to beaches that increases illness risk. Shallow water is warmer and hosts stronger wave action, two factors that increase exposure activity. Lotic systems and stormwater discharges located near beaches are popular swimming attractions. However, such inputs often contain high levels of pathogens from point and nonpoint sources of pollution, as described in the "Pollution sources" section.

Activities such as diving, triathlons and surfing are associated with a higher risk of infection compared to bathers due to more frequent and intense contact with contaminated water (Sanborn & Takaro 2013). Even partial water contact activities such as boating and fishing increase an individual's risks of acute gastrointestinal illness by 40–50 percent when compared to nonwater recreational activities (Sanborn & Takaro 2013).

Conclusion

Many pathogens have been identified in surface waters used for recreational activities. Pathogens may enter the water from a variety of sources, including urban/rural/agriculture stormwater runoff, partially or untreated wastewater, and industrial effluents. Humans are exposed to these pathogens during recreational activity. If the pathogens are ingested, they have the potential to cause illness, as demonstrated through multiple epidemiological studies at recreational beaches. However, scientists continue to develop improved understanding of relationships between pathogens, water quality and health. Past research has demonstrated recreational waters are highly dynamic and often require a site specific approach to adequately protect human health during recreational water activities.

Key recommended readings

1 Pond, K.R. 2005. Water recreation and disease: Plausibility of associated infections: acute effects, sequelae and mortality, Cornwall, UK. Available at : http://www.who.int/water_sanitation_health/bathing/recreadis.pdf. This article describes the pathogens, and associated diseases, encountered during recreational activities in marine, freshwater, hot tubs, spas and swimming pools.

2 Prüss, A. 1998. Review of epidemiological studies on health effects from exposure to recreational water. *International Journal of Epidemiology*, 27, 1–9. Available at: http://ije.oxfordjournals.org/content/27/1/1.full.pdf+html. This article summarizes 22

epidemiological studies from around the world that assess health risk caused by poor microbiological recreational water quality.

3 Hlavsa, M.C., Roberts, V.A., Kahler, A.M., Hilborn, E.D., Wade, T.J., Backer, L.C. and Yoder, J.S. (2014). Morbidity and mortality weekly report: Recreational water-associated disease outbreaks – United States, 2009–2010. *MMWR Surveillance Summary*, 63, 6–10. Available at: http://www.cdc.gov/mmwr/preview/mmwrhtml/mm6301a2. htm. This biennial article summarizes the Centers for Disease Control and Prevention's surveillance program results for recreational water-associated disease outbreaks.

References

Bolton, F.J. et al. 1999. Presence of *Campylobacter* and *Salmonella* in sand from bathing beaches. *Epidemiological Infections*, 122, 7–13.

Bonde, G.J. 1966. Bacteriological methods for estimation of water pollution. *Health Laboratory Science*, 3(2), 124–8. Available at: http://www.ncbi.nlm.nih.gov/pubmed/5326083 (accessed November 24, 2014).

Colford, J.M. et al. 2007. Water quality indicators and the risk of illness at beaches with nonpoint sources of fecal contamination. *Epidemiology*, 18(1), 27–35. Available at: http://www.ncbi.nlm.nih.gov/pubmed/17149140 (accessed October 24, 2010).

Coupe, S. et al. 2006. Detection of *Cryptosporidium*, *Giardia* and *Enterocytozoon bieneusi* in surface water, including recreational areas: a one-year prospective study. *FEMS Immunology and Medical Microbiology*, 47(3), 351–9. Available at: http://www.ncbi.nlm.nih.gov/pubmed/16872371 (accessed November 24, 2014).

Craun, G.F., Calderon, R.L. & Craun, M.F. 2005. Outbreaks associated with recreational water in the United States. *International Journal of Environmental Health Research*, 15(4), 243–62. Available at: http://dx.doi.org/10.1080/09603120500155716 (accessed November 8, 2014).

De Jonckheere, J.F. 2011. Origin and evolution of the worldwide distributed pathogenic amoeboflagellate *Naegleria fowleri*. *Infection, Genetics and Evolution: Journal of Molecular Epidemiology and Evolutionary Genetics in Infectious Diseases*, 11(7), 1520–8. Available at: http://www.sciencedirect.com/science/article/pii/ S1567134811002784 (accessed November 18, 2014).

Dufour, A.P. 1984. Bacterial indicators of recreational water quality. *Canadian Journal of Public Health*, 75(1), 49–56. Available at: http://www.jstor.org/stable/41990233?seq=2 (accessed November 24, 2014).

Fleisher, J.M. et al. 2010. The BEACHES Study: health effects and exposures from non-point source microbial contaminants in subtropical recreational marine waters. *International Journal of Epidemiology*, 39(5), 1291–8. Available at: http://www.pubmedcentral.nih.gov/articlerender.fcgi?artid=2984243&tool=pmcentrez&rendertype=abstract (accessed September 10, 2013).

Furness, B.W., Beach, M.J. & Roberts, J.M. 2000. Giardiasis Surveillance – United States, 1992–1997. *MMWR Surveillance Summaries*, 49, 1–13. Available at: http://www.cdc.gov/mmwr/preview/mmwrhtml/ ss4907a1.htm (accessed November 24, 2014).

Gerba, C.P. 2000. Assessment of enteric pathogen shedding by bathers during recreational activity and its impact on water quality. *Quantitative Microbiology*, 2(1), 55–68. Available at: http://link.springer.com/ article/10.1023/A:1010000230103 (accessed April 15, 2014).

Goodwin, K.D. & Pobuda, M. 2009. Performance of CHROMagar *Staph aureus* and CHROMagar MRSA for detection of *Staphylococcus aureus* in seawater and beach sand – comparison of culture, agglutination, and molecular analyses. *Water Research*, 43(19), 4802–11. Available at: at: http://www. cabdirect.org/abstracts/20093356937.html;jsessionid=9648379A10D8C66D6EF49C1D99C7889D (accessed January 9, 2014).

Griffith, J.F. & Weisberg, S.B. 2011. Challenges in implementing new technology for beach water quality monitoring: lessons from a California demonstration project. *Marine Technology Society Journal*, 45(2), 65–73.

Korajkic, A., Brownell, M.J. & Harwood, V.J. 2011. Investigation of human sewage pollution and pathogen analysis at Florida Gulf coast beaches. *Journal of Applied Microbiology*, 110(1), 174–83. Available at: http://www.ncbi.nlm.nih.gov/pubmed/21029275 (accessed January 9, 2012).

Lipp, E.K., Farrah, S.A. & Rose, J.B. 2001. Assessment and impact of microbial fecal pollution and human enteric pathogens in a coastal community. *Marine Pollution Bulletin*, 42(4), 286–93. Available at: http://www.ncbi.nlm.nih.gov/pubmed/11381749 (accessed May 15, 2015).

Maier, R.M., Pepper, I.L. & Gerba, C.P. 2009. *Environmental Microbiology*, Amsterdam: Academic Press. Available at: http://books.google.com/books?hl=en&lr=&id=A2zL8YBXQfoC&pgis=1 (accessed November 24, 2014).

Medema, G. 1997. Survival of *Cryptosporidium parvum*, *Escherichia coli*, faecal enterococci and *Clostridium perfringens* in river water: influence of temperature and autochthonous microorganisms. *Water Science and Technology*, 35(11–12), 249–52. Available at: http://www.sciencedirect.com/science/article/pii/S0273122397002679 (accessed October 30, 2014).

Pond, K.R. 2005. *Water Recreation and Disease: Plausibility of associated infections: Acute effects, sequelae and mortality*, London: IWA Publishing.

Prüss, A. 1998. Review of epidemiological studies on health effects from exposure to recreational water. *International Journal of Epidemiology*, 27(1), 1–9. Available at: http://www.ncbi.nlm.nih.gov/pubmed/9563686.

QMRA (n.d.) Quantitative Microbial Risk Assessment (QMRA) Wiki. http://qmrawiki.canr.msu.edu/index.php/Quantitative_Microbial_Risk_Assessment_(QMRA)_Wiki

Sanborn, M. & Takaro, T., 2013. Recreational water-related illness: office management and prevention. *Canadian Family Physician*, 59(5), 491–5. Available at: http://www.cfp.ca/content/59/5/491.short (accessed December 17, 2014).

Sinclair, R.G., Jones, E.L. & Gerba, C.P. 2009. Viruses in recreational water-borne disease outbreaks: a review. *Journal of Applied Microbiology*, 107(6), 1769–80. Available at: http://www.ncbi.nlm.nih.gov/pubmed/19486213 (accessed November 13, 2013).

Soller, J.A et al. 2010. Estimated human health risks from exposure to recreational waters impacted by human and non-human sources of faecal contamination. *Water Research*, 44(16), 4674–91. Available at: http://www.ncbi.nlm.nih.gov/pubmed/20656314 (accessed August 13, 2013).

15

WATER AND FOODBORNE CONTAMINATION[*]

Timothy R. Julian

GROUP LEADER, PATHOGENS AND HUMAN HEALTH, DEPARTMENT OF
ENVIRONMENTAL MICROBIOLOGY, EAWAG, DÜBENDORF, SWITZERLAND

Kellogg J. Schwab

PROFESSOR, ENVIRONMENTAL HEALTH SCIENCES AND THE HOPKINS
WATER INSTITUTE, JOHNS HOPKINS UNIVERSITY BLOOMBERG
SCHOOL OF PUBLIC HEALTH, BALTIMORE, MD, USA

Learning objectives

1 Describe the etiological agents, and their importance, that are primarily responsible for microbial foodborne disease due to inadequate or unsafe water access.
2 Describe the chemical contaminants in food that pose human health risks due to inadequate or unsafe water access.
3 Provide examples of appropriate interventions to reduce human health risks associated with foodborne contamination.

Introduction

Ensuring access to a sufficient quantity of safe water not only reduces exposure to biological and chemical waterborne contaminants, but also reduces exposures through other routes. One notable exposure route is via food. In the United States, a place with stringent food safety controls, there are an estimated 48 million cases of foodborne disease every year (Scallan et al., 2011a). Many of these cases result from inadequate hand hygiene; reliance on unsafe water for processing, cleaning, or preparing food; and/or hydrologic-scale water movement that contributes to contaminant transport and/or uptake (Motarjemi and Käferstein, 1999;

[*] Recommended citation: Julian, T.R. and Schwab, K.J. 2015. 'Water and foodborne contamination', in Bartram, J., with Baum, R., Coclanis, P.A., Gute, D.M., Kay, D., McFadyen, S., Pond, K., Robertson, W. and Rouse, M.J. (eds) *Routledge Handbook of Water and Health*. London and New York: Routledge.

Boxall et al., 2008). These processes are often compounded in low and middle income countries where access to insufficient quantities of safe water results in greater reliance on polluted water sources for irrigation (Raschid-Sally and Jayakody, 2009).

The term "foodborne disease" encompasses a number of biological and chemical contaminants that cause acute or chronic adverse health effects. Here, we focus our discussion on the subset of virus, bacteria, helminths, protozoa, mycotoxins, heavy metals, pesticides, and radionuclides whose presence in food is linked to the movement (or lack of movement) of safe water. Neither the list, nor the descriptions, is intended to be exhaustive. At the end of the chapter is a list of readings for further information.

Biological contamination

Description

Bacteria

Bacteria are single-celled microorganisms ubiquitous in both food and the environment. Pathogenic bacteria, which represent only a small fraction of total bacteria, can cause two distinct types of foodborne disease: infection and food intoxication (i.e., toxicosis; Cliver et al., 2011). Bacterial infection occurs when pathogenic bacteria are capable of colonizing and reproducing within a human host (Cliver et al., 2011). Symptomatic disease follows this process due to the growth of bacteria causing destruction of host cells, bacterial production of toxins impacting host cells (e.g., toxicoinfection), or the host immune system's response to disease (Cliver et al., 2011). Toxicosis is the second form of bacterial foodborne illness (Cliver et al., 2011). Toxicosis differs from infection in that human host exposure to a toxin produced by the bacteria, not the bacteria itself, induces disease. Toxicosis due to foodborne illness most commonly results from contamination of a foodstuff with a toxin-producing bacteria. When the bacteria grows, a toxin is produced either as a targeted response or as a metabolic byproduct. In both cases, the ability of bacteria to grow within food products prior to consumption increases the risk of subsequent infection and/or the concentration of toxins. This ability to grow within diverse matrices is unique to bacteria.

Infection due to contaminated food and water is typically caused by the following bacteria: *Salmonella* spp., *Shigella* spp., *Campylobacter* spp., *Listeria monocytogenes*, and pathogenic *E. coli* (Cliver et al., 2011). *Salmonella* spp. and *Shigella* spp. infections cause diarrhea, fever, cramps, blood loss, vomiting, and/or potentially death. *Salmonella* Typhi, a serovar of *Salmonella enterica*, can cause more severe complications, including jaundice, intestinal bleeding, myocarditis, and encephalothopy. *Campylobacter* spp. infection is characterized by acute diarrhea, abdominal cramping, fever, and bloody stools. *L. monocytogenes* infections are characterized by fever, diarrhea, sepsis, meningoencephalitis, and potentially abortion or fetal damage. Diarrheagenic *E. coli* (e.g., enteroaggregative, enterohemorragic, enteroinvasive, enteropathogenic, and enterotoxigenic) infections result in symptoms of diarrhea, vomiting, hemolytic-uremic syndrome, renal failure, and death. Notably, enterotoxigenic *E. coli* is one of the four etiological agents most responsible for moderate-to-severe diarrhea in children under five globally (Kotloff et al., 2013). Other important waterborne pathogens associated with foodborne outbreaks include *Vibrio cholerae*, *V. parahaemolytics*, and *V. vulnificus* (Cliver et al., 2011). *Vibrio* spp. infection symptoms include vomiting, diarrhea, abdominal pain, and potentially septic shock. Cholera infections may cause diarrhea and vomiting so severe that dehydration and electrolyte imbalance lead to death.

The most common toxin-producing bacteria are *Staphylococcus aureus* and *Clostridium botulinum* (Cliver et al., 2011). *S. aureus* produces a gastrointestinal toxin that causes nausea, vomiting, diarrhea, and stomach cramps. *Clostridium botulinum*, autochthonous to soil, produces one or more neurotoxins that cause difficulty swallowing, dry mouth, weakness, blurred vision, paralysis, and possibly death. In general, toxicosis is associated with foodborne, in contrast to waterborne, outbreaks as toxicosis requires a stable environment within which the toxin is capable of accumulating to sufficient levels to induce disease.

Viruses

Viruses are the smallest sized microbes and rely on invasion and infection of host cells to replicate. Viral infections can result in a number of adverse health effects caused by damage from both the viral replication processes (e.g., gastroenteritis, cirrhosis, dehydration) and from immune responses to infection (e.g., fever). The majority of foodborne viral infections are caused by gastrointestinal viruses (e.g., norovirus, astrovirus, human adenovirus, enterovirus, and rotavirus) (Stürchler, 2006). In developed countries, norovirus is particularly insidious and is the cause of an estimated 58 percent of all foodborne illnesses in the United States (Scallan et al., 2011b). In developing countries, rotavirus plays a larger role as it is one of the four major pathogens responsible for moderate-to-severe diarrhea in children across Africa and Asia (Kotloff et al., 2013), although modern vaccines are reducing this burden. Gastrointestinal viruses cause discomfort, nausea, diarrhea, vomiting, fever, dehydration, and/or death. Hepatitis A and E are two additional foodborne viruses that may be impacted by contaminated water supplies (Cliver et al., 2011). Symptoms of these diseases vary, but can include jaundice, fatigue, fever, diarrhea, and nausea.

Protozoa

Protozoa are heterotrophic, unicellar, eukaryotic organisms with motility. Like bacteria, protozoa are naturally found in the microbial flora of healthy animals. Although the most common pathogenic protozoa (e.g., *Cryptosporidium* spp. and *Giardia* spp.) may be transmitted primarily through contaminated drinking water, foodborne transmission as a result of contaminated water (as discussed below) is also common. *Cryptosporidum* spp., alongside the aforementioned enterotoxigenic *E. coli* and rotavirus, is one of the leading causes of diarrheal disease in children in Africa and Asia (Kotloff et al., 2013). Most protozoa transmitted via food (e.g., *Balantidium coli*, *Cryptosporidium* spp., *Cyclospora cayetanensis*, *E. histolytica*, *Giardia* spp., and *Isospora* spp.) are gastrointestinal pathogens that cause nausea, bloating, cramping, dehydration, diarrhea, and/or fever. Symptoms of other foodborne protozoa, such as *Sarcocystis* spp., include weight loss, anaemia, and/or fever often presenting without gastrointestinal symptoms. *Toxoplasma gondii*, which infects up to one third of the global population, is spread via contaminated water or vegetables. Although *T. gondii* may cause little-to-no symptomatic disease, infections can lead to complications that may include encephalitis, neurologic disorders, and adverse changes in behavior such as schizophrenia and suicidal tendencies.

Helminths

Helminths are eukaryotic, parasitic worms that receive nourishment and protection by living inside a host. Often, this arrangement is detrimental to the health and well-being of the host. Human hosts become infected with helminths via contact with contaminated food, water, soil, arthropods, or other infected humans. Helminths are of particular concern in the developing world as up to one quarter of the population in these regions is infected.

There are three groups of helminths that are parasitic to humans: tapeworms, roundworms, and flukes. Tapeworms (e.g., *Taenia* spp., *Hymenolepis* spp., *Echinococcus* spp.) reside in the intestinal tract and can cause discomfort, diarrhea, and/or anemia. Flukes parasitize blood (i.e., blood flukes) and/or organ tissue (i.e., tissue flukes). *Clonorchis sinesis* and *Paragonimus westermani* are two genera of foodborne tissue flukes that infect the liver and lungs, respectively. *Clonorchis sinesis* infection can cause nausea, diarrhea, weight loss, jaundice, bacterial infections, inflammation, and cancer. *Paragonimus westermani* symptoms may include abdominal pain, diarrhea, fever, and/or hives. In severe cases, *Paragonimus westermani* infection causes headaches, vomiting, seizures, and death if left untreated. Roundworms are distinguished from both tapeworms and flukes by the presence of a tubular digestive system. *Ascaris lumbricoides*, a type of roundworm, compromises nutritional status and can cause anemia, stunting, and cognitive deficiencies. A second type of roundworm, *Trichinella* spp., is one of the most common foodborne helminths. Symptoms of infection include nausea, vomiting, sweating, diarrhea, and/or fever. If left untreated, infections can lead to intense muscular pain, difficulty breathing, heart damage, kidney malfunction, and/or death as the parasite migrates from the site of first infection (typically the intestinal mucosa) to other host organs.

Fungi

Fungal contamination of food products is primarily a concern due to their potential production of mycotoxins. Sufficient exposure to mycotoxins leads to adverse health effects that can include growth and immune impairment, cancer, and death. Human exposure to mycotoxins generally occurs through ingestion of cereals (e.g., maize, rye, wheat) and nuts (e.g., peanuts, pistachios) contaminated with mycotoxin-producing fungi (Bryden, 2007). Fungi that produce mycotoxins include species within the *Aspergillus*, *Penicillium*, *Claviceps*, *Paecilomyces*, and *Fusarium* genera. Aflatoxin, fumonisin, ochratoxin, deoxynivalenol, and zearalenone are the most notable mycotoxins impacting human health (Bryden, 2007). Fungal contamination also has an economic impact as it may degrade food quality and thereby reduce food availability (Bryden, 2007).

The role of water in microbial contamination of foodstuffs

Microbial contamination frequently occurs due to reliance on unsafe water for growing, processing, cleaning, or preparing food (Steele and Odumeru, 2004). One useful classification for describing contamination events defines an event based on timing ("pre-harvest" or "post-harvest"). For example, crops grown using biosolids, animal manure, human manure, or untreated irrigation waters are at risk for contamination with pathogens pre-harvest (Steele and Odumeru, 2004). All classes of microorganisms are readily transmitted via manure and/or irrigation water to crops (Steele and Odumeru, 2004). *Salmonella* spp., *E. coli*, and *Clostridium* spp. for example, are bacteria that have readily been detected on crops (Steele and Odumeru, 2004). Similarly, helminths may be spread via crops as many spend a portion of their lifecycle in slugs and snails; contamination may occur due to the presence of gastropods or their residual mucus (Morgan et al., 2005).

One example outbreak due to pre-harvest contamination occurred in 2008 when over 1300 salmonellosis cases were reported in the United States. The outbreak was traced to the importation, from Mexico, of peppers irrigated with contaminated water (Barton Behravesh et al., 2011). Similarly, contaminated surface water used to irrigate spinach was implicated as a potential source of *E. coli* O157:H7 in a 2006 outbreak that infected over 250 people (Jay et al., 2007).

Pre-harvest contamination of meat products may occur due to infection of livestock via reliance on contaminated water. Viruses, helminths, and protozoa that infect livestock include hepatitis E virus, *Taenia* spp., *Trichenella* spp., *T. gondii*, and *Sarcocystis* spp. (Purcell and Emerson, 2010). One example is the association between use of private water supplies at pig farms and incidence of *T. gondii* infections in pigs bred in Italy (Villari et al., 2009).

When contaminated waters and/or crops are used as feed for livestock, meat and other animal products may also be contaminated pre-harvest. Pathogens commonly found in meat include *Campylobacter* spp., *Clostridium botulinum*, *Salmonella* spp., *Staphylococcus aureus*, *Listeria monocytogenes*, and pathogenic *E. coli*. Raw milk is also a reservoir for pathogens warranting pasteurization prior to consumption when possible, notably: *Brucella melitensis*, *Campylobacter* spp., *Coxiella burnetii*, hepatitis E, *Salmonella* spp., *S. aureus*, *Cryptosporidium* spp., and *T. gondii*. One example of the role of water in animal production contamination occurred in Finland in 1998, when *Campylobacter jejuni*-contaminated lake water was shown as the cause of infection in dairy cattle (Hänninen et al., 1998).

Another source of pre-harvest contamination occurs in molluscan shellfish (e.g., oysters, clams) exposed to contaminated environmental waters. Molluscan shellfish filter feed and inadvertently concentrate pathogens in environmental waters. Common pathogens in molluscan shellfish include bacteria (e.g., *V. vulnificus* and *V. parahaemolyticus*), virus (e.g., norovirus, hepatitis A), and helminths (e.g., *Paragonimus* spp., *A. cantonensis*, *Echinostoma* spp., and *Gnathostoma* spp.). The pathogens in molluscan shellfish primarily pose a threat because these foodstuffs are often served raw or undercooked. Helminths (e.g., *Echinostoma* spp., *Paragonimus* spp.) may also be transmitted via undercooked or raw fish.

Post-harvest contamination can occur during storage and/or transport, particularly in moist, warm environments. Fungal growth is particularly a concern for crops during storage and transport. Contaminated grain may also impact animal products, as mycotoxin in feed given to livestock may also pass into milk during dairy production. Bacteria are capable of impacting the food supply due to inadequate storage which allows for both growth (increasing risk of infection) and toxin production (increasing risk of toxinosis). Islam et al. (2012) demonstrated growth of *E. coli* in prepared foods left out at ambient temperature for hours prior to consumption. *Staphylococcus aureus*, which is naturally present on hands, and *Clostridium botulism*, which is naturally present in soils, secrete toxins during growth in food (Cliver et al., 2011).

But perhaps the most important role of water in foodborne contamination occurs during food preparation. When contaminated water is used to wash or prepare food, waterborne diseases may be transferred to the food directly (Motarjemi et al., 1993; Kirby et al., 2003). Indirectly, contaminants like *Clostridium botulism* may be introduced via contaminated water and then secrete toxins during growth due to poor preservation. Regions lacking safe, networked water supplies are particularly susceptible to food contamination via poor quality water (Marino, 2007). An example of this occurred in Sambalpur, India, in 2014 when over 1500 people were infected with hepatitis A. Although the cause of the outbreak was linked to a combination of poor drainage and damaged drinking water supply pipelines, the Food Safety Officer shut down roadside food stalls in an effort to slow transmission by improving food hygiene. Water treatment alone is not necessarily a complete barrier to water contamination. For example, oocysts (e.g., *Cryptosporodium* spp., *G. lamblia)* and helminth eggs are resistant to chlorine treatment and viruses may pass through common filtering devices (Galal-Gorchev, 1996).

In addition to contamination of water used to prepare food, inadequate quantity or quality of water for hand hygiene also contributes to foodborne illness. *Salmonella* spp., norovirus, hepatitis A, *Campylobacter jejuni*, *Giardia* spp., and helminths are examples of pathogens transmitted by infected food handlers who do not practice proper hygiene. Viral gastroenteritis transmission, in particular, is due largely to inadequate hygiene of food handlers (Acheson et al., 2002). Compounding the problem is that food handlers are often asymptomatic and are therefore not necessarily aware when consistent and effective hygiene is most essential (Idowu and Rowland, 2006). To improve hand hygiene compliance, sink accessibility with adequate soap and water is an important consideration (Whitby et al., 2007). Hand hygiene both reduces risk of food handlers transmitting own illnesses as well as risk of cross contamination between contaminated foodstuffs (e.g., raw meat and poultry) and uncooked foodstuffs (e.g., fruits, vegetables).

Control and remediation

Pre-harvest control and remediation techniques include interrupting pathways by which pathogens enter the food supply (see below "The role of water in contamination of the food supply"). Most notably, microbiologically safe water should be used for crop irrigation and livestock production. Post-harvest controls include screening food for contamination. Mycotoxin, in particular, is most effectively reduced when crops are properly screened for contamination (Binder, 2007). Once contamination of crops is identified, diets can be shifted away from highly contaminated crops, agricultural practices should be improved to reduce fungal infection (e.g., improved irrigation, use of fungicides, reducing plant density), or crop storage should be improved (e.g., reducing temperature and moisture, adding preservatives) (Wild and Gong, 2010; Chulze, 2010). However, screening for pathogens in the food supply is often too difficult. In regions with sufficient resources, a foodborne outbreak will trigger an epidemiological investigation to identify the food source responsible for the outbreak (Tauxe, 1997).

Proper food preparation and adequate hygiene (both hand and food) reduce foodborne contamination in the home. Milk should be pasteurized, and meat and shellfish should be thoroughly cooked. Other forms of meat preparation, including salting, drying, smoking, and microwaving, may be effective at reducing bacterial and viral contamination, but are insufficient to destroy helminths (Cross, 2000). Conversely, freezing meat for extended periods of time may destroy helminths but has little to no effect on bacteria and virus (Berry et al., 2008). Water-related interventions during food preparation include ensuring food is prepared with microbiologically safe water and/or cooking after water is added (Ehiri et al., 2001). For foods that are not intended to be cooked, like vegetables and fruits, thorough washing with clean water may further reduce surface contamination (Ehiri et al., 2001). Although rinsing or soaking in 1.5 percent bleach solutions has also been advocated in areas with high risk of disease, the high organic load may quench chlorine before it is sufficiently effective (Amoah et al., 2007). The association between water and food contamination was highlighted following a cholera outbreak in Latin American in the early 1990s. Nearly a million cases were reported with almost 9000 deaths. Case-control studies highlighted the importance of not only drinking water in spreading the disease, but also the consumption of unwashed vegetables (Guthmann, 1995).

Chemical contamination

Description

Heavy metals

The *Codex Alimentarius* Commission, established by the Food and Agricultural Organization (FAO) of the United Nations, established standards for six heavy metals (arsenic, cadmium, lead, mercury, methylmercury, and tin) to ensure food safety (*Codex Alimentarius* Commission, 1995). The adverse health effects caused by exposure to heavy metals vary by the specific element and whether the exposure is acute or chronic. For example, acute arsenic, lead, and tin exposure can cause nausea, abdominal pain, confusion, headaches, anemia, diarrhea, vomiting, seizures, comas, and/or death. Chronic lead exposure impacts cognitive and behavioral functions, and chronic arsenic or antimony exposure increases risks of heart disease, cancer, strokes, and death.

Radionuclides

Radionuclides are atoms with unstable nuclei that emit excess energy through radioactive decay. When radionuclides are ingested, radioactive decay causes cellular damage that may increase risk of cancer and/or organ damage if the radionuclides accumulate in essential organs (D'Mello, 2003). For example, exposure to ^{131}I can lead to thyroid cancer as iodine accumulates in the thyroid. There are over a hundred radionuclides, so to determine which, if any, present a health risk, the potential sources of contamination (e.g., industrial processing, radiological emergency, naturally high background levels) need to be considered. Generally, risks from food grown in areas with naturally high background levels of contamination are considered low.

Pesticides

Pesticides are chemicals intentionally applied in agricultural practice to control plants, animals, and/or microorganisms whose presence reduces productivity. Like the health effects caused by heavy metals, health effects from pesticides vary by the specific pesticide class and the exposure level. Here, we classify exposure as acute or chronic. Exposures can be due to one or more of the following: intentional ingestion; inadequate worker protection during production, application, and/or disposal of pesticides; ingestion of contaminated crops; ingestion of contaminated drinking water; and interaction with contaminated environments. Acute exposures, like those occurring due to inadequate worker protection, can lead to eye and skin irritation, headaches, dizziness, vomiting, delirium, seizures, and death (Thundiyil et al., 2008). Chronic exposures, like those occurring due to long-term ingestion of contaminated food, are linked to increased risk of cancers (e.g., non-Hodgkin lymphoma and leukemia), neurological conditions (e.g., Parkinson's disease), and impacts on fertility and fetal development (Ascherio et al., 2006; Sanborn et al., 2007; Bassil et al., 2007). Although there are over 24 pesticide classes, including fungicides, fumigants, herbicides, insecticides, insect repellants, and rodenticides (Thundiyil et al., 2008), many are restricted or banned due to adverse human or environmental health effects. Notably, some regions have limited capacity to enforce bans resulting in continued use of unsafe pesticides (Thuy et al., 2012).

The role of water in contamination of the food supply

Movement of water used in agriculture is a contributor to pesticide, heavy metal, and radionuclide contamination of the food supply. Irrigation water, or contaminated soils that leach into irrigation water, mobilize heavy metals which facilitate plant uptake even if water and soil concentrations are below permissible limits (Kumar Sharma et al., 2007). This is especially true in regions with naturally high levels of background contamination, where fertilizers are used, or where industrial processes have disturbed or contaminated the soils. Antimony, cadmium, lead, and radionuclides have all been detected in food from regions previously housing mining operations (Svoboda et al., 2006; Casado et al., 2007; Jibiri et al., 2007; Williams et al., 2009). In Northern France, for example, consumption of home grown vegetables is a risk factor for observed high blood lead levels due to high lead levels in soil (Pruvot et al., 2006). Radionuclide uptake in crops is dependent on a number of factors, including soil type, soil chemical characteristics, and specific plant uptake measures (Shanthi et al., 2010). Water soluble radionuclides (e.g., ^{238}U, ^{232}Th), arsenic, and other heavy metals may be transferred to the food supply via contaminated irrigation water. It should be noted that arsenic is especially pervasive in Bangladeshi groundwater; arsenic contamination of rice crops via irrigation is therefore a major public health concern in Bangladesh (Meharg and Rahman, 2003).

Water scarcity and food security are also inextricably linked. In regions with limited available clean water, poor quality water may be used for irrigation. In Marrakech City, Morocco, for example, wastewater is used to irrigate crops (Sedki et al., 2003). Once crops are contaminated, meat, diary, and fish products become contaminated by bioaccumulation following consumption of contaminated feed. In Marrakech, this manifests as high levels of cadmium in cattle fed alfalfa and corn leaves irrigated using the wastewater (Sedki et al., 2003). The linkages between water scarcity and food security are not limited only to foodborne contamination; other impacts (e.g., food scarcity) are outside the scope of this chapter (Fereres et al., 2011).

Control and remediation

Control and remediation of chemical contamination in the food supply typically requires shifting agricultural practices. The first step toward control is identifying the source of the contamination. Often, the regions impacted by chemical contamination have a history of industrial heavy metal and/or radionuclide sources. One example is the identification of 12 rural villages in Thailand where local farmers were exposed to cadmium because they were irrigating rice crops using water downstream of a gold mine (Swaddiwudhipong et al., 2012). To reduce disease burden, water-related interventions include shifting irrigation water to uncontaminated sources, though this may be cost-prohibitive.

When soil is identified as the source of contaminants, irrigation water may mobilize contaminants. Methods to remediate soil include metal extraction, immobilization, and phytoremediation, or the planting of contaminant-accumulating crops (Baker et al., 1994; Abumaizar and Smith, 1999). The high costs and technical expertise required render soil remediation impractical in low income regions. If soil remediation is not possible, chemical exposures due to the food supply may be controlled by reducing food crop production on contaminated lands (Sridhara Chary et al., 2008). Prioritizing non-food crops or food crops with low uptake of heavy metals or radionuclides may reduce exposure without losing land productivity.

Pesticide exposures are often driven by economics of increased crop yield. One method for the reduction of pesticide exposures is to employ resource-conserving farming methods that increase yield while reducing the reliance on pesticides (Pretty et al., 2006). Such methods are simultaneously effective at increasing water use efficiency; pesticide and water use reduction are not necessarily mutually exclusive. Alternative methods to reduce pesticide exposure include reducing superficial contamination by washing crops and prioritizing crops with limited adsorption of pesticides on edible components (Krol et al., 2000).

Regardless, shifting water sources and/or crops to reduce exposures may be difficult or impossible especially in developing countries with large populations, limited arable land, limited water resources, and a heavy reliance on subsistence farming. Geographic, geologic, and economic limitations may force continued use of contaminated lands and/or water supplies for production of food crops for local sustenance.

Conclusion

Globally, the causes of foodborne contamination are wide and varied. Efforts to reduce foodborne illness are driven largely by prevention: preventing contamination of the food supply pre-harvest and post-harvest. Chemical contamination, for example, occurs largely pre-harvest, and can be reduced by proper crop siting and limiting or controlling pesticide application. Biological contamination occurs predominately post-harvest, where production, distribution, and storage influence mycotoxin and bacterial toxins; consumer-level hygiene and food preparation influence microbial contamination. In all cases, clean water plays a role in helping to reduce contamination as irrigation water influences crop and livestock uptake of chemical contaminants and household access to water influences both hand hygiene and food preparation.

Key recommended readings

1 *Codex Alimentarius* Commission. 2009. Codex Stand 193-1995: Codex General Standard for Contaminants and Toxins in Food and Feed. http://www.fao.org/fileadmin/user_upload/livestockgov/documents/1_CXS_193e.pdf. The Codex General Standard for Contaminants and Toxins in Food and Feed is a comprehensive resource for identifying and understanding foodborne contamination. The document lists, describes, sets maximum levels for, and details sampling plans for contaminants and toxins found in internationally traded food products and feed products that are directly relevant to public health.

2 Food and Agriculture Organization of the United Nations. 2009. Microbiological Risk Assessment Series 17: Risk Characterization of Microbiological Hazards in Food. http://www.who.int/foodsafety/publications/micro/MRA17.pdf. A description of the microbiological risk assessment framework, with particular emphasis on risk characterization, for microbiological food contamination. The document discusses qualitative, semi-quantitative, and quantitative risk characterization; hazard characterization and exposure assessment; quality assurance; and economic analysis. Examples of assessment for foodborne pathogens (e.g., *Listeria*, *Vibrio vulnificus*, *E. coli*) are included.

3 Ercsey-Ravasz, M., Toroczkai, Z., Lakner, Z. & Baranyi, J. 2012. Complexity of the international agro-food trade network and its impact on food safety. *PLoS ONE*, 7 (5): e37810 DOI: 10.1371/journal.pone.0037810 The study explores the increasingly

connected global network of food distribution and highlights both the ease of contaminant transport along the network and the insufficiencies in tracking origins of contaminants. The study demonstrates that the seven countries that are most involved in international food trade each trade with over 77 percent of the world's countries, expanding on the interconnectedness of the developed and developing world with respect to food transport.

References

Abumaizar, R. J. & Smith, E. H. 1999. Heavy metal contaminants removal by soil washing. *Journal of Hazardous Materials*, 70, 71–86.

Acheson, D., Bresee, J. S., Widdowson, M.-A., Monroe, S. S. & Glass, R. I. 2002. Foodborne viral gastroenteritis: challenges and opportunities. *Clinical Infectious Diseases*, 35, 748–753.

Amoah, P., Drechsel, P., Abaidoo, R. & Klutse, A. 2007. Effectiveness of common and improved sanitary washing methods in selected cities of West Africa for the reduction of coliform bacteria and helminth eggs on vegetables. *Tropical Medicine & International Health*, 12, 40–50.

Ascherio, A., Chen, H., Weisskopf, M. G., O'Reilly, E., Mccullough, M. L., Calle, E. E., Schwarzschild, M. A. & Thun, M. J. 2006. Pesticide exposure and risk for Parkinson's disease. *Annals of Neurology*, 60, 197–203.

Baker, A., Mcgrath, S., Sidoli, C. & Reeves, R. 1994. The possibility of in situ heavy metal decontamination of polluted soils using crops of metal-accumulating plants. *Resources, Conservation and Recycling*, 11, 41–49.

Barton Behravesh, C., Mody, R. K., Jungk, J., Gaul, L., Redd, J. T., Chen, S., Cosgrove, S., Hedican, E., Sweat, D. & Chávez-Hauser, L. 2011. 2008 outbreak of *Salmonella* Saintpaul infections associated with raw produce. *New England Journal of Medicine*, 364, 918–927.

Bassil, K., Vakil, C., Sanborn, M., Cole, D., Kaur, J. & Kerr, K. 2007. Cancer health effects of pesticides: systematic review. *Canadian Family Physician*, 53, 1704–1711.

Berry, M., Fletcher, J., McClure, P. & Wilkinson, J. 2008. Effects of freezing on nutritional and microbiological properties of foods. In Evans J. (ed.) *Frozen Food Science and Technology*. Oxford: Blackwell Publishing.

Binder, E. M. 2007. Managing the risk of mycotoxins in modern feed production. *Animal Feed Science and Technology*, 133, 149–166.

Boxall, A., Hardy, A., Beulke, S., Boucard, T., Burgin, L., Falloon, P. D., Haygarth, P. M., Hutchinson, T., Kovats, R. S. & Leonardi, G. 2008. Impacts of climate change on indirect human exposure to pathogens and chemicals from agriculture. *Environmental Health Perspectives*, 117, 508–514.

Bryden, W. L. 2007. Mycotoxins in the food chain: human health implications. *Asia Pacific Journal of Clinical Nutrition*, 16, 95–101.

Casado, M., Anawar, H., Garcia-Sanchez, A. & Santa Regina, I. 2007. Antimony and arsenic uptake by plants in an abandoned mining area. *Communications in Soil Science and Plant Analysis*, 38, 1255–1275.

Chulze, S. 2010. Strategies to reduce mycotoxin levels in maize during storage: a review. *Food Additives and Contaminants*, 27, 651–657.

Cliver, D. O., Potter, M. & Riemann, H. P. 2011. *Foodborne Infections and Intoxications*. Burlington, MA: Academic Press.

Codex Alimentarius Commission 1995. Codex general standard for contaminants and toxins in food and feed. Available at: http://www.codexalimentarius.org/input/download/standards/17/CXS_193e2015.pdf [Accessed 19 May 2015].

Cross, J. H. 2000. Fish-and invertebrate-borne helminths. *Foodborne Disease Handbook: Volume 2: Viruses: Parasites: Pathogens, and HACCP*. New York: Dekker.

D'Mello, J. F. 2003. *Food Safety: Contaminants and Toxins*. Cambridge, MA: CABI.

Ehiri, J. E., Azubuike, M. C., Ubbaonu, C. N., Anyanwu, E. C., Ibe, K. M. & Ogbonna, M. O. 2001. Critical control points of complementary food preparation and handling in eastern Nigeria. *Bulletin of the World Health Organization*, 79, 423–433.

Fereres, E., Orgaz, F. & Gonzalez-Dugo, V. 2011. Reflections on food security under water scarcity. *Journal of Experimental Botany*, 62(12): 4079–4086.

Galal-Gorchev, H. 1996. Chlorine in water disinfection. *Pure and Applied Chemistry*, 68, 1731–1735.

Guthmann, J. 1995. Epidemic cholera in Latin America: spread and routes of transmission. *The Journal of Tropical Medicine and Hygiene*, 98, 419–427.

Hänninen, M. L., Niskanen, M. & Korhonen, L. 1998. Water as a reservoir for *Campylobacter jejuni* infection in cows studied by serotyping and pulsed field gel electrophoresis (PFGE). *Journal of Veterinary Medicine, Series B*, 45, 37–42.

Idowu, O. & Rowland, S. 2006. Oral fecal parasites and personal hygiene of food handlers in Abeokuta, Nigeria. *African Health Sciences*, 6, 160–164.

Islam, M., Ahmed, T., Faruque, A., Rahman, S., Das, S., Ahmed, D., Fattori, V., Clarke, R., Endtz, H. & Cravioto, A. 2012. Microbiological quality of complementary foods and its association with diarrhoeal morbidity and nutritional status of Bangladeshi children. *European Journal of Clinical Nutrition*, 66, 1242–1246.

Jay, M. T., Cooley, M., Carychao, D., Wiscomb, G. W., Sweitzer, R. A., Crawford-Miksza, L., Farrar, J. A., Lau, D. K., O'Connell, J. & Millington, A. 2007. *Escherichia coli* O157: H7 in feral swine near spinach fields and cattle, central California coast. *Emerging Infectious Diseases*, 13, 1908.

Jibiri, N., Farai, I. & Alausa, S. 2007. Estimation of annual effective dose due to natural radioactive elements in ingestion of foodstuffs in tin mining area of Jos-Plateau, Nigeria. *Journal of Environmental Radioactivity*, 94, 31–40.

Kirby, R. M., Bartram, J. & Carr, R. 2003. Water in food production and processing: quantity and quality concerns. *Food Control*, 14, 283–299.

Kotloff, K. L., Nataro, J. P., Blackwelder, W. C., Nasrin, D., Farag, T. H., Panchalingam, S., Wu, Y., Sow, S. O., Sur, D. & Breiman, R. F. 2013. Burden and aetiology of diarrhoeal disease in infants and young children in developing countries (the Global Enteric Multicenter Study, GEMS): a prospective, case-control study. *The Lancet*, 382(9888), 209–222..

Krol, W. J., Arsenault, T. L., Pylypiw, H. M. & Incorvia Mattina, M. J. 2000. Reduction of pesticide residues on produce by rinsing. *Journal of Agricultural and Food Chemistry*, 48, 4666–4670.

Kumar Sharma, R., Agrawal, M. & Marshall, F. 2007. Heavy metal contamination of soil and vegetables in suburban areas of Varanasi, India. *Ecotoxicology and Environmental Safety*, 66, 258–266.

Marino, D. D. 2007. Water and food safety in the developing world: global implications for health and nutrition of infants and young children. *Journal of the American Dietetic Association*, 107, 1930–1934.

Meharg, A. A. & Rahman, M. M. 2003. Arsenic contamination of Bangladesh paddy field soils: implications for rice contribution to arsenic consumption. *Environmental Science & Technology*, 37, 229–234.

Morgan, E. R., Shaw, S. E., Brennan, S. F., De Waal, T. D., Jones, B. R. & Mulcahy, G. 2005. *Angiostrongylus vasorum*: a real heartbreaker. *Trends in Parasitology*, 21, 49–51.

Motarjemi, Y. & Käferstein, F. 1999. Food safety, hazard analysis and critical control point and the increase in foodborne diseases: a paradox? *Food Control*, 10, 325–333.

Motarjemi, Y., Käferstein, F., Moy, G. & Quevedo, F. 1993. Contaminated weaning food: a major risk factor for diarrhoea and associated malnutrition. *Bulletin of the World Health Organization*, 71, 79.

Pretty, J. N., Noble, A. D., Bossio, D., Dixon, J., Hine, R. E., Penning De Vries, F. W. & Morison, J. I. 2006. Resource-conserving agriculture increases yields in developing countries. *Environmental Science & Technology*, 40, 1114–1119.

Pruvot, C., Douay, F., Hervé, F. & Waterlot, C. 2006. Heavy metals in soil, crops and grass as a source of human exposure in the former mining areas. *Journal of Soils and Sediments*, 6, 215–220.

Purcell, R. H. & Emerson, S. U. 2010. Hidden danger: the raw facts about hepatitis E virus. *Journal of Infectious Diseases*, 202, 819–821.

Raschid-Sally, L. & Jayakody, P. 2009. *Drivers and Characteristics of Wastewater Agriculture in Developing Countries: Results from a Global Assessment*. Colombo, Sri Lanka: IWMI.

Sanborn, M., Kerr, K., Sanin, L., Cole, D., Bassil, K. & Vakil, C. 2007. Non-cancer health effects of pesticides: systematic review and implications for family doctors. *Canadian Family Physician*, 53, 1712–1720.

Scallan, E., Griffin, P. M., Angulo, F. J., Tauxe, R. V. & Hoekstra, R. M. 2011a. Foodborne illness acquired in the United States—unspecified agents. *Emerging Infectious Diseases*, 17, 16.

Scallan, E., Hoekstra, R. M., Angulo, F. J., Tauxe, R. V., Widdowson, M.-A., Roy, S. L., Jones, J. L. & Griffin, P. M. 2011b. Foodborne illness acquired in the United States—major pathogens. *Emerging Infectious Diseases*, 17(11): 7–15.

Sedki, A., Lekouch, N., Gamon, S. & Pineau, A. 2003. Toxic and essential trace metals in muscle, liver and kidney of bovines from a polluted area of Morocco. *Science of the Total Environment*, 317, 201–205.

Shanthi, G., Thampi Thanka Kumaran, J., Gnana Raj, G. A. & Maniyan, C. 2010. Natural radionuclides in the South Indian foods and their annual dose. *Nuclear Instruments and Methods in Physics Research Section A: Accelerators, Spectrometers, Detectors and Associated Equipment*, 619, 436–440.

Sridhara Chary, N., Kamala, C. & Samuel Suman Raj, D. 2008. Assessing risk of heavy metals from consuming food grown on sewage irrigated soils and food chain transfer. *Ecotoxicology and Environmental Safety*, 69, 513–524.

Steele, M. & Odumeru, J. 2004. Irrigation water as source of foodborne pathogens on fruit and vegetables. *Journal of Food Protection*, 67, 2839–2849.

Stürchler, D. A. 2006. *Exposure: A Guide to Sources of Infections*. Hendon, VA: ASM Press.

Svoboda, L., Havlíčková, B. & Kalač, P. 2006. Contents of cadmium, mercury and lead in edible mushrooms growing in a historical silver-mining area. *Food Chemistry*, 96, 580–585.

Swaddiwudhipong, W., Limpatanachote, P., Mahasakpan, P., Krintratun, S., Punta, B. & Funkhiew, T. 2012. Progress in cadmium-related health effects in persons with high environmental exposure in northwestern Thailand: a five-year follow-up. *Environmental Research*, 112, 194–198.

Tauxe, R. V. 1997. Emerging foodborne diseases: an evolving public health challenge. *Emerging Infectious Diseases*, 3, 425.

Thundiyil, J. G., Stober, J., Besbelli, N. & Pronczuk, J. 2008. Acute pesticide poisoning: a proposed classification tool. *Bulletin of the World Health Organization*, 86, 205–209.

Thuy, P. T., Van Geluwe, S., Nguyen, V.-A. & Van Der Bruggen, B. 2012. Current pesticide practices and environmental issues in Vietnam: management challenges for sustainable use of pesticides for tropical crops in (South-East) Asia to avoid environmental pollution. *Journal of Material Cycles and Waste Management*, 14, 379–387.

Villari, S., Vesco, G., Petersen, E., Crispo, A. & Buffolano, W. 2009. Risk factors for toxoplasmosis in pigs bred in Sicily, Southern Italy. *Veterinary Parasitology*, 161, 1–8.

Whitby, M., Pessoa-Silva, C., Mclaws, M.-L., Allegranzi, B., Sax, H., Larson, E., Seto, W., Donaldson, L. & Pittet, D. 2007. Behavioural considerations for hand hygiene practices: the basic building blocks. *Journal of Hospital Infection*, 65, 1–8.

Wild, C. P. & Gong, Y. Y. 2010. Mycotoxins and human disease: a largely ignored global health issue. *Carcinogenesis*, 31, 71–82.

Williams, P. N., Lei, M., Sun, G., Huang, Q., Lu, Y., Deacon, C., Meharg, A. A. & Zhu, Y.-G. 2009. Occurrence and partitioning of cadmium, arsenic and lead in mine impacted paddy rice: Hunan, China. *Environmental Science & Technology*, 43, 637–642.

16

WATERBORNE ZOONOSES[*]

Victor Gannon

Ph.D., D.V.M., Laboratory for Foodborne Zoonoses,
Public Health Agency of Canada, Alberta, Canada

Chad R. Laing

Ph.D., Laboratory for Foodborne Zoonoses,
Public Health Agency of Canada, Alberta, Canada

Learning objectives

1 Understand the contribution of specific pathogens of animal origin to waterborne disease in humans.
2 Describe the routes of pathogen transmission and factors that influence the risk of waterborne disease.
3 Describe the role of watershed management and water treatment in the reduction of waterborne diseases associated with these pathogens.

Sources of exposure to waterborne zoonotic pathogens

Transmission of zoonotic pathogens to humans

While the term "zoonosis" may be used in a broad sense to refer to any infectious disease transmissible from one animal species to another, its most common usage is in reference to diseases which are transmitted from animals to humans. It is in the latter context that the term will be used through this chapter. Zoonoses represent a significant cause of persistent (endemic) infectious diseases and disease outbreaks in communities worldwide. Further, animals are thought to be the source of the majority (approximately 60 percent) of new, also known as "emerging", disease agents responsible for human illness, for example the Ebola virus and new strains of influenza viruses (Jones et al. 2008).

[*] Recommended citation: Gannon, V. and Laing, C.R. 2015. 'Waterborne zoonoses', in Bartram, J., with Baum, R., Coclanis, P.A., Gute, D.M., Kay, D., McFadyen, S., Pond, K., Robertson, W. and Rouse, M.J. (eds) *Routledge Handbook of Water and Health*. London and New York: Routledge.

There are typically multiple possible routes of exposure to zoonotic agents, with waterborne transmission being one of the most important. Therefore, a good understanding of this route of transmission is essential to the prevention and control of outbreaks and sporadic disease in humans. The drivers for disease transmission depend on interactions between the animal reservoir, the environment, human host and the pathogen. The characteristics (both quantitative and qualitative) of these interactions increase or decrease the risk of disease transmission events. Examples of these interactions are provided in Table 16.1.

At each step in zoonotic pathogen transmission there is a probability of survival of the agent that affects the ultimate levels of human exposure and probability of disease. In order for zoonotic pathogens to present a disease risk for humans, they must first establish themselves in animal populations. The reason for this is that human populations typically only sustain these pathogenic agents for a limited period of time. In terms of the source–sink evolutionary model, animal populations represent the "source" where the pathogen population is established, while humans are often "accidental" hosts, which provide a sink habitat in which pathogen populations are only transiently established but will eventually die out because of the marginal and inhospitable nature of the "sink" host.

The sustainability of pathogen populations in the animal reservoir species is dependent not only on the ability of the pathogen to find sufficient nutrients in this niche, but also on their ability to survive in harsh environments such as the gastrointestinal tract where they must

Table 16.1 Factors affecting the transmission dynamics of waterborne zoonotic pathogens

Compartment	Factors
Animal reservoir	Reservoir host species Age classes colonized Frequency of colonization Levels of acquired immunity and innate resistance to colonization Level of excretion of the pathogen Behavior/probability of contact with each other
Pathogen	Population numbers Ability to colonize and become established in animal reservoir and human host Virulence for humans Environmental persistence Resistance to physical and chemical stressors Competition with other microflora and predation
Environment	Reservoir species population density Exposure levels of the pathogen from food, water and other environmental sources Animal bio-security systems employed Degree of isolation from other herds or groups of the same and of different species including wildlife Animal wastes storage and disposal Chemical and biochemical agents such as acids, oxidizing agents, a variety of toxic organic chemicals, heavy metals, fungal and bacterial antimicrobial agents, and bacterial hormones, phage and protozoan predators Physical agents such as heat, cold, desiccation and ultraviolet (UV) light Topography, precipitation, climate
Human host	Host immune cells, immunoglobins, bactericidal peptides and intestinal microflora Levels of acquired immunity or innate resistance to colonization and disease

prevail against a large number of physical and chemical stressors (e.g. low pH and oxidizing agents). The pathogen must also be able to be transmitted from one animal to another.

The success of transmission is influenced by a large number of factors including the number of animals within a group which are colonized, the level of excretion of the pathogen by infected animals, the animal density and spatial proximity to one another, the likelihood of exposure to the excreta from herd or flock mates, and behaviors that promote the transmission of the agents. While these pathogens can be responsible for morbidity and mortality in the animal reservoir, frequently colonization of animals is asymptomatic with infections being cleared by the innate or acquired immune systems as the animal matures. It is for this reason that immature animals are frequently the driving force in the dissemination of these pathogens in the environment. Finally, the ability of the agent to persist in the external environment until it is acquired by a susceptible host plays a fundamental role in the transmission dynamics in the animal reservoir. It is for this reason that animal biosecurity measures and the proper removal, storage and treatment of animal excreta play a fundamental role in reducing levels of these organisms in the environment.

Many of the same factors that drive the transmission dynamics of these agents within the animal reservoir also drive the risks of human exposure to these pathogens. In the context of waterborne disease, not only the ability to survive in the environment but also factors affecting transport of the pathogen from animal excreta to surface or ground water play a significant role in determining human waterborne disease risks. It is for this reason that water source protection is fundamental in providing safe drinking water and that processes in nature that reduce pathogens in the environment, such as filtration, sedimentation, and chemical and UV treatment are exploited in modern water treatment systems.

Finally, the susceptibility of humans to waterborne pathogens is highly dependent on the level of exposure. The exposure dose is influenced by water treatment procedures and infrastructure, and by human behaviors such as choice of drinking water sources and recreational use of water. Following exposure, disease outcomes are significantly influenced by immunological and non-immunological defense mechanisms, with young children, the elderly and otherwise immune-compromised individuals being the most susceptible to disease. These high risk groups require the safest water and often act as sentinels alerting us to the presence of pathogenic agents in water.

While there are many zoonotic pathogens that have been implicated in human waterborne disease, this chapter will focus on three bacterial and two protozoan pathogens that have a global distribution and represent challenges for agencies tasked with ensuring water safety. Specific characteristics of enterohaemorrhagic *Escherichia coli*, *Salmonella enterica*, *Campylobacter*, *Cryptosporidium* and *Giardia* are given in Table 16.2.

Campylobacter

Campylobacter spp. are the most common cause of bacterial gastrointestinal illness throughout the world (Humphrey et al. 2007). While there are 21 different species of the organism, only 12 of these have been associated with human illness and the vast majority are caused by *C. jejuni* and *C. coli* (ca. 90 percent). Sporadic infections with these pathogens are the most common; however, outbreaks have been reported and there seems to be little immunity to subsequent infection with the organism following recovery. The organism is also associated with the Guillain-Barré syndrome, a debilitating disease characterized by an acute ascending neuropathy and the related Miller Fisher syndrome (Wassenaar and Blaser 1999). These conditions are thought to be a result of the host generating antibodies to specific carbohydrate

Table 16.2 Factors affecting the transmission of specific waterborne zoonotic pathogens

	Escherichia coli O157:H7 and other enterohaemorrhagic E. coli (EHEC)	*Non-typhoid Salmonella enterica*	*Campylobacter*	*Cryptosporidium*	*Giardia*
The animal reservoir	Carried largely asymptomatically in cattle and other ruminants. Most common in newborn and weaned calves. In temperate regions excretion is seasonal with individuals in groups excreting high levels (>106 CFU/g) in their faeces.	Many wild and domestic animals and species are colonized. Some serovars are associated with disease in the animal host. There is very high prevalence in young animals, specifically poultry.	Many animal species; very high prevalence in young poultry and ruminants.	Calves are the primary source of *C. parvum*. However, zoonotic infection occasionally occurs with *Cryptosporidium* spp. from other animal species. Individuals within groups excrete very high levels (>109 oocysts/g) in their faeces.	Many domestic animal and wildlife species carry zoonotic sub-assemblages.
The environment	Survives in un-composted animal wastes and soils for months. Short-lived in water. Survives in sediments. Resistant to low pH. Highly sensitive to chlorine and UV light.	Survives in un-composted animal wastes and soils for months. Resistant to low pH. Sensitive to chlorine and UV light.	Sensitive to desiccation and oxygen. Short-lived in water. Survives in sediments and beach sand for long periods. Sensitive to chlorine and UV light.	Survives in un-composted animal wastes and soils for months. Resistant to low pH. Resistant to chlorine but sensitive to UV light.	Survives in un-composted animal wastes and soils for months. Resistant to low pH. Resistant to chlorine but sensitive to UV light.
Pathogenicity	Infectious dose is low (fewer than 50 CFU). Certain serotypes and genetic lineages are associated with more severe disease.	Infectious dose is low for typhoid *Salmonella* (10⁵ CFU). Non-typhoid *Salmonella* have a high infectious dose. Certain serovars are more commonly associated with disease and more severe disease.	Infectious dose is low (fewer than 500 CFU). *C. jejuni*, *C. coli* and certain genotypes are more commonly associated with enteritis and the Guillian-Barré syndrome.	Infectious dose is low (fewer than 50 CFU). Certain species such as *C. parvum* are associated with zoonotic disease.	Infectious dose is low (fewer than 50 CFU). Certain assemblages and sub-assemblages are associated with zoonotic disease.

Nature of human disease	Large outbreaks and sporadic disease.	Large outbreaks and sporadic disease.	Few outbreaks, mostly sporadic disease.	Large outbreaks and sporadic disease.	Outbreaks and sporadic disease.
Foodborne / waterborne	Mostly foodborne but waterborne outbreaks occur. Regional hotspots but global in distribution.	Mostly foodborne but waterborne outbreaks and sporadic cases occur in developing world.	Mostly foodborne but waterborne outbreaks occur. Regional hotspots but global in distribution.	Waterborne outbreaks and sporadic cases occur.	Waterborne outbreaks and sporadic cases occur.
Disease	Children and elderly, higher rates in areas with high cattle density. Hemorrhagic colitis and sometimes fatal haemolytic uremic syndrome.	Highest rates in children in developing world. Acute diarrhea. Septicemia.	Highest rates in children in rural areas and developing world. Acute diarrhea. Guillain–Barré syndrome.	Highest rates in children in rural areas. Acute diarrhea.	Highest rates in children in rural areas. Acute and chronic diarrhea.
Treatment	Largely supportive, most antimicrobials contraindicated.	Largely supportive, antimicrobial use controversial.	Largely supportive, antimicrobials.	Largely supportive.	Anti-protozoans.
Control	Animal waste storage and treatment. Animal vaccination. Standard water treatment.	Animal waste storage and treatment. Bio-security. Depopulation of positive animals. Animal vaccination. Standard water treatment.	Animal waste storage and treatment. Standard water treatment.	Animal waste storage and treatment. Standard water treatment, high levels of chlorine and/or UV light.	Animal waste storage and treatment. Standard water treatment, high levels of chlorine and/or UV light.

antigens on the surface of certain genotypes of the pathogen which cross-react with antigens present on host nervous tissues.

The vast majority of *Campylobacter* infections in the developed world can be attributed to the consumption of undercooked poultry (Sheppard et al. 2009), with the disease burden being greatest in children greater than four years old in densely populated regions. In contrast, it has been shown that children less than four years of age are at greatest risk in less densely populated rural areas where animal agriculture is prominent (Strachan et al. 2009, Spencer et al. 2011). As might be expected, the genotypes of *Campylobacter* that predominate in children from cities are closely related to the genotypes isolated from poultry. Interestingly, the genotypes isolated from children in rural areas are more closely related to genotypes isolated from ruminants. Infection rates on a population adjusted basis are significantly higher in children than adults and higher in rural regions than cities. Average rates of infection in Europe are 45 cases per 100,000 (Silva et al. 2011) but can reach as high as 353 cases per 100,000 in New Zealand (Spencer et al. 2011). In countries in the developing world, infection rates in children less than four years of age are as high as 60,000 cases per 100,000 (Coker et al. 2002).

Campylobacters are highly sensitive to desiccation and atmospheric oxygen and die off quickly in the terrestrial environment. However, *C. jejuni* can be readily isolated from flowing surface waters and beach sand. A large number of genotypes abound in *Campylobacters* isolated from water, suggesting that there are multiple animal species that contribute to the abundance of this pathogen in water. Interestingly, the number of *Campylobacter* infections that have been attributed to drinking water is rather modest. Given the common occurrence of *Campylobacters* in surface water, the low rates of infection are likely related to their sensitivity to drinking water treatment procedures such as chlorination and UV light exposure. Craun et al. (2010) reported *Campylobacter* spp. were associated with 19 drinking water disease outbreaks as the only pathogen and in another 6 outbreaks where more than one agent was responsible between 1971 and 2006 in the USA. Waterborne outbreaks have also been reported in drinking water in Canada (Garg et al. 2006) and in drinking and recreational water in Sweden, Norway and Finland where a single genotype has predominated in each outbreak (Jakopanec et al. 2007, Kärenlampi et al. 2007).

Salmonella

Salmonella enterica is divided into six subspecies: *arizonae, diarizonae, enterica, houtenae, indica and salamae*. However, the species has over 2,500 serovars, with each serovar belonging to one of these distinct subspecies. This large amount of antigenic variety corresponds to some extent with an equally large variety of potential geographical and host species sources of the organism. *Salmonella enterica* can be isolated from most species of mammals, reptiles and birds. In some host species it is essentially a commensal, while in others it may result in high levels of morbidity and mortality. Most human infections are associated with *Salmonella enterica* subspecies *enterica* which includes more than 1,500 different serovars of the organism (Litrup et al. 2010). Certain serovars are found in a large number of host species while others appear to be highly host-adapted.

In the developed world, widespread treatment of municipal sewage and public water supplies has significantly reduced most human sources of waterborne enteric pathogens. Prior to this, human-derived pathogens were responsible for widespread and severe gastrointestinal and systemic illness, for example typhoid fevers and related paratypoid fevers are associated with *Salmonella* Typhi and *S.* Paratyphi (types A, B or C), respectively (Kothari et al. 2008). Today infections with these pathogens are rare in developed countries, but they are still common in many developing countries that lack the necessary water treatment infrastructure. In contrast, the *Salmonella* serovars most frequently associated with human disease in developed countries are now of animal origin (Majowicz et al. 2010). Further, most of the *Salmonella* of animal origin are associated with foodborne rather than waterborne disease. Human-derived *Salmonella* Typhi and *S.* Paratyphi also typically cause less severe gastrointestinal illness than animal-derived *Salmonella* serovars. Instead they enter immune cells lining the intestine, are carried into the bloodstream and are associated with severe systemic illness characterized by a persistently high fever. Infection with these human-derived serovars can also lead to the development of a long-term carrier state in humans. Animal-derived *Salmonella*, in contrast, are associated with mild to severe gastrointestinal illness and only occasionally are associated with systemic illness in immuno-compromised individuals (Raffatellu and Bäumler 2010).

The large number of serovars of *Salmonella* capable of colonizing a variety of animal species has resulted in a large diversity of *Salmonella* serovars associated with human and animal disease. Specific serovars seem to increase in numbers over a specific period of time in a given region and then decline in numbers again. However, other serovars such as *S.* Enteritidis and *S.* Typhimurium appear to be more persistent and widespread causes of gastrointestinal illness, and have become the target of national and international control programs. However, it is difficult to say if the high frequency of isolation of these serovars is related to intrinsic properties they possess such as increased virulence or if it is simply a matter of levels of human exposure. There is a tremendous diversity of *Salmonella* serovars isolated from water and some, but certainly not all of these, are commonly associated with human disease. According to Craun et al. (2010), *Salmonella* were the cause of only 20 drinking water-associated outbreaks in the USA between 1971 and 2006. It is likely that the risk of human disease from water sources is mitigated by chlorination of drinking water and the relatively low numbers typically found in untreated water may not constitute an infectious dose for most humans.

Escherichia coli O157:H7 and other enterohaemorrhagic E. coli

The term enterohaemorrhagic *E. coli* (EHEC) is used to describe a pathogroup of this species that produce one or more phage-encoded protein toxins (highly related to Shiga toxin of *Shigella dysenteria* type 1), and are associated with haemorrhagic colitis (HC) in humans. While EHEC have been isolated from a number of animal sources, ruminants and specifically cattle are considered the main animal reservoir of these pathogens. Young age classes of cattle have been shown to shed these organisms in their faeces in large numbers, and EHEC have been isolated from their carcasses at slaughter and from raw beef products such as ground beef. There is also a strong epidemiological link between both sporadic cases and outbreaks of human infections and the consumption of undercooked beef and other foods, including raw vegetables, unpasteurized milk and fruit juices, and water contaminated with ruminant faeces (Gyles 2007, Karmali et al. 2010). EHEC are associated with both systemic and enteric illnesses in humans which range in severity from mild diarrhea to the haemolytic uremic syndrome (HUS). During colonization of the intestine, EHEC release one or more antigenic variant of Shiga toxin (Stx). Once taken up into the bloodstream of the human host, this bipartite protein toxin attaches to and kills the endothelial cells in the small blood vessels of a number of organ systems. The endothelial cell damage promotes coagulation of blood and micro-thrombi formation in target organs such as the kidney. EHEC-associated

HUS is characterized by renal failure and hemolytic anemia. EHEC infections are most common in children and sometimes result in death.

While there are many serotypes of *E. coli* that produce Stx, only a few O serotypes, including O26, O45, O111, O103, O121, O145 and O157, are consistently associated with human disease worldwide. Among these, *E. coli* O157:H7 is the most common EHEC serotype associated with outbreaks and with severe infections. In addition to there being differences in virulence among different *E. coli* serotypes or "seropathotypes", certain lineages and genotypes of *E. coli* O157 are more commonly associated with human disease than others (Zhang et al. 2007, Manning et al. 2008, Karmali et al. 2010). While the reason for this difference in the frequency of association between specific EHEC and human disease is not known with certainty, it has been shown that the production of high levels of Stx2 may play a role (Zhang et al. 2010). However, it has also been postulated that specific genotypes of EHEC are excreted in the faeces of cattle in higher numbers than others and as a result represent a proportionally greater risk of human EHEC exposure and disease from these genotypes (Chase-Topping et al. 2008). As noted above, while EHEC have been isolated from a number of domestic and wild animal species, ruminants are considered the primary reservoir of this pathogen (Gyles 2007). However, this rigid view of the origin EHEC has recently been challenged by an outbreak of HC and HUS in Germany, associated with a presumably human-derived entero-aggregative *E. coli* O104 strain (Mellmann et al. 2011).

Most EHEC infections are foodborne and have been associated with ground beef and dairy products. However, waterborne outbreaks also occur (Smith et al. 2006, Craun et al. 2010) and are typically associated with untreated water from private wells. In the USA, 10 waterborne outbreaks associated with EHEC *E. coli* O157:H7 and O145:NM were reported between 1971 and 2006 (Craun et al. 2010). The largest outbreaks, reported in Swaziland in Africa with an estimated 40,000 cases and in Walkerton, Ontario, with 2,300 cases and seven deaths in Canada, were associated with heavy rainfall, contamination of run-off water with cattle wastes, and failure or lack of drinking water treatment (Effler et al. 2001, Garg et al. 2006). However, cases have also been associated with recreational use of water and consumption of raw fruits and vegetables. Waterborne outbreaks have occurred where standards for drinking water treatment procedures such as chlorination were not enforced or where there were failures in drinking water treatment associated with human error or negligence.

Cryptosporidium

Cryptosporidium are intracellular protozoan parasites belonging to the phylum Apicomplexa. There are 22 known species of the organism and most are thought to be reasonably host species-specific. *C. parvum* (formerly *C. parvum* genotype II) and *C. hominis* (formerly *C. parvum* genotype I) are thought to cause the majority of human infections. *C. hominis* is derived from humans, whereas certain lineages of *C. parvum* are zoonotic and infect both cattle and humans. In addition to these species, *C. meleagridis*, *C. felis*, *C. canis*, *C. muris*, *C. suis*, *C. andersoni*, *C. cuniculus* as well as several other species have also been associated with sporadic human infections. Among *Cryptosporidium* isolated from cattle are *C. parvum* which are isolated from calves less than two months of age, *C. ryanae* and *C. bovis* which are usually isolated from older calves, and *C. andersoni* which is isolated from adult cattle. However, *C. parvum* is associated with gastrointestinal illness in calves while the other species are not. More than 11 subtypes of *C. parvum* have been identified based on the sequence of the gene encoding the 60 kDa glycoprotein, and zoonotic subtypes have been assigned to families IIa and IId. Within the zoonotic family IIa, subtype A15G2R1 appears to be the most widespread in cattle populations worldwide (Feng et al. 2013).

It is thought that the underdeveloped immune systems of infants and adults with conditions which weaken the immune system such as HIV, and those undergoing cancer therapy and organ transplantation are particularly susceptible. In humans, there is a peak of infections in children 6–12 months of age. There are also more infections in the developing world than in the developed world. In much of the developing world cases are associated with human-derrived *C. hominis*, while in the developed world both *C. parvum* and *C. hominis* infections are observed. In tropical climates most infections occur during the rainy season suggesting that contamination of drinking water may be responsible. In temperate climates increases in temperature and rainfall are also associated with an increase in cases of cryptosporidiosis. Baldursson and Karanis (2011) reported that *Cryptosporidium* spp. were associated with 60 percent of 199 recorded waterborne disease outbreaks worldwide between 2004 and 2010. One of the largest waterborne disease outbreaks ever reported occurred in Milwaukee, Wisconsin, was caused by *Cryptosporidium*, and resulted in an estimated 400,000 people infected.

Oocysts of the organism are widely distributed in both the terrestrial and aquatic environments. It has a low infectious dose (as few as 10 oocycts) and is highly resistant to chlorine and other bactericidal agents used for disinfection and water treatment. The resistance of the oocycts to these agents explains how outbreaks of gastrointestinal illness associated with treated drinking water and swimming pools and waterparks occur. In the developed world, large outbreaks of waterborne cryptosporidiosis have been associated with failures of water treatment systems or high loads of oocysts that result from flooding or overt fecal contamination of water sources that overwhelm water treatment systems. Fortunately, cryptosporidia are sensitive to UV light disinfection procedures.

Giardia

Giardia duodenalis (also known as *Giardia lamblia* and *Giardia intestinalis*) is the cause of both acute and chronic diarrheal illness and dehydration worldwide. In the developed world the prevalence is thought to be between 0.4 and 7.5 percent, whereas in the developing world the prevalence is estimated to be between 8 and 30 percent (Feng and Xiao 2011). Baldursson and Karanis (2011) reported that *Giardia lamblia* (*duodenalis*) were associated with 35.2 percent of 199 recorded waterborne disease outbreaks worldwide between 2004 and 2010. While they are occasionally associated with foodborne disease they are also a leading cause of waterborne outbreaks and sporadic disease. Craun et al. (2010) reported that *G. intestinalis* was the sole pathogen in 123 (86.0 percent) of the 143 drinking water outbreaks associated with protozoa in the USA from 1971 to 2006. These outbreaks caused 28,127 cases of giardiasis with an average of 228 cases per outbreak. A large outbreak also occurred in Bergen, Norway, in 2004 resulting in 1,300 laboratory-confirmed cases of giardiasis. The cause of the outbreak is suspected to have been leakage of sewage pipes into a drinking water source. Interestingly, follow-up studies conducted three years after the outbreak found that a significantly higher proportion of those that had acute *Giardia* infections than controls suffered from irritable bowel syndrome (46.1 percent versus 14 percent) and chronic fatigue syndrome (46.1 percent versus 12 percent) (Wensaas et al. 2012).

Giardia cysts are moderately resistant to levels of chlorine normally used for drinking water treatment (0.2 to 1 mg/L) but are inactivated with higher concentrations of chlorine (4 mg/L), chlorine dioxide and UV disinfection. *Giardia* cysts are triggered to undergo excystation by the low pH of the stomach and enter the replicative trophozoite form in the duodenum. Bile in turn triggers the formation of cysts which are shed in the faeces. Like *Cryptosporidium* the infectious dose is small (10–100 cysts).

Giardia duodenalis is the only species that infects humans. However, members of this species are quite genetically diverse and, as a consequence, it is thought that *Giardia duodenalsis* really represents a collection of related species. While species names have been applied to these different subgroups, this nomenclature is not universally accepted and instead eight different assemblies (A to H) of the organism are recognized. However, only members of assemblages A (also known as *G. duodenalis*) and B (*G. enterica*) are thought to be zoonotic. Within each assemblage, species-specific subtypes termed sub-assemblies or multi-locus genotypes are recognized. Within assemblage A, sub-assemblage AI and AII and within assemblage B, sub-assemblage BIII and BIV are associated with human disease. Case control studies suggest that rates of giardiasis are higher in rural areas; however, pinpointing specific animal sources of the organism has been difficult. For example, zoonotic assemblage A can occasionally be isolated from several different domestic livestock species such as cattle and pigs and, interestingly, both of these host species more typically carry *Giardia* from non-zoonotic assemblage E. Assemblage A genotypes have also been isolated from dogs and wildlife species such as muskrats and beavers but like the livestock species mentioned above, they carry other non-zoonotic *Giardia* assemblages as well.

Watershed management in the prevention and control of zoonoses

Proper watershed management can play a fundamental role in decreasing the probability of source water contamination by zoonotic pathogens. This can be achieved through a number of means such as limiting the access of both domestic and wild animals to watersheds, through measures such as fencing waterways and providing alternative sources of drinking water to animals. While certain organisms such as *Cryptosporidium* oocysts are extremely resistant to environmental conditions, other infectious agents are sensitive to conditions encountered in the environment, and the rate of population decline is a critical factor in determining pathogen loads that enter water sources. Survival of these organisms is affected by a number of factors including cold, heat, desiccation, UV light, predation, competition from micro flora and lack of nutrients. Composting of animal wastes prior to application on land has also been shown to significantly reduce pathogen loads. The deposition of animal manure on cropland must also be limited to times of year with the least rain and/or snow melt run-off, in order to minimize overland transport of wastes to waterways. Other measures that are used to decrease pathogen transport include barriers to movement such as green strips. Natural and artificial wetlands are also effective in reducing nutrients, toxins and pathogens in run-off water.

Conclusions

With effective watershed management and the use of modern water treatment technologies in the developed world, risks of acquiring waterborne disease have diminished for most populations. The highest risk of infections with waterborne zoonotic pathogens occurs in children in rural regions with high animal densities and in developing countries where waste management and water treatment is inadequate. Immunologically compromised individuals such as those with HIV infections, those receiving immunosuppressive therapy and the elderly are also at risk of infections with these agents, and occupational and recreational exposure is also a concern. While a number of new technologies has improved the efficiency of water treatment procedures, there is always a risk that these systems will fail or simply be overwhelmed by pathogen loads.

Therefore, multi-barrier approaches and water safety plans have been implemented to ensure that water treatment systems are sufficiently robust to reduce risks associated with pathogens in the face of even the most adverse circumstances. However, further refinements are required as is the development of more cost-effective water treatment systems that will encourage more widespread use in smaller communities and in the developing world. In addition, improvements in technologies and policies related to the storage and disposal of animal excreta are required to prevent the transmission of zoonotic waterborne pathogens.

Key recommended reading

1 Dufour, A., Bartram, J., Bos, R., Gannon, V. (eds) 2012. *Animal Waste, Water Quality and Human Health*. IWA Publishing for World Health Organization. ISBN: 978-1-78040-123-2. This is a very comprehensive book on waterborne zoonotic pathogens which discusses a range of topics.
2 Atlas, R.M. and Maloy, S. (eds) 2014. *One Health: Animals, People and the Environment*. ASM Press, Washington, DC. ISBN 978-1-55581-842-5 (print); ISBN 978-1-55581-843-2 (electronic). This book is a compilation of chapters which promote a "One Health" collaborative approach in dealing with emerging and ongoing health issues arising from the human–animal–environment interface.
3 Dixon, M.A., Dar, O.A. and Heymann, D.L. 2014. Emerging infectious diseases: opportunities at the human–animal–environment interface. *Veterinary Record* 174: 546–551 The article is one in a series of five in the *Veterinary Record* promoting "One Health". The paper describes a number of recent zoonotic disease outbreaks, discusses immediate threats of other emerging zoonotic agents and argues that a better understanding of risk factors could prevent many zoonotic diseases from occurring.

References

Baldursson, S. and Karanis, P. (2011). Waterborne transmission of protozoan parasites: review of worldwide outbreaks – an update 2004–2010. *Water Res.* 45, 6603–6614.

Chase-Topping, M., Gally, D., Low, C., Matthews, L. and Woolhouse, M. (2008). Super-shedding and the link between human infection and livestock carriage of *Escherichia coli* O157. *Nat Rev Microbiol.* 6(12), 904–912.

Coker, A.O., Isokpehi, R.D., Thomas, B.N., Amisu, K.O. and Obi, C.L. (2002). Human campylobacteriosis in developing countries. *Emerging Infect. Dis.* 8(3), 237–244.

Craun, G.F., Brunkard, J.M., Yoder, J.S., Roberts, V.A., Carpenter, J., Wade, T., Calderon, R.L., Roberts, J.M., Beach, M.J. and Roy, S.L. (2010). Causes of outbreaks associated with drinking water in the United States from 1971 to 2006. *Clin. Microbiol. Rev.* 23, 507–528.

Effler, E., Isaäcson, M., Arntzen, L., Heenan, R., Canter, P., Barrett, T., Lee, L., Mambo, C., Levine, W., Zaidi, A. and Griffin, P.M. (2001). Factors contributing to the emergence of *Escherichia coli* O157 in Africa. *Emerging Infect. Dis.* 7(5), 812–819.

Feng, Y. and Xiao, L. (2011). Zoonotic potential and molecular epidemiology of *Giardia* and giardiasis. *Clin. Microbiology Reviews* 24, 110–140.

Feng, Y., Torres, E., Li, N., Wang, L., Bowman, D. and Xiao, L. (2013). Population genetic characterisation of dominant *Cryptosporidium parvum* subtype IIaA15G2R1. *Int. J. Parasitol.* 43(14), 1141–1147.

Garg, A.X., Marshall, J., Salvadori, M., Thiessen-Philbrook, H.R., Macnab, J., Suri, R.S., Haynes, R.B., Pope, J. and Clark, W. (2006). A gradient of acute gastroenteritis was characterized, to assess risk of long-term health sequelae after drinking bacterial-contaminated water. *J. Clin. Epidemiol.* 59(4), 421–428.

Gyles, C.L. (2007). Shiga toxin-producing *Escherichia coli*: an overview. *J. Anim. Sci.* 85, E45–E62.

Humphrey, T., O'Brien, S. and Madsen, M. (2007). *Campylobacters* as zoonotic pathogens: a food production perspective. *Int. J. Food Microbiol.* 117, 237–257.

Jakopanec, I., Borgen, K., Vold, L., Lund, H., Forseth, T., Hannula, R. and Nygård, K. (2008). A large waterborne outbreak of campylobacteriosis in Norway: the need to focus on distribution system safety. *BMC Infect. Dis.* 8, 128.

Jones, K.E., Patel, N.G., Levy, M.A., Storeygard, A., Balk, D., Gittleman, J.L. & Daszak, P. (2008). Global trends in emerging infectious diseases. *Nature* 451(7181), 990–993.

Kärenlampi, R., Rautelin, H., Schönberg-Norio, D., Paulin, L. and Hänninen, M.-L. (2007). Longitudinal study of Finnish *Campylobacter jejuni* and *C. coli* isolates from humans, using multilocus sequence typing, including comparison with epidemiological data and isolates from poultry and cattle. *Appl. Environ. Microbiol.* 73, 148–155.

Karmali, M.A., Gannon, V. and Sargeant, J.M. (2010). Verocytotoxin-producing *Escherichia coli* (VTEC). *Vet Microbiol.* 140(3–4), 360–370.

Kothari, A., Pruthi, A. and Chugh, T.D. (2008). The burden of enteric fever. *J. Infect. Dev. Ctries* 2, 253–259.

Litrup, E., Torpdahl, M., Malorny, B., Huehn, S., Christensen, H. and Nielsen, E.M. (2010). Association between phylogeny, virulence potential and serovars of *Salmonella enterica*. *Infect. Genet. Evol.* 10, 1132–1139.

Majowicz, S.E., Musto, J., Scallan, E., Angulo, F.J., Kirk, M., O'Brien, S.J., Jones, T.F., Fazil, A. and Hoekstra, R.M. (2010). The global burden of nontyphoidal *Salmonella* gastroenteritis. *Clin. Infect. Dis.* 50(6), 882–889.

Manning, S.D., Motiwala, A.S., Springman, A.C., Qi, W., Lacher, D.W., Ouellette, L.M., Mladonicky, J.M., Somsel, P., Rudrik, J.T., Dietrich, S.E., Zhang, W., Swaminathan, B., Alland, D. and Whittam, T.S. (2008). Variation in virulence among clades of *Escherichia coli* O157:H7 associated with disease outbreaks. *Proc. Natl. Acad. Sci. U.S.A.* 105, 4868–4873.

Mellmann, A., Harmsen, D., Cummings, C.A., Zentz, E.B., Leopold, S.R., Rico, A., Prior, K., Szczepanowski, R., Ji, Y., Zhang, W., McLaughlin, S.F., Henkhaus, J.K., Leopold, B., Bielaszewska, M., Prager, R., Brzoska, P.M., Moore, R.L., Guenther, S., Rothberg, J.M. and Karch, H. (2011). Prospective genomic characterization of the German enterohemorrhagic *Escherichia coli* O104:H4 outbreak by rapid next generation sequencing technology. *PLoS One* 6(7), DOI: 10.1371/journal.pone.0022751.

Raffatellu, M. and Bäumler, A.J. (2010). *Salmonella*'s iron armor for battling the host and its microbiota. *Gut Microbes* 1(1), 70–72.

Sheppard, S.K., Dallas, J.F., Strachan, N.J.C., MacRae, M., McCarthy, N.D., Wilson, D.J., Gormley, F.J., Falush, D., Ogden, I.D., Maiden, M.C.J. and Forbes, K.J. (2009). *Campylobacter* genotyping to determine the source of human infection. *Clin. Infect. Dis.* 48, 1072–1078.

Silva, J., Leite, D., Fernandes, M., Mena, C., Gibbs, P.A. and Teixeira, P. (2011). *Campylobacter* spp. as a foodborne pathogen: a review. *Front Microbiol.* 2, 200.

Smith, A., Reacher, M., Smerdon, W., Adak, G.K., Nichols, G. and Chalmers, R.M. (2006). Outbreaks of waterborne infectious intestinal disease in England and Wales, 1992–2003. *Epidemiol. Infect.* 134, 1141–1149.

Spencer, S.E.F., Marshall, J., Pirie, R., Campbell, D., Baker, M.G. and French, N.P. (2011). The spatial and temporal determinants of campylobacteriosis notifications in New Zealand, 2001–2007. *Epidemiol. Infect.* 63(6), 429–433.

Strachan, N.J., Gormley, F.J., Rotariu, O., Ogden, I.D., Miller, G., Dunn, G.M., Sheppard, S.K., Dallas, J.F., Reid, T.M., Howie, H., Maiden, M.C. and Forbes, K.J. (2009). Attribution of *Campylobacter* infections in northeast Scotland to specific sources by use of multi-locus sequence typing. *J. Infect. Dis.* 199, 1205–1208.

Wassenaar, T.M. and Blaser, M.J. (1999). Pathophysiology of *Campylobacter jejuni* infections of humans. *Microbes Infect.* 1, 1023–1033.

Wensaas, K.A., Langeland, N., Hanevik, K., Mørch, K., Eide, G.E., and Rortveit, G. (2012). Irritable bowel syndrome and chronic fatigue 3 years after acute giardiasis: historic cohort study. *Gut* 61, 214–219.

Zhang, Y., Laing, C., Steele, M., Ziebell, K., Johnson, R., Benson, A.K., Taboada, E. and Gannon, V.P.J. (2007). Genome evolution in major *Escherichia coli* O157:H7 lineages. *BMC Genomics* 8, 121.

Zhang, Y., Laing, C., Zhang, Z., Hallewell, J., You, C., Ziebell, K., Johnson, R.P., Kropinski, A.M., Thomas, J.E., Karmali, M. and Gannon, V.P.J. (2010). Lineage and host source are both correlated with levels of Shiga toxin 2 production by *Escherichia coli* O157:H7 strains. *Appl. Environ. Microbiol.* 76, 474–482.

PART III

Interventions
WHAT DO WE DO TO REDUCE EXPOSURE

17

INTRODUCTION TO INTERVENTIONS TO REDUCE WATER-RELATED DISEASE*

Katherine Pond

RESEARCH FELLOW, DEPARTMENT OF CIVIL AND ENVIRONMENTAL
ENGINEERING, UNIVERSITY OF SURREY, GUILDFORD, UK

Susan Murcott

RESEARCH SCIENTIST, DEPARTMENT OF URBAN STUDIES AND PLANNING,
MASSACHUSETTS INSTITUTE OF TECHNOLOGY, CAMBRIDGE, MASSACHUSETTS, USA

David M. Gute

PROFESSOR, CIVIL AND ENVIRONMENTAL ENGINEERING DEPARTMENT,
TUFTS UNIVERSITY, MEDFORD, MASSACHUSETTS, USA

Learning objectives

1 Identify the global achievements made in the implementation and impact of interventions in the field of water for drinking, recreation and sanitation.
2 Understand the key challenges faced in developing and implementing interventions to improve water quality.
3 Recognize the value of combining engineered and non-engineered interventions.

It has been estimated that, in 2012, 842,000 deaths from diarrhoea were caused by inadequate drinking water, inadequate sanitation and inadequate hand hygiene. In children under five years old, 361,000 deaths could have been prevented, representing 5.5 per cent of deaths

---

* Recommended citation: Pond, K., Murcott, S. and Gute, D.M. 2015. 'Introduction to interventions to reduce water-related disease', in Bartram, J., with Baum, R., Coclanis, P.A., Gute, D.M., Kay, D., McFadyen, S., Pond, K., Robertson, W. and Rouse, M.J. (eds) *Routledge Handbook of Water and Health*. London and New York: Routledge.

in that age group (Prüss-Ustün et al., 2014). Interventions in water supply, sanitation and hygiene contribute to reducing this disease burden and are therefore enormously important. Interventions in this context specifically refer to those related to engineering, behaviour and management to improve water quality.

Interventions targeted in development policy

Ensuring the availability of water, for drinking, for recreation and food, has preoccupied civilizations throughout history. However, this section reflects primarily on interventions since the 1980s, spanning the International Drinking Water Supply and Sanitation Decade (IDWSSD; 1981–1990) and the period covered by the Millennium Development Goals (2000–2015) (see Chapter 43). The aim of the IDWSSD was "to provide all people with water of safe quality in adequate quantity and basic sanitary facilities by 1990, if possible, according priority to the poor and less privileged' (UNDP, 1980). A challenge was to find low cost, low-tech interventions suitable for developing country settings. However, this statement did not adequately reflect what, in retrospect, was the real aim of improving health status through improvements in water supply and sanitation (O'Rourke, 1992).

Despite substantive achievements – at the end of the decade, around 1.3 billion additional people in developing countries had gained access to improved drinking water sources (an improved drinking water source is defined as one that, by nature of its construction or through active intervention, is protected from outside contamination, in particular from contamination with faecal matter (WHO and UNICEF, 2014)) – the decade raised a number of important questions. For example, in pursuit of the aims of the decade, were the most appropriate technologies considered? Is it always appropriate to change the interventions or technologies that already exist in a country or village? Is an appropriate intervention in one country or in one village necessarily appropriate to the circumstances of one in the next? (Howard, 1990 cited in O'Rourke, 1992).

With a goal of safe water supplies and adequate sanitation for all by the year 1990, it was estimated that 2.4 billion people out of 4.5 billion people, the equivalent of 660,000 people per day for 10 years, would need to be newly connected to a safe water supply (safe drinking water is defined by the World Health Organization (WHO) as water with microbial, chemical and physical characteristics that meet WHO guidelines or national standards on drinking

Table 17.1 Percentage of the world's population with improved water supplies, 1981–2012

	Percentage of world population with improved water supplies at the end of the stated period	
	International Water Supply and Sanitation Decade (1981–1990)	*Millennium Development Goal period (1990–2012)**
Urban	95%	96% (an increase of 1.2 billion people over the stated time period)
Rural	62%	82% (an increase of 851 million people over the stated time period)
Global average	76%	89%

Note: *figures as of 2012

Source: Adapted from WHO and UNICEF (2014).

water quality. This is achieved through an improved drinking water source: household connection; public standpipe; borehole; protected dug well; protected spring; rainwater (WHO and UNICEF, 2014)). This ambitious goal was not reached, although an estimated 370,000 people gained access to a safe water supply per day and almost 220,000 people per day gained basic sanitation (Table 17.1).

But population growth and rapid urbanization drastically reduced many countries' abilities to keep up with the need. (Population growth and urbanization increases the demand for food and water. Countries have to produce more food, with less water and a declining rural workforce.) At the end of the IDWSSD, it was recognized that a better system was needed to monitor progress at the country level, ideally one that could be used as a management tool for monitoring and influencing sector development. As a result, WHO and the United Nations International Children's Emergency Fund (UNICEF) launched the Joint Monitoring Programme for Water Supply and Sanitation (JMP) in 1990 (Bartram et al., 2014).

The JMP monitored progress towards targets, but lacking an international system to compile water quality data, they adopted a pragmatic indicator that focused on drinking water infrastructure improvement, and referred to this as access to an 'improved' source which was defined as a drinking water source that, by the nature of its construction and when properly used, adequately protects the source from outside contamination, particularly faecal matter.

The millennial year 2000 began with a UN Summit resolving

> To halve, by the year 2015, the proportion of the world's people whose income is less than one dollar a day and the proportion of people who suffer from hunger and, by the same date, to halve the proportion of people who are unable to reach or to afford safe drinking water.
>
> *(United Nations, 2000)*

To that end, 189 Member States adopted the Millennium Declaration and identified eight Millennium Development Goals (MDGs), ranging from halving extreme poverty to achieving universal primary education. These MDGs were a blueprint agreed to by all United Nations (UN) Member States and by all the leading development organizations. Goal 7 targeted environmental sustainability, calling on the international community to integrate the principles of sustainable development into country policies and programmes and reverse the loss of environmental resources. Target 7.C, once finalized was 'to halve, by 2015, the proportion of the population without sustainable access to safe drinking water and basic sanitation'. At the time of writing (2015) more than half the world's population now has a piped water supply to their home, with a further 33 per cent having access to other forms of improved drinking water supply such as a borehole or well. The most recent figures from the WHO/UNICEF Joint Monitoring Programme for Water Supply and Sanitation (WHO and UNICEF, 2014) state that:

- 55 per cent (3.9 billion people) have a piped drinking water supply on their premises;
- 89 per cent of the world's population (6.2 billion) are consuming water from an improved drinking water source;
- Only 4 per cent of urban populations rely on unimproved sources;
- Of the 2.1 billion people who gained access since 1990, almost two thirds, 1.3 million, live in urban areas.

However,

- 11 per cent of the world's population (0.8 billion) obtain their drinking water from an unimproved source, including 3 per cent (185 million) who rely on contaminated surface water sources to meet their daily drinking water needs.
- In four regions, sub-Saharan Africa, Oceania, southern Asia and southeastern Asia, less than one third of the population has a piped water supply.
- 83 per cent of the population drinking from an unimproved source live in rural areas. Indeed, Bain et al. (2014) has reported that drinking water in rural areas is found to be contaminated more often than in urban areas.

The following paragraphs introduce the intervention 'types' discussed further in this section: engineered interventions, behavioural interventions and management.

The implementation of engineered interventions has undoubtedly contributed to lowering the burden of disease related to poor water quality and inadequate quantity, allowing an improvement in hygiene. This is clearly illustrated by Wolf et al. (2014) who systematically analysed studies reporting on interventions examining the effect of drinking water and sanitation improvements in low- and middle-income settings published between 1970 and May 2013. Specific interventions, such as the use of water filters, provision of high-quality piped water and sewer connections, were found to be associated with greater reductions in diarrhoea compared with other interventions. The study also showed some evidence of a greater diarrhoea risk reduction from improving household water storage and combining the water intervention with hygiene education and/or improved sanitation than through the water intervention alone (Wolf et al., 2014).

Despite the progress that has been made in reducing the burden of disease related to poor water quality due to implementation of interventions, microbial contamination still affects all water sources including piped water supplies. Indeed, Bain et al. (2014) estimate that the number of people with an unimproved source or an improved source with $\geq$ 10 *E. coli* or thermotolerant coliforms (TTC) per 100ml is 1.3 billion.

There has been a re-focus on non-engineered interventions, in particular to the importance of behavioural interventions such as community-led total sanitation (an approach which aims to initiate a change in sanitation by a whole community rather than individual behaviours; WSP, 2007) and household water treatment and safe storage interventions such as boiling water, the use of solar disinfection and ceramic filters (see Clasen, 2009 and Chapter 21 for more detail). As reported by Freeman et al. (2014) it is estimated that the disease burden from inadequate hand hygiene amounts to 297,000 deaths each year. The same report estimates that only around 19 per cent of the world population wash hands with soap after contact with excreta (i.e. after using a sanitation facility or contact with children's excreta), emphasizing the importance of promotion and adoption of this simple intervention in conjunction with the engineered interventions. Indeed, inadequate drinking water and sanitation are estimated to cause 502,000 and 280,000 diarrhoea deaths, respectively. By improving water and sanitation provision *and* improving hand hygiene it is estimated that in children under five years old, 361,000 deaths could be prevented, representing 5.5 per cent of deaths in that age group (WHO, 2015).

Finally, the importance of management tools in conjunction with the interventions already discussed should be stressed. Water Safety Plans are central to ensuring whether the water supply chain as a whole, be it a large municipal water supply system or a small community supply, can deliver water of a sufficient quality. However, even if a sample taken from the supply indicates that it is free from contamination it does not mean that the supply is safe. Even well managed piped supplies do not present a negligible risk to health (Bain et al., 2014). This risk assessment/risk management process is key to identifying whether the implemented interventions are working adequately and, if not, how they can be fixed or improved. As noted above, engineered interventions will occasionally fail at some point – intermittent power, damaged pipes and so on are always a possibility. In addition, failures in the management, monitoring and control of piped water distribution systems can have catastrophic consequences, as shown by the outbreaks of waterborne disease in Walkerton, Canada, in 2000 and Milwaukee in 1993 (Chapters 69 and 70 respectively). By preventing these failures and having a system in place to check and correct the causes of the failures it is possible to reduce any compromises to the water quality.

As stressors such as rising population, increasing urbanization, growing water demands and climate change for example are interacting and adversely affecting water quality and quantity and thus human health, it has become clear that we need to start looking at more interdisciplinary interventions to account for the complexity of water-related health problems (Batterman et al., 2009).

This section discusses in further detail examples of key interventions to control human exposure to water-related hazards such as bacteria, viruses, protozoa, helminths, chemicals and personal physical factors at the individual, household and community levels. The individual chapters look at the traditional engineering interventions such as those designed to reduce the faecal–oral transmission of pathogens to provide an improved water supply (e.g. technologies such as piped household water connections, standpipes, or protected dug wells, springs or rainwater collection) (see Chapters 18, 19 and 20), and the non-engineered interventions such as water safety planning which considers the management of water from the source to the tap (WHO, 2009; see Chapter 23). The importance of behavioural interventions such as hand washing after toilet use and before the preparation of food is also discussed (Chapters 21 and 22). It is clear from this brief introduction that all of these interventions must be considered together rather than in isolation. As such we must continue to strive to develop more effective, integrated and sustainable interventions. Demonstrated experience and research clearly show that by combining engineered and non-engineered interventions considerable reductions in the burden of disease related to water can be achieved.

References

Bain, R., Cronk, R., Hossain, R., Bonjour, S., Onda, K., Wright, J., Yang, H., Slaymaker, T., Hunter, P., Prüss-Ustün, A. and Bartram, J. (2014). Global assessment of exposure to fecal contamination through drinking water based on a systematic review. *Tropical Medicine and International Health*, 19(8), 917–927.

Bartram, J., Brocklehurst, C., Fisher, M.B., Luyendijk, R., Hossain, R., Wardlaw, T. and Gordon, B. (2014). Global monitoring of water supplies and sanitation: history, methods and global challenges. *International Journal of Research and Public Health*, 11, 8137–8165.

Batterman, S., Eidenburg, J., Hardin, R., Kruk, M.E., Lemos, M.C., Michalak, A.M., Mukherjee, B., Renne, E., Stein, H., Watkins, C. and Wilson, M.L. (2009). Sustainable control of water-related infectious diseases: a review and proposal for interdisciplinary health-based systems research. *Environmental Health Perspectives*, 117(7), 1023–1032.

Clasen, T. (2009). *Scaling Up Household Water Treatment among Low-Income Populations*. Geneva: World Health Organization.

Freeman, M.C., Stocks, M., Cumming, O., Jeandron, A., Higgins, J., Wolf, J., Prüss-Ustün, A., Bonjour, S., Hunter, P.R., Fewtrell, L. and Curtis, V. (2014). Hygiene and health: systematic review of handwashing practices worldwide and update of health effects. *Tropical Medicine & International Health*, 19(8), 906–916.

O'Rourke, E. (1992). The international drinking water supply and sanitation decade: dogmatic means to a debatable end. *Water Science and Technology*, 26(7–8), 1929–1939.

Prüss-Ustün, A., Bartram, J. and Clasen, T. (2014). Burden of disease from inadequate water, sanitation and hygiene in low- and middle-income settings: a retrospective analysis of data from 145 countries. *Tropical Medicine & International Health*, 19(8), 894–905.

UNDP (United Nations Development Programme) (1980). *Decade Dossier, International Drinking Water Supply and Sanitation Decade 1981–1990*, New York: UNDP.

United Nations (2000). United Nations Millennium Declaration. http://www.un.org/millennium/declaration/ares552e.htm.

WHO (2009). Water safety plan manual: step-by-step risk management for drinking-water suppliers. Geneva: World Health Organization. http://www.who.int/water_sanitation_health/publication_9789241562638/en/

WHO (2015). Preventing diarrhoea through better water, sanitation and hygiene: exposures and impacts in low- and middle-income countries. http://www.who.int/water_sanitation_health/gbd_poor_water/en/

WHO and UNICEF (2014). *Progress on Sanitation and Drinking-Water 2013 Update: Joint Monitoring Programme for Water and Sanitation*. Geneva: WHO and UNICEF.

Wolf, J., Prüss-Ustün, A., Cumming, O., Bartram, J., Bonjour, S., Cairncross, S., Clasen, T., Colford, J.M., Curtis, V., De France, J., Fewtrell, L., Freeman, M.C., Gordon, B., Hunter, P.R., Jeandron, A., Johnston, R.B., Maüsezahl, D., Mathers, C., Neira, M. and Higgins, J. (2014). Assessing the impact of drinking water and sanitation on diarrhoeal disease in low- and middle-income settings: a meta-regression. *Tropical Medicine & International Health*, 19(8), 928–942.

WSP (ed.) (2007). *Community-Led Total Sanitation in Rural Areas. An Approach that Works*. Washington DC: Water and Sanitation Program.

18
DRINKING-WATER SUPPLY*

Jamie Bartram

DIRECTOR OF THE WATER INSTITUTE, UNIVERSITY OF
NORTH CAROLINA AT CHAPEL HILL, USA

Samuel Godfrey

WaSH CHIEF, UNICEF, ETHIOPIA

Learning objectives

1 Describe the different sources and technologies that supply drinking water to populations worldwide.
2 Explain the trends in drinking-water supply technologies and service levels.
3 Explore major challenges in both utility and community-managed supply.

Issues associated with household plumbing, with household water treatment and safe storage (HWTS; Chapter 21) and with the retail sale of water (packaged or otherwise) are not addressed.

Definition and scope of 'water supply'

'Drinking water' is 'required for all usual domestic purposes, including drinking, food preparation and personal hygiene' (WHO, 2011, p2) and is not limited to water consumed directly or indirectly (incorporated into beverages and foodstuffs). While the term 'domestic water' is more precise, in this chapter we use the conventional term 'drinking-water'.

Water supply is about ensuring access to safe water. According to General Comment 15 of the United Nations Committee on Cultural Economic and Social Rights the scope of water supply extends to: 'sufficient, safe, acceptable, physically accessible and affordable water for personal and domestic uses' (UNCESCR, 2002). A comprehensive definition would extend access to include all times and places of need – such as in schools and workplaces, health care and marketplaces; and in emergencies and disasters; and might also explicitly address the continuity and reliability of supply.

* Recommended citation: Bartram, J. and Godfrey, S. 2015. 'Drinking-water supply', in Bartram, J., with Baum, R., Coclanis, P.A., Gute, D.M., Kay, D., McFadyen, S., Pond, K., Robertson, W. and Rouse, M.J. (eds) *Routledge Handbook of Water and Health*. London and New York: Routledge.

Table 18.1 Water service levels, access measures and associated health-related needs met and levels of concern

Service level	Access measures	Needs met	Level of health concern
No access (quantity collected often below 5 l/c/d)	More than 1000 m or more than 30 minutes total collection time	Consumption – cannot be assured Hygiene – very restricted (bathing and laundry may be practiced at a source)	Very high
Basic access (quantity collected unlikely to exceed 20 l/c/d)	Between 100 and 1000 m or 5 to 30 minutes total collection time	Consumption – should be assured Hygiene – restricted: hand washing and basic food hygiene possible; laundry/bathing difficult to assure unless carried out at source	High
Intermediate access (average quantity used about 50 l/c/d)	Water delivered through one tap on-plot (to household) or within 100 m or less than 5 minutes total collection time	Consumption – assured Hygiene – all basic personal and food hygiene assured; laundry and bathing should also be assured	Low
Optimal access (quantity used at least 100 l/c/d)	Water supplied through one or more taps within the household, continuously	Consumption – all needs met Hygiene – all needs met	Very low

Source: Adapted from Howard and Bartram (2003)

How much water is 'sufficient' for all domestic purposes is explored in Howard and Bartram (2003, referenced in General Comment 15). They propose four broad categories of quantity and access, based on sufficiency for consumption and hygiene (Table 18.1).

General Comment 15 also requires that water be 'safe' and in doing so refers to the *Guidelines for Drinking-water Quality* of the World Health Organization (WHO). WHO defines safe drinking water as not presenting 'any significant risk to health over a lifetime of consumption, including different sensitivities that may occur between life stages' (WHO, 2011, p1). The *Guidelines* go on to describe safe water by specifying both water quality characteristics (such as tolerable concentrations of chemicals and faecal indicator bacteria) and safe practices that must be consistently achieved and applied for the water to be considered safe. It codifies water quality characteristics as Guideline Values, and good practices as sanitary inspections for small or community managed systems (WHO, 1997) or as Water Safety Plans for other system types, all under the umbrella of a Framework for Safe Drinking-water (WHO, 2011; and Chapter 23).

Water may not be *acceptable* to consumers, because of organoleptic characteristics (taste, odour and appearance), other cultural factors and so on, even if accessible. In these circumstances users may turn to preferable (more acceptable) sources – that may be unreliable, unsafe or less accessible – and would not reliably benefit from the intended water access.

Access to water may also be constrained for reasons of *physical access*. Around a fifth of humankind is physically disabled in some way and physical disability may restrict access. Physical access may also be restricted by means of caste and social exclusion.

The importance of water being *affordable* is reflected in international policy statements, such as the Millennium Declaration, and also in national policy in many countries through reference to concepts like 'equitable', 'fair' or 'reasonably priced' (Hutton, 2012). Affordability must reflect the ability and willingness to pay for the service, by users directly or by other means such as taxes or transfers. The apparently simple idea that water should be affordable is difficult to assess and implement: how much water (water for which purposes) must be affordable? What absolute amount or proportion of household income can reasonably be expended on water? How should costs of supply be covered when they exceed what the receiving population can afford? And if some members of a population cannot afford to cover their 'fair share' of costs, by what means are the costs to be covered? Hutton proposes that expenditure 'as a % of household disposable income is compared with a single affordability threshold'; notes that '[d]ifferent countries and international organizations have established a threshold that is considered appropriate, typically ranging from 3% to 6%'; and recommends that reporting be disaggregated to describe rural–urban, service level and wealth quintile differences as well as the situation of marginalized populations.

Status and trends in water supply

The situation of water supply is reported by WHO and the United Nations Children's Fund (UNICEF) through their 'Joint Monitoring Programme' (JMP). Since 2000, JMP reports have been based on data from censuses and nationally representative household surveys and are reliable sources of information on use of different forms of water source by households (Bartram et al., 2014). The global headlines (WHO and UNICEF, 2014) show 56 per cent of the world's population using piped water at the household (80 per cent in urban areas and 29 per cent in rural areas) and 33 per cent using 'other improved' sources like boreholes with handpumps, leaving 11 per cent using unimproved sources as of 2012. The highest proportions of the 'unserved' are in sub-Saharan Africa but the greatest numbers are in south Asia. Improved sources were a major focus of attention up to and during the Millennium Development Goal (MDG) period (1990–2015) and account for a large proportion of domestic supply in sub-Saharan Africa and southern and eastern Asia. Outside the developed nations most access to drinking water in rural areas is from 'improved sources', with rapid increase in piped-at-household rural water in Eastern Asia, Western Asia, Latin America and the Caribbean and North Africa. Globally the proportion of the urban population with piped water at-household has in fact declined since 1990 (from 81 to 80 per cent) despite having risen in most regions – the decline being largely due to deterioration in sub-Saharan Africa (from 42 to 34 per cent).

Data such as those from WHO and UNICEF, and the fact that the MDG target component to halve the proportion of the 'unserved' between 1990 and 2015 is reported as having been met, provide reason for optimism and indeed progress in this period was substantive. However, these data also present an overly positive picture of the situation as they do not account for important aspects such as safety and continuity of supply. Accounting for safety alone lowers estimates substantively (Bain et al., 2014a) and accounting for faecal contamination of 'improved sources' increases the number of people without access to safe water by over a billion (Onda et al., 2012; Wolf et al., 2013; Bain et al., 2014b). Another assessment of the data addresses multiple factors through the Water, Sanitation, and Hygiene (WaSH) Performance Index, which is a comparison of country performance in realizing universal WaSH. High performing countries include El Salvador, Mali, Tajikistan, Nepal, and Liberia (Cronk et al., 2015).

As noted above, a comprehensive perspective would also reflect access to safe water in other *settings*. Data from settings such as schools and health care settings are incomplete;

and from workplaces and markets are rare (Cronk et al., 2015). The available data from schools indicate that more than 30 per cent of schools in low and middle income countries lack access to improved water and sanitation (UNICEF, 2014); while data from health care settings indicate that 40 per cent of facilities lack access to an improved water source and 20 per cent lack access to sanitation (Bartram et al., 2015; WHO and UNICEF, 2015). International guidelines for such settings are available for schools (WHO, 2009) and health care settings (WHO, 2008).

Finally, in any given location, there may be households at very different levels of water service. The most consistent inequalities are between rural and urban areas (Bain et al., 2014c) and there are wide inequalities within urban areas, for example between wealthy neighbourhoods and slum and peri-urban settlements. As coverage with basic services has increased, so has concern for equality, and metrics have been proposed to compare levels of inequality and track progress towards equality (Luh et al., 2013). For example, as of 2010, coverage for access to total improved water was similar in Cape Verde (88 per cent) and Ghana (86 per cent), but Cape Verde is making no progress towards reducing the urban–rural gap, while Ghana is.

Sources and technologies for water supply

Water supply for households may be conveniently divided into two forms that reflect the data reported by WHO and UNICEF: households where water is collected from 'off-plot' shared community 'improved sources'; and households where piped water supply provides water on-plot or within-dwelling. Populations with neither of these largely collect water from unprotected sources such as rivers, lakes and unprotected wells (which are open to contamination, of which contamination by human or animal faeces is the principal health concern); or in urban areas purchase water from retail outlets or vendors.

While useful, this classification is based on the assumption that each household uses a single water source. In fact many households use multiple sources for water for domestic purposes. Furthermore water is put to diverse uses, including food production and economic purposes, the latter reflected in contemporary writing on 'multiple source use'.

In assessing a water source or supply system (as opposed to the service available to a population), from a health perspective the focus is typically on *safety* and *continuity*, whether a shared community 'improved source' or piped water supply.

Water *safety* assessment is based on a combination of water quality measures and assessment of the adequacy of the source or system in terms of preventing and, where treatment is applied, reducing contamination (Lloyd and Bartram, 1991). Water quality measurement is based on samples that are typically analysed for microbial safety (faecal indicator bacteria, Chapter 59) and chemical contaminants of health concern (Chapter 10). Microbial quality varies widely including seasonally (Kostyla et al., 2015). For sources from which water is collected, safety assessment is by 'sanitary inspection' of the hazards and protective measures at the source; whereas for piped systems a Water Safety Plan (Chapter 23) is advised.

Several related concepts determine the *continuity* of a source or supply and various authors use these in different and sometimes confusing ways. We view 'continuity' (or 'sustained functioning') as the proportion of time that water is available. Thus continuity relates to the functioning of a source or system over time, whereas 'sustainability' in the sense of sustainable development (which is readily confused with 'sustained functioning') is a prospective concept, looking at future viability. 'Functionality' is most commonly used as a cross-sectional measure to describe the proportion of sources working at a given time and is sometimes applied to a single source (working/not working). The notion of 'reliability' (or 'predictability') is distinct and relates to whether users can be confident in a source – not necessarily that it is continuously available, but that they know when it will be available.

Shared sources from which water is collected

People using off-plot 'improved sources' collect water from shallow or deep groundwater sources such as boreholes (tubewells) with handpumps and protected springs as well as rainwater harvesting systems that are constructed and maintained so as to reduce the likelihood of contamination with faeces.

Figures 18.1a and b illustrate the sanitary inspection of a borehole with handpump, the most common improved community water source type worldwide. Similar figures are available for other community-level water source types in WHO (1997).

While such wells and springs are typically designed with a projected working life of decades, limited evidence suggests that at any one time around a third of wells with handpumps in sub-Saharan Africa do not work (WaterAid, 2011).

Piped systems delivering water to households

The simplest piped water systems delivering water to households comprise a protected spring and simple branching network of pipes leading to households downhill of the source. Such systems rely on the relatively good quality of water from such sources (thereby avoiding the complexities of treatment); exploit the effect of gravity to move water to households (thereby avoiding the costs and complexity of pumping); and rely on copious water (sufficient flow to satisfy the needs of all households at all times) to maintain a sufficient, safe supply.

Systems this simple are uncommon. More often the flow of the source is insufficient to satisfy the maximum demand, or the source is remote from the community so that the cost of a large pipe to carry high flows of water is prohibitive. As a result it is common that a storage tank is constructed in or near the community to balance out variations in daily demand. Storage tanks are raised or positioned uphill so that water can flow from them to households by gravity.

Two further factors further complicate piped supply systems. Firstly the available source of water may be physically below the households that it is to supply – either on the ground surface (spring, river or lake) or underground (in the case of groundwater sources such as wells). In these circumstances pumping is required, with attendant costs, operational complexity and energy implications. Secondly many water sources are not reliably safe and treatment is required (Chapter 19). This may be as little as application of a disinfectant, or a more complex treatment chain that may often include coagulation, flocculation, filtration and chlorination.

Larger piped water supplies often exploit multiple water sources that may be similar (e.g. several wells) or diverse (e.g. several wells and a surface water source with associated water treatment). These multiple sources may be connected to households through a cross-connected (reticulated) system of pipes, often called a 'ring main', organized in hydraulic zones and providing multiple routes through which water may reach any given household from more than one source, with associated valves and regulation to ensure a controlled minimum water pressure, for example of 6–7.5m (Twort et al., 2009). Large systems may have multiple storage tanks and supplementary pumping. Supply may extend beyond directly-supplied households through standpipes (public taps accessible to many households) and retail sale (both formal and informal); and may also include other users including industry.

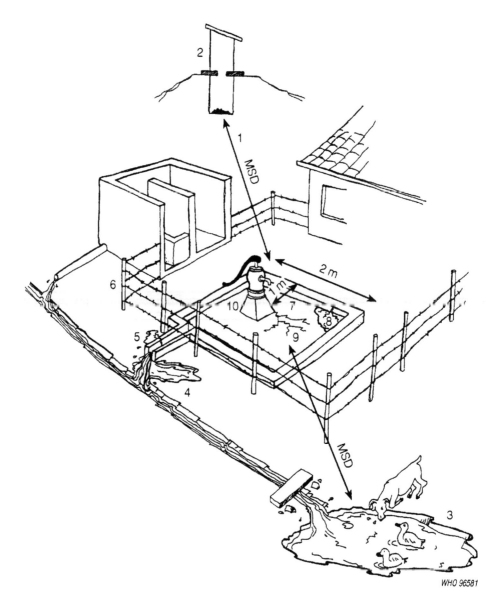

2

1

MSD

2 m

6

7

10

8

9

5

4

MSD

3

WHO 96581

Figure 18.1a Sanitary inspection of a borehole with hand pump

Source: WHO (1997, Figure A2.5)

I   Type of facility   TUBEWELL WITH HAND-PUMP

1. General information: Health centre ....
                          Village ....

2. Code no.—Address ....

3. Water authority/community representative signature ....

4. Date of visit ....

5. Water sample taken? .... Sample no. .... Thermotolerant coliform grade ....

II   Specific diagnostic information for assessment   Risk

1. Is there a latrine within 10 m of the hand-pump?   Y/N

2. Is the nearest latrine on higher ground than the hand-pump?   Y/N

3. Is there any other source of pollution (e.g. animal excreta, rubbish,   Y/N
   surface water) within 10 m of the hand-pump?

4. Is the drainage poor, causing stagnant water within 2 m of the
   hand-pump?   Y/N

5. Is the hand-pump drainage channel faulty? Is it broken, permitting
   ponding? Does it need cleaning?   Y/N

6. Is the fencing around the hand-pump inadequate, allowing animals in?   Y/N

7. Is the concrete floor less than 1 m wide all around the hand-pump?   Y/N

8. Is there any ponding on the concrete floor around the hand-pump?   Y/N

9. Are there any cracks in the concrete floor around the hand-pump which   Y/N
   could permit water to enter the well?

10. Is the hand-pump loose at the point of attachment to the base so that   Y/N
    water could enter the casing?

                                                    Total score of risks ..../10

Contamination risk score: 9–10 = very high; 6–8 = high; 3–5 = intermediate;
                          0–2 = low

III   Results and recommendations

The following important points of risk were noted: .... (list nos 1–10)
and the authority advised on remedial action.

Signature of sanitarian ....

Figure 18.1b Sanitary inspection of a borehole with hand pump

Source: WHO (1997, Figure A2.5).

The basic components of urban water systems (Cotton and Tayler, 2000) are therefore:

- Production facilities at source: intakes, treatment works, tubewells.
- Bulk supply mains: these carry water from production facilities to service reservoirs; they are only required when the source is remote from the supply area which in some large cities may be many tens or even hundreds of kilometres distant.
- Storage facilities: these allow variations in demand over the day to be balanced and may provide some back-up capacity in the event of a break in supply.
- Primary or trunk mains: these convey water in bulk from one part of the network to another; they are not normally required where the supply is from tubewells located in the supply area.
- Secondary mains: these link service reservoirs and trunk mains with service mains; they are normally of at least 150 mm in diameter and are laid to form a ring main or 'looped' system. Looped systems have the advantage of flexibility, allowing water to flow along different 'routes' around the system in response to varying demand, and operating pressures can be maintained at a more constant level throughout the network. It is also easier to extend ring mains in response to population growth and water demand.
- Tertiary or service mains: these are mains of typically 75 mm or 100 mm diameter that distribute water within a locality and may be laid in the form of either ring main/loops or branched systems.
- Service connections to houses and public standposts.

Piped water supply schemes, both large and small, are normally designed to supply water to all households continuously. However, discontinuous (or intermittent) supply is common and is sometimes considered the norm in south Asia. Discontinuous supply restricts effective access, introduces risks of contamination and leads to increased costs from coping measures adopted by households.

Disinfection

Water in both community sources and piped systems is often disinfected. This refers to reducing infectious particles (bacteria, viruses, protozoa) to concentrations at which they represent a negligible risk to health. It may be achieved through removal (filtration), chemical agents (such as chlorine) or physical agents (such as ultraviolet light). Water treatment is addressed in Chapter 19. The term 'residual disinfection' refers to treatments which leave some concentration of an oxidizing chemical, such as chlorine, in the water. This residual provides some degree of protection against contamination during distribution and storage. The concentration must be acceptable to consumers but also effective in microbial inactivation. WHO, for example, recommends that a minimum of 0.2 mg/l of chlorine be maintained throughout a piped distribution system, including its most remote parts, but that the concentration not exceed 0.5 mg/l to minimize the risk of infringing consumer acceptability (taste and odour) and thus drive consumers to use alternative, potentially more dangerous, sources.

Challenges at different scales and types of organization

In urban areas, the responsibility for and management of the water supply rests with a water utility or a department of local government. In rural areas, especially of less developed countries, such arrangements rarely exist.

WHO usefully distinguishes 'community managed' from other types of water supply, noting that:

> … while a definition based on population size or the type of supply may be appropriate under many conditions, it is often administration and management that set community supplies apart … The increased involvement of ordinary, often untrained and sometimes unpaid, community members in the administration and operation of water-supply systems is characteristic of small communities; this provides a ready distinction between community water supplies and the supply systems of major towns and cities.
>
> *(WHO, 1997, p1)*

A third type of supply is that to individual dwellings or very small settlements, sometimes referred to as 'private supply' or 'self-supply'. Such systems provide water to a significant population in developing and developed countries worldwide.

Community and private or self-supply systems generally have higher likelihood of contamination and lower levels of continuity and reliability than their urban counterparts.

Attitudes to community management of water supplies in developing countries have evolved over several decades (Figure 18.2), from a focus on provision of labour by community members, through increasing degrees of community influence in decision taking towards a 'demand responsive' approach. Notwithstanding this evolution in thinking, limited evidence suggests caution in interpreting the success of community management.

There are diverse claims concerning the cause and potential remedies for low rates of functionality that include technical, financial, social and community features, institutional and policy concerns, and environmental aspects (Lockwood et al., 2003; Clapham, 2004). However, there are few rigorous studies (e.g. Komives et al., 2008; Whittington et al., 2009; Marks et al., 2014) which generally point towards external support to community financial and technical management as critical (e.g. Harvey and Reed, 2007).

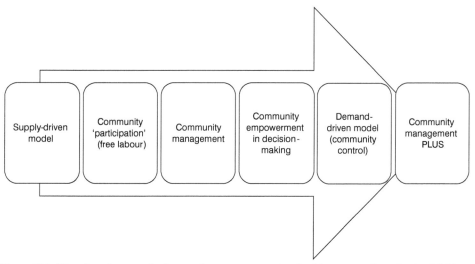

Figure 18.2 The changing terminology and concepts associated with community roles in drinking-water supply

Challenges encountered in urban water systems include high levels of leakage due to aging and under-maintained water assets, unplanned urban development resulting in inequitable supply and low continuity from high levels of demand, inappropriate domestic plumbing introducing risks of contamination and causing wastage, and microbial contamination due to inappropriate siting of sewer lines and with intermittent supply or low pressure in the reticulation system potentially allowing ingress of polluted water. In many towns and cities water utilities are challenged by inadequate investment in financial and billing systems which results in low cost recovery and thereby contributes to insufficient expenditure on operating expenses (AMCOW, 2011).

Water supply and health

The health consequences of water supply include both beneficial and adverse effects. The beneficial effects are primarily derived from the use of water for personal and domestic hygiene; and from the fact that water supplies tend to be safer than unprotected alternatives. The adverse effects are associated with contaminated water and the capacity of shared (community) sources and water distributed by piped systems to deliver contaminated water to large populations and cause common-source outbreaks of disease. These effects combine the water-borne and water-washed (or water access) categories of disease (Chapter 4).

The burden of disease associated with insufficient drinking-water supply is difficult to isolate from that of other causes – especially from the role of sanitation as a pollution source and the hygiene behaviours which must accompany water supply for effective disease prevention. WHO estimates that in 2012 around 502,000 deaths worldwide were due to diarrhea caused by insufficient water supply and a further 297,000 were due to diarrhea caused by inadequate hygiene (Prüss-Ustün et al., 2014). To these figures other disease outcomes preventable by sufficient safe water such as typhoid, infectious hepatitis, trachoma and scabies (Chapter 5) may be added.

Where water is collected and carried to the home there are further health risks. The act of carrying water, which is primarily undertaken by women and children, is associated with certain musculoskeletal disorders (Chapter 7), and collection of water, particularly from remote sources or at night, has been associated with violence against the collector.

Outbreaks of disease associated with contamination of drinking water, if detected and reported, often attract much attention. The efficiency of health surveillance systems in detecting water-borne disease outbreaks is low and it is likely that only a small proportion is recognized, even in countries with advanced surveillance systems. Outbreaks continue in countries at all stages of development but are best recognized and characterized in more industrialized nations. Evidence from such outbreaks has been collated and reviewed by Hrudey and Hrudey (2004) who highlight that the same basic problems tend to occur frequently. Such outbreaks affect both small and large communities (e.g. Walkerton, Canada, see Chapter 69; and Milwaukee, USA, see Chapter 70).

Future perspectives

The underlying challenges associated with water supply are threefold: extending access to the unserved and to unserved 'settings'; increasing service levels so that maximum benefits are secured by all; and maintaining the ever-increasing 'stock' of infrastructure.

While the first of these challenges is concentrated in least developed countries, the challenges of service improvement are more widespread, with deficiencies worldwide including rural areas of industrialized nations. Because of the greater stock of infrastructure and its increasing age, challenges associated with its maintenance and replacement are greatest in the upper and middle income countries. In the USA for example, the American Society for Civil Engineers routinely gives the country's water (and wastewater) infrastructure a 'D' grade, reflecting underinvestment that will transfer costs to future generations.

An additional challenge occurs with transition and change – for example where there is upgrading from community sources to piped supply, or where changes in supply are required to cope with demographic or climate change. Much water supply infrastructure is long term and where it is replaced before the end of its working life this increases inefficiencies.

It is widely perceived that slowing population growth will assist the endeavour of extending and improving water supply. However, the ongoing transition from community sources to piped water supply means that the number of households rather than the number of people is the principal determinant of the challenge. In fact in countries at all levels of development household number is increasing and globally will triple between 1990 and 2050, regardless of population projection scenarios (Bartram et al., 2012). This is accompanied by urbanization such that the global urban population will double between 2010 and 2050, whereas in Africa this doubling will be much more rapid and will take place by 2030 (HABITAT, 2008).

References

AMCOW, 2011. African Ministers Council on Water Country Status Overviews – Regional Synthesis Report. *Pathways to Progress: Transitioning to Country-Led Service Delivery Pathways to Meet Africa's Water Supply and Sanitation Targets*. Washington, DC: The World Bank/Water and Sanitation Program. http://www.wsp.org/wsp/content/pathways-progress-status-water-and-sanitation-africa.

Bain, R, Cronk, R, Wright, J, Yang, H, Slaymaker, T and Bartram, J 2014a. Fecal contamination of drinking-water in low and middle-income countries: a systematic review and meta-analysis. *PLoS Medicine*, 11(5). doi: 10.1371/journal.pmed.1001644.

Bain, R, Cronk, R, Hossain, R, Bonjour, S, Onda, K, Wright, J, Yang, H, Slaymaker, T, Hunter, P, Prüss-Ustün, A and Bartram, J 2014b. Global assessment of exposure to faecal contamination through drinking water based on a systematic review. *Tropical Medicine and International Health*, 19(8), 917–927. doi: 10.1111/tmi.12334.

Bain, R, Wright, J, Christenson, E and Bartram, J 2014c. Rural:urban inequalities in post-2015 targets and indicators for drinking water. *Science of the Total Environment*, 490, 509–513.

Bartram, J, Elliott, M and Chuang, P 2012. Getting wet, clean and healthy: why households matter. *Lancet*, 380(9837), 85–86. doi:10.1016/S0140-6736(12)60903-9.

Bartram, J, Brocklehurst, C, Fisher, M, Luyendijk, R, Hossain, R, Wardlaw, T, and Gordon, B 2014. Global monitoring of water supply and sanitation: a critical review of history, methods, and future challenges. *International Journal of Environmental Research in Public Health* 11, 8137–8165. doi:10.3390/ijerph110808137.

Bartram, J, Cronk, R, Montgomery, M, Gordon, B, Neira, M, Kelley, E, and Velleman, Y 2015. Lack of toilets and safe water in health-care facilities. *Bulletin of the World Health Organization* 93: 210. doi:10.2471/BLT.15.154609.

Clapham, D 2004. *Small Water Supplies: A Practical Guide*. London: Spon Press.

Cotton, A and Tayler, W 2000. *Services for the Urban Poor, Volume 3 Section 4b, Water Supply*, Water, Engineering and Development Centre, Loughborough University, Loughborough, pp 4.34–4.79. https://wedc-knowledge.lboro.ac.uk/details.html?id=11392.

Cronk, R, Bartram, J and Slaymaker, T 2015. Monitoring drinking water, sanitation, and hygiene in non-household settings: priorities for policy and practice. *International Journal of Hygiene and Environmental Health*, in press.

Cronk, R, Luh, J, Meier, B, Bartram, J 2015. *The Water, Sanitation, and Hygiene Performance Index: A Comparison of Country Performance in Realizing Universal WaSH*. UNC - Chapel Hill, Chapel Hill, NC. Available from: http://waterinstitute.unc.edu/wash-performance-index-report/.

HABITAT. 2008. *State of the African Cities Report*. Nairobi: UN-HABITAT.

Harvey, PA and Reed, RA 2007. Community-managed water supplies in Africa: sustainable or dispensable? *Community Development Journal*, 42(3), 365–378.

Howard, G and Bartram, J 2003. *Domestic Water Quantity, Service Level and Health*. Geneva: World Health Organization.

Hrudey, SE and Hrudey, EJ 2004. *Safe Drinking Water: Lessons from Recent Outbreaks in Affluent Nations*. London: IWA Publishing.

Hutton, G 2012. *Monitoring 'Affordability' of Water and Sanitation Services after 2015: Review of Global Indicator Options*. Geneva: United Nations Office of the High Commission for Human Rights.

Komives, K, Akanbang, B, Thorsten, R, Tuffuor, B, Wakeman, W, Larbi, E, Bakalian, A and Whittington, D 2008. Post-construction support and the sustainability of rural water projects in Ghana. 33rd WEDC International Conference, Accra, Ghana. http://wedc.lboro.ac.uk/resources/conference/33/Komvies_K.pdf.

Kostyla, C, Bain, R, Cronk, R and Bartram, J 2015. Seasonal variation of fecal contamination in drinking water sources in developing countries: a systematic review. *Science of the Total Environment*. http://www.ncbi.nlm.nih.gov/pubmed/25676921.

Lloyd, BJ and Bartram, J 1991. Surveillance solutions to microbiological problems in water quality control in developing countries. *Water Science and Technology*, 24(2), 61–75.

Lockwood, H, Bakalian, A and Wakeman, W 2003. Assessing sustainability in rural water supply: the role of follow up support to communities literature review and desk review of rural water supply and sanitation project documents. http://www.aguaconsult.co.uk/assets/Uploads/Publications/WorldBank-AssessingSustainability-2003.pdf

Luh, J, Baum, R and Bartram, J 2013. Equity in water and sanitation: developing an index to measure progressive realization of the human right. *International Journal of Hygiene and Environmental Health*, 216(6), 662–671.

Marks, SJ, Komives, K and Davis, J 2014. Community participation and water supply sustainability evidence from handpump projects in rural Ghana. *Journal of Planning Education and Research*. http://jpe.sagepub.com/content/early/2014/04/01/0739456X14527620.

Onda, K, LoBuglio, J and Bartram, J 2012. Global access to safe water: accounting for water quality and the resulting impact on MDG progress. *International Journal of Environmental Research and Public Health*, 9, 880–894.

Prüss-Üstün, A, Bartram, J, Clasen, T, Colford, JM Jr, Cumming, O, Curtis, V, Bonjour, S, Dangour, AD, De France, J, Fewtrell, L, Freeman, MC, Gordon, B, Hunter, PR, Johnston RB, Mathers, C, Mausezahl, D, Medlicott, K, Neira, M, Stocks, M, Wolf, J and Cairncross, S 2014. Burden of disease from inadequate water, sanitation and hygiene in low- and middle-income settings: a retrospective analysis of data from 145 countries. *Tropical Medicine and International Health*, 19(8), 894–905.

Twort, AC, Ratnayaka, DD and Brandt, MJ 2009. *Water Supply*. 6th ed. Oxford: Butterworth-Heinemann.

UNCESCR. 2002. *General Comment No. 15 on the Right to Water*, adopted by the UN Committee on Economic, Social and Cultural Rights at its twenty-ninth session in November 2002. https://www1.umn.edu/humanrts/gencomm/escgencom15.htm.

UNICEF. 2014. *Water, Sanitation, and Hygiene Annual Report 2013*. United Nations Children's Fund, New York.

WaterAid. 2011. *A Framework for Sustainable Water and Sanitation Services and Hygiene Behaviour Change*. WaterAid, London.

Whittington, D, Davis, J, Prokopy, L, Komives, K, Thorsten, R, Lukacs, H, Bakalian, A and Wakeman, W (2009) How well is the demand-driven, community management model for rural water supply systems doing? Evidence from Bolivia, Peru and Ghana. *Water Policy* 11(6), 696–718.

WHO. 1997. *Guidelines for Drinking-water Quality. Second edition Volume 3*. Geneva: World Health Organization.

WHO, 2008. *Essential Environmental Health Standards in Health Care*. World Health Organization, Geneva.

WHO. 2009. *Water, Sanitation and Hygiene Standards for Schools in Low-cost Settings*. Geneva: World Health Organization

WHO. 2011. *Guidelines for Drinking-water Quality. Fourth edition*. World Health Organization, Geneva.

WHO and UNICEF. 2014. *Progress on Drinking-water and Sanitation*. WHO, Geneva.

WHO and UNICEF. 2015. *Water, Sanitation and Hygiene in Health Care Facilities: Status in Low and Middle Income Countries and Way Forward*. WHO, Geneva.

Wolf, J, Bonjour, S and Prüss-Üstün, A 2013. An exploration of multilevel modeling for estimating access to drinking-water and sanitation. *Journal of Water and Health*, 11(1), 64–77. doi:10.2166/wh.2012.107.

19

DRINKING WATER
TREATMENT[*]

Donald Reid

ADJUNCT ASSISTANT PROFESSOR, DEPARTMENT OF CIVIL ENGINEERING, SCHULICH
SCHOOL OF ENGINEERING, UNIVERSITY OF CALGARY, ALBERTA, CANADA

Learning objectives

1 Understand the principle processes involved in producing drinking water fit for human consumption.
2 Appreciate the various disinfection practices used to ensure drinking water is microbiologically safe.
3 Distinguish between community (private) drinking water supplies and municipal (public) drinking water supplies.

The treatment of raw water (water derived from an environmental source) comprises a number of steps or processes called 'unit processes' (Gray, 1999a). The precise number and nature of the unit processes will depend on the qualities of the raw water source itself and the finished water quality objectives. The goal of drinking water treatment is to produce water that is fit for human consumption, that is, will not have harmful effects when consumed as drinking water. This requires that both microbiological and chemical (and on occasion radiological) parameters of concern are either reduced to levels below which they are likely to cause harmful effects or they are, in the case of microbiological contaminants, inactivated or removed through disinfection processes (Gray, 2008a).

In the discussions that follow, the ideal mode of operation (so-called 'steady-state' performance) is discussed. This will not always be the case as unit processes cycle throughout their operational life, for example rapid gravity filters need time to mature after they have been cleaned with a backwash cycle. These layers of operational complexity are beyond the scope of this chapter but, nevertheless, readers should be aware of their existence and appreciate that the descriptions given here are somewhat idealised when compared to the reality of a working drinking water plant.

[*] Recommended citation: Reid, D. 2015. 'Drinking water treatment', in Bartram, J., with Baum, R., Coclanis, P.A., Gute, D.M., Kay, D., McFadyen, S., Pond, K., Robertson, W. and Rouse, M.J. (eds) *Routledge Handbook of Water and Health*. London and New York: Routledge.

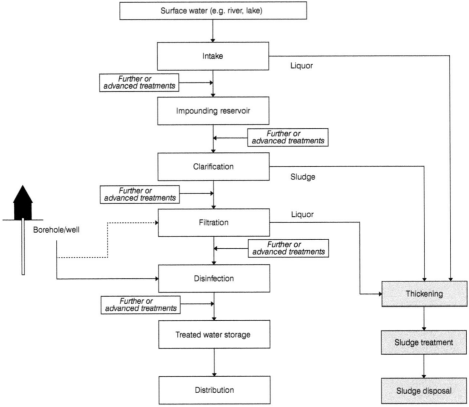

Figure 19.1 Drinking water treatment

Figure 19.1 illustrates the range of unit processes that contribute to drinking water treatment from surface water (such as rivers, lakes or reservoirs) or groundwater sources (boreholes or wells).

More material

http://www.who.int/water_sanitation_health/hygiene/om/linkingchap6.pdf
http://www.who.int/water_sanitation_health/dwq/S01.pdf

Intake

At an intake it is necessary to remove large-scale items, such as branches, stones or leaves as these may damage equipment such as pumps or block pipes, using screens (Hammer, 1986). At some intakes it will be necessary to remove fish before they enter into the treatment system; this is to preserve fish stocks as their progress through the treatment system will result in them being killed. Some instrumentation may also be located at the intake such as flow measurement to allow the operator to record how much water has been abstracted from the river (usually a condition of the abstraction rights granted to the drinking water plant) or pollution monitoring, for example for hydrocarbons.

Further or advanced treatments at the intake

Some chemical additions may be necessary at the intake. For example, some oxidants such as potassium permanganate may be injected at the intake to control algae (which can cause taste and odour problems). In some plants the disinfection process will start with some chlorine (or similar disinfectant chemical) being added. This practice was widespread in certain parts of the world, for example North America and parts of Europe, until the late 1990s when it started to be phased out due to the formation of disinfection by-products at levels that were higher than the regulated levels for these parameters.

More material

http://www.who.int/water_sanitation_health/hygiene/om/linkingchap3.pdf

Impounding reservoir

Some drinking water systems will have an impounding reservoir for the storage of the raw water. Impoundment reservoirs are artificial structures that allow the storage of water prior to the raw water entering the treatment plant proper (Foellmi, 2005). The advantages of impoundment reservoirs are that they provide storage during periods of low flow. The very act of storage provides some treatment capacity, for example ultraviolet (UV) light will act on the upper layers of the reservoir causing some inactivation of microorganisms, and the quiescent waters allow for some removal of solids through sedimentation by gravity. An old adage in drinking water treatment is that 'stored water is safe water'. Some disadvantages are that, under certain circumstances such as eutrophication, the reservoir may become 'contaminated' with blue-green algae (which can release cyanobacterial toxins and cause taste and odour problems in the final (treated) drinking water).

There is an increasing trend towards requiring reservoirs to be covered, for example in the United States. Such requirements are not universally applied across the globe and will vary depending on the jurisdiction.

Further or advanced treatments at the impounding reservoir

Further or advanced treatments undertaken at an impoundment reservoir are usually confined to dealing with eutrophication problems of algal blooms (Gray, 2008b). Such treatments may involve either chemical additions to control nutrients or kill the cyanobacterial cells; or may be physical in nature (aeration or ultrasonic probes) that aim to disrupt the formation of the blooms. There is a growing body of evidence suggesting that the phenomenon of 'quorum sensing' may play a role in the development, maintenance and ultimate loss of blue-green algal blooms.

More material

http://www.who.int/water_sanitation_health/resourcesquality/wqachapter8.pdf
http://www.sgmeet.com/jasm2014/viewabstract.asp?AbstractID=14365
http://www.nottingham.ac.uk/quorum/what.htm

Clarification

In most textbooks the next unit process is described as coagulation, flocculation and sedimentation but for simplicity we have termed these linked processes as 'clarification'. This process is the most commonly used one and is termed 'conventional treatment'. The purpose of clarification is to remove as much of the particulate material as possible through the use of gravity (Gray, 1999a). Larger particles such as sand will settle out as their mass is sufficient to allow gravity to overcome the opposing forces in the water column. Smaller particles will not have the same gravitation forces acting on them and so resist settling. This resistance is overcome by making the smaller particles larger by allowing them to stick together. This is not a simple proposition as most particles, when immersed in water, will assume a net negative charge and when they encounter another particle each will be repelled by electrostatic forces. A way around this is to provide a particle that retains its positive charge in water allowing the negatively charged smaller particles to attach and hence increase the mass of the resulting agglomeration or 'floc'. So coagulation is the addition and rapid mixing of a chemical (or chemicals) that acts as an aid to coagulation; flocculation is the gentle agitation of the water to ensure there is maximum interaction between particles and the coagulant to allow flocs (larger agglomerations of particles and coagulant) to occur; and the final stage, sedimentation, is when the flocs are allowed to settle out of the water column under gravity to provide a 'clarified' or clear water. Flocculants can include a range of aluminium and iron salts (aluminium sulphate, ferric sulphate etc.) but there are also chemicals known as 'polyelectrolytes' which form long strings when dissolved in water with multiple charges along their length (with positive and negative charges areas occurring on the sample molecule, hence 'poly' and 'electrolyte'). Polyeletrolytes can be used in addition to flocculants to enhance the formation of the floc and hence aid with sedimentation.

The water containing the flocs is allowed to enter a basin where the movement of the water is strictly controlled to allow gravity to act on the flocs and pull them out of the water column. In this stage we therefore have two products – the clarified (partially treated) water, which will proceed to the next steps of treatment, and the solids removed from the water column. These solids are a mixture of the particles contained in the raw water and the chemicals added to enhance their removal. The solids are termed 'sludge' and need to be treated before they can be disposed of. The treatment and disposal of solids is considered towards the end of this chapter.

Further or advanced treatments in clarification

Dissolved air flotation

Dissolved air flotation or DAF is a process that is used instead of sedimentation. It uses a stream of rising bubbles to lift the floc particles to the surface of the unit where the separated solids are then scraped off (think of the frothy top of a glass of beer being removed with a knife). DAF plants require compressed air to generate the column of bubbles (like blowing into a glass of water through a straw). The resulting solids are sometimes referred to as 'chocolate mousse' as they can resemble that dessert! DAF can be very effective at removing particles like algal cells or *Cryptosporidium* oocysts or with highly coloured waters.

More material

http://techalive.mtu.edu/meec/module03/WastewaterRegulations.htm
http://www.safewater.org/PDFS/knowthefacts/conventionalwaterfiltration.pdf
http://www.iwawaterwiki.org/xwiki/bin/view/Articles/
CoagulationandFlocculationinWaterandWastewaterTreatment
http://www.who.int/water_sanitation_health/dwq/en/watreatpath2.pdf

Filtration

There are two main types of filtration system – slow sand filtration and rapid gravity filtration. Despite the name applied to these unit processes, the removal of particles in the filtration stage is not by size exclusion but rather through physico-chemical adsorption (Crittenden et al., 2005).

Slow sand filtration is very old technology and works through the development of a complex biofilm (known as a schmutzdecke) on the top of the filter media. This biofilm traps particles, removing them from the water column. There is also some biological activity in the schmutzdecke where bioavailable nutrients are removed from the water column through incorporation into the biomass or removal through respiration processes, but the principle removal mechanisms are the trapping of materials in the polysaccharide biofilm. A slow sand filter uses very little energy and can be located (in temperate regions) outside in open tanks or inside (in more extreme environments). Once the filter performance starts to deteriorate, the filter is regenerated by mechanically removing the schmutzdecke. Periodically, the supporting sand material and the under-drain systems will need to be replaced and on large-scale slow sand processes the washing and grading of the sand material will be done on-site and stored ready for use.

Rapid gravity filtration uses layers of materials to provide the necessary physico-chemical conditions to allow the removal of particles. Most rapid gravity filters will have a dual media system with a layer of anthracite, coal or activated carbon on top of a layer of sand. The water will flow down through the layers to be collected in under-drains at the bottom of the filter. As the volume of material removed by the filter accumulates, the filter's performance will decrease and so it is necessary to periodically clean the filter. This cleaning is known as 'backwashing' and can be triggered by a number of factors, but typical control points are loss of filter head (that is, loss of pressure in the filter); when a certain level of turbidity[1] is reached in the filtered water; or on a time sequence. Backwashing involves reversing the flow of water to wash the collected material out of the filter matrix. In advanced systems the backwash will be assisted with an air scour where air is blown into the backwash water to enhance the agitation and physical removal of the trapped particles – rather like the difference when washing laundry of leaving clothes to soak or agitating them to help removed trapped dirt. Once the backwash cycle is completed the filter bed is allowed to re-establish before it is brought back into service. The lighter filter material, for example anthracite, will settle on top of the heavier sand allowing the dual media configuration to be re-established. The turbidity of the water running through the filter will be monitored, and until the required quality is achieved (know as filter maturation) the filtered water will be 'wasted' and either discharged as part of the waste stream or, more typically, will be returned to an earlier unit process to be retreated in the drinking water plant.

Further or advanced treatments for filtration

If the water entering the rapid gravity filter has not been chlorinated prior to this stage then the filter bed can act as a substrate for the growth of microorganisms as a biofilm that will cover the filter particles. The development of this biological component will turn the filter from a rapid gravity filter to a 'biofilter'. Biofilters work differently from slow sand filters as the microbial community in a biofilter develops to utilize dissolved materials such as nitrate, converting the dissolved substances into biomass and respiratory by-products rather than acting as a stick mat to trap particulate materials. Use of biofilters has become more widespread.

A more advanced method of filtration is provided by one of the different membrane technologies. Membranes come in a variety of types described on the basis of the pore size. Pore size is the spaces in the material which sieve out the particles or substances of concern and so membrane filtration does rely on size exclusion rather than physico-chemical forces to achieve separation of the solid and liquid phases. There are various degrees of size exclusion of particle removal which range from microfiltration (nominal pore sizes in the range 10 microns to 0.03 microns) through ultrafiltration (nominal pore sizes between 0.1 and 0.002 microns) to nanofiltration (nominal pore size 0.001 microns) and reverse osmosis which is capable of removing nearly all inorganic (and organic) contaminants from water. Generally, the water must be further treated to add minerals and to ensure that the water is adequately buffered after reverse osmosis treatment.

More information

http://www.who.int/water_sanitation_health/publications/ssf9241540370.pdf
http://www.cdc.gov/safewater/pdf/sand2011.pdf
http://iaspub.epa.gov/tdb/pages/treatment/treatmentOverview.do?treatmentProcessId=-1306063973
http://www.lifewater.org/resources/rws3/rws3o3.pdf
http://www.who.int/water_sanitation_health/hygiene/emergencies/fs2_14.pdf
http://www.who.int/water_sanitation_health/dwq/en/watreatpath2.pdf?ua=1
http://www.mrwa.com/WaterWorksMnl/Chapter%2019%20Membrane%20Filtration.pdf

Disinfection

The disinfection of drinking water ensures that harmful microorganisms (bacteria, viruses, protozoa and in some cases invertebrates) have been either killed or inactivated so that they can no longer pose a threat to human health when the water is ingested (CWWA, 1993; Lester and Birkett, 1999). Disinfection is one of the last unit processes, as the presence of other organic (and possibly inorganic) materials can react with the disinfectant reducing the efficacy of the disinfection process. The water being disinfected needs to be as pure as possible to allow disinfection to proceed optimally. It does need to be noted, however, that drinking water while disinfected is not sterile – there will still be some microbiological activity. For applications that require sterile water (medical, industrial etc.) alternative arrangements to treat the water need to be made which are beyond the scope of this chapter.

The most prevalent mode of disinfection is to use halogen-containing compounds to react with organic material such as bacterial cell walls. Amongst the halogens, iodine, bromine and chlorine have all been used for drinking water disinfection but by far the most common halogen for this purpose is chlorine.

When chlorine is dissolved in water it forms an equilibrium between the chlorine (Cl_2), hypochlorous acid (HOCl) and hydrochloric acid (HCl).

When chlorine is added to water some of the chlorine reacts with organic material (and inorganic material such as metals) that are present in the water and so it is unavailable for disinfection. This is termed the chlorine demand of the water. Chlorine that remains after the chlorine demand has been accounted for is called the total chlorine.

Total chorine is further divided into (a) combined chlorine which is the amount of chlorine that has reacted with organic nitrogen or ammonia to form chloramines (see below), and (b) free chlorine which is the chlorine available as either hypochlorous acid or the hypochlorite (OCl^-) ion or bleach to react with microorganisms, killing or inactivating them and making the water fit for consumption.

Total chlorine is thus the sum of free chlorine and combined chlorines and the level of total chlorine should always be higher than or equal to the level of free chlorine. Free chlorine is typically measured in drinking water systems using chlorine gas or sodium hypochlorite as the disinfectant; total chlorine is used to determine the amount of disinfection residual present in drinking water systems where chloramination is practised.

Chloramination

Chloramines are derivatives of ammonia by the substitution of one, two or three hydrogen atoms with chlorine atoms forming monochloramine (chloramine NH_2Cl), dichloramine ($NHCl_2$) and nitrogen trichloride (NCl_3) respectively, and their use in drinking water disinfection is termed 'chloramination'. The chemistry of the formation of chloramine is complicated and needs to be carefully controlled to ensure that monochloramine (chloramine), the most effective of the chloramines, is primarily formed and that as little as possible of the di- and tri-forms of chloramines are formed. The formation of dichloramine and nitrogen trichloride can result in taste and odour issues as well as concerns over their long-term health effects. Ammonia (NH_3) is reacted with chlorine (Cl_2). The chloramines speciation depends on the relative amounts of NH_3 and Cl_2 present:[2]

$NH_3 + Cl_2 \rightarrow NH_2Cl + HCl + Cl_2$ (favoured when the ratio of Cl_2:NH_3 is 3–5:1

     $NHCl_2 + HCl + Cl_2$ (favoured when Cl_2:NH_3 is 5–7:1

     $NCl_3 + HCl$ (favoured at higher Cl_2:NH_3 ratios)

Chloramines are a weaker disinfectant than chlorine but they are more stable and so their persistence in a distribution system is longer than chlorine. This stability allows for longer (in terms of time and distance) protection against microorganisms that may be present in the distribution pipes. As chloramines are weaker (less reactive) they tend to form fewer disinfection by-products such as the trihalomethanes (THMs) which are widely regulated. This reduction in the formation of THMs has led to an increasing trend of using chloramines instead of chlorine as a disinfectant, particularly where drinking water systems have very long water ages (large distribution systems) or where the removal of organic precursors that allow the formation of THMs is not good. Chloramines can produce disinfection by-products such as N-nitrosodimethylamine (NDMA). In addition, chloramine can degrade and release ammonia in the distribution system which can then lead to nitrification which, in turn, can cause other water quality problems.

Ozone

Ozone (O_3) is a colourless gas at room temperature and has a pungent odour. Ozone is a powerful oxidant which is used in water treatment for disinfection and also, in certain specific circumstances, for oxidation; for example it can be used for colour removal where the colour is due to the presence of humic acids. Ozone reacts with organic molecules either directly as an aqueous solution of molecular ozone or through the decomposition of the ozone molecule to form hydroxyl free radicals.

Ozone is very unstable and so has to be generated on-site just prior to use. The most common method of ozone generation for drinking water applications is by corona discharge. Corona discharge consists of an oxygen-containing gas being passed between two electrodes separated by a diaelectric and a discharge gap. When a very large voltage is applied an electron flow across the discharge gap occurs with the energy from the electrons ripping apart the oxygen molecules resulting in the formation of ozone.

Once the ozone gas has been generated it is transferred to water where the dissolved ozone reacts with organic matter (including microorganisms) and some inorganic constituents in a process known as 'ozone contactor'. Common methods for dissolving the ozone into water include bubble diffusers (which act like large aquaria 'air stones'), injectors and turbine mixers. Any ozone not dissolved needs to be collected and destroyed (thermal destruction) prior to venting of the air.

Contact time factor

The contact time (CT) is the time between the introduction of the disinfectant and when the water is first consumed (usually taken as the first customer to be supplied from a system after disinfection). A more robust measure is to calculate the CT factor. The CT factor is defined as the product of the residual concentration, C in mg/L, and the contact time T, in minutes, that residual disinfectant is in contact with the water. In practice the CT value used is the value that allows 90 per cent of the water to meet the required contact time (Lin, 2007). CT tables have been developed for a range of disinfectant types operating over a range of concentrations, pH and temperatures, with the rule of thumb being the higher the CT value the greater the proportion of microorganisms that will be killed or inactivated. The CT tables provide an estimate of the log inactivation for specific organisms (such as *Giardia*) or groups of organisms (such as viruses).

Further or advanced treatments for disinfection

The use of UV radiation has become more prevalent in some areas, particularly where concerns over the formation of disinfection by-products have resulted in a reluctance to use halogen-based disinfection practices. The UV damages the genetic code of the target organisms which prevents them from being able to replicate and divide their cells, a condition known as 'inactivated' as the cell is still 'alive' but can't replicate. Only certain wavelengths of UV are effective at causing this damage to the genetic material. In a typical drinking water system the UV is provided by UV-emitting light bulbs or 'tubes' which are mounted in a 'reactor' that allows water to pass through the light-path of the bulb, exposing the water to the UV rays. UV installations are popular with small systems and systems using groundwater (although some preliminary filtration using cartridge filters is usually required for groundwater systems). Increasingly UV installations are to be found on large-scale drinking water systems serving hundreds of millions of consumers. Regardless of the size, the principles remain the same – expose the water with potentially harmful microorganisms for a sufficient period of time that genetic damage will occur, inactivating the microorganisms.

UV treatment is also practised in developing countries where it is known as SODIS – SOlar ultraviolet water DISinfection. SODIS uses PET plastic bottles filled with water to be disinfected, which are left in direct sunlight. This is a free and effective method for decentralised water treatment at the household level.

More information

http://www.who.int/water_sanitation_health/dwq/en/watreatpath3.pdf?ua=1
http://water.epa.gov/drink/contaminants/basicinformation/disinfectants.cfm
http://en.wikipedia.org/wiki/Water_chlorination
http://waterquality.cce.cornell.edu/publications/CCEWQ-05-ChlorinationDrinkingWtr.pdf
http://www.who.int/water_sanitation_health/dwq/wsh0207/en/index4.html
http://www.sodis.ch/methode/index_EN
http://waterquality.cce.cornell.edu/publications/CCEWQ-10-UVWaterTrtforDisinfection.pdf
https://www.health.govt.nz/system/files/documents/publications/uv-disinfection-and-cartridge-filtration_0.pdf
http://www.epa.gov/ogwdw000/mdbp/pdf/alter/chapt_3.pdf

Treated water storage

Once the treated water has been disinfected it is transferred to large storage tanks where the time in storage will contribute to the CT of the drinking water (see above). Another function of the storage is to smooth out peaks in demand. Drinking water typically follows a daily demand curve with a peak in the morning as the population being served wakes up and starts to use water for showering, preparing breakfast and so on, and then again in the late afternoon and into the early evening as the population returns home and starts preparing their evening meals und using household appliances such as dishwashers or washing machines.

For many communities the treated water storage capacity also functions as storage for fire-fighting capability with hydrants being part of the distribution network. This can result in some storage tanks being significantly oversized for the population being served which can, in turn, lead to problems around the deterioration of the drinking water quality due to the age of the treated water, as the ability to replace or 'turnover' the stored water can be very limited.

More information

http://www.wpro.who.int/environmental_health/documents/docs/Household_Water_Treatment_Safe_Storage_PARTICIPANT.pdf

Distribution

The distribution network is the system of pumps, pipes and valves that allow the treated drinking water to flow from the treated water storage to the point of consumption, whether that is a tap (faucet) within a building or a communal stand-pipe or collection centre. The distribution network needs to be maintained to ensure that there are no leaks; that there is no build-up of microorganisms growing on the inside of the pipes (known as biofilm); and that all mechanical devices such as valves and backflow preventers are in proper working order. It is also important that the potential for cross-connections are rigorously controlled through by-laws and inspections, for example preventing the tapping of a wastewater discharge from a washing machine into the drinking water system. In terms of the capital asset of a drinking water system, the distribution system is typically the largest asset but it is usually one of the least maintained parts of a drinking water system, which can lead to deterioration of the water quality after it leaves the treatment plant. This potential for changing drinking water quality means that adequate monitoring of the drinking water quality delivered to consumers is a vital part of the whole system of quality assurance needed to keep consumer confidence in the quality of their drinking water (Gray, 1999b).

More information

http://www.who.int/water_sanitation_health/hygiene/om/linkingchap7.pdf
http://water.epa.gov/lawsregs/rulesregs/sdwa/tcr/distributionsystems.cfm
http://www.who.int/wsportal/dwflow/en/
http://www.epa.gov/safewater/pdfs/crossconnection/crossconnection.pdf

Solids handling

Figure 19.1 indicates the various points at which either liquor or sludge is withdrawn from the main drinking water process flow path. The ability of a drinking water treatment plant to deal with these 'waste' streams is vital – without this capability the various unit processes will quickly cease to be functional, for example if the plant cannot handle the sludge coming from the backwash on the filters then the filters will be out of operation and the plant will have to stop operating.

The solids removed from the various unit process are comprised of the particles contained in the raw water and the various chemicals added to the water to facilitate the removal of the solids. The use of aluminium or iron salts significantly increases the metals content of the solids and limits the routes of disposal. Where there is access to a sewer connection and the receiving wastewater treatment plant can accommodate the loads, some or all of the waste liquor and/or solids may be disposed to sewer. In areas where there is no capacity to dispose of the waste materials to sewer then the goal of the solids handling facility will be to concentrate the solids from the liquid waste stream to allow efficient transfer to a solid waste disposal facility (landfill).

There are various methods available to increase the solids content of the waste stream, ranging from simple sludge-drying beds where the evaporation of the liquid results in the formation of concentrated solids; through various 'sludge press' technologies such as filter presses or belt presses; through to the use of continuous centrifuges. The liquor derived from these mechanical solids concentrating process will need to be treated via some form of wastewater treatment.

Many of the techniques and processes used for solids handling in the drinking water plant are analogous to techniques and processes that are used for the treatment of solids derived from wastewater treatment.

More information

http://www.nesc.wvu.edu/pdf/dw/publications/ontap/2009_tb/water_treatment_DWFSOM49.pdf

http://www.google.ca/url?sa=t&rct=j&q=&esrc=s&frm=1&source=web&cd=4&ved=0CEgQFjAD&url=http%3A%2F%2Fwww.eea.europa.eu%2Fpublications%2FGH-10-97-106-EN-C%2Fdownload&ei=-BLAU-mEM6S3igLl3oFw&usg=AFQjCNHTEfD1eIs00W-ol4d7iAVqhLyaFA

Community (private) drinking water supplies versus municipal (public) drinking water supplies

There is an assumption that the quality of drinking water coming from a tap will be universally the same, particularly within the same country or area. This is not always the case. Where drinking water is supplied from a community or private supply, the quality is likely to be much more variable and hence of lower quality when compared to drinking water supplied by a large utility in an urban area. The problems of community or private supplies are myriad including lack of appropriate funding, lack of technical expertise and lack of management oversight to ensure operations are conducted appropriately. The problems and their solutions take up whole books and so this chapter will not be able to do adequate justice to the topic other than to raise it for the reader's awareness and recommend that they investigate further starting with the excellent book by David Clapham (Clapham, 2004).

Notes

1 Turbidity is a measure of the cloudiness of the water.
2 http://www.chloramine.org/articles_pdf/Chemicals_in_Drinking_Water_Chloramines.pdf

References

Clapham, D. (2004) *Small Water Supplies: A Practical Guide.* Spon Press, London and New York.

Crittenden, J.C., R.R. Trussell, D.W. Hand, K.J. Howe, G. Tchobanoglous. (2005) *Water Treatment: Principles and Design.* Chapter 11. 2nd edition. John Wiley & Sons Inc. New Jersey, USA.

CWWA (Canadian Water and Wastewater Association). (1993) Environmental Health Directorate, Health Protection Branch, Department of National Health and Welfare. Guidelines for Canadian Drinking Water Quality *Water Treatment Principles and Applications: A Manual for the Production of Drinking Water.* Ottawa: Canadian Water and Wastewater Association.

Foellmi, S.N. (2005) Chapter 4. In Baruth, E.E. (ed.) *Water Treatment Plant Design.* 4th edition. New York: McGraw Hill.

Gray, N.F. (1999a) Chapter 10. *Water Technology: An introduction for Environmental Scientists and Engineers.* London: Arnold.

Gray, N.F. (1999b) Chapter 11. *Water Technology: An introduction for Environmental Scientists and Engineers.* London: Arnold.

Gray, N.F. (2008a) Chapter 1. *Drinking Water Quality: Problems and Solutions.* 2nd edition. Cambridge: Cambridge University Press, Cambridge.

Gray, N.F. (2008b) Chapter 11. *Drinking Water Quality: Problems and Solutions.* 2nd edition. Cambridge: Cambridge University Press.

Hammer, M.J. (1986) Chapter 6. *Water and Wastewater Technology.* 2nd edition. Englewood Cliffs, NJ: Prentice-Hall Inc.

Lester, J.N. and J.W. Birkett. (1999) Chapter 20. *Microbiology and Chemistry for Environmental Scientists and Engineers.* 2nd edition. London: E & FN Spon.

Lin, S.D. (2007) Chapter 5. Part 18. *Water and Wastewater Calculations Manual.* 2nd edition. New York: McGraw Hill.

20

WASTEWATER TREATMENT[*]

Laura Sima

JOHNS HOPKINS SCHOOL OF PUBLIC HEALTH, BALTIMORE, MD, USA

Learning objectives

1 Describe wastewater sources.
2 Distinguish and understand the role of pretreatment, primary, secondary and tertiary steps for conventional wastewater treatment.
3 Describe membrane bioreactors, lagoons, wastewater irrigation and EcoSan.

The sanitation gap

The world is urbanizing, cities are growing, food production is intensifying, industries are producing a larger variety and quantity of chemical contaminants, and, as a result, we are producing more and more contaminated and polluted water and wastewater[1] every year (UNEP 2010). Untreated wastewater is a vector for pathogen exposure by recirculating pathogens in the environment. In a study of 27 South and Southeast Asian cities, the incidence of diarrhea was higher in cities with a larger share of liquid and solid human waste in untreated wastewater bodies (Asian Development Bank 2009). Release of untreated wastewater can also increase nutrient load to natural bodies, leading to de-oxygenated dead zones in some places. As of 2010, around 245,500 sq. km. of the world's marine ecosystems are already affected (UNEP 2010).

Wastewater can be collected at the point of degradation either through household pipes or through combined neighborhood drains. In urban areas, water is transported to a central location by segregated wastewater pipes, or along with rainwater in combined sewer systems. Sewer systems are covered in developed countries to prevent pathogen and pollutant exposure but these are most often open in developing cities. Sewer systems are relatively expensive to construct and maintain. Rural wastewater is treated in septic tanks, fishponds, irrigation systems and latrines or released directly into rivers or streams.

---

[*] Recommended citation: Sima, L. 2015. 'Wastewater treatment', in Bartram, J., with Baum, R., Coclanis, P.A., Gute, D.M., Kay, D., McFadyen, S., Pond, K., Robertson, W. and Rouse, M.J. (eds) *Routledge Handbook of Water and Health*. London and New York: Routledge.

In Jakarta, Indonesia, homeowners without access to municipal wastewater systems are required to operate household scale septic tank systems, but surveys show that most septic systems are in disrepair and that companies emptying the tanks often illegally discharge waste into city canals (Asian Development Bank 2009). DEWATS (the decentralized wastewater treatment systems) refer to decentralized, neighborhood simplified sewer systems connecting individual households, or community sanitation centers that are shared, coupled with passive anaerobic treatment (for more information, please see Key recommended reading, The Water Sanitation Program 2013). Though these systems have coupled one form of treatment with distribution, this does not need to be the case, and some DEWATS operate differently.

Wastewater treatment, or the process of enhancing wastewater quality, is critical to protecting ecosystems and public health from physical, microbial, chemical and high nutrient pollution. Wastewater treatment is nearly universal in developed countries, but around 90 percent of all wastewater in developing countries is discharged without treatment to rivers, lakes and oceans (UNEP 2010). Many people do not have access to any form of sanitation, such that, globally, 4.1 billion people lack access to improved sanitation, defined as sanitation including water treatment (Baum et al. 2013), and, depending on the enforcement of environmental standards, much of the wastewater produced by industry remains untreated in developing nations (Johnstone 2003). In densely populated South and Southeast Asia, only 30 percent of cities from one survey have wastewater treatment systems, but treatment remains a low government priority, with only 40 percent of cities reporting having a sanitation plan (Asian Development Bank 2009). This chapter describes domestic and commercial wastewater treatment. We direct the reader elsewhere for more background on processes to remove specific pollutants from specific industries (Wheeler and Pargal 1999; Johnstone 2003; International Finance Corporation 2007; Salmoaa and Watkins 2011).

Conventional municipal treatment

Conventional wastewater treatment includes pretreatment, primary and secondary treatment (see Figure 20.1). Each step is designed to speed up natural filtration and degradation processes, such that streams and other surface water receiving bodies are minimally impacted. Tertiary treatment is used in some cases to improve effluent quality.

Pretreatment involves the physical removal of substances from wastewater. This requires that water is passed through a bar screen to remove large debris and objects. Grit removal and/or fat and grease removal is accomplished via passage through a tank with skimmers as part of pretreatment (US Environmental Protection Agency 1986).

Primary treatment refers to the sedimentation stage, which, in conventional plants, is completed by passage of wastewater through large tanks, often referred to as clarifiers. Once

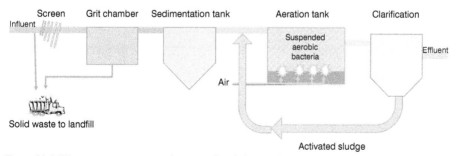

Figure 20.1 Wastewater treatment scheme with solid waste being transported to a landfill

the pretreatment is complete, water still contains suspended particles too small to settle on their own. These suspended particles can be removed through additional steps, such as sedimentation, gravity settling, chemical coagulation, or filtration. After this step, a mass of solids, called a primary sludge, is created, which is removed from plants.

Secondary treatment degrades dissolved organics and reduces biological oxygen demand. In secondary treatment, bacteria break down dissolved organics for their own consumption, as wastewater is passed through an attached or suspended-growth activated sludge chamber. When suspended growth bacteria are used, the wastewater is again separated from the bacteria in large settling basins. Finally, the liquid wastewater effluent can either be released into the environment or enter tertiary treatment.

Tertiary treatment can further improve effluent wastewater quality by additional filtration, nitrogen removal and/or phosphorous removal. In some countries, such as the U.S., tertiary treatment includes disinfection via ultraviolet (UV) light, ozone, chlorination, or other methods such as passage through wetlands (US Environmental Protection Agency 2012).

Membrane bioreactors

Membrane bioreactors (MBRs) combine biological processes with membranes for wastewater treatment, producing high quality water for recycling while using only a very small physical plant (Kimura 1991). Pretreatment mirrors the mechanisms described for conventional systems, but degradation by aerobic bacteria and clarification are combined into one step. Most reactor designs use a membrane to replace the sedimentation step in wastewater treatment (Stephenson et al. 2000). This allows membrane bioreactors to be much more compact than conventional plants. The use of membranes also allows these wastewater treatment systems to be designed to treat small flows with relatively infrequent monitoring.

Slightly over half of all operational MBR systems use membranes "submerged" in activated sludge, while the remainder have separated these processes (Judd 2008). Although the industrial market for MBR may already be considered mature, the municipal market continues to grow, especially as membrane costs continue to decrease (Lesjean and Juisjes 2007). The growth rate for MBR systems has been fastest in Japan and France, with North America adapting these systems at a slower rate (Yang et al. 2006). These systems have been shown to consistently remove viruses at higher rates than conventional systems (Sima et al. 2011; Figure 20.2).

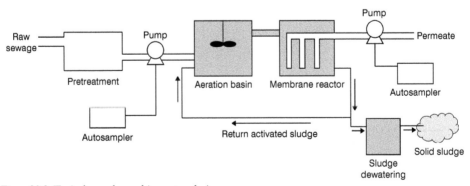

Figure 20.2 Typical membrane bioreactor design

Wastewater lagoons

A wastewater lagoon uses natural elements, such as sunlight, algae, oxygen and bacteria to improve water quality. The influent enters one end and travels through at least one slow-moving basin where pollutants can settle to the bottom and bacteria can degrade organics. Typical lagoons are designed such that water flows through several of these cells. During passage through a lagoon, total suspended solids and biological oxygen demand should be reduced. Lagoons are most effective at hotter temperatures and with plenty of sunlight, but tend to be less effective during winter. Lagoons require large land areas and may be inappropriate for densely populated cities, but they also treat water passively with no electric supply needs for minimal cost, making them ideal in some developing country scenarios.

Lagoons can be classified as aerobic, anaerobic or facultative lagoons (National Small Flows Clearing House 1997; Crites et al. 2014). Aerobic lagoons are typically 3–5 m deep and operate in a consistently aerated environment, ideal for nitrogen and chemical removal. These are able to stabilize organic matter at a faster rate than other lagoons (Crites et al. 2014), should have a retention time of 10–40 days (Yang and Wang 1990) and can be classified by oxygen saturation. Naturally aerobic lagoons are the most shallow with paved or lined bottoms to prevent weed growth on the bottom. These are often somehow mechanically mixed with the top being cleared to prevent algae growth. A partial-mix lagoon, one type of mechanically aerated lagoon, is supplied with oxygen by submerged diffusers, reducing the land required for a lagoon but increasing operating costs for oxygenation. The partial-mix lagoons require shorter retention times of 5–20 days. A completely mixed aerated lagoon is most common in the developed world and can produce excellent water quality. These are operated without access to light to prevent algal growth, and complete mixing to ensure oxygenation. These require energy but less land to operate. Anaerobic lagoons are 6 to 30 feet deep with no aerobic zones. These lagoons are typically used for industrial and distillery waste, and designed for 20–50 days of retention time. These work much like septic tanks and may produce strong odors (National Small Flows Clearing House 1997). Facultative lagoons, also called stabilization ponds, are 4–8 feet deep and consist of both aerobic and anaerobic zones. They are the most common lagoon design for small communities, with retention times of around 25 days. Algae on the top of the lagoons provide oxygen to aerobic bacteria on the top, while anaerobes degrade biodegradable matter on the bottom.

Land application

Controlled application of wastewater over land can also give time for physical, chemical and biological improvements to water quality through passage over soil. This type of treatment and water reuse is much less expensive than conventional water treatment and makes the most sense where land access is sufficient. Land treatment is particularity effective when water moves at slow rates and in areas with rapid infiltration, but ideally studies to understand site hydrology and potential impacts on groundwater, surface water and crops grown would be completed prior to use. The land application of pretreated municipal water for irrigation has been common, especially in drier areas; it is estimated that 7 percent (or 20 million hectares) of global agricultural land is irrigated with wastewater (WHO/FAO 2006). Wastewater should first be pretreated to meet regulations for end use safety based on the use of the land being irrigated and to protect the health of workers. Recent investigations have confirmed that there is a high risk for bacteria and intestinal worm infection for farmers applying untreated wastewater (Drechsel et al. 2010).

Slow rate infiltration, the most common form of land treatment in the U.S. for some time (US Environmental Protection Agency 1986), is the filtration of wastewater applied on a surface through the soil. This slow filtration through the soil facilitates microbial degradation and filtration while giving small particles the chance to attach to soil particles (Crites et al. 2014). Slow rate infiltration, like other types of land treatment, has a dual purpose: the irrigation of surface crops and percolation of treated water into the groundwater. To reduce the incidence of disease transmission, water may be disinfected prior to land application, depending on the crop and irrigation method (WHO/FAO 2006; US Environmental Protection Agency 2012).

Alternate wastewater solutions

Ecological sanitation, or EcoSan for short, is an approach to sanitation that seeks to close the nutrient resource loop with agriculture. Though the approach can be applied in a number of ways, including when resources are recovered from biosolids produced by municipal-scale wastewater treatment, it is most often associated with a urine diversion dry toilet, which does not use water for flushing and separates solid excreta from urine. Both the solid and liquid wastes in a urine diversion dry toilet may be used in agriculture and the solid excreta may be composted. This is also possible in other sanitation types and at a municipal scale.

Key recommended readings (open access)

1 The Water and Sanitation Program. (2013) "Review of Community Managed Decentralized Treatment Systems in Indonesia." Water and Sanitation Program: Technical Paper. Retrieved October 24th, 2014. From WPS: http://www.wsp.org/ sites/wsp.org/files/publications/WSP-Review-DEWATS-Indonesia-Technical-Paper. pdf. This report reviews the operation of numerous decentralized sanitation systems in dense urban settlements in Indonesia, and gives an overview of the challenge for sanitation in dense, unplanned urban settlements that is more universally applicable. The report explains successful non-governmental organization (NGO) involvement in planning these systems, and extrapolates and discusses challenges associated with growing the program into the future.
2 UNEP. (2010) "Sick Water: The Central Role of Wastewater Management in Sustainable Development." Birkeland Trykkeri AS, Norway. This report outlines the major challenges and possibilities for sanitation and wastewater treatment with a special focus on developing countries.
3 Asian Development Bank. (2009) *Asian Sanitation Data Book 2008: Acheiving Sanitation for All*. Manilla, Philippines: ADB. This study summarizes and presents detailed findings from a survey of 27 South and Southeast Asian cities, accounting for trends in wastewater services.

Note

1 Wastewater is an umbrella term for water degraded by human use with many distinct interpretations. For this chapter, we use the term broadly to describe domestic effluent, including black-water (excreta, urine and fecal sludge) and grey-water (kitchen and bathing water), industrial effluent, including from hospitals, commercial institutions and manufacturing plants, and agricultural and aquaculture effluent in either dissolved or suspended form (UNEP 2010).

References

Asian Development Bank. (2009) *Asian Sanitation Data Book 2008: Acheiving Sanitation for All.* Manilla, Philippines: ADB.

Baum, Rachel, Jeanne Luh, and Jamie Bartram. (2013) "Sanitation: A Global Estimate of Sewage Connections without Treatment and Resulting Impact on MDG Progress." *Environmnetal Science and Technology* 47, no. 4: 1994–2000.

Crites, R W, E J Middlebrooks, R K Bastian, and S C Reed. (2014) *Natural Wastewater Treatment Systems.* Boca Raton, FL: IWA Publishing.

Cunha Margues, Rui, and Pedro Simoes. (2010) *Regulation of Water and Wastewater Services: an International Comparison.* London: IWA Publishing.

Dalkmann, Philipp, et al. (2012) "Accumulation of Pharmaceuticals, Enterococcus, and Resistance Genes in Soils Irigated with Wastewater for Zero to 100 Years in Central Mexico." *PLOS One* 7, no. 9: e45397.

Drechsel, P, C A Scott, L Raschid-Sally, M Redwood, and A Bahri (eds). (2010) *Wastewater Irrigation and Health. Assessing and Mitigating Risk in Low-Income Countries.* London: IWMI-IDRC Earthscan.

Du Pisani, P L. (2005) "Direct Reclamation of Potable Water at Windhoek's Goreangab Reclamation Plant." In *Integrated Concepts of Water Recycling*, by S J Khan, M H Muston and A I Schafer. Wollongong: University of Wollongong Printing Services.

Environmental Protection Agency of Ireland. (1997) *Wastewater Treatment Manuals: Primary, Secondary and Tertiary Treatment.* Ardcavan, Wexford, Ireland: EPA.

International Finance Corporation. (2007) *Environmental, Health and Safety Guidelines for Water and Sanitation.* http://www.ifc.org/ehsguidelines (accessed June 11, 2014).

Johnstone, D W. (2003) "Effluent Discharge Standards." In *Handbook of Water and Wastewater Microbiology*, by D Mara and N Horan, Chapter 18. London: Academic Press Ltd.

Judd, Simon. (2008) "The Status of Membrane Bioreactor Technology." *Trends in Biotechnology* 26, no. 2: 109–116.

Kimura, S. (1991) "Japan's Aqua Renaissance '90 Project." *Water Science and Technology* 23, no. 7: 1573–1582.

Leader, Sheldon L and David M Ong. (2011) "Case Study of a Transnational, Non-State Actor Agreement: the EP." In *Global Project Finance, Human Rights and Sustainable Development*, 85–97. Cambridge: Cambridge University Press.

Lesjean, B and E H Juisjes. (2007) "Survey of European MBR Market: Trends and Perspectives." 4th IWA Conference on Membranes for Water and Wastewater Treatment. Harrogate, UK: IWA.

Mara, D. (2001) "Appropriate Wastewater Collection, Treatment and Reuse in Developing Countries." *Proceedings of ICE – Municipal Engineer* 145, no. 4: 299–303.

National Small Flows Clearing House (1997) "Lagoon Systems Can Provide Low-Cost Wastewater Treatment. Pipeline 8, No 2. Morgantown, WV". http://www.nesc.wvu.edu/pdf/WW/publications/pipline/PL_SP97.pdf retrieved Aug 2014.

PUB. (2012) *Every Drop More than Once: Determination Drives Innovation.* http://www.pub.gov.sg/annualreport2013/article04_01.html (accessed June 6, 2014).

Salmoaa, Eila, and Gary Watkins. (2011) "Environmental Performance and Compliance Costs for Industrial Wastewater Treatment – An International Comparison." *Sustainable Development* 19, no. 5: 325–336.

Sima, L, J Schaeffer, JC Le Saux, S Parnadudeau, M Elimelech, and S Le Guyader. (2011) "Calicivirus Removal in a MBR Treatment Plant." *Applied and Environmental Microbiology* 77, no. 15: 5170–5177.

Stephenson, T, S Judd, B Jefferson and K Brindle. (2000) *Membrane Bioreactors for Wastewater Treatment.* London: IWA Publishing.

UNEP. (2010) "Sick Water: The Central Role of Wastewater Management in Sustainable Development." Oslo: Birkeland Trykkeri AS.

US Environmental Protection Agency. (1986) *Design Manual: Municipal Wastewater Disinfection.* Cincinnati, OH: National Service Center for Environmental Publications.

US Environmental Protection Agency. (2012) *Guidelines for Water Reuse.* Washington, DC: National.

Wheeler, David, and Sheoli Pargal. (1999) "Informal Regulation of Industrial Pollution in Developing Countries: Evidence from Indonesia." *Policy Research Working Papers*, http://dx.doi.org/10.1596/1813-9450-1416.

WHO/FAO. (2006) *Guidelines for the Safe Use of Wastewater Excreta and Grey Water in Agriculture and Aquaculture, Third Edition, Vol 1-4.* Geneva: World Health Organization.

World Health Organization. (2000) *Global Water Supply and Sanitation Assessment.* World Health Organization. http://www.who.int/water_sanitation_health/monitoring/jmp2000.pdf

Yang, P. Y., and M. L. Wang. (1990) *Biotechnology Applications in Wastewater Treatment.* Environment and Sanitation Information Centre Paper No. 29, Bangkok, Thailand: AIT.

Yang .Wenbo, Nazim Cicek. and John Ilg. (2006) "State-of-the-art of Membrane Bioreactors: Worldwide Research and Commercial Applications in North America." *Journal of Membrane Sciences* 270, no. 1–2: 201–211.

21

HOUSEHOLD WATER TREATMENT AND SAFE STORAGE[*]

Maria Elena Figueroa

ASSISTANT SCIENTIST, JOHNS HOPKINS UNIVERSITY BLOOMBERG SCHOOL OF PUBLIC HEALTH, DEPARTMENT OF HEALTH, BEHAVIOR AND SOCIETY, BALTIMORE, MD, USA

D. Lawrence Kincaid

ASSOCIATE SCIENTIST, JOHNS HOPKINS UNIVERSITY BLOOMBERG SCHOOL OF PUBLIC HEALTH, DEPARTMENT OF HEALTH, BEHAVIOR AND SOCIETY, BALTIMORE, MD, USA

Learning objectives

1 Describe how the adoption of new practices and products occurs over time through the concept of innovation and the theory of diffusion of innovations (DoI).
2 Present a model of communication to guide new research for the design and evaluation of household water treatment and safe storage (HWTS[1]) interventions.
3 Illustrate the adoption of HWTS under three different communication strategies.

Using theory for research and design of HWTS interventions

Beyond childhood mortality, the negative impact of exposure of children to fecal bacteria includes long-term consequences on children's physical and mental development, as indicated by undernutrition, poor weight–height ratios, stunting and disability adjusted life years (DALYs) (Guerrant et al., 2002). In addition, recent studies on environmental enteropathy show that chronic exposure to fecal contamination negatively affects the intestinal absorptive and immunologic functions that limit the impact of nutritional interventions and oral vaccines in the developing world (Korpe and Petri, 2012).

---

[*] Recommended citation: Figueroa, M.E. and Kincaid, D.L. 2015. 'Household water treatment and safe storage', in Bartram, J., with Baum, R., Coclanis, P.A., Gute, D.M., Kay, D., McFadyen, S., Pond, K., Robertson, W. and Rouse, M.J. (eds) *Routledge Handbook of Water and Health*. London and New York: Routledge.

While the public health community has, since 1958, identified the fecal–oral pathways of disease transmission and the appropriate interventions to break the fecal–oral cycle (Wagner and Lanoix 1958; Prüss-Üstün and Corvalán, 2006), changing the behaviors necessary for these interventions to work – HWTS, hand washing with soap and proper disposal of feces – has proven to be difficult. For example, the research findings of the first study on household water treatment behavior almost six decades ago (Wellin, 1955) identified challenges that continue to trouble HWTS interventions today. Maximum positive public health impact on water, sanitation and hygiene-related diseases happens when interventions change the behavior of a majority of people to consistently treat drinking water, wash hands with soap and properly dispose of fecal matter. Particularly for HWTS, studies and literature reviews conclude that adherence to HWTS intervention is low and more research is needed to identify strategies to increase uptake and adherence (Lantagne et al., 2006; Kols, 2012; Boisson et al., 2013). Such research proves most effective when guided by behavior change theories and frameworks to help identify social, cultural and structural constraints and facilitators that are the critical determinants of behavior, and to help develop effective communication strategies and interventions to address these determinants. In this chapter we discuss a theory of how innovations diffuse over time and a conceptual framework based on theories that complement the DoI and that can be used for designing successful behavioral interventions to increase and sustain HWTS and related sanitation and hygiene practices.

HWTS as an innovation that still lacks widespread adoption

Research shows that the majority of people who live in diarrhea-prone areas do not treat their drinking water, do not wash their hands with soap at critical times to prevent diarrhea (Freeman et al., 2014) and do not use an appropriate sanitation facility for disposal of feces. Furthermore, they may perceive diarrhea as a normal part of growing up and developing resistance to disease (Nichter, 1988) and thus not a reason to treat drinking water or follow other hygiene practices. Those who follow these practices may be doing so for reasons other than disease prevention (Wellin, 1955; Curtis et al., 2009; Figueroa and Kincaid, 2010; see also Chapter 22). To develop successful HWTS, sanitation and hygiene interventions, one has to understand the concept of innovations and how an innovation spreads over time. This chapter focuses on HWTS but the discussion can also be applied to other behaviors.

What is an innovation?

An innovation is an idea, practice or object that is *perceived* as new by a person or other unit of adoption (Rogers, 2003). The unit of adoption is a person or group that is expected to adopt and use the innovation (such as HWTS). According to the DoI theory, the rate of adoption of an innovation over time is determined by five factors, each with an independent contribution affecting that rate. If all of them are positive and work in the same direction, the innovation (HWTS) will spread quickly; otherwise it will take longer to be adopted, if it is adopted at all.

How does DoI apply to HWTS and how can it help develop successful interventions?

Interventions that promote HWTS introduce a complex innovation with the intention of quick diffusion. For these interventions to be effective, people must make significant

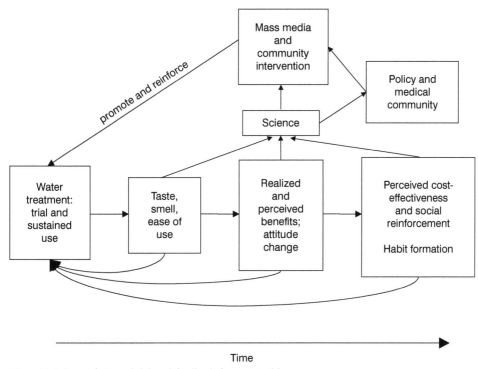

Figure 21.1 Immediate and delayed feedback for sustainable water treatment

changes to their current way of thinking and practices in at least three ways: 1) accept a new idea (drinking water needs to be treated); 2) agree to buy a new product (water treatment technology); and 3) start and then maintain the new practice (use the treatment technology to treat water consistently before drinking). Some technologies may also require the allocation of extra labor, financial resources and valuable time. In addition, HWTS interventions expect all members in the household to accept the new idea, the product and its practice. The five components of the DOI theory provide insights on how to improve formative research[2] to inform the design of the HWTS intervention and its monitoring and evaluation so that the requirements for a successful diffusion are more likely to be met. In the following, we address each of these components.

Attributes of the innovation

The more a new idea, practice or product *is perceived* as superior in the five DoI attributes listed below, the more likely it will be adopted and spread quickly. But as Figure 21.1 indicates, potential users have to perceive *for themselves* that the innovation is superior. It is not enough to just tell them that research has proven it is good for their health. The role of science is to establish what is known about water treatment technology, users' experiences with it and users' preferences. Sharing this information with policy makers, the medical community and those conducting interventions using mass media and community approaches will make these more effective.

Results of the consumers' initial trials with new technology will have immediate feedback (curved arrows) in terms of taste, smell and ease of use; intermediate feedback in terms of benefits actually realized and changed attitudes; and long-term feedback in terms of cost-effectiveness and social reinforcement. Positive experiences and feedback at each of these moments will reinforce the trial behavior and lead to sustained use over time (habit formation) (Wood and Neal, 2007). The order of these feedback loops suggests that the consumer may not make it to later stages if immediate taste and ease of use (the experience) lead to a premature decision to stop using it. If the intervention (top box) is science-based and if it includes policy support, HWTS promotion and consumer experience can become successful.

How does HWTS perform in terms of the five classic DoI attributes?

The following examples are from Figueroa and Kincaid (2010).

1 *Relative advantage:* Will people *perceive* the new HWTS idea, practice and product as better than the current ones?

    In some contexts the idea of HWTS creates concern that if one starts treating water, and then shifts to untreated water, it will actually be worse for people's health. Likewise, the practice of HWTS may imply changes in household tasks or priorities, and increases in expenditures that may not be easy to bear. And people may not realize a benefit themselves if the practice is incorrect and inconsistent (e.g., low chlorine dose, malfunctioning filter, contamination due to lack of proper hand washing).

2 *Compatibility:* Will HWTS be *perceived* as compatible with the cultural context?

    The idea of adding something (such as chlorine) to the water or changing the "natural" essence of the water may be at odds with some local cultures. Also, containers that are of materials other than what people traditionally use (such as plastic versus clay) may be seen as incompatible.

3 *Complexity:* How complicated is HWTS to perform?

    Some HWT technologies may not be easy to use in some contexts (for example, water purification packets[3]). The actual task of treating water may also become problematic if no household member is available to perform it regularly.

4 *Trialability:* Can people try HWTS themselves without any big commitment?

    Availability of products at minimum or no cost is one way to allow consumers to try a new idea or product. Likewise, product demonstrations allow potential users to try it.

5 *Observability:* Can people *see what happens to others* when they try the new idea, practice or product?

    Observability is often difficult to accomplish with HWTS because unless others visit inside the household, many people will not see the new practice. This slows down the process of social modeling[4] and influence, thus decelerating the rate of becoming the social norm. One way to overcome this obstacle is to place signs prominently in front of homes that adopt HWTS as is often done when home security systems or solar panels are installed on a home.

Just for comparison, how does cell phone use perform in each of the five attributes above? We can talk with anyone we want at any time without the need for a landline (high relative advantage); we like to talk to friends and family (compatibility); we can find cell phones that are simple to use (complexity is low); we can try a phone without having a long-term plan (trialability has become easier); we often see others talking on the street or carrying the phone around in a belt holster or purse as a sign of prestige (high observability). Of course cell phone use receives high scores on all attributes, which helps explain why the number of mobile phone subscriptions by the end of 2013 was higher than 6 billion (http://www.itu.int/en/ITU-D/Statistics/Pages/stat/default.aspx) in a world of 7 billion people.

Locus of decision

Who the adopter of the innovation is will influence how rapidly an innovation spreads. Innovations that require group or community decision-making take longer to spread than those that require individual decision-making. Typically, the adopter of HWTS is not one person but a whole household. In addition to adults, the household often includes children who will benefit the most from drinking safe, potable water as well as older relatives who may have different ideas and beliefs about household tasks than younger members.

Household dynamics and the relative power of household members will also play a role in whether a HWTS intervention is adopted or not. In many contexts women do not have enough decision-making power and/or do not have access to economic resources. In these instances, interventions that address only women will have very limited success. Understanding household dynamics will help develop strategies focused on the household members who can support or hinder the decision about whether HWTS is adopted in the household. Likewise, research that involves these various household members will help identify ideational orientations and concerns that each may have regarding the new idea, the practice and the product.

The model of communication for water treatment, sanitation and hygiene behaviors presented in Figure 21.2 can be used to guide new research to design and evaluate HWTS interventions (Figueroa and Kincaid, 2010). Reading from right to left, column four in the model lists the health outcomes that are expected from improved HWTS and other sanitation and hygiene behaviors listed in column three. Column two identifies a set of theory-based intermediate factors that predict behavior. These factors are the *most important* to identify through in-depth qualitative research[5] with various household and community members such as men and women, older parents, parents-in-law, community leaders, members of health committees and health providers, among others. Through focus group discussions, in-depth interviews and other qualitative methods, research can explore individual/ideational, household and community level intermediate factors (also known as predictors of behavior) listed in column two to identify motivations and barriers in the study setting that the intervention would need to address to be successful.[6] Column one lists a set of communication/promotion interventions that could be used to fit the results of the research.

A majority of HWTS interventions lack behavioral research related to the factors in column two of Figure 21.2. Most of the time, interventions assume that all that is needed is to increase knowledge, and thus the behavioral component is limited to providing such information, mostly to mothers (who may lack power to exert change), about the negative effects of consuming contaminated water and the importance of treating it. This message, while accurate, has been proven since 1955 to have limited effect. We need to acknowledge that the promise of diarrhea prevention that most interventions use to promote HWTS is either not believable or not realized (as shown in Figure 21.1) after various trials because of the role other factors play, such as lack of latrines and hand washing with soap at critical times, shown in column three of the model. The alternative is to invest in formative research to identify the ideational orientations for HWTS in each local context, to use that information to design the content of messages (not necessarily health-related) and to use multimedia strategies to promote HWTS (see communication channels below).

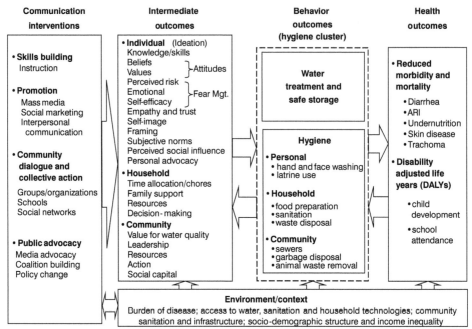

Figure 21.2 A model of communication for water treatment, sanitation and hygiene behaviors

Source: Adapted from Figueroa and Kincaid (2010).

Nature of the social system

Existing social norms, levels of authority and type of relations that are expected of people can influence the success of HWTS interventions. In situations where women are not allowed out of their homes, authoritarian leaders oppose change and/or community factions constrain community interaction, the diffusion of HWTS will face difficult challenges. Contextual assessments can be conducted to provide information about the conditions of the local system, the existence of social networks that can increase the "social contagion effect,"[7] and to help design strategies to overcome challenges and enhance opportunities. For example, the intervention may include talking first to leaders to find out what they think about HWTS and to convince them to support HWTS. Leaders that are centrally located in a social network can become multipliers of the HWTS message and adopters themselves of the new practice – the best possible message if visible and known (Rogers and Kincaid, 1981). The intervention may also include creating opportunities for women to learn about HWTS without the need to leave their home or to join efforts with existing local groups to improve health conditions.

Communication channels

The increasing availability of mass media and mobile communication technologies are making it possible to reach large numbers of people with HWTS messages. The problem today is not access to media but the effectiveness of the message content. Effective messages speak to the cognitive and emotional components of the HWTS behavior (column 2 in Figure 21.2). Generally, the content of successful messages comes from the findings of the formative research that invoke ideas and concepts people are familiar with. Some non-health factors that have been identified by research elsewhere include being a good mother, pride in creating safe water for the family. and taste or freshness of the water produced.

Interventions need to be strategically designed to take advantage of specific communication channels. Each channel has attributes that when used appropriately and designed to interact can create synergy for increased effectiveness. The basic attributes of *mass media* are: 1) it is dynamic and can be used to *validate* a new idea, practice or product; 2) it can also reach large numbers of people with a standard message; and 3) it can provide multiple voices (community members, health providers, government officials, personalities) and spark debate about HWTS (set the agenda[8]). *Community* level approaches are more effective when they provide opportunities for members and their leaders to express *their* priorities and *their* concerns about HWTS and help them identify their water-related problems *themselves* as well as their plans to address them (Kincaid and Figueroa, 2009). *Interpersonal* communication by external change agents is most effective if it is directed to clarify doubts about HWTS and if it involves health care providers who can credibly inform about the health benefits of HWTS. *Print materials* such as billboards, posters and flyers are most effective when they are used as reminders of access to products or services and benefits (such as healthy and happy children).

Since these channels complement each other, they are highly effective when used jointly as part of an integrated, multimedia campaign. Regardless of the channel, however, the content of the message itself needs to be *credible*, *relevant* and *timely*. This is why formative research to identify the most effective content, and pretesting to confirm the message is understood and provides the expected response, are so important. Since diarrhea as malaria tend to be seasonal diseases, health campaigns should also be timed accordingly.

Change agents

The credibility and empathy of the source of the message will substantially influence how an innovation is received. If the message of the HWTS intervention is that it prevents disease, a credible source will be someone that *the potential user* identifies as knowledgeable about health and disease such as a representative from the health sector. This may explain why HWTS interventions that link to health services provision or that market HWTS products after a cholera epidemic have seen some success (Olembo et al., 2004).

Lack of credibility, on the other hand, may explain why the health promise of HWTS delivered by non-health sources has not worked as intended. In the classic study in Peru, the health message became credible only after an authorized medical doctor spoke of germs and water treatment (Wellin, 1955). In 2006 in Indonesia, the Aman Tirta project to promote safe drinking water through use of a chlorine-based solution initially floundered before receiving the needed endorsement of the Ministry of Health. Only after the project messages showed the approval of the Ministry of Health did the population accept the new treatment technology, and the ministry went on to develop and implement a policy for HWTS in the country that expanded water treatment options beyond boiling, which had been the only acceptable method in Indonesia to make water safe to drink (http://ccp.jhu.edu/household-water-treatment-and-safe-storage-hwts-from-household-policy-to-national-policy/).

In summary, when an innovation such as HWTS scores high on its attributes; when the intervention takes into account adopters' situations, preferences and concerns and addresses them through promotional messages that resonate with adopters; and when these are widely disseminated through multiple channels by credible change agents, the innovation will have a much faster rate of adoption than will interventions lacking these features.

Making HWTS a social norm

The DoI theory proposes that the rate of adoption of an innovation will be self-sustaining when enough individuals have adopted the innovation (i.e., it reaches a critical mass, a tipping point) and thus have higher chances to become a new norm. This implies that even when an intervention cannot start at scale, research and program design can be done with full-scale implementation in mind so that a critical mass can be reached. Research and situational assessments to inform the HWTS intervention should cover the five components proposed by the DoI theory with emphasis on the following issues:

1 Assessment of appropriate technology for the local context (e.g., specific water contaminants) and consumer preferences.
2 Formative research on the motivations and barriers regarding HWTS and other hygiene behaviors (factors in column 2 of model in Figure 21.2).
3 Assessment of the local context to identify local organizations and former or ongoing HWTS and other hygiene efforts (learning from experiences of others and creating collaborations).
4 Media consumption patterns and other culturally appropriate communication channels (e.g., local markets and sporting events).

As behavioral research is interdisciplinary, the above research and the design of HWTS interventions would benefit from the involvement and collaboration of experts in the various fields of water quality and technology development, behavioral sciences, social marketing and communication sciences.

Product promotion versus practice promotion

Given the low practice of HWTS in most contexts, an intervention will need to address two objectives: to increase HWTS as a behavior–practice; and to inform about HWTS technologies available to potential consumers – product promotion. So far, most interventions have focused on product promotion to induce behavior. While technology does have a role in inducing behavior as discussed above, the focus of the interventions should not be at the expense of addressing people's needs and preferences, and understanding how it would fit into their existing lifestyle. Product promotion can supplement practice but it cannot replace it. This strategy of focusing on both practice and product promotion was tested in Pakistan in 2004 (Figueroa and Hulme, 2008).

Adoption scenarios

Public health benefit occurs when a substantial portion of poor households adopt HWTS and other hygiene-related behaviors. A societal change of this magnitude requires: (1) the wide-scale, collaborative effort of multiple agencies (public and private sectors, non-governmental organizations (NGOs), local communities and schools); (2) a series of well-researched and strategically designed promotional campaigns, each one improved by what is learned from the previous ones (Kincaid et al., 2013); (3) an increase in accessibility to suitable and affordable treatment and hygiene products; (4) an increase in public advocacy and awareness of water treatment and hygiene practices as community norms; (5) the tangible realization by adopters of the promised benefits of these behaviors; and (6) the sharing of those experiences with others (personal advocacy).

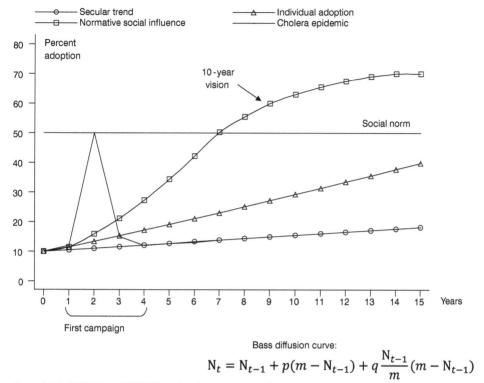

Bass diffusion curve:

$$N_t = N_{t-1} + p(m - N_{t-1}) + q\frac{N_{t-1}}{m}(m - N_{t-1})$$

Figure 21.3 Diffusion of HTWS under three strategies for change

Figure 21.3 shows a hypothetical diffusion of household water treatment under three different strategies for change, assuming a low initial 10 percent level of adoption for purposes of illustration. The normal secular trend assumes a low level of sustained promotion and negligible social influence, leading to a gradual increase of less than 1 percentage point per year and reaching only 18 percent of HWTS adoption after 15 years.[9] Occasionally, this trend may be interrupted by a natural catastrophe or epidemic, creating a temporary spike in water treatment that falls back to the pre-existing secular trend once the threat disappears.

With an effective mass media intervention, the rate of adoption would be expected to double to 2 points per year, with HWTS reaching approximately 38 percent within 15 years – not yet a social norm. If social change processes also operate, then the rate of change can accelerate substantially. If the intervention promotes *community adoption* as well as individual or household adoption, emphasizing that HWTS is "increasing and becoming the social norm," then we would expect the resulting "contagion" or "normative social influence" process to accelerate the rate of diffusion. This normative social influence can also be *reinforced* by featuring successful community projects in the mass media. By the third year with this strategy, adoption would double to 21 percent and, if continued, exceed 50 percent after eight years, becoming a social norm for the intended population. This increase in the rate of change is indicated by the inflection point in the curve where social influence "kicks in," popularly called the "tipping point." Diminishing returns would be expected to set in by the seventh year, gradually slowing this rate of diffusion.

How can social contagion be used to accelerate change?

Fortunately for HWTS interventions, at least a minority in most developing countries does treat its water on a regular basis. These *trendsetters*, if given the chance, could help lead a movement to make water treatment a social norm.[10] This can happen through bounded normative influence – the tendency of social norms to influence behavior within relatively bounded local subgroups of a social system where the new practice is already the norm, if not consensus, rather than in the system as a whole (Kincaid, 2004, 2009). New members are recruited and over time the new practice grows until it becomes the norm for the system. An example of this type of bounded group strategy was applied in Zimbabwe (Waterkeyn and Cairncross, 2005). The process can accelerate when these subgroups are influential in the community's social network and HWTS interventions can identify trendsetters and help create and support such subgroups.

The next generation of WASH interventions

There is no magic bullet to changing health behavior, whether for HWTS, sanitation, hand washing with soap, bed net use, HIV prevention, increased exercise or decreasing tobacco use. A combination of elements such as those discussed here need to be present to make change possible and sustainable. Commercial companies rely on regular and carefully designed consumer research that provides valuable information on consumers' needs and preferences that are then used to craft messages and products that fit the findings. HWTS, sanitation and hygiene (WASH) interventions have valuable theory-based tools that can be used to design formative research for a more comprehensive understanding of people's preferences and lifestyles but also of whether water, sanitation and hygiene practices cluster (Figueroa and Kincaid, 2010). The next generation of WASH interventions will benefit from these tools to enhance their success and contribute to reducing the public health burden of water- and hygiene-related diseases.

Key recommended readings (open access)

1   Figueroa, M.E., & Kincaid, D.L. (2010). *Social, cultural and behavioral correlates of household water treatment and storage.* Center Publication HCI 2010-1: Health Communication Insights, Baltimore, MD: Johns Hopkins Bloomberg School of Public Health, Center for Communication Programs. http://ccp.jhu.edu/wp-content/uploads/Household-Water-Treatment-and-Storage-2010.pdf. This publication comprises a comprehensive literature review of water treatment interventions and uses a behavior change model to identify behavior-related variables addressed by the HWTS interventions. It makes recommendations for HWTS research and practice.

2   Kols, Adrienne. (2012). Sparking demand for household water treatment products: Lessons from commercial projects in four countries. Project Brief. Seattle, WA: PATH. http://www.path.org/publications/files/TS_swp_demand_br.pdf. This project brief summarizes PATH's demand generation experiences in encouraging HWTS in Cambodia, Kenya, India and Vietnam.

3   Waterkeyn, J., & Cairncross, S. (2005). Creating demand for sanitation and hygiene through community health clubs: A cost-effective intervention in two districts in Zimbabwe. *Social Science & Medicine, 61*(9), 1958–1970. http://www.africaahead.org/wp-content/uploads/2011/01/2005_SSM.pdf.   doi:10.1016/j.socscimed.2005.04.012.

This study describes the success of a hygiene and sanitation intervention that changed various key hygiene practices through the influence of Community Health Clubs in Zimbabwe and positively influenced hygiene norms.

Notes

1 Household water treatment refers to treatment that happens at the point of water collection or use, rather than at a large, centralized location; it improves water quality and reduces diarrheal disease in developing countries. Safe water storage consists of storing treated water in a container that protects it from recontamination (http://www.cdc.gov/safewater/pdf/sws-overview-factsheet508c.pdf).
2 Formative research takes place before a program is designed with the purpose of identifying the elements of behavior, generally known as motivations and barriers, that the intervention needs to address to increase the chances that the intervention will be successful. See Rossi et al. (2004) for a comprehensive discussion of evaluation research.
3 Previously called PuR, the P&G water purification packet is a powder that treats 10 liters of water and removes pathogenic microorganisms and suspended matter, making previously contaminated water safe to drink (www.csdw.org/csdw/pur-packet-technology.shtml). See Figueroa & Hulme (2008) for examples of using this purification method in Haiti.
4 Social learning theory (Bandura, 1986) posits that people learn by observing the behaviors of others (social modeling) and the outcomes/consequences of those behaviors.
5 Qualitative research focuses on gaining in-depth understanding of human behavior and the reasons for such behaviors. It uses data collection methods and approaches that allow investigating the why and how of the behaviors of interest. See Denzin and Lincoln (2005) for a thorough discussion of qualitative research as a field of inquiry.
6 An example of formative research (Rimbatmaja et al., 2007) that explored ideational and other intermediate factors for the design of a water, sanitation and hygiene (WASH) intervention in Indonesia can be found here: http://ccp.jhu.edu/wp-content/uploads/Indonesia-ESP-formative-research-health-and-hygiene.pdf
7 Social contagion refers to the process of adoption that takes place through interpersonal and peer networks (Rogers, 2003; Blanchet, 2013).
8 The agenda-setting function of the mass media was first proposed in 1972 by McCombs and Shaw. See also McCombs (2005).
9 The Bass diffusion curve was used to calculate the rate of diffusion under each condition (Bass, 1969; Mahajan et al., 1990), where m is the market size, p is the coefficient of individual innovation, and q is the coefficient of social contagion. For the three strategies, the following coefficients were used: p = .005, .015, .02, and q = .01, .05, 1.8, respectively for the secular trend, individual adoption and normative social influence (up to year 7).
10 We prefer the positive term "trendsetters" over "positive deviants" because it implies leadership for others rather than negative social sanctions from others.

References

Bandura, A. (1986). *Social foundations of thought and action: A social cognitive theory*. Engelwood Cliffs, NJ: Prentice-Hall.

Bass, F. (1969). A new product growth model for consumer durables. *Management Science, 15*(5), 215–227. doi:10.1287/mnsc.15.5.215

Blanchet, K. (2013). How to facilitate social contagion? *International Journal of Health Policy and Management, 1*, 189–192. doi: 10.15171/ijhpm.2013.35

Boisson, S., Stevenson, M., Shapiro, L., Kumar, V., Singh, L.P., Ward, D., & Clasen, T. (2013). Effect of household-based drinking water chlorination on diarrhoea among children under five in Orissa, India: A double-blind randomised placebo-controlled trial. *PLOS Medicine, 10*(8). doi:10.1371/journal.pmed.1001497

Curtis, V.A., Danquah, L.O., & Aunger, R.V. (2009). Planned, motivated and habitual hygiene behaviour: An eleven country review. *Health Education Research, 24*(4), 655–673. doi: 10.1093/her/cyp002

Denzin, N.K., & Lincoln, Y.S. (eds). (2005). *The Sage handbook of qualitative research* (3rd ed.). Thousand Oaks, CA: Sage.

Figueroa, M.E., & Hulme, J. (2008) *Experiences in Haiti, Ethiopia, and Pakistan: Lessons for future water treatment programs*. Washington, DC: Safe Drinking Water Alliance. http://www.ehproject.org/ PDF/ehkm/gda2010.pdf

Figueroa, M.E., & Kincaid, D.L. (2010). *Social, cultural and behavioral correlates of household water treatment and storage*. Center Publication HCI 2010-1: Health Communication Insights, Baltimore, MD: Johns Hopkins Bloomberg School of Public Health, Center for Communication Programs. http://ccp.jhu.edu/wp-content/uploads/Household-Water-Treatment-and-Storage-2010.pdf

Freeman, M.C., Stocks, M.E., Cumming, O., Jeandron, A., Higgins, J.P.T., Wolf, J., Prüss-Ustün, A., Bonjour, S., Hunter, P.R., Fewtrell, L., & Curtis, V. (2014). Hygiene and health: Systematic review of handwashing practices worldwide and update of health effects. *Tropical Medicine and International Health*, *19*(8), 906–916. doi: 10.1111/tmi.12339

Guerrant, R.L., Kosek, M., Lima, A.A.M., Lorntz, B., & Guyatt, H.L. (2002). Updating the DALYs for diarrhoeal disease. *TRENDS in Parasitology*, *18*(5), 191–193. doi:10.1016/S1471-4922(02)02253-5

Kincaid, D.L. (2004). From innovation to social norm: Bounded normative influence. *Journal of Health Communication*, 9: 37–57. doi: 10.1080/10810730490271511

Kincaid, D.L. (2009). Convergence theory. In S.W. Littlejohn & K.A. Foss (eds), *Encyclopedia of communication theory* (pp. 188–191). Thousand Oaks, CA: Sage.

Kincaid, D.L., & Figueroa, M.E. (2009). Communication for participatory development: Dialogue, collective action, and change. In L. Frey & K. Cissna (eds), *Handbook of applied communication* (pp. 506–531). Mahwah, NJ: Lawrence Erlbaum Associates.

Kincaid, D.L., Delate, R., Figueroa, M.E., & Storey, D. (2013). Closing the gaps in practice and in theory: Evaluation of the scrutinize HIV campaign in South Africa. In R. Rice & C. Atkin (eds), *Public communication campaigns, 4th edition.* (pp. 305–319). Thousand Oaks, CA: Sage.

Kols, A. (2012). Sparking demand for household water treatment products: Lessons from commercial projects in four countries. Project Brief. Seattle, WA: PATH. http://www.path.org/publications/ files/TS_swp_demand_br.pdf

Korpe, P.S., & Petri, W.A. (2012). Environmental enteropathy: Critical implications of a poorly understood condition. *Trends in Molecular Medicine*, *18*(6), 328–336. doi:10.1016/j. molmed.2012.04.007

Lantagne, D.S., Quick, R., & Mintz, E.D. (2006). Household water treatment and safe storage options in developing countries: A review of current implementation practices. In M. Parker, A. Williams & C. Youngblood (eds), *Water stories: Expanding opportunities in small-scale water and sanitation projects* (pp. 17–38). Washington, DC: Woodrow Wilson International Center for Scholars. http://www. wilsoncenter.org/sites/default/files/WaterStoriesHousehold.pdf

Mahajan, V., Muller, E., & Bass, F.M. (1990). New product diffusion models in marketing: A review and directions for research. *Journal of Marketing*, *54*(1), 1–26. doi: 10.2307/1252170

McCombs, M. (2005). A look at agenda-setting: Past, present and future. *Journalism Studies*, *6*(4), 543–557. doi: 10.1080/14616700500250438

McCombs, M., & Shaw, D. (1972). The Agenda-Setting Function of Mass Media. *Public Opinion Quarterly*, *36*(2), 176–187. http://links.jstor.org/sici?sici=0033-362X%28197222%2936%3A2%3C 176%3ATAFOMM%3E2.0.CO%3B2-5

Nichter, M. (1988). From aralu to ORS: Sinhalese perceptions of digestion, diarrhea, and dehydration. *Social Science & Medicine*, *27*(1), 39–52. doi:10.1016/0277-9536(88)90162-1

Olembo, L., Kaona, F.A.D., Tuba, M., and Burnham, G. (2004). Safe water systems: An evaluation of the Zambia CLORIN program (final report). US Agency for International Development through the Environmental Health Project. Available from http://www.ehproject.org/pdf/others/ zambia%20report%20format.pdf

Prüss-Üstün, A., & Corvalán, C. (2006) *Preventing disease through healthy environments: Towards an estimate of the environmental disease burden*. Geneva: World Health Organization. http://www.who. int/quantifying_ehimpacts/publications/preventingdisease.pdf

Rimbatmaja, R., Figueroa, M.E., Semiarto, A., Pooroe-Utomo, N., Amirudin, A., Lestari, A., & Amini, F (2007). *Health and hygiene behaviors in Indonesia: Results of the formative research.* Jakarta, Indonesia: USAID Environmental Services Project (ESP). http://ccp.jhu.edu/wp-content/ uploads/Indonesia-ESP-formative-research-health-and-hygiene.pdf

Rogers, E.M. (2003). *Diffusion of innovations.* (5th ed.). New York: Free Press.

Rogers, E.M., & Kincaid, D.L. (1981). *Communication networks: Toward a new paradigm for research*. New York: Free Press.

Rossi, P.H., Lipsey, M.W., & Freeman, H.E. (2004). *Evaluation: A systematic approach*. (7th ed.) Thousand Oaks, CA: Sage.

Wagner E.G., & Lanoix, J.N. (1958). Excreta disposal for rural areas and small communities. *World Health Organization Monograph Series, 39*, 1–182. http://apps.who.int/iris/handle/10665/41687

Waterkeyn, J., & Cairncross, S. (2005). Creating demand for sanitation and hygiene through community health clubs: A cost-effective intervention in two districts in Zimbabwe. *Social Science & Medicine, 61*(9), 1958–1970. doi:10.1016/j.socscimed.2005.04.012

Wellin, E. (1955). Water boiling in a Peruvian town. In B. Paul (ed.), *Health, culture and community* (pp. 71–103). New York: Russell Sage Foundation.

Wood, W., & Neal, D.T. (2007). A new look at habits and the habit–goal interface. *Psychological Review, 114*(4), 843–863. doi: 10.1037/0033-295X.114.4.843

22
WATER FOR HYGIENE*

Aidan A. Cronin and Therese Dooley

UNICEF WATER, SANITATION AND HYGIENE (WASH) PROGRAM

Learning objectives

1 Understand the importance of hygiene and sanitation to achieve the full public health impact of water.
2 Understand the evidence base for hygiene and sanitation and their impact on health.
3 Describe how sanitation and hygiene interventions are complementary and behaviour change is central to sustaining both.

Introduction

Context and definitions

WASH (water, sanitation and hygiene) approaches aim to minimize faecal–oral disease transmission. The health aspects of water are dealt with in other chapters of this handbook while this chapter deals with how the other two pillars of WASH, sanitation and hygiene, interact with water. This is specifically in the context of their relationship with water and public health (focusing primarily on faecal–oral transmission but also referring to broader impact where appropriate) and the essential nature of water for good hygiene and sanitation.

Many broad definitions of sanitation encompass a wide range of public interventions including safe disposal of excreta, solid waste management, liquid waste management, proper drainage, indoor air pollution, vector control, etc. However, for the purposes of this chapter, sanitation shall focus on the management of human faeces. This entails 'hardware' aspects (i.e. toilets and their components to ensure the safe separation of faeces from human or water contact) but also in parallel are the critically important 'software' components – often referred to as good hygiene practices – that encompass changing or adapting human behaviour to reduce the risk of faecal–oral disease transmission.

There are many hygiene behaviours which interrupt faecal–oral disease transmission such as the safe disposal of child stools, the management of water during its transportation, storage and use (for example the use of ladles to take drinking water from the storage container), vector control (such as flies, mosquitoes), food hygiene and behaviours linked to food safety

---

* Recommended citation: Cronin, A.A. and Dooley, T. 2015. 'Water for hygiene', in Bartram, J., with Baum, R., Coclanis, P.A., Gute, D.M., Kay, D., McFadyen, S., Pond, K., Robertson, W. and Rouse, M.J. (eds) *Routledge Handbook of Water and Health*. London and New York: Routledge.

(storage, reheating, covering of food, etc.) or general hygiene (personal washing, laundry, cleaning, etc.), but the main hygiene behaviour this chapter will deal with due to its critical public health importance is handwashing with soap (HWWS). Water plays a key role for HWWS and the full public health benefit of improved water access (be it quality, quantity, sustainability) cannot be fully realized without good hand hygiene. It is also central to menstrual hygiene management (MHM); not in terms of the faecal–oral transmission route but because of its important role in reduction of urinal tract infections and its important influence on quality of life for women and girls.

Linking waterborne disease with sanitation and hygiene

Sanitation and hygiene are intrinsically linked to all water 'related' diseases; the mere provision of water without the accompanying hygiene and sanitation behaviours will not have the desired public health impacts and the faecal–oral transmission routes will not be adequately interrupted. In communities where facilities for the safe disposal of faeces are inadequate or where hygiene behaviours are poor, infectious agents can be transmitted through many routes – fingers, faeces, flies, foods, fluids, surfaces, etc. – and contamination with faecal organisms can be spread either directly or indirectly. The famous F-diagram is often used to display this (Wagner and Lanoix, 1958) – see also Chapter 3. Here we summarise the broader WASH disease categories, the linkages and interventions (Table 22.1).

Table 22.1 Categories of disease linked to WASH (columns one and two adopted and expanded from Montgomery and Elimelech, 2007) with the hygiene and sanitation intervention points for each category identified

Category	Disease type	Hygiene / sanitation intervention
Water-borne	Caused by the ingestion of water contaminated by human or animal excreta or urine containing pathogenic bacteria or viruses; includes cholera, typhoid, amoebic and bacillary dysentery and other diarrheal diseases.	Sanitation, handwashing with soap, safe water storage.
Water-based	Caused by parasites found in intermediate organisms living in water; includes dracunculiasis, schistosomiasis and some other helminths.	Sanitation, safe hygiene and water handling practices.
Water-related	Caused by microorganisms with life cycles associated with insects that live or breed in water; includes dengue fever, lymphatic filariasis, malaria, onchocerciasis and yellow fever.	Breeding site control; drainage, waste management.
Excreta-related	Caused by direct or indirect contact with pathogens associated with excreta and/or vectors breeding in excreta; includes trachoma and most waterborne diseases. There is a growing evidence base also on the linkage with increased risk of child stunting.	Sanitation, personal hygiene; handwashing with soap.
Water collection and storage	Caused by contamination that occurs during or after collection, often because of poorly designed, open containers.	Proper hygiene in handling / transport of water.
Toxin-related	Caused by toxic bacteria, such as cyanobacteria, which are linked to eutrophication of surface-water bodies; causes gastrointestinal and hepatic illnesses.	–

Impact of HWWS on diarrhoea

From a meta-analysis of 42 studies reporting handwashing prevalence, Freeman et al. (2014) estimate that only approximately 19 per cent of the world population practices HWWS after contact with excreta (i.e. use of a sanitation facility or contact with children's excreta). This is an extremely low figure for such an important public health action. Several HWWS meta-analyses on the reduction in diarrhoea morbidity in children under five have been undertaken; these have shown strong reduction rates in diarrhoea morbidity, ranging from 37 per cent to 48 per cent (Fewtrell et al., 2005; Waddington et al., 2009; Cairncross et al., 2010; Freeman et al., 2014). Prüss-Ustün et al. (2014) estimate that, correcting for bias, a figure of 23 per cent may be more appropriate but that this still amounts to 297,000 deaths in 2012 from diarrhoeal disease linked to inadequate hand hygiene and is actually slightly higher than the estimate for sanitation (280,000 deaths). A review of several studies shows that handwashing in institutions such as primary schools and day-care centres reduces the incidence of diarrhoea by an average of 30 per cent (Ejemot et al., 2009).

The broader impact and cost-effectiveness of HWWS

In addition, HWWS can reduce the incidence of acute respiratory infections (ARIs), according to systematic review estimates, by 16–21 per cent (Rabie and Curtis, 2006; Aiello et al., 2008) and so HWWS is also a core intervention in the control of pneumonia, the biggest cause of mortality in children under five. A systematic review of water and sanitation and maternal mortality found an association of increased risk with inadequate water access and higher rates of maternal mortality (Benova et al., 2014) while Rhee et al. (2008) found in one setting that HWWS by birth attendants and mothers significantly increased newborn survival rates by up to 44 per cent (Rhee et al., 2008). Recent analyses also suggest the supportive role of sanitation and HWWS on reducing the risk of stunting though noting the limitations of self-reporting in handwashing studies (Rah et al., in press). Though Haller et al. (2007) urge that caution should be taken with interpreting cost-effectiveness figures as they are not applicable across different settings, it is still clear, however, that hygiene promotion is ranked among the most cost-effective public health interventions (e.g. Curtis et al., 2011), though more evidence may exist on the cost–benefit of sanitation and water (e.g. Haller et al., 2007; Cronin et al., 2014).

The role of water in sanitation and hygiene

Water linking with sanitation and hygiene and the impact on diarrhoea

All the key components of WASH are inextricably linked – piped water into a home that is then contaminated by dirty hands doesn't fulfil the public health objective. Likewise sanitation and hygiene require water in most settings to wash oneself and maintain a clean latrine and healthy household environment. Advanced technology that reduces and reuses water alongside tariff incentives has managed to reduce the volume of water required for sanitation in many high- and middle-income countries but water is still central to good sanitation and hygiene, while sanitation and hygiene are essential if the full public health benefits of improved water are to be realized.

It is clear that when all three pillars of WASH complement each other and where the hardware and software are planned and implemented in sync then significant public health gains can be made. There have been challenges in terms of quantifying the impact of WASH, in particular sanitation, though the devastating impact of poor WASH is clear (Schmidt, 2014). Wolf et al. (2014) found large potential reductions in diarrhoeal disease risk through improvements to both water and sanitation in low- and middle-income settings while stressing the need for proper containment and treatment of excreta to avoid diarrhoeal disease being passed downstream from the point of collection.

Studies that assessed the impact of water quantity improvements on diarrhoea morbidity found median disease reductions ranging from 20 to 25 per cent (Esrey et al., 1991; Fewtrell et al., 2005; Waddington et al., 2009), most likely driven by better hygiene with more available water. Distance to the water source determines ease of water availability and positively impacts on hygiene (Cairncross and Feachem, 1983; see also Chapter 18). In Burkina Faso, a tapstand in the yard resulted in a threefold increase in hygienic behaviours in comparison with those using wells outside the compound and a twofold increase in hygienic behaviours over mothers who used either public standpipes or wells within the yard (Curtis et al., 1995). Pickering and Davis (2012) analysed demographic and health survey data from 26 countries and determined that a 15-minute decrease in the one-way walk time to water source was associated with a 41 per cent average relative reduction in diarrhoea prevalence, improved anthropometric indicators of child nutritional status, and a 11 per cent relative reduction in under-five child mortality. It is easy to relate to this quote made in a focus group discussion on WASH from a mother in Eastern Indonesia. 'I will not let my kids wash their hands unless we have enough water as getting water is difficult and our priority is for drinking and cooking' (UNICEF, 2014b).

Water linking with sanitation and hygiene and the broader health impact

Improved personal hygiene, with an adequate quantity of water, could conceivably contribute to reductions in ringworm, louse-borne illnesses and potentially scabies, although the size of the protective effect remains to be determined (DFID, 2013). Soil-transmitted helminths are strongly impacted by WASH interventions, especially sanitation, though there is still need for high quality research (e.g. Strunz et al., 2014). In specific local settings there may be other aspects of behaviour that deserve attention, for example measures to prevent schistosomiasis or guinea worm disease (Esrey et al., 1991) where water is part of the transmission cycle – here good hygiene may focus more on the proper management of water resources. Knowledge gaps still remain around the most effective times for HWWS (e.g. Luby et al., 2011) and also the relative effectiveness of soap as opposed to other commonly used substances like ash and mud (e.g. Bloomfield and Nath, 2009) as well as methodological challenges around blinding and self-reporting (e.g. Schmidt, 2014).

Water and sanitation are also required for MHM (see also Chapter 52 of this volume) which impacts on the health, developmental and societal development of girls and women. Water is needed for bathing and cleaning during menstruation, but also a toilet or private bathing area is needed for the changing of pads and cloths which then needs an appropriate and effective solid waste management system to dispose of them (House et al., 2012).

Changing behaviours

WASH behaviours

The reasons why individuals or communities change their sanitation and hygiene practices are still not fully or universally understood. There is some evidence that people reduce hygiene- and sanitation-related health risks for reasons that are not always health driven (e.g. Curtis et al., 2011; Guiteras et al., 2014). Biran et al. (2014) suggest emotional drivers, notably nurture (the desire for a happy, thriving child), disgust (the desire to avoid contamination), affiliation (the desire to fit in with peers), status (or aspiration) and habit, may play a more important role than rational health beliefs. Even where health has been the core motivating factor (e.g. Waterkeyn and Cairncross, 2005) other drivers such as social norms, aspiration, habit formation etc. have also been present.

A major formative survey in Eastern Indonesia found that both quantitative and qualitative results identified that there is a relatively high knowledge of the environmental and health risks of poor hygiene and sanitation but that this knowledge does not necessarily change practice; key drivers for change were convenience and practicality, health and community environment, in that order (UNICEF, 2014b). Formative research in Kerala state in India suggests that people want to be hygienic for reasons of comfort, to remove smells, to demonstrate love for children and for social acceptability (Scott et al., 2003), while research from Ghana highlights that motives for hygiene behaviour can be classified in desires to nurture, to avoid disgust and the desire to gain social status (Scott et al., 2007); drivers for sanitation may be different and social norms (see 'The role of social norms and habit formation' below) may play a larger role.

Conventional top-down education and supply-driven interventions have struggled to bring about or sustain large-scale behavioural change, even when there has been an increase in knowledge levels (Webb and Sheeran, 2006; Marteau et al., 2012). This has also been found for HWWS (e.g. Rabbi and Dey, 2013) and has driven social marketing campaigns to help drive HWWS behaviour change at scale (Curtis et al., 2011). However, concerns have been raised about a dilution of impact with increasing scale, a divergence between efficacy and effectiveness (Curtis et al., 2011). Rates of HWWS around the world are low (Freeman et al., 2014). Observed rates of HWWS at critical moments – that is, before handling food and after using the toilet – range from 0 per cent to 34 per cent (Scott et al., 2003). It is clear even in developed countries with well-funded public health systems, where there is widespread availability of soap and handwashing locations, increasing HWWS for behaviour change at scale remains a challenge (e.g. Judah et al., 2009; Guiteras et al., 2014).

Water is the most important starting point for hygiene and sanitation but sustainable interventions need to go further into the theory of behaviour change and social marketing, recognizing that availability of water alongside knowledge may not be enough. There is growing recognition that the underlying elements to sustainable behaviour change may be multi-faceted and need to be better understood for success at scale. (For further details on such drivers, models of behaviour change and the associated issues for sanitation and HWWS see also Coombes and Devine, 2010; de Hooge et al., 2011, UNICEF, 2013, Aunger and Curtis, 2014; Biran et al., 2014; Sigler et al., 2014.)

The role of social norms and habit formation

WASH sector thinking, and in particular approaches to achieve open defecation free communities, has, in the past few years, become strongly influenced by social norm theory (UNICEF, 2014a). The term 'norm' can refer to a variety of behaviours and accompanying expectations. Social norms are social constructs that can describe behaviour; they entail obligations and are supported by normative expectations. Not only do we expect others to conform to a social norm, we are also aware that we are expected to conform. Social norms are not simple behavioural regularities (e.g. wearing shoes), nor legal norms (e.g. inheritance laws, traffic rules), nor moral norms (e.g. do no harm) or conventions (e.g. money, language). Indeed, four (individually) necessary and (jointly) sufficient conditions exist to define a social norm (Bicchieri, 2005):

1 Contingency (i.e. individuals know that the rule exists and when it applies, e.g. I should use a toilet).
2 Individuals prefer to conform to the rule when they (a) expect a sufficiently large part of the population to conform to the rule (empirical expectation – i.e. others use toilets), and (b) believe that a sufficiently large part of the population think they ought to conform and may sanction their behaviour (normative expectations, i.e. others believe I should use a toilet).
3 Individuals may follow a social norm in the presence of the relevant expectations, but may not follow the norm in the absence of such expectations (e.g. if others are not open defecating I won't; if they don't follow this rule then I won't; i.e. my compliance to the rule is determined also by the actions of the others in the group). This is known as conditional preference.

These points are important to reflect on as they have a number of implications for scaling up and sustaining changed sanitation and hygiene practices. Given the collective nature of changing social norms, the interventions have to reach the entire group so that behaviour change is adopted across the whole community and so the community themselves must realize that sustained usage of sanitation facilities and good hygiene is their own responsibility and collectively pledge to achieve it. Hence, all must know the new norms and all must know that all others know this new way forward. There may be some who do not adopt the new norm but the expectation among all is that is now the accepted behaviour.

Trust in leaders and messengers as well as collective deliberation on common values by the community are important in the process, that is, does WASH service provision complement these values and become a source of pride to a community while addressing individual needs around privacy, increased social status and improved health and economic needs (Dooley, 2011) and do most people / everyone know that most other people / everyone else knows that everyone aspires to those changes; potential values to build on have been mapped out (Figure 22.1).

Based on the identified values, there is a need to harmonize the drivers of change associated with legal, moral and social norms (Table 22.2).

Table 22.2 provides a structural outline of how various strategies may be brought together for a WASH programme. While the moral norm may exist it may be reinforced by religious leaders and trusted personalities in the communities, while the authorities may also support it by banning practices like open defecation with a viable alternative in place. For social norms the starting point may be one that necessitates community dialogue around the values

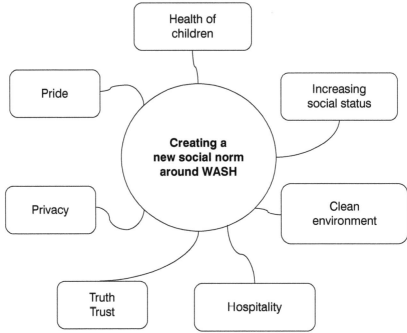

Figure 22.1 Mapping of potential values to build a new social norm on WASH

Source: Cronin (2011).

Table 22.2 WASH from the perspectives of various types of norms

Norm component	Legal norm	Moral norm	Social norm
Intervention	Introduce accountability for WASH (open defecation free, operation and maintenance) at community and household level	Community may host guests and offer a safe latrine and clean water as per hospitality and religious traditions – such values may be shared but may also be at the individual level	Taking responsibility for using / maintaining your toilet and water source, even when shared – perhaps within a larger rights discourse
Intended positive reaction in the community of the intervention	Authorities give WASH services legitimacy and instigate redressal systems with clear outlines of responsibilities and expectations of all actors	Community proud and have good conscience	Enhance community esteem around services and the environment, health, privacy
Intended negative reaction in the community of the intervention	Prosecution and fine – at least to start to drive to legal–social–legal ratchet	Villagers neglect their duty or are at odds with hospitality / religious teaching on sharing	Sanctions may include gossip and other non-stigma-related censure
Emotion in violator	Fear	Guilt	Shame to pride

Source: Adapted from Mackie (2011)

that the community lives to and what the vision is that they aspire to. This may be focused on WASH as an entry point or it may be that deeper human rights discussions are required. Transforming sanitation behaviour, in particular through a rushed initiative, has proven very often to be not sustainable as it takes time to build trust around these new concepts in the community (Dyalchand et al., 2011). Such approaches, based on social norms theory, have progressed sanitation and hygiene uptake globally through community-led total sanitation (CLTS; Kar and Chambers, 2008) or community approaches to total sanitation (CATS; UNICEF, 2010, 2014a). These approaches emphasize the sustainable use of safe, affordable, user-friendly sanitation facilities and promote HWWS via a solid process and trust and rely on community mobilization and behaviour change to improve sanitation and integrate hygiene practices. This process must be inclusive of the entire community if all are to be reached and behaviours sustained. This is also essential if impact is to lead to improved gender and equitable results.

A major recent evaluation of CATS (UNICEF, 2014a) emphasized the importance of social norm theory and that this has helped CATS in successfully contributing to shift the sanitation sector towards demand-driven and not directly subsidized approaches. It found that social norms form part of CATS programme planning but may not be explicitly used across the range of CATS implementing partners. It also recommended further work around developing and embedding social norms indictors into monitoring and evaluation frameworks, while at the same time needing to be in balance with supply mechanisms in order to achieve the desired change to social norms. There is also a need for further research on sustainability of open defecation status and social norm approaches.

Different behaviour theory may be relevant to different behaviours (Sigler et al., 2014), especially at the individual, household or community level. Neal et al. (2014) postulate that HWWS may need a dual approach targeting both the basal ganglia in interaction with neocortex and the neocortex parts of the brain. The former, system 1, is driven by habits and heuristics while the latter, system 2, is driven by attitudes, gaols, social norms and emotions. Traditionally HWWS programmes target system 2 and not enough system 1, while the latter may have much longer lasting effects with stronger habit formation. Neal et al. (2014) present a seven-step framework to help support habit formation to scale up HWWS; these basically emphasize the need for immediate availability of soap and water, piggy-backing on existing habits, keeping the steps of HWWS simple, creatimg cues in the environment and group theory.

Conclusions and looking forward

All the pillars of WASH together are key to improved public health and nutritional outcomes, and the availability of good service provision around water, sanitation and hygiene can lead to each area of intervention becoming mutually reinforcing of the others. Improved quality and quantity of water must be complemented by safe disposal of excreta and good personal and community hygiene. Integrating the pillars of WASH is being increasingly recognized as critical to impact and sustainability; HWWS and sanitation have clear impact on diarrhoea and broader health benefits. However, they require behaviour change, which has proven challenging to sustain and at scale. Work on the drivers of change at individual and community level is helping to better understand the components in interventions to this end.

Though the benefits associated with improved hygiene are significant, there was no specific target for hygiene in the Millennium Development Goals. The discourse has evolved though and proposals for the post-2015 Sustainable Development Targets advocate for an integrated approach to WASH in order to maximize positive impacts on the health, welfare and productivity of populations and in doing so call specifically for the inclusion of hygiene in post-2015 WASH targets (JMP/WSSCC, 2014). This is a very positive step in advocating for further mainstreaming and prioritizing hygiene into the WASH and health sectors as well as the push to eliminate open defecation by 2030, still practised by 1 billion people globally.

Key recommended reading

1 Freeman, M.C., Stocks, M., Cumming, O., Jeandron, A., Higgins, J., Wolf, J., Prüss-Üstün, A., Bonjour, S., Hunter, P.R., Fewtrell, L., Curtis, V. (2014) Hygiene and health: systematic review of handwashing practices worldwide and update of health effects. *Tropical Medicine and International Health*, 19(8): 906–916. This review provides a systematic overview of the current global prevalence and impact of handwashing.
2 Biran, A., Schmidt, W.P., Varadharajan, K.S., Rajaraman, D., Kumar, R., Greenland, K., Gopalan, B., Aunger, R., Curtis, V. (2014) Effect of a behaviour-change intervention on handwashing with soap in India (SuperAmma): a cluster-randomised trial. *Lancet Glob. Health*, 2: e145–154. This study describes the approach and impact of a handwashing intervention programme in India, along with the key drivers for behaviour change.
3 UNICEF (2013) Handwashing Promotion Monitoring and Evaluation Module, prepared by Jelena Vujcic, MPH and Pavani K. Ram, MD of University at Buffalo for UNICEF, 120 pages. This module walks you through planning and implementing monitoring and evaluation (MandE) for your handwashing promotion programme, including indicators for impact.

Disclaimer

The opinions expressed are those of the authors and do not necessarily reflect the views of UNICEF or the United Nations.

References

Aiello, A., Coulborn, R.M., Perez, V., and Larson, E.L. (2008) Effect of hand hygiene on infectious disease risk in the community setting: a meta-analysis. *Am J Public Health*, 98(8): 1372–1381.

Aunger, R., and Curtis, V. (2014) The evo-eco approach to behaviour change. In *Applied Evolutionary Anthropology*, ed. by David Lawson and Mhairi Gibson. New York: Springer.

Benova, L., Cumming, O., and Campbell, O.M.R. (2014) Systematic review and meta-analysis: association between water and sanitation environment and maternal mortality. *Tropical Medicine and International Health*, 19(4): 368–387.

Bicchieri, C. (2005) *The Grammar of Society, The Nature and Dynamics of Social Norms.* Cambridge: Cambridge University Press.

Biran, A., Schmidt, W.P., Varadharajan, K.S., Rajaraman, D., Kumar, R., Greenland, K., Gopalan, B., Aunger, R., and Curtis, V. (2014) Effect of a behaviour-change intervention on handwashing with soap in India (SuperAmma). a cluster-randomised trial. *Lancet Glob Health*, 2. e145–154.

Bloomfield, S.F., and Nath, K.J. (2009) Use of ash and mud for handwashing in low income communities. An IFH expert review. http://www.ifh-homehygiene.org

Cairncross, S., and Feachem, R.G. (1983) *Environmental Health Engineering in the Tropics: An Introductory Text.* Chichester: John Wiley.

Cairncross, S., Hunt, C., Boisson, S., Bostoen, K., Curtis, V., Fung, I.C.H., and Schmidt, W.P. (2010) Water, sanitation and hygiene for the prevention of diarrhoea. *Int J Epidemiol* 39: i193–i205.

Coombes, Y., and Devine, J. (2010) Introducing FOAM: A Framework to analyze Handwashing Behaviors to Design Effective Handwashing Programs. WSP Working Paper. Washington, DC: World Bank.

Cronin, A.A. (2011) Changing Social Norms around the Sustainability of WASH Services in India, University of Pennsylvania Social Norms paper series, 16 pages (unpublished).

Cronin, A.A., Ohikata, M., and Kumar, M. (2014) Social and economic cost benefit analysis of sanitation in Orissa, India. *Journal of Water, Sanitation and Hygiene for Development*, 4(3): 521–531.

Curtis, V., Kanki, B., Mertens, T., Traore, E., Diallo, I., Tall, F., and Cousens, S. (1995) Potties, pits and pipes: Explaining hygiene behaviour in Burkina Faso. *Soc Sci Med*, 41(3): 383–393.

Curtis, V., Schmidt, W., Luby, S., Florez, R., Touré, O., and Biran, A. (2011) Hygiene: new hopes, new horizons. *Lancet Infect Dis.*, 11: 312–321.

de Hooge, I.E., Zeelenberg, M., and Breugelmans, S.M. (2011) A functionalist account of shame-induced behaviour. *Cognition and Emotion*, 2 (5), 939–946.

DFID. (2013) Water, Sanitation and Hygiene, Evidence paper, May, 128 pages. London: UK Department for International Development.

Dooley, T. (2011) Lecture on Creating a New Social Norm – Open Defecation Free Communities, July 12.

Dyalchand, A., Khale, M., and Vasudevan, S. (2011) Institutional arrangements and social norms influencing sanitation behavior in rural India. In Lyla Mehta and Synne Movik (eds) *Shit Matters – The Potential of CLTS*. Rugby: Practical Action.

Ejemot, R.I., Ehiri, J.E., Meremikwu, M.M., and Critchley, J.A. (2009) Hand washing for preventing diarrhoea. *Cochrane Database of Systematic Reviews*, Issue 1. Art. No. CD004265. doi: 10.1002/14651858. CD004265.pub2

Esrey, S.A., Potash, J.B., Roberts, L., and Shiff, C. (1991) Effects of improved water supply and sanitation on ascariasis, diarrhoea, dracunculiasis, hookworm infection, schistosomiasis, and trachoma. *Bull World Health Organ*, 69: 609–621.

Fewtrell, L., Kaufmann, R.B., Kay, D., Enanoria, W., Haller, L., and Colford, J. (2005) Water, sanitation, and hygiene interventions to reduce diarrhoea in less developed countries: a systematic review and meta-analysis. *The Lancet Infectious Dis*eases, 5(1): 42–52.

Freeman, M.C., Stocks, M., Cumming, O., Jeandron, A., Higgins, J., Wolf, J., Prüss-Üstün, A., Bonjour, S., Hunter, P.R., Fewtrell, L., and Curtis, V. (2014) Hygiene and health: systematic review of handwashing practices worldwide and update of health effects. *Tropical Medicine and International Health*, 19(8): 906–916.

Guiteras, R.P., Katun-e-Jannat, K., Levine, D.I., Luby, S.P., Polley, T.H., and Unicomb, L. (2014). Disgust and Shame: Motivating Contributions to Public Goods. Working Paper Series No. WPS-034. Center for Effective Global Action. University of California, Berkeley.

Haller, L., Hutton, G., and Bartram, J. (2007) Estimating the costs and health benefits of water and sanitation improvements at global level. *Journal of Water and Health*, 5(4): 467–480.

House, S., Mahon, T., and Cavill, S. (2012) Menstrual hygiene matters: a resource for improving menstrual hygiene around the world. WaterAid/SHARE toolkit.

JMP/WSSCC. (2014) WASH post-2015: proposed targets and indicators for drinking-water, sanitation and hygiene. Recommendations from International Consultations Comprehensive Recommendations. Updated April 2014. Factsheet by WSSCC based on consultations of Joint Monitoring Program of WHO/UNICEF. http://www.wssinfo.org/fileadmin/user_upload/resources/post-2015-WASH-targets-factsheet-12pp.pdf

Judah, G., Aunger, R., Schmidt, W.P., Michie, S., Granger, S., and Curtis, V. (2009) Experimental pretesting of hand-washing interventions in a natural setting. *Am J Public Health*, 99(suppl 2): S405–S411.

Kar, K., and Chambers, R. (2008) *Handbook on Community-Led Total Sanitation*. London: PLAN.

Luby, S.P., Halder, A.K., Huda, T., Unicomb, L., and Johnston, R.B. (2011) The effect of handwashing at recommended times with water alone and with soap on child diarrhea in rural Bangladesh: an observational study. *PLoS Medicine*, 8(6). doi: 10.1371/journal.pmed.1001052.

Mackie, G. (2011). Lecture on Mockus, Mayor of Bogota, July 12. University of Pennsylvania Social Norms Course.

Marteau, T.M., Hollands, G.J., and Fletcher, P.C. (2012) Changing human behavior to prevent disease: the importance of targeting automatic processes. *Science*, 337(6101): 1492–1495.

Montgomery, M., and Elimelech, M. (2007) Water and sanitation in developing countries: including health in the equation. *Environmental Science and Technology*, 41(1): 17–24.

Neal, D., Vujcic, J., Hernandez, O., and Wood, W. (2014) Handwashing and the Science of Habit. Oral presentation at the 2014 UNC Water and Health Conference, October 13.

Pickering, A.J., and Davis, J. (2012) Freshwater availability and water fetching distance affect child health in Sub-Saharan Africa. *Environ. Sci. Technol.* 46(4): 2391–2397.

Prüss-Üstün, A., Bartram, J., and Clasen, T. (2014) Burden of disease from inadequate water, sanitation and hygiene in low- and middle-income settings: a retrospective analysis of data from 145 countries. *Tropical Medicine and International Health*, 19(8): 894–905.

Rabbi, S.E., and Dey, N.C. (2013) Exploring the gap between hand washing knowledge and practices in Bangladesh: a cross-sectional comparative study. *BMC Public Health*, 13: 89.

Rabie, T., and Curtis, V. (2006) Handwashing and risk of respiratory infections: a quantitative systematic review. *Tropical Medicine and International Health*, 1 (3): 258–267.

Rah, J.H., Cronin, A.A., Badgaiyan, B., Aguayo, V.M., Coates, S.J., and Ahmed, S. (2015) Household sanitation and personal hygiene practices are associated with child stunting in rural India. *BMJ Open.* 5: e005180. doi:10.1136/bmjopen-2014-005180

Rhee, V., Mullany, L.C. Khatry, S.K., Katz, J., Le Clerq, S.C., Darmstadt, G.L., and Tielsch, J.M. (2008) Maternal and birth attendant: hand washing and neonatal mortality in southern Nepal. *Arch Pediatr Adolesc Med*, 162(7): 603–608.

Schmidt, W.P (2014) The elusive effect of water and sanitation on the global burden of disease. *Tropical Medicine and International Health*. doi:10.1111/tmi.12286.

Scott, B., Curtis, V., and Rabie, T. (2003) Protecting children from diarrhoea and acute respiratory infections: the role of handwashing promotion in water and sanitation programmes. *WHO Regional Health Forum,* 7, 42–47.

Scott, B., Curtis, V., Rabic, T., and Garbrah-Aidoo, N. (2007) Health in our hands, but not in our heads: understanding hygiene motivation in Ghana. *Health Policy and Planning*, 22(4): 225–233.

Sigler, R., Mahmoudi, L., and Graham, J.P. (2014) Analysis of behavioral change techniques in community-led total sanitation programs. *Health Promotion International*, doi:10.1093/heapro/dau073

Strunz, E.C., Addiss, D.G., Stocks, M.E., Ogden, S., Utzinger, J., and Freeman, M.C. (2014) Water, sanitation, hygiene, and soil-transmitted helminth infection: a systematic review and meta-analysis. *PLoS Med.* 11(3): e1001620. doi:10.1371/journal.pmed.1001620

UNICEF. (2010) Community Approaches to Total Sanitation, UNICEF Field Notes: UNICEF Policy and Programming in Practice, New York: UNICEF

UNICEF. (2013) Handwashing Promotion Monitoring and Evaluation Module, prepared by Jelena Vujcic, MPH and Pavani K. Ram, MD of University at Buffalo for UNICEF.

UNICEF. (2014a) Evaluation of the WASH Sector Strategy 'Community Approaches to Total Sanitation' (CATS); Summary of key findings, conclusions and recommendations. UNICEF WASH Section, New York (unpublished).

UNICEF. (2014b) KAP Survey on WASH in Eastern Indonesia. UNICEF (unpublished report).

Waddington, H., Snilstveit, B., White, H., and Fewtrell, L. (2009) *Water, sanitation and hygiene interventions to combat childhood diarrhoea in developing countries.* New Delhi: International Initiative for Impact Evaluation.

Wagner, E.G., and Lanoix, J.N. (1958) Excreta Disposal for Rural Areas and Small Communities. World Health Organization monograph series, no. 39. World Health Organization, Geneva.

Waterkeyn, J., and Cairncross, S. (2005) Creating demand for sanitation and hygiene through community health clubs: a cost-effective intervention in two districts in Zimbabwe. *Social Science & Medicine*, 61: 1958–1970.

Webb T.L., and Sheeran, P. (2006) Does changing behavioral intentions engender behavior change? A meta-analysis of the experimental evidence. *Psychol Bull*, 132(2): 249–268.

Wolf, J., Prüss-Üstün, A., Cumming, O., Bartram, J., Bonjour, S., Cairncross, S., Clasen, T., Colford, J.M., Curtis, V., De France, J., Fewtrell, L., Freeman, M.C., Gordon, B., Hunter, P.R., Jeandron, A., Johnston, R.B., Maüsezahl, D., Mathers, C., Neira, M., and Higgins, J. (2014) Assessing the impact of drinking water and sanitation on diarrhoeal disease in low- and middle-income settings: systematic review and meta-regression. *Tropical Medicine and International Health*, 19(8): 928–942.

23

WATER SAFETY PLANS[*]

Katrina Charles

DEPARTMENTAL LECTURER, COURSE DIRECTOR ON MSc IN WATER
SCIENCE, POLICY & MANAGEMENT, UNIVERSITY OF OXFORD, UK

Learning objectives

1 Understand the Water Safety Plan process.
2 Appreciate the benefits of Water Safety Plans.
3 Understand the key challenges in developing, implementing and sustaining effective Water Safety Plans.

Introduction

Water Safety Plans (WSPs) are a risk-based water supply management intervention that aims to "consistently ensure the safety and acceptability of a drinking water supply" (Bartram et al., 2009, p1). WSPs were developed to provide a proactive and comprehensive risk assessment and risk management approach that included all stages of water supply from catchment to consumer.

International guidance for WSPs were first articulated in the third edition of the World Health Organization's (WHO) *Guidelines for Drinking-Water Quality* in 2004 after evaluation in a series of expert review meetings. The WSP methodology was further developed by WHO into a practical manual (Bartram et al., 2009) designed to guide implementers through the process.

This chapter will provide an introduction to the WSP approach (based on the WSP manual by Bartram et al. (2009)), and review the developing evidence base for WSPs, the challenges of implementation and provide some examples of the increasingly broad range, with respect to geography and system type, of application.

Water Safety Plans: an overview

WSPs provide a systematic methodology for the proactive identification and control of hazards and risks in drinking water supply systems, including all stages of the water supply:

---

[*] Recommended citation: Charles, K. 2015. 'Water Safety Plans', in Bartram, J., with Baum, R., Coclanis, P.A., Gute, D.M., Kay, D., McFadyen, S., Pond, K., Robertson, W. and Rouse, M.J. (eds) *Routledge Handbook of Water and Health*. London and New York: Routledge.

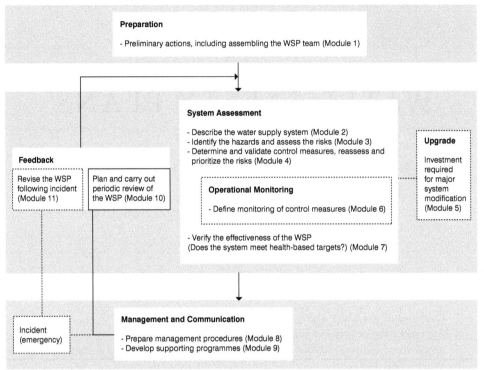

Figure 23.1 How to develop and implement a Water Safety Plan

Source: Bartram et al. (2009).

catchment, treatment, distribution and the consumer. The methodology is designed to be adaptable to all different types of water sources, and scales and types of water supply. While the aim is for safe water, the process reflects the challenges in providing this water, promoting prioritized improvements to make the water consumed of acceptable quality, considering its intended use, and the costs and issues associated with achieving that quality.

The WSP manual published by WHO (Bartram et al., 2009) summarizes the process into eleven steps or modules (Figure 23.1). These are summarized in Table 23.1.

Implicit in the aim of WSPs is that they are not just about quality, but that the drinking water supply should aim to provide sufficient water and continuous service to ensure users are not relying on supplementing water from sources that may potentially be unsafe or storing water in a manner that might contribute to degradation of the quality.

Why Water Safety Plans?

Water supply systems managed by effective WSPs provide an improved water service and better quality water, helping reduce the risk from waterborne diseases. Understanding WSPs, and the benefits of them, will increase the engagement of staff and therefore improve implementation (e.g. Bartram et al., 2009; Summerill et al., 2010). Communicating the benefits will also aid in achieving support from management and stakeholders. This section will summarize firstly the need for WSPs, and secondly the research that has been reported on the benefits of WSP implementation, including health, reductions in noncompliance and financial benefits.

Table 23.1 Summary of the WSP approach presented in Bartram et al. (2009)

Module	Task	Key actions	Typical challenges	Outputs
1	Assemble the team	Engage senior management; secure financial and resource support; identify who will be on the team, their roles, who will lead and the time frame.	Finding skilled personnel. Managing the workload. Engaging external stakeholders. Communicating effectively.	An experienced, multidisciplinary team.
2	Describe the water supply system	Develop a detailed description of the water supply system, considering water quality standards, source/s, water quality changes, land use, storages, treatment, distribution systems, materials, uses, users and training.	Lack of up-to-date information, e.g. on land use/management in catchments; on industry; on procedures; and distribution system maps. Staff time to undertake fieldwork.	A detailed description of the water supply system, including a flow diagram; an understanding of water quality; and identification of users and uses of water.
3	Identify hazards and hazardous events and assess the risks[1]	Identify, for each step of the process flow diagram, what could go wrong at what point in terms of hazards and hazardous events. Assess the risk associated with each hazard in terms of likelihood and severity.	Missing new hazards and hazardous events which arise. Lack of data and knowledge leading to uncertainty in assessment of risks. Avoiding subjective assessments.	A description of what could go wrong and where, with an assessment of risks expressed in an interpretable and comparable manner.
4	Determine and validate control measures, reassess and prioritize risks	Identify existing controls in place to protect drinking water quality, and validate their effectiveness. Use this information to reassess and prioritize the risks.	Identifying responsibilities for undertaking the field work. Ensuring controls are cost-effective and sustainable.	Identification of existing controls. Validation of the effectiveness of the controls. Identification and prioritization of insufficiently controlled risks.
5	Develop, implement and maintain an improvement/upgrade plan	Draw up and implement a plan for short, medium and long term mitigation or controls for each significant risk.	Ensuring the WSP is kept up to date. Accessing resources and expertise. Ensuring improvements don't introduce new risks.	Development, implementation and monitoring of a prioritized improvement/upgrade plan for each significant uncontrolled risk.
6	Define monitoring of the control measures	Define monitoring of controls and responses through understanding of what is monitored (and analyzed), where, how, how often and by whom.	Lack of resources to carry out monitoring and analysis and corrective actions. Inadequate or absent evaluation of data.	A plan for ongoing assessment of the performance of control measures and corrective actions.

continued …

Table 23.1 continued

Module	Task	Key actions	Typical challenges	Outputs
7	Verify effectiveness of the WSP	Compliance monitoring. Internal and external auditing of the WSP. Consumer satisfaction.	Lack of appropriate auditors, laboratories, resources and knowledge of consumer satisfaction.	Confirmation that the WSP is appropriate, and water quality targets are met. Evidence that the WSP is being implemented effectively.
8	Prepare management procedures	Prepare clear management procedures (e.g. standard operating procedures) for normal conditions and incidents.	Keeping the procedures up to date and staff aware of changes.	Management procedures for normal and incident conditions.
9	Develop supporting programmes	Develop and/or revise programmes that indirectly support water safety, e.g. training, research and development.	Resources. Equipment. Not identifying procedures and processes as part of the WSP.	Programmes and activities embed the WSP approach in the water utility's operations.
10	Plan and carry out periodic review of the WSP	Keep WSP up to date and continuously improve the WSP. Convene regular review meetings.	Reconvening the team. Ensuring continued support. Managing transitions. Keeping records.	An up-to-date and effective WSP.
11	Revise the WSP following an incident	Determine the cause of an incident (including emergencies and near misses) and revise the WSP as necessary.	Obtaining open and honest appraisals of causes of incidents. Focusing and acting on positive lessons learned, rather than apportioning blame.	Comprehensive review of incident causes and responses. Incorporation of the lessons learned into WSP.

Prior to WSPs, the WHO (2004) *Guidelines for Drinking-Water Quality* advocated a reactive approach to water supply management of water quality surveillance and control. This approach assessed health risk based primarily on detection of indicator bacteria in drinking water. However, indicator bacteria have limitations as indicators of health risk as they are not as robust as some pathogens, and results are only available after the water has been distributed (also see Chapter 59), and while the level of disease associated with contaminated water has decreased with improved treatment and management, outbreaks still occur from all types of water systems. WSPs represent a shift from this reactive approach to a proactive approach, based on comprehensive risk assessment and risk management that included all stages of water supply from catchment to consumer, which addresses the sources of hazards.

The most comprehensive work on the benefits of WSPs to date was done in Iceland, where risk assessments were legislated in 1995 resulting in over 80 per cent of the population being served by a water utility with a WSP by 2008. It has been reported that, based on 10 years of data, areas where WSPs had been implemented have shown significant improvements in water quality (based on heterotropic plate counts) and a statistically significant reduction of 14 per cent in the clinical cases of diarrhea (Gunnarsdottir et al., 2012b). The efficacy of WSPs have also been reported for a hospital water supply system, where WSP implementation resulted in a reduction in the number of water-related infections in immunosuppressed and other at-risk patients based on three years of data (Dyck et al., 2007). For more information on the health benefits of individual interventions see the other chapters in this part.

The second major benefit of successful WSPs reported is the reduction in noncompliance with regulatory requirements. In Iceland, WSP implementation reduced the mean noncompliance of the five water utilities studied with Icelandic Drinking Water Regulation on heterotropic plate counts and *E. coli* concentrations in source and tap water by 80 per cent (Gunnarsdottir et al., 2012b). Reductions in noncompliance with WSP implementation have also been reported for Águas do Algarve in Portugal, which supplies drinking water for 1.5 million people (Sancho et al., 2012), and in a hospital in Germany (Dyck et al., 2007).

Financial benefits resulting from WSP implementation have also been reported. Development of a WSP requires financial commitment to ensure there are sufficient resources to undertake the work involved; however, there are also financial benefits from implementation of WSPs. Sancho et al. (2012) reported a reduction in overall costs due to two key factors: the risk-based approach allowed them to reduce laboratory analyses, including being allowed to reduce mandatory analyses; and a decrease in the cost of maintenance due to the implementation of operational asset management programmes. Gunnarsdottir et al. (2012a) reported that there was the perception of financial gain from WSP implementation in Iceland from better asset management; however, no actual cost–benefit analyses had been undertaken.

In less developed water supplies the direct financial benefits might not be realized, and ongoing technical and financial assistance may be required to achieve a safe water supply (Chang et al., 2013). However, the benefit to the communities from improved water quality has been estimated to be high; for example for the community in Palau, the expected net benefits of a WSP for the K-A Water Treatment Plant which serves the states of Koror and Airai were estimated at US$ 1.1 million, based on decreases in health care costs, reduced loss of productivity due to disease and reductions in the purchase of bottled water, providing a return of US$5.90 on each US$1.00 invested (Gerber, 2010).

Steps to successful WSP implementation

There are many steps required for successful WSP implementation, some of which have been addressed above. This section will expand on three key aspects to successful implementation: the risk assessment, implementation and achieving long term success.

At the core of the WSP method is the assessment of the risks to drinking water quality and prioritization of control measures to mitigate the risks. This is based on the identification of hazards and hazardous events that might affect water quality, thereby causing an impact on public health, or on aesthetics, continuity, quantity or utility reputation. Assessing the risks allows managers to prioritize the most significant risks for immediate action. The risk assessment process can use quantitative, semi-quantitative or qualitative approaches, depending on the resources, expertise and context. These are outlined with examples in Module 3 of the WSP manual (Bartram et al., 2009), and there are many examples of different approaches for large utility systems to small community supplies available via the WSPortal (see 'Key recommended readings'). What is important, irrespective of the approach, is that the methodology used is clear and consistent, enabling a valid comparison of the risks involved. Two of the key challenges in undertaking the risk assessment process are: firstly, the WSP team must have the expertise and experience to understand and assess risks throughout the water supply, and to understand the issues surrounding implementation of improvements. For a large utility, this will usually involve engagement of external stakeholders who work within the water supply catchment or industry or are users of the water. Secondly, there needs to be a commitment to spend the time (and budget) necessary to gain a thorough understanding of the water supply system through site visits.

The initial implementation of improvements prioritized in the WSP, the move from planning to successfully delivering outcomes, is a major challenge. This stage can include significant infrastructure investments, changes in management procedures and regulation, and training programmes (e.g. Reid et al., 2014), and will require financial resources. Setting interim goals can help maintain interest, and enhance the confidence of policy makers and stakeholders, throughout the process (Kot et al., 2015). WSPs can be adapted to small water utilities and utilities in low resources settings; the latter have successfully linked WSPs to donor programmes, to ensure that WSP priorities are considered by donors. Additionally, implementation is done by staff working with the water supply and therefore it is important that they are included in the development of the WSP to support effective implementation.

For the long term success of a WSP, one of the important challenges is the ongoing revision. A WSP needs to be dynamic to respond to changes that may arise in the water supply and in scientific understanding of the water supply. There is also a need to respond to emergencies and incidents. To ensure that the plan is up to date, the work of the WSP team does not finish with the plan but is crucial in the ongoing processes of revising the plan and the supporting programmes. This continued engagement is particularly important in a crisis as institutional relationships need to be strong to ensure the response is appropriate and timely (Jalba et al., 2010). A culture within the organization of continual improvement can support this process, and can act as an enabling factor for WSP implementation, just as complacency can act as a barrier to implementation (Summerill et al., 2010). This ability to respond to rare or even unforeseen hazards is becoming an increasingly important challenge for water supplies with changes in climate and extreme weather events. While climate change is not listed in the manual as an explicit hazard, it is implicit in the issues of ensuring reliability of supply through flooding and droughts and damage to infrastructure.

Applications of WSPs

Since their conception over a decade ago, WSPs have grown to have a wide ranging application in the provision of drinking water worldwide. This section provides an introduction to the different places and systems, beyond the references already provided, where WSPs have been applied, to help the reader find case studies that will enable them to further understand the challenges and benefits of managing risk through WSPs.

The WSP approach is typically most advanced in countries where it has been incorporated in the regulatory framework for water supplies, including Australia, Iceland, New Zealand, Nigeria, the Philippines, Singapore, Uganda and the United Kingdom. Case studies illustrating challenges from each WSP module from Australia, the UK and Latin America and the Caribbean are provided in the WSP manual (Bartram et al., 2009). The WSPortal provides case studies from these countries and more, particularly from Africa, Asia, and the Latin American and Caribbean region.

While much of the evidence discussed relates to large municipal supplies, WSPs have been successfully applied to small water supplies, including small piped supplies, rainwater harvesting systems, hand-pumps, hand dug wells and protected springs. Guidance on applying WSPs to the small community supplies has been developed by the World Health Organization (WHO, 2012) and Tearfund (Greaves & Simmons, 2011). One example of application of WSPs to community water supplies from Nepal developed innovative tools to support development and implementation including a well hazard mapping exercise, a WSP game to encourage WSP knowledge building, and a household WSP checklist to address common hazards (Barrington et al., 2013). Additional guidance has been developed for country-level implementation of WSPs (WHO and IWA, 2010). The WSP approach is also being adapted by WHO for sanitation safety planning.

Summary

Water Safety Plans offer an adaptive approach to management of drinking water supplies based on proactively identifying and managing risks. The adaptability of the risk-based approach has enabled it to be applied to an increasing number of different types and sizes of water supplies in countries across the world. There is a small body of evidence that this approach has significant positive impacts for public health, and can have financial benefits for more developed drinking water supplies; this body of evidence is expected to grow in the coming years from the work of ongoing WHO programmes that are gathering further evidence on the effectiveness and benefits of WSPs.

Key recommended readings (open access)

1 Bartram, J., L. Corrales, A. Davison, D. Deere, D. Drury, B. Gordon, G. Howard, A. Rinehold and M. Stevens (2009) Water Safety Plan manual: step-by-step risk management for drinking-water suppliers. World Health Organization. Geneva. This manual summarizes the WSP process, providing examples of application in Australia, Latin America and the Caribbean, and the United Kingdom (England and Wales).

2 WSPortal. www.wsportal.org. This website is operated by the World Health Organization and the International Water Association (IWA) and contains a database of tools and case studies provided by water experts.

3 WHO and IWA (2010) Think big, start small, scale up. A road map to support country-level implementation of water safety plans. World Health Organization. Available at

http://www.who.int/water_sanitation_health/dwq/thinkbig_small.pdf. A guide to scaling up WSPs based on the experience of WHO and IWA's work.

Note

1 Hazards are defined as "physical, biological, chemical or radiological agents that can cause harm to public health". Hazardous events are defined as "an event that introduces hazards to, or fails to remove them from, the water supply". An example is heavy rainfall (a hazardous event) which may introduce pathogens (hazards) into the water supply (Bartram et al., 2015).

References

Barrington, D., K. Fuller and A. McMillan (2013). "Water safety planning: adapting the existing approach to community-managed systems in rural Nepal." *Journal of Water, Sanitation and Hygiene for Development* 3(3): 392–401.

Bartram, J., L. Corrales, A. Davison, D. Deere, D. Drury, B. Gordon, G. Howard, A. Rinehold and M. Stevens (2009) *Water Safety Plan Manual: Step-By-Step Risk Management for Drinking-Water Suppliers.* Geneva: World Health Organization.

Chang, Z. K., M. L. Chong and J. Bartram (2013). "Analysis of water safety plan costs from case studies in the Western Pacific Region." *Water Science & Technology: Water Supply* 13(5): 135–136.

Dyck, A., M. Exner and A. Kramer (2007). "Experimental based experiences with the introduction of a water safety plan for a multi-located university clinic and its efficacy according to WHO recommendations." *BMC Public Health* 7(1): 34.

Gerber, F. (2010). An Economic Assessment of Drinking Water Safety Planning Koror-Airai, Palau, SOPAC Technical Report 440. SOPAC, Fiji.

Greaves, F. and C. Simmons (2011). Water Safety Plans for Communities: Guidance for Adoption of Water Safety Plans at Community Level. Teddington: Tearfund.

Gunnarsdottir, M., S. Gardarsson and J. Bartram (2012a). "Icelandic experience with water safety plans." *Water Science & Technology* 65(2) 277–288.

Gunnarsdottir, M. J., S. M. Gardarsson, M. Elliott, G. Sigmundsdottir and J. Bartram (2012b). "Benefits of water safety plans: microbiology, compliance, and public health." *Environmental Science & Technology* 46(14): 7782–7789.

Jalba, D. I., N. J. Cromar, S. J. Pollard, J. W. Charrois, R. Bradshaw and S. E. Hrudey (2010). "Safe drinking water: critical components of effective inter-agency relationships." *Environment International* 36(1): 51–59.

Kot, M., H. Castleden, and G. A. Gagnon (2015). "The human dimension of water safety plans: a critical review of literature and information gaps." *Environmental Reviews* 23(1): 24–29.

Reid, D., K. Abramowski, A. Beier, A. Janzen, D. Lok, H. Mack, H. Radhakrishnan, M. Rahman, R. Schroth and R. Vatcher (2014). "Implementation of Alberta's drinking water safety plans." *Journal of Canada* 49(1): 5–9.

Sancho, R., L. Costa and A. Calvinho (2012). Benefits obtained due to Water Safety Plan implementation – a case study, Águas do Algarve, SA. Water Safety Conference, November 13–15, 2012. Kampala, Uganda, IWA/WHO.

Summerill, C., S. J. Pollard and J. A. Smith (2010). "The role of organizational culture and leadership in water safety plan implementation for improved risk management." *Science of the Total Environment* 408(20): 4319–4327.

WHO (2004). *Guidelines for drinking-water quality, third edition, volume 1: recommendations.* World Health Organization, Geneva. Downloadable from http://www.who.int/water_sanitation_health/dwq/guidelines/en/.

WHO (2012). Water safety planning for small community water supplies. Step-by-step risk management guidance for drinking-water supplies in small communities. Geneva: World Health Organization. Downloadable from http://www.who.int/water_sanitation_health/publications/2012/water_supplies/en/.

WHO and IWA (2010). *Think big, start small, scale up. A road map to support country-level implementation of water safety plans.* Geneva: World Health Organization.

24

SYSTEM MAINTENANCE AND SUSTAINABILITY*

Neil S. Grigg

PROFESSOR, CIVIL AND ENVIRONMENTAL ENGINEERING, COLORADO
STATE UNIVERSITY, FORT COLLINS, COLORADO, USA

Learning objectives

1 To impart appreciation for the importance of effective and sustained maintenance of water and sanitation systems and services.
2 To learn general procedures for maintenance management of any water system or service and be able to apply them to specific systems at different scales and community levels.
3 To understand and explain what is meant by the institutional arrangements required to support effective maintenance management systems.

Introduction

Effective maintenance of water-handling systems is required to sustain the barriers that protect people from disease-causing agents. People are exposed to these agents primarily through paths involving: 1) ingestion of water, 2) coming into contact with contaminated water, and 3) being exposed to water-dependent vectors. These are referred to as water-borne, water-washed, and water-based pathways and lead to gastrointestinal disease, exposure to parasitic threats such as schistosomiasis, and diseases such as malaria with mosquitos as vectors (White et al., 1972).

The barriers to prevention of water-related illnesses rely on constructed and environmental systems that require effective operations to achieve performance goals and effective maintenance to sustain the systems. Constructed systems include water and wastewater facilities of diverse types. The basic strategy for safe drinking water in community systems is to apply multiple barriers of safe water sources, treatment equipment, and distribution systems. Environmental systems are places where water exists in nature such as in wetlands or any place where water is found outside of directly-managed infrastructure systems.

---

* Recommended citation: Grigg, N.S. 2015. 'System maintenance and sustainability', in Bartram, J., with Baum, R., Coclanis, P.A., Gute, D.M., Kay, D., McFadyen, S., Pond, K., Robertson, W. and Rouse, M.J. (eds) *Routledge Handbook of Water and Health*. London and New York: Routledge.

Billions of people globally are exposed to threats from environmental water, especially where safeguards of organized and formal systems are not provided.

The focus on safety of drinking water is explained in Chapter 23, which describes Water Safety Plans. According to Chapter 23, Water Safety Plans were first codified in the World Health Organization's 2004 *Guidelines for Drinking-Water Quality*. As explained there, a Water Safety Plan will require a management team with expertise and experience to assess risks and a commitment to devote resources toward gaining a thorough understanding of systems. These plans are called by different names, such as the "sanitary survey" in the U.S., which is a comprehensive on-site review of a water system source, facilities, equipment, operation, and maintenance (U.S. Environmental Protection Agency, 2014).

The principles of system maintenance seem simple, but they are actually complex because of the multiplicity of situations and the many failures due to institutional problems involving diverse groups of roles and responsibilities. Although the importance of maintenance is obvious, it is often under-appreciated and easy to ignore. For example, water system developers often value new construction and facilities but not maintenance, political leaders leave maintenance off their radar screens, and the public is normally unaware of the importance of taking care of existing systems. While maintenance procedures are straightforward, getting them organized and sustained is not, particularly in challenging situations where services are not well organized or supported, which is common in many rural areas and urbanized areas with informal settlements. The informal settlements are, of course, a major issue in peri-urban and rural areas, which lack the organizational and funding capacity of larger cities.

This chapter identifies the water-handling systems that require maintenance to protect health, and outlines the need for effective institutional arrangements and technical procedures for effective and sustainable maintenance programs. Maintenance procedures are presented in a general framework, and a strategy to improve management systems is outlined in the conclusion. A selection of key references for further study is provided at the end.

Water-handling systems and environmental water

The water-handling systems required to protect people from disease-causing agents are at the interfaces where people are in contact with water for any purpose. The overall picture can be seen from a classification in six basic systems:

1 *Drinking water*: people ingest water and require adequate hydration.
2 *Sanitation*: people contact water while washing or incidentally during waste disposal.
3 *Agricultural water*: people ingest water or waterborne contaminants in food.
4 *Water bodies*: people contact water during recreation or other times.
5 *Standing water*: people are exposed to insects that breed in standing water.
6 *Floods*: people are at risk from flooding.

While reliable statistics are not available, it is apparent that drinking water problems are the greatest health risk to humans, along with gastrointestinal diseases. However, given the large number of water-health exposure paths, all of the systems shown above are significant in requiring maintenance attention.

The nature of the water-handling systems will vary according to the scale and to whether they occur in rural or urban settings, which will be different in more- and less-developed regions. Scale refers to the scope of organized systems, such as a city's water utility, versus individual systems, such as household plumbing or water supply from wells. Examples of

Table 24.1 Examples of maintenance scenarios

Type of water-handling system	Urban examples	Rural examples
Drinking water	Organized water utilities	Self-supplied systems
Sanitation	Organized utilities	Self-supplied facilities
Agricultural water	Lawns and gardens	Irrigated crops
Body contact with water	Showers, swimming pools	Swimming in rivers
Standing water	Drainage near buildings	Wetlands, swamps
Floods	Urban floods	Rural flooding

systems and situations will show a range of scenarios where maintenance is required, such as in these urban and rural cases (Table 24.1).

While these examples illustrate wide ranges of situations, the need for maintenance is similar across classes of organized and individual water-handling systems, including drinking water and sanitation systems from urban to household scales; on-site sanitation; site drainage, stormwater, and flood systems; and irrigation and farm systems.

Many health problems with water contact stem from environmental water, where human contact occurs during recreation or other exposure to streams, lakes, wetlands, or coastal waters. These do not normally involve organized water-handling systems, but maintenance of water extraction and treatment-disposal systems affects the quality of water in these situations. For example, if bathers are exposed to contaminated water at a river or beach, the solution will be to solve the source of contamination rather than to treat the environmental water itself. Another water-handling example would be improvement of flows in streams or lakes to flush or eliminate over-enrichment or other pollution problems.

Types of equipment that require maintenance differ among the categories given above and include components such as shown in Table 24.2 (Grigg, 2005).

How the different types of systems fit into the overall picture of water management is illustrated in Figure 24.1. The environmental flow systems provide the inputs for public supplies, which depend on services for the different uses. Discharges to environmental waters determine the quality of the supplies to the parties downstream. The infrastructures and systems require maintenance by the appropriate parties, depending on the scale and context (Grigg, 2011). The additional features shown on the figure are the informal settlements, which may lack organized services, and the environmental water, which may create conditions to breed mosquitos and otherwise create health threats. Not shown are the many drainage paths that enable the environmental water to reach the stream. The informal settlement near Community B is shown near the stream, which is often the case where people depend on the nearby water for domestic purposes. Also note that Community B is affected by Community A's water-handling actions.

Table 24.2 Examples of infrastructure components requiring maintenance

Drinking water	Dams, wells, diversions, canals, pipes, tunnels, meters, treatment plants, pumps, tanks, valves, plumbing
Wastewater	Sewer pipes, manholes, treatment plants, pumps, sludge disposal, reuse, plumbing and other sanitation facilities
Agricultural water	Dams, wells, diversions, canals, distribution systems, sprinkler systems, drip and subsurface irrigation, drains
Standing water and stormwater	Collectors, culverts, headwalls, pipes, ponds, tanks, outfalls
Floods	Dams, ponds, channels, pipes, pumps, bridges, levees,

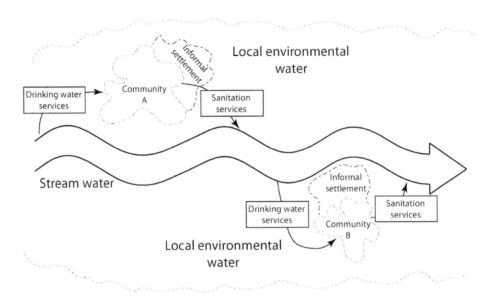

Figure 24.1 Water services and environmental water

Exposure and maintenance scenarios

A correct perspective on the exposure of people and their reliance on different water-handling systems can help governance authorities and donors target their efforts on maintenance. The number of people worldwide who are served by or dependent on different types of water-handling systems varies according to more- or less-developed regions and whether they live in urban or rural areas. Also, the cultural roles of men and women differ across regions, leading to different degrees of exposure, including to children. Therefore, a wide difference in exposure and safety exists between citizens in more-developed regions with modern systems and those in rural areas of less-developed regions with poorly-organized services. This disparity leads to the long-standing international concern about water, sanitation, and hygiene and to current estimates that some 2.5 billion people lack improved sanitation facilities and some 768 million people use unsafe drinking water sources (UNICEF, 2013).

While these estimates identify the order-of-magnitude of the problem, the numbers of people threatened by poor management of constructed systems is much greater (Biswas, 2013). To understand the full scale of the problem of poor maintenance we must also count the populations who have some services but are threatened by breakdowns. To see the scale of this issue, consider the percentages of urban and rural populations in more and less developed regions, as shown in Table 24.3 (United Nations Department of Economic and Social Affairs, 2013a).

Table 24.3 Percentage urban population

	1950	2010	2050
More-developed regions	54.5	77.5	85.9
Less-developed regions	17.6	45	64.1

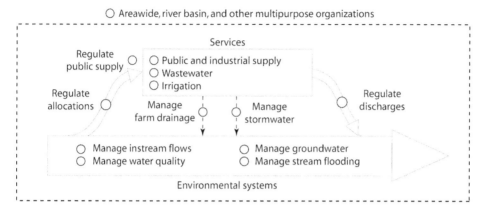

Figure 24.2 Scenarios of water services

With estimates of 82 percent of the world's current total of some 7 billion living in less-developed regions, it is apparent that between 5 and 6 billion people are actually at significant risk from poor system management in urban areas where services are not well-organized, as well as in rural areas where access to effective services is lacking (United Nations Department of Economic and Social Affairs, 2013b). The differences between urban and rural scenarios are important because in rural areas people are often required to maintain their own self-supplied or small community systems, whereas in urban areas the people may be dependent on authorities with uncertain performance records.

While the range of scenarios is wide, they can be classified for purposes of maintenance effectiveness by size and income of community (Figure 24.2). In this simplified classification, four basic situations are shown, but many other situations occur between them. For example, in developed regions, many smaller low-income communities experience similar maintenance challenges to those of communities in less-developed regions. Alternatively, higher-income communities in less-developed regions may resemble situations in wealthier regions, even while they are surrounded by lower-income rural areas.

These four simple scenarios can be used to create maintenance archetypes for the purpose of illustration as: urban areas in more-developed regions; urban areas in less-developed regions; rural areas in more-developed regions; and rural areas in less-developed regions. In the more-developed regions, organized utilities are expected to be adequately financed and to have access to modern methods and equipment. Even in higher-income rural areas small communities and individual families can access needed methods and technologies. Deferred or ineffective maintenance can challenge systems in these areas, but the challenges are much greater in less-developed regions and in low-income areas of more-developed regions. Regardless of the scenario or type of water-handling system, adequate institutional arrangements must be in place and required procedures must be followed. As would be expected, the greater challenge is to assure the necessary institutional arrangements.

Institutional arrangements for maintenance management

Many investigations have shown the importance of institutional arrangements for maintenance and operations since the advent of the Water Supply and Sanitation Decade of the 1980s. These arrangements create incentives, controls, and empowerment for system maintenance. For example, during the 1980s the US Agency for International Development

(1990) organized a Water and Sanitation for Health project with numerous projects in less-developed countries. Lessons learned included that technical assistance is most successful when it helps people learn to do things for themselves and that local institution-building is the key to transferring sustainable skills, such as maintenance of systems. The elements of institutional development have also been outlined, and they point to arrangements such as autonomy in management, effective leadership and management, administrative procedures and policies, sound financial practices, and supportive organizational cultures (Interagency Task Force, 1992).

The challenge to creating effective institutional arrangements such as these is caused by different patterns of culture, social organization, and socio-political status which the billions of people at risk experience. The distinction between maintenance scenarios where people are at greatest risk in small communities of less-developed regions and in the large and rapidly-urbanizing cities of the same regions was explained by Naughton (2013). She wrote that the cost per capita of water systems is higher in rural than in urban areas, and this leads to higher system operating costs but with fewer users. Also, no institution may be available to take responsibility, leading to poor collection of fees and inadequate operation and maintenance. While there are no simple solutions, she outlined three projects with promise in varied contextual arrangements. In Vietnam four provinces piloted rural enterprises to manage water and sanitation programs. Another project in Senegal was managed by a water user association with participation by the government and involving a private operator. The third project was for rural water supply and sanitation in Haiti involving local entrepreneurs selected by the community to operate, maintain, and manage the systems. The lessons offer promise for use of rural enterprises and water user associations and underline the importance of local initiatives.

Maintenance procedures

The principles of maintenance management can be applied to all of the systems described above and at different scales. In its basic form, maintenance management is a set of tasks to sustain the condition and integrity of equipment and systems. These begin with the preventive maintenance system, which involves care-taking and minor repair procedures for equipment such as pumps, valves, hydrants, and treatment components. The other part of the management system is for corrective maintenance, which involves more difficult repairs, rehabilitation, and sometimes replacement of components.

The range of application of these tasks extends from the most basic to highly-advanced and complex systems practiced by large utilities. At the basic level, an example of procedures is available through the Water Supply & Sanitation Collaborative Council (2013), which sponsors an Operation and Maintenance Network (2013) and publishes training aids for maintenance of water and wastewater systems. The tools offered by the network range from practical knowhow to advanced concepts, such as strategic asset management. At the advanced level, networks of utilities cooperate to perform research on best practices and publish the results and tools for methods such as strategic asset management and failure diagnosis for broken pipes (UKWIR, 2007; Water Research Foundation, 2013).

By arraying these parts of an overall management system, how they apply to systems for water-handling as outlined above becomes clear. Table 24.4 illustrates tasks and responsibilities for the five categories of system as they vary among urban and rural cases.

The need for effective organization and control is evident across these many types of systems and responsibilities. In all cases, it is important that water users and managers

Table 24.4 Maintenance tasks and responsibilities for different systems

	Urban	Rural
Drinking water	Water utility maintenance for source, treatment, and distribution; quality control of household plumbing systems	Maintenance of wells, small treatment systems, and distributed or small community water systems
Agricultural water	Use of clean water sources to avoid contact with pathogens	Use of clean water sources to avoid contact with pathogens
Body contact with water	Maintenance of plumbing systems for public showers, swimming pools, and related facilities	Programs identify and remediate sources of pathogens and pesticides/insecticides in rivers and swimming waters
Standing water	Maintenance of formal and informal building drainage systems	Maintenance of formal and informal building drainage systems. Monitoring of wetlands and swamps
Floods	Maintenance of urban stormwater and flood control systems	Programs to assure safety from flood damages in rural areas

understand their responsibilities to perform the required tasks, whether they involve basic facility maintenance at the level of the household or high-level system responsibilities in major utilities.

Progress and gaps in system maintenance and sustainability

While different types of systems are involved, their maintenance and sustainability require durable infrastructure, effective management processes, and institutional support. Failures can stem from shortcomings in any of these, but sustaining systems requires an effective overall approach. Studies from the field show that maintenance depends on local people, but often their participation is not considered adequately by system designers and donors. Also, infrastructure that is appropriate for high-income countries may not work in lower-income situations. Another issue is lack of replacement parts, which plagues maintenance workers in rural areas (Pickford, 1987).

The problem of failed systems is also endemic in urban areas, where there is a paradox that great demands for infrastructure occur at the same time that existing systems fall into disrepair before the end of their design lives. While maintenance is identified as the key to sustainability of assets, responsible stakeholders do not take actions within the context of governance to ensure that it occurs. Case studies from India, Pakistan, and Sri Lanka showed that findings from community involvement can benefit communities to improve maintenance, even when traditional centralized management systems fail. Neither communities nor governments can ensure sustainability unless effective partnerships are implemented. Required are availability of management information and clear delineation of roles and responsibilities. In turn, effective operation and maintenance of services is a key determinant of citizen satisfaction and leads to better governance (Sohail, et al., 2005).

These findings about the effectiveness of maintenance in sustaining water systems in rural and urban areas can be scaled across the spectrum of income and development situations. Whereas organized utilities are non-existent in many rural areas, they may exist in urbanized settings but not work well. Therefore, community involvement may be required to initiate reforms in urban and rural cases. The lesson from research is that a combination of good condition in technical systems and effective governance is required no matter the context involved.

Summary and conclusions

No matter where people live, work, or recreate, effective maintenance of water-handling systems is required to protect them from disease-causing agents. These systems provide direct water-related services to people and help to ensure that environmental water bodies do not become polluted or incubators for disease-causing vectors such as mosquitos. However, while the procedures and importance of maintenance are well-known, they are frequently neglected because of the multiplicity of situations and institutional problems.

The intentional or unintentional neglect of maintenance increases the exposure of people to multiple disease threats. The numbers of people exposed varies according to more- or less-developed regions and urban or rural areas as well as by gender roles and the special case of vulnerable children. The populations at greatest risk are in rural areas and in urbanizing areas of less-developed regions. While published estimates of the populations without access to safe water and sanitation show one aspect of the problem, many more people are threatened by poor management of constructed systems.

The water-handling systems that require maintenance operate where people are in contact with or near water for drinking, sanitation, agriculture, body contact, standing water, and flooding.

Drinking water problems create the greatest health risk through gastrointestinal diseases, but all water-handling systems are significant for creating barriers between human contact with water and disease-causing agents.

Regardless of the exposure or type of water-handling system, required maintenance procedures are important, but the greatest challenge is in the institutional arrangements. Experience with many water supply and sanitation projects in urban and rural situations shows that these are the key to success but they differ among patterns of culture, forms of social organization, and socio-political scenarios which create exposure to risk for billions of people.

When institutional arrangements support effective approaches, the principles of maintenance management can be applied through regular tasks for sustaining the condition and integrity of equipment and systems. These begin with preventive maintenance, which involves care-taking and is critical for Water Safety Plans, along with minor repair procedures and corrective maintenance for more difficult repairs, rehabilitation, and replacement of components.

Experience shows that reducing the incidence of water-related diseases requires improved understanding of the links between barriers, systems, and effective maintenance. Perspective on the importance of maintenance can help governance authorities and donors target their efforts in the many situations around the world where people are at risk.

Key recommended readings (open access)

1 Rural Water Supply Network. 2013. Guidelines and Tools for Rural Water Supplies: RWSN IFAD Rural Water Supply Series: Volume 3. http://www.rural-water-supply.

net/en/resources/details/398. A set of tools for operation and management of rural water systems, showing the range of tasks which must be followed to ensure integrity in system condition.

2 U.S. Environmental Protection Agency. 2013. Water & Wastewater Utility Operation and Maintenance Training for Small, Rural Systems. http://water.epa.gov/type/watersheds/wastewater/smallsystemsoperatortraining.cfm An explanation of the range of training required for operators of small water and sanitation systems.

3 World Health Organization. 2013. Operation and Maintenance. http://www.who.int/water_sanitation_health/hygiene/om/en/. An introduction to the resources and networks related to system operation and management and provided by the World Health Organization.

References

Biswas, Asit. 2013. Future of the World's Water. Water Institute Distinguished Lecture 2013, University of Waterloo. http://www.youtube.com/watch?v=Axs3FzgKj2U&feature=youtube. Accessed June 24, 2013.

Grigg, Neil S. 2005. *Water Manager's Handbook*. Fort Collins, CO: Aquamedia Publications.

Grigg, Neil S. 2011. *Water, Wastewater and Stormwater Infrastructure Management*. Boca Raton, FL: CRC Press.

Interagency Task Force. 1992. Protection of the Quality and Supply of Freshwater Resources, Country Report, USA. International Conference on Water and the Environment, January.

Naughton, Meleesa. 2013. 3 Innovative Ways to Manage Rural Water Supply. The Water Blog. World Bank. http://blogs.worldbank.org/water/3-innovative-ways-manage-rural-water-supply. June 25, 2013. Accessed September 1, 2013.

Operation and Maintenance Network. 2013. Key Focus Areas. http://www.operationandmaintenance.net/.

Pickford, John. 1987. Water, maintenance, and people. *Water International*. 12: 1.

Sohail, M., Cavill, S., Cotton, A. P. (2005) Sustainable operation and maintenance of urban infrastructure – Myth or reality? *Journal of Urban Planning and Development*, 131(1), pp.39–49.

UKWIR. 2007. A Road Map of Strategic R&D Needs to 2030. ukwir.forefront-library.com/reports/07-rg-10-3/91770/90055/91293/91293. Accessed September 7, 2013.

UNICEF. 2013. Water, Sanitation and Hygiene. http://www.unicef.org/wash/. Accessed August 31, 2013.

United Nations Department of Economic and Social Affairs. 2013a. World Urbanization Prospects, the 2011 Revision. http://esa.un.org/unup/CD-ROM/Urban-Rural-Population.htm. Accessed August 30, 2013.

United Nations Department of Economic and Social Affairs. 2013b. World Population Prospects: The 2012 Revision. http://esa.un.org/wpp/. Accessed August 20, 2014.

U.S. Agency for International Development. 1990. Lessons Learned from the WASH Project. USAID, Water and Sanitation for Health Project, Washington, DC.

U.S. Environmental Protection Agency. 2014. Sanitary Survey. http://water.epa.gov/learn/training/dwatraining/sanitarysurvey/. Accessed August 20, 2014.

Water Research Foundation. 2013. About the Foundation. http://www.waterrf.org/the-foundation/Pages/default.aspx. Accessed September 7, 2013.

Water Supply & Sanitation Collaborative Council. 2013. Sanitation, Hygiene and Water for All. http://www.wsscc.org/. Accessed March 31, 2015.

White, G. F., Bradley, D. J. and White, A. U. 1972. *Drawers of Water: Domestic Water Use in East Africa*. Chicago, IL: University of Chicago Press.

25

MANAGING CHEMICAL HAZARDS[*]

Jacqueline MacDonald Gibson

Department of Environmental Sciences & Engineering, Gillings School of Global Public Health, University of North Carolina at Chapel Hill

Nicholas DeFelice

Department of Environmental Health Sciences, Mailman School of Public Health, Columbia University

Learning objectives

1 Learn different approaches used by international health organizations and national agencies to prioritize chemical contaminants in drinking water for risk management.
2 Learn to apply alternative approaches for prioritizing chemical contaminants.
3 Understand the range of options for managing chemical contaminants in drinking water.

Historically, concerns about water quality have been driven by the need to reduce risks of exposure to waterborne pathogens. Indeed, microbiologically unsafe drinking water still poses major risks in many parts of the world. For example, the World Health Organization (WHO) estimates that 9 percent of the 6.3 million deaths of children under age five globally are caused by diarrhea, much of it attributable to waterborne pathogens.[1,2] However, in their efforts to reduce this infectious disease burden, well-intentioned donor organizations have sometimes inadvertently introduced new risks into communities by failing to account for potential chemical contamination. An example is programs initiated by the United National Children's Fund in the 1970s to reduce childhood diarrheal mortality rates by replacing local surface water supplies with groundwater extracted from shallow tube wells. The wells were not tested for arsenic, but many were highly contaminated, leading to what one author called "the largest poisoning of a population in history".[3] Thus, even though eliminating or

---

[*] Recommended citation: MacDonald Gibson, J. and DeFelice, N. 2015. 'Managing chemical hazards', in Bartram, J., with Baum, R., Coclanis, P.A., Gute, D.M., Kay, D., McFadyen, S., Pond, K., Robertson, W. and Rouse, M.J. (eds) *Routledge Handbook of Water and Health*. London and New York: Routledge.

decreasing exposure to pathogens remains the foremost goal of many water supply programs, attention to potential chemical exposures also is critical.

Because water is a universal solvent, a wide variety of chemical contaminants may be present in drinking water supplies. Contaminants may originate from natural sources, such as geologic formations rich in arsenic or radioactive compounds, or from man-made sources, such as agricultural fertilizers and industrial wastes. With more than 84,000 known chemicals in commerce,[4] and with the variety of potential natural contamination sources, identifying contaminants of potential concern in and chemical risk management strategies for a given water supply can be a daunting challenge.

This chapter first discusses approaches for prioritizing potential chemical contaminants for monitoring and control. Then, it discusses alternative approaches for managing identified chemical hazards. The feasibility of pursuing the priority-setting and management options discussed in this chapter will vary with the available resources (for example, resources to support drinking water quality monitoring or the construction of treatment plants). Nonetheless, the fundamental principles of prioritizing and managing chemicals are the same regardless of setting. Throughout the chapter, we provide example approaches suitable for a variety of settings, from those with no pre-existing data on concentrations of chemicals of concern to those with well-established monitoring programs.

Prioritizing chemicals for risk management

Confronted with the many possible chemical drinking water contaminants, as described in the chemical hazards chapter (Chapter 10), figuring out which chemicals may be of concern in any given water supply can be daunting. This section discusses example approaches for prioritizing chemicals for risk management. The first approach requires no pre-existing data, while the subsequent approaches build in the complexity of their data requirements.

Priority-setting example 1: essential priority chemicals

Experience around the world shows that, while the prevalence of contaminants varies geographically, a relatively short list of contaminants tends to dominate chemical risks in many geographically distant parts of the world. On this basis, for settings with little or no pre-existing data on the chemical quality of drinking water and with scarce resources for collecting new data, the WHO recommends establishing an initial list of four "essential priority chemicals":[5, 6]

1 arsenic,
2 fluoride,
3 selenium,
4 nitrate.

These contaminants are singled out because of their common global occurrence and high health risks (see Chapter 10). Along with these, the WHO recommends prioritizing the control of iron and manganese in groundwater sources because of their common occurrence and the tendency of consumers to seek other (potentially unsafe) water supplies when high manganese and iron levels cause turbidity and discoloration of the water. Initial chemical risk management programs therefore should investigate the potential presence of these chemicals through a combination of assessment of local geologic conditions and water quality monitoring. Geologic assessment can be particularly useful in characterizing the risks of arsenic and fluoride exposure.

Because arsenic, fluoride and selenium arise mainly from geogenic sources, collecting the data needed to assess exposure risk requires expertise in geology. Amini et al. developed a regression model that predicts arsenic occurrence globally based on readily available digital data on soil, geology, climate and elevation to predict the probability of arsenic occurring in groundwater at concentrations exceeding the WHO guideline value of 10 μg/liter.[7] Amini et al.'s results could be used as a starting point for deciding on the need to consider the potential for arsenic exposure, where monitoring data and geologic expertise are limited.

Priority-setting example 2: water safety plan risk scoring matrix

In situations where expertise in potential local activities that may lead to chemical pollution risks is available but where monitoring data are scarce, water managers could expand the priority list beyond the four priority chemicals by developing a risk priority matrix. The WHO document *Guidelines for Drinking-water Quality* provides an example matrix form (Table 25.1). The guidelines recommend the matrix be developed by an expert team appointed to gather available data on potential contamination sources in the catchment area of the water supply. Potential sources can be identified through local knowledge and field surveys. The expert team then can identify the potential chemical contaminants that may be associated with each source, in order to prioritize chemicals and target sources for action. Table 25.2 shows part of a risk matrix from a water safety plan in Manila; in this case, chemical risks other than manganese were not considered, but the matrix was used to prioritize potential contamination sources for action. In the Manila example, the likelihood and severity of each potential risk were ranked from 1 to 5, with 5 indicating the highest likelihood or severity, and then the numbers were multiplied to yield a risk score for each potential source.

A major limitation of the qualitative risk matrix approach is that it provides no information on the relative magnitudes of the risks posed by chemicals or sources of those chemicals in

Table 25.1 Example qualitative risk scoring matrix

Likelihood	Severity of consequences				
	Insignificant (no impact or not detectable)	*Minor (potentially harmful to small population)*	*Moderate (potentially harmful to large population)*	*Major (potentially lethal to small population)*	*Catastrophic (potentially lethal to large population)*
Almost certain (once/day)	5	10	15	20	25
Likely (once/week)	4	8	12	16	20
Moderately likely (once/month)	3	6	9	12	15
Unlikely (once/year)	2	4	6	8	10
Rare (once/5 years)	1	2	3	4	1

Note: The matrix is adapted from WHO *Guidelines for Drinking-water Quality*.[25] The original WHO matrix assigns variable shading rather than numerical scores to matrix elements, but in practical applications water safety plans often use numerical scores such as those shown above.

Table 25.2 Example application of qualitative risk scoring matrix from a water safety plan

Hazardous event/cause of contamination	Hazard	Likelihood	Severity	Risk
La Nina	High turbidity	1	5	5
Land slide	High turbidity	1	5	5
Chemical (manganese)	Manganese	1	5	5
Forest fire	Color, taste, odor	2	5	10
Illegal logging	High turbidity	5	2	10
Human access (Dumagat squatters)	Microbial contamination, turbidity	5	2	10
Security threats (terrorist act)	Chemical toxic substance	1	5	5
Illegal settling/ intrusion into the watershed	Microbial contamination, turbidity	5	4	20

Source: Maynilad Water Services, Inc., Manila, Philippines.

the different cells of the matrix. For example, consider using the matrix to compare two chemicals with the following likelihood and severity:

- Chemical 1: daily exposure to a population of 10; probability of mortality from daily exposure is 0.001
- Chemical 2: exposure once per week to a population of 10,000; probability of mortality from weekly exposure is 0.001/7.

According to the approach in Tables 25.1 and 25.2, Chemical 1 would be scored as 5 for likelihood and 4 for severity, while Chemical 2 would have likelihood and severity scores of 4 and 5, respectively. Hence, both would receive scores of 4 x 5 = 20. However, Chemical 1 would be expected to lead to 10 x 0.001 = 0.01 premature deaths, while Chemical 2 is expected to lead to 10,000 x 0.001/7 = 1.4 premature deaths. Hence, considered on a population scale, Chemical 2 should be regarded as 140 times as risky as Chemical 1, not equal in risk to Chemical 1. Qualitative matrices based solely on expert judgment also may be subject to the well-known human cognitive bias to under-estimate high risks and over-estimate low risks; and the bias, especially strong among experts, to be over-confident in their risk judgments.[8] Nonetheless, where no data on chemical occurrence are available, a qualitative matrix may be the best option.

Priority-setting example 3: dimensionless risk index

Researchers in the United Kingdom's Department of Environment, Food and Rural Affairs have developed an approach for prioritizing pesticides and their transformation products in drinking water supplies that addresses some of the limitations of a qualitative approach by considering chemical and soil properties relevant to environmental mobility and quantitative toxicity information. This prioritization approach could be extended to other anthropogenic contaminants and applied even in settings with no water quality monitoring data, although data on the extent of use (or extent of natural sources) of each chemical are needed.[9] The approach involves calculating a dimensionless exposure index (E)—the product of the relative amount of chemical in use, the fraction likely to end up in water (determined from the chemical's physical properties) and the fraction lost through degradation (determined from degradation studies). The exposure index is divided by the chemical's acceptable daily intake (or another relevant health risk metric) to provide an overall risk index.

The first step in using this prioritization scheme is to define a geographic area of interest, such as a state, watershed, or water supply service area. Then, for each potential pollutant of concern, E is calculated as the product of three terms representing the amount (A) of chemical in use, the fraction (F) mobilized into water and the persistence (P) of the compound. Each term is estimated from chemical properties commonly available in compendia or published research studies plus properties of soil and water. The equation for calculating A is

$$A = \frac{U}{U_{max}} f \tag{25.1}$$

where A is the "amount" index, U is the total amount of the chemical in use in kg/year (or, for chemical transformation products, the amount of the parent product in use), U_{max} is the total amount of the highest used chemical of concern in the geographical area (kg/year) and f is the maximum mass of transformation product mass formed per unit mass of the parent compound (relevant only when considering transformation products and otherwise equal to 1).

The mobility index (F) is computed from

$$F = \frac{1}{1 + K_d r_{sw}} \tag{25.2}$$

where K_d (cm³/g) is the distribution coefficient for adsorption of the contaminant to soil, and r_{sw} (g/cm³) is the solid-phase mass to the aqueous-phase volume of the compartment (either the soil overlying groundwater or, for contaminants discharged to surface waters, the surface water body). K_d values can be estimated from the organic carbon partition coefficient (K_{OC}), commonly available in the literature along with information about the fraction of the soil mass comprised of organic carbon (f_{oc}): $K_d = f_{OC} {}^* K_{OC}$. Where the latter data are unavailable, the authors of the approach recommend using $f_{oc} = 0.02$. For chemicals applied to soil (e.g., agricultural pesticides applied to crops or pesticides and fertilizers applied to lawns and golf courses), the authors recommend using $r_{sw} = 7.5$ g/cm³, whereas for applications or discharges to water they recommend using $r_{sw} = 0.005$ (both derived from typical volume fractions of solids in soil and sediments).

The third exposure index term, the persistence index (P), is calculated using degradation half-lives of the chemical in soil and water:

$$P = e^{-(\ln 2/DT_{50w})t} \times e^{-(\ln 2/DT_{50s})t} \tag{25.3}$$

where DT_{50w} and DT_{50s} are the half-lives of the chemical in water and soil, respectively, and t is the residence time in the environmental compartment. The authors recommend using $t = 40$ days for surface water and potentially larger values for groundwater sources.

The risk index (RI) for a particular chemical considers both the potential for human exposure (represented as $E = A {}^* F {}^* P$) and contaminant's acceptable daily intake (ADI):

$$RI = \frac{E}{ADI} = \frac{A \times F \times P}{ADI} \tag{25.4}$$

Table 25.3 Data for example risk index calculation for aldicarb

Variable	Value for aldicarb	Source	Notes
U (kg/year)	118,887	Sinclair et al.[9]	Year 2003
U_{max} (kg/year)	13,000,000	Sinclair et al. [9]	Year 2003
f (dimensionless)	1		Aldicarb is a parent compound (not a transformation product), so $f=1$
K_d (cm³/g)	0.16	U.S. EPA[10]	K_{oc} is reported as 8-37; $K_d=0.02\star 8$ (conservative estimate)
r_{sw} (g/cm³)	0.005	Sinclair et al.[9]	Conservative estimate
DT_{50w} (days)	1,300	U.S. EPA[10]	Longest half-life reported in *in situ* groundwater
DT_{50s} (days)	15	U.S. EPA[10]	Longest half-life reported in *in situ* soil
t (days)	40 days	Sinclair et al.[9]	
ADI (mg/kg-day)	1×10^{-3}	U.S. EPA[11]	

The ADI is a measure of the amount of the chemical that can be ingested daily over a lifetime without risk of adverse health effects; ADI values are available from a number of international and national organizations, for example the WHO, Food and Agricultural Organization (FAO) and U.S. Environmental Protection Agency (EPA; for example, in the Integrated Risk Information System, which expresses ADI as reference dose, or RfD, available at http://www.epa.gov/iris/).

As an example, consider the calculation of *RI* for aldicarb potential contamination of groundwater sources of drinking water in California. Data from the California Department of Pesticide Regulation indicated that, in 2003, 118,887 kg of aldicarb were applied to crops in California and total pesticide use in the state was approximately 13 million kg. Table 25.3 shows additional data required to compute the *RI* for aldicarb. Using the data in Table 25.3, the *RI* for aldicarb can be computed as follows:

$$A = \frac{U}{U_{max}} f = \frac{118,887 \, \text{kg}}{13,000,000 \, \text{kg}} \times 1 = 0.0091 \tag{25.5}$$

$$F = \frac{1}{1 + K_d r_{sw}} = \frac{1}{1 + \dfrac{0.16 \, \text{cm}^3}{\text{g}} \times \dfrac{0.005 \, \text{g}}{\text{cm}^3}} = 0.99 \tag{25.6}$$

$$P = e^{-(\ln 2/DT_{50w})t} \times e^{-(\ln 2/DT_{50s})t} \tag{25.7}$$

$$= e^{-\left(\frac{\ln 2}{1300 \, \text{days}}\right)40 \, \text{days}} \times e^{-\left(\frac{\ln 2}{15 \, \text{days}}\right)40 \, \text{days}} = 0.15$$

$$RI = \frac{A \times F \times P}{ADI} = \frac{0.0091 \times 0.99 \times 0.15}{1 \times 10^{-3}} = 1.4 \tag{25.8}$$

Table 25.4 shows the results of a case study ranking of pesticide transformation products for Great Britain based on pesticide usage data from the Pesticide Usage Survey Group of the Central Science Laboratory. As shown from this ranking, several of the compounds have very low *RI* estimates, while a few (e.g., breakdown products of isoproturon, propachlor and diclofob-methyl) stand out as having risk indexes that are orders of magnitude larger than those of many of the other compounds. This ranking exercise also included an additional metric

Table 25.4 Results of chemical risk priority setting exercise in Great Britain using risk index [9]

Transformation product	Parent pesticide(s)	Pesticide usage (U) (kg/year)	Formation fraction (f)	Sorption (K_d) (cm^3/g)	Persistence in soil (DT_{50w}) (days)	Persistence in water (DT_{50w}) (days)	Data availability classification	Pesticide ADI (mg/(kg-day))	Risk index
3,5,6-trichloro-2-pyridinol	chlorpyrifos/triclopyr	67,684/38,295	0.32/0.26	0.53	279	383	A	0.003/0.005	0.69
thifensulfuron acid	thifensulfuron-methyl	16,786	0.25	0.138	365	109	A	0.01	0.066
kresoxim-methyl acid	kresoxim-methyl	94,944	0.84	0.34	131	383	A	0.4	0.019
O-desmethyl thifensulfuronmethyl	thifensulfuron-methyl	16,786	0.19	0.68	15.3	51	A	0.01	0.002
CGA-321113	trifloxystrobin	76,011	1	0.96	350	289	B	0.038	0.091
carbendazim	thiophanate-methyl/benomyl	18,633/922	0.76/1	0.45	320	743	B	0.02/0.03	0.066
	fluquinconazole/tebuconazole	22,443/12,6787	0.161/0.06					0.006/0.03	
1,2,4-triazole	tetraconazole/propiconazole	4,906/10,760	0.1/0.43	0.86	98	190	B	0.004/0.04	0.044
	myclobutanil	3,877	1					0.1	
CL 153815	picolinafen	3,103	1	3.2	77	31.4	B	0.014	0.001
diclofop acid	diclofop-methyl	33,683	0.9	0.2	63	105	B	0.001	2.653
ethyl-m-hydroxyphenyl carbamate	desmedipham	4,349	0.16	0.2	27	26	B	0.0018	0.008
triazine amine A	tribenuron-methyl	4,074	0.91	0.2	240	105	B	0.12	0.004
BTS 27919	amitraz	642	0.35	0.2	150	21	B	0.0025	0.004
desmethylisoptoturon	isoproturon	2,260,278	0.14	1.07	65	300	B	0.015	0.615
deethylatrazine	atrazine	139,758	0.19	1.7	121	300	B	0.006	0.277
deisopropyl-atrazine	simazine/atrazine	139,758	0.11/0.1	0.16	17	300	B	0.005/0.006	0.201
thiophene sulfonamide	thifensulfuron-methyl	16,786	0.29	0.052	96.6	300	B	0.01	0.106

propachlor oxanilic acid	propachlor	138,592	0.33	0.03	300	300	C	0.009	1.539
propachlor ethane sulfonic acid	propachlor	138,592	0.19	0.03	300	300	C	0.009	0.883
4-hydroxy-2,5,6-trichloroisophthalonitrile	chlorothalonil	479,833	0.32	0.2	43	300	C	0.018	0.722
triazamate metabolite II	triazamate	1,020	0.91	0.28	300	300	C	0.0003	0.367
1-methyl-3-(4-isopropyl phenyl)-urea	isoproturon	2,260,278	0.16	0.2	300	300	D	0.015	3.458
TCPSA	tri-allate	372,093	0.18	0.2	300	300	D	0.005	1.916
3-carbamyl-2,4,5-trichlorobenzoic acid	chlorothalonil	479833	0.25	0.2	300	300	D	0.018	0.98
methiocarb sulfoxide	methiocarb	38,567	0.3	0.2	300	300	D	0.002	0.851

pertinent to whether the *RI* was based mostly on default values (data availability classification value in Table 25.4 of C–D), required one default value (data classification B, with subscripts denoting the type of default value), or required no default assumptions (A). This approach allows decision-makers to consider not only the relative potential for exposure and adverse health effects but also the quality of the data upon which the estimates are based.

Priority-setting example 4: quantitative chemical risk assessment

When data on chemical concentrations in the water supply of interest are available, a quantitative risk or, equivalently, burden of disease assessment can be used to estimate the probability of each contaminant leading to health risks. Such quantitative assessments provide the most accurate measure of risk differences among chemicals, given current knowledge of chemical toxicity (which, for many chemicals, is still sparse). Where public health data on incidence rates of diseases linked to chemical pollutants are available (for example, bladder cancer mortality rates), a burden of disease approach is preferred to a typical risk assessment, since the latter does not consider observed illness rates in the exposed population but rather relies exclusively on data from other populations or animal toxicity studies. Chapter 57 explains the details of how to carry out such a burden of disease assessment.

DeFelice used a burden of disease approach to prioritize risks of cancer-causing chemicals in community drinking water supplies in North Carolina, United States.[12] For each community water system and each of North Carolina's 100 water systems, DeFelice estimated the number of premature cancer deaths that could be attributable to chemical contaminants per year. The North Carolina analysis showed that among 20 regulated carcinogenic chemicals, 2 (total trihalomethanes (TTHMs) and arsenic) dominated overall risks: 53 annual premature cancer deaths (95 percent confidence interval (CI) 28–78) were attributed to TTHMs and 1.4 (95 percent CI 0.28–5.2) to arsenic, only only 1 premature death every 10 years could be attributed to all other 18 chemicals combined. By computing these risks for each of the 2,120 water systems in the state, DeFelice also was able to assess the spatial distribution of risks and identify water systems and counties where risks were elevated, in comparison to other systems and counties (Figure 25.1). Box 25.1 illustrates the process by which the risks in each community were computed.

Managing chemical risks

Since the earliest water distribution systems were built in ancient times, water engineers have recognized that the most effective step for controlling risks of exposure to pollutants is selection of the cleanest water source available. In ancient Rome, for example, waters distributed via aqueducts were separated by quality, with the best-quality water reserved for potable uses and lower-quality water diverted to non-potable uses.[13] In his description of the Roman water supply, Roman Water Commissioner Sextus Julius Frontinus wrote,

> It was ... determined to separate [waters from different sources] and then to arrange so that Marcia [the aqueduct delivering the cleanest source] should serve wholly for drinking purposes, and that the others should be used for purposes adapted to their special qualities. For example, it was ordered for several reasons, that Old Anio should be used for watering the gardens, and for the more dirty uses of the city, because the further from its source its waters are drawn, the less wholesome they are.[13]

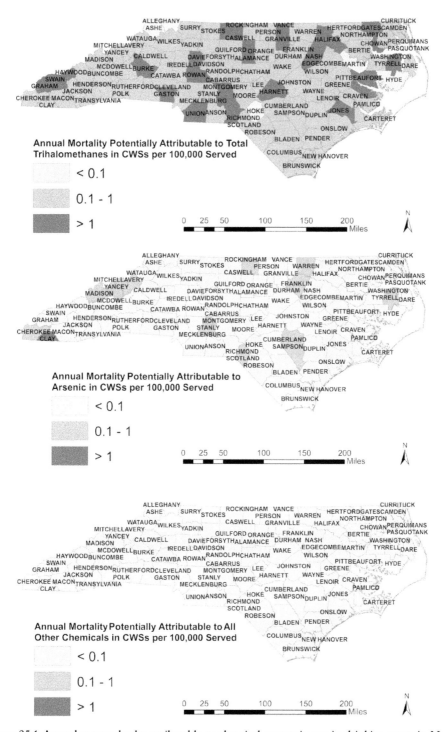

Figure 25.1 Annual cancer deaths attributable to chemical contaminants in drinking water in North Carolina, USA (cases per 100,000). A comparison of these three figures shows that risks from triholomethanes (THMs) and arsenic far exceed those from all other regulated chemical carcinogens combined. CWS = community water system.

Box 25.1 **Example quantitative burden of disease calculation for chemical hazards priority-setting**

DeFelice used quantitative chemical burden of disease assessment to prioritize carcinogenic chemicals for further analysis in North Carolina (NC) community water systems. This example illustrates the process of computing the annual number of premature cancer deaths attributable to one chemical (trichloroethylene, TCE) in one community water system.

Quantitative chemical burden of disease assessment follows four steps: (1) hazard identification (identifying the chemicals of concern); (2) exposure assessment (quantifying the number of people exposed and exposure concentrations); (3) dose–response assessment (estimating the fraction of observed cases attributable to each contaminant); and (4) risk characterization (quantifying the potential burden of cancer at population scale).

For this example, the hazard identification step found that TCE was detected in a community water system serving 100,000 people. According to the EPA 2005 *Guidelines for Carcinogen Risk Assessment,* TCE is classified as a human carcinogen and has been connected with liver cancer, kidney cancer and non-Hodgkin lymphoma[11, 22].

The exposure assessment step used regulatory data collected by the NC Division of Environment and Natural Resources to estimate the exposure concentration. For this example, TCE was detected in 30 percent of monthly water samples over a six-year period, and the concentration in positive samples was found to fit a lognormal probability distribution with mean and standard deviation of 1.02×10^{-2} and $9.6 \times 10^{-}$ mg/liter, respectively.

Using these concentration data, the next step was to determine the lifetime average daily dose (*LADD*, mg/kg-day). This example considers only exposure by ingestion; however, to fully quantify *LADD* from drinking water, one may also need to account for other exposure pathways (e.g., inhalation of volatilized vapors and dermal contact). The following equation quantifies the *LADD* due to ingestion:

$$LADD_{ingest} = \frac{C \times IR}{BW} \tag{25.9}$$

where C is the contaminant concentration (mg/liter) in tap water, IR is the daily volume of tap water ingested (liters/day) and BW is body weight (kg).

DeFelice computed the *LADD* using Monte Carlo simulation, in which values for each variable in equation 25.9 were sampled at random from their corresponding probability distributions, and the calculation was repeated 1,000 times for 1,000 different combinations of variables, yielding a distribution of *LADD*. In each simulation run, C was estimated as pX, where p represents a Bernoulli random variable with parameter equal to the contaminant detection frequency (0.3 for this example) and X represents a lognormal variable of the contaminant concentration in positive samples. Table 25.5 summarizes values of all variables in equation 25.9, which were derived from the EPA *Exposure Factors Handbook* and other literature sources[23, 24]. From the simulations, the lifetime average daily dose for an individual exposed in this town is 5.4×10^{-5} (mg/kg-day) (95 percent CI: $0 - 4.3 \times 10^{-4}$).

Table 25.5 Parameters used in TCE burden of disease assessment example

Parameter	Distribution or discrete value	Data source
Exposure concentration (*C*) (mg/liter)	$p \star X$, where $p \sim$ Bernoulli (0.3), $X \sim$ Lognormal (mean: 1.02×10^{-2}, sd: 9.6×10^{-3})	North Carolina Department of Environment and Natural Resources
Body weight (*BW*) (kg)	Normal (mean: 69, sd: 17.8)	U.S. EPA *Exposures Factors Handbook* [23]
Intake rate (*IR*) (liters/day)	Lognormal (mean: 1.129, sd: 0.67)	Roseberry and Burmaster, 1992 [24]
Cancer slope factor (*SF*$_{oral}$) (kg-day/mg)	0.0462	U.S. EPA Integrated Risk Information System [11]
National lifetime probability of death from liver cancer, kidney cancer, or non-Hodgkin lymphoma (I_u)	Normal (mean: 0.0193, sd: 0.0001)	U.S. National Cancer Institute [25]
Annual deaths from liver cancer, kidney cancer and non-Hodgkin lymphoma in the case study community	39	North Carolina Division of Public Health

The excess cancer risk (*ECR*) was calculated by multiplying the *LADD* for the exposure pathway by the appropriate slope factor (*SF* (mg/kg-day)$^{-1}$) for the chemical exposure pathway:

$$ECR = SF_{oral} \times LADD_{ingest} \tag{25.10}$$

The *ECR* from exposure to a chemical is defined as the probability of developing cancer by age 70 due to a lifetime of exposure. The slope factor for TCE is 0.0462 (mg/kg-day)$^{-1}$. Thus the excess cancer risk was estimated as 2.5×10^{-6} (95 percent CI: $0 - 2.0 \times 10^{-5}$).

The following equation was used to convert the *ECR* to an attributable fraction (*AF*):

$$AF = \frac{ECR}{ECR + I_u} \tag{25.11}$$

In equation 25.11, the calculated *ECR* is added to the background probability of cancer for the unexposed population (I_u) in order to determine the total probability of cancer for the exposed group in the denominator. The background cancer probabilities (I_u) for each cancer associated with TCE exposure was determined from the national average lifetime risk of dying from cancer in the U.S. (SEER Table 1.18)[6] In this example, I_u is the sum of liver cancer (0.68 percent, 95 percent CI: 0.67–0.69), kidney cancer (0.47 percent, 95 percent CI: 0.47–0.48) and non-Hodgkin lymphoma (0.78 percent, 95 percent CI: 0.77–0.78) lifetime risks, so I_u is equal to 1.93 percent (95 percent CI: 1.91–1.95).

Using the estimated *AF*s plus observed cancer case counts (I_o) in the community, the number of cancer cases potentially attributable (*AC*) to TCE was calculated as:

$$AC = AF \times I_o \tag{25.12}$$

The observed incidence rates for the case study community were 11, 13 and 15 deaths due to liver cancer, kidney cancer and non-Hodgkin lymphoma, respectively. Thus I_o for this community were 39 deaths due to liver cancer, kidney cancer and non-Hodgkin lymphoma. TCE has an attributable fraction of 0.013 percent (95 percent CI: 0–0.10 percent) resulting in 0.005 (95 percent CI: 0–0.04) annual deaths that were potentially attributable to TCE. This estimate can be interpreted as meaning that one death every 200 years ($=1/0.005$) in the community of 100,000 could be attributed to TCE exposure via the community water supply.

To compare and prioritize chemical carcinogens, DeFelice carried out this type of analysis for each of the 2,120 community water systems in North Carolina and for each of 20 regulated carcinogens occurring at concentrations above the maximum allowable concentration under the U.S. Safe Drinking Water Act in at least one of the systems. The results were then aggregated to the county level to produce the maps in Figure 25.1 and the state-wide burden of disease estimates described in the main text.

Frontinus also described the influence of the watershed on source water quality, quoting Emperor Nerva's order that the New Anio aqueduct stop drawing from the Anio River and instead

> to take from the lake lying above the Sublacensian Villa of Nero, there where the Anio is the clearest; for inasmuch as the source of Anio is above Treba Augusta, it reaches this lake in a very cold and clear condition, be it because it runs between rocky hills and because there is but little cultivated land around that hamlet, or because it drops its sediment in the deep lakes into which it is taken.[13]

An ancient Roman law charged steep fines to those polluting source waters: "No one shall with malice pollute the waters where they issue publicly; should any one pollute them his fine shall be 10,000 sestertii".[13]

In more modern times, New York City recognized early on the importance of source water selection and in 1840 began constructing a network of aqueducts to transport water from distant, pristine watersheds. Those watersheds remain the source of the 1.3 billion gallons per day that the city currently provides to its customers.[14] As a result of the pristine nature of the water supply and an extensive program to protect the watersheds around the city's source waters, the U.S. EPA granted a waiver from the requirement to filter most of its water.

For settings in which new water supplies are being considered, an inventory of potential contamination sources in the catchment area of each supply should be conducted. The types of pollutants originating from each potential source should be cataloged. Ideally, schematics or maps will be drawn indicating potential pollution sources. Then, data on the pollutant mass released by each source and chemical and geological characteristics describing its potential to be transported into the water source will be gathered. Figure 25.2 and Table 25.6 show the results of a pollution source inventory for a small water supply well in Gaston County, North Carolina.[15] From information on these sources combined with hydrogeologic characteristics of the well and properties of each potential chemical pollutant, the researchers

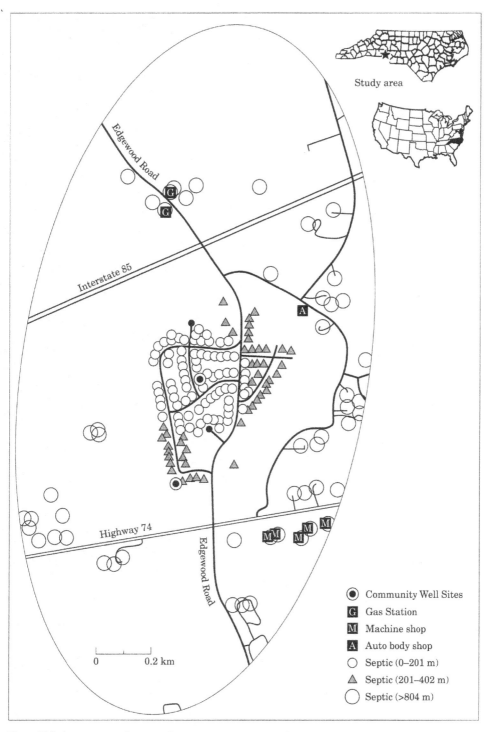

Figure 25.2 Assessment of potential contamination sources for a drinking water well serving Gaston, North Carolina, United States[15]

Table 25.6 Results of assessment of potential contamination sources for a drinking water well serving Gaston, North Carolina, United States [15]

Name	Category	Potential contaminants
Machine shop	Container storage and material transfer	Chloroform, carbon tetrachloride, organics mix 2, benzene, methyl ethyl ketone
Body shop	Container storage and material transfer	Benzene, acetic acid, sulfuric acid, organics mix 1 5, organics mix 2 6
Gas station A	Medium underground storage tank (UST), gasoline	Benzene
	Small UST, waste oil	Benzene
	Small UST, waste oil	Organics mix 1
	Small UST, waste oil	Methanol
Gas station B	Medium UST, gasoline	Benzene
Septic systems	Septic systems	Nitrates
Edgewood Road	Material transport	Benzene, acetic acid, sulfuric acid, organics mix 1 5, organics mix 2 6
Highway 74	Material transport	Benzene, acetic acid, sulfuric acid, organics mix 1 5, organics mix 2 6, chromium mix 3, lead, chloroform, carbon tetrachloride, organics mix 4, organics mix 1, methyl ethyl ketone, methanol, organics mix 2, toluene
Interstate 85	Material transport	Benzene, acetic acid, sulfuric acid, organics mix 1 5, organics mix 2 6, chromium mix 3, lead, chloroform, carbon tetrachloride, organics mix 4, organics mix 1, methyl ethyl ketone, methanol, organics mix 2, toluene

Note: Organics mix 1: acetone ethyl ketone; organics mix 2: 1, 2 dichlorobenzene; metals mix 4: zinc, lead, cadmium and nickel; chromium mix 3: chromium and methanol; organics mix 1 5, methyl ethyl ketone, cresol, acetone; organics mix 4: 1, 1, 1 trichloroethane; organics mix 2 6, xylene, napthalene, toluene and 1, 1, 1 trichloroethane.

conducting this inventory were able to rank the sources and accompanying pollutants using a method similar to the dimensionless risk index approach described above. The importance of inventorying potential chemical pollution sources cannot be overstated, as illustrated by the catastrophic failure to consider arsenic as a potential pollution source in programs to develop alternative water sources for Bangladesh (described earlier in this chapter).

In situations where many of the contaminants of concern are from natural sources (but leaching into groundwater has been accelerated by human land and groundwater use practices) a catchment survey for sources of pollution may give an incomplete picture of groundwater contaminant hazards. In addition to catchment survey data, geologic information and expertise are required. When considering very small supplies in such regions, a first step is to assess whether rainfall harvesting and capture is a viable alternative to groundwater use (largely eliminating the need for costly chemical analysis). If this is not possible, simple chemical screening of groundwater using preliminary pH tests would help to determine the hazards that need to be assessed: pH < 4.5 – assess arsenic, lead (corrosion of pipes), other metals; pH > 8 higher risk of arsenic, uranium, fluorine, selenium in groundwater. In the absence of pH testing equipment, dyes from local natural foods can be used as pH indicators. Appleyard et al. describe an example in which a community used red cabbage dye to screen wells for potential arsenic contamination associated with sulfide oxidation.[16]

Source water protection

Beyond selecting the best water source available, watershed and wellhead protection problems are the most cost-effective way to prevent most kinds of chemical contamination. Such protection programs can prevent not only contamination from industrial sources and those associated with human settlements but also with disinfection byproducts and harmful algal blooms. For example, the main precursor to disinfection byproduct formation is often derived from the decomposition of algae, the production of which can be reduced by preventing runoff of nitrogen and/or phosphorus from agricultural and other human activities into the watershed. Similarly, preventing over-fertilization of water sources is key to controlling harmful algal blooms that can produce chemical toxins. Watershed and wellhead protection programs can consist of laws restricting development or certain kinds of activities in the vicinity of the source water, economic incentives to discourage such activities, and/ or education programs to encourage voluntary measures to protect the watershed or zone of influence around a groundwater supply well.

Egypt's program to delineate protected areas around 1,500 water supply wells in the densely populated Nile Delta region is an example of a legal approach to source water protection.[17] To take charge of this task, Egypt in the 1990s established a new government agency, the Groundwater Sector (GWS). To determine the size of the protected areas, the GWS used hydrogeologic modeling to develop a simplified spreadsheet that estimates the size of the protection area based on readily available well and hydrogeologic parameters (pumping rate, aquifer thickness, conductivity, clay thickness, aquifer porosity, clay porosity and hydraulic gradient). For areas without the required hydrogeologic data, they developed default values based on previous surveys.[17] Activities that might pollute wells are restricted in the resulting wellhead protection areas.

An economic approach known as "payment for ecosystem services" (PES) is increasingly being used around the world to protect downstream water users from upstream activities that may pollute the water supply. Under this approach, downstream users pay upstream polluters to forego activities that degrade water quality. Some implementations of this approach also require that upstream polluters pay downstream users if downstream pollution rises above pre-determined levels. China has experimented with the latter type of PES approach to improve water quality in eight watersheds.[18] Early research in one of these watersheds, the Shaying River, has shown that the PES has improved water quality. Under this PES program, downstream prefecture-level governments pay upstream governments a fee established by the provincial-level environmental agency if concentrations of indicator pollutants are below provincial standards, whereas upstream governments pay the downstream governments when pollutant concentrations exceed the standards.[18]

An example program that includes a mix of voluntary, regulatory and economic approaches is the Clean Water Initiative of the Tennessee Valley Authority (TVA) in the United States.[19] The TVA oversees the conservation and development of the Tennessee River, which drains 106,000 km² of land area covering portions of seven states and includes 30 major reservoirs. Deteriorating water quality in the system caused by increasing development led the TVA to search for innovative solutions to improve water quality throughout this river system.

As part of the solution, the TVA appointed 12 River Action Teams, one for each of the 12 subwatersheds of the river system. The teams were charged with inventorying the quality of tributaries in the subwatersheds through analysis of the health of aquatic species and then inventorying sources of contamination through field visits and aerial photography. The teams then identify the key stakeholders with a role in reducing pollution from the identified sources; these may include regulatory agencies, community leaders, businesses, universities and industries. Working with these stakeholder groups, the River Action Teams develop and implement mitigation plans to reduce pollution from high-priority sources. Plans vary from community to community but often include education of farmers and other landholders on land management practices that can reduce pollution loads to the watershed and persuasion of these landholders to implement these practices.

In the United States, perhaps the most ambitious watershed protection program is that of New York City.[14] As previously mentioned, when developing its water supply during the nineteenth century, New York sought pristine water sources distant from the city. Today, New York transports 90 percent of its water from the Catskill and Delaware watersheds via two aqueducts, one 92 miles and the other 100 miles long (with the remaining water drawn from the Croton watershed, closer to the city). Due to an agreement with U.S. EPA to invest $300 million over 10 years in a watershed protection program, the city continues to be exempted from the requirement to filter water from the Catskill and Delaware watersheds.[20] The city adopted this program in large part because it was determined to be more cost-effective than installing a filtration system, which was estimated to cost $8 billion.[20] The watershed protection program includes an array of activities designed to control both nonpoint and point sources of pollution, including extensive land acquisition, payments to watershed communities to compensate for economic losses of foregone development opportunities, requiring setback distances between water's edge and various land uses, programs to limit nutrient runoff, siting and technology requirements for wastewater treatment plants and septic systems, and enhanced source water quality monitoring.[14]

Treatment

With population pressure and development, selecting pristine water sources is becoming an increasing challenge, and water treatment may be the only way to manage chemical contaminants. Even in communities with relatively clean water sources, treatment may be needed as an additional barrier to prevent exposure. Treatment technologies vary with the type of chemical contaminant. The WHO *Guidelines for Drinking-water Quality* lists treatment options for chemical contaminants for which guideline values exist. The U.S. EPA also provides on-line summaries of treatment technology options for all contaminants regulated in the United States (available from contaminant-specific links at http://water.epa.gov/drink/contaminants/).

Unfortunately, conventional treatment technologies are ineffective in removing many of the chemical contaminants discussed in this chapter. Conventional groundwater treatment systems often employ only disinfection and therefore do not remove chemical contaminants. Conventional surface water treatment—coagulation/flocculation, sedimentation, sand filtration and disinfection—also has limited capability to remove chemical contaminants. Design of systems to remove chemical contaminants is complex due to differences in source water characteristics and generally requires bench and pilot testing prior to implementation. The complexity and high costs of chemical contaminant removal reinforce the importance of prevention. For example, a U.S. EPA study of wellhead protection programs in seven communities found that treating chemical contamination problems after they arose cost, on average, 27 times as much as wellhead protection programs that could have prevented the contamination.[21]

Additional resources and readings

1 Smith AH, Lingas EO, Rahman M. 2000. Contamination of drinking-water by arsenic in Bangladesh: a public health emergency: RN – *Bull. WHO*, 78:1093–1103. http://www.ncbi.nlm.nih.gov/pmc/articles/PMC2560840/

2 U.S. National Research Council. 2000. Watershed Management for Potable Water Supply. Washington, D.C. http://www.nap.edu/catalog/9677/watershed-management-for-potable-water-supply-assessing-the-new-york

3 World Health Organization. 2004. *Guidelines for Drinking-water Quality*, Third Edition, Volume 1: Recommendations. Geneva: WHO. http://www.who.int/water_sanitation_health/dwq/guidelines/en/

References

1. World Health Organization. 2014. Child Mortality and Causes of Death. World Health Organization (accessed 2015 Jan 19). Available from: http://www.who.int/gho/child_health/mortality/en/.

2. Prüss-Üstün, A, Bos, R, Gore, F, and Bartram J. 2008. Safer water, better health: costs, benefits and sustainability of interventions to protect and promote health. Geneva: World Health Organization.

3. Smith, AH, Lingas, EO, and Rahman, M. 2000. Contamination of drinking-water by arsenic in Bangladesh: a public health emergency: *Bull. WHO*, 78:1093–1103.

4. U.S. Environmental Protection Agency. 2014. Toxic Substances Control Act Chemical Substance Inventory. 2014 (accessed 2015 Jan 19). Available from: http://www.epa.gov/oppt/existingchemicals/pubs/tscainventory/basic.html.

5. Thompson, T, Fawell, J, Shoichi, K, Jackson, D, Appleyard, S, Callan, P, Bartram, J and Kinston, P 2007. *Chemical Safety of Drinking-Water: Assessing Priorities for Risk Management*. Geneva: WHO.

6. World Health Organization. 2004. *Guidelines for Drinking-water Quality*, Third Edition, Volume 1: Recommendations. Geneva: WHO.

7. Amini, M, Abbaspour, KC, Berg, M, Winkel, L, Hug, SJ, Hoehn, E, Yang, H, and Johnson, CA. 2008. Statistical modeling of global geogenic arsenic contamination in groundwater. *Environ. Sci. Technol.* 42:3669–3675.

8. Lichtenstein, S, Slovic, P, Fischhoff, B, Layman, M, and Combs, B. 1978. Judged frequency of lethal events. *J. Exp. Psychol. Hum. Learn. Mem.* 4:551–578.

9. Sinclair, CJ, Boxall, ABA, Parsons, SA, and Thomas, MR. 2006. Prioritization of pesticide environmental transformation products in drinking water supplies. *Environ. Sci. Technol.* 40:7283–7289.

10. U.S. Environmental Protection Agency. Technical Factsheet on Aldicarb and Metabolites. Washington, DC.

11. U.S. Environmental Protection Agency National Center for Environmental Assessment. Integrated Risk Information System (IRIS). (accessed 2014 Oct 17). Available from: http://www.epa.gov/iris/.

12. DeFelice, NB. 2014. *Drinking Water Risks to Health 40 Years After Passage of the Safe Drinking Water Act: A County-by-County Analysis in North Carolina*. University of North Carolina at Chapel Hill, NC.

13. Frontinus, SJ. 1913. *The Two Books on the Water Supply of the City of Rome*. New York: Longmans, Green and Co.

14. U.S. National Research Council. 2000. Watershed Management for Potable Water Supply. Washington, DC.

15. Harman, WA, Allan, CJ, and Forsythe, RD. 2001. Assessment of potential groundwater contamination sources in a wellhead protection area. *J. Environ. Manage.* 62:271–282.
16. Appleyard, SJ, Angeloni, J, and Watkins R. 2006. Arsenic-rich groundwater in an urban area experiencing drought and increasing population density, Perth, Australia. *Appl. Geochemistry.* 21:83–97.
17. Fadlelmawla, AA, and Dawoud, MA. 2006. An approach for delineating drinking water wellhead protection areas at the Nile Delta, Egypt. *J. Environ. Manage.* 79:140–149.
18. Lu, Y, and He, T. 2014. Assessing the effects of regional payment for watershed services program on water quality using an intervention analysis model. *Sci. Total Environ.* 493:1056–1064.
19. Ungate, CD. 1996. Tennessee Valley Authority's clean water initiative: building partnerships for watershed improvement. *J. Environ. Plan. Manage.* 39:113–122.
20. DePalma, A. 2007. City's Catskill water gets 10-year approval. *New York Times.* Apr.
21. Job, CA. 1996. Benefits and costs of wellhead protection. *Gr. Water Monit. Remediat.* 16: 65–68.
22. U.S. Environmental Protection Agency. 2005. Guidelines for Carcinogen Risk Assessment. Washington, DC.
23. U.S. Environmental Protection Agency. 2011. *Exposure Factors Handbook: 2011 Edition.* Washington, DC.
24. Roseberry, AM, and Burmaster, DE. 1992. Lognormal distributions for water intake by children and adults. *Risk Anal.* 12:99–104.
25. U.S. National National Center for Health Statistics. 2013. Surveillance, Epidemiology, and End Results (SEER) Stat Database.

PART IV

Implementing interventions

26

INTRODUCTION: SETTINGS-BASED APPROACHES[*]

Laura Linnan

DIRECTOR OF THE CAROLINA COLLABORATIVE FOR RESEARCH ON WORK & HEALTH, PROFESSOR, DEPARTMENT OF HEALTH BEHAVIOR, UNC GILLINGS SCHOOL OF GLOBAL PUBLIC HEALTH, CHAPEL HILL, NC, USA

Anna Grummon

DOCTORAL STUDENT, DEPARTMENT OF HEALTH BEHAVIOR, UNC GILLINGS SCHOOL OF GLOBAL PUBLIC HEALTH, CHAPEL HILL, NC, USA

Learning objectives

1 Discuss the differences between issue, population and setting or venue-based approaches to addressing public health problems.
2 Describe the social ecological framework and how a water-based intervention might include strategies at the intrapersonal, interpersonal, organization, community and policy levels of influence.
3 Identify two strengths and two limitations of the social ecological framework.

By focusing direct attention to the wide array and complexity of considerations associated with gaining access to clean, safe, sufficient amounts of water to sustain life, and maintaining access over time, we can categorize interventions as taking an *issue-based approach*, where *water* is the issue. In public health, we often take an issue-based approach to concentrate needed attention on a particular disease or condition or concern. This *Handbook* represents a comprehensive exploration of water as an important public health issue. Alternatives to an issue-based approach in public health are *population-based approaches* or *settings-based approaches*. To illustrate population-based approaches, consider that many health disparities clearly exist among lower-income, lower-educated individuals and among certain groups by race/ethnicity. If we organized an exploration of the many potential health concerns/issues for

[*] Recommended citation: Linnan, L. and Grummon, A. 2015. 'Introduction: settings-based approaches', in Bartram, J., with Baum, R., Coclanis, P.A., Gute, D.M., Kay, D., McFadyen, S., Pond, K., Robertson, W. and Rouse, M.J. (eds) *Routledge Handbook of Water and Health*. London and New York: Routledge.

each subgroup, this would be an example of a *population-based approach*. Another approach, and one which represents the focus of this section of the *Handbook*, is a *settings-based approach*. An example of a settings-based approach would be addressing health concerns in different settings such as schools, health-care centers, and workplaces.

These different approaches (issue, population and settings) are often used in combination to plan and implement public health interventions. For example, while addressing water as a priority health issue (issue-based approach), we turn to certain public health settings (settings-based approach) where water problems exist, and then plan appropriate interventions. In this *Handbook*, water issues have been explored from the perspective of hazards, sources of exposure, interventions to reduce exposures, policy implementation, distal influences, investigative tools and measures, and historic cases. And, in this part, we will add a settings-based focus to our consideration of priority water-related issues.

A settings-based approach is consistent with an important tenet of the World Health Organization, Ottawa Charter (1986, p.2) which emphasized that "Health is created and lived by people within the settings of their everyday life: where they learn, work, play and love." Further study and evidence of the effectiveness of a settings-based approach has been offered in several thoughtful books and manuscripts (McLeroy et al., 1988; Mullen et al., 1995; Poland et al., 2008). Here we first summarize some of the strengths, and then discuss the limitations and challenges, of using a settings-based approach to address important water-related issues.

Strengths of the settings-based approach

There are many types of settings that can be mobilized to address water-related issues. Educational institutions, health-care organizations and workplaces are three important venues that are prevalent in most communities. By activating water interventions in just these three locations you would reach youth, adults and those who may be at high risk to other health co-morbidities in most geographic locations. Adding other important community settings such as churches, senior centers/senior housing, barbershops/beauty salons and retail locations would permeate several other subgroups of the population who may be more difficult to reach in a single location. As a result of water activities directed to these specific settings, one might expect to reach many individuals by intervening in a relatively small number of setting categories – creating effort and cost efficiencies.

Another strength of the settings-based approach is that each setting has a unique set of structures, norms, policies and environmental supports that can be engaged to support intervention efforts. For example, a health-care organization may have a set of rules or policies which influence patients who visit and employees who work there. Norms about water use may be formal and informal, and are likely to influence individual behavior both directly and indirectly. By fully appreciating and mobilizing them in support of water interventions, the setting can exert a powerful influence on both individual and organizational change.

A third strength of settings-based approaches is that they typically include multi-level and multidisciplinary strategies to produce change. While this can also be a challenge, the power in having contributions from, say, educators, statisticians, engineers, policy-makers and administrators all thinking about the same issue, lends important insights and expertise that extends beyond single-minded, expert advice. If one were to change policies about water use in a school setting, it would be most successfully implemented if it involved school teachers, the principal, parents, the school engineer and a representative from the local water department so that all key stakeholders had an opportunity to be involved in the planning and decision-making process.

In addition to broad reach, a settings approach also provides an opportunity to target certain high-risk or high-influence target populations or potential beneficiaries of a water-based intervention. For example, if prioritized interventions were designed to reach parents and children, then a school-based focus is likely to be more effective than trying to reach parents/children in some other community location. Finally, specific organizational settings are often in the "middle ground" between individual and larger social change. They are places where an intervention can be tested in a more defined population, which is often a more practical way to approach intervention efforts for both practitioners and researchers.

Limitations and challenges of setting-based approaches

Despite a number of important strengths of the setting-based approach, several challenges are noteworthy. First, some health issues are broad enough to cross virtually all settings. Water is one such issue. As a result, "missing" a particular setting would potentially be quite problematic. Second, one must be careful when using a settings-based approach to avoid "using" a particular setting to "reach" a particular audience. A somewhat subtle, but important, acknowledgement of this limitation points out that "one size does not fit all" when it comes to planning and implementing water interventions in a workplace setting. For example, the types of interventions to improve access to water among employees in a corporate office is quite different than efforts required for workers on an oil rig, or workers on an assembly line in a manufacturing worksite. Assumptions about the homogeneity of the individuals (e.g. workers) or organizations (e.g. workplaces) in a settings-based approach may seriously mask potentially important differences that demand consideration of culture and context.

There is also the challenge of getting too specialized in a settings focus and thereby missing important potential synergies across different places or settings. Consider how an individual moves through a typical day. Let's consider that Juan wakes up and uses some type of transportation to get to work at a landscaping business; works for approximately eight hours; leaves work and stops at a grocery store to pick up dinner supplies; travels home and eats dinner with his family; attends a school play or the soccer field after dinner; stops to visit briefly with a neighbor; and turns in for the evening. In any given day, Juan moves between home, work, school, neighborhood and family "settings" – each of which provides opportunities for receiving some type of water-focused intervention. The point is that settings are permeable, with defined but not necessarily hard boundaries. As people move in/through venues, they may have pre-existing social relations in each that will influence intervention implementation and outcomes. Moreover, because settings have historical, social, political and environmental imperatives, processes and contexts, change may take time and considerable effort. Within any given setting, power makes a difference. Sometimes it is a challenge to understand who is in power, and who is not. To marshal support and engage powerbrokers within a given setting, one must often consider both informal and formal brokers. To fully succeed with a settings-based approach, these important challenges must be appreciated and integrated into an intervention planning effort. One way to begin to understand the strengths and challenges of planning and implementing water-related interventions using a settings-based approach is to organize intervention efforts within the social ecological framework.

Social ecological framework – an organizing heuristic for planning and implementing settings-based water interventions

The social ecological (SE) framework provides an overarching way to conceptualize the dynamic relationships among personal and environmental factors affecting human development and behavior (McLeroy et al., 1988; Stokols, 1996). The SE framework emphasizes that factors at multiple levels work to influence individual health behavior. Similar to McLeroy (1988), we conceptualize individual behavior as being determined by five levels of influence, organized in decreasing proximity to the individual (see Figure 26.1):

1 Intrapersonal – characteristics of the individual including attitudes, beliefs, knowledge, intentions, motivations, personality traits and skills.
2 Interpersonal – relationships between the individual and others, including, but not limited to, family members, friends, neighbors, co-workers, supervisors and other acquaintances.
3 Organizational/institutional – characteristics of relevant settings, including structures, policies, norms, processes and other rules.
4 Community – relationships that exist between organizations, institutions and other informal networks.
5 Society/Public Policy – the historical, cultural, economic and political context in which an individual exists including social and political policies, structures and other realities.

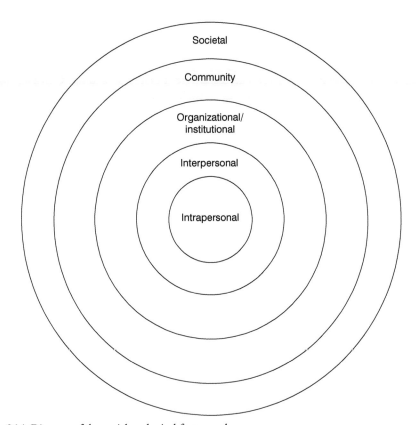

Figure 26.1 Diagram of the social ecological framework

The SE framework is particularly useful when considering settings-based intervention planning. The SE framework assumes that multiple levels of influence exist; and, if they are all engaged, they are more likely to prove successful with producing individual behavior change. For example, Table 26.1 clarifies how a SE framework would organize intervention strategies to influence water access for Juan, the worker at a landscape firm we mentioned previously.

In this example, it becomes clear that enacting this multi-level array of intervention activities is more likely to be effective than only enacting a single level effort. This is a strong assumption of the SE framework. If Juan receives information about the importance of water breaks, and how much he should consume, these intrapersonal interventions may increase his water consumption. However, if his manager is encouraging him to take water breaks (interpersonal), and the employer provides free water on the job (organization), and his company becomes an "employer of choice" for landscaping because they were recognized by the local media for supporting the health and safety of its employees through these policies/practices (community), then intervention efforts not only have a better chance for ensuring appropriate water consumption among workers, but are also more likely to be sustained

Table 26.1 Using the SE approach to organize a worksite setting intervention to improve water access

SE level	Target of the intervention	Intervention strategies
Intrapersonal	Worker	Educational materials in multiple languages/appropriate reading levels which emphasize the amount and timing of appropriate water breaks during typical workday
	Supervisor	Materials included in training of managers to understand appropriate water breaks and how to manage that
Interpersonal	Employee–supervisor	Interactions and/or team meetings to discuss appropriate water breaks
Organizational/institutional	Work environment	Water is made available to workers by employer on all landscape jobs
	Manager training and evaluation policies/procedures	Manager trainings all include information about water breaks
		Policy in place about taking appropriate water breaks
		Managers evaluated based on performance checks regarding appropriate water breaks in the field
Community	Chamber of Commerce	Recognition program (local media) for employers who have and appropriately enforce water break policy
	Department of Labor/health department	Random inspections at landscape job sites to assess whether state policy is being complied with by employer
Society/public policy	National and/or state regulations or policy on occupational safety and health	Policy in place (and enforced) to make sure workers get the amount and correct timing of water on the job site

as institutionalized programs and policies. Further, if managers/supervisors who do not encourage employees to take water breaks are reprimanded, or if the employer receives a OSHA fine because they are not following a policy about water breaks at the national level (public policy), this further embeds the program as the "normal way of doing things". The levels of influence do not work alone; they have a synergy and dynamism that support, reinforce and strengthen over time.

The SE framework is increasingly used for implementing interventions in public health research and practice (Sallis et al., 2008; Golden and Earp, 2012). SE models are not theories per se, but rather an organizing heuristic, helpful for conceptualizing potential factors affecting behavior at several levels of influence. The focus on multiple levels is a key strength of ecological models, as it broadens intervention options. The SE framework does not, however, specify which level of influence is most important; nor does it clarify the mechanisms by which factors at each level are expected to influence behavior. Practitioners must use additional theories or models specific to each level of influence to guide them in developing effective interventions (Linnan et al., 2001; Baron et al., 2013). SE frameworks are also behavior-specific, and therefore practitioners should tailor intervention strategies for each outcome of interest. For example, Table 26.1 specifies an intervention designed to increase the appropriate, timely access to water breaks at work. If more than one outcome was of interest, for example to use water with limited chemicals as part of the landscape business, then a new set of behaviorally-specific strategies would be planned. Obviously, this planning process is more demanding, logistically challenging and resource-intensive than focusing on a single level, or using a generic (versus behaviorally-specific) intervention. Yet it makes good sense that the intensity of multi-level interventions all focused on one behavior could produce more successful outcomes.

Multi-level settings-based water interventions

In this *Handbook*, contributing authors of the settings-based approach have identified multi-level strategies in their chapters. We summarize these strategies (see Table 26.2) for each of the following venues: schools, household, workplaces, health-care organizations, rural settings and urban settings. In each chapter, we see examples of interventions that target individual behavior among students, workers, residents or health-care professionals (intrapersonal); interventions that address the relationships between key individuals (interpersonal); or, interventions that operate on the organizational, community or social/policy levels. The SE framework can categorize the different levels of influence, identify where gaps in intervention efforts exist and consider potential synergies across levels of influence.

While there are many strengths of using the SE framework, planning multi-level interventions often becomes complex, and implementation and evaluation efforts may be more demanding and time-consuming than a single-level intervention. There are also analytic challenges to evaluating multi-level interventions. Specifically, disentangling the independent effects of a particular intervention requires multi-level or hierarchical analyses that may be beyond the skill set of many practitioners and even some researchers. If additional biostatistical assistance is required, evaluation costs and time demands will increase. Yet, these apparent challenges are outweighed by the strengths of using the SE framework to organize settings-based interventions by considering their potential synergies and inter-dependencies.

Given the importance and complexity of this important public health issue, water-related interventions that are organized within a SE framework, including a settings-based approach, are in a position to produce real, effective and sustained change that occurs at the individual,

Table 26.2 Matrix of multi-level water interventions using the SE framework in a settings-based approach to water interventions

Setting (author)	Intrapersonal	Interpersonal	Institution/organization	Community	Public policy/societal
Schools (Freeman, Chapter 28)	Hygiene behaviors (p. n)	"Approaches to improve WASH behaviors among pupils…requires systematic engagement by teachers and parents" (p. n) "…(some) evidence that pupils can act as agents of change, influencing their siblings and parents to change their own wash practices" (p. n)	"…ministries of education typically have policy and budgetary purview over schools" (p. n)	"Changes to social norms… may prove a more effective approach" (p. n)	"Sustained provision of WinS requires capital expenditures…funds for maintenance…allocation for low-cost consumables (e.g. soap)…" (p. n)
Households (Lantagne and Gute, Chapter 27)	"Changing behavior requires the individual to invest energy…" (p. n) Description of an intrapersonal-focused randomized, controlled trial (RCT) in Liberian refugee camp (p. n) provided participants with materials, training, demonstrations	NA	"…school-based deworming (chemotherapy delivered to populations to suppress the burden of parasitic worms) was found to increase school attendance" (p. n)	NA	"Adoption of healthy behaviors can be, and often is, facilitated by a governmental central authority, through mechanisms such as…" (p. n) "…those living without central authority also are less likely to have the resources, support, and education that would assist them in using household interventions effectively" (p. n)

continued…

Table 23.1 continued

Setting (author)	Intrapersonal	Interpersonal	Institution/organization	Community	Public policy/societal
Workplaces (Rogers and Randolph, Chapter 29)	Amount of water an individual needs in workplace will depend on rate of sweating, which is affected by wearing protective equipment (as well as work conditions (organizational level)) (p. n)	NA	Job characteristics/work conditions – "…those who work …outdoors…may be at significant risk because they work in the direct sunlight, perform strenuous and prolonged work, often wear protective or impermeable suits, and may not always have ready and quick access to water" (p. n) Policies about safe water use during international travel	NA	Water-related OSHA regulations (e.g., must have potable water available to all employees at all times, employers must provide place for employees to wash hands)
Health care (Exner, Chapter 30)	Hand washing for individual health-care workers	"The water and sanitation safety group must be implemented as a subgroup of the infection control committee as a multidisciplinary group" (p. n)	Water and sanitation safety group	NA	Regulations for surveillance of bacteria and chemicals is not always routine in health-care facilities in many developed and developing countries

Intervention					
Rural community (Marks and Schwab, Chapter 31)	NA	Users of water systems have preferences about quality/taste, reliability, proximity to home, and cost of water (p. 5) "Water sector professionals widely cite the need to instill a sense of ownership for the project among water users" (p. n) Hygiene education has been found to be an effective means for reducing diarrheal disease in a systematic review and meta-analysis. "To fully realize the health benefits of a water supply project, it is important to provide hygiene education…" (p. n)	NA	In the "Demand responsive approach" (DRA), the water system involves community members in planning, operating, and maintaining water projects (p. n)	Low access to clean water in rural areas explained by barriers including lack of institutional support, unreliable/non-existent supply chains, intermittent power supplies (p. n) Projects that emphasize "gender- and poverty-sensitive decision-making" perform better over the long term. Varied modes in which women/poor individuals were engaged "suggests that [project teams must] allocate sufficient resources to assessing gender roles, power divisions, and caste or class disparities within each community" (p. n) Financing and management decisions such as cost sharing, tariff structure, and subsidies (p. n)
Urban interventions (Vairavamoorthy et al., Chapter 32)	NA	NA	Authors argue that the main barriers to achieving Integrated Urban Water Management are "institutional" and state that the "reform agenda should include coordination of … divisions…" (p. n)	"Multiple stakeholders need to be coordinated / involved in the planning, decision, and implementing process for IUWM" (p. n)	See example given for institutional – management may take place at a state or local level Authors mention the need to think about specific policy structures to ensure the quantity and quality of water in urban areas

organization, community and societal levels. By mobilizing support at all of these levels, we assume that reach and reinforcement of behaviorally-specific intervention strategies will produce lasting effects for targeted populations. By addressing the nuances of each specific setting via changes in norms, procedures, policies, structures and environmental support, contextually-relevant interventions can address specific subgroups of the population who are at highest risk, as well as the general population. Water is essential for life, and settings-based, multi-level interventions are critical for producing successful strategies for positive public health change.

References

Baron, S.L., Beard, S., Davis, L., Delp, L., Forst L., Kidd-Taylor A., Liebman A.K., Linnan L., Punnet L., and Welch L. (2013). Promoting an integrated approach to reducing health inequities among low-income workers: Applying a social ecological framework. *Am J Industrial Medicine.* doi10.1002 ajim22174.

Golden, S.E., and Earp, J.L. (2012). Social ecological approaches to individuals and their contexts: Twenty years of *Health Education & Behavior* health promotion interventions. *Health Educ Behav.* 39(3): 364–372.

Linnan, L., Sorensen, G., Colditz, G., Klar, N., and Emmons, K.M. (2001). Using theory to understand the multiple determinants of low participation in worksite health promotion programs. *Health Education and Behavior* 8(5): 591–607.

McLeroy, K.R., Bibeau, D., Steckler, A., and Glanz, K. (1988). An ecological perspective on health promotion programs. *Health Educ Quarterly* 15(4): 351–377.

Mullen, P., Evans, D., Forster, J., Gottlieb, N., Kreuter, M., Moon, R., O'Rourke, T., and Strecher, V. (1995). Settings as an important dimension in health education/promotion policy, programs and research. *Health Education and Behavior* 22: 329–345.

Poland B., Green L., and Rootman I. (eds). (2000). *Settings for Health Promotion: Linking Theory and Practice.* Thousand Oaks, CA: Sage.

Sallis, J.F., Owen, N., and Fisher, E.B. (2008). Ecological models of health behavior. In K. Glanz, B.K., Rimer, and K. Vishwanath (eds), *Health Behavior and Health Education: Theory, Research, and Practice* (4th ed., pp. 465–485). San Francisco, CA: Jossey-Bass.

Stokols, D. (1996). Translating social ecological theory into guidelines for community health promotion. *Am J Health Promotion* 10(4): 282–298.

World Health Organization. (1986). *The Ottawa Charter for Health Promotion.* Geneva, Switzerland: WHO. Retrieved from http://www.who.int/healthpromotion/conferences/previous/ottawa/en/index.html

27

HOUSEHOLD-FOCUSED INTERVENTIONS*

Daniele Lantagne

Usen Family Career Development Assistant Professor, Tufts University School of Engineering, Medford, MA, USA

David M. Gute

Professor, Tufts University School of Engineering, Medford, MA, USA

Learning objectives

1 Understand the factors that lead to successful public health programs and societal health change.
2 Understand the benefits and drawbacks of implementing interventions at the household level, and the appropriate times and contexts to implement household-based interventions.
3 Understand how best to evaluate and implement household interventions.

Introduction

The adoption of healthy behaviors can be, and often is, facilitated by a governmental central authority, through mechanisms such as:

- *Engineering*, such as the installation of water and sanitation infrastructure and seat belts in cars.
- *Subsidization*, as in the provision of subsidies to grocery stores to stock more healthy foods or providing no-interest loans to landlords and families to de-lead residences.
- *Legislation*, such as legal sanctions for failure to comply with seat belt or bicycle helmet use or international standards for attaching child safety seats in cars.
- *Taxation*, such as "sin taxes" on cigarettes and alcohol.

* Recommended citation: Lantagne, D. and Gute, D.M. 2015. 'Household-focused interventions', in Bartram, J., with Baum, R., Coclanis, P.A., Gute, D.M., Kay, D., McFadyen, S., Pond, K., Robertson, W. and Rouse, M.J. (eds) *Routledge Handbook of Water and Health*. London and New York: Routledge.

- *Education*, such as "Smoking Kills" labeling on cigarette packages, or posters about disease transmission from unclean hands above bathroom sinks.

In the twentieth century, huge strides were made in public health in the United States and other industrializing nations through programs implemented by central authorities. For example, in 1900 only slightly more than half of United States cities had any public water supply (Melosi, 1999), and the first chlorination of public water supplies was implemented in 1908. By 2000 – due to civil infrastructure engineering, subsidization at the local and national level, and legislation – greater than 90 percent of Americans had piped, disinfected water flowing into their homes.

In another example, a combination of engineering design, legislation based on research, and individual behavior change caused seatbelt use in the United States to increase from 14 percent to 84 percent in the relatively short time frame between 1983 and 2011 (Wikipedia, 2014). Although seat belt use depends on individual behavior change, seatbelt design improvements in cars and the "Click It or Ticket" legislative programs encouraged the majority of the population – excepting the approximately 15 percent of laggards/resistors – to use seatbelts regularly.

In an example from the United Kingdom, Conservative Prime Minister Margaret Thatcher established one of the world's first needle-exchange programs using legislation, not because of political ideals, but because it met the bottom line of reducing costs of HIV clinical care in the National Health Service (Szalavitz, 2008). This program was highly successful, with the United Kingdom maintaining a <1 percent HIV infection rate in people who inject drugs in the late 1980s, compared to an almost 50 percent infection rate in some US cities during the same time period. The reason for this success is that the incentives between legislation (to reduce costs and provide health care) and individual adopters (to legally use clean needles) were favorably aligned, and this led to high adoption of needle-exchange programs by individual injectors.

Although these successes have significantly improved public health in the United States and other countries, more work remains. While adult smoking rates have declined from 41 percent in 1944 to 20 percent in 2011 (Gallup, 2014), smoking remains the leading cause of preventable death in the United States, accounting for one out of every five deaths (CDC, 2014). Strategies to encourage people to not smoke and to stop smoking are continuously promoted. More recently, there is active debate on how best to address the emerging public health challenges of an increasingly obese population and gun violence. These current public health challenges in the United States, to date, have been less able to be effectively addressed through the centralized mechanisms of engineering, legislation and subsidization, and even significant taxation has not been effective at reducing cigarette purchases. Education to encourage individual behavior change has been the main mechanism used, to date, to address these public health challenges.

As citizens, educational messages encouraging us to improve our own health appear pervasive: *wear your seatbelt, wash your hands, use a condom, stop smoking, wear a bicycle helmet, exercise more, drink less alcohol, don't reuse needles, eat healthier, eat more fruits and vegetables, store your gun safely*. These messages are intended to motivate individuals to take action – to change their individual behavior – to prevent societally costly adverse health outcomes.

Changing behavior requires the individual to invest energy; in most cases, it is *easier* and/or more *fun* to eat the cookies, drink too much, not use a condom, or not bother with clicking your seatbelt, washing your hands, or exercising each day. In contrast to needle-exchange programs, sometimes the incentives are not aligned between those interested in protecting public health (who are working to combat obesity) and the individual (who wants to eat the cookie). Thus, the educational mechanisms that central authority can use to address these public health challenges are not always effective, as evidenced when someone smokes a cigarette from a package labeled "Smoking Kills!" Contrary to popular opinion, knowledge of health impacts is often not the primary driver of behavior change (Jenkins and Scott, 2006). Cultural and sociological factors often play a larger role; the desire for privacy while defecating encourages latrine use, wanting to extend one's life in order to play with grandchildren encourages smoking cessation, and social pressure encourages handwashing in public bathrooms.

The activation energy and commitment required to adopt healthy behaviors (that are sometimes counter to basic human tendencies) is reached when individuals determine that the perceived benefits outweigh the perceived inconveniences, as described in Chapter 21. There are a number of behavior change theories which address how health promotion activities can lead to behavior change.

While all individuals must take some responsibility for protecting their own health, in societies where there is not an effective functional central governmental authority, the individual must take on additional responsibility for protecting their own health. Many times this is by using products meant for the individual to use at the household level, such as: treating water with a filter in the home instead of receiving centrally treated water piped to prevent diarrhea; or, sleeping under a bednet instead of your community being sprayed with pesticides to prevent malaria. Ironically, this places the greatest burden for maintaining their own health on some of the world's most vulnerable populations: those living without central authority also are less likely to have the resources, support and education that would assist them in using household interventions effectively. The success of household interventions, by definition, depends on individual behavior change.

Because of this added burden, household level health interventions should be used as a last resort, promoted when centralized or community-managed options are not available, appropriate or possible. For example, handwashing is an efficacious intervention that depends on household/individual behavior change. However, handwashing can still be centrally supported by providing educational messages, providing handwashing stations, and providing soaps and sanitizers. Even with a household intervention without a way to provide it centrally (like handwashing), central support can still remove barriers to correct use.

In areas where it is not possible to currently implement centralized or community-managed options, there is extensive programming and interest in household interventions to reduce the burden of the three largest causes of death in children under five: improved cookstoves to prevent respiratory infections; household water treatment and handwashing to reduce diarrhea; and, mosquito bednets to reduce malaria.

There is active debate on how effective these commonly promoted household interventions actually are. Randomized, controlled trials (RCTs) are the gold standard for research for evaluating development interventions in general, and household interventions specifically. The intent of such research is to document the *efficacy* of the intervention – measuring the ability of the intervention to produce the desired result in observed populations. If programmatic impact is documented during an RCT, the intervention is then made available to more people during "scale-up". Using a community-based example, after school-based deworming (chemotherapy delivered to populations to suppress the burden of parasitic worms) was found to increase school attendance in one study in Western Kenya, the project was expanded throughout Kenya and 26 other countries (DTW, 2014).

RCTs often provide important evidence in understanding the efficacy and impact of interventions in the specific context being studied, where there is well-managed program implementation. For example, in an RCT of a flocculant/disinfectant sachet for household water treatment in Liberian refugee camps, participants were provided with all materials necessary to use the product, including replacement of materials stolen during the trial. Additionally, the primary caretaker received an extensive initial training, which included visual instructional materials and live demonstrations of the products. Lastly, weekly active diarrheal disease surveillance and water quality testing was conducted in households. Overall, a 91 percent reduction in diarrheal disease was documented (Doocy and Burnham, 2006).

A smooth transition from the well-supported RCT context to a scale-up program cannot be assumed. Adoption rates generally decrease in scale-up programs because there is less funding to support the provision of materials, training and ongoing effectiveness monitoring. *Effectiveness*, as distinct from *efficacy*, is whether the intervention is successful in real-world programs outside a trial situation. In the actual emergency distribution of the flocculant/ disinfectant sachet (where product was handed out in a non-food-item distribution box with only one minimal training), less than 5 percent of recipients knew how to use the sachet correctly and less than 5 percent had water treated with the sachet (Lantagne and Clasen, 2012). As can be seen in Figure 27.1, the recipients of the flocculant/disinfectant were busy with the tasks of everyday life. The woman shown is grinding corn and nursing her child sitting next to drinking water she collected and a house that she built. She has less than a third grade education. It is unreasonable to expect her to be able to treat her own water after one minimal training using a five-step flocculant/disinfectant treatment process.

Figure 27.1 Recipient of household water treatment and safe storage products in cholera emergency in Turkana, Kenya

Source: D. Lantagne.

Additionally, interventions that are efficacious or effective in one context may not be in another. A recent review of all the relevant literature found that deworming alone had no effect on growth, cognitive ability or school attendance (*BMJ*, 2013). While deworming was efficacious in increasing school attendance in Kenya, it was not, on average, efficacious across all studies. Thus, indiscriminate scaling-up of interventions without sufficient data documenting transferability is generally unwarranted.

Thus, the potential impact of a household level intervention depends strongly on a thoughtful and well-designed implementation strategy. Logistics such as transportation chains, availability of product and cost all impact access to, and acceptance of, products. Local traditions and beliefs affect willingness to use. Local conditions vary, and modifications to the product are often necessary. There is, generally, no one product appropriate for all locations in the world – even Coca Cola franchises modify the sweetener added to better match local taste preferences.

With regards to household water treatment products, research has shown that successful programs – those where the microbiological quality of user's household water is improved and the burden of diarrheal disease is reduced – distribute an *appropriate* product in a *cost-effective* manner to a trained population (Lantagne and Clasen, 2012). Two examples of successful real-world, long-term household water treatment implementation programs are presented in the following paragraphs.

A ceramic filter program in Cambodia reached over 2,000 households in the four years before the evaluations – where households that received a ceramic filter 0–44 months before the study were compared to matched controls – were conducted. In a cross-sectional evaluation, 156 of 506 (30.8 percent) households were still using the ceramic filter at the time of the visit, and the geometric mean *E. coli* concentration in filtered water was 14 colony forming units (CFU)/100 mL, as compared to 474 CFU/100 mL in non-filtered water (Brown et al., 2007). The main reason for filter disuse was breakage, with a 2 percent breakage rate per month being reported. A concurrent longitudinal diarrheal disease evaluation demonstrated a reduction in self-reported diarrheal disease of 46 percent in ceramic filter using households versus non-using households (Brown et al., 2008).

A liquid chlorine program in rural Haiti began in 2002 and subsequently reached more than 4,000 families over 10 years (Harshfield et al., 2012). In 2010, a total of 201 program participants were randomly selected for a cross-sectional survey and compared to controls. Participants had been enrolled in the program an average of 53 months at the time of the survey. Overall, 56 percent of participants (versus 10 percent of controls) had stored household water with free chlorine residuals between 0.2 and 2.0 mg/L, indicating effective disinfection and sufficient chlorine to ensure the water is both clean and would not be recontaminated. Additionally, significantly fewer children in participant households had an episode of diarrhea in the previous 48 hours with a 57 percent reduction in odds of diarrhea. When economics tools were used to complete additional analysis it was found that respondents who self-reported always treating (67 percent reduction in odds), had a bottle of chlorine in their home (52 percent reduction in odds), had a record of chlorine bottle sales (48 percent reduction in odds) or had free chlorine residual in their drinking water (50 percent reduction in odds); all had increased disease reduction compared to the study population (including adults and children) as a whole (39 percent reduction in odds) (Table 27.1).

Both of these long-term, successful household water treatment programs had extensive oversight and technical assistance by highly qualified program managers over the long-term nature of the program who could ensure the appropriate products were distributed to the

Table 27.1 Treatment on treated analysis example on the odds of diarrhea in the past 48 hours among sub-groups of program participants compared with controls, Haiti

Program participant sub-group	All ages				Children <5			
	OR	95% CI	p	N[1]	OR	95% CI	p	N[1]
All participants	0.61	0.45–0.82	0.001	3,148 (614)	0.43	0.26–0.70	0.001	328 (240)
Self-report always chlorinating	0.33	0.21–0.50	<0.001	2,462 (481)	0.30	0.15–0.60	0.001	268 (195)
Record of technician visit	0.68	0.47–0.99	0.043	2,620 (512)	0.44	0.22–0.89	0.021	270 (199)
Record of chlorine purchase in past year	0.52	0.26–1.06	0.071	2,228 (433)	0.46	0.12–1.75	0.252	233 (172)
Presence of bottled chlorine in household	0.48	0.32–0.72	<0.001	2,731 (534)	0.29	0.15–0.57	<0.001	286 (207)
Presence of free chlorine in water	0.50	0.31–0.79	0.003	2,619 (511)	0.30	0.14–0.64	0.002	268 (196)

Notes:

Standard errors adjusted for clustering at the household level.

1: Individuals (households)

Source: Adapted from Harshfield (2012).

trained population. While both of these programs are product cost-recovery (as users pay for the product), it is unclear whether they would be sustainable in the absence of the extensive oversight (which is not fully accounted for in product cost). These results raise concerns over program sustainability.

To translate these program successes into a theoretical framework (from Rehfuess and Bartram, 2014), the success of environmental health programs comes from the confluence of five factors: 1) direct (intrinsic) impact as measured by an RCT; 2) behavior change to use the product by the user ("compliance" in Rehfuess terminology); 3) delivery; 4) programming; and 5) policy measures. These two successful intervention programs delivered an efficacious intervention to populations that wanted to use the product with programmatic systems to encourage delivery that were supported by national policy. The confluence of these five factors leads to effective household-based environmental health interventions.

In summary, household interventions decentralize the management of health from a centralized authority to the individual level, and inherently place the greatest burden on those with the least resources. They are not simple to implement as time, training and support are needed to design and launch programs that lead to the household behavior change which yields the desired beneficial health outcomes. In spite of these difficulties, there is evidence that household interventions can be successful if well implemented, and as such they may be the preferred option in areas where centralized or community options are not available.

Key recommended readings

1   Harshfield, E., Lantagne, D., Turbes, A., Null, C. 2012. "Evaluating the sustained health impact of household chlorination of drinking water in rural Haiti." *Am J Trop Med Hyg* 87(5), 786–95. In this manuscript, data on the impact of a real-world household chlorination program in Haiti is presented and described. Significantly fewer children in participant households had an episode of diarrhea in the previous 48 hours with 59 percent reduced odds, and treatment-on-treated estimates of the odds of diarrhea indicated larger program effects for participants who met more stringent verifications of participation.

2   Lantagne, D., Clasen, T. 2012. "Use of household water treatment and safe storage methods in acute emergency response: case study results from Nepal, Indonesia, Kenya, and Haiti." *Environ Sci Technol* 46(20), 11352–60. In this manuscript, data on the effectiveness of household water treatment options in the acute emergency situation is presented. Effective use: the percentage of the targeted population with contaminated household water who used the household water treatment and safe storage (HWTS) method to improve stored drinking water microbiological quality to internationally accepted levels ranged from 0 to 67.5 percent; more successful programs provided an effective HWTS method, with the necessary supplies and training provided, to households with contaminated water who were familiar with the method before the emergency.

3   Lantagne, D., Quick, R. E., Mintz, E. 2006. "Household water treatment and safe storage options in developing countries: a review of current implementation practices." Woodrow Wilson International Center for Scholars' Environmental Change and Security Program. http://www.wilsoncenter.org/sites/default/files/WaterStoriesHousehold. pdf. In this article, five of the most common household water treatment options – chlorination, filtration (Biosand and ceramic), solar disinfection, combined filtration/ chlorination, and combined flocculation/chlorination – are described, including

common implementation strategies for each option. Implementing organizations are identified; the successes, challenges, and obstacles they have encountered in their projects are described; sources of funding and the potential to distribute and sustain each option on a large scale are presented; and, goals for future research and implementation are proposed.

References

BMJ (2013). Deworming debunked. http://www.bmj.com/content/346/bmj.e8558

Brown, J., Sobsey, M. D., amd Loomis, D. 2008. "Local drinking water filters reduce diarrheal disease in Cambodia: a randomized, controlled trial of the ceramic water purifier." *Am J Trop Med Hyg* 79(3), 394–400.

Brown, J., Sobsey, M., and Proum, S. 2007. *Use of Ceramic Water Filters in Cambodia*. Cambodia: Water and Sanitation Program of the World Bank/UNICEF.

CDC. 2014. "Current Cigarette Smoking Among Adults in the United States." http://www.cdc.gov/tobacco/data_statistics/fact_sheets/adult_data/cig_smoking/

Doocy, S., and Burnham, G. 2006. "Point-of-use water treatment and diarrhoea reduction in the emergency context: an effectiveness trial in Liberia." *Trop Med Int Health* 11(10), 1542–52.

DTW. 2014. Deworm the World Initiative. http://www.dewormtheworld.org/

Gallup. 2014. "One in Five U.S. Adults Smoke, Tied for All-Time Low." http://www.gallup.com/poll/156833/one-five-adults-smoke-tied-time-low.aspx

Harshfield, E., Lantagne, D., Turbes, A., and Null, C. 2012. "Evaluating the sustained health impact of household chlorination of drinking water in rural Haiti." *Am J Trop Med Hyg* 87(5), 786–95.

Jenkins, M., and Scott, B. 2006. "Behavioral indicators of household decision-making and demand for sanitation and potential gains from sanitation marketing in Ghana." London: London School of Hygiene and Tropical Medicine.

Lantagne, D., and Clasen, T. 2012. "Use of household water treatment and safe storage methods in acute emergency response: case study results from Nepal, Indonesia, Kenya, and Haiti." *Environ Sci Technol* 46(20), 11352–60.

Melosi, M. 1999. *The Sanitary City: Urban Infrastructure in America from Colonial Times to the Present*. Baltimore, MD: The Johns Hopkins University Press.

Rehfuess, E., and Bartram, J. 2013. "Beyond direct impact: evidence synthesis towards a better understanding of effectiveness of environmental health interventions." *International Journal of Hygiene and Environmental Health* 217, 155–9.

Szalavitz, M. 2008. "Needle exchange and the new drug czar." STATS, Statistical Assessment Service, Arlington, VA, USA. http://stats.org/stories/2008/needle_exchange_drug_czar_dec03_08.html

Wikipedia. 2014. "Seat belt use rates in the United States." http://en.wikipedia.org/wiki/Seat_belt_use_rates_in_the_United_States.

28

WATER IN SCHOOLS*

Matthew C. Freeman

ASSISTANT PROFESSOR OF ENVIRONMENTAL HEALTH,
EMORY UNIVERSITY, ATLANTA, GEORGIA, USA

Learning objectives

1 Understand existing evidence on the health and educational impact of access to water in schools.
2 Understand the contextual considerations for provision of water in low-income settings.
3 Understand the links between water access and sanitation and hygienic conditions and behaviors in schools.

Introduction

Children spend considerable time at school, and the condition of the learning environments is critical for educational attainment and to prevent illness. While the focus of this handbook is on water, any discussion of access to safe and sustainable water in schools necessarily includes how these conditions impact other hygienic conditions and behaviors in the school. This is because water, sanitation, and hygiene (WASH) are complementary for achieving health and educational outcomes and because they are typically budgeted together at the government and school level. As discussed in Chapter 26 in this handbook, schools provide an opportunity to target a vulnerable population – children – who may not receive education about WASH in other contexts. In addition, schools are seen throughout the world as trusted institutions that promote learning, which provides the opportunity to reach marginalized populations in low-income settings that may be missed by other development approaches. School-based management of water resources, or WASH infrastructure and behavior change in general, may serve as an entry point or model for community-based approaches.

Access to safe and sustainable water, basic sanitation, materials for hygiene and associated hygiene education are seen as critical for health and educational attainment of school children. It is estimated that 1.9 billion school days would be gained if the international targets for

---

* Recommended citation: Freeman, M.C. 2015. 'Water in schools', in Bartram, J., with Baum, R., Coclanis, P.A., Gute, D.M., Kay, D., McFadyen, S., Pond, K., Robertson, W. and Rouse, M.J. (eds) *Routledge Handbook of Water and Health*. London and New York: Routledge.

water and sanitation were achieved (Hutton and Haller, 2004). Schools were not included in the original Millennium Development Goal (MDG) targets, but have been proposed as part of the international development initiatives from 2015 onwards. This inclusion may provide some additional attention to provision of water in schools from national governments and international donors.

To meet the requirements of school children, WASH in schools (WinS) should include the necessary hardware and education to facilitate hygienic behaviors and recurrent costs for maintenance and consumables. In most countries, funds for the provision of WinS are provided by some combination of the ministry of education, parents, and school management committee. The largest capital expenditure is on water supply and, where water is piped, on monthly water bills. In many low-income settings, responsibility for maintenance and management of these facilities falls to the teachers and community members, but also pupils and, if schools employ them, janitors. The primary purpose of WinS programming is to provide or enable sustained WASH access at school (Adams et al., 2009). While this handbook's focus is on the provision of water, it is imprudent to talk about water without discussing the associated facilities and behaviors that a sufficient quantity of safe water supports, which include:

1 Sufficient quantities of safe water for drinking, handwashing and personal hygiene, and cleaning, and when appropriate water for cooking, flushing toilets, and school farms.
2 Child-friendly, gender-specific, culturally appropriate, private, and well maintained toilet facilities.
3 Personal hygiene materials, such as toilet paper, water, soap, and menstrual pads.
4 Hygiene education.
5 Safe disposal of solid waste.
6 Control measures to reduce transmission and morbidity of WASH-related illnesses.

The standards and approaches needed for WinS vary widely by country and context, just as schools themselves vary by target population, governance structure, and location. Targeting primary-school children will require different behavior change approaches and more child-friendly designs as compared to secondary schools. Private schools have different funding and governance mechanisms, as compared to public schools, just as religious schools may have different cultural needs as compared to secular institutions. An example is the provision of water needed for personal hygiene differs between Muslim and non-Muslim populations within East Africa and South Asia. In low-income settings, urban schools may have access to water and sewer infrastructure, while rural schools often rely on on-site systems, which require different technologies, financing, and approaches to maintenance. Differences in climate and geography, such as the water table depth, soil type and temperature, all impact the types of water systems required and toilet technologies that can and should be constructed at school. Finally, boarding schools require bathing facilities and additional water, as compared to day schools. Schools are regulated and funded by a variety of governance structures and may be subject to differing policies at the local and national levels. Gender needs are of critical importance for the provision of sanitation and personal hygiene related to WinS, and special attention must be paid to the needs of girls who represent a marginalized population in many parts of the world. Importantly, schools educate children with a variety of language skills, mental and physical abilities, and of all ages. These myriad populations, governance structures, and conditions require context-specific technical solutions for provision of WinS hardware and behavior change approaches.

While there are many similarities between community and school WASH infrastructure and behavior change approaches, there are a number of key differences, including the type of hardware needed to support use by school-aged children, financing and governance, approaches to behavior change needed to target school-age children, and the maintenance of facilities. Until recently there has been little attention to the importance of WinS to create an enabling environment for school learning, and as a necessary complement for other school education, health, and nutrition programs. An understanding of the most appropriate facilities and behavior change strategies are needed.

Global coverage data on WinS are limited. A number of countries have developed educational management informational systems (EMIS) to track WinS coverage, as well as other school parameters; access to those data are limited as well. The best estimates of provision of water supply and sanitation come from the 2013 UNICEF WASH annual report (UNICEF, 2014), which found that 47 percent of schools in least developed countries (69 percent in all reporting UNICEF countries) provide water, while 46 percent provide sanitation (67 percent in all reporting UNICEF countries). There are several limitations of these data, including (i) they rely on self-reported data from UNICEF country offices, (ii) are not systematically collected within each country with consistent definitions of "adequate" access, and (iii) they are unweighted, meaning that they treat countries like Ghana and China the same in analysis.[1] Inclusion of WinS indicators as part of the 2015 Sustainable Development Goals may lead to better monitoring and increased attention for WinS programming. WinS is not merely a problem in low- and middle-income settings as school conditions can also impact school attendance in high-income settings (Durán-Narucki, 2008). While the challenges and potential solutions differ in high-income settings, they do necessitate attention.

Evidence

Adequate provision of WASH in schools is considered a fundamental right, and some evidence exists that demonstrates WinS to positively impact the health and education of school-aged children. However, compared to the large body of evidence demonstrating the health impacts of improved WASH access and behaviors in the household (Fewtrell et al., 2005), relatively few rigorous studies have been conducted (Birdthistle et al., 2011; Jasper et al., 2012). There are no rigorous studies available purely assessing the impact of improving water quantities or water quality independent of hygiene or sanitation improvements; as such we discuss all aspects of WinS below. The primary outcomes assessed in WinS impact studies include:

1 Increased attendance, enrollment, and grade progression.
2 Reduced prevalence of infectious disease, including soil-transmitted helminth infection, diarrheal diseases, and acute respiratory infection.
3 Gender and socio-economic equity.
4 Enabling access for children with disabilities.

Figure 28.1 is a conceptual model that details the behaviors, health impacts, and educational impacts associated with WinS. The majority of studies assessing the impact of WinS on pupil absenteeism due to diarrheal disease have been conducted in high-income settings. These studies have focused on establishing a link between handwashing with hand sanitizers (either alcoholic or non-alcoholic) and absence, and have found reductions in absence of between

20 and 51 percent (Hammond et al., 2000; Dyer, 2001; White et al., 2001; Guinan et al., 2002; Morton and Schultz, 2004; Sandora et al., 2008). However, with the exception of Hammond and colleagues (2000), "these studies relied on data from selected classrooms in five schools or fewer and are of questionable rigor." A recent study from New Zealand found no effect of the provision of hand sanitizer, though the control group also included hygiene education (Priest et al., 2014).

In low-income settings, there have been few gold-standard randomized trials of WinS on pupil health and educational attainment. Bowen and colleagues (2007) assessed the impact hygiene improvements have on absenteeism in Chinese primary schools. Students that received a hygiene education program – which included education, soap provision, and enlistment of student hand washing "champions" – reported 42 percent fewer absence episodes and lower median duration of absences compared to the control group. Schools that received only education, but not soap provision, had no difference in absence relative to controls.

In Kenya, a comprehensive school hygiene promotion, water treatment, and sanitation intervention was shown to reduce absenteeism for girls between 21 percent and 37 percent (Freeman et al., 2012a). In a non-randomized study, O'Reilly and colleagues (2008) found that absence was 35 percent lower among pupils that attended schools that received handwashing and water treatment promotion, while it increased by 5 percent in nine nearby control schools over the same time period. A program that promoted a flocculant-disinfectant for water treatment reduced absence by 26 percent (Blanton et al., 2010). In India, a national school-based sanitation construction initiative increased enrollment among all students, and generally impacted girls more (Adukia, 2014), while in Kenya, Garn and colleagues found that a comprehensive school WASH program increased enrollment of girls by 4 percent relative to the control schools (Garn et al., 2013).

Regarding health, while the main driver of global WASH-related morbidity associated with diarrhea is children under five years, over 20 percent of morbidity for school-aged

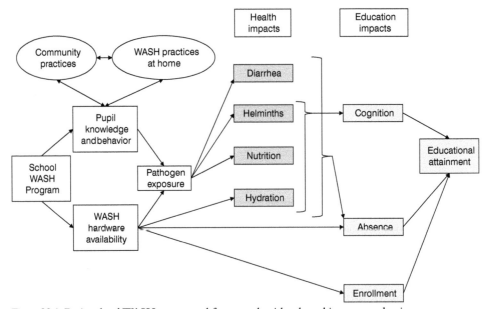

Figure 28.1 Basic school WASH conceptual framework with selected impact mechanisms

children can be linked to unsafe WASH (Prüss-Üstün et al., 2008). A study in Kenya found that a comprehensive WASH package was shown to reduce diarrhea among school children and their siblings, though diarrhea was not reduced for an intervention that just promoted hygiene and sanitation (Freeman et al., 2013a; Dreibelbis et al., 2014). School-aged children are the population at greatest risk of soil-transmitted helminth infections (STHs), a set of pervasive parasitic worms. The same study in Kenya showed that school sanitation and hygiene improvements in conjunction with mass drug administration (MDA) can reduce some STH infections compared to MDA alone (Freeman et al., 2013b). Similar to studies in children under five, improvements to handwashing have been shown to reduce acute respiratory illness (Lopez-Quintero et al., 2009; Talaat et al., 2011), but not others (Bowen et al., 2007). In addition to the indirect impact of WASH improvements on educational attainment associated with absence and health, there is some evidence that reducing dehydration improves pupils' cognitive abilities in high-income settings (Edmonds and Burford, 2009; Edmonds and Jeffes, 2009; Benton and Burgess, 2009).

School absenteeism may serve as a proxy for health status among children in developed countries (Houghton, 2003), but is associated with WinS in ways beyond pathogen control, especially for girls (Pearson and McPhedran, 2008). In schools without adequate water supply, the burden of collecting water often falls on the students, who report this chore both as the most demanding in time and energy and as the most dangerous (Hemson, 2007). Fetching and carrying water often causes children to miss school or arrive late, especially among students who must make more than one trip to collect water per day (Fisher, 2004; Hemson, 2007; Nauges and Strand, 2013). Fetching water also detracted from children's time to study, and decreased their alertness in class and morale (Hemson, 2007). Schools without adequate WinS and limited support structures are ill-equipped to support girls during menstruation (see Chapter 52). Children with disabilities, who are frequently unable to attend school, are marginalized further if adequate WASH facilities don't exist (Erhard et al., 2013). Further, WinS may also differ based on geography, age, or religion (Freeman et al., 2012a; Fehr et al., 2013). Teachers may also miss school due to the necessity of carrying water. And schools may have difficulty hiring and retaining good teachers in schools that do not have adequate WASH facilities (Fisher, 2004).

Provision of water and the link to sanitation and hygiene

Provision of water at school is critical to ensure that children have access to safe drinking water and sufficient water for personal hygiene throughout the day. Water in schools is typically programmed and budgeted for within the context of sanitation and hygiene behavior change and even menstrual hygiene management (Chapter 52). As discussed in the introductory chapter of this part on settings-based approaches (Chapter 26), one could consider the socio-ecological model when considering the provision of WinS. When considering WinS, the key domains are:

1   Policy: policies at the national level, including ministerial oversight, teacher training, and budgetary provision, will differ depending on the country and location.
2   Community: local and contextual factors that influence water provision, including the environment, access to water at the community level, and prioritization of the education environment.
3   Institutional: governance of school resources, engagement of parents, and roles and responsibilities at the local level.

4 Interpersonal: learning context at the school level, motivation and opportunity to engage in peer-to-peer learning, and behaviors among models, such as teachers and parents.
5 Intrapersonal: behaviors, attitudes, and motivation of individual students.

Sustained provision of WinS requires capital expenditures for hardware (e.g. water supply, sanitation facilities, handwashing stations), funds for maintenance of facilities (e.g. spare parts, cleaning supplies, and the time of school staff and children), allocation for low-cost consumables (e.g. soap, anal cleansing materials), and behavior change programs. In many countries, there are established standards for capital expenditures associated with WinS. The ministries of education are largely responsible for allocation of WinS resources, and the focus of these resources are typically on capital expenditures, and perhaps paying utility bills, not maintenance of facilities or consumables, such as soap. The construction standards for facilities are typically set by the government though, in many countries, these standards may not meet the needs of all children in all contexts. WinS is part of an enabling learning environment that ensures the health, safety, and dignity of children, and facilitates other essential education, nutrition, and public health programs. However, in many low-income countries, the resources dedicated to provision and maintenance of facilities is insufficient.

Water provision

To meet the needs of school children, provision of water should be both safe for drinking and in sufficient quantities for drinking, practicing personal hygiene (including anal cleansing in certain settings), and cleaning of facilities. There is no empirical evidence for the number of liters per child that should be provided, but international standards are for 5L per child per day for day schools and 20L liters for boarding schools (Adams et al., 2009). Ideally, water will also be available for bathing (especially for boarding schools) and productive purposes, such as school gardens. Where water is not provided on school grounds, children, specifically girls, may be required to fetch water. Many schools do have rainwater catchment facilities, but those are only viable during certain times of year. The fetching of water by girls, at times forcing them to miss class, represents a considerable barrier to educational equity. To prevent water-borne disease, water should be free of fecal pathogens, and proper water handling practices can be promoted to prevent contamination. Water provided at school should be tested to ensure that it meets World Health Organization (WHO) and national standards for chemical contamination. There is little data on the frequency of water testing at school water sources in low-income settings. In rural areas of many low-income settings, water systems at school are managed by the community and provide access to the community in addition to the school children. The sustainability of service delivery has been poor using this model of community management. In high-income settings, one challenge is encouraging drinking water over consumption of sugary beverages in order to reduce calorie intake, reduce cavities, and improve cognition (Patel and Hampton, 2011). Lead and other chemical contamination is also of critical importance in schools with aging infrastructure (Bryant, 2004).

Basic sanitation

Water supply is critical for the maintenance of sanitary facilities in schools. In many schools throughout the world, sanitation facilities require piped water networks for flushing or water access for pour flush sanitation designs. The proper construction and use of school sanitation facilities, as well as the management of effluent, can influence contamination of ground and surface water. While all households have some access to water, that is not the case for all schools. The access to water at school will greatly influence the types and conditions of sanitation facilities for pupils. Though standards vary between contexts, there is general agreement that school sanitation facilities should be gender-separated, clean, easy to maintain and safe (Adams et al., 2009), and the dirtiness of facilities may deter use, especially for girls (Garn et al., 2014). Few rigorous, empirical studies have been conducted that describe pupil preferences regarding sanitation, though it has been shown that improved pupil to latrine ratios do lead to an increase in toilet use (Garn et al., 2014). Whether the facilities are pour flush, dry sanitation, or ecological sanitation depends on soil type, local materials and resources, and government standards. One of the key differences between community sanitation and school sanitation is that school sanitation requires a level of durability – both in terms of the superstructure and capacity – not needed for a typical family. Whether those capital expenditures are met by the government, community or outside donors, there is frequently little money for maintenance of the facilities or funds for recurrent costs associated with cleaning. Moreover, cleaning burden may fall disproportionately on girls, as has been reported in Mexico (Snel and Shordt, 2005).

Hygiene

Access to sufficient quantities of water in school is required for maintaining hygienic conditions at school and for handwashing with soap at key times. Hygiene, as related to water supply within the context of school-based provision, is mostly focused on promotion of handwashing with soap at key times, specifically after using the toilet and before eating. Water, along with soap, should be provided close to toilet facilities to ensure that handwashing takes place following defecation. In schools where there are school feeding programs, having a safe and reliable access to water is necessary for food safety. There are other aspects of hygiene that could be considered and are often neglected, including provision of water or tissue for anal cleansing (McMahon et al., 2011; Jasper et al., 2012); materials, infrastructure, and education on menstrual hygiene management (Caruso et al., 2013; Haver et al., 2013; Long et al., 2013); education of food vendors and food preparers for school feeding programs; and bathing facilities for personal hygiene. Without proper hygiene behaviors, provision of school sanitation may actually lead to increased risk of fecal pathogen exposure (Greene et al., 2012). One of the key challenges of hand hygiene is provision of water and sustainability of soap provision. Tippy taps are a simple, low-cost technology that has been used widely in sub-Saharan Africa to establish handwashing stations at schools. The use of soap solution, as opposed to reliance on bar soap, has seen some promising results in promoting sustained provision of soap (Saboori et al., 2010; Saboori et al., 2013).

Behavior change

Approaches to improve WASH behaviors among pupils both at school and at home require systematic engagement by teachers and parents. Key water-related behaviors include safe water handling to prevent microbiological contamination (Chapter 21) in places where water is stored, water treatment at the point of use, and the use of water for personal hygiene and general cleaning of the school compound and sanitation facilities. There are several well-known behavior change approaches that have been used to promote WASH in schools, though evidence on the effectiveness of these approaches is limited. Few behavior change approaches target water provision or safe handling of water specifically, and none do so without also discussing sanitation and hygiene-related behavior change. Many of these behavior change approaches use knowledge generation around disease risk (specifically diarrhea) to promote WASH behavior change, such as the Personal Hygiene and Sanitation Education (PHASE) approach. Participatory Hygiene and Sanitation Transformation (PHAST) and School-led Total Sanitation (SLTS) both rely on participatory approaches to increase demand and take action for changing the environment.

Changes to social norms and habit formation may prove to be a more effective approach (Sidibe, 2007). Hygiene promotion can facilitate interactions among school children and reinforce social norms within the school that can lead to sustained behavior change and habit formation (Dutton et al., 2011; Chittleborough et al., 2012; Le Thi Thanh Xuan et al., 2013). There is (some) evidence that pupils can also act as agents of change, influencing their siblings and parents to change their own wash practices (Onyango-Ouma et al., 2005). However, impact depends on continued access to functional WASH services at home and systematic engagement either through specific activities or peer-to-peer learning.

Monitoring WinS

In general, the monitoring of water and WinS in low-income settings is not particularly robust. It is frequently the case that a ministry of water will be responsible for monitoring and managing water supply, though, in many parts of the world, communities manage school-based supplies. Monitoring may extend only to the reliable provision of water, but lack microbiological or chemical contamination, even if policies exist requiring normal testing. Since ministries of education typically have policy and budgetary purview over schools, monitoring of WinS in general, and water specifically, may be included in teacher or school performance measures. Due to the intersectoral nature of WinS, many countries have adopted inter-ministerial working groups. Typically at the local level, teachers, parent-led school management committees, or local education officials take responsibility for managing and monitoring school facilities. Due to the many competing priorities, this may mean that WinS is neglected. In many contexts, students are engaged in maintenance and teachers are assigned to ensure that water is available and that WinS facilities are clean and well-maintained, though no formal monitoring processes may be in place. Few countries have robust monitoring systems that hold teachers, policy makers, or other stakeholders accountable for inadequate water access or poor WinS conditions. The 2015 Sustainable Development Goals will target universal access to water in schools, which may change what we know about global access and inform a more robust monitoring system. A number of countries have recently, or are developing, electronic education management information systems (EMIS) that collect WASH information, among other data, and the key recommendations from the AfricaSan conference held in 2014 was to set minimum standards and improve monitoring through EMIS systems (Cross and Coombes, 2014). While there are not yet standard indicators for EMIS, UNICEF has developed a toolkit that can serve as a starting place for monitoring WinS access.

Acknowledgements

I wish to acknowledge Bethany Caruso and the knowledge group of UNCIEF, including Christie Chatterly, Leslie Moreland, Nick Chudeau, and Peter van Maanen for their contributions.

Note

1   UNICEF has supported an effort to gather additional data on WinS coverage (http://www.washinschoolsmapping.com/maps.html).

Key recommended readings (open access)

1   Water, Sanitation and Hygiene Standards for Schools in Low-cost Settings: http://www.who.int/water_sanitation_health/publications/wash_standards_school.pdf. This set of standards developed by the WHO is designed to provide a way for policy makers to prioritize standards within their country.
2   Toward Effective Programming on WASH in Schools: http://www.unicef.org/wash/schools/files/rch_effective_programming_2010.pdf. This manual, developed by IRC Netherlands, discusses some of the key components of school WASH.
3   WASH in Schools Monitoring Package: http://www.unicef.org/wash/files/WASH_in_Schools_Monitoring_Package_English.pdf. This document provides guidance on monitoring WinS programs.
4   What impact does the provision of separate toilets for girls at school have on their primary and secondary school enrolment, attendance and completion? A systematic review of the evidence: http://www.3ieimpact.org/en/evidence/systematic-reviews/details/70/ Systematic review of the evidence on the impact of WinS.

References

Adams, J., Bartram, J., Chartier, Y. & Sims, J. (eds) 2009. *Water, Sanitation and Hygiene Standards for Schools in Low-cost Settings*. Geneva: WHO. http://www.who.int/water_sanitation_health/publications/wash_standards_school.pdf

Adukia, A. 2014. *Sanitation and Education*. Cambridge, MA: Harvard University.

Benton, D. & Burgess, N. 2009. The effect of the consumption of water on the memory and attention of children. *Appetite*, 53, 143–146.

Birdthistle, I., Dickson, K., Freeman, M. & Javidi, L. 2011. What impact does the provision of separate toilets for girls at schools have on their primary and secondary school enrolment, attendance and completion? Social Science Research Unit, Institute of Education, University of London.

Blanton, E., Ombeki, S., Oluoch, G., Mwaki, A., Wannemuehler, K. & Quick, R. 2010. Evaluation of the role of school children in the promotion of point-of-use water treatment and handwashing in schools and households–Nyanza Province, Western Kenya, 2007. *American Journal of Tropical Medicine and Hygiene*, 82, 664–671.

Bowen, A., Ma, H., Ou, J., Billhimer, W., Long, T., Mintz, E., Hoekstra, M. & Luby, S. P. 2007. A cluster-randomized controlled trial evaluating the effect of a handwashing-promotion program in Chinese primary schools. *Am J Trop Med Hyg*, 76, 1166–1173.

Bryant, S. 2004. Lead-contaminated drinking waters in the public schools of Philadelphia. *Clinical Toxicology*, 42, 287–294.

Caruso, B. A., Fehr, A., Inden, K., Sahin, M., Ellis, A., Andes, K. L. & Freeman, M. C. 2013. *WASH in Schools Empowers Girls' Education in Freetown, Sierra Leone: An Assessment of Menstrual Hygiene Management in Schools*. New York, United Nations Children's Fund.

Chittleborough, C. R., Nicholson, A. L., Basker, E., Bell, S. & Campbell, R. 2012. Factors influencing hand washing behaviour in primary schools: process evaluation within a randomized controlled trial. *Health Education Research*, 27, 1055–1068.

Cross, P. & Coombes, Y. 2014. *Sanitation and Hygiene in Africa. Where Do We Stand? Analysis from AfricaSan Conference, Kigali, Rwanda*. London: IWA. http://www.iwaponline.com/wio/2013/wio2013RF9781780405421.pdf

Dreibelbis, R., Freeman, M. C., Greene, L. E., Saboori, S. & Rheingans, R. 2014. The impact of school water, sanitation, and hygiene interventions on the health of younger siblings of pupils: a cluster-randomized trial in Kenya. *Am J Public Health*, 104, e91–e97.

Durán-Narucki, V. 2008. School building condition, school attendance, and academic achievement in New York City public schools: a mediation model. *Journal of Environmental Psychology*, 28, 278–286.

Dutton, P., Peschiera, R. F. & Nguyen, N. K. 2011. The power of primary schools to change and sustain handwashing with soap among children: the cases of Vietnam and Peru. World Bank's Water and Sanitation Program, Lima.

Dyer, D. L. 2001. Does use of an instant hand sanitizer reduce elementary school illness absenteeism? *The Joural of Family Practice*, 50, 64.

Edmonds, C. J. & Burford, D. 2009. Should children drink more water?: The effects of drinking water on cognition in children. *Appetite*, 52, 776–779.

Edmonds, C. J. & Jeffes, B. 2009. Does having a drink help you think? 6–7-Year-old children show improvements in cognitive performance from baseline to test after having a drink of water. *Appetite*, 53, 469–472.

Erhard, L., Degabriele, J., Naughton, D. & Freeman, M. C. 2013. Policy and provision of WASH in schools for children with disabilities: a case study in Malawi and Uganda. *Global Public Health*, 8, 1000–1013.

Fehr, A., Sahin, M. & Freeman, M. C. 2013. Sub-national inequities in Philippine water access associated with poverty and water potential. *J WASH Dev*, 3, 638–645.

Fewtrell, L., Kaufmann, R. B., Kay, D., Enanoria, W., Haller, L. & Colford Jr, J. M. 2005. Water, sanitation, and hygiene interventions to reduce diarrhoea in less developed countries: a systematic review and meta-analysis. *The Lancet Infectious Diseases*, 5, 42–52.

Fisher, J. 2004. The EDUCATION Millennium Development Goal: What water, sanitation and hygiene can do. WELL Briefing Note. Leicestershire: WEDC.

Freeman, M., Greene, L., Dreibelbis, R., Saboori, S., Muga, R., Brumback, B. & Rheingans, R. 2012a. Assessing the impact of a school-based water treatment, hygiene, and sanitation program on pupil absence in Nyanza Province, Kenya: a cluster-randomized trial. *Trop Med Int Health*, 17, 380–391.

Freeman, M. C., Erhard, L., Fehr, A. & Ogden, S. 2012b. Equity of access to WASH in schools: a comparative study of policy and service delivery in Kyrgyzstan Malawi, Philippines, Timor-Lest, Uganda, and Uzbekistan. New York: UNICEF and Emory University.

Freeman, M., Clasen, T., Dreibelbis, R., Saboori, S., Greene, L., Brumback, B., Muga, R. & Rheingans, R. 2013a. The impact of a school-based water supply and treatment, hygiene, and sanitation programme on pupil diarrhoea: a cluster-randomized trial. *Epi Infect*, 1136: 80–91..

Freeman, M. C., Clasen, T., Brooker, S. J., Akoko, D. O. & Rheingans, R. 2013b. The impact of a school-based hygiene, water quality and sanitation intervention on soil-transmitted helminth reinfection: a cluster-randomized trial. *Am J Trop Med Hyg*, 89, 875–883.

Garn, J., Greene, L. E., Dreibelbis, R., Saboori, S., Rheingans, R. & Freeman, M. C. 2013. A cluster-randomized trial assessing the impact of school water, sanitation, and hygiene improvements on pupil enrollment and gender parity in enrollment. *Journal of Water, Sanitation, and Hygiene for Development*, 3(4): 592–601.

Garn, J. V., Caruso, B. A., Drews-Botsch, C. D., Kramer, M. R., Brumback, B. A., Rheingans, R. D. & Freeman, M. C. 2014. Factors associated with pupil toilet use in Kenyan primary schools. *International Journal of Environmental Research and Public Health*, 11, 9694–9711.

Greene, L. E., Freeman, M. C., Akoko, D., Saboori, S., Moe, C. & Rheingans, R. 2012. Impact of a school-based hygiene promotion and sanitation intervention on pupil hand contamination in western Kenya: a cluster randomized trial. *Am J Trop Med Hyg*, 87, 385–393.

Guinan, M., Mcguckin, M. & Ali, Y. 2002. The effect of a comprehensive handwashing program on absenteeism in elementary schools. *American Journal of Infection Control*, 30, 217–220.

Hammond, B., Ali, Y., Fendler, E., Dolan, M. & Donovan, S. 2000. Effect of hand sanitizer use on elementary school absenteeism. *Am J Infect Control*, 28, 340–346.

Haver, J., Caruso, B. A., Ellis, A., Sahin, M., Villasenor, J. M., Andes, K. L. & Freeman, M. C. 2013. *WASH in Schools Empowers Girls' Education in Masbate Province and Metro Manila, Philippines An Assessment of Menstrual Hygiene Management in Schools*. New York: United Nations Children's Fund.

Hemson, D. 2007. "The toughest of chores": policy and practice in children collecting water in South Africa. *Policy Futures in Education*, 5, 315–326.

Houghton, F. 2003. The use of primary/national school absenteeism as a proxy retrospective child health status measure in an environmental pollution investigation. *Journal of the Royal Institute of Public Health*, 117, 417–423.

Hutton, G. & Haller, L. 2004. *Evaluation of the Costs and Benefits of Water and Sanitation Improvements at the Global Level*. Water Sanitation and Health: Protection of the Human Environment. Geneva: WHO.

Jasper, C., Le, T.-T. & Bartram, J. 2012. Water and sanitation in schools: a systematic review of the health and educational outcomes. *International Journal of Environmental Research and Public Health*, 9, 2772–2787.

Le Thi Thanh Xuan, T. R., Hoat, L. N., Dalsgaard, A. & Konradsen, F. 2013. Teaching handwashing with soap for schoolchildren in a multi-ethnic population in northern rural Vietnam. *Global Health Action*, 6, 1–12.

Long, J., Caruso, B. A., Lopez, D., Vancraeynest, K., Sahin, M., Andes, K. L. & Freeman, M. C. 2013. *WASH in Schools Empowers Girls' Education in Rural Cochabamba, Bolivia: An Assessment of Menstrual Hygiene Management in Schools*. New York: United Nations Children's Fund.

Lopez-Quintero, C., Freeman, P. & Neumark, Y. 2009. Hand washing among school children in Bogota, Colombia. *American Journal of Public Health*, 99, 94.

McMahon, S., Caruso, B. A., Obure, A., Okumu, F. & Rheingans, R. D. 2011. Anal cleansing practices and faecal contamination: a preliminary investigation of behaviours and conditions in schools in rural Nyanza Province, Kenya. *Tropical Medicine & International Health*, 16, 1536–1540.

Morton, J. L. & Schultz, A. A. 2004. Healthy hands: use of alcohol gel as an adjunct to handwashing in elementary school children. *Journal of School Nursing*, 20, 161–167.

Nauges, C. & Strand, J. 2013. *Water Hauling and Girls' School Attendance: Some New Evidence from Ghana*. World Bank Policy Research Paper 6443. Washington, DC: World Bank.

Onyango-Ouma, W., Aagaard-Hansen, J. & Jensen, B. B. 2005. The potential of schoolchildren as health change agents in rural western Kenya. *Soc Sci Med*, 61, 1711–1722.

O'Reilly, C., Freeman, M., Ravani, M., Migele, J., Mwaki, A., Ayalo, M., Onbeki, S., Hoekstra, R. & Quick, R. 2008. The impact of a school-based safe water and hygiene programme on knowledge and practices of students and their parents: Nyanza Province, western Kenya. *Epidemiol Infect*, 136, 80–91.

Patel, A. I. & Hampton, K. E. 2011. Encouraging consumption of water in school and child care settings: access, challenges, and strategies for improvement. *American Journal of Public Health*, 101, 1370.

Pearson, J. & McPhedran, K. 2008. A literature review of the non-health impacts of sanitation. *Waterlines*, 27, 48–61.

Priest, P., McKenzie, J. E., Audas, R., Poore, M., Brunton, C. & Reeves, L. 2014. Hand sanitiser provision for reducing illness absences in primary school children: a cluster randomised trial. *PLoS Medicine*, 11, e1001700.

Prüss-Üstün, A., Bos, R., Gore, F. & Bartram, J. 2008. *Safer Water, Better Health: Costs, Benefits and Sustainability of Interventions to Protect and Promote Health*. Geneva: World Health Organization.

Saboori, S., Mwaki, A. & Rheingans, R. 2010. Is soapy water a viable solution for handwashing in schools? *Waterlines*, 29, 329–336.

Saboori, S., Greene, L. E., Moe, C. L., Freeman, M. C., Caruso, B. A., Akoko, D. & Rheingans, R. D. 2013. Impact of regular soap provision to primary schools on hand washing and *E. Coli* hand contamination among pupils in Nyanza Province, Kenya: a cluster-randomized trial. *The American Journal of Tropical Medicine and Hygiene*, 80(4): 696–708.

Sandora, T. J., Shih, M.-C. & Goldmann, D. A. 2008. Reducing absenteeism from gastrointestinal and respiratory illness in elementary school students: a randomized, controlled trial of an infection-control intervention. *Pediatrics*, 121, e1555–e1562.

Sidibe, M. A. 2007. Can hygiene be cool and fun? Understanding school children's motivations to use their school toilets and wash their hands with soap in Dakar, Senegal. DrPH. http://esa.un.org/iys/docs/san_lib_docs/Can_Hygiene_be_cool_and_fun_-_Senegal__2007.pdf

Snel, M. & Shordt, K. 2005. The evidence to support hygiene, sanitation and water in schools. *Waterlines*, 23, 2–5.

Talaat, M., Afifi, S., Dueger, E., El-Ashry, N., Marfin, A., Kandeel, A., Mohareb, E. & El-Sayed, N. 2011. Effects of hand hygiene campaigns on incidence of laboratory-confirmed influenza and absenteeism in schoolchildren, Cairo, Egypt. *Emerging Infectious Disease*, 17, 619–625.

UNICEF. 2014. *Water, Sanitation and Hygiene Annual Report 2013*. New York: UNICEF.

White, C., Shinder, F. S., Shinder, A. L. & Dyer, D. L. 2001. Reduction of illness absenteeism in elementary schools using an alcohol-free instant hand sanitizer. *Journal of School Nursing*, 17, 258–265.

29

WATER AND HYDRATION IN THE WORKPLACE*

Bonnie Rogers

DIRECTOR, NC OCCUPATIONAL SAFETY AND HEALTH AND EDUCATION AND
RESEARCH CENTER AND OHN PROGRAM, GILLINGS SCHOOL OF GLOBAL PUBLIC
HEALTH, UNIVERSITY OF NORTH CAROLINA, CHAPEL HILL, NC, USA

Susan Randolph

CLINICAL ASSISTANT PROFESSOR, OCCUPATIONAL HEALTH NURSING
PROGRAM AND DEPUTY DIRECTOR, NC OCCUPATIONAL SAFETY AND HEALTH
EDUCATION AND RESEARCH CENTER, GILLINGS SCHOOL OF GLOBAL PUBLIC
HEALTH, UNIVERSITY OF NORTH CAROLINA, CHAPEL HILL, NC, USA

Learning objectives

1 Discuss the importance of water and hydration in the workplace.
2 Describe hot environment jobs and related health hazards.
3 Describe water-related health risks in the work environment (e.g., mold, contamination).

Water is essential for life; however, only a small percentage of fresh water is available for human life. It is estimated that nearly 2 billion people in low and middle income countries lack access to safe water for drinking, personal hygiene, and domestic use (Onda et al. 2012) and nearly 4 billion lack access to adequate sanitation facilities (Baum et al. 2013). Lack of access to clean drinking water can result in water-related disease such as diarrheal illnesses estimated at more than 500,000 deaths per year worldwide (World Health Organization 2014a). Available water resources vary markedly across countries and, as reported by the U.S. Agency for International Development (USAID), projections are that by the year 2025 two-thirds of the world's population could be living in severe water stress condition (USAID

---

* Recommended citation: Rogers, B. and Randolph, S. 2015. 'Water and hydration in the workplace', in Bartram, J., with Baum, R., Coclanis, P.A., Gute, D.M., Kay, D., McFadyen, S., Pond, K., Robertson, W. and Rouse, M.J. (eds) *Routledge Handbook of Water and Health*. London and New York: Routledge.

2013). However, there is no discussion about the impact of water availability and sanitation related to the workplace. As pointed out by the World Health Organization (2014a), when water comes from improved and more accessible sources, less time and effort is spent physically obtaining it, meaning that people can be productive in other ways. This includes having safe, accessible water in the work environment.

Water in the workplace is essential in order to maintain adequate hydration of workers and for good hygiene. While water treatment is available in many developed countries, such as the United States, pathogen contamination remains a public health threat from existing, new and resistant pathogens (e.g., *Cryptosporidium*, Ebola). Water/dampness can also pose health hazards in terms of mold growth, or if drinking water becomes contaminated with foreign (biological, chemical) agents. In addition, industrial/agricultural contamination in the workplace and surrounding areas from various elements (e.g., waste water toxins, pesticides) continues to be prevalent (Foran 2011). This chapter will discuss the importance of water and hydration in the workplace, hazards in the work environment related to water needs and exposure (e.g., heat, mold, water-related hypothermia, wastewater exposure), and work-related interventions such as hygiene and decontamination. Pertinent U.S. Department of Labor (US DOL) Occupational Safety and Health Administration (OSHA) regulations for potable water are addressed.

Hydration

Water is an essential and necessary nutrient for good hydration and should not be overlooked in the workplace. Water's importance for prevention of nutrition-related noncommunicable diseases has emerged because of the shift toward fluids from caloric beverages (Popkin et al. 2011). There is disagreement about what the average person needs to consume per day to survive in a moderate climate with average activity including work and ranges from 2 to 5 liters per day (Forrester 2006; Gleick et al. 2009). Good hydration is important to workers' health and safety. Even mild levels of dehydration adversely affect both physical and mental performance, but these effects can be made worse by the physical demands of the job, a hot work environment, or the need to wear protective clothing (Burke 2000). The issue of high caffeine-related beverages is particularly important as energy drinks (generally high in caffeine) are often offered and consumed at the workplace. In a sample of 18 adult males aged 24–39, investigators reported no significant differences in hydration, as measured by weight change from various combinations of beverages, including caffeinated drinks. However, much of the data reported related to diet and weight were self-reported, lacking control, and the age group was severely limited. Investigators indicated a more controlled study would be beneficial (Grandjean et al. 2000). The Institute of Medicine (2004) has also reported that caffeine effects are transient with regards to hazards to hydration effects. However, more recent studies show significant intracellular dehydration, hypertension, fatigue, higher rates of illness, and mental health disorders among workers who have consumed energy drinks (Abbott 2010; Toblin et al. 2012). Industries that employ mostly male populations, such as construction, mining, agriculture, and utilities, are at greater risk because men consume more energy drinks than women (Substance Abuse and Mental Health Services Administration, Center for Behavioral Health Statistics Quality 2013). Workers who have jobs with high physical demands, long hours, tedious work, or interruption of circadian rhythms or those who work more than one job look for ways to stay alert. When energy drinks are consumed, workers may be at risk from the ill effects such as dehydration that can directly affect their fitness for duty, especially when combined with exposures to other workplace hazards (Dennison et al. 2013).

We know about dehydration at sports events, school practice, and in hot work environments when workers are water depleted, but the effects of dehydration, although evident in the workplace, are not widely promoted. When a worker on the job is doing physical work, sweat can be a greater output than the water intake. This is the cause of dehydration and as the activity is heightened or the heat goes up, the dehydration can accelerate. Even the type of clothes worn can impact the dehydration cycle making it difficult to keep up with proper hydration.

Some of the effects that can be observed from dehydration include decreased productivity, increase in work-related accidents, decreased cognitive performance, decreased visual motor tracking, reduced short-term memory and attention, dizziness, cramping, fatigue, and mobility and standing issues all of which put the worker at significant health risk, can threaten the safety of co-workers, and can be costly for the employer. What is becoming clear is that dehydration is a key source of widespread problems on the job. Operations involving high air temperatures and humidity, radiant heat sources, direct physical contact with hot objects, or strenuous physical conditions have the potential for inducing dehydration and heat stress in workers engaged in job functions in specific industries, such as construction, firefighting, kitchen work, certain manufacturing plants, mines, and military (Rogers et al. 2007).

Hazards and the work environment

Hot work environments

For the human body to maintain a constant internal temperature, the body must rid itself of excess heat. This is achieved primarily through varying the rate and amount of blood circulation to the outer layers of the skin and the release of fluid onto the skin by the sweat glands. The evaporation of sweat cools the skin, releasing large quantities of heat from the body. As area temperatures approach normal skin temperature, cooling of the body becomes more difficult. If air temperature is as warm or warmer than the skin, blood brought to the body surface cannot lose its heat, and sweating becomes the primary means of maintaining a constant body temperature. Sweating does not cool the body unless the moisture is removed from the skin by evaporation. Under conditions of high humidity, the evaporation of sweat from the skin is decreased and the body's efforts to maintain acceptable body temperature may be significantly impaired (Texas Department of Insurance [TDI], Division of Workers' Compensation Workplace Safety 2008). Sweat evaporates very rapidly in low humidity, which can lead to severe dehydration if one does not drink enough water throughout the day.

Working in hot environments leads to water loss from both sweating and increased respiration if water is not replaced. Dehydration can occur along with a loss of electrolytes and plasma volume, further increasing the core temperature (Popkin et al. 2011). According to Clap et al. (2002), the rate of sweating varies among individuals and will be affected by work conditions and wearing protective equipment that can lead to sweating rates of as much as 2.25 liters per hour. For example, those who work in hot environments or outdoors, including construction and agricultural workers, baggage handlers, electrical power transmission and control workers, sports professionals and trainers, and landscaping and yard maintenance workers, may be at significant risk of heat stress because they work in the direct sunlight, perform strenuous and prolonged work, often wear protective or impermeable suits, and may not always have ready and quick access to water (US DOL, OSHA 2013). Care must be taken with the older worker who, with advancing age, experiences a reduction in total body water, a decline in renal function, and decreased responses to heat (Grandjean et al. 2000; Popkin et al. 2011). With the combination of high heat and often high humidity and the lack of fluid replacement, workers can suffer not only heat exhaustion or stroke but can die on the job. When heat stroke occurs, the core body temperature exceeds 105°F, resulting in complications damaging most body organ systems, central nervous system failure, and death (Karriem-Norwood 2014).

There are a variety of heat-related health problems and illnesses (TDI 2008), including heat cramps, heat rash, heat exhaustion, and heat stroke.

Heat cramps may occur alone or simultaneously with other heat-related illnesses. Heat cramps are painful muscle spasms caused by sweating while performing hard physical labor in a hot environment. The cramps may be caused by either too much or too little salt. Tired muscles are very susceptible to heat cramps.

Heat rash (also known as prickly heat) often occurs in hot, humid environments where sweat does not easily evaporate from the skin. The sweat ducts become clogged, resulting in a rash. Taking frequent breaks in a cool place during the work day and bathing and drying the skin regularly can help to prevent heat rash.

Heat exhaustion is caused by the loss of large amounts of fluid by sweating, sometimes with excessive loss of salt.

Heat stroke is the most serious heat-related illness and occurs when the body's temperature-regulating system fails and sweating becomes an inadequate way of removing excess heat. A comparison of signs/symptoms and treatment modalities for heat exhaustion and heat stroke is shown in Table 29.1.

In 2011, U.S. DOL, OSHA (2013) established a Heat-Illness Prevention Campaign for those working in hot environments. Part of the campaign advises that employers should establish a complete heat illness prevention program to prevent heat illness. This includes providing workers with water, rest and shade; gradually increasing workloads and allowing more frequent breaks for new workers or workers who have been away for a week or more to build a tolerance for working in the heat (acclimatization); modifying work schedules as necessary; and monitoring workers for signs of illness, and planning for emergencies and training workers about the symptoms of heat-related illnesses and prevention.

In order to prevent heat-related illness and fatalities, workers should drink water every 15 minutes, even if not thirsty, and employers should provide cold water and disposable cups in convenient locations close to work areas. In addition, workers should rest in the shade to cool down when needed and wear a hat and light-colored clothing. It is vital to know the signs of heat illness and what to do in an emergency, and keep an eye on fellow workers for signs or symptoms of heat-related stress. "Easy does it" on the first days of work in the heat in order to get used to it should be encouraged. If workers are new to working in the heat or returning from more than a week off, and for all workers on the first day of a sudden heat wave, a work schedule should be implemented to allow them to get used to the heat gradually.

The U.S. National Oceanographic and Atmospheric Administration (US NOAA) developed the heat index system (Figure 29.1). The heat index combines both air temperature and relative humidity into a single value that indicates the apparent temperature in degrees Fahrenheit, or how hot the weather will feel. The higher the heat index, the hotter the weather will feel, and the greater the risk that outdoor workers will experience heat-related illness. U.S. NOAA issues heat advisories as the heat index rises. It is critical that water replacement be easily available and accessible to all affected workers.

Table 29.1 Comparison of heat exhaustion and heat stroke signs, symptoms, and treatment

Heat exhaustion		Heat stroke	
Signs/symptoms	*Treatment*	*Signs/symptoms*	*Treatment*
Headache	Move the person to a cool, shaded area	Dry pale skin (no sweating)	Call for emergency help
Dizziness	Provide cool water to drink	Hot red skin	Move person to a cool, shaded area
Weakness	Cool the person by fanning them	Mood changes (irritable, confused)	Provide cool water to drink if conscious
Mood changes (confused or irritable)	Cool the skin with a wet cloth	Seizures/fits	Lay person on back and be sure airway is clear
Feeling sick to stomach	Lay person on back and raise legs 6 to 8 inches if they are dizzy	Collapse/ unconsciousness	Lay the person on side if nausea occurs
Vomiting	Lay the person on side if nausea occurs		Loosen and remove heavy clothing
Decreased and dark-colored urine	Loosen and remove heavy clothing		Remove any objects close by if person has a seizure
Light-headedness or fainting	Stay with the person		Place ice packs under armpits and in the groin area and cool skin with a wet cloth
Pale clammy skin	*Emergency help should be called if no improvement. Untreated heat exhaustion may advance to heat stroke.*		Stay with the person

Temperature (°F)

	80	82	84	86	88	90	92	94	96	98	100	102	104	106	108	110
40	80	81	83	85	88	91	94	97	101	105	109	114	119	124	130	136
45	80	82	84	87	89	93	96	100	104	109	114	119	124	130	137	
50	81	83	85	88	91	95	99	103	108	113	118	124	131	137		
55	81	84	86	89	93	97	101	106	112	117	124	130	137			
60	82	84	88	91	95	100	105	110	116	123	129	137				
65	82	85	89	93	98	103	108	114	121	128	136					
70	83	86	90	95	100	105	112	119	126	134						
75	84	88	92	97	103	109	116	124	132							
80	84	89	94	100	106	113	121	129								
85	85	90	96	102	110	117	126	135								
90	86	91	98	105	113	122	131									
95	86	93	100	108	117	127										
100	87	95	103	112	121	132										

Relative Humidity (%)

Likelihood of Heat Disorders with Prolonged Exposure and/or Strenuous Activity
  Caution   Extreme Caution   Danger   Extreme Danger

Figure 29.1 National Oceanic and Atmospheric Administration National Weather Service Heat Index

Source: US National Oceanic and Atmospheric Administration National Weather Service (n.d.)

Dampness and mold in buildings

Workers may be exposed to mold hazards from dampness in buildings. Dampness results from water incursion either from internal sources (e.g., leaking pipes) or external sources (e.g., rainwater). Flooding causes dampness. Dampness becomes a problem when various materials in buildings (e.g., rugs, walls, ceiling tiles) become and remain wet for extended periods of time. Excessive moisture in the air (i.e., high relative humidity) that is not properly controlled with air conditioning can also lead to excessive dampness. Dampness is a problem in buildings because it provides the moisture that supports the growth of bacteria, fungi (i.e., mold), and insects (Centers for Disease Control and Prevention [CDC] 2013b).

In the presence of damp building materials the source of water incursion is often readily apparent (e.g., leaks in the roof or windows or a burst pipe). However, dampness problems can be less obvious when the affected materials and water source are hidden from view (e.g., wet insulation within a ceiling or wall; excessive moisture in the building foundation due to the slope of the surrounding land).

Mold is a fungal growth that forms and spreads on various kinds of damp or decaying organic matter. There are many different mold species that come in many different colors. They are found both indoors and outdoors in all climates, during all seasons of the year. Outdoors, molds survive by using plants and decaying organic matter such as fallen leaves as a source of nutrition. Indoors, molds need moisture to grow as well as a carbon source from building materials or building contents.

Recent media attention has increased public awareness and concern over exposure to molds in the workplace. While this may seem to be a new problem, exposure to molds has actually occurred throughout history. In fact, the types of molds found in office buildings are not rare or even unusual. It is important to understand that no indoor space is completely free from mold spores – not even a surgical operating room. Molds are everywhere, making our exposure to molds unavoidable, whether indoors or outdoors, at home or at work. Mold growing in buildings indicates that there is a problem with water or moisture. Health problems associated with excessive damp conditions and mold include allergies, hypersensitivity pneumonitis, and asthma.

When workers suspect their health problems are caused by exposure to building-related dampness or mold, they should report concerns immediately to supervisors or those persons responsible for building maintenance. Management must respond to these concerns and provide maintenance and monitoring including:

- Regularly inspect building areas for evidence of dampness; take prompt steps to identify and correct the causes of any dampness problems found.
- Prevent high indoor humidity through the proper design and operation of heating, ventilation, and air conditioning (HVAC) systems.
- Dry any porous building materials that have become wet from leaks or flooding within 48 hours.
- Clean and repair or replace any building materials that are moisture-damaged or show evidence of visible mold growth.
- Encourage occupants who have developed persistent or worsening respiratory symptoms while working in the building to see a health care provider.
- Follow health care provider recommendations for relocation of occupants diagnosed with building-related respiratory disease.

NIOSH has developed an observational assessment tool for dampness and mold in buildings. The tool can be used to evaluate signs of dampness, water damage, mold growth, and musty odors in rooms and areas throughout a building.

Cold water environments

"The safety and security of life at sea, protection of the marine environment and over 90% of the world's trade depends on the professionalism and competence of seafarers" (International Maritime Organization 2014a, para 1). Workers face a wide variety of hazards when working in, on, or near bodies of water. Bodies of water include retaining pools, rivers and streams, ponds and lakes, marshes, and the ocean. "Shipping is perhaps the most international of all the world's great industries—and one of the most dangerous" (International Maritime Organization 2014b, para 1). Hazards range from drowning and cold water immersion-related injuries in maritime-based occupations, aquaculture, and commercial fishing; slips and falls on wet floors; falling overboard; electrocution from power tools that come in contact with water; to exposure to marine animals and snakes. Workers, such as seafarers, police, and so on, may be exposed to frigid water while performing emergency water rescues or performing work (American Red Cross [ARC] n.d.a.; International Maritime Organization 2014b).

Commercial fishing and shipping are known for hazardous working conditions such as strenuous labor, long work hours, and harsh weather (European Maritime Safety Agency 2010; NIOSH 2014b). Falling overboard is the second leading cause of death among commercial fishermen in the United States (National Institute for Occupational Safety and Health [NIOSH] 2014b). In 2010, 61 seafarers lost their lives on commercial vessels in and around European Union (EU) waters, compared to 52 in 2009 and 82 in both 2008 and 2007 (European Maritime Safety Agency 2010). Vessels may also be struck by high waves, capsize, or sink, which contributes to cold water exposures.

Occupational hazards of aquaculture can vary based on the types of operation, scale of production, and fish species (Cole et al. 2008). Electrocution and high-voltage electrical injuries, drowning, and long-term exposure in extreme environments of sunlight, wind, cold, and water are a few of the hazards. Workers can also have exposures to toxic chemicals such as hydrogen sulfide and fish pathogens.

Hypothermia occurs from prolonged exposure to cold water temperatures or cold air which causes the core, internal body temperature to drop below 95 degrees F (35 degrees C). Early symptoms of hypothermia are shivering, fatigue, loss of coordination, and confusion and disorientation. Late symptoms include no shivering, blue skin, dilated pupils, slowed pulse and breathing, and loss of consciousness (NIOSH 2014a). As the core body temperature decreases, physical and mental processes slow down and death may result. Survival time in cold water varies based on water temperature. As the water temperature decreases, so does the expected survival time (US Search and Rescue Task Force, n.d.). Cold water immersion causes immersion hypothermia which develops more quickly than hypothermia as water conducts heat away from the body 25 times faster than air (National Institute for Occupational Safety and Health 2014a). An international hypothermia registry was developed as survival after accidental deep hypothermia (core temperature, <28 ⁰C) and prolonged cardiac arrest is rare, and the long-term survival rate is 47 percent higher than previously reported (Walpoth et al. 1997).

Besides hypothermia, workers exposed to high winds and cold temperatures can suffer from frostbite and trench foot which are exacerbated by wet conditions (National Institute for Occupational Safety and Health 2014a). Frostbite develops when ice crystals form in the fluids and underlying soft tissues of the skin, causing tissue destruction. Exposed areas, such as ears, nose, mouth, cheeks, fingers, and toes, are most commonly affected. Frostbitten skin is hard, pale, cold, and lacks feeling. When the affected area thaws, it becomes red and painful and may cause permanent damage to body tissues. Symptoms of frost bite are reduced blood flow to hands and feet, numbness, tingling, throbbing sensation, and bluish or pale waxy skin. Trench foot or immersion foot is an injury to the feet caused by prolonged exposure to wet and cold conditions. Skin tissue will die because of a lack of oxygen and nutrients to the feet, and buildup of toxic products. Symptoms include reddened skin, numbness, leg cramps, edema, tingling pain, blisters, bleeding under the skin, and gangrene.

Where the danger of drowning exists, employers should provide approved life jackets or buoyant work vests to all workers exposed to water hazards. These personal floatation devices (PFDs) can be approved by the U.S. Coast Guard, the International Convention for the Safety of Life at Sea (SOLAS), or flag state regulations. The PFDs and immersion (exposure) suits can help to prevent serious injury or death. "Vessels operating beyond coastal waters (beyond three mile limit) when the waters are 'cold' (meaning the monthly water temperature is 59 degrees F or less) must be equipped with at least one Coast Guard Approved Immersion Suit for each person on board" (Rosecrans et al. 2008). At least one employee should be trained in cardiopulmonary resuscitation, first aid, and emergency response skills. Workers must be encouraged or required to know how to swim if their jobs require them to work around water.

When cold/wet environments cannot be avoided, workers should wear appropriate clothing to keep warm. Several layers of loose clothing should be worn as layering provides better insulation than one layer. Tight clothing should be avoided as it reduces blood circulation and may restrict movement. The face, ears, hands, and feet must be protected in cold and wet environments. Boots should be waterproof and insulated, and hats help to reduce the amount of body heat that escapes from the head. Employers should monitor workers who are at risk of cold stress.

Landscapers

Landscapers often work around a variety of bodies of waters and face an increased risk of drowning. Waterscape work and other landscaping jobs performed near water involve use of bulky equipment, unstable ledges, and electrical wiring and pumps (Professional Landcare Network 2010).

Water-related jobs should be assessed for any slip and trip hazards, and types of tools or equipment that will be used near water. Employees should be prepared for a water emergency event. Life preservers, grab lines or poles should be available and placed within reach if a worker falls into water and needs help.

Wastewater workers

Water used by industries, businesses, and homes must be treated before being released back to the environment. Wastewater is used water and includes human waste, food scraps, oils, soaps, and chemicals (US Geological Survey 2014). The presence of toxic chemicals and organisms in sewage, in sludge, and in the air can pose hazardous situations affecting the health of workers. In 2006, the World Health Organization (WHO) published Guidelines for the Safe Use of Wastewater, Excreta and Greywater in Agriculture and Aquaculture (WHO 2006a, 2006b). The guidelines "propose a flexible approach of risk assessment and risk management linked to health-based targets that can be established at a level that is realistic under local conditions" (WHO n.d., para 3).

Drowning is a serious risk to wastewater workers due to extreme currents and process equipment. Guard railings should be in place around all process tankage and pits. PFDs should be worn by workers who are inside the railing area. Falls are common in wastewater facilities from wet and slippery surfaces so walkways should be as clear and dry as possible. Electrocution due to energizing circuits on equipment being repaired or serviced is a concern near water sources. Employee contact with raw sewage and wastewater can be a health hazard, and personal protective equipment (PPE), such as rubber gloves, liquid-repellent coveralls, rubber boots, goggles, protective face masks or splash-proof face shields, are necessary to prevent wastewater contact. For example, exposure to infectious agents such as hepatitis in wastewater can occur. Proper hand washing and hygiene practices are essential to help prevent exposure. Workers should be trained on basic hygiene practices, use and disposal of PPE, proper handling of human waste or sewage, signs and symptoms of infectious disease, and vaccine recommendations.

Agricultural workers

Agricultural workers, particularly migrant and seasonal farmworkers, may not have access to sufficient drinking water, sanitation facilities, or hand washing facilities while working in the field. Workers exposed to hot and humid conditions can suffer heat stroke and heat exhaustion without adequate intake of potable water (see section 'Hot work environments' above).

Agricultural workers may also be exposed to pesticides and other chemicals from spills or splashes, if their drinking fluids become inadvertently contaminated from pesticide exposure, or "if they drink from, wash their hands, or bathe in irrigation canals or holding ponds, where pesticides can accumulate" (US DOL, OSHA n.d.a., para 17). These workers become exposed because they perform hand-labor tasks in areas that have been treated with pesticides. The International Labour Organization (ILO 2001) has C184 – Safety and health in agriculture convention, 2001, which covers agricultural and forestry activities. Sound management of chemicals is addressed in Articles 12–14, and welfare and accommodation facilities are addressed in Article 19.

The OSHA Field Sanitation standard (1928.110) applies to any agricultural establishment where 11 or more workers are engaged on any given day in hand-labor operations in the field. OSHA standards require covered employers to provide toilets, potable drinking water, and hand washing facilities to hand-laborers in the field; to provide each worker reasonable use of the above; and to inform each worker of the importance of good hygiene practices (US DOL, OSHA n.d.a., para 15). The ILO also has international labor standards on migrant workers which "provide tools for both migrant sending and receiving countries to manage migrant flows and ensure adequate protection for this vulnerable category of workers" (ILO, n.d., para 1).

Water contamination

While water can become contaminated in the community after weather-related events such as flooding, tornados, hurricanes, and the like, or accidental release of chemicals into groundwater or water supply, water may also become contaminated with bacteria, parasites, and chemicals in the workplace which can cause disease. Sources where temperatures allow bacteria to live include hot water tanks, cooling towers, and evaporative condensers of large air-conditioning systems, commonly found in hotels and large office buildings. Legionnaires' disease, which first occurred in 1976 at a convention of the American Legion at a hotel in Philadelphia, is a prime example of inhaled aerosolized water contaminated with *Legionella* bacteria (CDC 2013c).

Box 29.1 Water decontamination process

1. Filter the water using a piece of cloth or coffee filter to remove solid particles.
2. Bring it to a rolling boil for about one full minute.
3. Let it cool at least 30 minutes. Water must be cool or the chlorine treatment described below will be useless.
4. Add 16 drops of liquid chlorine bleach per gallon of water, or 8 drops per 2-liter bottle of water. Stir to mix. Sodium hypochlorite of the concentration of 5.25–6 percent should be the only active ingredient in the bleach. There should not be any added soap or fragrances. A major bleach manufacturer has also added sodium hydroxide as an active ingredient, which they state does not pose a health risk for water treatment.
5. Let stand 30 minutes.
6. If it smells of chlorine, you can use it. If it does not smell of chlorine, add 16 more drops of chlorine bleach per gallon of water (or 8 drops per 2-liter bottle of water), let stand 30 minutes, and smell it again. If it smells of chlorine, you can use it. If it does not smell of chlorine, discard it and find another source of water.

Source: American Red Cross (n.d.b., para 2). Readers are requested to go to the website redcross.org and search "water treatment" for the most current information.

The ARC recommends that all water of uncertain purity should be treated or decontaminated before use. This may be difficult in the workplace but if facilities are available, water should be treated for consumption and hygiene, using the steps recommended by the ARC n.d.b.) (Box 29.1).

Interventions

OSHA water regulations

OSHA requires that clean and sanitary water, also known as potable water, should be available to all employees in adequate supply at all times for personal use in the workplace. This includes drinking water, water to wash hands, body and clothes, and water to wash food, cooking equipment, environments, cups, and utensils. It must be readily accessible at suitable places and marked by appropriate signage (Standards OSHA n.d.). Drinking water taps should not be placed where contamination is likely, such as where lead is handled, or in toilet facilities or washrooms (Forrester, 2006). Additionally, Standards OSHA (n.d.) has the following water requirements:

- Water used for purposes other than consumption (e.g., fire, industrial purposes) should be clearly marked as to its use and that it is not for consumption.
- Sanitary washing facilities must be available in all workplaces with tepid or hot and cold running water.

Water and hand washing

Hand washing is an important public health activity that decreases exposure to a variety of workplace hazards, such as heavy metals, biological hazards such as Ebola, and chemicals, and removes these harmful contaminants from the skin. Basic hand hygiene can be achieved by hand washing with adequate quantities of clean (ideally running) water and soap or hand rubbing with an alcohol-based hand rub solution. Workers should wash hands after removing gloves or other PPE, after touching contaminated surfaces, before eating, after using the bathroom, and so on.

Several OSHA standards require appropriate hand washing. For example, the OSHA Bloodborne Pathogens Standard (29 CFR 1910.1030) (US DOL, OSHA 2001) requires employers to provide hand washing facilities which are readily accessible to employees. Employees must wash their hands with soap and water immediately or as soon as feasible after removal of gloves or other PPE. Employees should follow universal precautions to prevent contact with blood and other bodily fluids. Universal precautions is "an approach to infection control to treat all human blood and certain human body fluids as if they were known to be infectious for HIV, HBV, and other bloodborne pathogens" (US DOL, OSHA n.d.b., para 5). The ILO also emphasizes good hand hygiene to protect workers' health.

Ebola has significance here, particularly for the health care worker. Ebola viruses, first discovered in 1976, are found mostly in several African countries (WHO 2014b). The natural reservoir is unknown although animals are the most likely host source. Ebola spreads through contact with infected blood and body fluids. Thus, health care workers caring for patients infected with Ebola are at significant risk. Use of PPE and good hygiene with soap and water are of paramount importance.

The OSHA lead standard (29 CFR 1910.1025) specifies several control measures to reduce exposure to lead, a toxic metal. Shower facilities must be provided for employees when the airborne exposure to lead is above the permissible exposure limit (PEL), without regard to use of respirators. In addition, an adequate number of bathrooms must be provided so employees can wash their hands and face prior to eating, drinking, smoking, or applying cosmetics (US DOL, OSHA n.d.c.). Workplace hygiene is essential in preventing excessive lead exposure at work and at home, particularly if young children are in the home.

As previously mentioned, OSHA regulations require the workplace to have running water available. All health care professionals must wash their hands after evaluating and treating workers. In addition, any first aiders or workers who provide first aid to injured or ill workers must wash their hands after any blood or body fluid exposure.

Water and international travel

Business travelers should take several precautions when traveling abroad, particularly relating to water. Travelers should drink at least eight ounces of water per waking hour to prevent dehydration in the dry atmosphere of airplanes (Tompkins 2008).

Food-borne and water-borne illnesses, such as travelers' diarrhea, may occur from eating or drinking food or beverages contaminated by bacteria, parasites, or viruses (CDC 2013a). For travelers to high-risk areas, several approaches may reduce this risk. Travelers should not drink tap or well water and should avoid ice made with tap or well water. Other drink options include water and sodas that are bottled and sealed, water that has been disinfected, ice that has been made with bottled or disinfected water, hot coffee or tea, and pasteurized milk (CDC 2013a).

If safe water is not available to wash hands, alcohol-based hand sanitizers can be used to clean hands before eating. Travelers should also use bottled water when brushing their teeth, and avoid swallowing water while showering or swimming.

The World Business Council for Sustainable Development is engaged in the global Safe Water, Sanitation, and Hygiene (WASH) initiative and applies this to the workplace saying, "many businesses have operations, employees, contractors and customers in countries lacking access to safe water, sanitation and hygiene" (World Business Council for Sustainable Development n.d.) Furthermore, investing in safe WASH for employees leads to:

- Healthier and more productive workforce: Adequate access to safe WASH is associated with decreased absenteeism due to water-related diseases, and thus improved productivity.
- Demonstration of leadership in supporting global objectives: Ensuring safe WASH at the workplace contributes to the achievement of internationally recognized objectives such as the universal realization of the human right to water and sanitation.

Conclusion

Water is an essential substance in the workplace. In order for workers to function properly, they must be well hydrated. Care should be taken to make certain that access to drinking water is provided to all employees at all times as well as when water is needed for hygiene purposes. Specific situations such as working in hot environments must be carefully monitored to be certain that worker health is not compromised. Where water-related issues pose a threat, such as with increased mold exposure, management needs to take steps to mitigate or eliminate the risks through clean-up and repair.

References

Abbott, S 2010, "Assessing the effects of energy drinks upon firefighter health and safety". Available from www.usfa.fema.gov/pdf/efop/efo45842.pdf (18 September 2014).

American Red Cross (ARC) n.d.a., "Basic water rescue". Available from http://www.watersafetyguy.org/wp-content/uploads/2012/07/Basic_Water_Safety_Rescuesb0606.pdf (20 September 2014).

American Red Cross (ARC) n.d.b., "Water treatment: Ensuring that your water is safe". Available from http://www.redcross.org/prepare/disaster/water-safety/water-treatment (20 September 2014).

Baum, R, Luh, J, & Bartram, J 2013, "Sanitation: A global estimate of sewerage connections without treatment and the resulting impact on MDG progress", *Environmental Science and Technology*, vol. 47, no. 4, pp. 1994–2000.

Burke, ER 2000, "Health hydration", *Occupational Health and Safety*, vol. 69, no. 5, pp. 52–54.

CDC 2013b, Indoor environmental quality: Dampness and mold in buildings. Available from http://wwwnc.cdc.gov/niosh/topics/indoorenv/mold.html (18 September 2014).

CDC 2013c, Legionella (Legionnaires' disease and Pontiac fever). Available from http://www.cdc.gov/legionella/about/history.html (18 September 2014).

Centers for Disease Control and Prevention (CDC) 2013a, Food and water safety: Travelers' health. Available from http://wwwnc.cdc.gov/travel/page/food-water-safety (4 February 2014).

Clap, AJ, Bishop, PA, Smith, JF, Lloyd, LK & Wright, KE 2002, "A review of fluid replacement for workers in hot jobs", *AIHA Journal*, vol. 63, pp. 190–198.

Cole, DW, Cole, R, Gaydos, SJ, Gray, J, Hyland, G, Jacques, ML, Powell-Dunford, N, Sawhney, C & Au, WW 2008, "Aquaculture: Environmental, toxicological, and health issues", *International Journal of Hygiene and Environmental Health*, vol. 212, pp. 369–377.

Dennison, K, Rogers, B & Randolph, SA 2013, "Energy drinks and worker health risks", *Workplace Health & Safety*, vol. 61, no. 10, p. 468.

European Maritime Safety Agency 2010, Maritime accident review 2010. Available from http://emsa.europa.eu/publications/technical-reports-studies-and-plans/download/1388/1219/23.html (23 November 2014).

Foran, J 2011, "Water contamination and wastewater treatment" in *Occupational and Environmental Health*, eds B Levy, D Wegman, S Baron & R Sokas. Oxford University Press, New York, pp. 154–169.

Forrester, H 2006, Wise up on water. Available from www.water.org.uk/home/resources-and-links/water-for-health/ask-about (4 February 2014).

Gleick, PH, Cooley, H, Cohen, MJ, Morikawa, M, Morrison, J & Palaniappan, M 2009, *The World's Water, 2008–2009*. The Island Press, Washington, DC.

Grandjean, AC, Reimers, KJ, Bannick, KE & Haven, MC 2000, "The effect of caffeinated, non-caffeinated, caloric and non-calorie beverages on hydration", *Journal of the American College of Nutrition*, vol. 19, no. 5, pp. 591–600.

Institute of Medicine 2004, Hydration guidelines. Available from http://www.beverageinstitute.org/article/hydration-guidelines/ (24 November 2014).

International Labour Organization (ILO) 2001, C184 – Safety and health in agriculture convention, 2001 (No. 184). Available from http://www.ilo.org/dyn/normlex/en/f?p=NORMLEXPUB:12100:0::NO::P12100_ILO_CODE:C184 (24 November 2014).

International Labour Organization n.d., International labour standards on migrant workers. Available from http://ilo.org/global/standards/subjects-covered-by-international-labour-standards/migrant-workers/lang--en/index.htm (24 November 2014).

International Maritime Organization 2014a, Human element. Available from http://www.imo.org/OurWork/HumanElement/Pages/Default.aspx (23 November 2014).

International Maritime Organization 2014b, Maritime safety. Available from http://www.imo.org/OurWork/Safety/Pages/Default.aspx (23 November 2014).

Karriem-Norwood, V 2014, Heat stroke: Symptoms and treatment. Available from www.m.webmd.com (3 December 2014).

National Institute for Occupational Safety and Health (NIOSH) 2014a, Cold stress. Available from http://www.cdc.gov/niosh/topics/coldstress/ (20 September 2014).

NIOSH 2014b, Commercial fishing safety. Available from http://www.cdc.gov/niosh/topics/fishing/ (20 September 2014).

Onda, K, LoBuglio, J & Bartram, J 2012, "Global access to safe water: Accounting for water quality and the resulting impact on MDG progress", *International Journal of Environmental Research and Public Health*, vol. 9, no. 3, pp. 880–894.

Popkin, BM, D'Anci, KE & Rosenberg, IH 2010, "Water, hydration and health", *Nutrition Reviews*, vol. 68, no. 8, pp. 439–458.

Professional Landcare Network 2010, Safety sense: Working near water. Available from https://www.landcarenetwork.org/riskmgmt/ssense/Jul10.pdf (30 September 2014).

Rogers, B, Stiehl, K, Borst, J, Hutchins, S & Hess, A 2007, "Heat related illnesses: Role of the occupational and environmental health nurse", *AAOHN Journal*, vol. 55, no. 7, pp. 279–288.

Rosecrans, MM, Kemerer, J & Dzugan, J 2008, Commercial fishing vessel safety digest. Available from http://www.uscg.mil/d1/prevention/CommFishSafetyDigest-20081.pdf (20 September 2014).

Standards OSHA n.d., Sanitation – potable water. 1910.141. https://www.osha.gov/pls/oshaweb/owadisp.show_document?p_table=interpretations&p_id=22932 (30 September 2014).

Substance Abuse and Mental Health Services Administration, Center for Behavioral Health Statistics Quality 2013, The DAWN Report. Update on emergency department visits involving energy drinks: A continuing public health concern. Available from www.samhsa.gov/data/2k13/DAWN126/sr126-energy-drinks-use.pdf (18 September 2014).

Texas Department of Insurance (TDI), Division of Workers' Compensation Workplace Safety 2008, Heat stress. Available from http://www.tdi.texas.gov/pubs/videoresource/stpheatst.pdf (18 September 2014).

Toblin, RL, Clarke-Walper, K, Kok, BC, Sipos, ML & Thomas, JL 2012, "Energy drink consumption and its association with sleep problems among U.S. service members on a combat deployment: Afghanistan, 2010", *Morbidity and Mortality Weekly Report*, vol. 61, no. 44, pp. 895–898.

Tompkins, OS 2008, "Business traveler fitness", *AAOHN Journal*, vol. 56, no. 6, p. 272.

US Agency for International Development (USAID) 2013, Water and development strategy 2013–2018. Available from http://www.usaid.gov/sites/default/files/documents/1865/USAID_Water_Strategy_3.pdf (4 February 2014).

US Department of Labor, Occupational Safety and Health Administration (US DOL, OSHA) 2001, Bloodborne pathogens standard, 29 CFR 1910.1030. Available from https://www.osha.gov/pls/oshaweb/owadisp.show_document?p_id=10051&p_table=STANDARDS (4 February 2014).

US DOL, OSHA 2013, Occupational heat exposure. Available from https://www.osha.gov/SLTC/heatillness/index (20 September 2014).

US DOL, OSHA n.d.a., Agricultural operations. Available from https://www.osha.gov/dsg/topics/agriculturaloperations/hazards_controls.html (4 February 2014).

US DOL, OSHA n.d.b., Healthcare wide hazards, (lack of) universal precautions. Available from https://www.osha.gov/SLTC/etools/hospital/hazards/univprec/univ.html (4 February 2014).

US DOL, OSHA n.d.c., Lead standard, 29 CFR 1910.1025. Available from https://www.osha.gov/pls/oshaweb/owadisp.show_document?p_table=standards&p_id=10030 (4 February 2014).

US Geological Survey 2014, Wastewater treatment, water use. Available from http://water.usgs.gov/edu/wuww.html (4 February 2014).

US National Oceanic and Atmospheric Administration, National Weather Service n.d., Heat index chart. Available from http://www.crh.noaa.gov/images/unr/preparedness/heatindex.pdf (4 February 2014).

US Search and Rescue Task Force n.d. Cold water survival. Available from http://www.ussartf.org/cold_water_survival.htm (4 February 2014).

Walpoth, BH, Walpoth-Aslan, BN, Mattle, HP, Radanov, BP, Schroth, G, Schaeffler, L, Fischer, AP, von Segesser, L. & Althaus, U 1997, "Outcome of survivors of accidental deep hypothermia and circulatory arrest treated with extracorporeal blood warming", *New England Journal of Medicine*, 337, pp. 1500–1505. DOI: 10.1056/NEJM199711203372103.

World Business Council for Sustainable Development n.d., Implementation case studies. Available from http://www.wbcsd.org/WASHatworkplace.aspx (24 November 2014).

World Health Organization 2006a, Guidelines for the Safe Use of Wastewater, Excreta and Greywater. Volume 2: Wastewater Use in Agriculture. Available from http://www.who.int/water_sanitation_health/wastewater/wastewateruse2/en/ (22 May 2015).

World Health Organization 2006b, Guidelines for the Safe Use of Wastewater, Excreta and Greywater. Volume 3: Wastewater and Excreta Use in Aquaculture. Available from http://www.who.int/water_sanitation_health/wastewater/wastewateruse3/en/ (22 May 2015).

World Health Organization n.d., Wastewater use. Available from http://www.who.int/water_sanitation_health/wastewater/en/ (24 November 2014).

World Health Organization, 2014a Water. Available from http://www.who.int/media centre/factsheets/fs391/en/ (July 2014).

World Health Organization, 2014b. International travel and health West Africa – Ebola virus disease. Available http://who.int.ith/updates/2012042/en/

30

HEALTH CARE SETTINGS*

Martin Exner

PROFESSOR OF HYGIENE AND PUBLIC HEALTH, UNIVERSITY OF BONN, BONN, GERMANY

Learning objectives

1 Understand the risks and challenges related to water in health care facilities (hospitals, primary health care centres, isolation camps, burn-patient units, feeding centres and others).
2 Understand the most important water associated nosocomial infections in health care facilities.
3 Describe the major points of prevention and control strategies associated with water, plumbing systems, sinks and wastewater drainage in health care facilities.

Health care facility risks

Clean water is necessary for many purposes in health care facilities and patient care beyond drinking, including washing patients, washing hands of health care workers, cleaning instruments and surfaces, and preparation of solutions and disinfectants. In intensive care units, patients may be washed completely two to three times per day, necessitating ample clean water. Many patients have open skin, wounds, catheters, and so on which provide pathways for pathogens to enter the body that are not usually penetrated.[1, 2] Water is also essential to operate machines and other devices. For example, cleaning of endoscopes after disinfection is necessary to flush out disinfectants. Shower heads and water taps can produce aerosols by which *Legionella* can disseminate and reach the lungs.[3]

Many health care facility patients may also be at greater risk of infection by waterborne pathogens due to predisposed factors, including immunosuppression. Chronic wounds or burns also enhance the vulnerability of infection for patients. Decreased clearance of the respiratory tract or the bedridden situation of patients increases the risk of aspiration pneumonia after cleaning their mouths with contaminated drinking water.[1, 2, 4]

---

* Recommended citation: Exner, M. 2015. 'Health care settings', in Bartram, J., with Baum, R., Coclanis, P.A., Gute, D.M., Kay, D., McFadyen, S., Pond, K., Robertson, W. and Rouse, M.J. (eds) *Routledge Handbook of Water and Health*. London and New York: Routledge.

Antibiotic therapy changes the microbioma of the patients and eliminates the physiological microbial flora. Pathogens with a high intrinsic antibiotic resistance like Gram-negative bacteria such as *Pseudomonas aeruginosa* or *Acinetobacter* can colonize the respiratory tract and enhance the risk of nosocomial pneumonia. The most important waterborne pathogens causing nosocomial outbreaks can be found in Table 3.1. These pathogens are shed into sinks, drainage paths, toilets and wastewater systems where they find an ideal biotope with biofilms and appropriate temperatures in which they can persist and even develop antibiotic resistance. Therefore wastewater systems or sanitation systems must be included in a holistic system of prevention and control especially of Gram-negative nosocomial pathogens. Given these risks, and lack of research in health care facilities specifically, there is great potential to improve the prevention and control of health care related infections and disease from water-related pathogens.[1, 2, 4, 5]

Controlling the risks

It is necessary to distinguish between the different areas of reservoirs in the water and wastewater system in health care facilities. Waterborne pathogens can be introduced by the central water system from outside the health care facilities when there is an insufficient protected water catchment area, insufficient water treatment or insufficient disinfection especially in situations of heavy rainfalls.[6–8]

Pathogens can be introduced by drinking water from outside – because drinking water is not sterile – into plumbing systems in low concentrations. These pathogens can proliferate in the plumbing systems of health care facilities under specific conditions or on materials which enhance the growth of bacteria like rubber and some plastic materials.

Waterborne pathogens like *Legionella*, *Pseudomonadaceae*, *Acinetobacter* spp. and Enterobacteriaceae can also grow independently from the other parts of the plumbing system in water taps and shower heads.

Sinks with overflow and sink drainages are ideal reservoirs for pathogens to proliferate and be splashed into the environment, on the water taps and on the hands of health care workers.[9, 10] *Klebsiella oxytoca* was isolated in concentrations of up to 1×10^7/ml from water sink drainages of neonatal wards during the source tracking of a *Klebsiella oxytoca* outbreak in neonates. After identification and sanitization of one distinct sink in which neonates had been bathed, the outbreak was brought under control.

In the last few years more attention has also been drawn to the sewage system as an important reservoir for *Pseudomonas aeruginosa*, *Acinetobacter* spp. and antibiotic resistant Enterobacteriaceae like *Citrobacter, Klebsiella and Enterobacter*.[11–13] In the case of obstruction of the sanitation or sewage water system, there can be a reflow into the sinks with high concentrations of nosocomial waterborne pathogens.

Waterborne and wastewater pathogens

Waterborne and wastewater pathogens from outside the health care facility

Waterborne pathogens coming from the centralized community water system may enter and be transmitted throughout health care facilities with severe consequences. Pathogens that may enter health care facilities through community water systems have varying degrees of health significance. These pathogens should be prioritized in health care facilities based upon their persistence and significance (see Table 3.1[14]).

Water and wastewater-borne pathogens with properties to proliferate in the plumbing and wastewater system

In the plumbing system of hospitals – water taps and shower heads, sink drains and sanitation systems – only bacterial pathogens can proliferate; viruses and parasites are not able to proliferate. Some of the most important plumbing system associated waterborne pathogens are: *Acinetobacter* spp., *Klebsiella* spp., *Legionella* spp., non-tuberculosis *Mycobacteria*, *Pseudomonas aeruginosa*, *Burkholderia cepacia*, Enterobacteriaceae (*E. coli*, *Serratia* spp., *Enterobacter* spp., *Citrobacter*, *Raoultella* spp.), *Raoultella* spp., *Stenotrophomonas maltophilia*, *Sphingomonas* spp., *Ralstonia pickettii*, fungi (*Aspergillus* spp., *Fusarium* spp.) and amoeba-associated bacteria (*Leegionella anisa*, *Bosea massiliensis*).[1, 2, 4]

Acinetobacter spp.

Acinetobacter spp. are ubiquitous inhabitants of soil, water and sewage environments. *Acinetobacter* has been isolated from 97 per cent of natural surface water samples in numbers of up to 100/ml.

In health care facilities, infection is most commonly associated with contact with wounds and burns, major surgery, and weakened immune systems, such as in neonates and elderly individuals. *Acinetobacter* is an opportunistic pathogen that may cause urinary tract infections and pneumonia. *Acinetobacter* is frequently found in sink drains or in sewage systems. From these points, water taps and shower heads can also be contaminated. Outbreaks of infection have been associated with water baths and humidifiers. Injection is not a usual source of infection. In cases of sporadic and nosocomial infection, and especially in nosocomial outbreaks, water outlets, sink drains and sanitation systems must be taken into account as possible sources.

Klebsiella spp.

Klebsiella spp. are Gram-negative bacilli that belong to the family of Enterobacteriaceae. The genus *Klebsiella* consists of a number of species including *Klebsiella pneumoniae* and *Klebsiella oxytoca*. *Klebsiella* spp. has been identified as colonizing hospital patients. *Klebsiella* spp., notably *Klebsiella pneumoniae* and *Klebsiella oxytoca*, may cause serious nosocomial infections such as pneumonia, wound infections and sepsis.

Klebsiella spp. is a natural inhabitant of many water environments. In drinking water distributions, they are known to colonize water taps. Like other bacteria and non-ferment Gram-negative pathogens, they are generally biofilm organisms. In well-treated plumbing systems, *Klebsiella* spp. can be found. But in sink drains and sanitations systems *Klebsiella* spp. are often isolated in high concentrations. Especially in nosocomial outbreaks, water, water outlets, and sink drains must be regarded as infection reservoirs. The same is true for *Enterobacter* spp. and *Citrobacter* spp. In a recent outbreak in Germany a broad spectrum of antibiotic-resistant bacteria including *Klebsiella* spp., *Enterobacter* spp., *E. coli*, *Citrobacter* spp. and *Raoultella* were found as colonizing pathogens in more than 140 patients. All of the different Enterobacteriaceae had a KPC2-wearing plasmid. An infection reservoir in the sanitation system of the hospital with 525 beds was identified. After breaking the transmission pathway the outbreak could be brought under control.[13]

Legionella

The genera *Legionella* has more than 50 species that are Gram-negative and are found in a wide range of water environments and can proliferate at temperatures from 25° C up to 50° C (Chapter 8).

In health care facilities, plumbing systems are well-described infection reservoirs of a broad range of different *Legionella* species (not only *Legionella pneumophila*).

In some countries like Germany, the Netherlands, France and the United Kingdom, *Legionella* species or *Legionella pneumophila* must be controlled in a quantitative way. This is based on the European Council Directive 98/83/EC of 3 November 1998 on the quality of water intended for human consumption. For the purposes of the minimum requirements of this directive, water intended for human consumption shall be wholesome and clean if it is free from any microorganisms and parasites and from any substances which, in numbers or concentrations, constitute a potential danger to human health. In plumbing systems in Germany, water samples must be tested every year, and in intensive care units and other areas with special risks like wards for immunosuppressive patients, water must be tested every six months.

Atypical mycobacteria

In contrast to typical species of mycobacteria such as *Mycobacterium tuberculosis*, the non-tuberculose or atypical species of mycobacteria are natural inhabitants of a variety of water environments. Examples include the species *M. gordonae, M. kansasii, M. marinum, M. xenopi, M. intracellulare, M. avium, M. chelonae* and *M. fortuitum*. The diseases caused by atypical mycobacteria include skin and soft tissue diseases as well as respiratory, gastrointestinal and genitourinary tract diseases. Atypical *Mycobacterium* spp. multiply in a variety of suitable water environments, notably in biofilms. High numbers of atypical *Mycobacterium* spp. may occur in distribution systems after events that dislodge biofilms, such as flushing or flow reversal. They are relatively resistant to treatment and disinfection and have also been detected in well-operated and well-maintained drinking water supplies. Principal routes of infection appear to be inhalation, contact and ingestion of contaminated water.

Detections of atypical mycobacteria in drinking water and identified routes of transmission suggest that drinking water supplies are a plausible source of infection. Because they are relatively resistant to disinfection, disinfection itself is not a safe way to control atypical mycobacteria. Control measures that are designed to minimize biofilm growth, including treatment to optimize organic carbon removal, restriction of the residence time of water in distribution systems and maintenance of disinfectant residuals, could result in less growth of these organisms.

A routine surveillance of atypical mycobacteria in drinking water is not regulated in most countries. However, in the case of sporadic nosocomial infections and source tracking, plumbing systems must be considered as a potential reservoir for atypical mycobacteria.

Pseudomonas aeruginosa

Pseudomonas aeruginosa is a member of the family Pseudomonadaceae and is an aerobic Gram-negative rod. *Pseudomonas aeruginosa* is one of the most important nosocomial pathogens with a high intrinsic antibiotic resistance. *Pseudomonas aeruginosa* can cause a range of infections but rarely causes serious illness in healthy individuals without some predisposing factor. It predominantly colonizes damaged sites such as burn and surgical wounds, the respiratory tracts of people with underlying diseases and physically damaged eyes. From these sites, it may invade the body, causing destructive lesions or septicaemia and meningitis. Cystic fibrosis and immunocompromised patients are prone to colonization with *P. aeruginosa*, which may lead to serious progressive pulmonary infections.

Pseudomonas aeruginosa is a common environmental organism and can be found in water and sewage. It can multiply in many water environments and it is a recognized cause of hospital-acquired infection with serious complications. It has been isolated from a range of most moist environments such as sinks, water baths, hot water systems, showers and spa pools. It can be isolated in water flow meters, hydrants and water softeners in the plumbing system of hospitals. *Pseudomonas aeruginosa* can be introduced from the centralized water utility system to the distribution system of health care facilities where it can grow in niches like water flow meters, water softener, dead ends of plumbing systems on plastic and rubber materials and also especially in water taps in biofilms. In the biofilm environment they are able to survive even against high chlorine concentrations.

In some countries like France, Germany and the United Kingdom, *Pseudomonas aeruginosa* must be controlled in drinking water by regular drinking water monitoring. *Pseudomonas aeruginosa* should not be isolated in 100 ml. Well managed plumbing systems without chlorine are free of *Pseudomonas aeruginosa* even up to 1 l. In the case of sporadic infections, and especially in *Pseudomonas aeruginosa* outbreaks, water sinks and sinks drainage must be excluded as reservoirs of *Pseudomonas aeruginosa*. The evidence of transmission from sink drainage system to the hands of health care workers when washing their hands leads to the development of automatic disinfected sink drainage systems by which the rate of colonization and infection could be significantly reduced.

Prevention and control strategies

Prevention strategies should be implemented to proactively identify and mitigate incidents or outbreaks associated with the water and sanitation systems in health care facilities before they happen. Control strategies should be designed in the case of an incident or outbreak associated with the water and sanitation system.

Prevention strategies

Implementation of a water and sanitation safety group

The water and sanitation safety group (WSSG) must be implemented as a subgroup of the infection control committee and as a multidisciplinary group. The WSSG should be composed of at least the director of infection prevention and control, the technical director of the health care facility team and the infection control nurse.

The task of the WSSG is to develop the water and sanitation safety plan and to advise on the remedial action required on water and sanitation systems in the case of incidents and outbreaks associated with water and sanitation systems. It must be regulated that all monitoring results are immediately recorded by the responsible person of the WSSG, analysed, evaluated and assessed, and ensured that the remedial action will be implemented.[1, 2, 14–17]

The WSSG must implement the water and sanitation safety plan, which includes information from where water travels into the plumbing system of the health care facility. They must assess the risk points of the distribution system and install a plan of verification and reaction. The WSSG will need to ensure that decisions affecting the safety and integrity of the water system do not go ahead without being agreed by them.

Where estate and facility provided services are part of a contract, it is essential that these providers participate in all aspects of estate and facility management that can affect patients. This includes responding to specific requests from the infection prevention and control (IPC) team and WSSG, which may be in addition to relevant guidance and documentation. The WSSG is also responsible for the risk assessment and coordination of the management of the sanitation and wastewater system. The WSSG should always act in an appropriate and timely manner.

The WSSG is responsible for developing the water and sanitation safety plan which provides a risk-management approach to microbiological safety of water and sanitation and establishes good practices in local water usage distribution supply and sanitation management.

The first step in the development of a water safety plan (Chapter 23) is to gain a comprehensive understanding of the whole water system including the external supply of water, the catchment area, the ward treatment and the system of the water supplier, and describe the plumbing system from the point of entrance, critical control points.

The WSSG has to define the monitoring of control measures, acceptable performance and how the control measures are monitored; verify that the water safety plan is working effectively and will meet the health-based targets; and develop supporting programmes by teaching hygienic practices and preparing management procedures, including corrective actions for normal and incident conditions. A plan for the sampling and microbiological testing in identified risk units must be done.

The water and sanitation safety plan should identify potential microorganisms which indicate a risk and ecological aspects of microbiological growth by *Legionella*, *Pseudomonas aeruginosa* and other opportunistic pathogens, and then consider practical and appropriate control measures. The implementation of the water and sanitation safety plan should be coordinated by the responsible person of the WSSG.

The risk assessments that inform the water and sanitation safety plan should identify potential microbiological hazards caused by *Legionella*, *P. aeruginosa* and other opportunistic pathogens, and the hazardous events and risks that may arise during storage, delivery and use of water in augmented care settings. The sampling points and the interval of verification must be identified.

The risk assessment should be done under the coordination of the hospital hygienist or chair of the infection control team. They should consider:[16, 17]

- The susceptibility of patients from each type of water use.
- Clinical practices where water may come into contact with patients in their invasive devices.
- The cleaning of patient equipment.
- The disposal of blood, bodily fluids and patients' washing water.
- The maintenance and cleaning of wash hand basins and associated tubs, specialist baths and other ward outlets.
- Change in use (for example, clinical area changed to office accommodation or vice versa) due to refurbishment or operational necessity.
- Other devices that increase/decrease the temperature of water (for example, ice-making machines, water chillers) which may not be appropriate in augmented care settings.
- Engineering assessment of water systems, including correct design installation, commissioning, maintenance and verification of the effectiveness of control measures.

- Underused outlets.
- Flushing policy.
- Sampling, monitoring and testing programme that needs to be put in place.
- Design of sinks.
- Design of sink drainage system.
- Education and training.
- Microbiological environmental sampling and verification.

After identification of the critical control points, the water safety plan must specify the sampling points, the frequency of environmental sampling, the volume and the method of taking water samples. When the risks have been identified, an action plan needs to be developed with defined roles and responsibilities, and agreed timescales to minimize these risks.

In detail, it must be regulated which criteria must be fulfilled in other areas where instruments are cleaned or sterilized. The design of water tubs, shower heads, water basins and sink trays must be considered and regulated. Considering the contamination potential of electronic faucets with *Legionella* and *Pseudomonas aeruginosa* in Germany, these faucets are not recommended. In patient areas, sinks should not have an overflow. The water must not flow directly into the sink drainage to avoid highly contaminated splashes into the environment of the sink. Sink drainage must be considered as high-risk areas. Therefore in high-risk areas like hemato-oncological wards or intensive units, special sink drains should be regularly cleaned and disinfected. The cleaning and disinfection regime of water sink outlets and sink drains must be regulated in the water and sanitation safety plan especially after the discharge of patients and before a new patient is transferred to the room.

All patient equipment should be stored clean, dry and away from potential water and drainage splashing. All preparation areas for aseptic procedures for injection and infusion and preparation of parenteral food and any associated sterile equipment should not be located where there is a risk of splashing contamination from water outlets and sinks. In the central kitchen a contamination of the kitchen environment and foods with sewage water must also be excluded. All taps that are used infrequently on augmented care units should be flushed regularly (at least daily in the morning for one minute).

If the outlet is fitted with a point of use filter, the filter should not be removed in order to flush the tap unless the manufacturer's instructions advise otherwise.

Sampling and testing for indicator pathogens in health care facilities

The WSSG must install a sampling protocol. The time of sampling has significance. The first water to be delivered from the outlet (pre-flush sample) should be collected to assess the microbial contamination in the outlet. If water flows over a biofilm containing waterborne pathogens located at or near the outlet, planktonic bacteria arising from that biofilm will be diluted and a subsequent sample will give low bacterial counts. If contamination is upstream in the system, this will not affect bacterial counts. The sample obtained after allowing water to flow from an outlet is referred to as a "post-flush" sample. Comparison of counts from pre- and post-flush samples can help locate the source of the waterborne pathogens. If a pre-flush sample gives a high count, subsequent paired pre- and post-flush samples should be tested to help locate the source of the contamination.

Associated outbreaks in health care facilities are well described in the literature whereas sanitation system associated outbreaks are described only in a few publications. In the last few years, there have been well documented nosocomial outbreaks in which the sanitation system and its controls were identified as infection reservoirs, especially for *Pseudomonas aeruginosa* and antibiotic-resistant carbapenemase-producing Enterobacteriaceae.[11-13] Therefore it is of utmost importance to also take into account the sanitation system when there is an increase in those pathogens. Even today, official guidelines for prevention and control of health care associated infections to control carbapenemase-producing Enterobacteriaceae do not take into account water and sanitation systems as a reservoir. All that is proposed is the screening and training of health care workers in hand hygiene and isolating or grouping infected or colonized patients.

Water system associated with incidents and outbreaks

In outbreaks of *Legionella*, *Pseudomonas aeruginosa* and other waterborne pathogens, it is of great importance to also screen the water system for these pathogens. We recommend not only collecting 100 ml but collecting up to 1 l of water to enhance the chance of detecting pathogens. It must be verified whether there is a systemic or an isolated contamination of the plumbing system beginning at the entrance point from the community water system into the health care facility.

In Germany in 2014, there was a recall of water flow meters contaminated with *Pseudomonas aeruginosa*. The water flow meters were contaminated at the time of testing to verify their integrity during the manufacturing of the water flow meters. Thousands of water flow meters had to be exchanged by the water utility companies.

It is important to describe the kinds of contamination when answering the question of whether the pathogens have been introduced from the water utilities, or have colonized the plumbing system of the health care facilities or if only an isolated contamination of taps or shower heads exists. The pathogens should be preserved to enable analysis and identification of them by molecular typing systems. After demonstrating the kind of contamination, a plan for sanitizing the system should be developed. The first step must be to exchange identified contamination points of the plumbing system such as water flow meters, water softeners or parts of the plumbing system. Only then would a disinfection be successful. In the case of a contamination of the plumbing system coming from the community water system, it is proposed that an ultraviolet-disinfection device is integrated between the water flow meters and the plumbing system to protect the system from the contamination introduced from the community of the central water system.

Control of the sanitation system

As mentioned before, only a few publications have identified the sanitation system as a responsible reservoir for nosocomial pathogens. In cases of outbreaks, it is proposed that the sink drains, toilets and other parts of the sanitation system are investigated and the isolated pathogens typed. For the future, there must be more investigation into the significance of the sanitation system where pathogens can find unique niches to exchange antibiotic resistance and to persist over a long time. People have also begun to investigate the central wastewater system of health care facilities to establish whether there are antibiotic resistant pathogens. The sink water drainage must be constructed in such a manner that splashes of highly antibiotic-resistant pathogens in the environment are avoided. Perhaps with these changes, the dramatic increase of antibiotic-resistant pathogens can be better controlled.

References

1 Exner, M, Kramer, A, Kistemann, T, Gebel, J, and Engelhart, S. 2007. [Water as a reservoir for nosocomial infections in health care facilities, prevention and control] *Bundesgesundheitsblatt Gesundheitsforschung Gesundheitsschutz* 50(3):302–11.

2 Exner, M, Kramer, A, Lajoie, L, Gebel, J, Engelhart, S, and Hartemann, P. 2005. Prevention and control of health care-associated waterborne infections in health care facilities. *Am J Infect Control* 33(5 Suppl 1):S26–40.

3 Schoen, ME, and Ashbolt, NJ. 2011. An in-premise model for *Legionella* exposure during showering events. *Water Res* 45(18):5826–36.

4 Anaissie, EJ, Penzak, SR, and Dignani/ MC. 2002. The hospital water supply as a source of nosocomial infections: a plea for action. *Arch Intern Med* 162(13):1483–92.

5 Trautmann, M, Halder, S, Lepper, PM, and Exner, M. 2009. [Reservoirs of Pseudomonas aeruginosa in the intensive care unit. The role of tap water as a source of infection] *Bundesgesundheitsblatt Gesundheitsforschung Gesundheitsschutz* 52(3):339–44.

6 Auld, H, MacIver, D, and Klaassen, J. 2004. Heavy rainfall and waterborne disease outbreaks: the Walkerton example. *J Toxicol Environ Health A* 67(20–22):1879–87.

7 MacKenzie, WR, Hoxie, NJ, Proctor, ME, Gradus, MS, Blair, KA, Peterson, DE, et al. 1994. A massive outbreak in Milwaukee of *Cryptosporidium* infection transmitted through the public water supply. *N Engl J Med* 331(3):161 7.

8 Nichols, G, Lane, C, Asgari, N, Verlander, NQ, and Charlett, A. 2009. Rainfall and outbreaks of drinking water related disease and in England and Wales. *J Water Health* 7(1):1–8.

9 Doring, G, Ulrich, M, Muller, W, Bitzer, J, Schmidt-Koenig, L, Munst, L, et al. 1991. Generation of *Pseudomonas aeruginosa* aerosols during handwashing from contaminated sink drains, transmission to hands of hospital personnel, and its prevention by use of a new heating device. *Zentralbl Hyg Umweltmed* 191(5–6):494–505.

10 Engelhart, S, Saborowski, F, Krakau, M, Scherholz-Schlosser, G, Heyer, I, and Exner, M. 2003. Severe *Serratia liquefaciens* sepsis following vitamin C infusion treatment by a naturopathic practitioner. *J Clin Microbiol* 41(8):3986–8.

11 Longtin, Y, Troillet, N, Touveneau, S, Boillat, N, Rimensberger, P, Dharan, S, et al. 2009. *Pseudomonas aeruginosa* outbreak in a pediatric intensive care unit linked to a humanitarian organization residential center. *Pediatr Infect Dis J* 29(3):233–7.

12 La Forgia, C, Franke, J, Hacek, DM, Thomson, RB, Robicsek, A, and Peterson, LR. 2009. Management of a multidrug-resistant *Acinetobacter baumannii* outbreak in an intensive care unit using novel environmental disinfection: a 38-month report. *Am J Infect Control* 38(4):259–63.

13 Carstens, A, Kepper, U., Exner, M., Hauri, A., Kaase, M., and Wendt, C. 2014. Plasmid-vermittelter Multispezies-Ausbruch mit Carbapenem-resistenten Enterobacteriaceae. [Plasmid-mediated multispecies-outbreak with Carbapenem-resistant Enterobacteriaceae.] *Epidemiologisches Bulletin* 47:455–9.

14 WHO. 2011. *Guidelines for Drinking-water Quality*. http://whqlibdoc.who.int/publications/2011/9789241548151_eng.pdf.

15 Dyck, A, Exner, M, and Kramer. A. 2007. Experimental based experiences with the introduction of a water safety plan for a multi-located university clinic and its efficacy according to WHO recommendations. *BMC Public Health* 7:34.

16 Department of Health. 2012. Water sources and potential *Pseudomonas aeruginosa* contamination of taps and water systems – advice for augmented care units. http://www.his.org.uk/files/8113/7088/0902/8_DH_Water_sources_and_potential_

Pseudomonas_aeruginosa_contamination_of_taps_and_water_systems_Advice_for_
augmented_care_March_2012.pdf

17 Department of Health. 2013. Water systems Health Technical Memorandum 04-01:
Addendum *Pseudomonas aeruginosa* – advice for augmented care units. https://www.
gov.uk/government/uploads/system/uploads/attachment_data/file/140105/Health_
Technical_Memorandum_04-01_Addendum.pdf

31

WATER SUPPLY IN RURAL COMMUNITIES[*]

Sara J. Marks

Senior Scientist, Department of Water and Sanitation in Developing
Countries (Sandec) at Eawag, Dübendorf, Switzerland

Kellogg J. Schwab

Professor, Environmental Health Sciences Johns Hopkins
Bloomberg School of Public Health; Director, Johns
Hopkins Water Institute, Baltimore, MD USA

Abbreviations

DRA – demand-responsive approach
FLOW – Field Level Operations Watch
JMP – Joint Monitoring Programme
NGO – non-governmental organization
SWE – small water enterprise

Learning objectives

1　Understand the nature of the challenges associated with water service provision in rural communities.
2　Be familiar with the key definitions and technology options for rural water supply schemes.
3　Critically analyze the evidence regarding successful and failed approaches to implementing rural water projects.

---

[*] Recommended citation: Marks, S.J. and Schwab, K.J. 2015. 'Water supply in rural communities', in Bartram, J., with Baum, R., Coclanis, P.A., Gute, D.M., Kay, D., McFadyen, S., Pond, K., Robertson, W. and Rouse, M.J. (eds) *Routledge Handbook of Water and Health*. London and New York: Routledge.

Introduction

Nearly half the world's population lives in rural communities where water service providers often face unique challenges. In industrialized countries such as the United States, access to piped water in the home is virtually universal. Yet even in such settings rural communities may experience inconsistent water quality and deteriorating infrastructure.

In developing countries the nature of the challenges differ. Access to improved water sources in rural areas is low, especially in sub-Saharan Africa (Table 31.1). The WHO/UNICEF Joint Monitoring Programme (JMP) defines an improved water source as "one that, by nature of its construction or through active intervention, is protected from outside contamination, in particular from contamination with faecal matter" (JMP 2014). To allow for international comparability, the JMP classifies the following sources as improved:

- Piped water into dwelling
- Piped water to yard/plot
- Public tap or standpipe
- Tubewell or borehole
- Protected dug well
- Protected spring
- Rainwater.

Expanding access to improved water is slowed by several barriers, including a mismatch between technology options and customer demand, lack of institutional support, unreliable or non-existent supply chains, intermittent power supplies, and an inability to recover the

Table 31.1 Water supply coverage by region as defined by the WHO/UNICEF Joint Monitoring Programme, 2012 estimates

Region	Urban			Rural		
	Total improved (%)	*Piped on premises (%)*	*Other improved (%)*	*Total improved (%)*	*Piped on premises (%)*	*Other improved (%)*
Caucasus and Central Asia	96	86	10	78	29	49
Developed countries	100	98	2	98	83	15
Eastern Asia	98	95	3	85	46	40
Latin America/Caribbean	97	94	3	83	66	16
Northern Africa	95	91	4	89	74	15
Oceania	94	74	21	45	11	34
Southern Asia	96	54	42	89	15	74
South-eastern Asia	95	50	44	85	13	71
Sub-Saharan Africa	85	34	51	53	6	47
Western Asia	96	92	4	79	66	13
Total	*96*	*80*	*16*	*82*	*30*	*52*

Source: JMP (2014).

recurrent cost of service provision. As a result, delivering adequate and sustainable water services has proven difficult in rural areas throughout many developing countries.

This chapter reviews and critically analyzes strategies for implementing rural water supply projects, with a focus on the evidence regarding successful and failed approaches. We focus on the water systems most common to rural communities, where the hardware and delivery scheme must be suited to the unique realities of the setting. These include non-networked water points such as borewells equipped with handpumps and protected wells; piped networks delivering water to public kiosks or homes; and small water enterprises such as cart or truck vendors.

The chapter is organized as follows: first, we present definitions and a description of water delivery options. Next, we describe eight key considerations for implementing water supply schemes. We then conclude with a discussion of innovations for tackling sustainability challenges in the water sector.

Definitions and water delivery options

What is a community?

Schouten and Moriarty (2003) define a community as a complex, dynamic, and interconnected collection of households composed of diverse sub-groups. Communities are often linked to external groups such as businesses, government agencies, schools, and clinics. The line where one community ends and another begins may be unclear. Further, communities are not homogenous; they consist of a diverse and heterogeneous collection of sub-groups. As the authors explain: "Diversity and unclear boundaries are characteristics of communities and they are arguably the characteristics that have the most important impact on community [water supply] management." The sub-groups within a community (not necessarily mutually exclusive) most relevant to water supply projects include:

- *Water users* – The water supply system may not serve every community member. Among those using the system, the level of service enjoyed might not be equal. Project teams must consider how the water project might differentially impact households, for water users as well as those without access to the system.
- *Women and girls* – As the primary managers of water within the home, women and girls face the heaviest time and labor burden due to water collection, and they are therefore disproportionately affected by rural water supply projects. By one estimate, women in rural sub-Saharan Africa spend an average of 15–17 hours per week fetching water, including walking to and from the source and queuing time (UNDP 2006).
- *Leaders* – Decisions related to the project are often made by one or more community leaders. These leaders may be elected, appointed, or they may have inherited their positions.
- *Elite* – Similar to leaders, wealthy elites may hold considerable power and exert influence over project-related decisions.

Water delivery options

Water systems encompass a wide array of designs to match varying levels of demand and resource availability. The list below is not meant to be all-encompassing, but rather it includes the technologies and service levels often found throughout rural areas of low income countries. In general, water supplies can be categorized as networked systems or non-networked (point) sources. Networked systems deliver water to public kiosks, yard taps, or household taps. They can be categorized as either:

- *Pumped schemes* – Groundwater is accessed through a borehole or well and pumped to a storage tank feeding a piped distribution network. The distribution network consists of water mains and sub-mains, and it may be looped, branched, or a combination of both. Pumping may be powered by diesel generators, solar photovoltaic arrays, wind-powered turbines, or manually (Fraenkel and Thake 2006).
- *Gravity-fed schemes* – Surface water or spring sources are collected at elevation and distributed through a piped network. Such water schemes harness energy due to gravity to maintain pressure throughout the system.

Non-networked water points provide water to customers at the location of the source. Common technology options include:

- *Handpump* – A manually operated pump typically fitted to a deep borehole or well. Many designs are available, with most using check valves for positive displacement of water.
- *Protected spring* – Groundwater reaching the surface is accessed at a protective structure and/or stored or piped to another access point.
- *Rainwater harvesting* – Rainwater is collected from the surface of the household's roof and stored nearby.

Not fitting easily into the above categories are *water vendors* who sell water to rural households via tanker trucks, carts, stationary shops, or hand carriers. Although the WHO/UNICEF JMP does not recognize it as an "improved" water supply, water vending nonetheless constitutes a major source of domestic water for some rural dwellers, especially in more densely populated areas (Opryszko et al. 2009).

Planning and implementation: key considerations

Planning and implementing rural water supply projects involves technical, financial, and management-related considerations. Premature failure of rural water systems has been linked to inappropriate design, poor management of water resources, rent-seeking behavior, and limited institutional capacity (Brookshire and Whittington 1993, Lovei and Whittington 1993, Singh et al. 1993). Moreover, community members often struggle to operate and maintain the water system over the useful life of the infrastructure (Kleemeier 2000, Schouten and Moriarty 2003). Whereas these challenges are not unique to rural water projects, the planning and implementation solutions used in rural communities may differ as compared to urban or peri-urban settings.

This section reviews and analyzes the literature regarding eight key implementation issues for rural water supply projects. These issues include: assessing household demand; determining the level of service; achieving sound financial management; instilling community members' sense of ownership for the project; protecting local water resources; ensuring health and hygiene benefits; considering gender- and poverty-sensitivity of the project; and establishing institutional support networks.

Household demand

Demand-oriented planning refers to a flexible approach that prioritizes user control over choices about technology features and the level of service received (Garn 1997). Salient features of water supplies include quality/taste, quantity, reliability, proximity to home, and cost of the water (Hope 2006). Multiple studies have demonstrated that a demand-responsive approach (DRA) to planning improves the sustainability of water projects as compared to centrally planned and operated projects, which in many instances failed prematurely (Therkildsen 1988, Sara and Katz 1998, Whittington et al. 2009). Through DRA, community members typically make upfront cash and labor contributions toward the technology of their choice, attend planning meetings, and participate in decisions about the project (Narayan 1995). In addition, the community is often responsible for covering 100 percent of the cost of operating and maintaining the project after installation.

Level of service

In the DRA framework, community members typically choose amongst a suite of water supply designs delivering various levels of service. Evidence suggests that such decision-making improves outcomes for projects (Marks and Davis 2012), but more so if water users are engaged in management-related decisions rather than technical decisions (Khwaja 2009, Marks et al. 2014). Other studies have shown that water supply projects that are exceptionally large or complex are prone to operational challenges and eventual failure in the absence of external support (Kleemeier 2000, Schouten and Moriarty 2003). On the other hand, projects that are overly-simplified and only offer a low level of service may not meet the water users' needs and be neglected over time (Singh et al. 1993).

Financing and management

For financing, key considerations include determining the level of capital investment and any cost sharing arrangements; designing a tariff structure for covering ongoing operation, maintenance, and replacement costs; and deciding whether subsidies or loans will be made available to customers. Assessing households' willingness and ability to pay for various service levels through demand-oriented planning is a critical step for determining such financial arrangements (Whittington et al. 1998). Management considerations include determining the water system's hours of operation, who will perform key technical and managerial functions (e.g., collecting fees, bookkeeping, making repairs, emptying and cleaning tanks, etc.), any rules and enforcement mechanisms, and the roles and responsibilities of supporting non-governmental organizations (NGOs) or government agencies.

Sense of ownership

Following installation of the infrastructure, rural communities are often responsible for operating and maintaining their water system. Water sector professionals widely cite the need to instill a sense of ownership for the project among water users, since this is thought to ensure their willingness to pay for and maintain the infrastructure. A study of 50 rural communities in Kenya found that water users' sense of ownership for the communal system was linked to their participation in planning and implementing the project. In particular, users who had made decisions about the level of service received and made a meaningful investment (typically toward a private yard tap) were most likely to report the strongest feelings of ownership for the infrastructure (Marks and Davis 2012).

Source protection

A water system's design must balance households' demand for water with the availability of local water resources – both in terms of quantity and quality. The system's capacity should be sufficient to meet current as well as future demand, based on population growth estimates and potential changes in water uses. Strains on the system may arise from illegal connections, new water uses within the catchment area, and seasonal or climatic patterns. A water system that is not designed to respond to such inevitable fluctuations will likely experience serious technical problems and potential depletion of the source. Implementation teams must conduct a careful assessment of the available water resources, create a plan for managing and protecting the source and catchment area, and design the system to meet current and projected water demand. A study of water user associations in the Philippines found that cooperative management among users was most likely if the water resource was neither too scarce nor overly abundant (Araral 2009).

Hygiene education

To fully realize the health benefits of a water supply project, it is important to provide hygiene education and to promote consistent handwashing. Fewtrell and colleagues conducted a systematic review and meta-analysis to compare the relative effectiveness of water, sanitation, hygiene, and combined interventions for reducing illness. Hygiene interventions were found to be the most effective means for reducing diarrheal disease incidence (Fewtrell et al. 2005). These results are consistent with previous reviews, which found that handwashing and other hygiene interventions are associated with a 43 percent and 33 percent reduction in illness, respectively (Esrey et al. 1991, Curtis and Cairncross 2003).

Gender and poverty sensitivity

Women and the poor are disproportionately affected by water supply projects, yet they are often the least involved in planning and implementing the project (Schouten and Moriarty 2003). In a 15-country study, projects that emphasized gender- and poverty-sensitive decision-making during implementation performed better over the long term, in terms of both technical and financial outcomes (Gross et al. 2001). Despite this generalized finding, the modes by which women and the poor were engaged in project planning were remarkably site-specific. This suggests that successful projects are those which allocate sufficient resources to assessing gender roles, power divisions, and caste or class disparities within each community.

External support

Left entirely to their own devices, rural communities will almost always face challenges with operating their water system. The sector is increasingly recognizing that some form of post-construction support (PCS) is essential for achieving sustainable service delivery. PCS may be provided in the form of circuit riders delivering spare parts or training on bookkeeping and financial management of the project. Evidence from a multi-country study shows that

PCS was critical for ensuring the functionality of rural water supplies four to eight years after construction (Whittington et al. 2009). In Bolivia, management-oriented PCS was most strongly associated with sustainable outcomes for piped systems (Davis et al. 2008).

Toward sustainability: innovations and policies

Despite the significant progress made in recent decades in improving access to water services globally, the sector still faces serious challenges with sustainability issues, particularly in rural areas. Projects still fail to deliver services that match what water users want and need, many communities still lack access to spare parts or assistance, the roles and responsibilities regarding the system's management often remain unclear, and few water supply programs consider replacement or upgrading of the infrastructure in their planning agenda. In recent years there has been a renewed enthusiasm for tackling sustainability problems, as evidenced by the broad acceptance of a sustainability charter (http://www.washcharter.org/) and the increasing number of NGOs, donors, and governments putting sustainability at the forefront of their water supply programs.

Innovations for sustainability

Monitoring and real-time mapping

Using open-source software and phone-based data collection devices, monitoring and real-time mapping efforts have enabled program managers to provide affordable and effective support to rural communities throughout several countries. Examples include Akvo's Field Level Operations Watch (FLOW) platform, which monitors water supplies in Malawi, and WaterAid's Water Point Mapper program that is used throughout east and southern Africa.

Small water enterprises

Private sector involvement in the installation and operation of small water enterprises (SWE) has been shown to improve the sustainability record of water services in rural communities. In Ghana, for example, SWEs delivered higher quality water (Opryszko et al. 2013) and were better maintained (Opare 2011) than water systems that were managed by communities alone.

Water for multiple uses

Water supply projects that are designed to support domestic and productive activities improve household income and encourage entrepreneurship, especially among women (Marks and MacDonald 2014). Moreover, water-based productive activities are associated with households' satisfaction and willingness to pay for their water service, as well as financial health of the system (Moriarty et al. 2004).

From infrastructure to services

As evidenced by each of the innovations above, there is a shift within the development community to re-frame water supply planning around delivery of services, rather than simple installation of infrastructure. The service delivery approach emphasizes the *attributes* of water services that are most valued by customers, including water quality and quantity, taste, convenience, safety of accessing, and the cost of delivery. A key innovation in support of the service-oriented approach is the establishment of circuit rider programs, which connect rural communities (especially those in remote areas) to supply chains for spare parts (Rosenberg 2011).

Key recommended readings (open access)

1 For further reading and forum discussion on multiple-use water services, see: www. musgroup.net
2 The WASH Sustainability Charter can be found at: http://www.washcharter.org/
3 Information about the FLOW tool by Akvo and Water For People: http://www. waterforpeople.org/flow-mapping/
4 Information about the Water Point Mapper tool: http://www.waterpointmapper.org/
5 Additional readings and discussion topics on rural water supply planning are found on the Rural Water Supply Network: http://www.rural-water-supply.net/en/
6 "The Real Future of Clean Water" by David Bornstein: http://opinionator.blogs. nytimes.com/2013/08/21/the-real-future-of-clean-water/?hp&_r=0

References

Araral, E. (2009) What explains collective action in the commons? Theory and evidence from the Philippines. *World Development*, 37(3), pp. 687–697.

Brookshire, D. and Whittington, D. (1993) Water resources issues in the developing countries. *Water Resources Research*, 29(7), pp. 1883–1888.

Curtis, V. and Cairncross, S. (2003) Effect of washing hands with soap on diarrhoea risk in the community: A systematic review. *The Lancet Infectious Diseases*, 3(5), pp. 275–281.

Davis, J., Lukacs, H., Jeuland, M., Alvestegui, A., Soto, B., Lizárraga, G., Bakalian, A. and Wakeman, W. (2008) Sustaining the benefits of rural water supply investments: Experience from Cochabamba and Chuquisaca, Bolivia. *Water Resources Research*, 44, p. W12427.

Esrey, S. A., Potash, J. B., Roberts, L. and Schiff, C. (1991) Effects of improved water supply and sanitation on ascariasis, diarrhoea, dracunculiasis, hookworm infection, schistosomiasis, and trachoma. *Bulletin of the World Health Organization*, 69(5), pp. 609–621.

Fewtrell, L., Kaufmann, R. B., Kay, D., Enanoria, W., Haller, L. and Colford, J. M. (2005) Water, sanitation, and hygiene interventions to reduce diarrhoea in less developed countries: A systematic review and meta-analysis. *The Lancet Infectious Diseases*, 5(1), pp. 42–52.

Fraenkel, P. and Thake, J., eds. (2006) *Water Lifting Devices: A Handbook for Users and Choosers*, Third Edition. Rome, Italy: FAO/IT/Practical Action.

Garn, H. (1997) Lessons from large-scale rural water and sanitation projects: Transition and innovation. Urban Environmental Sanitation Working Papers. Washington, DC: Water and Sanitation Program (WSP).

Gross, B., van Wijk, C. and Mukherjee, N. (2001) *Linking Sustainability with Demand, Gender and Poverty*. Washington, DC: World Bank Water and Sanitation Program (WSP).

Hope, R. A. (2006) Evaluating water policy scenarios against the priorities of the rural poor. *World Development*, 34(1), pp. 167–179.

JMP (2014) *Progress on Drinking Water and Sanitation: 2014 Update*. Geneva/New York (available online at: http://www.wssinfo.org/): World Health Organization / United Nations Children's Fund Joint Monitoring Programme.

Khwaja, A. I. (2009) Can good projects succeed in bad communities? *Journal of Public Economics*, 93(7–8), pp. 899–916.

Kleemeier, E. (2000) The impact of participation on sustainability: An analysis of the Malawi rural piped scheme program. *World Development*, 28(5), pp. 929–944.

Lovei, L. and Whittington, D. (1993) Rent-extracting behavior by multiple agents in the provision of municipal water supply: A study of Jakarta, Indonesia. *Water Resources Research*, 29(7), pp. 1965–1974.

Marks, S. J. and Davis, J. (2012) Does user participation lead to sense of ownership for rural water systems? Evidence from Kenya. *World Development*, 40(8), pp. 1569–1576.

Marks, S. J. and MacDonald, L. (2014) Households' motivations for investing in multiple-use water services in rural Burkina Faso (Available online at: http://water.jhu.edu/magazine/households-motivations-for-investing-in-multiple-use-water-services-in-rura).

Marks, S. J., Komives, K. and Davis, J. (2014) Community participation and water supply sustainability: Evidence from handpump projects in rural Ghana. *Journal of Planning Education and Research*, 34(3), pp. 276–286.

Moriarty, P., Butterworth, J. and van Koppen, B., eds. (2004) Beyond domestic: Case studies on poverty and productive uses of water at the household level. IRC Technical Paper Series 41, Delft, the Netherlands: IRC International Water and Sanitation Centre.

Narayan, D. (1995) The contribution of people's participation: Evidence from 121 rural water supply projects. Environmentally sustainable development occasional paper series, no. 1., Washington, DC: The World Bank.

Opare, S. (2011) Sustaining water supply through a phased community management approach: Lessons from Ghana's "OATS" water supply scheme. *Environment, Development and Sustainability*, 13(6), pp. 1021–1042.

Opryszko, M. C., Huang, H., Soderlund, K. and Schwab, K. J. (2009) Data gaps in evidence-based research on small water enterprises in developing countries. *Journal of Water and Health*, 7(4), pp. 609–622.

Opryszko, M. C., Guo, Y., MacDonald, L. H., MacDonald, L., Kiihl, S. and Schwab, K. J. (2013) Impact of water-vending kiosks and hygiene education on household drinking water quality in rural Ghana. *American Journal of Tropical Medicine and Hygiene*, 88(4), pp. 651–660.

Rosenberg, T. (2011) Keeping the water flowing in rural villages. *New York Times*, December 8. http://opinionator.blogs.nytimes.com/2011/12/08/keeping-the-water-flowing-in-rural-villages/?_r=0

Sara, J. and Katz, T. (1998) Making rural water supply sustainable: Report on the impact of project rules. Washington, DC: World Bank Water and Sanitation Program (WSP).

Schouten, T. and Moriarty, P. (2003) *Community Water, Community Management: From System to Service in Rural Areas*. London: ITDG Publishing.

Singh, B., Ramasubban, R., Bhatia, R., Briscoe, J., Griffin, C. C. and Kim, C. (1993) Rural water supply in Kerala, India: How to emerge from a low-level equilibrium trap. *Water Resources Research*, 29(7), pp. 1931–1942.

Therkildsen, O. (1988) *Watering White Elephants? Lessons From Donor Funded Planning and Implementation of Rural Water Supplies in Tanzania*. Uppsala: Scandinavian Institute of African Studies.

UNDP (2006) *Human Development Report 2006. Beyond Scarcity: Power, Poverty, and the Global Water Crisis*. New York: United Nations Development Programme.

Whittington, D., Davis, J. and McClelland, E. (1998) Implementing a demand-driven approach to community water supply planning: A case study of Lugazi, Uganda. *Water International*, 23(3), pp. 134–145.

Whittington, D., Davis, J., Prokopy, L., Komives, K., Thorsten, R., Lukacs, H., Bakalian, A. and Wakeman, W. (2009) How well is the demand-driven, community management model for rural water supply systems doing? Evidence from Bolivia, Peru and Ghana. *Water Policy*, 11(6), pp. 696–718.

32

INTEGRATED URBAN WATER MANAGEMENT*

Kalanithy Vairavamoorthy, Jochen Eckart, Kebreab Ghebremichael and Seneshaw Tsegaye

PATEL COLLEGE OF GLOBAL SUSTAINABILITY, UNIVERSITY
OF SOUTH FLORIDA, TAMPA, FLORIDA, USA

Learning objectives

1 Describe the principles for the integrated provision of water and sanitation services in urban areas.
2 Analyze the benefits of integrated urban water management for public health.
3 Discuss how integrated urban water management can be applied in specific case studies.

Water-related diseases are major public health concerns in growing urban centers of developing countries and are mainly attributed to the lack of proper access to improved water supply and sanitation (Lee and Schwab, 2005; Corcoran et al., 2010). The provision of water and sanitation services based on linear, sectoral and centralized systems (where water is used once and then discharged) has resulted in significant improvements to public health. Nevertheless, worldwide almost 750 million people in urban and rural areas still do not have access to an improved drinking water source and some 2.5 billion, almost half the population of the developing world, do not have access to improved sanitation (WHO and UNICEF, 2014). According to Biswas (2004) this is caused by the increasing multi-dimensional, multi-sectoral and multi-interest nature of urban water problems. Isolated approaches of the different sub-sectors' water supply, sanitation and drainage cannot capture those interactions and tend to be insufficient to further improve public health in many cities throughout the developing world.

In contrast to the linear, sectoral and centralized urban water systems, integrated urban water management (IUWM) adopts a holistic view of all components of the urban water cycle (water supply, sanitation and drainage) including feedback loops. The concept of IUWM was developed from the broader framework of integrated water resource management (IWRM)

---

* Recommended citation: Vairavamoorthy, K., Eckart, J., Ghebremichael, K. and Tsegaye, S. 2015. 'Integrated urban water management', in Bartram, J., with Baum, R., Coclanis, P.A., Gute, D.M., Kay, D., McFadyen, S., Pond, K., Robertson, W. and Rouse, M.J. (eds) *Routledge Handbook of Water and Health*. London and New York: Routledge.

which gained popularity in the 1990s. Both approaches attempt to overcome the limitations of the sectoral perspective on the urban water cycle. While IWRM encompasses a whole catchment, IUWM works within the boundaries of a city while considering the interactions with the wider catchment. Because of the difference in scales, IWRM is more focused on the integration of the different water users (agriculture, domestic, industry and ecosystem) while IUWM aims to integrate the different parts and flows of the urban water cycle (water supply, sanitation and drainage).

Guiding principles for integrated urban water management

IUWM is not a method but rather a way of thinking that can be described by guiding principles for the planning and design of water and sanitation systems (Jacobsen et al., 2012).

Apply an integrated framework

The provision of water and sanitation should be contextualized within an integrated urban water framework. The integrated framework allows the relationship between the various components of the urban water system to be analyzed and articulated, viewing water supply, stormwater and wastewater as components of an integrated physical system (van der Steen and Howe, 2009). The integrated framework can be applied at two different spatial scales: neighborhood/cluster scale and urban/city scale.

The neighborhood/cluster scale allows the articulation of detrimental as well as directional relationships between the different components of the urban water cycle. Figure 32.1 presents an integrated framework for water systems as a stock and flow model from sources to sinks. The framework is tailored for low-income neighborhoods in developing countries illustrating typical elements such as different water uses, onsite sanitation and harmful interactions that include cross-contamination of water sources or leakage. The framework facilitates a systematic exploration of the possible interactions between the different elements of the urban water cycle. An example of a harmful interaction is the cross-contamination of treated water by dysfunctional or poor sanitation systems. In several cities in developing countries, although adequate water treatment is provided at the treatment plant, the water distribution pipes may go through foul water bodies exposing them to the risk of contamination. Furthermore, intermittent supplies and low pressures in the distribution system encourage contaminant ingress from the surrounding environment and stagnancy may lead to declining microbiological and chemical water quality. The resulting contamination of the potable water creates public health risks.

At the urban/city scale, water flows are described between different neighborhoods within an urban area. The integrated perspective helps to explore the benefits of providing water and sanitation services for all including low-income communities. An example of such a holistic approach to urban improvement in which slums are seen as an integral part of the city is the Indore's Slum Network Project (SNP) (Verma, 2000). Typically the unwillingness of the municipality to provide drainage systems in slums may impact the performance of drainage systems in the entire city. Similarly, lack of sanitation services in slums results in water quality degradation of water sources in other areas of the city. In this regard the SNP aims to provide concerted improvements of water, sanitation and drainage services in slums, to achieve sustainable improvements to both the slums and the whole city.

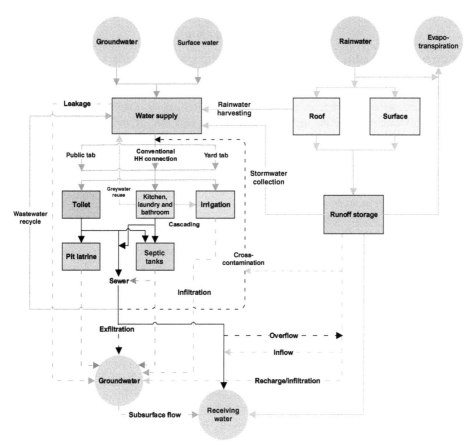

Figure 32.1 Integrated urban water framework for low-income urban areas

Source: Vairavamoorthy et al. (2012).

Provide water fit for purpose

When considering water demands for different purposes, it is important to match water of a certain quality for its intended use. This perspective avoids the need for the highest water quality for all purposes, hence reducing water treatment costs (Muller, 2010). The application of this principle highlights the importance of alternative sources of water that can be safely used for different purposes. For example, greywater can be used for toilet flushing or gardening and non-potable grade quality can be utilized by the industrial sector (Vairavamoorthy et al., 2012; Jacobsen et al., 2012). Cascading water use maximizes the benefits from each drop of water by utilizing water multiple times from higher to lower-quality needs. This can be achieved by linking different water users in the city. A common example in many low-income cities is the use of domestic wastewater for urban agriculture. Nevertheless untreated wastewater used for irrigation can result in public health risks since pathogens and heavy metals in the wastewater may end up in the food chain. In order to minimize health risks there is a need to provide low cost treatment systems such as waste stabilization ponds or constructed wetlands to facilitate a safe reuse of wastewater for irrigation purpose. Guidelines for the safe reuse of water are provided by the World Health Organization (WHO, 2006).

Portfolio of water sources

To increase the reliability of available water it is important to diversify water sources rather than depend on limited options of water sources (Vairavamoorthy et al., 2009; Gleick, 2009; Bahri, 2012). There is a need to consider all potential water sources (surface water, groundwater, water transfer, desalination, wastewater reuse, rainwater harvesting, etc.) on local and catchment scale and combinations of end-use, as well as system efficiency measures that would provide additional water sources and/or reduce water demand (Sharma and Vairavamoorthy, 2009). For example the reduction of high leakage rate in water supply systems, a common problem in many low-income cities, provides a great opportunity to improve system efficiency. Leakage refers to physical water loss in the distribution system. As leakage management programs are costly, an optimum leakage level can be calculated where the cost of reducing leakage and the value of water saved are balanced. Compared to other options, leakage management and water conservation are some of the cheapest and the largest sources of water available within cities. Reducing the average water losses of 35 percent by half could be used to increase water supply to an additional 90 million people in low-income countries (Kingdom et al., 2006). In addition, leakage reduction improves public health by reducing the risk of cross-contamination as described above.

Maximize the benefits from wastewater

Besides the improvement of public health, wastewater treatment can also help to maximize the benefits harvested from wastewater (Vairavamoorthy et al., 2012). This perspective of beneficiation highlights the potential of reclaimed water, renewable energy and nutrients which can be extracted from wastewater. This can provide additional incentives for the provision of sanitation services. An example is the recovery of nutrients (phosphorus and nitrogen) from wastewater. Mineral phosphorus is a finite resource that may be depleted in 50–100 years (Cordell et al., 2009) and at the same time the demand is projected to increase. Alternative and more sustainable sources such as mineral recovery from wastewater are becoming increasingly important. Treating phosphorus as a finite resource shifts the management paradigm from mitigating a noxious substance causing eutrophication to recovering a precious element. In addition, treated wastewater needs to be viewed as potential water sources for irrigation, environmental flows, non-potable uses or even as a new drinking water source. The use of wastewater can help to ensure the long-term reliability of sufficient water supply, guarantees wastewater treatment, and fosters the integrated provision of water supply and sanitation with associated public health benefits (Bieker et al., 2010).

Combination of urban and catchment scale

Water management at the city scale has to include consideration of the interactions with the wider catchment. An integrated perspective aims to capture the interactions between cities and the catchments in which they are embedded. Activities at the catchment scale determine a city's access to adequate quality water and also provide flood protection. For example upstream changes in land-use patterns or water allocation may change the local hydrology and available water resources and can result in the necessity of watershed protection plans or water allocation strategies. On the other hand, the city's impact on watershed has to be considered. This may refer to the efficient use of the water resources within cities as well as the impact of cities on downstream uses through the discharge of wastewater and stormwater. Hence IUWM should be combined with IWRM strategies on the catchment scale (GWP, 2010).

Involving all the players

Critical to the success of IUWM is the early and continuous integration of all relevant stakeholders in the planning, decision and implementation process. The stakeholder participatory process is required not only to understand the priorities of potential users, but also to take account of the prevailing institutional context, to undertake the IUWM strategy in partnership with implementers and other key stakeholders, and to communicate results and emerging innovations effectively (Butterworth et al., 2011; Bahri, 2012). Many challenges of the provision of water supply and sanitation in cities in developing countries are often more a function of unfair distribution than an absolute shortage of resources. For example politics plays a major role in determining how water and sanitation services are provided to different potential users. Therefore stakeholder engagement has to include consideration of the politics of the provision of water and sanitation services and should be seen as part of the wider political process.

Institutional environment

Many barriers to achieving IUWM are institutional in nature because of a highly fragmented institutional division of responsibilities and tasks. There is a need to ensure the coordination of the institutional divisions of responsibilities and tasks that exist during the entire process of planning and implementation (GWP, 2010; Bahri, 2012). The institutional structure should allow institutions to deal with water supply, wastewater, stormwater and solid waste management collectively, enabling them to coordinate their plans and actions in a way that recognizes and reconciles the important interrelations and interdependencies that exist between each of these subsystems

Example for the application of IUWM principles

Although the guiding principles of IUWM are straightforward, their application on the ground is challenging. Since 2000 the concept of IUWM has been successfully applied on the city scale with documented applications from Brazil (Braga, 2000), Australia (Mitchell et al., 2004), Singapore (Tan et al., 2009), Europe (Howe et al., 2011) and African cities (Jacobsen et al., 2012). As an example, the application of IUWM principles for the city of Arua, Uganda, is presented (Jacobsen et al., 2012).

Arua, a rapidly emerging town located in northern Uganda, is experiencing a critical shortage of water. The main water source, the Enyau River, is affected by the increasing water demands of upstream users, exacerbating the low flow conditions during the dry season. The current water supply was not sufficient to meet the demand in year 2010, and with an estimated population growth of up to 200 percent in the next 20 years, the water supply problem will only get worse. In addition to water shortage, Arua also lacks adequate sanitation provisions, with dysfunctional pit latrines, open defecation and untreated wastewater disposal posing both health and water pollution risks. In order to cope with these challenges, a feasibility study for future water supply and sanitation was developed by applying IUWM principles. As an emerging urban area without extensive inherited water infrastructure systems, Arua provides the opportunity to implement water system configurations, where surface water, groundwater and stormwater together are considered as potential sources. In addition, solutions could be applied that allow source separation of wastes and implementation of reclamation schemes. The feasibility study proposed that Arua surface water, groundwater, artificial aquifer recharge and recycled greywater were considered as potential water sources, resulting in increased water security and the provision of sanitation services. The decentralized wastewater treatment is viewed from the dual perspective of providing improved sanitation and generating additional water sources. The allocation of the different water resources is prioritized from a cost–benefit perspective. The feasibility study estimates that for Arua the average unit costs for the proposed IUWM scenario would be US\$0.57 per cubic meter, while the unit costs for the traditional approach of using water from conventional surface water sources (20 km away from the city) would be US\$0.74 per cubic meter. The feasibility report concluded that building a decentralized system for wastewater recycling using innovative options, such as a decentralized wastewater treatment system (DEWATS) and soil aquifer treatment (SAT), can both improve sanitation and provide sufficient water resources to meet increasing demands in the next 20 years.

Critical discussion of IUWM principles

The approach of IUWM has been promoted in practice since the 2000s. Nevertheless no information is available to indicate whether this integrated approach has led to an improvement of water supply and sanitation services that would not have occurred under conventional management practice, and without any explicit use of the integrated principles. The principles have also been criticized for being unrealistic. For example, Biswas (2004) and several of those who responded to his article pointed out the definitions of IUWM and IWRM appear overly broad and that there is no clear idea about what exactly the concept means in operational terms. The critique was that the absence of both an operational definition and measurable criteria makes it difficult to identify what constitutes an IUWM approach, or how water should be managed so that the system remains inherently integrated on a long-term basis. The guiding principles for IUWM were developed in response to this critique in order to make the general concepts more manageable.

Another major criticism of IUWM is that it ignores politics (Watson and Wester, 2003; Gyawali et al., 2006; Molle, 2008), which is one of the main mechanisms in society for organizing decision making. The principles of stakeholder engagement and institutional reform are criticized for being simplistic in relation to politics. This critique does not only apply to IUWM but also to many technical planning approaches. While IUWM does not provide a framework which always results in developing political viable solutions, it can still be helpful in promoting integrated thinking in terms of systems and processes.

Besides the criticism, the presented principles of IUWM can help to identify the necessary steps needed to solve many water-related public health concerns, especially in low-income countries. The integrated perspective improves the understanding of pollution pathways (e.g. cross-contamination between water supply and sanitation system) and can support the provision of sufficient water resources (e.g. exploring the full spectrum of water resources). Furthermore the IUWM principles highlight additional drivers for the provision of water supply and sanitation such as the recovery of benefits from wastewater that go beyond the traditional improvement of public health.

Key recommended readings (open access)

1 Jacobsen, M., Webster, M. and Vairavamoorthy, K. (2012) The Future of Water in African Cities: Why Waste Water? World Bank Publication. The publication provides an overview on the urban water management challenges in African cities. In order to address these challenges, principles of IUWM are presented. The application of these principles in three African cities is documented.
2 Butterworth, J., McIntyre, P., da Silva, C. (eds) (2011) *SWITCH in the City: Putting Urban Water Management to the Test*. Delft. The book gives an overview on the IUWM approach developed in the SWITCH (Sustainable Water Management Improves Tomorrow's Cities' Health) project and presents the application of the approach in 12 demonstration cities. In addition the book provides detailed guidance on the required stakeholder engagement for IUWM.
3 Bahri, Akiça (2012) Integrated Urban Water Management. Global Water Partnership TEC Background Papers No. 16. The paper provides a background to water resource and urbanization and the need for IUWM. It explores how to create an enabling environment for IUWM.

References

Bahri, Akiça (2012) Integrated Urban Water Management. Global Water Partnership TEC Background Papers No. 16. Retrieved from: http://www.gwp.org/Global/The%20Challenge/Resource%20material/GWP_TEC16.pdf

Bieker, S., Cornel, P. and Wagner, M (2010) Semicentralised supply and treatment systems: integrated infrastructure solutions for fast growing urban areas. *Water Science & Technology*, 61(11):2905–2913.

Biswas, A. K. (2004) Integrated water resources management: a reassessment: a water forum contribution. *Water international*, 29(2):248–256.

Braga Jr, B. P. (2000) The management of urban water conflicts in the metropolitan region of Sao Paulo. *Water International*, 25(2):208–213.

Butterworth, J., McIntyre, P., & da Silva Wells, C. (2011) *SWITCH in the City: Putting Urban Water Management to the Test*. The Hague: IRC International Water and Sanitation Centre.

Corcoran, E., Nellemann, C., Baker, E., Bos, R., Osborn, D., and Savelli, H. (eds) (2010) Sick Water? The Central Role of Wastewater Management in Sustainable Development. A Rapid Response Assessment. United Nations Environment Programme, UN-HABITAT, GRID-Arendal, International Atomic Energy Agency. Retrieved from http://www.unep.org/pdf/SickWater_screen.pdf

Cordell, D., Drangert, J.-O. and White, S (2009) The story of phosphorus: global food security and food for thought. *Global Environmental Change*, 19(2):292–305.

Gleick, P. H. (2009) Doing more with less: improving water use efficiency nationwide. *Southwest Hydrology*, 8(1):20–21.

Global Water Partnership (GWP) (2010) Towards a water secure world: What is Integrated Water Resource Management? http://www.gwp.org/The-Challenge/What-is-IWRM/ (access date: August 20, 2011).

Gyawali, D., Allan, J. A. & Antunes, P. (2006) EU-INCO water research from FP4 to FP6 (1994–2006): a critical review. Retrieved from: http://ec.europa.eu/research/water-initiative/pdf/incowater_fp4fp6_rapport_technique_en.pdf

Howe, C. A., Vairavamoorthy, K. & van der Steen, N. P. (2011) Sustainable water management in the city of the future. Retrieved from: http://www.switchurbanwater.eu/outputs/pdfs/Switch_-_Final_Report.pdf

Jacobsen, M., Webster, M. and Vairavamoorthy, K. (2012) *The Future of Water in African Cities: Why Waste Water?* World Bank Publication. Retrieved from: http://water.worldbank.org/sites/water.worldbank.org/files/publication/iuwm-full-report.pdf

Kingdom, B., Liemberger, R. and Marin, P. (2006) *The Challenge of Reducing Non-Revenue Water (NRW) in Developing Countries*. Washington, DC: The World Bank.

Lee, E. and Schwab, K. (2005) Deficiencies in drinking water distribution systems in developing countries. *Journal of Water and Health*, 3(2):109–127.

Mitchell, V. G., Bui, E., Cleugh, H., Diaper, C., Grant, A., Gray, S. R., Sharma, A. and Toze, S. (2004) Delivering planning and performance assessment tools for integrated urban water management, CSIRO Water for a Healthy Country Flagship. Melbourne, Australia: CSIRO.

Molle, F. (2008) Nirvana Concepts, narratives and policy models: insights from the water sector. *Water Alternatives*, 1(1):131–156.

Muller, M. (2010) Fit for purpose: taking integrated water resource management back to basics. *Irrigation and Drainage Systems*, 24(3–4):161–175.

Sharma, S. K. and Vairavamoorthy, K. (2009) Urban water demand management: prospects and challenges for the developing countries. *Water and Environment Journal*, 23(3):210–218.

Tan, Y. S., Lee, T. J. & Tan, K. (2009) *Clean, Green and Blue: Singapore's Journey Towards Environmental and Water Sustainability*. Singapore: Institute of Southeast Asian Studies.

Vairavamoorthy, K., Tsegaye, S. and Eckart, J. (2012) Urban water management in cities of the future – emerging areas in developing countries. *On the Water Front – Selections from the 2011 World Water Week in Stockholm*. Stockholm: SIWA. Retrieved from: http://www.worldwaterweek.org/documents/Resources/Best/2011/OntheWaterFront2011_final.pdf

Vairavamoorthy, K., Zhou, Y. and Mansoor, M. (2009) Urban water systems and their interactions. *Desalination*, 251:402–409.

van der Steen, P. and Howe, C. (2009) Managing water in the city of the future: strategic planning and science. *Rev Environ Sci Biotechnol*, 8:115–120.

Verma, G. D. (2000) Indore's Habitat Improvement Project: success or failure? *Habitat International*, 24(1):91–117.

Watson, N. & Wester, P. (2003) Inter-organizational Domains, Collaboration and the Uncertain Quest for Integrated Water Resources Management. Paper presented at the 2nd IAHS International Symposium on Integrated Water Resources Management (IWRM): Towards Sustainable Water Utilisation in the 21st Century, 22–24 January 2003, Stellenbosch, South Africa.

WHO (2006) WHO Guidelines for the Safe Use of Wastewater, Excreta and Greywater. Vol. I: Policy and Regulatory Aspects. Vol. II: Wastewater Use in Agriculture. Vol. III: Wastewater and Excreta Use in Aquaculture. Vol. IV: Excreta and Greywater Use in Agriculture. Geneva: World Health Organization.

WHO and UNICEF (2014) Progress on Drinking and Sanitation – 2014 Update. Geneva: WHO. Retrieved from: http://www.unicef.org/publications/files/JMP_report_2014_webEng.pdf

PART V

Distal influences

33

INTRODUCTION:
DISTAL INFLUENCES*

David M. Gute

PROFESSOR, DEPARTMENT OF CIVIL AND ENVIRONMENTAL
ENGINEERING, TUFTS UNIVERSITY, USA

Learning objectives

1 Appreciate how a wide array of factors that you may not associate closely with the provision of the adequate supply of safe water and sanitation services around the world influence the distribution of these assets.
2 Appreciate the broad array of disciplines required to assess and provide commentary on the elements covered in this section.
3 Understand the complexity of arriving at solutions of sufficient "uptake" to ensure both sustainability and effectiveness.

Overview

This part of the *Handbook* covers factors relating to water and health that have been termed "distal" to the all too present outcomes that result from the lack of access to water of sufficient quantity or quality. Distal is an elastic term having both a popular meaning as well as a precise clinical usage. It is instructive to reflect upon the clinical interpretation by comparing distal (far) with proximal (near). Through this lens we can reflect upon the array of factors we are about to review which include: water scarcity, climate change, poverty, emergencies and disasters, demographics, water reuse, and war and conflict.

Some of the relative "distance" between these factors and the outcomes reported in other parts of the *Handbook* may be quite different depending upon the disciplinary orientation of the beholder. For instance an observer strongly imbued in the science of climate change might see the closeness of this set of factors relative to vector borne disease quite differently than an observer who is not aware of the evidence base. In a similar manner an observer dedicated to social justice issues may well move the influence of poverty closer to the negative health outcomes attributable to the lack of safe water than an observer well grounded in ecological effects and impacts.

* Recommended citation: Gute, D.M. 2015. 'Introduction: distal influences', in Bartram, J., with Baum, R., Coclanis, P.A., Gute, D.M., Kay, D., McFadyen, S., Pond, K., Robertson, W. and Rouse, M.J. (eds) *Routledge Handbook of Water and Health*. London and New York: Routledge.

Many of the elements of this part of the *Handbook* address issues that are central to both international planning and policy frameworks. Within the planning category, the present Millennium Development Goals (MDGs) (United Nations, 2014) adopted by the United Nations as quantitative targets (and which are about to sunset in 2015) cover topics addressed in the present part of the *Handbook*. (See Chapter 43 ("International Policy") within the *Handbook* for additional detail on the MDGs directly related to water and sanitation.) However, the full set of MDGs also cover issues such as the eradication of extreme poverty, the promotion of environmental sustainability, the preservation of biodiversity, global partnership for development and other topics of specific relevance to this part of the *Handbook*.

Social determinants of health

The conceptual thrust of issues raised by authors in this part of the *Handbook* have been utilized by such policy review activities as the World Health Organization's (WHO) Commission on Social Determinants of Health (CSDH) as chaired by Dr. Michael Marmot. Preceding his chairmanship, Marmot effectively extended the well established tradition within the United Kingdom of evaluating the impact of social class on population health in his 2005 book entitled *Social Inequities in Health* (Marmot and Wilkinson, 2005) by making the simple statement, "In the Scottish city of Glasgow, people living in the most deprived districts have life expectancy 12 years shorter than those living the most affluent" (Marmot and Wilkinson, 2005). Of interest here is the lack of major differences in terms of access to health care given the existence of the National Health Service. Dr. Marmot further disseminated his approach to addressing social inequities within the official activities of the CSDH which was established by WHO in March 2005 to support countries and global health partners in addressing the social factors leading to ill health and health inequities. The CSDH completed its work in 2008.

The yawning differences seen even in high income nations such as Scotland become even more pronounced when such examinations are extended to low income nations. Sepulveda and Murray (2014) write, "A child born in western sub-Saharan Africa is almost 30 times more likely to die by the age of 5 than a child born in Western Europe (under 5 mortality rate (U5MR) of 114.3 versus 3.9 per 1000 live births)." The impression should not be left that such examination focused on social determinants of health and the inequitable distribution of resources is only found in low income nations. Dr. Marmot, aided by other experts, recently applied the same approach to the countries comprising Western Europe to elucidate actionable policy initiatives (Marmot et al., 2012).

In addition to the public health community, the necessity of viewing the issues covered in this part of the *Handbook* as being of great importance is also of great interest to other disciplines and classes of professionals. Engineers are recognizing that the scale of the problems affecting both the social and the physical infrastructure may well require alterations in the training of these professionals.

The transformation needs to lie not in the technical areas, as engineers have known how to produce safe drinking water and how to build toilets in high income nations for more than 100 years. Professor Bernard Amadei, founder of Engineers Without Borders, put it perhaps best, "What the world needs now is a new kind of engineer," he said. "My goal is to develop a new generation of engineers for the 21st century who are not just providers of technical solutions, but are social entrepreneurs, community builders and peacemakers." "If all the problems in the world were technical, we would have solved them by now" (Macmillan, 2008).

A tension runs through any technical discipline in trying to balance innovation with achievability, engineering certainly included in this set. This was admirably framed in the following passage:

> Competitions such as the Gates Foundation's Reinvent the Toilet Challenge reflect the kind of integrative thinking that must occur to create a next-generation toilet that can not only manage water but also harvest water and energy resources…Although such competitions highlight important challenges, funders often solicit solutions with a high degree of technological innovation. An unintended consequence of this premium on innovation can be to complicate downstream implementation efforts. It is time for the engineering and international aid communities to adopt approaches that can improve global health in ways that can be sustained.
>
> *(Niemeier et al., 2014)*

Sustainability has been focused upon in Chapter 24 of the *Handbook*. Sustainability also must comprehend the "uptake" or acceptance of the proffered intervention or technology. As an example of the critique being expressed regarding the unsuitability of certain approaches, Jason Kass, an environmental engineer and the founder of the organization Toilets for People, wrote an op-ed in the *New York Times* (Kass, 2013) on the urgent need for sustainable toilets in the developing world. He essentially upbraided developers of the designs of new toilets of essentially satisfying the needs of the developed world rather than those of low income nations. Keeping the needs and requirements of the populations that interventions are intended for must become a defining element of innovation, combined with simplicity, albeit with sufficient functionality, at the point of delivery.

Organization of this part: distal influences

The chapters included in this part explore a set of drivers of great importance to the understanding of how to proceed in designing and implementing interventions directed at global populations. The narratives reflect the disparate backgrounds of the authors.

Sarah Bell at University College London is a public health professional extremely interested in protective interventions which can be provided for young people. With regards to a WASH perspective Ms. Bell, at the School of Social and Community Medicine, is particularly interested in the benefits of hand washing and authored the water scarcity chapter. The provision of sufficient quantities of water is a rate limiting influence on hygiene. Dr. Katrina Charles, a Departmental Lecturer at Oxford University, is an environmental engineer with an interest in improving access to safe drinking water and sanitation globally and authored the climate change chapter. In this chapter Dr. Charles explores the many ramifications that will affect the regions of the globe that will suffer deficits in precipitation, and outlines adaptation approaches. The co-authors of the poverty chapter are: Dr. Leo Heller who is a researcher at the Oswaldo Cruz Foundation in Brazil and the current UN Special Rapporteur on the Human Rights to safe drinking water and sanitation. Alexander (Sand) Cairncross is a Professor at the London School of Hygiene and Tropical Medicine. Dr. Heller holds a BA in Civil Engineering, a MSc. in Sanitary Engineering and a PhD. in Epidemiology. Dr. Cairncross is an Epidemiologist who is also trained in soil mechanics. Both are very interested in the impact of the Millennium Development Goals as a driver to lifting portions of the global population out of poverty and other social and economic deficits. They see this as a key determinant in achieving greater progress in the welfare of global poor.

Andy Bastable and Ben Harvey co-authored the chapter on emergencies and disasters based upon their rich set of field experiences gained in over 20 years of practical emergency and development work in the water and sanitation sector. Mr. Bastable is the Head of Water and Sanitation for Oxfam while Mr. Harvey serves as an independent WASH consultant. Carl Haub has been a senior demographer at the Population Reference Bureau located in Washington, D.C. since 1979. A specialist in the compilation and analysis of demographic data and dissemination, he applied his wide analytic expertise to the authoring of the chapter on demographics. Professor Ong Choon Nam is a Professor in the Department of Community, Occupational and Family Medicine at the National University of Singapore (NUS). He is currently the Director for the Centre for Environmental and Occupational Health at NUS, an advisor to the National Water Research Institute (US) and was the chairperson of the expert panel which advised PUB independently on the NEWater Study from 2001 to 2003. He brought this flexible interdisciplinary approach to his authoring of the chapter on water reuse. This is a topic of particular importance to Singapore given its population requirements and available natural resource assets.

Finally, the last chapter in this part is co-authored by two physicians, Drs Barry Levy (Tufts University) and Victor Sidel (Montefiore Medical Center and the Albert Einstein College of Medicine) on the topic of war and conflict. These authors have written a book on this subject that will likely be of interest to the readers of this *Handbook*.

Summary

The issues covered in this part of the *Handbook* map into broader movements such as environmental justice and social justice that exert significant ramifications both domestically and internationally. A common thread linking many of the issues explored in this part is that the effects of these factors will not be distributed equally across different global populations. The reality is that the influences of climate change, war and conflict, and natural disasters will have a greater impact on those populations which are poor, rural or peri-urban and of minority status. Imbalances in global health care research and development investment by the Secretariat of the Global Forum for Health Research (2004) are chronicled in the 10/90 report series which held that less than 10 percent of the global health research funds are devoted to research into the health problems that account for 90 percent of the global disease burden (measured in disability-adjusted life years or DALYs). Much progress has been made in reversing this imbalance partly as a result of simply making the situation more widely known (Secretariat of the Global Forum for Health Research, 2004). Part of the rationale for re-focusing investment dollars is similar to lines of argument that are explored in this part of the *Handbook*. There is a growing desire on the part of researchers, practitioners and policy makers alike to identify and work on the "root cause" of the numerous problems plaguing 21st century society. Recent attempts to facilitate the allocation of funds and the dissemination of appropriate technologies have relied upon transparent communication through the formation of a "global observatory" as a means of promoting non-commercial areas of research for neglected tropical diseases (NTDs). This was the subject of a meeting convened by the Wellcome Trust in February 2013 (Terry et al., 2014).

Some elements are currently operational. Emory University manages a database called the Global Health Primer (www.globalhealthprimer.org/default.aspx), which tracks products in the pipeline for 25 neglected tropical diseases and charts the progress of the development of health technologies (diagnostics, vaccines and drugs) for NTDs. In a similar vein (NTDs) Policy Cures operates the G-FINDER survey (https://g-finder.policycures.org/gfinder_

report) which tracks the investment profile of more than 800 global research funders, not covered by confidentiality clauses, for 31 tropical diseases.

The ability to marshal quantitative data and to disseminate information regarding the funding of these global scourges is a positive development that has contributed to a partial realignment of funding and research intensity. It is hoped that such examples of progress will translate into a more cohesive international response to the "distal" factors present in this part.

Key recommended readings

1 Toole, M. and Waldman, R.J. (1990) Prevention of excess mortality in refugee and displaced populations in developing countries. *JAMA* 263(24), 3296–3302.
   Early seminal research on factors which produce heightened risk among populations at increased susceptibility in low income nations which lead to poor health outcomes.
2 Braverman, P., Cubbin, C., Egerter, S., Williams, D.R. and Pamuk, E. (2010) Socioeconomic disparities in health in the United States: what the patterns tell us. *Am J Public Health* 100, S186–S196. doi:10.2105/AJPH.2009.166082.
   Groundbreaking assessment of similar influences at work even in high income nations such as the United States.
3 *10/90 Report on Health Research 2003–2004, Global Forum for Health Research, ISBN 2-940286-16-7. http://www.globalforumhealth.org/about/1090-gap/
   This document reports on the progress being made on ameliorating the 10/90 gap. The 10/90 gap was first written about in the mid-1980s. These reports noted that only 5 percent of global medical research funds were being directed at diseases which accounted for over 90 percent of the preventable deaths. Progress on rectifying this imbalance are presented.

References

Kass, J. (2013) Bill Gates Can't Build a Toilet. *New York Times* November 18, 2013.
Macmillan, L. (2008) Engineering for humanity. *Tufts Journal* November 19, 2008. http://tuftsjournal.tufts.edu/2008/11_2/features/03/. Last accessed November 2, 2014.
Marmot, M. and Wilkinson, R. (eds) (2005) *Social Inequities in Health*, Oxford: Oxford University Press.
Marmot, M., Allen, J., Bell, R., Bloomer, E. and Goldblatt, (2012) WHO European review of social determinants of health and the health divide. *Lancet* 380(9846), 1011–1129. doi: 10.1016/S0140-6736(12)61228-8.
NHS Health Scotland (2004) *Public Health Institute for Scotland*. Glasgow: Public Health Institute for Scotland
Niemeier, D., Gombachika, H. and Richards-Kortum, R. (2014) How to transform the practice of engineering to meet global health needs? *Science* 345, 1287–1290.
Secretariat of the Global Forum for Health Research (2004) *10/90 Report on Health Research 2003–2004*. http://announcementsfiles.cohred.org/gfhr_pub/assoc/s14789e/s14789e.pdf. Last accessed November 3, 2014.
Sepulveda, J. and Murray, C. (2014) The state of global health in 2014. *Science* 345, 1275–1278.
Terry, R.F., Salm, Jr. J., Nannei, C. and Dye, C. (2014) Creating a global observatory for health R&D. *Science* 345, 1302–1304.
United Nations (2014) Millennium Development Goals. http://www.un.org/millenniumgoals/. Last accessed October 29, 2014.

34

WATER SCARCITY*

Sarah Bell

Senior Lecturer, University College London, UK

Learning objectives

1 Identify common methods for defining water scarcity and understand the key differences between them.
2 Understand the concept of water poverty and its importance to human health.
3 Outline key technical, institutional and economic responses to water scarcity.

Introduction

Water scarcity is one of the most worrying manifestations of the global environmental crisis. Water is a renewable but finite resource. Water is essential for life – not only for meeting immediate needs for drinking and hygiene, but also to produce food, electricity and industrial products. Water scarcity can be a source of conflict within local communities and regions, and between nations. Water scarcity can have direct impacts on human health; however, lack of access to safe water for drinking and sanitation is usually the result of inadequate infrastructure provision, rather than absolute scarcity of water. Measurement and prediction of water scarcity and its impact on human health are therefore more complex than simply running out of water for people to meet their daily needs.

Throughout the twentieth century, global water use grew at twice the rate of population growth (UNCSD 1997). This was primarily due to irrigation for agricultural production, but also reflects growing industrial and domestic consumption. Agriculture accounts for approximately 70 per cent of global water use, with industry consuming 19 per cent and domestic users consuming 11 per cent (Gleick 2008, FAO 2010). Demand for water is predicted to continue to grow to meet demands for food from larger populations, as well as increased domestic and industrial use.

The 'nexus' between demands for water, food and energy is well established. Water and fossil fuels are important inputs to industrial agriculture, water is used to cool power stations, and water and wastewater treatment and pumping require energy. In 2009 Sir John Beddington, the then Chief Scientific Advisor to the UK Government, described

* Recommended citation: Bell, S. 2015. 'Water scarcity', in Bartram, J., with Baum, R., Coclanis, P.A., Gute, D.M., Kay, D., McFadyen, S., Pond, K., Robertson, W. and Rouse, M.J. (eds) *Routledge Handbook of Water and Health*. London and New York: Routledge.

the interconnections between food, water, energy and climate change as the 'perfect storm' brewing on the horizon of global events (Beddington 2009). Growing population and increasing per capita consumption as a result of economic development propel trends towards a 45 per cent increase in demand for energy, and a 40 per cent increase in demand for water and food by 2030 (IEA 2008, Shen et al. 2008, Alexandratos and Bruinsma 2012). With food production dependent on water resources, climate change likely to result in changing precipitation patterns, and energy demand most likely to be met from fossil fuel sources and requiring water for cooling, these trends intersect and interact in complex and dangerous ways.

This chapter provides an introduction to some of the important issues associated with water scarcity as it affects development. It begins by reviewing the different ways that water scarcity is defined and measured, and summarizes key studies about the future predictions of global water scarcity. The concept of water poverty is then introduced and is used to highlight the reasons why people lack access to safe water, even under conditions of plentiful water resources. Key responses to water scarcity are summarized, including the idea of virtual water as a means of understanding the connections between global trade in food and goods and its impacts on local water resources. The chapter concludes by emphasizing the need for a critical and nuanced understanding of water resources, water usage and its relation to human health and development, to move from worrisome statistics to sustainable solutions.

Defining water scarcity

Water scarcity is the absence of sufficient water to meet human and ecological needs. Defining water scarcity in a way that can be assessed quantitatively, meaningful in different contexts and based in sound science and good quality data, is not simple. Water resources in different parts of the world include surface water, groundwater, soil moisture, snow and glaciers, all of which vary within and between years. Water resources are most often assessed on a catchment or watershed scale, which rarely correspond to political boundaries. Demand for water is determined by population, climate, lifestyles and economic activity. Agreeing upon a definition of water scarcity that reasonably reflects the seriousness of the problem of over-exploitation of water resources, without diverting attention away from problems of water distribution and economic development, has been the basis of considerable and ongoing debate amongst water professionals and policy makers since the 1980s (see for example, Falkenmark et al. 1989, Arnell 2004, Molden et al. 2007 and Vörösmarty et al. 2000).

One of the first and most enduring definitions of water stress was devised by Marin Falkenmark and colleagues (Falkenmark et al. 1989). The purpose of the 'Falkenmark Index' was to identify the point at which countries were at risk of not being self-sufficient in food, and were likely to experience considerable difficulties with managing competing demands for water. The index was originally conceived as the 'number of people per flow unit', but is more often referred to in terms of the inverse 'cubic meters of water per person per year'. Countries are considered to be under moderate water stress if the available resources are between 1,000 and 1,700 m³/person/year, high stress between 500 and 1,000 m³/person/year, and extreme stress if available resources are less than 500 m³/person/year (Arnell 2004). Available resources are considered to be surface water runoff, usually based on long-term average but in some cases water scarcity may be assessed in terms of a dry year average runoff.

Although the Falkenmark Index is widely used, its attention to food self-sufficiency as defining thresholds for water stress mean that it does not adequately consider trade in food and water intensive industrial products. A catchment with high per capita water availability may still experience water stress if water withdrawals for agriculture or other export industries exceed local hydrological limits. Conversely a catchment with low water availability may be able to overcome water scarcity by importing food and products from more water rich regions. More recent measures of water scarcity have been developed to incorporate a more holistic view of water use and availability, although often at the expense of greater complexity and uncertainty in the data used in their formulation.

An alternative to per capita availability of water resources is to define water scarcity in terms of the water withdrawals relative to available water. Once more, water availability is defined in terms of surface water runoff, but water withdrawal is based on hydrological data reflecting the actual demand for water in a river basin, region or country. Countries or river basins are defined as suffering from moderate water stress if withdrawals are between 0.1 and 0.2 of the available water, medium–high water stress between 0.2 and 0.4, and high water stress above 0.4 (Vörösmarty et al. 2000). Alternatively, countries are considered to be subject to physical water scarcity when more than 75 per cent of available water is withdrawn, with little or no water scarcity if withdrawals are less than 25 per cent of available water, and are considered to be 'approaching water scarcity' in between (Molden et al. 2007). This indicator enables more nuanced evaluation of the state of water resource, but its reliability depends on the quality of hydrological and economic data available, which varies between countries, a problem that is amplified when predicting future water scarcity.

Current assessments of water scarcity based on withdrawals as a proportion of water available estimate that about 1.2 billion people live in river basins facing physical water scarcity (Molden et al. 2007). A further 1.6 billion people are thought to live under conditions of economic water scarcity, where less than 25 per cent of available water resources are withdrawn, but malnutrition exists due to the lack of access to water (Molden et al. 2007).

The use of surface water runoff as the key indicator of water resources availability has been criticized for underestimating the available resources, thereby exaggerating the scale of the problem of water scarcity (Taylor 2009). Runoff does not take account of water storage that is available for use, such as groundwater, glaciers and soil moisture. Where these storages are renewable, such as through groundwater recharge, they can be considered a sustainable, available water resource. On the other hand, exploitation of so-called 'fossil' groundwater resources in some regions has provided short-term relief from water scarcity that is not sustainable in the long term. For instance, Saudi Arabia was able to achieve self-sufficiency in wheat production between the 1970s and 1990s by irrigating crops using groundwater, but as groundwater resources have been depleted this system of food production is no longer viable and wheat production dropped by more than two thirds between 2007 and 2010 (Brown 2011).

Predicting water scarcity

Future projections of water scarcity are based on changes in population, water demand and climate, each of which brings their own uncertainties. An often quoted United Nations report (UN 1997) forecast that by 2025 two thirds of the world's population will be living under conditions of water stress, defined as more than 20 per cent of available resources being abstracted for human use.

A study by Vörösmarty et al. in 2000 considered increasing demand due to population growth and changing resource availability due to climate change to forecast that by 2025 one third of the world's population will be under conditions of water stress due to domestic and industrial demand exceeding 20 per cent of available resources, and 49 per cent will be under water stress due to agricultural demands on water resources. This study concluded that population growth will have a more significant impact on increasing water stress than economic development or climate change.

A more detailed consideration of the impact of climate change and population on water scarcity in 2004 used the simpler indicator of water resource availability per capita, to avoid uncertainty in economic forecasts about changing per capita demands that underpin forecasts of water withdrawals (Arnell 2004). This study used UN climate change carbon emissions and population scenarios to estimate changes in available surface water runoff, and confirmed previous findings that population growth is likely to have a bigger impact on water scarcity than climate change. The study estimated that by the 2050s the different climate change scenarios have relatively little impact on resource availability, but different population scenarios have a significant impact on forecasts for water scarcity. Without the impact of climate change, by 2025 34–53 per cent of the global population may be living in water stressed watersheds, with runoff of less than 1,700 m3/per person/year, compared to a 1995 baseline of 7 per cent. By 2055 this is forecast to increase to 39–68 per cent of the global population living under moderate to high water stress.

Water poverty

Measures of water scarcity that focus on water resource availability, withdrawals and the impact of population growth have been criticized for underemphasizing the role of economic, environmental, institutional and social factors that determine access to water. Lack of access to water for the world's poor, particularly the urban poor, is rarely the result of physical or absolute water scarcity, and is more often the result of political, economic and institutional failure. Lack of infrastructure is more often to blame for poor access to water than physical scarcity of water. Development professionals and academics have argued that an emphasis on water scarcity supports policies that favor market based allocation of resources and large engineering solutions, rather than directly addressing the causes of lack of access to safe water in ways that meet the needs of the poor (Allen et al. 2006, Chenoweth 2008). An alternative approach focuses on the concept of water poverty, rather than water scarcity, to emphasize the economic, social and institutional barriers to access to safe and reliable water, rather than resource scarcity as the driver for policy and development.

The Water Poverty Index (WPI) has emerged as an alternative measure to the conventional indicators of water scarcity, as a way of incorporating a wider range of factors into determining policy priorities and actions (Sullivan et al. 2006). The WPI includes assessment of resources, access, use, capacity and environment. Assessment of each of these elements involves a series of indicators, requiring a wider selection of data input than conventional indicators. The calculation of the final WPI score is also based on weighted averaging of each of the components, requiring agreement on the relative importance of each factor in the overall evaluation by the stakeholders and researchers involved in the assessment. The WPI aims to provide a more holistic indication of the status of water within communities, catchments and countries, but at the cost of a more complicated methodology than conventional measures of water scarcity. The complexity of the method for calculating the WPI and the judgments required to agree on the weighting of individual components have limited its application as a measure of comparing water poverty in different parts of the world. Whilst it is valuable to have stakeholder engagement in defining measures of water poverty in local catchments, this limits the international comparability and widespread adoption of the WPI as an alternative to less comprehensive indices of water scarcity.

Responses

Water scarcity is considered by many to be one of the greatest threats facing global economic, environmental and political systems. Using Saudi Arabia as an example, authors such as Lester Brown warn that short-term exploitation of groundwater resources has created 'food bubbles' in many regions of the world, particularly in South Asia and the Middle East, and as major aquifers are depleted global food production looks set to decline sharply in coming decades (Brown 2011).

Whilst forecasts of catastrophic water scarcity must be approached with critical awareness of the complexity of water resources and the role of water in development, there is little doubt that in coming decades meeting the needs of a growing global population for food, water and sanitation will become more complex. Water resources must be managed to meet a range of needs, for agriculture, domestic and industrial uses, and to ensure the health and vitality of ecosystems that also depend on water to thrive. Ecosystems provide many services to human society and economies, such as flood protection, water purification, fisheries, limiting soil erosion and salinization, and pest management. Ensuring sufficient water is available for natural systems is important for human health and economic growth. Policies based on integrated water management are being applied in urban and rural catchments to attempt to address the needs of different stakeholders and to ensure maximum benefit from the use of available water resources. Allocation of water as an economic good, that can be traded in newly established water markets, is criticized by many who contend that this undermines the status of water as a public good, but is increasingly applied as a means of achieving economic efficiency in meeting competing demands for a scarce resource.

Technological innovation has an important role to play in integrated water resources management, but technology alone is unlikely to be able to solve the complex problems of water scarcity. In evaluating technological options for water resources it is important to consider the energy requirements for treating and pumping water, environmental impacts, social acceptability, reliability, and economic costs and benefits. Water scarcity has driven technological innovation to improve water efficiency in homes, businesses, industry and agriculture. Improving the efficiency of irrigation and ensuring appropriate crops are cultivated in water scarce regions is key to producing food for a growing population without continuing to increase the water use. Recycling wastewater is of increasing importance in addressing water scarcity, but can also increase demand for energy, depending on the level of treatment required to meet the desired quality of water for the end use. Water reuse can be as simple as households watering their gardens with dishwashing water, or as complex as treating discharge from wastewater treatment plants to a very high level of purity before recycling through the drinking water supply system. Recycled wastewater is an important component of water used for agriculture in many water scarce countries, including Israel. Desalination is a new source of water that is increasingly important for domestic and industrial use, but it is more expensive and requires more energy than conventional water sources, thus intensifying the water–energy–food nexus (Cooley and Wilkinson 2012).

Global trade in agricultural and industrial products can be both a cause of and a solution to local water shortages. The concept of 'virtual water' was developed by Tony Allen and colleagues to demonstrate the amount of water needed to produce the food and materials that are traded and consumed around the world (Allen 2011). Virtual water can be used to highlight the water resource requirements of different food and consumer products. For instance, depending on the local conditions of production, a 150-gram beef burger requires 2,400 liters of water to produce grain and pasture, and raise and slaughter the cattle, and a cotton T-shirt requires 2,000 liters of water to grow the cotton and process it into fabric (Hoekstra and Chapagain 2007). The water footprint or virtual water content of different commodities and products will depend on the production system and local climate, and the hydrological and ecological impact of water use is also highly dependent on local conditions (Ridoutt and Pfister 2010). However, the concept of water foot printing can be useful to draw attention to the positive and negative effects of global trade on water scarcity. Water scarce countries can import 'virtual water' as food or other commodities as a means of overcoming local hydrological constraints. At the same time, consumption of imported food or other products might contribute to local water scarcity in the source regions of these goods. Water resources used to produce food, fiber and manufactured goods for export from water rich to water scarce regions may spur economic growth and development, but must be carefully managed in order to avoid local water scarcity or water poverty in the longer term.

Conclusions

Water is a renewable but limited resource that is essential for life. Water scarcity is one of the most alarming problems confronting development. Sharp declines in availability of water could have catastrophic impacts on global food production, and could jeopardize the ability of populations to meet their basic needs for health and hygiene (please see Chapters 5 and 22 for additional detail). However, access to water and the sustainable management of water resources are as much political and economic problems as they are technical and environmental problems. How water scarcity is measured can shape the political and institutional debates which inform policy decisions. Presenting water as a scarce resource can divert attention from political and economic failure to deliver water to the people who need it. Global trade in agricultural, industrial and consumer goods can alleviate local water scarcity, but exacerbate it in other places. A clear understanding of the complexity of water resources, water use and their role in trade, development and everyday life is vital in developing fair and sustainable policies and technologies for agriculture, industry and domestic water use.

Key recommended readings (open access)

1 The United Nations Water Development Report http://www.unesco.org/new/en/ natural-sciences/environment/water/wwap/wwdr/. This report is released every year, focusing on a different aspect of water and development, and providing a comprehensive overview of the state of the world's water resources.

2 Molden D. (2007) *Water for Food, Water for Life: A Comprehensive Assessment of Water Management in Agriculture*. London: Earthscan, and Colombo: International Water Management Institute. This report addresses the agricultural uses of water and available water resources. Individual chapters are available to download from http:// www.iwmi.cgiar.org/Assessment/.

3 Food and Agriculture Organization (2011) *The State of the World's Water and Land Resources for Food and Agriculture*. London: Earthscan. This report is summary of the trends in water and land resources for agriculture, and assessment of technical and institutional options for addressing water scarcity and land degradation. It is available to download from http://www.fao.org/nr/solaw/solaw-home/en/.

References

Alexandratos, N., and Bruinsma, J. (2012) *World Agriculture towards 2030/2050. The 2012 Revision*. ESA Working Paper No. 12-03. Rome: FAO.

Allen, A., Dávila, J, and Hofmann, P. (2006) The peri-urban water poor: citizens or consumers? *Environment and Urbanization* 18: 333–335.

Allen, T. (2011) *Virtual Water*. London: I.B. Tauris.

Arnell, N. (2004) Climate change and global water resources: SRES emissions and socio-economic scenarios. *Global Environmental Change* 14: 31–52.

Beddington, J. (2009) *Food, Energy, Water and the Climate: A Perfect Storm of Global Events?* Government London: Office for Science. http://www.bis.gov.uk/assets/goscience/docs/p/perfect-storm-paper.pdf

Brown, L. (2011) *World on the Edge*. London: Earthscan.

Chenoweth, J. (2008) A re-assessment of indicators of national water scarcity. *Water International* 33: 5–18.

Cooley, H., and Wilkinson, R. (2012). *Implications of Future Water Supply Sources for Energy Demands*. Alexandria, VA: Water Reuse Research Foundation.

Falkenmark, M., Lundqvist, J., and Widstrand, K. (1989) Macro-scale water scarcity requires micro-scale approaches. *Natural Resources Forum* 13: 258–267.

FAO. (2010) Aquastat. http://www.fao.org/nr/water/aquastat/main/index.stm (accessed 26 November 2012).

Gleick, P. (2008) *The World's Water: 2008–09*. Washington, DC: Island Press.

Hoekstra, A., and Chapagain, A. (2007) Water footprints of nations: water use by people as a function of their consumption pattern. *Water Resources Management* 21: 35–48.

IEA. (2008) *World Energy Outlook 2008*. Paris: IEA.

Molden, D., Frenken, K., Barker, R., de Fraiture, C., Mati, B., Svendsen, M., Sadoff, C., and Finlayson, C. (2007) Trends in Water and Agricultural Development. In Molden D (ed.) *Water for Food, Water for Life*. London: Earthscan.

Ridoutt, B., and Pfister, S. (2010) A revised approach to water footprinting to make transparent the impacts of consumption and production on global freshwater scarcity. *Global Environmental Change* 20: 113–120.

Shen, Y., Oki, T, Utsumi, N., Kanae, S., and Hanasaki, N. (2008) Projection of future world water resources under SRES scenarios: water withdrawal. *Hydrological Sciences* 53: 11–33.

Sullivan, C., Meigh, J., and Lawrence, P. (2006) Application of the water poverty index at different scales: a cautionary tale. *Water International* 31:; 412–426.

Taylor, R. (2009) Rethinking water scarcity: the role of storage. *EOS* 28: 237–248.

UNCSD (1997) *Comprehensive Assessment of the Freshwater Resources of the World*. New York: UN Division for Sustainable Development.

Vörösmarty, C., Green, P., Salisbury, J., and Lammers, R. (2000) Global water resources: vulnerability from climate change and population growth. *Science* 289: 284–288.

35

CLIMATE CHANGE[*]

Katrina Charles

Departmental Lecturer, Course Director on MSc in Water
Science, Policy and Management, University of Oxford, UK

Learning objectives

1 Understand the basics of climate change.
2 Explain the effect of climate change on the hydrological cycle.
3 Relate climate change impacts to water-mediated health outcomes.

Introduction

Climate change is increasingly affecting the availability and quality of freshwater supplies. The latest report, the Fifth Assessment Report (AR5) from the Intergovernmental Panel on Climate Change[1] (IPCC, 2014) has highlighted how climate change is affecting the quantity and quality of water, and the distribution of water-borne illnesses and disease vectors. It is important for people working in the area of water and health to understand how changes in the climate are likely to impact on water, its provision and quality, as well as how to adapt to those changes. This chapter will introduce the science of climate change and climate forecasting, followed by a summary of how these observed and forecast changes will impact on health.

Climate change: the science

Climate is defined as average weather, and is typically averaged over 30 years. Climate change is defined by the IPCC as a "change in the state of the climate that can be identified (e.g., by using statistical tests) by changes in the mean and/or the variability of its properties, and that persists for an extended period, typically decades or longer" (Cubasch et al., 2013, p126). Changes in the climate are recognized to occur through natural and anthropogenic (caused by human activities) means, however, it is extremely likely[2] that human influence has been the main cause of the observed increase in global temperatures from 1951 to 2010, with the major contributor being the increase in greenhouse gas concentrations (IPCC, 2013). Greenhouse gases absorb infrared radiation from the earth, and emit it in all directions; the

---

[*] Recommended citation: Charles, K. 'Climate change', in Bartram, J., with Baum, R., Coclanis, P.A., Gute, D.M., Kay, D., McFadyen, S., Pond, K., Robertson, W. and Rouse, M.J. (eds) *Routledge Handbook of Water and Health*. London and New York: Routledge.

net effect being that less radiation is emitted to space but is retained in the earth's atmosphere resulting in an increase in temperature.

Observed changes in the hydrological cycle with climate change

The ability of the atmosphere to hold water increases with rising temperature at a rate of about 7 per cent per 1°C. In the past 40 years, there has been an increase of about 3.5 per cent in tropospheric[3] water vapor, which is consistent with the observed 0.5°C temperature increase (IPCC, 2013). This is expected to bring with it an increase in precipitation, albeit a smaller change than is experienced in the water vapor content. These increasing temperatures and water vapor content have resulted in changes in the observed amount, intensity, frequency and type of pattern of precipitation. These results for precipitation are not uniformly distributed on a global basis; there are regional differences in the changes in precipitation and in the level of confidence in the ability to identify a change due to a lack of data, the latter is a particular issue for understanding changes in extreme events.[4] For more details on the regional differences and the degree of confidence see the IPCC AR5 report. The main trends observed in the hydrological cycle are:

- Global precipitation has increased. However, there is significant variation between different regions.
- The number of heavy precipitation events and their frequency or intensity has increased. These changes can be seasonal, for example in Europe where much of the increase in frequency or intensity is experienced in winter.
- Fewer snowfall events are occurring where winter temperatures are increasing, the extent of the snow cover has decreased, and glaciers are shrinking.
- Global mean sea level rose 0.19m between 1901 and 2010.

A summary of changes in extreme events is provided in Figure 35.1.

There is also the potential for impacts to have synergistic effects. For example, in areas where total precipitation is decreasing, increases in the intensity of precipitation will result in longer dry periods between rain events.

Our understanding of the role of climate change in the hydrological cycle is still developing. As more data becomes available, spatially and temporally, our understanding changes. Two key areas where our overall understanding has changed since AR4 are in river discharge and droughts. For river discharge, in AR4 it was concluded that global runoff increased during the 20th century. However, more recent work has identified many areas where streamflow has decreased making any overall trend hard to distinguish. This is further complicated by the anthropogenic changes that occur along rivers such as development of storages and withdrawals, and changes in the cryosphere.[5] Shrinking glaciers and reduced snowfall are changing the seasonality of flows with the former increasing the spring and summer melt flows, and the latter contributing to lower spring melt flows.

The second change is regarding droughts. In AR4 it was concluded that there had been a global increase in droughts since the 1970s. A drought is defined as "a period of abnormally dry weather long enough to cause a serious hydrological imbalance", but it is a relative term. This differs from water scarcity which is related to demand for water (see Chapter 34). The latest research suggests that on a global scale there is not sufficient data to have confidence that there are defined trends; however, there is good evidence that since the 1950s the Mediterranean and West Africa have experienced more intense and longer droughts.

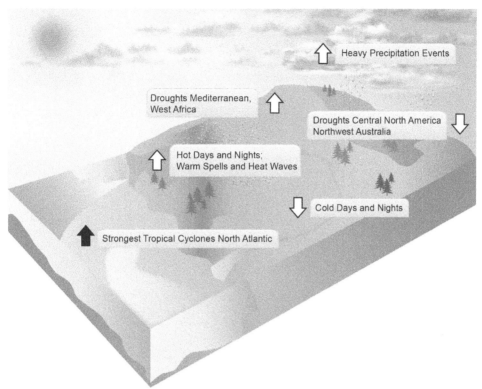

Figure 35.1 Trends in the frequency (or intensity) of various climate extremes (arrow direction denotes the sign of the change) since the middle of the 20th century (except for North Atlantic storms where the period covered is from the 1970s)

Source: IPCC (2013).

Climate forecasts

The IPCC predictions for future climate change are based on a range of Atmosphere-Ocean General Circulation Models and Earth System Models, the latter of which includes biogeochemical cycles relating to the carbon cycle, and others. Since AR4 there have been many modelling advancements that have improved our understanding of the impacts of climate change. The models include projections of future emissions or concentrations of greenhouse gases and other climate drivers. They are based on scenarios, or Representative Concentration Pathways (RCPs), defined by the scientific community. In AR5 four key RCPs are used which include a mitigation scenario (RCP2.6), two stabilization scenarios (RCP4.5 and RCP6), and one scenario with very high greenhouse gas emissions (RCP8.5).

The forecasts indicate that global mean temperatures will continue to rise, with a temperature increase by the end of the 21st century (relative to 1850–1900) of more than 1.5°C likely for RCP4.5, and more than 2°C for RCP6 and RCP8.5, with the latter projected to increase temperatures by 4.3°C. The general impacts associated with this increase in temperature are summarized in Figure 35.2. For the hydrological cycle, the projection is that globally the already observed large-scale patterns will continue. However, at the regional scale, the impacts will be strongly affected by human activities and natural internal variability. Extreme precipitation events are likely to become more intense and more frequent at the mid-latitudes and wet tropical regions, and the area subject to monsoons will increase

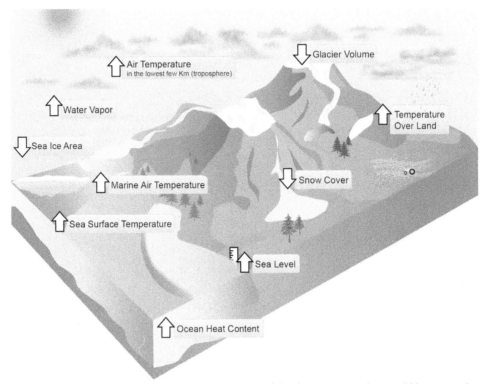

Figure 35.2 Independent analyses of many components of the climate system that would be expected to change in a warming world exhibit trends consistent with warming (arrow direction denotes the sign of the change)

Source: IPCC (2013).

with monsoons forecast to become longer and to have more intense precipitation. There are several areas forecast to be subject to large-scale drying including the Mediterranean, southwest USA, and southern Africa and it is considered likely that they will see an increase in the risk of agricultural drought (lack of rainfall in the growing season).

Human interactions with the hydrological cycle can make it hard to discern the impact of climate change. Groundwater is one of these areas. The direct impacts of climate change on groundwater recharge will vary depending on the geology of the area, as well as the human and climate factors that dictate the water surplus available for recharge. It is likely that the indirect impacts of climate change will have a significant impact on groundwater as pumping increases to meet the rising irrigation water demand, a particular concern in areas forecast to experience more agricultural droughts (Taylor et al., 2013). This demand for water, and more specifically for irrigation water, will be further increased with population growth and rising living standards. In coastal areas, it is likely to be over-extraction rather than sea level rise which will have the biggest impact on saline intrusion.

How will climate change impact on water and health?

These observed climate change impacts have had a significant influence on the hydrological cycle, and through this an impact on health. This will continue with the forecasted impact of climate change.

Climate change will have global impacts, but the risks are generally greater for disadvantaged people and communities. In developing countries, the lack of resources and institutional capacity are likely to limit their ability to adapt and therefore result in the impacts of climate change being more disruptive. The low levels of access to safe and reliable water supplies, and the prevalence of water-related diseases likely to be affected by climate change will also increase the impact on health.

It is difficult to attribute the impact of climate change on health. Climate change was estimated to be responsible for approximately 2.4 per cent of worldwide diarrhea cases in 2000 (WHO, 2002). This is expected to rise as climate change increasingly impacts on water and sanitation provision, especially in developing countries, as almost 58 per cent of diarrheal deaths are attributable to lack of access to adequate drinking water, sanitation and hygiene (Prüss-Üstün et al., 2014). However, diarrhea is not the only health impact. This section will focus on how climate change will affect health through changes in the distribution of diseases, direct health impacts, water scarcity, and reduced access to water supply and sanitation facilities (Figure 35.3). As these topics have been addressed in some detail in Chapters 4, 16 and 48, the focus here is on how climate change will impact them. More detailed discussion of these issues can be found in the WHO (2009) *Vision 2030* reports.

Changes in disease distribution and occurrence

Changing environmental conditions, as well as human behavior, can affect the distribution and occurrence of insect vector-borne diseases (see Chapter 6). Increasing temperatures and changes in rainfall pattern will change the area of land with climates suitable for disease transmission, including mosquito-borne diseases such as malaria and dengue. Increased rainfall is more likely to result in standing water for mosquitoes to breed in, while household water storage during periods of water scarcity can also provide breeding grounds. Small changes in warming may lead to large increases in transmission of malaria if conditions are suitable (Smith et al., 2014), and there is potential for spread to areas not currently affected, especially if health and governance systems are not strong.

Direct health impacts

Direct impacts of climate change on health, with respect to water, include deaths or injuries caused by flooding. Floods represented six of the ten biggest natural disasters in 2010, in terms of the number of people affected, and four of the ten biggest natural disasters in terms of economic damages (Guha-Sapir et al., 2011). The intensity and frequency of flooding is likely to be affected by changes in rainfall with climate change; however, the development of river basins and increases in populations living in flood prone areas are likely to have a significant impact on the frequency and consequences of flooding. Storm surges and sea level rise will also increase the populations at risk of flooding events on small islands and along low-lying coastal areas.

Water scarcity

Water scarcity increases reliance on poor quality drinking water sources, and reduces water availability for hygiene purposes and for agriculture, affecting food availability, and potentially resulting in increases in waterborne disease, water-washed disease and malnutrition. Water scarcity will be affected by changes in drought frequency and intensity, seasonality of

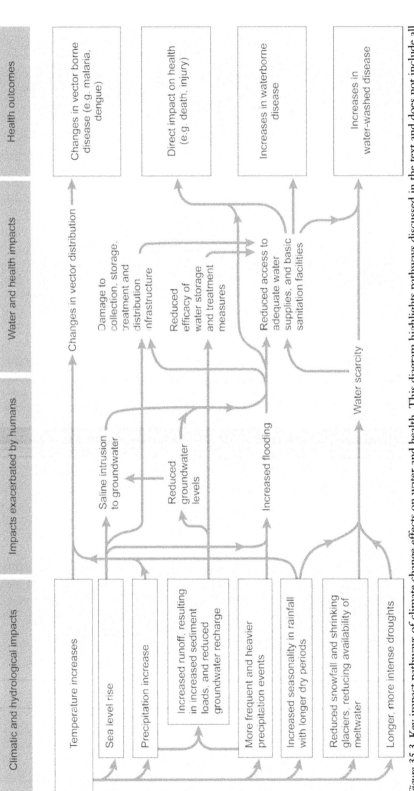

Figure 35.3 Key impact pathways of climate change effects on water and health. This diagram highlights pathways discussed in the text and does not include all pathways. Climatic and hydrological impacts will vary regionally

rainfall, reduced snowfall, as well as changes in the efficacy of water storage facilities to harvest increased volumes runoff over shorter time periods (Figure 35.3). Melting glaciers will contribute to water shortages in the long term, although may increase water availability in the short term. Falling water levels in groundwater and surface water may reduce access to water as wells and other infrastructure become nonproductive; the resulting reliance on fewer functional systems can increase wear and tear (WHO, 2009). Water scarcity will be exacerbated by population changes, with rising demand associated with both population growth and rising affluence (Bates et al., 2008).

Reduced access to adequate water supplies and basic sanitation facilities

Reduced access incorporates how both physical access to existing facilities and the quality of access, in terms of safety or quantity, may be affected, and can be temporary or more permanent. Climate change will affect both water quality as well as the water supply and sanitation facilities that help to mitigate the influence of climate change on health. Following are examples illustrating the different pathways whereby flooding reduces access to adequate water supplies and basic sanitation facilities:

- In January 2013, within the space of a week, floods forced the closure of two water treatment works, one in Santiago, Chile (BBC, 2013), and one in Brisbane, Australia (*Brisbane Times*, 2013). The water treatment works were reported to have stopped treatment due to the high sediment loads in the floodwater. This reduced the availability of water for residents, with 2 million residents without water for a day in Santiago.
- Floods in Mozambique in 2000 caused a reduction in access through several pathways (Cairncross and Alvarinho, 2006). There was large-scale destruction of onsite water and sanitation infrastructure with up to 100,000 pit latrines destroyed, and 300 out of 800 rural water points (predominantly handpumps) destroyed in one province. Access to safe water was further compromised from pollution, including overflow from 3,000 septic tanks, and damage to piped water supplies in eight smaller towns. The number of diarrheal disease cases reported were 11,399, including cholera and dysentery. In addition to this, an estimated 500,000 people were displaced, many of whom likely experienced a reduction in access.
- In 2007, the Mythe water treatment works, Gloucester, UK, was inundated with up to half a meter of flood water. The treatment works was closed for 17 days, leaving 140,000 households without water (EA, 2007).

Physical access will be affected temporarily by flooding, with onsite water supply and sanitation facilities in particular likely to have reduced accessibility. More permanent reductions in physical access will result from damage to infrastructure used in the collection, storage, treatment or distribution of water and sewage (including onsite sanitation) resulting from rising sea level, flooding, thawing permafrost and extreme events (WHO, 2009). In addition, changes in climate and subsequent impacts on livelihoods may result in migration, which in turn may affect access.

The quality of access will be affected by changes in water quality, with contamination potentially leading to outbreaks of diarrheal diseases, as well as longer term impacts from chemical contamination and related non-communicable diseases which are not addressed here. Impacts to water quality include increased sediment loads, lack of dilution of pollution during low flows, and overloading of, or damage to, sanitation facilities and sewage systems during heavy rainfall and flood events resulting in overflows of untreated sewage into the environment. These will contribute to poorer water quality overall, especially in surface waters, with reduced efficacy of treatment systems likely to result from the increased pollution and sediment load.

Changes in quality that affect the organoleptic properties of the water or suitability for agriculture, such as increases in salinity, may result in reduced access as people choose to use more acceptable, but potentially contaminated, alternatives. Saline intrusion of groundwater resulting from rising sea level in coastal areas can have more direct health impacts, such as the seasonal hypertension observed in pregnant women in Bangladesh (Khan et al., 2011). The quality of access can also be affected by reductions in available water quantity, such as from sediments filling up storages, or the reduced productivity of groundwater wells.

Summary

In the coming decades climate change will increasingly affect the location, intensity and frequency of rainfall events; sea level will rise and temperatures will increase. These impacts will affect the ability of populations to access reliable, safe drinking water sources, as well as affecting the maintenance of sanitation and wastewater treatment systems. Increasing temperatures and changes in rainfall will affect the spread of water-related diseases, such as malaria and dengue fever. The impacts of climate change are not fully understood yet, either globally or regionally. This chapter draws heavily on the latest reports from the IPCC; however, these results, especially the regional results, may change over time as improved monitoring systems and data are available.

Key recommended readings

1 IPCC (2013). Summary for Policymakers. In: Stocker, T.F. et al. (eds.) *Climate Change 2013: The Physical Science Basis. Contribution of Working Group I to the Fifth Assessment Report of the Intergovernmental Panel on Climate Change.* Cambridge University Press, Cambridge, UK and New York. This report provides the basic introduction to the current state of knowledge on climate science, including key statistics, the level of certainty of the knowledge, and modelling of future climate. A glossary of key terms is provided in Annex III of the main report.

2 WHO (2009). *Summary and Policy Implications. Vision 2030: The Resilience of Water Supply and Sanitation in the Face of Climate Change.* World Health Organization, Geneva. http://www.who.int/water_sanitation_health/publications/9789241598422_cdrom/en/. This document, and the associated technical reports, provides a review of how climate change will influence drinking water supply and sanitation facilities in the coming decades, and recommendations on the technologies and approaches for adaptation.

3 Bates, B.C., Z.W. Kundzewicz, S. Wu and J.P. Palutikof (eds) (2008). Climate Change and Water. Technical Paper of the Intergovernmental Panel on Climate Change, IPCC Secretariat, Geneva. This report was the first to deal with water and climate change in depth. While some of the science and the predictions have now been improved upon, it still provides an overview of the expected impacts regionally.

Notes

1 The Intergovernmental Panel on Climate Change (IPCC) was established in 1988 to provide clarity on the current state of knowledge in climate change and its potential environmental and socio-economic impacts. Since then, they have produced reports every seven years that review and assess scientific, technical and socio-economic information relevant to the understanding of climate change, with the most recent being the Fourth Assessment Report (AR4) in 2007 and the Fifth Assessment Report (AR5) in 2013/14. The IPCC is an intergovernmental body with 195 member countries, and thousands of scientists, engaged in this process of developing and reviewing reports. Review is an essential part of the IPCC process and includes review by scientists and governments prior to publication of the Assessment Reports.
2 Extremely likely is equivalent to a 95–100 per cent certainty.
3 The troposphere is the lowest part of the atmosphere, where clouds and weather phenomena occur.
4 An extreme weather event is an event that is rare at a particular place and time of year.
5 The cryosphere includes all regions on and beneath the surface of the earth and ocean where water is in solid form, including sea ice, lake ice, river ice, snow cover, glaciers and ice sheets, and frozen ground (which includes permafrost).

References

Bates, B.C., Z.W. Kundzewicz, S. Wu and J.P. Palutikof (eds) (2008). *Climate Change and Water.* Technical Paper of the Intergovernmental Panel on Climate Change, Geneva: IPCC Secretariat,

BBC (2013). Chile water shortage hits Santiago from Maipo river. http://www.bbc.co.uk/news/world-latin-america-21148402. Last accessed 25 September 2014.

Brisbane Times (2013). Brisbane suburbs risk running out of drinking water. http://www.brisbanetimes.com.au/queensland/brisbane-suburbs-risk-running-out-of-drinking-water-20130129-2dij8.html. Last accessed 25 September 2014.

Cairncross, S. and M. Alvarinho (2006). The Mozambique floods of 2000: health impact and response. In: Few R., Matthies F., (eds.) *Flood Hazards and Health: Responding to Present and Future Risks.* London: Earthscan.

Cubasch, U. et al. (2013). Introduction. In: Stocker, T.F. et al. (eds) *Climate Change 2013: The Physical Science Basis. Contribution of Working Group I to the Fifth Assessment Report of the Intergovernmental Panel on Climate Change.* Cambridge and New York: Cambridge University Press.

EA (2007). *Review of 2007 Summer Floods.* Bristol: Environment Agency.

Guha-Sapir, D., F. Vos and R. Below (2011). *Annual Disaster Statistical Review 2010. The Numbers and Trends.* Brussels: Center for Research on the Epidemiology of Disasters.

IPCC (2013). Summary for policymakers. In: Stocker, T.F. et al. (eds) *Climate Change 2013: The Physical Science Basis. Contribution of Working Group I to the Fifth Assessment Report of the Intergovernmental Panel on Climate Change.* Cambridge and New York: Cambridge University Press.

IPCC (2014). *Climate Change 2014: Synthesis Report. Contribution of Working Groups I, II and III to the Fifth Assessment Report of the Intergovernmental Panel on Climate Change.* Core Writing Team, R.K. Pachauri and L.A. Meyer (eds). Geneva: IPCC.

Khan, A.E. et al. (2011). Drinking water salinity and maternal health in coastal Bangladesh: implications of climate change. *Environmental Health Perspectives* 119(9), 1328–1332.

Prüss-Üstün, A. et al. (2014). Burden of disease from inadequate water, sanitation and hygiene in low- and middle-income settings: a retrospective analysis of data from 145 countries. *Tropical Medicine & International Health* 19(8), 894–905. doi: 10.1111/tmi.12329

Smith K.R. et al. (2014). Human health: impacts, adaptation, and co-benefits. In: Field, C.B. et al. (eds) *Climate Change 2014: Impacts, Adaptation, and Vulnerability. Vol I: Global and Sectoral Aspects. Contribution of Working Group II to the Fifth Assessment Report of the Intergovernmental Panel on Climate Change.* Cambridge, UK and New York: Cambridge University Press, Chapter 11.

Taylor, R.G. et al. (2013). Ground water and climate change. *Nature Climate Change* 3(4), 322–329.

WHO (2002). *World Health Report 2002: Reducing Risks, Promoting Healthy Life.* Geneva: World Health Organization.

WHO (2009). *Summary and Policy Implications. Vision 2030: The Resilience of Water Supply and Sanitation in the Face of Climate Change.* Geneva: World Health Organization. http://www.who.int/water_sanitation_health/publications/9789241598422_cdrom/en/

36
POVERTY[*]

Léo Heller

RESEARCHER, OSWALDO CRUZ FOUNDATION, BRAZIL, UN SPECIAL RAPPORTEUR
ON THE HUMAN RIGHTS TO SAFE DRINKING WATER AND SANITATION

Sandy Cairncross

LONDON SCHOOL OF HYGIENE AND TROPICAL MEDICINE, UK

Learning objectives

1 Understand that hazardous environmental conditions are more often present and more dangerous in the places where the poorest live, work or visit.
2 Appreciate that of those who suffer from environmentally determined disease, the poorest are most likely to suffer illness and death as a result; in other words, the poor die young.
3 Be aware that given the same intensity of environmental exposure, the poorest run the greatest risks of poor health as a result.

Introduction

According to Amartya Sen, awarded the Nobel Prize in Economics in 1998, "poverty must be seen as the deprivation of basic capabilities rather than merely the lowness of incomes" (Sen 1999). Sen sees every individual as endowed with certain capacities, and the inability to realise them fully represents a state of "un-freedom" which prevents the individuals from overcoming their poverty. A consequence of this definition is that poverty is no longer solely perceived as an attribute of the individual, but rather becomes a population level factor of the society in which it is found, and which necessitates an economic social, political and cultural environment such that poverty is eliminated. In other words, the "poor" does not refer simply to those who failed to take advantage of opportunities, but to an outcome which is the result of a society which provided an insufficient set of opportunities to fully develop the livelihoods of all its citizens.

---

[*] Recommended citation: Heller, L. and Cairncross, S. 2015. 'Poverty', in Bartram, J., with Baum, R., Coclanis, P.A., Gute, D.M., Kay, D., McFadyen, S., Pond, K., Robertson, W. and Rouse, M.J. (eds) *Routledge Handbook of Water and Health*. London and New York: Routledge.

The production or attenuation of poverty is related to the adoption of specific economic or social policies which interfere in the way in which people get access to goods and services. There are many examples of how socio-political measures have led to an increase in poverty, from the monetarist fiscal policies in the early years of Mrs Thatcher's Britain, to the transition of the Eastern European countries after 1989, when observers noted "dramatic declines in income, the reappearance of diseases long forgotten, growing poverty and great uncertainty" (Milanovic 1998). Even in times of rapid economic growth and in spite of increases in income level, poverty can grow with increasing income inequality, as observed by Mújica et al. (2014) in four of the BRICS countries between 1990 and 2010 – the Russian Federation, India, China and South Africa and with the exception of Brazil.

On the other hand, poverty affects the relationship between water and health in ways which are determined by various factors, particularly the characteristics of the community. Factors such as ethnicity (see UNDP 2006) and migration are closely involved, not least because the poorest people are often heavily represented among migrants and among particular ethnic groups. Educational level has an ambiguous role in this area; poor children usually reach only the lowest levels of attainment. As a result, they learn least about diseases, and least about the hygiene measures to prevent them. Gender also (Saleth et al. 2003) is a mediating factor, in the sense that among the poor, women are often the most vulnerable. As a positive approach, the human rights framework for water services could be a way to prioritize the access to water and sanitation of the most disadvantaged, marginalized and vulnerable, attenuating the influence of poverty on adverse health impacts from inadequate water and sanitation services (Winkler and Roaf in this volume).

The poverty of a society, or of groups in that society, is a fundamental mediating (or facilitating) factor affecting the outcome of environmental conditions on human health. Three premises – which are linked to the learning objectives of the chapter – support that statement, and will be explored in this chapter:

1 Hazardous environmental conditions are more often present and more dangerous in the places where the poorest live, work or visit.
2 Among those who suffer from environmentally determined disease, the poorest are most likely to suffer illness and death as a result; in other words, the poor die young.
3 Given the same intensity of environmental exposure, the poorest run the greatest risks of poor health as a result.

The poor are most affected by environmental hazards or change

An extensive literature considers how environmental impacts do not affect everyone uniformly. Martinez-Allier (2002) used the phrase, "the environmentalism of the poor", to express a political position regarding the interpretation on how the environment affects the global poor. The author sets this perspective against the two prevailing rationales: the "cult of wilderness" – representing the defence of an immaculate Nature – and the "gospel of eco-efficiency" – which focuses on the modernization of the economy and the compatibility between economic growth and acceptable environmental and health impacts. The "environmentalism of the poor" blames economic development for producing increased environmental impacts, with geographical displacements from North to South, of "sources and sinks." This position is perfectly compatible with the tenets of environmental justice, which holds that the worldwide burden of environmentally-caused disease is disproportionately present in poor people and that, globally, much of the environmental damage is caused by the richer countries (Schlosberg 2007; Stephens 2007). Moreover it can be argued that poverty, especially urban or peri-urban poverty, is not a significant contributor to environmental degradation, but that on the contrary, urban environmental hazards are major contributors to urban poverty. It accordingly follows that well-conceived environmental policies could be important instruments to propel poverty reduction forward (Satterthwaite 2003).

Applying the concept of environmental justice to water issues would require water to be regarded simultaneously as a natural or material and social good, as well as an explicit acceptance of water problems as increasingly being contested. Water justice includes, but transcends, questions of the distribution of services by including both cultural recognition and political participation (Zwarteveen and Boelens 2014). In addition, the institutional arrangements we employ for governing water must address issues of democratization, human welfare and ecological conditions (Perreault 2014). The concept of access to water resources and water services is thus broadened, as Sen (1982) did for access to food, by introducing the concept of entitlement.

Specifically regarding access to water services, the disparity between the developed and the developing countries is very clear, as shown in Figures 36.1 and 36.2.

There are various reasons for the striking differences in the proportion of the population with access to these services. They range from a history of inadequate and irregular investment in the sector to shortcomings in public policy in the poorer countries, which fail

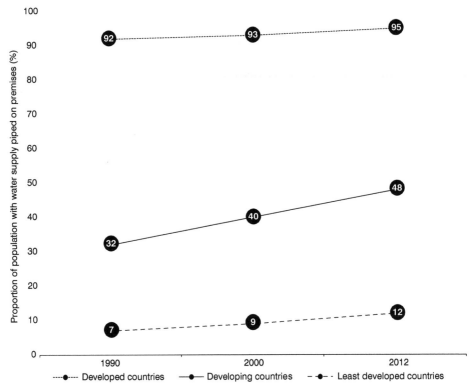

Figure 36.1 Proportion of population with access to water supply piped on premises. Developed, developing and least developed countries (1990, 2000, 2012)

Source: Elaborated by the authors on data from WHO/UNICEF (2014) Joint Monitoring Programme.

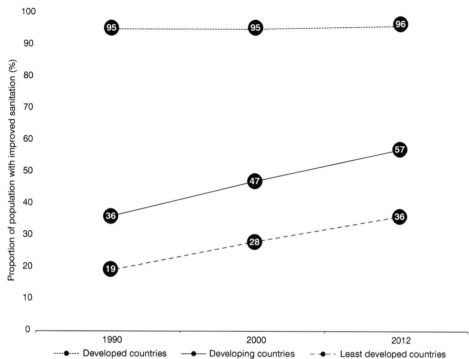

Figure 36.2 Proportion of population with improved sanitation. Developed, developing and least developed countries (1990, 2000, 2012)

Source: Elaborated by the authors on data from WHO/UNICEF (2014) Joint Monitoring Programme.

to provide for their populations a planned expansion of access to facilities and the means for their sustainable use thereafter.

"The poor die young"

This was the title of a book published more than two decades ago (Hardoy et al. 1990), which brought together evidence of much higher levels of ill health and premature death in the poorest urban populations in developing countries, compared with the better-off residents in the same cities. The book analysed housing and living conditions, including relations with water supply, sanitation and urban drainage, as the principal explanatory factors accounting for these differences. This finding was supported by the analysis of a World Bank economist (Listorti 1996), also considering the urban environment, who concluded that most environmental health problems are man-made and therefore preventable. He calculated that infrastructure improvements, by improving urban environmental health, could prevent up to 44 per cent of the burden of disease in the cities of the developing world. This can be compared with the World Bank estimate (World Bank 1993) that basic health services could prevent only 32 per cent of the disease burden, and at greater cost to the economy.

The social determinants of human health have received increasing international attention, particularly since 2004 when the WHO established the Commission on Social Determinants of Health, with a view to promoting policies to reduce inequalities in health within and between countries. At the launch of the Commission it was suggested that,

"The gross inequalities in health that we see within and between countries present a challenge to the world. That there should be a spread of life expectancy of 48 years among countries and 20 years or more within countries is not inevitable." It was also noted that mortality of children under five varied from 316 to 3 per 1000 live births among countries (Marmot 2005).

The study of the social determinants of health has for several decades been the concern of the discipline known as social epidemiology, the early development of which was strongly rooted both in Latin America, where it was known as critical epidemiology (Breilh 2008), and also in the North (Berkman and Kawashi 2000; Galea and Link 2013). The different schools of thought considered social factors "in the causal chain of factors that lead to the production of health" (Galea and Link 2013), or else adopted a radically different explanatory model of the health–disease process in which social exclusion by capitalist society plays a key role (Breilh 1991). In different ways, these theoretical streams start from social factors at the individual, collective or macro level, to explain the process of health and disease and health inequalities.

The controlled effect of income inequality on health has been demonstrated in quantitative studies and in meta-studies, such as the meta-analysis developed by Kondo et al. (2009), showing that each increase of 0.05 in a country's Gini coefficient (a measure of income inequality), is associated with an increase in the odds of dying by a factor of 1.08 (95 per cent confidence interval (CI); 1.06 to 1.10) for cohort studies, 1.04 (95 per cent CI; 1.02 to 1.06) in cross sectional studies, and stronger associations in studies conducted on countries with a Gini coefficient higher than 0.3. In other words, social equality is good for all our health, benefiting society as a whole and not only the poor.

In the same environmental conditions, the poorest run the greatest health risks

"The Bangladeshi who is drinking water with 50 micrograms per litre of arsenic and has poor nutrition may have worse health than a well-nourished person drinking the same water" (*Bulletin of the World Health Organization* 2008). This statement, from a Bangladeshi researcher commenting on the tragic contamination of many groundwater sources in that country with naturally occurring arsenic, illustrates the premise of this section. Not only are the poorest frequently subjected to the most hazardous environmental health conditions; even if they are in the *same* environmental health conditions, there are various features of their poverty which amplify the health risks to which their environment exposes them. Specifically in the case of Bangladesh, the asymmetry of the burden occurs even though the tubewells were installed as an environmental health improvement. These asymmetric outcomes can be explained by various factors: individual factors (nutrition, immunity, interaction with other diseases...), collective factors (the insanitary living environment) and social factors (access to health care services, presence of social capital).

Thus, poverty exposes to a greater risk of disease, groups who are already vulnerable because of other factors: their age (children and the elderly); gender; demography (families in crowded, unhygienic conditions); occupation (relative to certain high-risk occupations); socio-economic (the poorest of the poor); as well as their environment, their access to health care, among others.

Poverty, environment and health: how they connect

The three premises above help us to understand *why* poverty is a fundamental potentiating factor to the links between environment and health. Because of this relationship, several processes can arise, explaining *how* environmental exposures are particularly important to the poor.

Poor people live where environmental health conditions are worst. This is no accident. Poor people cannot afford to live on spacious, well-drained land away from busy roads and with access to clean air, good water supply and sanitation, and effective solid waste management. Those who can afford to pay for sufficient environmental services generally do so. The poor cannot afford these, and so live with the consequences of a dirty and unhealthy environment.

The poor cannot afford good housing. Besides, it would not be sensible for them to invest their scant resources in housing improvements when they are squatters, and so liable to eviction, or damage by flooding, fire or landslides. That means that poor communities do not have pumps, like their wealthier counterparts, to provide water from a piped system when the pressure is low. Millions do not even have their own toilet. So they suffer more from diarrhoea, intestinal worms and other diseases associated with poor hygiene.

The burden of environmental disease falls more harshly on the poor. The poor are vulnerable not only because of where they live, but also the work they do, the greater risks they run, and their lower resistance to infection. In poor families, it is not only those who are ill who suffer; the others have to make sacrifices to care for them or to work in their place, especially if it is the breadwinner or the provider of child care who is ill or injured. When poor adults die, their children often die soon afterwards (Over et al. 1992).

The poor already pay more for environmental health services. In low-income urban areas, many people buy their water from vendors, who sell it for 10 to 20 times more than the water tariff charged by the formal water utility to people who can afford house connections. Other costly items are mosquito control, latrine emptying and rubbish collection when purchased from the informal sector. These items can amount to over 20 per cent of a poor household's income (Zaroff and Okun 1984; Cairncross and Kinnear 1992), and the money for them comes out of the food budget; according to Engel's law (Engel 1857) there is nowhere else it could come from. Affordable environmental health services therefore enable people to both live and eat better (M'Gonigle and Kirby 1936)!

Disease contributes to poverty. When poor people fall ill, they lose income and often lose their jobs. When the illness is serious or affects the breadwinner, desperate relatives spend their savings on treatment, often on inappropriate cures prescribed by charlatans. Impoverished families can often trace the origin of their predicament to the cost of a health disaster affecting one of their members (Xu et al. 2003). Environmentally caused disease affects their prospects in other ways, too; children with intestinal worms or exposed to environmental lead may be stunted in their growth or handicapped in their intellectual performance (Nokes et al. 1992; Walker et al. 2011).

Environmental health is about more than health. In a number of studies, "environmental health interventions" (improved water supply, drainage, sanitation, roads) rank highly among the poor as signs of progress, as ways of knowing that life has improved. But for the poor, health is only a side benefit of these improvements. The main benefits are often about:

- Saving precious time.
- Reducing the burdens of daily life.
- Lowering the basic cost of living.

- Emancipation of women.
- Increasing dignity, self-respect and safety.
- Creating a more pleasant and ordered living space.
- Increasing income through service provision and sound resource management (recycling rubbish, building latrines, etc.).

These are not "wrong" reasons to want environmental health improvements; rather, they are additional benefits. Moreover, they mean that people are often willing to pay for the services (Bartram and Cairncross 2010).

Implications of poverty for the link between water and health

Implications for research and evaluation

Studies of the relation of water and public health often fail to identify the mediating role in the relationship played by the condition of poverty. Poverty is frequently treated simply as a confounder in epidemiological studies, which effectively reduces the dimensions of this complex factor to only one. One of the limitations often encountered is the difficulty of characterizing the state of poverty, as the variables commonly used to "measure" poverty are inaccurate and unreliable. Some of the indicators, such as the *ownership of goods*, may show the amount of goods accumulated by the patriarch of the extended family at a particular stage in the household development cycle, without accurately measuring the current productive capacity of other individuals and families. Other variables, such as *wages*, can be unreliable where poor people are concerned as they often do not have a steady job or a constant monthly wage.

Studies of poverty at the collective level, such as those which compare countries, can easily fail to map the full extent of poverty in its various forms, leading to underestimates of the magnitude of the problem, even in the apparently objective statistics collected and published by international initiatives like the WHO/UNICEF Joint Monitoring Programme (Satterthwaite 2014). One of us (SC) organized the reform of that monitoring programme in 2000, so we state this advisedly.

Thus any scientific study of the links between poverty, water and health needs to take special care to avoid a number of pitfalls. First, the condition of poverty needs to be considered at various levels of aggregation, from the individual to the nation, whether or not the study is epidemiological. Second, the mediating role of poverty in the relationship between water and health deserves careful study, bearing in mind that these relations are not necessarily linear. Limited access to water produces poverty, just as poverty explains limited access; poor health produces poverty, just as poverty explains poor health. There are many settings in which poverty is more than a confounding variable, and may even be a determinant, or a driving force of ill health, and needs to be considered as such. Figure 36.3 shows an example of how environmental improvements can weaken this link between poverty and ill health. In Figure 36.3, the width of each vertical bar shows the proportion of diarrhoea risk attributable to socio-economic status and mediated by the intermediate variables shown. The two figures show conditions respectively (a) before and (b) after implementation of a major sanitation project. The project was associated with a 21 per cent reduction in diarrhoea citywide, and 42 per cent in the high incidence areas. Socio-economic status accounted for 23 per cnet of the variance in diarrhoea rates before the project, but afterwards the strength of that link (corresponding to the total thickness of

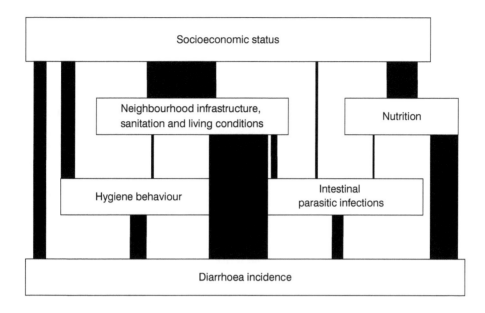

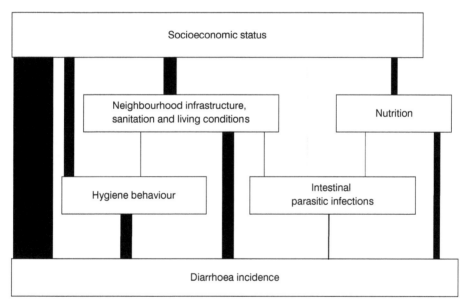

Figure 36.3 Determinants of diarrhoea in El Salvador, Brazil, 1997–2004; results of a hierarchical effect decomposition analysis (a: before and b: after implementation of a major sanitation project)

Source: Bartram and Cairncross (2010).

all four bars near the top of the figure) had been halved, to 11 per cent. The proportion of that association mediated by intermediate variables, particularly sanitation, was also greatly diminished.

Implications for policy

Clearly, the issues discussed in this chapter have public policy implications. On one hand, adequate integrated water management can contribute to poverty reduction along the four dimensions adopted in the poverty reduction framework of the Poverty-Environment Partnership (PEP 2006):

- Enhanced livelihoods security, assured by the provision of improved water services and the consequent opportunities for livelihoods improvement, with greater security and sustainability.
- Reduced health risks, through the mitigation of factors related to water management that put the poor and most vulnerable at risk from various diseases, poor nutrition and untimely death.
- Reduced vulnerability, by means of the reduction of threats from water hazards, such as floods, droughts and major storms; the rise in sea levels due to climate change; and the impact of water pollution, including saline intrusion.
- Pro-poor economic growth, facilitated by the improved management of water resources, especially where changes in water management are part of a wider development strategy aiming to create opportunities for poor people.

On the other hand, there is a debate, among those concerned with public policy for poverty elimination between universalist approaches, in which the entire population is the beneficiary of social benefits – and targeting – through the identification of groups to be benefited, based on eligibility criteria (Mkandawire 2005; Fischer 2010), with the balance of opinion favouring universalist policies as a means of poverty reduction. However, there is evidence that at least in some cases, targeting can offer sound health impacts to the beneficiaries (Rasella et al. 2013). Once it is accepted that poverty can reduce the health benefits which accrue to populations enjoying improved access to water and of initiatives toward integrated water resources management, this can have important policy implications. For public policies in the water sector to realize their full potential benefit to health, they should preferably be linked to wider social programmes, whether universalist or targeted, which are designed to alleviate poverty.

Key recommended readings

1 Berkman, L.F., Kawashi, I. (eds) (2000) *Social Epidemiology*. Oxford University Press, Oxford. An advanced reading, containing 16 chapters on the social determinants of health, from different perspectives, including socio-economic position, income inequality and social capital.
2 Hardoy, J.E., Mitlin, D., Satterthwaite, D. (2001) *Environmental Problems in an Urbanizing World*. Earthscan Publications Ltd., London. Emphasizes the environmental hazards in the cities, including the lack of water and sanitation, and how the urbanizing process affects the environment.

3   Martinez-Allier, J. (2002) *The Environmentalism of the Poor: A Study on Ecological Conflicts and Valuation*. Edward Elgar Publishing Limited, Cheltenham, UK and Northampton, MA. A reference book for the social movements of environmental justice, bridging the fields of political ecology and ecological economics to explain how and why environmental burdens fall disproportionately on the poor.

References

Bartram, J., and Cairncross, S. (2010) Hygiene, sanitation, and water: forgotten foundations of health, *PLoS Med*, vol 7, no 11: e1000367. doi:10.1371/journal.pmed.1000367

Berkman, L. F. and Kawashi, I. (ed) (2000) *Social Epidemiology*, Oxford: Oxford University Press

Breilh, J. (1991) Componente de metodología: la construcción del pensamiento en medicina social, in S. Franco, E. Nunes, E. Duarte, J. Breilh, E. Granda, J. Yépez, P. Costales, and A. C. Laurell (eds) *Debates em medicina social*, Quito: Organización Panamericana de la Salud

Breilh, J. (2008) Latin American critical ('social') epidemiology: new settings for an old dream, *International Journal of Epidemiology*, vol 37, no 4, pp 745–750

Bulletin of the World Health Organization (2008) Bangladesh's arsenic agony. An interview with Mahmuder Rahman, *Bulletin of the World Health Organization*, vol 86, no 1, pp 11–12

Cairncross, S. and Kinnear, J. (1992) Elasticity of demand for water in Khartoum, Sudan, *Social Science and Medicine*, vol 34, no 2, pp 183–189

Engel, E. (1857) Die productions und consumtionsverhältnisse des königreichs sachsen, *Zeitschrift Des Statistischen Bureaus Des Königlich Sachsischen Ministerium Des Inneren*, vol 8, no 9, pp 28–29

Fischer, A. M. (2010) Towards genuine universalism within contemporary development policy, *Institute of Development Studies Bulletin*, vol 41, no 1, pp 36–44

Galea, S. and Link, B. G. (2013) Six paths for the future of social epidemiology, (Commentary), *American Journal of Epidemiology*, vol 178, no 6, pp 843–849

Hardoy, J. E., Cairncross, S., and Satterthwaite, D. (eds) (1990) *The Poor Die Young*, London: Earthscan

Kondo, N., Sembajwe, G., Kawachi, I., van Dam, R. M., Subramanian, S. V., and Yamagata, Z. (2009) Income inequality, mortality, and self rated health: meta analysis of multilevel studies, *British Medical Journal*, vol 339. doi: 10.1136/bmj.b4471

Listorti, J. A. (1996) Bridging Environmental Health Gaps. AFTES Working Papers nos 20–22. Urban Environmental Management, Africa Technical Department, Washington, DC: World Bank

Marmot, M. (2005) Social determinants of health inequalities, *The Lancet*, vol 365, pp 1099–1104

Martinez-Allier, J. (2002) *The Environmentalism of the Poor: A Study on Ecological Conflicts and Valuation*, Cheltenham: Edward Elgar

M'Gonigle, G. C. M. and Kirby, J. (1936) *Poverty and Public Health*, London: Gollancz

Milanovic, B. (1998) *Income, Inequality, and Poverty during the Transition From Planned to Market Economy*, Washington, DC : World Bank

Mkandawire, T. (2005) Targeting and universalism in poverty reduction, United Nations Research Institute for Social Development, Social Policy and Development Programme Paper Number 23. Available: http://www.unrisd.org/80256B3C005BCCF9/(httpAuxPages)/955FB8A594EEA0B0C12 570FF00493EAA/$file/mkandatarget.pdf. Access: 05 August 2014

Mújica, O. J., Vázquez, E., Duarte, E. C., Cortez-Escalante, J. J., Molinab, J., and Silva Junior, J. B. (2014) Socioeconomic inequalities and mortality trends in BRICS, 1990–2010, *Bulletin of the World Health Organization*, vol 92, pp 405–412

Nokes, C., Grantham-McGregor, S. M., Sawyer, A. W., Cooper, E. S., and Bundy, D. A. (1992) Parasitic helminth infection and cognitive function in school children, *Proc Biol Sci.*, vol 247, no 1319, pp 77–81

Over, M., Ellis, R. P., Huber, J. H., and Solon, O. (1992) The consequences of adult ill-health. In R. G. A. Feachem, T. Kjellstrom, C. Murray, M. Over (eds) *The Health of Adults in the Developing World*, Oxford: Oxford University Press for the World Bank

PEP – Poverty-Environment Partnership (2006) Linking poverty reduction and water management, SEI and UNDP. Available: http://www.who.int/water_sanitation_health/resources/povertyreduc2. pdf. Access: 03 July 2014

Perreault, T. (2014) What kind of governance for what kind of equity? Towards a theorization of justice in water governance, *Water International*, vol 39, no 2, pp 233–245

Rasella, D., Aquino, R., Santos, C. A. T., Paes-Sousa, R., and Barreto, M. L. (2013) Effect of a conditional cash transfer programme on childhood mortality: a nationwide analysis of Brazilian municipalities, *The Lancet*. Published online May 15, 2013 http://dx.doi.org/10.1016/S0140-6736(13)60715-1

Saleth, R. M., Samad, M., Molden, D., and Hussain, I. (2003) Water, poverty and gender: an overview of issues and policies, *Water Policy*, vol 5, pp 385–398

Satterthwaite, D. (2003) The links between poverty and the environment in urban areas of Africa, Asia, and Latin America, *The Annals of the American Academy*, vol 590, no 1, pp 73–92

Satterthwaite, D. (2014) Guiding the goals; empowering local actors, *SAIS Review of International Affairs*, vol 34, no 2, pp 51–61. doi: 10.1353/sais.2014.0025

Schlosberg, D. (2007) *Defining Environmental Justice: Theories, Movements, and Nature*, New York: Oxford University Press

Sen, A. (1982) *Poverty and Famines: An Essay on Entitlement and Deprivation*. Oxford and New York: Clarendon Press and Oxford University Press

Sen, A. (1999) *Development as Freedom*, New York and Toronto: Alfred A. Knopf and Random House of Canada Limited

Stephens, C. (2007) Environmental justice: a critical issue for all environmental scientists everywhere (Editorial), *Environ. Res. Lett.*, vol 2, pp 1–2

UNDP – United Nations Development Programme (2006) *Human Development Report 2006. Beyond Scarcity: Power, Poverty and the Global Water Crisis*, Palgrave Macmillan, Houndmills, Basingstoke, Hampshire

Walker, S. P., Wachs, T. D., Grantham-McGregor, S., Black, M. M., Nelson, C. A., Huffman, S. L., Baker-Henningham, H., Chang, S. M., Hamadani, J. D., Lozoff, B., Gardner, J. M., Powell, C. A., Rahman, A., and Richter, L. (2011) Inequality in early childhood: risk and protective factors for early child development, *The Lancet*, vol 378, no 9799, pp 1325–1338. doi: 10.1016/S0140-6736(11)60555-2

WHO/UNICEF (2014) Progress on drinking-water and sanitation 2014 Update. Available: http://www.wssinfo.org/fileadmin/user_upload/resources/JMP-report2014Table_Final.pdf. Access: 03 July 2014

World Bank (1993) *World Development Report 1993: Investing in Health*, Oxford: Oxford University Press for the World Bank

Xu, K., Evans, D.B., Kawabata, K., Zeramdini, R., Klavus, J., and Murray, C. J. L. (2003) Household catastrophic health expenditure: a multicountry analysis, *The Lancet*, vol 362, no 9378, pp 111–117

Zaroff, B. A., and Okun, D. (1984) Water vending in developing countries, *Aqua*, vol 5, pp 289–295

Zwarteveen, M. Z., and Boelens, R. (2014) Defining, researching and struggling for water justice: some conceptual building blocks for research and action, *Water International*, vol 39, no 2, pp 143–158

37

EMERGENCIES
AND DISASTERS[*]

Andy Bastable

HEAD OF WATER AND SANITATION, OXFAM GB, OXFORD, UK

Ben Harvey

INDEPENDENT WASH ADVISOR

Learning objectives

1 Provide an overview of recent global emergencies and disasters.
2 Describe what typical water, sanitation, and hygiene interventions look like during emergencies and disasters and explain the rationale for water, sanitation, and hygiene interventions in emergency response.
3 Present a broad understanding of current issues facing the emergency water, sanitation, and hygiene sector.

Note: It is generally recognized that the health gains of water provision alone are small compared with a holistic emergency response program that combines water, sanitation, and hygiene promotion to improve the public health situation of an emergency-affected community. It is for this reason that during this chapter we talk about the importance of water, sanitation, and hygiene as an emergency response sector rather than focusing on water provision alone.

Introduction

WASH is defined as water, sanitation, and hygiene promotion and incorporates:

- Water supply: for human consumption, handwashing, bathing, cleaning, cooking, sanitation, and laundering.

---

[*] Recommended citation: Bastable, A. and Harvey, B. 2015. 'Emergencies and disasters', in Bartram, J., with Baum, R., Coclanis, P.A., Gute, D.M., Kay, D., McFadyen, S., Pond, K., Robertson, W. and Rouse, M.J. (eds) *Routledge Handbook of Water and Health*. London and New York: Routledge.

- Sanitation: excreta disposal, solid waste management, drainage, and vector control.
- Hygiene promotion: community mobilization, health data monitoring, information, education and communication (IEC), and hygiene kit distribution.

During and directly after an emergency the affected population is largely traumatized, and may have received insufficient food or clean water for some time and is therefore more susceptible to opportunistic diseases. As people are often displaced to overcrowded camps or centers, the risk of disease epidemics is greatly increased. In most emergencies and disasters, WASH interventions form part of the immediate first-phase, life-saving response in conjunction with shelter, nutrition, protection, and healthcare interventions. Since 2010 the right to safe water and sanitation has become a human right, enshrined in international humanitarian law that must be immediately addressed in both times of peace and emergencies (UNHCHR, 2010).

This chapter aims to provide a brief overview of recent global emergencies and disasters; describes what typical water, sanitation, and hygiene promotion interventions look like during emergencies and disasters, and explains why they are so important; and finally describes current issues facing the emergency WASH sector – in particular the need for more innovation, improved preparedness, better response times, increased monitoring, enhanced coordination, and greater accountability.

Defining emergencies and disasters

Disasters can be generally defined as:

> Events that occur when significant numbers of people are exposed to hazards to which they are vulnerable, with resulting injury and loss of life, often combined with damage to property and livelihoods.
>
> *(WHO, 1999)*

Emergencies can be defined as:

> Situations that arise out of disasters, in which the affected community's ability to cope has been overwhelmed, and where rapid and effective action is required to prevent further loss of life and livelihood.
>
> *(WHO, 1999)*

Complex emergencies can be defined as:

> Situations of disrupted livelihoods and threats to life produced by warfare, civil disturbance and large-scale movements of people, in which any emergency response has to be conducted in a difficult political and security environment.
>
> *(WHO, 1999)*

The overall capabilities of communities and authorities to efficiently respond to emergencies vary considerably across the world, and not all disasters will result in an emergency if there is sufficient preparedness or absorption capacity. However, in many developing countries, high rates of population growth and urban migration continue to force populations to live in locations where they may be more vulnerable to disasters. UN-

HABITAT (2003) estimates that approximately one of every three urban inhabitants in the world lives in a slum where many are not able to afford the necessary levels of building reinforcement to protect against large magnitude earthquakes or cyclones. In addition, many of the world's poor are reluctant to leave high disaster risk areas such as flood plains or coastal regions as they depend on these areas for their livelihoods.

While the number of people affected by emergencies, disasters, and complex emergencies varies tremendously from year to year, the general trend is upwards. Since 1988, the Centre for Research on the Epidemiology of Disasters (CRED) has maintained EM-DAT, a worldwide database on emergencies and disasters. The criteria for a disaster or emergency to be entered into the EM-DAT database are that: ten or more people reported killed; 100 people or more reported affected; declaration of a state of emergency; or a call for international assistance (CRED, 2013). According to the EM-DAT online database, during the period 2004–2014 there were 4,335 disasters related to natural hazards resulting in 979,697 deaths and 26.8 million persons being made homeless. An analysis of the occurrence, types, and number of deaths from disasters during the period 2004 to 2014 can be found in Figure 37.1.

Whilst millions of persons have been affected by emergencies resulting from natural causes, during the last decade the number of persons displaced by conflict has also increased dramatically. According to the most recent UNHCR Global Trends Report (UNHCR, 2013) there are currently 16.7 million refugees worldwide and 33.3 million internally displaced persons (IDPs), meaning people forced to flee their homes as a result of conflict but still in their own country. These totals alone are the highest UNHCR has seen since 2001. In addition, more than half of the refugees under UNHCR's care (6.3 million) had at end 2013 been in exile for more than five years. Overall, the biggest refugee populations under UNHCR care and by source country are Afghans, Syrians, and Somalis, together accounting for more than half of the global refugee total.

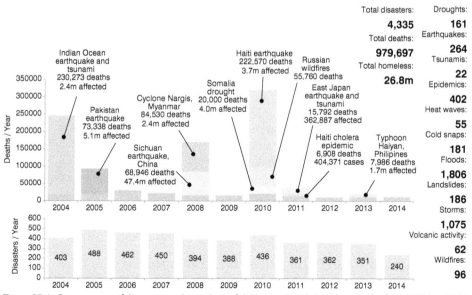

Figure 37.1 Occurrence of disasters and number of disaster related deaths per year from 2004 to 2014

Source: EM-DAT.

The importance of WASH interventions in emergencies

During emergencies and disasters, stress, fatigue, and malnourishment, coupled with unsanitary living conditions such as substandard sanitation, inadequate water supplies, inadequate shelter, overcrowding, and poor hygiene, make the population especially vulnerable to disease epidemics (Johns Hopkins, 2008). In addition to putting the population at increased risk from epidemics, natural disasters can cause power failures, damage to water supply and sanitation infrastructure, transportation difficulties, and shortage of personnel. Flooding may result in an excess of excreta being washed out of septic tanks, sewers, and latrine pits into the public environment, or a proliferation of disease vector breeding. Complex emergencies can result in damage to critical public infrastructure (theft, looting, or conflict damage to critical water and sanitation infrastructure) or more importantly mass movements of the civilian population (Adams and Wisner, 2002).

The effectiveness of WASH interventions in reducing exposure to pathogens in non-emergency settings is well known with significant health gains reported by Esrey et al. (1991), Curtis and Cairncross (2003), and Fewtrell et al. (2005). Findings generally showed that:

- Using a pit latrine can reduce diarrheal diseases by 36 percent or more (Fewtrell et al., 2005).
- Handwashing with soap and water at critical times can reduce diarrheal diseases by 44 percent or more (Curtis and Cairncross, 2003).
- Source water treatment can reduce diarrheal diseases by 11 percent or more (Fewtrell et al., 2005); however, improving the quality of water at the household level can reduce risk of diarrheal diseases by 39 percent or more (Fewtrell et al., 2005).
- Increasing the quantity of available water can reduce risk of diarrheal diseases by 25 percent or more (Fewtrell et al., 2005).

It is well documented that the most common causes of death following an emergency are from acute respiratory infection, diarrheal diseases, measles, malaria, and malnutrition (Lambert and Davis, 2002; WHO, 2005). The most at risk during emergencies are typically infants, the elderly, and the chronically ill. Outbreaks of diarrheal diseases, including dysentery and cholera, are common in emergencies and may account for more than 40 percent of deaths in the acute phase of an emergency, with greater than 80 percent of deaths in children under two years of age (Connolly et al., 2004). Following the arrival of 800,000 Rwandan refugees into the Democratic Republic of the Congo in 1994, 85 percent of the first month's 50,000 deaths were due to cholera and shigellosis (Connolly et al., 2004). The influx of 86,000 Bhutanese refugees into eastern Nepal during 1992 and 1993 created many problems, especially in crowded camps where uncontrolled defecation polluted the groundwater leading to high levels of diarrheal infection, cholera, and typhoid. Studies showed that 87 percent of refugees had never used a latrine before (Puddifoot, 1995). Important risk factors and diseases related to WASH in emergencies have been summarized in Table 37.1.

Typical WASH interventions during an emergency response

In the first days and weeks of an emergency, it is critical to meet survival needs as quickly as possible (Johns Hopkins, 2008). WASH related diseases can be greatly minimized through the immediate provision of basic interventions that include (Adams et al., 2002):

Table 37.1 Summary of important risk factors and diseases related to WASH in emergencies

Emergency type		WASH related risks	WASH related diseases
Flooding Population is generally living in or near their homes.		**Risk factors** Flooded latrines. Contaminated water used for drinking and bathing. Loss of household items for domestic and hygiene use. Vectors breeding in water.	**Diarrheal diseases** Cholera, shigellosis, typhoid, and hepatitis A. **Water-washed diseases** Scabies (skin infections). Louse borne typhus. Infectious skin and eye diseases. **Vector based diseases** Malaria, dengue.
Displacement Either due to conflict or natural disasters such as hurricanes, cyclones, earthquakes, tsunamis etc.	**Host communities**	**Risk factors** Contaminated water at source or due to poor handling or storage. Open defecation. Loss of household items for domestic and hygiene use. Poor hygiene practices.	**Diarrheal diseases** Cholera, shigellosis, typhoid, and hepatitis A. **Water-washed diseases** Louse borne typhus. Infectious skin and eye diseases.
	Crowded camps	**Risk factors** Unsafe handling or storage of water. Open defecation. Poor hygiene practices.	**Diarrheal diseases** Cholera, shigellosis, typhoid, and hepatitis A. **Water-washed diseases** Scabies (skin infections). Louse borne typhus. Infectious skin and eye diseases. **Vector based diseases** Malaria, dengue.
Drought Population is either living in or near their homes or seeking food aid/assistance in camps.		**Risk factors** Dehydration. Lack of water for personal hygiene, i.e. handwashing, bathing, and laundering.	**Diarrheal diseases** Cholera, shigellosis, typhoid, and hepatitis A. **Water-washed diseases** Scabies (skin infections). Louse borne typhus. Infectious skin and eye diseases.

- The provision of an adequate quantity of water.
- The consumption of water of an adequate quality.
- The safe containment of excreta.
- The provision of safe water management and hygiene materials, in particular clean water storage containers and soap.
- The promotion of hygienic practices, in particular handwashing with soap at critical times.
- The efficient control of disease vectors, in particular the reduction of vector breeding sites, clean up campaigns, and the efficient management of domestic solid waste.

The provision of an adequate quantity of water following an emergency is essential not only for drinking but to facilitate basic personal hygiene, handwashing, bathing, laundering, cleaning, cooking, and toilet flushing. SPHERE (2011) recommends that emergency response agencies should aim to provide at least 15 liters per person per day, with the maximum distance from any household to the nearest water point no more than 500 meters, and queuing times of no more than 30 minutes. The importance of sufficient water quantity is supported by Cronin et al. (2008) who compared water and sanitation provision in refugee camps with selected health and nutrition indicators and observed that households reporting diarrhea within the previous 24 hours had a mean 26 percent less water available. Spiegel et al. (2002) also compared mortality rates to water availability and concluded that refugee camps with lower than 15 liters of water per person per day had significantly higher under-five mortality rates.

The consumption of water of an adequate quality is another essential requirement following an emergency or disaster. SPHERE (2011) recommends that all water supplies during the emergency response should be fit for human consumption and advocates that supplies are tested to ensure they are free from chemical and radiological contamination and that there are no fecal coliforms per 100ml of water at the point of delivery. In addition it advises that all water supplies at times of risk of diarrheal epidemics should be treated with chlorine so that there is a residual of at least 0.5mg/l. While it is generally considered best practice to use bulk water treatment methods in camp based settings, there is increasing evidence that point-of-use (POU) water treatment interventions can play an important role in reducing disease risk in non-camp settings. Lantagne and Classen (2011) undertook an extensive study of POU interventions in emergencies and concluded that they are most likely to succeed when the products that are being distributed are familiar to the affected population. Not only is it important that water is of good quality at the point of distribution, but it is also essential that it is protected from contamination during transportation and storage. SPHERE (2011) recommends that water collection and storage containers have narrow necks and/or covers to prevent post-delivery contamination. The protective effects of safe water storage containers against diarrheal diseases have been confirmed by Roberts (2001) who found 31 percent less self-reported diarrhea in households with improved water storage containers in a refugee camp in Malawi. Kunii et al., (2002) also found safe water storage to be protective against diarrheal disease following the 1998 Bangladesh floods. During emergencies it is typical to find families displaced with limited possessions, therefore SPHERE (2011) recommends verifying that each household affected by the emergency has at least two clean water collecting containers of 10–20 liters, one for storage and one for transportation.

The safe containment of excreta is an essential part of the primary barrier to fecal–oral pathogen transmission. Safe containment is generally classified as any system that breaks the fecal–oral transmission routes via groundwater contamination, flies, or direct contamination of the environment. SPHERE (2011) advises that the environment in which the emergency affected population lives must be free from human feces. It recommends that all toilets are at least 30 meters away from any groundwater source and the bottom of any latrines are at least 1.5 meters above the water table. It also advises that in emergency settings there should be no more than 20 persons using each toilet. Technical options for excreta management in emergencies have been extensively documented (Harvey et al., 2002; Lambert and Davis, 2002; Adams and Wisner, 2002; ACF, 2005; Harvey, 2007; MSF, 2010); however, despite this, problems with safe excreta management were particularly evident in Haiti (Bastable and Lamb, 2012) due to a number of factors including the urban environment, private land ownership, high water tables, and problems with desludging operations monopolized by the private sector costing up to US$33 to empty a single porta-loo.

The provision of soap and the promotion of hygienic practices, in particular handwashing with soap, safe water storage, and safe food practices, are important interventions to reduce the risk of disease epidemics following an emergency. Handwashing with soap has been well documented as being the single most effective and lowest-cost intervention for reducing diarrheal diseases in non-emergency settings, with a systematic review of the evidence base showing that the simple act of handwashing with soap at critical times could reduce diarrhea risk by 47 percent and shigellosis by 59 percent (Curtis and Cairncross, 2003). A study carried out by Peterson et al. (1998) in Nyamithuthu Refugee Camp in Malawi found 27 percent less episodes of diarrhea in households when soap was present compared to when no soap was present. SPHERE (2011) recommends that 250g of bathing soap and 200g of laundry soap are available per person per month, and that there are private laundering and bathing areas available for women with sufficient water. In addition it recommends that relief organizations systematically provide information on hygiene-related risks and preventive actions using appropriate channels of mass communication.

Finally, emergency WASH programs are typically responsible for ensuring that within any displaced settlements that are established for the emergency affected population there is adequate domestic waste management and sites are free from wastes, wastewater, and stormwater, in addition to disease vectors, and disease vector breeding sites.

Current issues facing the emergency WASH sector

Despite decades of experience, recent emergencies have revealed that the emergency WASH sector still faces challenges when responding to large or complex emergencies. A recent major gap analysis exercise consulting a range of emergency WASH practitioners from over 45 different organizations in 40 countries identified 12 significant gaps in emergency WASH provision (Bastable and Russell, 2013). Of the 12 significant gaps, it was generally agreed that the most critical was the need to continue to develop rapidly deployable sanitation kits for difficult emergency response settings including: high water tables, urban settings, and unstable soil situations. The Haiti earthquake response in particular showed that agencies are still poorly equipped to deal with the rapid provision of safe excreta disposal in a large-scale urban emergency (Bastable and Lamb, 2012). Other notable gaps identified included: improved water treatment, particularly improving the adherence rates of POU treatment in non-camp emergency settings; and improved hygiene promotion, in particular the importance of understanding context, including socio-anthropology issues, the need for improved rapidly deployable handwashing hardware, and greater accountability to the affected population (Bastable et al., 2013).

Other non-technical issues facing the emergency WASH sector are related to improving monitoring, evaluation, and coordination of WASH responses. While great strides have been achieved in setting WASH indicators and standards for emergency response, there remains disparity in the abilities of WASH agencies to monitor and report on progress against these indicators, in addition to their abilities to rapidly take their WASH programs to scale. Despite these problems, recent monitoring initiatives carried out by the REACH Initiative using Geographical Information Systems (GIS) in refugee camps in Jordan, Iraq, and following Typhoon Haiyan in the Philippines have shown how effective monitoring can be achieved. Finally, great advances have been made in terms of emergency sector coordination through the Inter-Agency Standing Committee (IASC) Humanitarian Reforms agreed by United Nations (UN) agencies in 2006 and in particular the 'cluster approach'. However, there are still great needs for improving coordination in emergencies as recent emergencies have shown that the success of the coordination efforts are highly variable, and depend heavily on the individual experience, skills, and personalities of the emergency WASH Sector Coordinator.

Conclusions

WASH interventions form a major and essential part of the first wave of response during most large emergency and disasters and are essential in reducing the risk of disease and ensuring the affected population has access to basic needs and life with dignity. Based on experience from the last decade with 4,335 disasters resulting in 979,967 deaths and 26.8 million persons being made homeless it is clear that large-scale emergencies and disasters requiring WASH assistance will continue to emerge in the near future. Despite decades of experience, the emergency WASH sector still faces many challenges which require continued innovation (particularly in terms of safe excreta disposal in urban or difficult environments), improved preparedness, better response times, increased monitoring, enhanced coordination, and greater accountability to the affected population and other stakeholders.

References

ACF (2005), *Water, Sanitation and Hygiene for Populations at Risk*. Second edition. Paris: ACF. http://www.actioncontrelafaim.org/sites/files/pubs/fichiers/wsh_acf_0.pdf

Adams, J. and Wisner, B. (2002), 'Environmental health in emergencies and disasters: a practical guide'. Geneva: World Health Organization (WHO). http://whqlibdoc.who.int/publications/2002/9241545410_eng.pdf

Bastable, A. and Lamb, J. (2012), 'Innovative designs and approaches in sanitation for challenging and complex humanitarian urban contexts'. *Waterlines* 31(1). DOI: 10.3362/1756-3488.2012.007

Bastable, A. and Russell, L. (2013), 'Gap analysis in emergency water, sanitation, and hygiene promotion'. Humanitarian Innovation Fund Report. http://www.humanitarianinnovation.org/sites/default/files/hif_wash_gap_analysis_1.pdf

Brown, J., Jeandron, A., Cavill, S., and Cumming, O. (2012), 'Evidence review and research priorities: water, sanitation, and hygiene for emergency response'. London: Department for International Development (DFID). http://r4d.dfid.gov.uk/pdf/outputs/sanitation/Evidence_review__WASH_for_emergency_response_March_2012.pdf

Chalinder, A. (1994), 'Water & sanitation in emergencies: a good practice review'. London: Overseas Development Institute (ODI). http://www.odihpn.org/download/gpr1pdf

Connolly, M. A., Gayer, M., Ryan, M. J., Spiegel, P., Salama, P. and Heymann, D. L. (2004), 'Communicable diseases in complex emergencies: impact and challenges'. *The Lancet* 364(9449), 1974–1983. http://www.thelancet.com/journals/lancet/article/PIIS0140673604174813/abstract

CRED (2013) 'What are the EM-DAT disaster criteria?' http://www.emdat.be/frequently-asked-questions

Cronin, A., Shrestha, D., Cornier, N., Abdalla, F., Ezard, N. and Aramburu C. (2008), 'A review of water and sanitation provision in refugee camps in association with selected health and nutrition indicators – the need for integrated service provision'. *Journal of Water and Health* 06.1. http://www.unhcr.org/4add71179.pdf

Curtis, V. and Cairncross, S. (2003), 'Effect of washing hands with soap on diarrhoea risk in the community: a systematic review'. *The Lancet* 3(5): 275–281. http://www.ncbi.nlm.nih.gov/pubmed/12726975

Esrey, S., Potash, J. B., Roberts, L. and Shiff, C. (1991), 'Effects of improved water supply and sanitation on ascariasis, diarrhoea, dracunculiasis, hookworm infection, schistosomiasis, and trachoma'. *Bulletin of the World Health Organization* 69(5), 609–621. http://www.ncbi.nlm.nih.gov/pmc/articles/PMC2393264/pdf/bullwho00050-0101.pdf

Fewtrell, L. et al. (2005), 'Water, sanitation, and hygiene interventions to reduce diarrhoea in less developed countries: a systematic review and meta-analysis'. *The Lancet Infectious Diseases.* http://www.thelancet.com/pdfs/journals/laninf/PIIS1473-3099(04)01253-8.pdf

Harvey, P. A. (2007), 'Excreta disposal in emergencies: a field manual'. Loughborough: WEDC, Loughborough University. http://www.unhcr.org/4a3391c46.html

Harvey, P. A., Baghri, S. and Reed, R. A. (2002) 'Emergency sanitation – assessment and programme design'. Loughborough: WEDC, Loughborough University. http://reliefweb.int/lou-water-02.pdf

IFRC (2014), 'World disasters report: focus on culture and risk'. Geneva: International Federation of Red Cross and Red Crescent Societies (IFRC). http://www.ifrc.org/Global/Documents/Secretariat/201410/WDR%202014.pdf

Johns Hopkins (2008), *Public Health Guide for Emergencies,* Second edition. Geneva: International Federation of Red Cross and Red Crescent Societies (IFRC). http://www.jhsph.edu/research/pubs/_CRDR_ICRC_Public_Health_Guide_Book/doc.pdf

Kunii, O., Nakamura, S., Abdur, R. and Wakai. S. (2002), 'The impact on heath and risk factors of the diarrhoea epidemics in the 1998 Bangladesh floods'. *Public Health* 116: 68–74. http://www.publichealthjrnl.com/article/S0033-3506(02)00506-1/abstract

Lambert, R. and Davis, J. (2002), *Engineering in Emergencies*, Second edition. London: Register of Engineers for Disaster Relief (RedR)

Lantagne, D. and Clasen, T. (2011), 'Assessing the implementation of selected household water treatment and safe storage (HWTS) methods in emergency settings'. UNCEF/Oxfam GB/ PhD London School of Hygiene and Tropical Medicine. http://blogs.washplus.org/drinkingwaterupdates/wp-content/uploads/2012/01/Oxfam-LSHTM-acute-emergency-report-final.pdf

MSF (2010) *Public Health Engineering in Precarious Situations*, Second edition. Brussels: MSF. http://refbooks.msf.org/msf_docs/en/public_health_en.pdf

Oxfam GB (2011), 'Urban WASH lessons learned from post earthquake response in Haiti'. Oxfam GB, Oxford, UK. http://oxfamlibrary.openrepository.com/oxfam/bitstream/10546/136538/8/tbn20-wash-urban-lessons-learnt-haiti-06052011-en.pdf

Peterson, E. A., Roberts, L., Toole, M. J. and Peterson. D. E. (1998), 'The effect of soap distribution on diarrhoea: Nyamithuthu Refugee Camp'. *International Journal of Epidemiology* 27(3): 520–524. http://www.ncbi.nlm.nih.gov/pubmed/9698146

Puddifoot, J. (1995), 'Pit latrines in Nepal – the refugee dimension'. *Waterlines* 14(2): 30–32

Roberts, L., Chartier, Y., Chartier, O., Malenga, G., Toole, M., and Rodka, H. (2001), 'Keeping clean water clean in a Malawi refugee camp: a randomized intervention trial'. *Bulletin of the World Health Organization* 79(4): 280–287. WHO, Geneva, Switzerland. http://www.ncbi.nlm.nih.gov/pubmed/11357205

SPHERE (2011) 'Humanitarian charter and minimum standards in disaster response'. http://www.sphereproject.org/resources/download-publications

Spiegel, P., Sheik, M., Gotway-Crawford, C. and Salama P. (2002), 'Health programmes and policies associated with decreased mortality in displaced people in post emergency phase camps: a retrospective study'. *The Lancet* Dec 14 (360): 1927–1934. http://www.ncbi.nlm.nih.gov/pubmed/12493259

UN-HABITAT (2003), 'The challenge of slums: global report on human settlements'. London: United Nations Human Settlements Programme. http://www.unhabitat.org.jo/pdf/GRHS.2003.pdf

UNHCHR (2010), 'Legal obligations of the rights to water and sanitation'. Human right to water and sanitation UN special rapporteur. Geneva: Office of the United Nations High Commissioner for Human Rights, UNHCHR. http://www.ohchr.org/Documents/Issues/Water/LegalObligations_en.pdf

UNHCR (1992), 'Water manual for refugee situations'. Geneva: UNHCR. http://www.unhcr.org/3ae6bd100.pdf

UNHCR (2007), *Handbook for Emergencies*, Third edition. Geneva: UNHCR. http://www.unhcr.org/472af2972.html

UNHCR (2013), 'Global refugee trends, statistical overview of populations of refugees, asylum-seekers, internally displaced persons, stateless persons, and other persons of concern to UNHCR'. Geneva: UNHCR. http://www.unhcr.org/5399a14f9.html

WHO (2005), 'Communicable disease control in emergencies: a field manual'. Geneva: World Health Organization. http://whqlibdoc.who.int/publications/2005/9241546166_eng.pdf?ua=1

38

POPULATION AND DEMOGRAPHICS*

Carl Haub

SENIOR DEMOGRAPHER, POPULATION REFERENCE BUREAU, WASHINGTON, DC, USA

Learning objectives

1 Understand the importance of how rising world population increases pressure on water resources.
2 Have an appreciation of common sources of demographic data.
3 Grasp the implications of the greatly varying population increases and decreases within high and low income nations.

The demographic background

Past population growth

Global population growth accelerated greatly in the past century, placing obvious demands on all natural resources, including water. In 1900, world population totaled 1.6 billion, a figure that had taken about 50,000 years of human history to attain. Merely 100 years later, in 2000, it stood at 6.1 billion, the two numbers simply reversing. In 2013, the global total stands at 7.1 billion (PRB, 2013) and growth continues at about 86 million a year with nearly all of that occurring in the poorest countries.

The rapid growth observed in the past 100 years results from the different way in which the "demographic transition" evolved in the developed and developing countries (the United Nations defines the developing countries as those in Africa, Asia (excepting Japan), Latin America, and Oceania (excepting Australia and New Zealand). All of the rest are considered developed; UNPD, 2013). In the developed countries, improvements in public health, food distribution, and medical care contributed to slowly rising life expectancy and gradually decreasing infant mortality. At the same time, the process of industrialization and urbanization often reduced the need and desire for traditionally large families. No small role was played by rising female literacy. As a result, birth and death rates declined roughly in

---

* Recommended citation: Haub, C. 2015. 'Population and demographics', in Bartram, J., with Baum, R., Coclanis, P.A., Gute, D.M., Kay, D., McFadyen, S., Pond, K., Robertson, W. and Rouse, M.J. (eds) *Routledge Handbook of Water and Health*. London and New York: Routledge.

parallel so that, while population growth rates did increase, they remained at a modest level in the developed countries, rarely exceeding one percent per year.

The demographic transition process was entirely different in the developing countries. Declines in the death rate occurred much more swiftly, particularly after the Second World War. In Indonesia, for example, life expectancy at birth stood at only 38 years in the early 1950s, but had risen to 66 years by the end of the century (UNPD, 2013). Such gains had taken centuries in the developed countries. The improvements in public health were then comparatively quickly "exported" to the developing countries. But the improvement in mortality was often not accompanied by a decrease in the birth rate as the basic makeup of societies had not changed. Many remained fundamentally rural and agrarian with the traditional desire for large families remaining unchanged. Population growth rates rose to much higher levels than they ever had in the developed countries, sometimes as high as three percent per year and more. Ultimately, many countries adopted policies to lower birth rates and slow growth but the actual progress achieved has been quite uneven (UNPD, 2014).

Future population growth

The projections of future population growth are rather speculative. Population projections which offer a variety of scenarios necessarily require that assumptions on the future course of birth and death rates first be made. Those assumptions are often revised to take the current observed trends into account. Consider projections issued by two population organizations over the past few years in Table 38.1. The differences in Table 38.1 are due to different assumptions regarding fertility. The Population Reference Bureau has higher estimates for some countries based on more recent surveys.

The increase in projected population size has been rather substantial and may be expected to continue. A clearer understanding of why that is the case can be gained by a brief overview of projections for selected regions of the world.

Future population growth in the developed countries

The projection of the total population of the developed countries in 2050, 1.3 billion, has remained essentially constant for some time, due to their very low birth rates which show little or no sign of rising. Women in developed countries average less than two children each, often much less. For some, population decline is likely while others should maintain population growth through immigration as well as higher birth rates among immigrants.

The most significant demographic consequence of low birth rates is the unprecedented aging of their populations. The projected aging of Japan is a dramatic example. Today one fourth of Japan's population is in the age group 65 and over; that is projected by the Japan

Table 38.1 Projections of world population for 2050 (billions)

Year of projection	United Nations Population Division	Population Reference Bureau
2008	9.15	9.35
2010	9.31	9.49
2012	9.55	9.62
2013	n.a.	9.73

Sources: UNPD (2013) and PRB (2013).

government to rise to 39 percent by 2050 (JNIPSSR, 2014). This "medium" projection assumes that the birth rate will remain constant at 1.4 children per woman and that a modest amount of immigration will lessen population decline to some extent. Total population is projected to decline from 127 million currently to 97 million by 2050, with the decrease accelerating rapidly after that. While aging in Japan is the most extreme, projections for other developed countries are similar. By 2050, one third of the population of Germany, Italy, and Spain, among others, is projected to be aged 65 and over. Countries with higher birth rates, such as France and the US, should experience less aging, rising to one fifth to one fourth of their populations of 65 and over by 2050.

Projections can, of course, prove wrong in the long run but one aspect of low birth rates adds to the robustness of the outlook. Long periods of low birth rates have reduced the number of young people who are the parents of tomorrow. The age group 0–4 years, for example, is often little more than half the size of their parents' age group. Only a sharp, and generally unexpected, rise in birth rates could forestall deaths from exceeding births, something that has already happened in some developed countries.

Future population growth in the developing countries

As can easily be seen in Figure 38.1, future world population growth will be confined to the developing countries. The graph also distinguishes the growth projections the 43 United Nations (UN) defined "least developed countries" primarily located in sub-Saharan Africa. Much of the growth results from population "momentum" in the developing countries, which are assumed to experience continued birth rate decline. Momentum results when a country reaches the "replacement" level of two children per woman but continues to grow. That, in turn, occurs due to the comparatively large number of young people born during periods of higher birth rates. Population growth, then, continues for decades after reaching

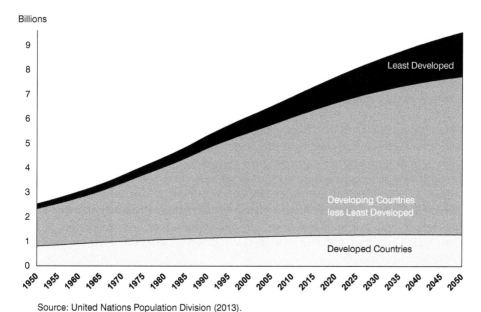

Source: United Nations Population Division (2013).

Figure 38.1 Population growth worldwide, 1950–2050

the benchmark two-child family. While it is true that the age at childbearing tends to rise over time, the trend in the total fertility rate makes a much larger difference in population size.

Despite having the largest population of any world region, 4.3 billion, Asia will likely not have the largest amount of population growth. That region is expected to be Africa, primarily due to growth in the 49 countries of sub-Saharan Africa. Those countries are projected to rise from 1.1 billion in 2013 to 2.4 billion in 2050. The amount of that increase can be appreciated when we consider that Asia's much larger population is projected to increase by one billion over the same period. Latin America/Caribbean projected growth is far less, from 606 million to 780 million. That region has reached replacement level fertility and grows solely from momentum, its growth also slowed by emigration.

Future regional population growth in the developing countries

Africa's future population growth is the largest question mark as it is the region with the highest fertility, at 4.8 children per woman. Its projected increase of 1.3 billion in the short span of 37 years will certainly place unprecedented pressure on the need for water. And, in the UN projections, growth continues for many decades after 2050. Again, projections are simply a result of the assumptions from which they result. For all countries with fertility above replacement, the assumption is usually made that fertility will decline smoothly – and without interruption – to the replacement level of two children per woman and even below.

In the absence of a reliable vital statistics enumeration system to count births and deaths in most developing countries, nationally representative sample surveys are conducted on a regular basis. One such global program is the Demographic and Health Surveys (DHS) conducted by national governments in cooperation with international aid agencies, such as the United States Agency for International Development (USAID). Similar surveys are conducted by the United Nations International Children's Emergency Fund (UNICEF). These programs are key to our understanding of possible future population trends.

In sub-Saharan Africa, many surveys show that birth rates are not coming down or are coming down much more slowly than projections have customarily assumed. Currently, women in the region average 5.2 live births. Surveys also show that the use of modern contraceptive methods is often low and the use of temporary methods, used for spacing the number of children rather than limiting their number, predominates. In the projections cited in Table 38.1, sub-Saharan Africa accounts for about 80 percent of the increase in world population by 2100. It is likely that the 2050 projected population for the region will continue to rise.

Fertility in Asia is now below the replacement level, but not if the large statistical effect of China is removed. Outside China, women in Asia average 2.5 children. Overall, national policies to reduce the birth rate have met with considerable success in Asia. In India, the region's second largest country, women now average 2.4 children but several states in the north with very large populations, such as Bihar and Uttar Pradesh, have lagged behind, leaving the country's future population size in some doubt. China, where women average 1.5 children, has recently announced a change in its stringent population policy to allow couples where only one spouse was an only child (previously both had to be) to have a second child. The change will undoubtedly result in some increase in the birth rate, but it is uncertain just how much.

As noted earlier, Latin America now has low fertility, in part due to its two largest countries, Brazil and Mexico, which account for half the region's population and who report fertility of 1.8 and 2.2, respectively. In a number of other countries, fertility decline has slowed and appears not to be declining to the replacement level.

Table 38.2 Change in renewable water supply per capita, 1988–1992 to 2008–2012 and projected population increase, selected countries

	Total renewable water resources per capita 10⁹ m³year		*Projected 2050 population as a multiple of 2013*
	1988–1992	*2008–2012*	
Afghanistan	4,222,000	1,956,000	1.8
Australia	28,068,000	21,467,000	1.5
Brazil	63,260,000	41,505,000	1.2
China	2,362,000	2,051,000	1.0
Egypt	971,000	683,000	1.7
Germany	1,918,000	1,878,000	0.9
India	2,100,000	1,519,000	1.3
Niger	4,054,000	2,022,000	3.9
Nigeria	2,794,000	1,718,000	2.5
Saudi Arabia	140,000	84,000	1.6
Uganda	3,485,000	1,853,000	3.1

Note: Water supply is estimated to be constant in both time periods.

Sources: FAO (2014) and PRB (2013).

The coming demand for water

In its simplest terms, population growth will exert varying constraints on water supply, looking ahead to the year 2050. Such constraints are likely to be felt most in sub-Saharan Africa where the availability of renewable water per capita is low but projected population growth is the highest. However, the increase in population masks the projected growth in the number of households. In France, for example, the number of households will increase by 80 percent between 1990 and 2050 (Bartram et al., 2014). This is due to the fact that, as populations grow, average household size decreases leading to an increase in the overall number of households. By contrast, in the Dominican Republic, the number of households is projected to rise 280 percent, as it transitions from community-level to household-level. But demographic projections change and, with them, the forecast outlook for the need for water. Will the demand for water be greater or less than the projections indicate? Policy makers are searching for perfect information. Not likely in this sector of data gathering.

Key recommended readings

1 Bongaarts, J., Sathar, Z., and Mahmood, A. (2013). *Population Trends in Pakistan*. New York: The Population Council.
   Excellent country-specific discussion of the effects of different future fertility trends on population growth. http://www.popcouncil.org/pdfs/2013_PakistanDividend/Chapter2.pdf
2 MEASURE Evaluation, University of North Carolina Chapel Hill Population Center. (2013). Population Analysis for Planners. Lesson 8: The Cohort Component Population Projection Method. Chapel Hill: MEASURE Evaluation.
   Detailed description of the cohort-component method of making population projections. Explains the importance of age–sex structure and its interaction with fertility,

mortality, and migration. Free online course. https://training.measureevaluation.org/non-certficate-courses/pap/lesson-8

3 ★Population Reference Bureau. (2013). Family Planning Worldwide 2013 Data Sheet. Washington, DC: Population Reference Bureau.

Wallchart providing very recent survey data and the use of family planning for the world's countries. Very useful in evaluating the likelihood of fertility decrease in developing countries. http://www.prb.org/Publications/Datasheets/2013/family-planning-worldwide-2013.aspx

References

Bartram, J., Elliot, M., and Chuang, P. (2014). Getting wet, clean, and healthy: why households matter. London: *The Lancet*. Available at http://www.thelancet.com/journals/lancet/article/PIIS0140-6736(12)60903-9/fulltext

FAO (United Nations Food and Agriculture Organization) (2014). AQUASTAT. FAO's Information System on Water and Agriculture. Rome: United Nations Food and Agriculture Organization

JNIPSSR (Japan National Institute of Population and Social Security Research) (2014). Population Projection for Japan. Tokyo: Japan National Institute of Population and Social Security Research. Available at: http://www.ipss.go.jp/site-ad/index_english/esuikei/gh2401e.asp

PRB (Population Reference Bureau) (2013). 2013 World Population Data Sheet. Washington, DC: Population Reference Bureau. Available at: http://www.prb.org/Publications/Datasheets/2013/2013-world-population-data-sheet/data-sheet.aspx

UNPD (United Nations Population Division) (2013). World Population Prospects, the 2012 Revision. New York: United Nations Population Division. Available at: http://esa.un.org/unpd/wpp/index.htm

UNPD (United Nations Population Division) (2014). World Population Policy Database. New York: United Nations Population Division. Available at: http://esa.un.org/PopPolicy/about_database.aspx

39

WATER REUSE[*]

Choon Nam Ong

PROFESSOR AND DIRECTOR, NUS ENVIRONMENTAL RESEARCH
INSTITUTE, NATIONAL UNIVERSITY OF SINGAPORE

Learning objectives

1 Better understanding of the major types of water reuse.
2 Outline the main technologies that can be used for water recycling.
3 Understand the health and water quality issues of water reuse.

Freshwater is a limited resource

Freshwater is a precious resource. More than 97 percent of the earth's water is seawater and, of the remainder, more than two thirds is frozen glaciers and ice caps (cryosphere), leaving just 1–2 percent of all the water on our planet as fresh. Over the last several decades, water shortages have become increasingly common globally. As climate change and population growth will further strain our freshwater resources, we may experience more water shortages in the future.

Water reuse is the practice of using water that has already been used. The terms "reclaimed water," "recycled water" and "reused water" are used interchangeably. Water reuse can be defined as, "Wastewater treated or processed to a certain standard suitable for reuse." This recovered water can serve as a source for many different applications; however, its quality must be tailored to the requirements of the end usage. The main objective of this chapter is to review the current technologies for water reuse and their applications in various parts of the world. The health and water quality issues of wastewater reuse and the future of water reuse are also discussed.

Types of water reuse

There are two major types of water reuse: direct reuse and indirect reuse. Direct reuse refers to the introduction of reclaimed water from a water reclamation plant to a distribution system, for example treating used water and then piping it to a waterworks for an industrial

---

[*] Recommended citation: Ong, C.N. 2015. 'Water reuse', in Bartram, J., with Baum, R., Coclanis, P.A., Gute, D.M., Kay, D., McFadyen, S., Pond, K., Robertson, W. and Rouse, M.J. (eds) *Routledge Handbook of Water and Health*. London and New York: Routledge.

403

center or a housing estate for use. Indirect reuse is the use of water, usually treated and then placed back into a water supply source such as a lake, river or aquifer, retrieved to be used again. A planned reuse system involves treatment of wastewater to careful standards to be reused with predetermined purposes.

Both direct and indirect reuse can be suitable for drinking (potable) and non-drinking purposes. Indirect potable reuse (IPR) typically refers to wastewater, planned or unplanned, that is discharged into a stream, lake, river or aquifer where it is then reused for drinking purposes, often after advanced treatment. Direct potable reuse (DPR) is planned or intentional usage of treated wastewater as a drinking water supply. In most parts of the world, some form of treated or untreated wastewater is discharged into rivers, streams and lakes where it is later used for many purposes, including drinking. An unplanned reuse system involves inadvertent reuse of wastewater. For example, the City of London receives 20 percent of its drinking water from a tributary of the Thames and all the cities along the Rhine, Danube, Ohio, Colorado, Mississippi, Yangtze and Huangpu and other major rivers use water that has been used by residents upstream, put back into the river, and then reused, in most cases many times. Thus, the quality of the water is much more important than the type of the source of the water.

Today, with advanced treatment technologies increasingly available, we can treat water to a quality following stringent guidelines for the intended use. The recycled water can be used in numerous applications to satisfy most water demands, depending on the level of treatment. Presently, recycled water is more commonly used for non-drinking purposes, such as agriculture, landscape, horticulture and garden/park irrigation. Other major applications include cooling water for buildings, power plants and oil refineries, toilet flushing, dust control, construction activities, concrete mixing and artificial lakes.

Technologies for recycling water

Currently, there are a variety of treatment processes that can be used to recycle/reclaim water. The four core stages of wastewater treatment are primary, secondary, tertiary and advanced treatments. The number of treatment steps varies depending on how the water will be eventually used. Most recycled water, however, will undergo some form of disinfection.

Advances in reuse technology since the 1960s have made reuse a feasible option for treating wastewater for various beneficial purposes. An important development is the "multiple-barrier" concept, which provides several barriers, including biological, physical and chemical processes, to prevent the passage of contaminants into the final effluent. Table 39.1 shows the types of *advanced* treatment processes, and suggested *usage* at each level of treatment. In uses where there is a greater chance of human exposure to the water, more treatment is required. As for any water source that is not properly treated, health problems could arise from drinking or being exposed to recycled water if it contains disease-causing organisms or chemical contaminants of sufficient concentration to engender disease.

Primary treatment is a physical process that removes a portion of suspended solids and organic matter from wastewater by allowing it to settle, as well as removing floating materials. This process does not remove microbiological pathogens and thus is not recommended as a single process for potentiating reuse. Secondary treatment is usually a combined biological and chemical process that removes biodegradable organic matter and suspended solids, through the use of microorganisms. Traditionally, secondary treatment has been sufficient for most non-potable applications and for environmental release. Nevertheless, this process reduces but does not eliminate pathogens. Tertiary treatment is usually more complex, and

Table 39.1 Suggested water recycling treatment and uses

Increasing levels of human exposure; increasing levels of treatment

Primary treatment Sedimentation	**Secondary treatment** Biological oxidation, disinfection	**Tertiary treatment** Chemical coagulation, filtration, disinfection	**Advanced treatment** Microfiltration, reverse osmosis membrane and disinfection
No uses recommended at this level	Agriculture and horticulture; surface irrigation of orchards and vineyards Non-food crop irrigation Restricted landscape usage Groundwater recharge of non-potable aquifer Wetlands, wildlife habitat, stream augmentation Industrial cooling processes	Landscape and golf course irrigation Toilet flushing Vehicle washing Food crop irrigation Unrestricted recreational impoundment	Indirect potable reuse: groundwater recharge of potable aquifer and surface water reservoir augmentation

Note: Suggested uses are based on Guidelines for Water Reuse, developed by the US EPA. Recommended level of treatment is site-specific (US EPA, 2012b).

is also specifically designed to remove residual suspended solids and pathogens. Primary and secondary treatment technologies have not changed significantly, but tertiary and advanced treatment technologies have resulted in higher efficiency, increased reliability, and, in some cases, better quality. Advanced treatment is tailored for more specific desired purpose. This can include removal of total dissolved solids, pathogens and trace constituents (Asano et al. 2007; US EPA, 2012a).

If the intended application is IPR, treated wastewater needs to be subject to more stringent and well proven purification processes, including microfiltration (MF) and ultrafiltration (UF), which use a low-pressure membrane that removes small suspended particles and other materials out of the water. During the last decade, synthetic membranes have been regarded as one of the best available technologies for the production of high quality recycled water for IPR. Among the processes used for wastewater treatment, membrane bioreactor (MBR) technology is also advancing rapidly worldwide. Further, membranes that can stand pressure using reverse osmosis (RO) have been widely used for treatment and the production of high-quality water, suitable for potable usage. Today more wastewater treatment facilities are using membrane technologies, and this number is on the rise as the technology offers unparalleled capability in meeting rigorous requirements (National Research Council, 2012). Membranes are currently being used as a tertiary or advanced treatment for the removal of dissolved species; organic compounds; colloidal and suspended solids; and human pathogens, including bacteria, protozoan cysts and viruses. Also, recent research has shown that micro-constituents, such as pharmaceuticals and personal care products, can be removed by high-pressure RO membranes. A typical advanced treatment process that is used for treating wastewater for IPR is illustrated in Figure 39.1.

Disinfection must accomplish the dual objective of inactivating pathogenic organisms while not harming water users (human or environmental) or plant workers. Traditionally, chlorine has been used as a disinfectant. However, ozone and ultraviolet (UV) radiation have

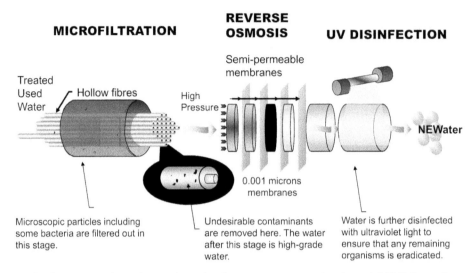

MICROFILTRATION **REVERSE OSMOSIS** **UV DISINFECTION**

Figure 39.1 Singapore reclaimed water through a three-stage process: going through MF, RO membrane and UV disinfection for the production of NEWater

Source: Courtesy of PUB, Singapore.

rapidly become alternative methods of disinfection. Ozone and UV radiation are both more effective disinfectants than chlorine, but both are energy intensive and expensive, although prices are decreasing. To achieve an adequate disinfection performance, the combination of UV with other disinfection agents is usually used to provide greater reliability and higher efficiency for inactivation of different types of microorganisms. Ozone is as effective as UV and chlorine but like chlorine it may also lead to formation of secondary disinfection byproducts (DBPs) (Plewa et al., 2011).

Water reuse in different parts of the world

Water reuse is a growing practice in many parts of the world. Countries and regions in which water reuse is on the rise include Australia, Israel, Europe, Singapore and the USA.

According to a recent review (Australian Academy of Technological Sciences and Engineering, 2004), in Australia a total of 166.2 Gl/y (43.91 billion gal/y) was reused in 2001–2002, up from 112.9 Gl/y (29.83 billion gal/y) during 1996–1999. The proportion of effluent recycled also increased during this period from 7.3 percent during 1996–1999 to 9.1 percent in 2001–2002. As Australia continues to battle the drought which has been going on for the last decade, reclaimed effluent is becoming a popular and necessary option. Two major cities, Adelaide and Brisbane, have already committed to adding reclaimed effluent to supplement their dwindling water stocks maintained by dams. Brisbane has been seen as a leader in this trend, and other cities and towns such as Melbourne are considering building a reclaimed effluent process.

Europe, over the last two decades, has witnessed growing water stress, both in terms of water scarcity and quality deterioration, which has prompted many municipalities to look for a more efficient use of water resources, including a more widespread acceptance of water reuse practices. According to Bixioa et al. (2006), currently there are more than 200 water reuse projects in Europe while many others are in an advanced planning phase. The second largest waste reclamation program in the world is in Spain, where 12 percent of the nation's waste is treated (Bixioa et al., 2006). Nevertheless, only one water reuse project referred to as "Europe" has been identified for potable water production. The project was set up to reduce the extraction of natural groundwater for potable water production and to hold back the saline intrusion at the Flemish coast of Belgium (Bixioa et al., 2006). On the other hand, indirect or even unplanned potable reuse occurs in most of the major European cities. In Europe there is a growing interest in artificial groundwater recharge with reclaimed wastewater to prevent saline intrusion in coastal aquifers.

As of 2010, Israel leads the world in the proportion of water it recycles (Futran, 2013). Israel treats 80 percent of its sewage, and 100 percent of the sewage from the Tel Aviv metropolitan area is treated and reused as irrigation water for agriculture and public works. The recycled water allows farmers to plan ahead and not be limited by water shortages. However, plans to implement the usage of reclaimed water for drinking are complicated by the psychological hurdles of the public regarding the quality of reclaimed water, and the fear of its origin. As of today, all the reclaimed sewage water in Israel is used for agricultural and land improvement purposes.

As an island nation, Singapore in the past depended on Malaysia for a large part of its water supply. The amount of imported water, however, was subject to ongoing negotiations. This lack of security regarding future water resources was considered a significant risk by Singapore. Although Singapore's annual rainfall is more than double the global average, collecting and storing enough water is difficult because open land is scarce. Even with reservoirs covering half the country's land area, Singapore only has enough water for 60 percent of its daily needs. In order to reduce dependence on imported water, Singapore invested in a highly regulated desalination and water reclamation system. Singapore tested the safety of reclaimed water for two years in a demonstration plant before inaugurating a state-of-the-art wastewater reclamation plant in 2002. Please see Figure 39.1 for a schematic of this system. This plant treats domestic wastewater and storm water to produce reclaimed water, referred to as "NEWater" (Lim and Seah, 2013). The NEWater plants use secondary treatment membrane UF, RO, UV disinfection, and chlorination to treat reclaimed water to potable quality. This water is then tested on over 200 different quality parameters. The quality of the NEWater exceeds drinking water standards set by the US Environmental Protection Agency (US EPA) and the World Health Organization. According to the Singapore national water agency, PUB, the recycled water treated with advanced technologies is in fact cleaner than the water that flows to taps in Singapore. After treatment, Singapore's reclaimed wastewater is blended with raw water in reservoirs, and then the mixture of reclaimed and raw water is treated for domestic use via the potable water distribution system. NEWater is used for wafer industry which requires high purity water, but it also comprises about 3 percent of the potable water system. The national water agency has plans to increase this share of indirect potable water.

In the US, the practice of recycling/reclaiming water is a growing enterprise. It is estimated that 2.6 billion gallons per day (bgd) are reused in the US. However, this is only a fraction of the total volume of wastewater generated. According to the US EPA, a total of 34.9 bgd is produced (National Research Council, 2012). Thus, the reclaimed water amounts to only 7.4 percent, suggesting future potential for the reclaiming of wastewater in the USA. Florida and California were the first two states to approve the use of reclaimed water for in-building piping and in flushing toilets. Similar efforts to stretch limited water resources are also in progress in other states, although to a lesser extent. Some municipalities are adapting existing wastewater treatment plants for this purpose. For example, San Diego

County recently opened a pilot wastewater recycling plant that uses a method similar to Singapore's. The county's example is indicative of the difficulties of introducing recycled wastewater into the water supply and may serve as a model for future attempts. Like other counties in Southern California, San Diego imports much of its water from inland regions, an expensive proposition made increasingly tenuous by drought and population growth that increase upstream demand. Thus, the water authorities in different parts of the world that reuse their water have turned to community and environmental groups to help educate the public about the safety of recycled water (O'Connor et al., 2007; PUB Singapore, 2014). Orange County, California, serves as another example of early adoption by employing RO and advanced oxidation processes to generate as much as 265 million liters of clean water from wastewater each day, sufficient to supply 20,000 average US households (Water Reuse Association, 2012). In Orange County, this resource replenishes supplies that have been severely depleted by drought and population growth, and it helps to prevent seawater from the Pacific Ocean from intruding and contaminating the freshwater. Recently, the US National Research Council reported that expanding the reuse of municipal wastewater for irrigation, industrial uses and drinking water augmentation could significantly increase the US total available water resources. The panel also indicated that the risk of exposure to certain microbial and chemical contaminants from drinking reclaimed water is not any higher than from drinking water from current water treatment systems and, in some cases, may be orders of magnitude lower (National Research Council, 2012).

Health and water quality issues of wastewater reuse

As population pressures grow, the demand for water reclamation will increase as a result of the need for more potable water and better ways to dispose of wastewater. Future success in water reuse will depend on whether this can be carried out without adverse effects on human health and the environment. Reclaimed water can be highly engineered for safety and reliability so that the quality of the water is more predictable than many existing surface and groundwater sources. Reused water is thus usually considered safe when treated and use dappropriately. Reclaimed water planned for use in recharging aquifers or augmenting surface water supplies should receive adequate and reliable treatment before mixing with naturally occurring water and undergoing natural restoration processes.

For each defined usage, the water quality should be driven by a number of health, safety, socio-psychological and technical-economic criteria. As a rule, water quality objectives are set by guidelines and regulations, which in turn determine the treatment technology to be used. Assurance of treatment reliability and appropriate operation of water recycling and reuse systems are the major water quality control measures. A thorough knowledge and appropriate monitoring of water quality is needed to protect public health and minimize the negative impacts on the receiving water bodies and the environment.

For IRP, it is also important to note that the advancements in treatment technology have been accompanied by a simultaneous increase in the number and chemical complexity of wastewater effluent pollutants. A growing concern in all drinking water risk management is the appearance of "emerging contaminants" that have only recently been discovered and are generally unregulated and lack monitoring. These include disinfection byproducts, pharmaceutically active chemicals, personal care products, and suspected endocrine disruptors. Emerging contaminants are present in raw wastewater, but they are also found in natural sources (Fawell and Ong, 2012). This lack of regulation and increased occurrence present an additional challenge for wastewater treatment and reuse (National Academy of Sciences, 2012).

Water treatment processes, especially the use of RO in water reuse, can affect mineral concentrations and hence affect intake of essential elements such as calcium and/or magnesium. Therefore, treatment and stabilization practices for water reuse should ensure that the overall process does not significantly reduce these minerals in the drinking water (WHO, 2009). During the past two decades, numerous studies have been conducted on the potential health effects of potable and non-potable reuse of treated wastewater and most studies have concluded that clean and safe effluents can be produced from municipal wastewater treatment through a combination of approaches. The direct and indirect reuse of treated wastewater for potable uses can thus be assured. As previously mentioned, it is important to note that there is a potential health risk from little-known or unknown pathogens that might be found in wastewater and may not be eliminated by treatment. A recent study by the US National Academy of Sciences offered a series of recommendations to operators of water-supply systems considering potable reuse of treated wastewater. The most important aspects are (1) reuse systems should use a combination of advanced physical treatment processes and strong chemical disinfectants as the main defense against most microbial contaminants; (2) facilities should assess and report the effectiveness of their treatment processes in removing microbial pathogens; and (3) research should be conducted on the detection of emerging pathogens in environmental samples and regarding the effectiveness of various water or wastewater treatment processes and disinfectants in removing or inactivating these pathogens (National Research Council, 2012).

The future of water recycling

With recent advances in technology, treating municipal wastewater and reusing it could significantly increase the total available stock of water resources and also offer significant additional potential supplies to help meet future needs (National Research Council, 2012). Non-potable reuse is a widely accepted practice and will continue to grow. Advances in wastewater treatment technology and health studies of IPR have led many to believe that planned IPR will become more common. Moreover, new analyses suggest that the possible health risks of exposure to chemical contaminants and disease-causing microbes from wastewater reuse do not exceed, and in some cases may be significantly lower than, the risks of existing water supplies. Thus far, water reuse has been considered as a viable, long-term solution to the challenges of population growth, industrial and agricultural demands for water. Today, it is technically feasible to produce water of virtually any quality desired. The emerging issue of climate change is also expected to favor water recycling, and reuse development as water recycling is widely recognized as a proven response to water scarcity which enhances sustainability by providing drought-proof alternative resources. Therefore, water reuse should be viewed as one of several alternative sources of new supplies of water or as an important part of any water resource management planning activity, especially in water stressed regions.

Key recommended readings (open access)

1 National Research Council (2012) *Water Reuse: Potential for Expanding the Nation's Water Supply through Reuse of Municipal Wastewater*. This document presents a portfolio of treatment options in reclaimed water along with new analysis and the risks of exposure to certain contaminants.
2 US EPA (2012) Guidelines for Water Reuse. EPA/600/R-12/618, September. This document's primary purpose is to facilitate further development of water reuse by

serving as an authoritative reference on water reuse practices. http://nepis.epa.gov/Adobe/PDF/P100FS7K.pdf

3 WHO (2011) *Guidelines for Drinking Water Quality* (4th Ed.). World Health Organization. WHO produces international norms on water quality and human health in the form of guidelines that are used as the basis for regulation and standard setting, in developing and developed countries worldwide. It is a useful reference for water regulators, academia and water suppliers.

References

Asano, T, Burton, F, Leverenz, H, Tsuchihashi, R, and Tchobanoglous, G. (2007) *Water Reuse: Issues, Technologies, and Applications*. Wakefield, MA: Metcalf & Eddy.

Australian Academy of Technological Sciences and Engineering (2004) *Water Recycling in Australia*. Parkville: AATSE.

Bixio, D, Thoeye, C, De Koning, J, et al. (2006) Wastewater reuse in Europe, *Desalination*, 187: 89–101.

Fawell, J, and Ong, CN. (2012) Emerging contaminants and the implications for drinking water. *Int. J. Water Res Dev*. 28: 247–263.

Futran, V. (2013) Tackling water scarcity: Israel's wastewater recycling as a model for the world's arid lands. Discussion Paper 1311. Global Water Forum, Canberra, Australia.

Lim, MH, and Seah, H. (2013) Role of Water Reuse for the City of Future (Singapore). In *Milestones in Water Reuse*. Lazarova, V, Asano, T, Bahri, V, and Anderson, J. London: International Water Association.

National Research Council (2012) *Water Reuse: Potential for Expanding the Nation's Water Supply through Reuse of Municipal Wastewater*. Washington, DC: National Academies Press.

O'Connor, GA, Elliott, HA, and Bastian, RK. (2007) Degraded water reuse: an overview. *J. Environ. Qual*. 37: S157–S168.

Plewa, MJ, Wagner, ED, and Mitch, WA. (2011) Comparative mammalian cell cytotoxicity of water concentrates from disinfected recreational pools. *Environ Sci Technol*. 45(9): 4159–4165.

PUB, Singapore (2014) *NEWater – The Third National Tap*. http://www.pub.gov.sg/water/newater/Pages/default.asp.x

US EPA (2012a) Water Recycling and Reuse: The Environmental Benefits, Water Division Region IX – EPA 909-F-98-001. http://www.epa.gov/region9/water/recycling/

US EPA (2012b) *Guidelines for Water Reuse*. EPA/600/R-12/618, September. http://nepis.epa.gov/Adobe/PDF/P100FS7K.pdf

Water Reuse Association (2012) Orange County Chapter. https://www.watereuse.org/sections/california/orange-county

WHO (2009) *Calcium and Magnesium in Drinking Water – Public Health Significance*. Geneva: World Health Organization.

40

WAR AND CONFLICT*

Barry S. Levy

MD, MPH, Adjunct Professor of Public Health, Tufts
University School of Medicine, Sherborn, MA, USA

Victor W. Sidel

MD, Distinguished University Professor of Social Medicine Emeritus,
Montefiore Medical Center and Albert Einstein Medical College; Adjunct
Professor of Medical Ethics in Medicine; and Adjunct Professor of Health
Care Policy and Research at Weill Cornell Medical College, New York, USA

Learning objectives

1 Describe how war reduces access to safe water.
2 Describe how decreased access to safe water can lead to conflict.
3 Provide selected specific examples of what can be done to ensure access to safe water.

Introduction

War and conflict are associated with water in two ways. First, reduced access to freshwater is increasingly a cause of conflict, including armed conflict. Second, armed conflict frequently limits access to water by non-combatant civilians, such as by damaging and destroying water treatment facilities, water supply systems, and other critical segments of civil and related societal infrastructure. In this chapter, we discuss these two associations between conflict and water and what can be done to resolve water-related conflict and to preserve access to water during conflict.

Reduced access to safe freshwater: a cause of conflict

There is far less freshwater in the world than one may think. About 97.5 percent of water globally is either salt water or polluted water. Almost 70 percent of the rest is frozen in the polar ice caps and in glaciers. Less than 0.01 percent of all water globally is available in lakes,

---

* Recommended citation: Levy, B.S. and Sidel, V.W. 2015. 'War and conflict', in Bartram, J., with Baum, R., Coclanis, P.A., Gute, D.M., Kay, D., McFadyen, S., Pond, K., Robertson, W. and Rouse, M.J. (eds) *Routledge Handbook of Water and Health*. London and New York: Routledge.

rivers, reservoirs, and aquifers that can be easily accessed – and therefore available for human use.

Potential causes of conflict are the sharing of river basins between or among countries and the sharing of transboundary groundwater reserves (aquifers) between or among countries. About 60 percent of all river water in the world is shared by at least two countries. This river water is in 263 river basins that are located in 145 countries, where about 40 percent of people live worldwide (Wolf et al., 1999). Therefore, many countries are dependent on water that originates beyond their territory. As examples, 76 percent of water resources in Pakistan and 34 percent in India originate outside these countries (Renner, 2010). A more complex example is the Nile River Basin, which is shared by 11 countries.

People generally require 100 to 200 liters of water daily to successfully meet all of their basic needs (for all individual and community uses of water, not limited to drinking, cooking, and bathing). If one considers all of the water needs in society, including industry, agriculture, and energy generation, the average freshwater requirement per capita each year is about 1,000 cubic meters (Gleick, 1993). About 25 years ago, there were already 11 countries in Africa and the Middle East that had less than 1,000 cubic meters available per person (Gleick, 1993). Given high population growth rates in each of these countries, each of these 11 countries will have substantially less water per person available as time goes on.

Freshwater is increasingly scarce in many parts of the world. One billion people do not have access to safe water. With population growth and climate change, scarcity of freshwater will become an increasingly severe problem, especially in arid and semi-arid countries in Africa and the Middle East (Klare, 2001).

Over the past five decades, the number of conflicts over water, both within and between countries, has been sharply increasing. For example, between 1900 and 1959 there were 22 global water conflicts, an average of 0.37 per year. From 1960 to 1989, there were 38, an average of 1.27 per year. Between 1990 and 2007, there were 83 conflicts, an average of 4.61 per year. Of these 83 conflicts, at least 61 were violent or in the context of violence (Gleick, 2009). The Pacific Institute provides an online database of water-related conflict during the past 5,000 years, which can be analyzed by time, place, and other parameters (Pacific Institute, 2014). While "conflict" is not specifically defined, it includes the following types of conflicts: military tool (state actors); military target (state actors); terrorism, including cyber-terrorism (non-state actors); and development disputes (state and non-state actors).

Underlying reasons for water-related conflict include (a) low rainfall, an inadequate water supply, or dependency on only one major water source; (b) a high population growth rate and a high rate of urbanization; (c) industrialization and modernization; and (d) a prior history of armed conflict and tensions between countries or among groups within countries (Gleick, 2009). It is important to note that scarcity of water, in and of itself, is not often the cause of armed conflict over water. Factors that immediately precipitate armed conflict include political, social, or economic tensions; disputes related to large-scale projects, such as dams and reservoirs; and disputes over resources and environmental issues (Gleick, 2009). For example, the conflict in Darfur, which began in 2003, started when residents revolted against the central government, which then fought back by unleashing the Janjaweed (Sudanese Arabs, whom the United States claims killed 200,000 to 400,000 Darfur residents); however, the roots of this conflict dated back to the mid-1980s when a severe drought and famine affected Sudan and the entire Horn of Africa (Borger, 2007).

Violent conflict over water, like other armed conflict, can have disastrous health consequences for individuals and populations, including not only death, injury, illness, and long-term physical and mental impairment, but also destruction of the health-supporting infrastructure of society, including systems that provide freshwater; forced internal or external migration, which generally decreases access to freshwater; and diversion of human and financial resources, including resources to maintain and improve access to freshwater (Levy and Sidel, 2008).

Conflicts over water are not limited to low-income countries. They also occur in the United States and other high-income countries, as demonstrated in recent years by water-related disputes in the southwest, where seven states in the Colorado River Basin share its water, and the southeast of the country, where three states are in dispute over water flow from the Apalachicola-Chattahoochee-Flint and the Alabama-Coosa-Tallapoosa river basins.

Reduced access to safe water: a consequence of conflict

Armed conflict often results in damage to, or destruction of, water treatment facilities and water supply systems, thereby reducing access to safe water. In addition, destruction of other components of the health-supporting infrastructure of society, such as transportation networks as well as power generation and supply systems, contributes to this reduced access.

For example, as a result of the Persian Gulf War (1990–1991) and related United Nations sanctions, there were major shortages of safe water in Iraq. During the war, there was much damage to and destruction of water treatment plants. Because of the sanctions that were imposed from 1990 to 1998, some water treatment facilities and water supply systems that had been damaged or destroyed were not repaired for long periods of time. War-related damage to electrical grids, which powered water-pumping stations, contributed to water shortages. Shortages of safe drinking water were a contributing factor to the estimated 400,000 excess deaths in young children that occurred during the war and when sanctions were in force (Ali et al., 2003).

Forced migration is another contributing factor to reduced access to safe water supplies during war and its immediate aftermath. As people are uprooted from their homes and communities, they are less likely to access safe water, at least in the short term. Those who are able to migrate to other countries are likely to have their basic needs met, including access to safe water. However, people who are internally displaced within their own countries during war often cannot access safe water – and other essentials for life. Multiple factors force people to migrate during armed conflict, including fear, death of loved ones, damage to their homes and communities, and reduced access to safe food and water.

Climate change can also be a contributing factor to water shortages. While some areas of the world will become wetter, especially in the tropics and in high latitudes, arid and semi-arid areas, especially in temperate zones, are likely to become drier, thereby worsening existing water shortages.

Prevention of water-related conflicts

There are two major ways of preventing water-related conflicts: increasing the availability of water and resolving water conflicts before they "boil over."

Increasing the availability of water

The availability of water can be increased in several ways. For example, the use of water can be reduced, such as by decreasing leaks in the distribution system and wasteful uses and by increasing the efficient use of water. In addition, availability of clean water can be increased, such as reducing industrial pollution and sewage contamination of water, improving sewage and wastewater treatment, and improving watershed management. Yet other ways of increasing the availability of water include establishing and maintaining new groundwater wells, designing and implementing improved methods of desalinization, and expanding use of "greywater" (wastewater from wash basins, showers, and baths, which can be recycled for uses such as toilet flushing, irrigation, and constructed wetlands).

Resolving water conflicts before they "boil over"

There are a variety of ways to resolve conflicts over water. For example, laws and regulations at the local, state or provincial, national, or international level can promote the resolution of conflicts at an early stage. Various approaches to mediation and arbitration can also help to resolve conflicts over water. There have been more than 3,800 unilateral, bilateral, or multilateral declarations or conventions concerning water, including 286 treaties (United Nations, 2006). The Program in Water Conflict Management and Transportation at Oregon State University provides a searchable online database of summaries and/or the full text of more than 400 international, freshwater-related agreements from 1820 to 2007 (http://transboundarywater. geo.orst.edu/database/interfreshtreatdata.html. Accessed on October 29, 2014).

In addition, proactive cooperation among nations or among states (or provinces) within nations can prevent conflicts over water and potentially resolve them before they become significant. Such cooperation can improve public health, maintain food security, and promote social, economic, and environmental stability. There are a number of good examples of such cooperation, including the Good Water Neighbors Project in the Middle East (Kramer, 2008), the Nile Basin Initiative in Africa (Nile Basin Initiative, 2010), and the cooperation between Bolivia and Peru in working together to manage water resources related to Lake Titicaca (Wolf and Newton, 2009).

A framework has been developed and applied for resolving transboundary water conflicts that occur when people, states or provinces, or countries that share a resource that crosses legal or political jurisdictions disagree about the use of that resource (Rowland, 2005). It is a model for equitable use of shared water resources throughout the world.

What you can do

There are a number of things that individuals (and the organizations to which they belong) can do to help prevent water-related conflicts. These include the following:

1 Raising awareness about the importance of access to freshwater.
2 Documenting conflicts over water and their adverse health effects.
3 Promoting activities for preventing contamination of water, to conserve it, and use it more efficiently.
4 Promoting nonviolent approaches to resolving conflicts over water.
5 Promoting proactive cooperation among countries or groups within countries.
6 Supporting global, national, and local programs to improve access to safe drinking water and basic sanitation.

By helping to prevent conflicts over water, you can help to ensure a basic requirement for the public's health, adequate access to freshwater, and sustainable peace.

Key recommended readings

1 Levy, B.S. and Sidel, V.W. (eds). (2008) *War and Public Health (Second Edition)*. New York: Oxford University Press. This book comprehensively describes the adverse consequences of war on public health, provides examples from several recent wars, and describes ways in which public health professionals can help to prevent war and minimize its adverse health consequences.

2 Levy, B.S. and Sidel, V.W. (2014) Collective violence caused by climate change and how it threatens health and human rights. *Health and Human Rights Journal*. 16. pp. 32–40. This paper reviews the association between climate change and collective violence, based on a systematic literature review. It also describes preventive measures, research issues, and the legal and moral obligations of countries to protect human rights.

3 Pacific Institute. The World's Water: "Water Conflict" (Online database). Available from: http://worldwater.org/water-conflict/. (Accessed: October 23, 2014.)
This online database provides detailed information on the relationship of water and conflict over the past several decades.

References

Ali, M., Blacker, J., and Jones, G. (2003) Annual mortality rates and excess deaths of children under five in Iraq, 1991–98. *Population Studies: A Journal of Demography*. 57. pp. 217–226.

Borger, J. (2007) Scorched. *The Guardian*. (Online) April 27. Available from: http://www.theguardian.com/environment/2007/apr/28/sudan.climatechange. (Accessed: September 26, 2014.)

Gleick, P.H. (1993) Water and conflict. *International Security*. 18(1). pp. 90, 100–101.

Gleick, P.H. (2009) Water conflict chronology. In *The World's Water, 2008-2009: The Biennial Report on Freshwater Resources*. Washington, DC: Island Press. pp. 151–196.

Klare, M.T. (2001) *Resource Wars: The New Landscape of Global Conflict*. New York: Henry Holt and Company.

Kramer, A. (2008) *Regional Water Cooperation and Peacebuilding in the Middle East*. Berlin: Adelphi Research.

Levy, B.S. and Sidel, V.W. (eds.). (2008) *War and Public Health* (Second Edition). New York: Oxford University Press.

Nile Basin Initiative. (2010) Nile Basin Initiative (Online) Available from: http://www.nilebasin.org. (Accessed: November 20, 2014.)

Pacific Institute. (2014) Water conflict chronology timeline. (Online). Available from: http://www2.worldwater.org/conflict/timeline/. (Accessed: September 26, 2014.)

Renner, M. (2010) Troubled waters: Central and South Asia exemplify some of the planet's looming water shortages. *Water World*. May/June. pp. 14–20.

Rowland, M. (2005) A framework for resolving the transboundary water allocation conflict conundrum. *Ground Water*. 43. pp. 700–705.

United Nations. (2006) From water wars to bridges of cooperation: Exploring the peace-building potential of a shared resource. 10 stories the world should hear more about. (Online) Available from: http://www.un.org/events/tenstories/06/story.asp?storyID=2900. (Accessed: September 26, 2014.)

Wolf, A.T. and Newton, J.T. (2009) Autonomous binational authority of Lake Titicaca. In: Delli Priscoli, J. and Wolf, A.T. (eds) *Managing and Transforming Water Conflicts*. Cambridge: Cambridge University Press, pp. 211–214.

Wolf, A.T., et al. (1999) International river basins of the world. *International Journal of Water Resources Development*. 15. pp. 387–427.

PART VI

Policies and their implementation

41

INTRODUCTION TO POLICIES AND REGULATIONS ON WATER AND HEALTH*

Michael J. Rouse

INDEPENDENT INTERNATIONAL ADVISOR AND DISTINGUISHED
RESEARCH ASSOCIATE, UNIVERSITY OF OXFORD

Introduction

The subject of water in relation to health is very broad. Health matters are influenced by quantity as well as quality aspects, as the management of water resources involves allocation between the needs for irrigation, domestic and industrial water supplies, and critically the requirement to manage the water environment for current and future generations. Water taken by various users has to be used efficiently and responsibly and returned to the environment in a condition which provides for sustainability. New water resources and so-called wastewaters should be considered as integrated water resources. Rainfall does not arrive in convenient quantities equally spread over time, but intermittently with an increasing number of extreme events. This can and does result in droughts with the need for various forms of water storage and water demand management by all users. Equally it can result in excess quantities at one time resulting in flooding and consequent risks to health, both physically and from faecal contamination. Such flooding can be regarded as the mismanagement of a water resource with the need for integration of water resource and flood management including the use of wetlands for water storage. All aspects can impinge on health. The challenges go much further than scientific understanding with the need for integrated and coherent policies which are progressed through legislation, and implemented through effective governance covering institutional frameworks, regulations and delivery management. The roles of the various parts of government are discussed under joined-up government.

It is instructive to consider the scale of the water, sanitation and hygiene (WASH) challenges. The various international initiatives such as the First UN Water Decade of

---

* Recommended citation: Rouse, M.J. 2015. 'Introduction: policies and regulations on water and health', in Bartram, J., with Baum, R., Coclanis, P.A., Gute, D.M., Kay, D., McFadyen, S., Pond, K., Robertson, W. and Rouse, M.J. (eds) *Routledge Handbook of Water and Health*. London and New York: Routledge.

1981–90, and the Millennium Development Goals (2000–2015) may not have fulfilled their ambitions, but they resulted in international focus on water and access to improved water supplies for additional billions of people. The post-2015 sustainable development goals with the expected target of achieving universal coverage of both water and sanitation by 2030, will provide international and national focus for policies and their implementation for the next 15 years. The challenge of achieving such goals is immense; making assumptions on population growth, and using definitions of successful delivery provided by the United Nations (UN) handbook on the human rights of water and sanitation[1] (HRWS), it can be estimated that additional provision of water supply services will be required to be made to around 2.7 billion people worldwide by 2030. That is an average of around half a million people each day until 2030 and more for sanitation. Different assumptions result in different figures but all produce challenging numbers.

Joined-up government

Political commitment is vital, and the development and implementation of effective policies will require joined-up government, as many aspects of government need to be involved. Figure 41.1 shows the functions of government with important inputs; the government ministries have been given generic names for illustrative purposes. Water is often part of a Ministry for Environment, but can be part of a Ministry for Agriculture and Irrigation, or some other aspect of government depending on policy priorities for a country. Wherever it is located it should be the lead/coordination unit for all aspects of water policy. Clearly the Ministry of Finance has a key role in policy on the provision of finance for capital projects, and on charging policies and use of subsidies. It has responsibility for financial stewardship and will wish to see evidence of value for money, with good documentation and effective monitoring of construction projects and operations. Subsidy policy is important, both in connection with pro-poor policy, but also in relation to incentives for efficient service delivery and sustainability. With effective briefing Ministries of Finance[2] are in the best position to recognise the financial benefits of investment in WASH and to support progress. A close working relationship between the Ministries of Water and Finance is essential. Ministries of Finance often oversee the bodies responsible for economic regulation and tariff determinations. The Ministry of Social Services may be responsible for interventions on pro-poor provisions and should have a role in the implementation of the HRWS. Such provisions would be expected to be integrated with other aspects of social welfare such as housing. The Ministries of Health are consulted on water-related health matters, particularly on drinking water and often have the role of monitoring drinking water quality. The quality regulation functions, both for drinking water and wastewater, can be within government departments, but increasingly are carried out by government agency regulators. There are many approaches to economic regulation,[3] from municipal self-determination to economic regulatory agencies. However this aspect is managed it is critical that the function is seen in the wider picture of cost recovery to allow the necessary capital provision and for sustainability. There are other functions, such as monitoring and reporting of international goals; this might be the responsibility of the ministry which is responsible for international affairs. This function might also involve negotiations on rivers which cross country boundaries in relation to integrated water resources management (IWRM). Establishing these connections within national governments is a challenge in itself, but necessary to achieve the targeted progress.

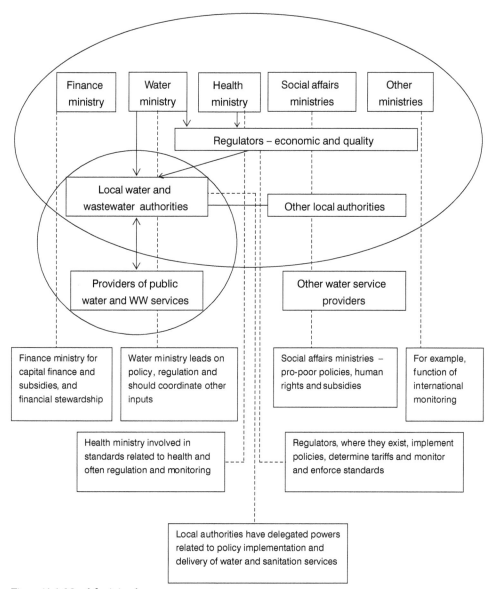

Figure 41.1 Need for joined-up government

Depending on national structures local governments may have responsibility for policy but more often have the role of local implementation with the responsibility for the service delivery provision. Many local authorities have the delivery function within the municipality, although increasingly steps are being taken to separate the local policy and delivery functions. Figure 41.1 has been drawn as these being separate organisations as the functions within the 'oval' make up the content of this part of the book.

The inclusion of other aspects of local authority functions is important particularly in relation to urbanisation. IWRM is very important for water provision to cover all water requirements including that of irrigation, but predicted rapid urbanisation requires integration of water and sanitation planning with city planning. This requires joined-up local government policy and planning but also is a factor in national policies. With the effect of the increase in the world's population being experienced in urban areas, due to the migration from rural areas, many small local towns will become the cities of the future. To avoid the problems of slum development in peri-urban areas, with associated lack of safe water and gross pollution, a paradigm shift[4] in system provision is required, and should be seen as an opportunity to plan for sustainability. This important aspect is covered in Chapter 32 on Integrated Urban Water Management

Concepts and tools

There are a number of key concepts and tools which both educate policy and define its implementation: precautionary principle, polluter pays, risk management and cost–benefit analysis. None of these stands alone, with there being interplay between them.

The precautionary principle concept was first defined in Germany.[5] There are various ways of defining it.

> One description [Wikipedia] is the precautionary principle or precautionary approach to risk management states that if an action or policy has a suspected risk of causing harm to the public or to the environment, in the absence of scientific consensus that the action or policy is not harmful, the burden of proof that it is not harmful falls on those taking an action.

In economics,[6] the precautionary principle has been analysed in terms of the effect on rational decision making of the interaction of irreversibility and uncertainty. In relation to the environment the most influential reference to the precautionary approach was the Rio Declaration[7] of 1992. It noted: 'In order to protect the environment, the precautionary approach shall be widely applied by States according to their capabilities. Where there are threats of serious or irreversible damage, lack of full scientific certainty shall not be used as a reason for postponing cost-effective measures to prevent environmental degradation.' Two key words arise from these definitions: they are risk and irreversibility. Actions which impact on the earth's atmosphere causing global warming may not be irreversible in the foreseeable future. In contrast, although the consequences of degradation of the water environment may be severe in the short term, they are not irreversible. Note the clean-up of the Great Lakes in North America, and the rebirth of the dead rivers in the United Kingdom. So in terms of water, perhaps we should be more concerned with the time frames for reversibility and the costs of allowing deterioration of the water environment and water assets. Cost–benefit analysis is used in decision making. Primarily it is used by businesses[8] to consider whether an investment is sound or to provide a basis for comparing investment options. It is also used by governments and is valuable in comparing options within relatively short business-type horizons. It considers future costs to be less important and uses a discount factor to calculate total costs and benefits over time. To a degree this is sensible due to the greater uncertainties in predicting future costs and benefits, but this poses risks and brings issues back to the precautionary principal. Risk is associated with a 'stitch in time saves nine' in which, for example, delayed maintenance of water and sanitation supply systems can result in expensive refurbishment in the future and pose risks to health. This relates to the folly of not establishing sustainable cost recovery in that cost savings for the current generation are to the detriment of future generations.

Risk management is used widely in water to determine priorities for investment and management focus. It is the foundation of water safety planning which is the subject of Chapter 23. Equivalent thinking is now being applied to sanitation with an approach[9] being trialled in Portugal. Safety planning has to be considered alongside standards and regulation, but is vital in establishing operational management procedures and operator training.

In environmental law, the polluter pays principle is enacted to make the party responsible for producing pollution responsible for paying for the damage done to the natural environment. It came about when governments, often under pressure from environmental groups, were required to think how to control pollution rather than to pay for its clean-up. There had been the mistaken belief that imposing such costs on the manufacturing industry would adversely affect the economy, whereas today generally there is recognition that a polluted environment is detrimental to progress. In practice, manufacturers have realised that recovery of valuable substances, as opposed to 'dumping' them, is sound economics. Dealing with historical pollution is often difficult as the polluters may be difficult to identify, or may no longer be in business. In the United States the 'Superfund' Law required polluters to pay but, if not locatable, provides funds for clean-up. The polluter pays principle is also known as 'extended producer responsibility' with far greater reach. OECD describe[10] it as

> a concept where manufacturers and importers of products should bear a significant degree of responsibility for the environmental impacts of their products throughout the product life-cycle, including upstream impacts inherent in the selection of materials for the products, impacts from manufacturers' production process itself, and downstream impacts from the use and disposal of the products. Producers accept their responsibility when designing their products to minimise life-cycle environmental impacts, and when accepting legal, physical or socio-economic responsibility for environmental impacts that cannot be eliminated by design.

Polluter pays sits along environmental standards. If the discharge of specific substances is banned, as in European law, the potential polluter pays through the cost of the processes used to remove the substances before discharge. If a polluter does discharge such substances illegally there will be a prosecution and associated penalties, and perhaps loss of licence to operate. In cases where discharges are allowed into a sewer, but increase the organic load on sewage treatment works, there are formulae for levying increased charges to cover the increased wastewater treatment costs. In the UK the Mogden[11] formula is used, which takes into account the increase in both oxygen demand and suspended solids. A wastewater company may choose that the effluent is treated on site, with the 'polluter' having the choice of making that investment or paying the wastewater company to install and manage the treatment processes. In a regulated environment the 'polluter' pays in one way or another.

Importance of information, communication and transparency

The ability to establish sound policies, to regulate and to operate efficiently is highly dependent upon the quality of information. Service providers require the most detailed information. Much of it has to be linked with physical locations in order to focus resources on priorities related to operational risks, efficiency and health safety. It is highly beneficial to be able to compare performance and methods to learn from 'best in class'. This requires

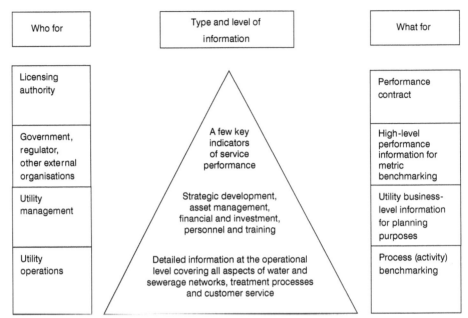

Figure 41.2 Hierarchy of information

Source: Michael J. Rouse.

'standard' performance measures. There are web-based guidance/comparison networks such as IBNET[12] on benchmarking and IWA Water Safety Portal[13] on water safety planning.

The information requirements of regulators and policy makers are integrations of the raw operational data, and financial information. Figure 41.2 illustrates the information relationships.

Modern data handling systems such as management information systems (MIS), geographic information systems (GIS) and web-based search systems make information handling easier, but it is critical that the data displayed is accurate and auditable. There is the risk that the 'gloss' of computer-based presentation systems give greater credence to the data being presented. Data sources should always be questioned and verified.

Another important customer for information is the general public. Such data has to be understandable in lay terms, honest and not misleading. This requires integrity and skills of communicating often complex issues in plain language. Chapter 50 'Information on Water and Health' provides essential advice on informing the public.

Openness in reporting is part of the essential transparency[3] to gain the trust and support of the general public whether that be in relation to financial matters, or drinking water quality data, or in the reporting of incidents which are a risk to health. The public have greatest concern if they feel that information is being withheld. Open reporting of problems and what is being done to manage events provides confidence in the authorities.

The chapters

The chapters in this part cover the enabling elements of policies and interventions related to health of humans, not specifically to the water environment, although they are closely related. To give structure, the part is sub-divided into different aspects and there are many interconnections between those elements. For example, although the human rights framework is in the aspect 'Managing policies and regulations' it is also very much part of international policy.

Within the part there is recognition of the balance to be made between the use of economic instruments and human rights considerations. The chapter on demand driven approaches recognises that the needs of the poor must be met, and equally the chapter on the human rights framework recognises the need for cost recovery for sustainability. The challenge is in the use of tariff systems and subsidies which achieve affordability for all yet do not discourage a drive for greater quality and efficiency of the services provided.

Regulations have an important role in giving definition to policy, especially on quality aspects, and in providing the basis for monitoring. With the emphasis of the book on water and health, there are chapters covering all aspects of human contact with water, both water for consumption and water for recreation.

In addition to the hard aspects of regulation, the social aspects are critical. Public consultation and information provides for understanding and trust, and facilitates recognition of necessary cost recovery and associated service charges. That trust is enhanced by the knowledge of the application of health impact assessments. In developing countries, gender issues, in particular, require specific attention, with recognition of menstrual hygiene management requirements of women and its association with girls' education opportunities. All of these aspects are embraced in implementation of human rights. The UN guidance on implementation provides definition to the key elements of the availability, safety and affordability. This part provides the tools for policies and their implementation.

References

1. UN General Assembly (2010) Resolution 64/292. The human right to water and sanitation. 3 August.
2. Outcomes of a Meeting of Senior Finance Ministry Officials to Discuss Decision-making for WaSH. UNC Water Institute Policy Brief May 2013
3. Rouse, M. J. (2013) *Institutional Governance and Regulation of Water Services: The Essential Elements*, second edition. London: IWA Publishing.
4. Novotny, V. and Brown, P. (eds) (2007) *Cities of the Future: Towards Integrated Sustainable Water and Landscape Management*. London: IWA Publishing.
5. Christiansen, S. B. (1994). 'The precautionary principle in Germany: enabling government'. In Tim O'Riordan and James Cameron (eds) *Interpreting the Precautionary Principle*, London: Earthscan.
6. Arrow, K. J. and Fischer, A. C. (1974). 'Environmental preservation, uncertainty and irreversibility'. *Quarterly Journal of Economics* 88 (2): 312–319.
7. UNEP (1992) The Rio Declaration on Environment and Development. Agenda 21 United Nations Conference on Environment and Development. New York: UNEP.
8. Cellini, S. R. and Kee, J. E. (2010) 'Cost-effectiveness and cost-benefit analysis'. In Kathryn E Newcomer (ed.) *Handbook of Program Evaluation*. Upper Sadle River, NJ: Wiley.
9. IWA (2014) Sanitation Safety Plans, Water 21 International Water Association, December.
10. OECD (2006) 'Extended producer responsibility'. Project Fact Sheet. Paris: OECD Environment Directorate.
11. Merrett, S. (2005) *The Price of Water: Studies in Water Resource Economics and Management*. London: IWA.
12. IBNET (n.d.) 'The international benchmarking network for water and sanitation utilities' http://www.ib-net.org/
13. IWA (n.d.) 'Water safety portal'. www.wsportal.org/wspmanual

42

INTEGRATED WATER RESOURCES MANAGEMENT[*]

Kebreab Ghebremichael, Jochen Eckart, Krishna Khatri and Kalanithy Vairavamoorthy

PATEL COLLEGE OF GLOBAL SUSTAINABILITY, UNIVERSITY OF SOUTH FLORIDA, USA

Learning objectives

1 Describe the concept and principles of Integrated Water Resources Management (IWRM).
2 Describe the planning process for IWRM.
3 Analyze how IWRM enhances public and environmental health.

Introduction

In the face of climate change, economic growth, growing demand for water, and diminishing availability of fresh water sources, many regions are becoming increasingly challenged to achieve sustainable water resources management. For example, the current average available annual per capita fresh water resources of 750 m^3 is expected to reduce to about 450 m^3 by 2050, even without taking into account climate change effects (UNESCO-WWAP, 2006). This will result in more than 80 percent of the countries worldwide to be water stressed.[1] As the challenge of water scarcity is becoming more prevalent, we need to look at effective ways of allocating and using our limited water resources. As a result water resources management practices have been undergoing a change moving from a mainly supply-oriented, engineering based approach to towards a more comprehensive demand-oriented, multi-sectoral, and sustainability focused approach, often labeled as Integrated Water Resources Management (IWRM).

IWRM strives to address the concerns of conventional water resources management by considering social, economic, and environmental aspects as well as quantitative,

---

[*] Recommended citation: Ghebremichael, K., Eckart, J., Khatri, K. and Vairavamoorthy, K. 2015. 'Integrated Water Resources Management', in Bartram, J., with Baum, R., Coclanis, P.A., Gute, D. M., Kay, D., McFadyen, S., Pond, K., Robertson, W. and Rouse, M.J. (eds) *Routledge Handbook of Water and Health*. London and New York: Routledge.

qualitative, supply, and demand issues. The integrated approach ensures water resources are well managed across different water users considering the different dimensions of entire hydrological cycle, water users, spatial scale, and temporal scale. The IWRM approach has received wider acceptance internationally as the way forward for efficient, equitable, and sustainable development and management of water resources and to cope with conflicting demands of various users.

This chapter will discuss the concept of IWRM, its principles, framework for implementation, and best practice examples and its challenges. It will also highlight how the IWRM approach could be beneficial to enhance public and environmental health.

Concept and principles of IWRM

There are different definitions of IWRM in the literature; however, the most commonly used is the definition proposed by the Global Water Partnership (Cap-Net, GWP and UNDP 2005):

> ...a process which promotes the coordinated development and management of water, land and related resources in order to maximize economic and social welfare in an equitable manner without compromising the sustainability of vital ecosystems.

A similar definition by USACE states that,

> IWRM aims to develop and manage water, land, and related resources, while considering multiple viewpoints of how water should be managed (i.e. planned, designed and constructed, managed, evaluated, and regulated). It is a goal-directed process for controlling the development and use of river, lake, ocean, wetland, and other water assets in ways that integrate and balance stakeholder interests, objectives, and desired outcomes across levels of governance and water sectors for the sustainable use of the earth's resources.
>
> *(USACE, 2010)*

These definitions recognize that IWRM is a coordinated and participatory process that acknowledges the entire water cycle with all its natural aspects and brings together stakeholders and focuses on economic development, equity, and ecosystem protection. IWRM offers a guiding conceptual framework with a goal of sustainable management of water resources. It seeks integration between *natural systems* (fresh and salt water; surface and ground water), between *human systems* (institutions, policy, governance), and *across different water users* (domestic, industry, agriculture, ecosystem) (GWP, 2000). For example, integration in terms of policy would imply that water resources policy must be integrated with national economic and sectoral policies. Conversely, economic and social policies need to take into account water resources implications. Cross-sectoral integration is an overarching component that ties the different water use sectors together.

IWRM aims to achieve three key strategic objectives in water resources management that include efficiency in water use, equity, and environmental sustainability. Efficiency in water use recognizes that water resource is finite and vulnerable and that we need to improve water use efficiency to meet the increasing demand. Equity emphasizes the basic rights of people to have access to water of adequate quantity and quality for the sustenance of human well-being. Similarly, environmental sustainability acknowledges aquatic ecosystems as one of the water users and requires that adequate water (quantity and quality) is allocated to sustain their natural functioning. IWRM does not provide a specific blueprint for a given water management problem, but rather it is a broad set of principles, tools, and guidelines which must be tailored to the specific context.

The concept of IWRM evolved as an outcome of continuous research feedback, national and international dialogues, conferences, and United Nations (UN) summits. IWRM is based on four guiding principles that were outlined during the 1992 International Conference on Water and the Environment in Dublin, Ireland (ICWE, 1992). The four principles state that:

1 Fresh water is a finite and vulnerable resource, essential to sustain life, development, and the environment.
2 Water development and management should be based on a participatory approach, involving users, planners, and policymakers at all levels.
3 Women play a central part in the provision, management, and safeguarding of water.
4 Water has an economic value in all its competing uses and should be recognized as an economic good.

The fourth principle was highly debated and opposed by water professionals, particularly from the developing world that argued no water development initiatives could be sustainable if water is considered an economic good without considering the issues of equity and poverty. This issue is being addressed by declaring water as a human right. For example, in 2010, the UN recognized the human right to water and sanitation and called upon states and organizations to provide financial support and build capacity to provide safe drinking water (and sanitation) for all (UN, 2010).

Successful implementation of IWRM requires a framework with the necessary tools and instruments. GWP has developed a framework that includes three components: enabling environment, institutional roles, and management instruments (GWP, 2000).

Enabling environment, supported by the right policies and legislations, provides mechanisms for stakeholders' engagement and allows them to play their respective roles in the development and management of water resources. The enabling environment will require both top-down and bottom-up approaches to ensure successful stakeholder engagement from the national level down to the village or municipality. *Institutional roles* involve the creation of effective coordination mechanisms between different agencies. It focuses on developing an organizational framework and building institutional capacity. Finally, an IWRM framework should be supported by *management instruments* that include tools and methods that enable and help decision makers apply rational and informed decisions. GWP has developed an IWRM toolbox that contains the three components of the framework that include 59 tools, which serve as guidelines for implementing IWRM.

IWRM planning cycle

IWRM requires a well-crafted planning process to develop long-term actions. The IWRM planning cycle is process-oriented and follows a logical sequence of iterative phases that is driven by continuous management support and consultation. The main phases of the planning cycle are shown in Figure 42.1 and include: Initiation, Work plan, Vision and policy, Situation analysis, Strategy development, IWRM plan, Implementation, and Evaluation.

The planning cycle is often initiated by internal or external drivers, such as international agreements, pollution concerns, water scarcity, and conflicting demands and it requires

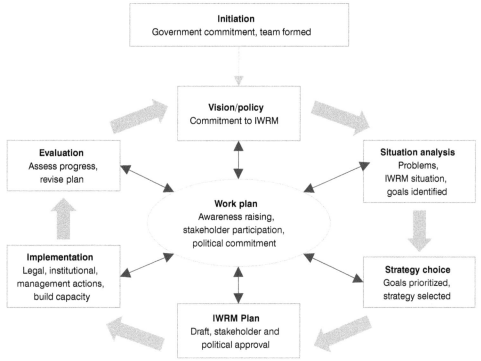

Figure 42.1 IWRM planning cycle

Source: Cap-Net, GWP and UNDP (2005).

government endorsement and commitment. A long-term strategic plan is developed that takes into consideration the outcome of situation analysis. The process results in a plan that informs stakeholders and decision makers on sustainable water resources management. The plan is continuously refined through periodic evaluation and reformulation.

Role of IWRM in enhancing public and environmental health

IWRM is not only about managing water resource from a quantity perspective but, also, it is about improving the water quality. In this respect, IWRM plays an important role in maintaining the objectives of public and environmental health. There are several ways by which IWRM contributes to these objectives.

- IWRM uses watershed as a unit of management that is key to balancing upstream and downstream water related interests. This is important in reducing pollution and over-abstraction by upstream users and in ensuring availability of sufficient water of good quality for downstream users. It is also important in maintaining minimum environmental flow requirements for a healthy ecosystem that provides needed ecosystem services for the well-being of societies and reduces remediation or treatment costs.
- IWRM aims to ensure equity in terms of access to water sources by all social groups including the poor. It recognizes the need for adequate quantity and quality of water for the sustenance of healthy communities (Hunter et al., 2010).

- In most developing countries women are critical in water management and maintaining the health of families. IWRM recognizes this and strives to empower women in water management and operation. By doing so, it helps to promote public health aspects at household and community levels.
- In urban areas an integrated approach to water management helps to identify negative interactions between water supply, wastewater, and drainage systems and helps reduce potential cross-contamination between water supply and sewer systems. Additionally, an integrated approach incorporates catchment protection into the management practice and helps to exploit reuse potential of wastewater and stormwater, thus reducing the pollution risks of water sources and enhancing environmental and public health.

IWRM examples and challenges of implementation

During the past 15 years or so, IWRM has been heavily promoted to solve water problems all over the globe. The challenges and solutions vary according to variability in physical, economic, social, and environmental conditions. Most of the successful case studies set out to solve particular water-related problems or achieve development goals by embracing a holistic approach within larger physical and development contexts. Some examples that embrace components of IWRM include large projects such as: California's Integrated Regional Water Management Plan, the IWRM plan for management of the transboundary Guarani Aquifer System in South America, and the IWRM project in Guiyang in China.

The California Integrated Regional Water Management is designed to address the mismatch of the state's water supply and demand where most of the water sources are in the northern part of the state while much of the water demand is concentrated in the south. In 2002 California's Integrated Regional Water Management Act set forth a new way of thinking about water by authorizing the development of Integrated Regional Water Management plans (IRWMPs) to increase collaboration between local agencies. The Act is guided by overarching principles at the state level, but includes the flexibility to develop a bottom-up, truly local approach to water management. Today, IRWM regions encompass 87 percent of the area of the state, and about 99 percent of the population is represented by a planning region. Some of the key strategies that have been introduced within the plan include: reducing water demand by improving efficiency for water uses, improving operational efficiency and water transfer in the region, increasing water supply through different options including conjunctive surface and groundwater management, desalination, precipitation enhancement, recycled municipal water, and enhancing resources stewardship (AWRA, 2012). While this plan has helped to improve water management, integration and holistic management is evolving slowly due to some challenges such as the issue of insufficient and uncoordinated information storage and exchange.

The IWRM project in Guiyang, China, is another example. Guiyang is a major industrial hub for Guizhou province with a population of about 4 million, as well as an ecotourism destination in China. Guiyang does not have enough water storage capacity to satisfy the demand of its growing population and economy where only 33 percent of rainfall and runoff in the area can be captured and stored because of variation in rainfall patterns, its mountainous terrain, karst deposits, and high evaporation. In order to address this challenge, Guiyang is applying the country's first large IWRM project at the municipal level. With a $150m million loan from the Asian Development Bank, the project will improve and build urban water supply, reservoirs, dams, transmission pipelines, and a treatment plant. In the rural areas, over 40 small reservoirs will be established, dilapidated irrigation systems will be restored, and soil conservation measures applied. At the farm level, over 100,000 small water storage tanks will be built to collect spring and rainwater. An innovative feature of the project is the greater community involvement in water resources management and the introduction of a payment for the ecosystem services scheme (ADB, 2014).

It is hard to disagree with the concept of IWRM; however, translating this concept into practice is definitely a challenge. Implementation has been hindered by the lack of a consistent, operational definition with measurable criteria that has resulted in slow uptake of the approach despite the noted benefits (AWRA, 2012). Complex political, social, and physical factors also make its practice difficult, which requires tailored approaches to each specific condition and challenge. Some believe that the interpretation of the IWRM concept and its implementation are issues for its slow uptake and, in some cases, for failure (Biswas, 2008). The key to successful IWRM implementation is the integration of local resources in the local context. But integration requires that the ground conditions, such as institutions, government, policies, and commitment from stakeholders, are in place and conducive. GWP acknowledges the challenges by stating that "IWRM is a challenge to conventional practices, attitudes and professional certainties. It confronts entrenched sectoral interests and requires that the water resource is managed holistically for the benefits of all." However, GWP also states that although addressing the challenges will not be easy, "it is vital that a start is made now to avert the burgeoning crisis" (Cap-Nat GWP amd UNDP, 2005).

Indeed, it is important that first steps are taken to implement IWRM and more refinements and adjustments are made based on lessons learned from successful and failing experiences.

Key recommended readings

1 Rahaman, M.M. and Varis, O. (2005). Integrated Water Resources Management: evolution, prospects and future challenges, *Sustainability: Science, Practice and Policy*, 1, 1. http://sspp.proquest.com/archives/vol1iss1/0407-03.rahaman.html

2 Biswas, A. (2008). Integrated Water Resources Management: is it working? *Water Resources Development*, 24(1), 5–22.

3 GWP (2000). Integrated Water Resources Management, Technical Advisory Committee (TAC) Background paper No. 4. http://dlc.dlib.indiana.edu/dlc/bitstream/handle/10535/4986/TACNO4.PDF?sequence=1

Note

1 The commonly used definition ('water stress index', or 'Falkenmark indicator') describes water scarcity or water stress based on the amount of renewable freshwater that is available per person per. A country or region can be: water stressed if the amount is below 1,700 m^3; water scarce if the amount is 1,000 m^3; and absolute water scarce if the amount falls below 500 m^3.

References

ADB – Asian Development Bank (2014) ADB report series. http://www.adb.org/search?keyword=38594&id=38594

AWRA (American Water Resources Association) (2012) Case Studies in Integrated Water Resources Management: From Local Stewardship to National Vision. http://www.awra.org/committees/AWRA-Case-Studies-IWRM.pdf (retrieved on May 29, 2014).

Biswas, A. (2008) Integrated Water Resources Management: is it working? *Water Resources Development*, 24(1), 5–22.

Cap-Net, GWP and UNDP (2005) *Integrated Water Resources Management Plans: Training Manual and Operational Guide.* http://www.cap-net.org/documents/2014/06/iwrmp-training-manual-and-operational-guide.pdf

GWP (2000) Integrated Water Resources Management, Technical Advisory Committee (TAC) Background Paper No. 4. http://dlc.dlib.indiana.edu/dlc/bitstream/handle/10535/4986/TACNO4.PDF?sequence=1

Hunter, P.R., MacDonald, A.M. and Carter, R.C. (2010) Water supply and health. *PLoS Med* 7(11). e1000361. doi:10.1371/journal.pmed.1000361

ICWE (1992) The Dublin statement and report of the conference. International Conference on Water and the Environment: Development Issues for the 21st Century, 26–31 January, Dublin.

UN (2010) Resolution Adopted by the General Assembly on 28 July 2010, A/RES/64/292, Sixty Fourth Session, agenda item 48. http://www.un.org/en/ga/64/resolutions.shtml

UNESCO-WWAP (2006) *Water: A Shared Responsibility.* The UN World Water Development Report 2. http://www.unesco.org/bpi/wwap/press/

USACE (United States Army Corps of Engineers) (2010) *National Report: Responding to National Water Resources Challenges, Building Strong Collaborative Relationships for a Sustainable Water Resources Future.* Washington, D.C. http://www.building-collaboration-for-water.org/ Documents/nationalreport_final.pdf

43

INTERNATIONAL POLICY*

Jamie Bartram

THE WATER INSTITUTE, UNIVERSITY OF NORTH CAROLINA
AT CHAPEL HILL, CHAPEL HILL, NC, USA

Georgia Kayser

THE WATER INSTITUTE, UNIVERSITY OF NORTH CAROLINA
AT CHAPEL HILL, CHAPEL HILL, NC, USA

Bruce Gordon

WORLD HEALTH ORGANIZATION, GENEVA, SWITZERLAND

Felix Dodds

THE WATER INSTITUTE, UNIVERSITY OF NORTH CAROLINA
AT CHAPEL HILL, CHAPEL HILL, NC, USA

Learning objectives

1 Provide a brief history of international water and health policy.
2 Identify key international water and health policy actors and institutions.
3 Discuss the drivers of policy development and change.

International water and health policy derives from a series of influences including human development/poverty alleviation, sustainable development, health and human rights constituencies. International policy on water and health is dominated by bilateral and regional arrangements. The first of these influences is principally manifest through a series of decades focused on human development and similar initiatives that began in the 1960s and most recently took the form of the Millennium Development Goals (MDGs); and the second

* Recommended citation: Bartram, J., Kayser, G., Gordon, B. and Dodds, F. 2015. 'International policy', in Bartram, J., with Baum, R., Coclanis, P.A., Gute, D.M., Kay, D., McFadyen, S., Pond, K., Robertson, W. and Rouse, M.J. (eds) *Routledge Handbook of Water and Health*. London and New York: Routledge.

since the 1970s through sustainable development conferences and summits, the Bruntland report, and the UN Commission on Sustainable Development (CSD). The third has been principally under the aegis of the World Health Organization (WHO). Most recently, recognition of the human right to water and sanitation has provided a complementary policy perspective. Each of these brings a different outlook, with distinct values. Current negotiations towards proposed Sustainable Development Goals have the potential, and to some extent reflect the intent, to bring these perspectives together.

Water presents specific challenges to policy in general and international policy in particular, in part because water management spans ministerial, sectoral and disciplinary barriers; while *health* policy is divided between health services – which tend to be consolidated in a single ministry (the WHO at global level); while wider responsibility for and impact on public health also spans conventional separations, whether its oversight is consolidated into an authority charged to oversee public health or otherwise. Even though health benefits from the prevented diseases and averted health care costs arising from sound policies and actions on water, there is little collective planning between health and water sectors, either at national level or in international policy. Beyond setting standards for the safety of drinking-water or running hygiene campaigns, ministries of health rarely engage substantively on preventive environmental actions to reduce water-related disease. In consequence neither national nor international policy on *water and health* has a single core "node" but is found in policies concerned with development, water, health, environment, human rights and elsewhere such as consumer protection. The one exception is the UNECE-WHO/Europe Protocol on Water and Health, adopted in 1999.

International policy on water and health has spanned a period of great change: since 1970, the global population has nearly doubled and 54% of the world's population lives in urban areas (United Nations, 2013; United Nations, DESA, 2014):

> The number of people using improved sources of drinking water expanded from 2.4 billion in 1970 to 6.2 billion in 2012 (64% to 89% coverage), while the number using basic sanitation increased from 1.3 to 4.4 billion (36% to 64%). More than half of the world's population now gets water from a piped source in the home.
>
> *(WHO and UNICEF, 2014)*

Today, the population is increasingly dominated by an urban existence, a longer lifespan and improved access to basic services. Similarly, ongoing climate change; problems associated with water availability and quality; as well as disease emergence are garnering attention from policy makers at all levels, driving policy responses related to establishing resilient water and health systems. Diarrheal disease mortality declined between 2000 and 2015 and some of the causes may include: water and sanitation infrastructure, improvements in case detection and treatment, water governance, demographic change, and improvements in health and development.

For international policy and governance purposes the subject matter of water spans from international water courses and international lakes (1992 UN Convention on the Protection and Use of Transboundary Watercourses and International Lakes) and international trade in water whether directly as a commodity (bottled water) or as "virtual water" (e.g., food, clothing); to water for domestic purposes – intrinsically a local/national issue but the subject of policy from human development and human rights perspectives. Figure 43.1 provides timelines of international policy in water and health divided into thematic areas – development, sustainable development, health and human rights.

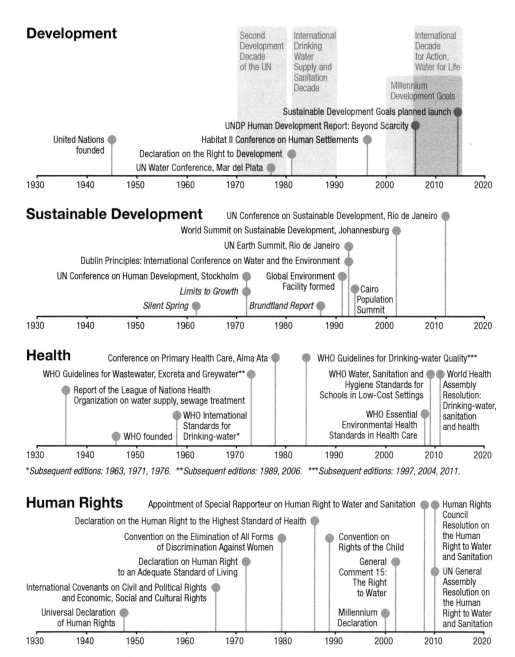

Development

Second Development Decade of the UN

International Drinking Water Supply and Sanitation Decade

International Decade for Action, Water for Life

Millennium Development Goals

Sustainable Development Goals planned launch ●

UNDP Human Development Report: Beyond Scarcity ●

United Nations founded ●

Habitat II Conference on Human Settlements ●

Declaration on the Right to Development ●

UN Water Conference, Mar del Plata ●

1930   1940   1950   1960   1970   1980   1990   2000   2010   2020

Sustainable Development

UN Conference on Sustainable Development, Rio de Janeiro ●

World Summit on Sustainable Development, Johannesburg ●

UN Earth Summit, Rio de Janeiro ●

Dublin Principles: International Conference on Water and the Environment ●

UN Conference on Human Development, Stockholm ●

Global Environment Facility formed ●

Limits to Growth ●

● Cairo Population Summit

Silent Spring ●

Brundtland Report ●

1930   1940   1950   1960   1970   1980   1990   2000   2010   2020

Health

Conference on Primary Health Care, Alma Ata ●

● WHO Guidelines for Drinking-water Quality***

WHO Guidelines for Wastewater, Excreta and Greywater** ●

WHO Water, Sanitation and ● ● World Health Hygiene Standards for Assembly Schools in Low-Cost Settings Resolution: Drinking-water, sanitation and health

● Report of the League of Nations Health Organization on water supply, sewage treatment

● WHO International Standards for Drinking-water*

WHO Essential ● Environmental Health Standards in Health Care

● WHO founded

1930   1940   1950   1960   1970   1980   1990   2000   2010   2020

*Subsequent editions: 1963, 1971, 1976. **Subsequent editions: 1989, 2006. ***Subsequent editions: 1997, 2004, 2011.*

Human Rights

Appointment of Special Rapporteur on Human Right to Water and Sanitation ● ● Human Rights Council Resolution on the Human Right to Water and Sanitation

Declaration on the Human Right to the Highest Standard of Health ●

Convention on the Elimination of All Forms ● of Discrimination Against Women

● Convention on Rights of the Child

Declaration on Human Right ● to an Adequate Standard of Living

General ● Comment 15: The Right to Water

● UN General Assembly Resolution on the Human Right to Water and Sanitation

International Covenants on Civil and Political Rights ● and Economic, Social and Cultural Rights

Universal Declaration ● of Human Rights

Millennium ● Declaration

1930   1940   1950   1960   1970   1980   1990   2000   2010   2020

Figure 43.1 Water and health in development policy timeline

Water-and-health in development policy

Water (or "drinking-water", meaning water for domestic purposes), often combined with sanitation and sometimes hygiene, has been a component of international development policy since 1936, with the Report of the League of Nations Health Organization (the predecessor of the WHO) on water supply and sewage treatment under its rural hygiene programme. This coincided with the LNHO emphasis on "social medicine" (i.e., public health) in the 1930s.

In the 1970s, water was reflected in the Second Development Decade of the United Nations (UN): "each developing country will endeavor to provide an adequate supply of potable water to a specified proportion of its population, both urban and rural, with a view to reaching a minimum target by the end of the Decade" (UN, 1970).

In 1977 at the UN Water Conference at Mar del Plata, Argentina, it was resolved that "... member states should be encouraged to prepare programmes to provide all people with water of safe quality and in adequate quantity and basic sanitary facilities by 1990, according priority to the poor and less privileged" (Anon, 1977). The conference initiated the International Drinking Water Supply and Sanitation Decade (IDWSSD or "Water Decade", 1981–1990), with the declared objective to achieve "Substantial improvement in ... drinking water and sanitation by 1990." In 1983, the World Bank took the operational role of implementing drinking water supply programs from WHO and irrigation development from the Food and Agriculture Organization (FAO). This switched the focus from water for health to water for development The Water Decade coincided with the third UN Development Decade and was associated with the Mar del Plata Action Plan which improved awareness of the problem around the lack of access to basic water and sanitation and increased the allocation of resources to the water sector (Biswas, 2004). It has been linked to the establishment of national water and river basin authorities as well (Falkenmark, 1997). Despite this, it has been criticised for providing a road map that was not followed (Falkenmark, 1997).

The New Delhi Global Consultation on Safe Drinking Water and Sanitation for the 1990s, which concluded the Water Decade, provided governments, international organizations (IOs), non-government organizations (NGOs) and multilateral agencies opportunity to reflect on lessons learned. They included the need to:

- Focus on poverty, especially serving the un-served.
- Build capacity, the responsibility of the government.
- Meet demand by understanding what people want and are willing to pay for.
- Share costs and provide appropriate pricing as a means to improve sector performance.
- Technical innovation to provide a range of options to meet demand.
- Focus on women, the primary water users and caretakers of the sick.
- Monitor the extended coverage and create achievable goals that can be tracked.
- Coordinate between organizations to build national and international collaborative networks (Anon, 1990).

All of these have subsequently been addressed in some national activities. This is especially the case with meeting demand, sharing cost and technical innovation, reflected in the trend towards a demand-driven approach (Gibbons et al., 1996). The need to emphasize the role of women has also been a focus of development work in the water, sanitation and hygiene (WaSH) (Sims, 1994; IRC, 1995). Less activity occurred around the role of government, monitoring and national coordination, until recently (Kayser et al., 2013).

In 1990, the World Summit for Children, with its associated Declaration and Plan of Action (UN, 1990), called for universal access to safe drinking water (and sanitary excreta disposal) by 2000. The Summit resulted in ratification by 192 countries of the Convention on the Rights of the Child.

When first formulated in the Millennium Declaration, adopted by world leaders in 2000, a target was "To halve, by the year 2015, the proportion of people who are unable to reach or to afford safe drinking water" (UN, 2000). This reflected a continuation of the rate of decline in "unserved" populations that had been achieved in the preceding period. The target wording was repeatedly modified (Bartram et al., 2014), until adopted in its final form as Target 7C: to halve, between 1990 and 2015, "the proportion of the population without sustainable access to safe drinking water and basic sanitation". A perception (Cumming et al., 2014) that progress in sanitation was lagging drinking-water led to both the addition at the World Summit on Sustainable Development in 2002 of sanitation to the initially water-only MDG target and to a series of regional sanitation conferences in the period 2000–2015, which catalyzed the political will to spend and achieve more on sanitation. Despite the ongoing MDG process, the policy importance of water was further confirmed through the announcement of the Decade 2006–15 as the International Decade for Action, "Water for Life".

The 2006 UNDP Human Development Report identified some of the shortcomings of the past and contributed to creating a space for Sanitation and Water for All (SWA) and the UN-Water global analysis and assessment of sanitation and drinking-water (GLAAS). The SWA alliance was formed in 2009. It is a partnership of 90 developing country governments, donors, civil society organizations and development partners that works to improve leadership, action and accountability for improving access to safe water and sanitation. The objective of GLAAS is "to monitor the inputs required to extend and sustain water, sanitation and hygiene systems and services" (UN-Water, 2012).

Over eight decades, much action has been driven by a need to meet international targets centered on increasing the number of people with access to water and sanitation. This led to impacts in terms of number served – for example two billion people gained access to an improved source of drinking water between 1990 and 2015 – but this focus has led to insufficient attention to sustaining services, improving their quality and pursuing equity.

The sustainable development movement

The roots of sustainable development may be traced to rising concerns about the impact of human activities on the environment and planet in the 1960s and 1970s. Influential publications which reflected emerging concerns in the industrialized nations and captured public imagination included Rachel Carson's *Silent Spring* (1962), *Limits to Growth* (Meadows et al., 1972), and the work of Lester Brown.

The warnings about pesticide use and ecosystem destruction in Rachel Carson's book *Silent Spring* sparked concern about water quality among other issues, and launched the environmental movement in the USA. *Limits to Growth* (Meadows et al., 1972) modeled the interactions among population, industrialization, pollution, food production and resource depletion and their consequences on the earth and human systems, projecting scenarios for future development that varied between "overshoot and collapse" of the global system in the 21st century and a "stabilized world". In *The Twenty-ninth Day* (1978), Brown warned of "the various dangers arising out of our manhandling of nature ... by overfishing the oceans, stripping the forests, turning land into desert."

On the international policy stage, these concerns were first captured in the UN Conference on the Human Environment in 1972 in Stockholm. The conference, attended by 114 states, approved the Stockholm declaration – 26 principles on the management of the global environment – and the WHO was recommended to support governments in achieving the principles, especially those related to health and water and sanitation. It was the precursor to the UN's third development decade (1980–1990) and responsible for the formation of the United Nations Environment Programme (UNEP).

The maturation of these concerns into the notion of "sustainable development", with its three pillars of economic, social and environmental sustainability, was advanced in *Our Common Future* (World Commission on Environment and Development, 1987), commonly referred to as the Brundtland report after its chair, former Prime Minister of Norway, Dr Gro Harlem Brundtland, and later Director-General of WHO. The report coined a definition of sustainable development, "development that meets the needs of the present without compromising the ability of future generations to meet their own needs", that has endured.

The 1990s had been the decade of UN conferences and summits starting with the Children's Summit (1990), followed by the Rio Earth Summit (1992), the Cairo Population Conference (1993), the Social Summit (1994), the Beijing Women's Summit (1995), the Istanbul Habitat II Conference (1996) and the Food Summit (1996). Each of these created action plans that mention water. By 2000, however, governments were having problems in their implementation. The Millennium Summit in September 2000 tried to simplify this by agreeing on eight MDGs and assorted targets. This would focus the development community, in particular, to try to deliver a small set of goals and targets that could be achieved by 2015.

At national level, the creation of ministries of environment, environmental protection agencies and similar entities often began by carving off the "environmental sanitation" role of health agencies. This had the unintended consequence of distancing health authorities from environmental determinants of health at a time when such concerns were evolving rapidly under the influences of urbanization and industrialization. These new agencies sometimes sought to differentiate themselves by focusing on ecological health, reflected at international level in the Global Environment Facility, which excluded health concerns from its funding decisions, initially. UNEP was strengthened by the 2012 Conference on Sustainable Development. The Commission on Sustainable Development (CSD), replaced by the High Level Political Forum (HLPF), have over the years sought to increase coherence and inter-organizational commitment. Within the field of water alone, UN-Water, the body charged with increasing coherence and coordination of UN system actions on water, has a membership of 27. The fragmentation means that resources for water may not be adequately allocated between sectors and aspects are missed. Another perspective is that water permeates everything and can only be managed in a participatory manner, and not by centralizing it in a single entity.

The distance that emerged between ecological/green and narrowly-defined health perspectives left environmental health largely a local government-driven basic inspectorate. This distance creates a present-day opportunity to reflect and value the health impacts (beneficial and detrimental) of environmental management in decision-making, for example through tools such as environmental impact assessments and health impact assessments. There is a conundrum in that the absence of a lead agency at national or international levels responsible for water and health is perceived as a weakness; yet, the need to integrate measures across ministerial, sectoral and disciplinary boundaries suggests that it may provide an opportunity if new ways of working across traditional boundaries can be further developed. For example, a lead ministry or inter-ministerial coordination mechanism could be proposed with responsibility for integration of water policy issues and integrated planning. Interestingly, the value of environment to health protection and disease prevention is under-appreciated because of the difficulties in valuing the benefits of environment and environmental health services. The result, for example, is in assessment of proposed hydropower dams and irrigation projects, an environmental impact assessment is required; however, rarely is a health impact assessment also conducted (Lerer, 1999).

Water in health policy

The WHO serves as the UN lead agency on health issues and provides the locus of international policy on health. Established in 1948, it replaced the Health Organization of the League of Nations (LNHO), which had its own roots in earlier international agreements and concern for international transmission of disease and quarantine. The WHO constitution defines health as a "state of complete physical, mental and social well-being and not merely the absence of disease or infirmity" and identifies a WHO function as "to promote, in cooperation with other specialized agencies where necessary, the improvement of nutrition, housing, sanitation, recreation, economic or working conditions and other aspects of environmental health" (WHO, 1948).

WHO comprises a governance structure – the World Health Assembly – made up of representatives of each of its member states, which provides policy orientation and adoption of formal measures; and its secretariat – comprising staff who implement its policies and programs by: providing leadership in global health; developing guidelines; reporting on good practices and evidence-based policy options; providing technical country support; and monitoring progress.

Unlike other UN specialized agencies, WHO's decentralized structure provides independent governance for each of its six regions (regional committees, comprising representatives of the corresponding regional states) and their associated regional secretariats.

The normative functions of WHO are especially relevant to the theme of this book. They include establishing guidelines (quasi-standards) and, in collaboration with United Nations Children's Fund (UNICEF) and UN-Water, providing global monitoring related to water.

WHO's guidelines are intended to support governments as they develop national standards and regulations that reflect risk management strategies for hazards of concern to human health. Like the national standards they inform, they may include descriptions of quality (such as tolerable concentrations – guideline values – for chemicals in water) and minimum requirements of safe practices. Current guidelines related to water include: *Guidelines for Safe Drinking Water* (WHO, 2011); *Essential Environmental Health Standards in Health Care* (WHO, 2008); *Guidelines for the Safe Use of Wastewater, Excreta, and Greywater in Agriculture and Aquaculture* (WHO, 2006a); *Water, Sanitation and Hygiene Standards for Schools in Low-Cost Settings* (WHO, 2006b); and *Guidelines for Safe Recreational Water Environments*, volumes 1 and 2 (WHO, 2003 and 2006c).

The importance of water for health and development was also reflected in the International Conference on Primary Health Care (Alma-Ata, USSR (now Kazakhstan)) in 1978 mainstreaming universal basic health care, equity in health services and a focus on disease prevention. However, a subsequent shift towards "selective primary health care" (Walsh and Warren, 1979) marginalized much of primary prevention, including actions on water. Only belatedly was it appreciated that its underlying cost–benefit analysis ascribed all costs of water service provision but accounted only health benefits, making water incorrectly appear a poor investment.

WHO produced landmark reports quantifying the health and economic impacts of inadequate water, sanitation and hygiene. The 2002 World Health Report estimated that 88 percent of all cases of diarrhea were attributable to water, sanitation and hygiene (WHO, 2002). A 2007 report on global costs and benefits of water supply (updated in 2012) demonstrated a US$ 1 of spending has a US$ 5.5 return on every dollar invested(WHO, 2012). Both were widely cited figures that drove home the significant benefits of establishing sound water and health policies at all levels.

World Health Assembly resolution 64.24 (2011), the first WaSH resolution adopted by the Assembly in 20 years, recalled the human right to water and sanitation. It urged Member States to develop national public health strategies to include water and sanitation; and urged strengthening of intersectoral coordination between relevant ministries and stakeholders in order to put in place water and sanitation risk management practices. Finally, it called for improved and expanded national and international monitoring relating to water and health.

The human right to water and sanitation

The 1948 Universal Declaration of Human Rights, states that "[e]veryone has the right to a standard of living adequate for the health and wellbeing of himself and his family" (United Nations, 1948), but makes no direct reference to "water". Rights-based language that specifically references water can be traced to 1977 and the UN Conference on Water in Mar del Plata, Argentina, where it was declared that "all peoples, whatever their stage of development and their social and economic conditions, have the right to have access to drinking water in quantities and of a quality equal to their basic needs". This established the basic parameters for judging the adequacy of water supply services and it is important that emphasis was given to accessibility and equal importance placed upon quality and quantity of water supplies.

State obligations around water were mentioned in the Human Right to an Adequate Standard of Living in 1972, the Human Right to Development in 1986, and the Human Right to the Highest Standard of Health in 1989 (Meier et al., 2013).

The 1992 "Dublin Principles" from the International Conference on Water and the Environment, declared in one of four guiding principles that: "it is vital to recognize first the basic right of all human beings to have access to clean water ... at an affordable price" (Anon, 1992). However, the rights-based vocabulary was not carried forward into the UN Conference on Environment and Development (UNCED or the "Earth Summit", Rio de Janeiro, 1992). While such a right was also referred to in the 2000 Millennium Declaration, human rights in general and the right to water specifically were absent from the MDGs.

More formal and specific international recognition of the human right to water began in 2002 with General Comment 15 of the UN Committee on Economic, Social and Cultural Rights (CESCR, 2002), which further clarified the parameters by which adequacy of drinking water and sanitation was to be judged; and through resolutions in the UN General Assembly (UN, 2010) and Human Rights Council (UN Human Rights Council, 2010).

A human rights perspective on water (Chapter 51) creates an obligation on governments as primary duty bearers. It does not create a right that each individual can demand to be immediately satisfied; rather, a human right creates state obligations to respect, protect and fulfil. *Respect* means taking no action that would adversely affect anyone currently enjoying the right; to *protect* prevents infringement by others; while "fulfil" creates an obligation to "progressively realise" and to do so with the maximum of the means at one's disposal. The concept of "progressive realisation" means that rights are measured by degree and not through a have and have not perspective and that compliance with the right is not measured by country status but by its rate of improvement (Luh et al., 2013). Countries can then be benchmarked against the maximum rate that a country can achieve, allowing for a measure of rate of improvement rather than a level of achievement (Luh et al., 2013).

Importantly, the right has no role in determining the means by which a government may realize it. Thus despite claims that water should be "free" and invocation of rights language in debate surrounding the privatization of water services, a human rights perspective simply means that everyone is entitled to sufficient, safe, accessible water and sanitation without discrimination. The recognition of water as a human right and the adoption of a rights-based approach are neutral as to the means of realization.

Future Sustainable Development Goals

At the time of writing this chapter final political negotiations were ongoing about "Sustainable Development Goals" (SDGs) to supersede the MDGs and the sustainable development framework and with the potential to address human health, international development and human rights concerns.

A series of public consultations, scientific and political processes connected by a loose "theoretical" framework supported the development of SDG proposals but were undertaken separately, with different stakeholder groups and methods of work. Water and health generally featured as central, with strong political support for a target related to universal access to safe drinking-water. The most influential report was that of the "Open Working Group", which was adopted by the UN General Assembly as the main basis for formulating SDGs. Water and health featured in a few of the 17 goals as well as Goal 6, to ensure sustainable management of water and sanitation for all.

Proposed targets under the water goal address improving drinking water, sanitation, hygiene access, quality and quantity, wastewater management, reducing water pollution, increasing water efficiency, implementing integrated water resources management, protecting water ecosystems, safely managing and reusing wastewater and expanding international cooperation and local participation (Table 43.1). Specific indicators to track these targets are still being negotiated. As part of the development process WHO and UNICEF organized working groups and discussions to propose drinking-water and sanitation goals and targets, and analyze the associated monitoring challenges.

One of the concerns raised by governments was the linkages between different goal areas. The German government had addressed this in their Nexus Conference on Water–Energy–Food in 2011. The conference identified a number of drivers that would have an impact on food, water and energy availability. In a background paper for the conference the Stockholm Environment Institute noted that increased urbanization, population and economic growth by 2030 would require an additional demand of 40 percent for energy and between 30 and 50 percent for food resources and that would contribute to a 40 percent shortage of water. This was followed by a Declaration on Water–Energy–Food and Climate produced in 2014 at a conference bearing the same name organized by the Water Institute at the University of North Carolina. It added climate change to the nexus themes because of the impacts that climate change will have on water and food and recognized that water is a major requirement for energy provision. The nexus will be an increasing important area over the next 15–20 years as we try to balance water demand in these different areas.

Table 43.1 Draft targets for the Sustainable Development Goal 6. Ensure availability and sustainable management of water and sanitation for all

Targets	
6.1	By 2030, achieve universal and equitable access to safe and affordable drinking water for all
6.2	By 2030, achieve access to adequate and equitable sanitation and hygiene for all and end open defecation, paying special attention to the needs of women and girls and those in vulnerable situations
6.3	By 2030, improve water quality by reducing pollution, eliminating dumping and minimizing release of hazardous chemicals and materials, halving the proportion of untreated wastewater and increasing recycling and safe reuse by [x]% globally
6.4	By 2030, substantially increase water-use efficiency across all sectors and ensure sustainable withdrawals and supply of freshwater to address water scarcity and substantially reduce the number of people suffering from water scarcity
6.5	By 2030, implement integrated water resources management at all levels, including through transboundary cooperation as appropriate
6.6	By 2020, protect and restore water-related ecosystems, including mountains, forests, wetlands, rivers, aquifers and lakes
6.6a	By 2030, expand international cooperation and capacity-building support to developing countries in water- and sanitation-related activities and programmes, including water harvesting, desalination, water efficiency, wastewater treatment, recycling and reuse technologies
6.6b	Support and strengthen the participation of local communities in improving water and sanitation management

The act of bringing together human development, sustainable development, health and human rights concerns into a unified SDG framework seems at face value positive. To date, however, on water, the discussion is largely combining and not adding value from the combination and reflects the silo approach, characteristic of the MDG period. One example is that sanitation and reduction of environmental contamination were not combined during technical preparations but rather during the political negotiations.

Global monitoring of water

Monitoring has policy relevance if it is both able to identify gaps and opportunities and support refinement of policies and practices; and also by enhancing accountability, encouraging critical comparison and providing evidence for advocacy.

International monitoring concerned with water and health is dominated by three processes: monitoring of the status and trends in drinking water and sanitation coverage (by WHO and UNICEF through their Joint Monitoring Programme (JMP) (Bartram et al., 2014)); monitoring of the associated financing, institutional arrangements, capacities and constraints (through the GLAAS of UN-Water, also implemented by WHO); and global water information system (by FAO's AQUASTAT, and water quality of surface and groundwater by GEMStat).

JMP monitoring is based on responses from household members about their principal water source, collected through censuses and nationally representative household surveys. It provides disaggregated data on different levels of water service. The approach taken is low cost and minimizes duplication of efforts. Despite recognized deficiencies, it is an internationally representative monitoring activity that provides consistent and comparable information. Criticisms of it fall into three groups – omissions, apparent inconsistencies and future relevance. Two important omissions are the failure to reflect water safety and system sustainability – dimensions that require approaches other than household surveys. Criticisms related to apparent inconsistency, with evaluations or assessments made by others largely reflect definitional differences (for example using a higher or lower "benchmark" in determining access; accounting or not accounting for water safety and so on). They also concern the real world complexity of household use of multiple water sources. Future relevance issues overlap in part with these, but also include necessary improvements to the modeling approach taken, especially for the purpose of projection.

GLAAS uses existing reports and country data collected from questionnaires filled out by government officials. The data expands on the national statistics from JMP. The GLAAS report includes information on water and sanitation, policies, monitoring and evaluation, budgeting equity, outputs, sustainability and human resources (UN-Water, 2012).

AQUASTAT is the FAO's geographical water information system that provides data on water resources, water uses, irrigation, dams and water-related institutions. GEMstat is a global water quality database for surface and groundwater quality with 3000 monitoring stations worldwide.

While it may be perceived that monitoring is undertaken in response to policy demands, in the case of international monitoring, the relationship is less simple and bi-directional. The timelines required to establish procedures and baselines make a simple monitoring response to a new policy direction challenging, and in fact the availability of data collection capacities and information on baseline conditions influence the formulation of policies and targets.

Conclusion

There is value in looking "back", as if from some future date to critique present decisions. We can be confident that the world of 2030 will be very different from that of today but less confident about what it will be like. Drivers that appear important today include human population demographics; environmental limits and constraints; greater frequency and intensity of extreme events; ongoing human development; increases in demand for water from larger and wealthier populations; improved local, national and international governance accompanied by concerns for security; and the ongoing fruits of human ingenuity whether in new technology and communications or in access to basic infrastructure and management of service delivery.

Over recent decades the pace of development and mushrooming of governmental, non-governmental, private and social sector actors has led to increasing engagement and coordination mechanisms (such as UN-Water which provides an early and unusual example of UN engagement with "non-traditional partners"). Clearly the need for enhanced coordination is unlikely to decrease and many opportunities for improved performance seem to call for inputs that cross traditional ministerial, sectoral and disciplinary divides.

One might ask whether policy in general and international policy in particular really matters. Some work suggests that the impacts of simple manifestations of policy – such as the relative flow of international aid funding – cannot be detected (Bain et al., 2013). Other work concludes that pace of progress in WaSH may in fact not be primarily determined by gross national income (GNI) per capita, government effectiveness, female education, foreign assistance or inequality (Luh and Bartram, accepted), which suggests that potentially deliberate policy may have a determining impact. It is clear, however, that targets and timetables are critical to international policy implementation (Chasek et al., 2014) and these targets must be implemented, monitored and enforced within sovereign nations so that the health, economic and human rights-related benefits can be realized (Kayser et al., 2013).

Key recommended readings

1   Bartram, J., C. Brocklehurst, M. Fisher, R. Luyendijk, R. Hossain, T. Wardlaw and B. Gordon. (2014). Global monitoring of water supply and sanitation: history, methods and future challenges. *Int. J. Environ. Res. Public Health* 11, 8137–8165; doi:10.3390/ijerph110808137. This paper describes and reviews the methods used by the JMP to track access and progress to water and sanitation, globally.
2   Kayser, G.L., P. Moriarty, C. Fonseca and J. Bartram. (2013). Domestic water service delivery indicators and frameworks for monitoring, evaluation, policy & planning: a review. *International Journal of Environmental Research and Public Health* 10(10), 4812–4835; doi:10.3390/ijerph10104812. This article provides a review of indicators and frameworks used in WaSH monitoring, evaluation, and policy, globally.
3   Bain, R., S. Gundry, J. Wright, H. Yang, S. Pedley and J. Bartram. (2012). Accounting for water quality in monitoring access to safe drinking-water as part of the Millennium Development Goals: lessons from five countries. *Bulletin of the World Health Organization* 90(3), 228–235. This article estimates how accounting for water quality might affect progress on the 2015 MDGs in five countries.

References

Anon. (1977). Report on the United Nations Conference on Water, Mar del Plata, 14–25 March 1977, New York: United Nations.

Anon. (1990). Global Consultation on Safe Water and Sanitation for the 1990s, Background Papers. Secretariat for the Global Consultation on Safe Water and Sanitation for the 1990s, New Delhi.

Anon. (1992). International Conference on Water and the Environment: Development Issues for the 21st Century. The Dublin Statement and Report of the Conference, World Meteorological Organization, Geneva, Switzerland.

Bain, R., R. Luyendijk and J. Bartram. (2013). Universal access to drinking-water: the role of aid. WIDER Working Paper no. 2013/088. http://www.wider.unu.edu/publications/working-papers/2013/en_GB/wp2013-088/ (accessed 18 May 2015).

Bartram, J., C. Brocklehurst, M. Fisher, R. Luyendijk, R. Hossain, T. Wardlaw and B. Gordon. (2014). Global monitoring of water supply and sanitation: a critical review of history, methods, and future challenges. *International Journal of Environmental Research and Public Health* 11, 8137–8165.

Biswas, A. (2004). From Mar del Plata to Kyoto: an analysis of global water policy dialogue. *Global Environmental Change* 14 (Supplement), 81–88.

Brown, L. (1978). *The Twenty-Ninth Day*. Toronto: World Water Institute.

Carsons, R. (1962). *Silent Spring*. New York: Houghton Mifflin Company.

Chasek, Pamela S., David L. Downie, and Janet Welsh Brown. (2014). *Global Environmental Politics*. Sixth edition. Boulder, CO: Westview Press.

Committee on Economic, Social and Cultural Rights (CESCR). (2002). General Comment 15. Geneva: United Nations.

Cumming, O., M. Elliott, A. Overbo and J. Bartram. (2014). Does global progress on sanitation really lag behind water? An analysis of global progress on community- and household-level access to safe water and sanitation. *PLoS ONE* 9(12), e114699. Doi:10.1371/journal.pone.0114699

Falkenmark, M. (1997). Analytic summary *Mar Del Plata, 20 year anniversary seminar*. SIWI, Stockholm Sweden. http://www.siwi.org/documents/Resources/Reports/Report1_Mar_del_Plata_1997.pdf

Gibbons, G., L. Edwards, A. Gross, S.M. Lee and B. Noble. (1996). *The UNDP-World Bank Water and Sanitation Programme, Annual Report July 1994 – June 1995*. Washington, DC: International Bank for Reconstruction and Development.

IRC. (1995). *Water and Sanitation for All: a World Priority. Vol 2: Achievements and Challenges.* The Hague: Ministry of Housing, Spatial Planning and the Environment.

Kayser, G.L., P. Moriarty, C. Fonseca and J. Bartram. (2013). Domestic water service delivery indicators and frameworks for monitoring, evaluation, policy & planning: a review. *International Journal of Environmental Research and Public Health* 10(10), 4812–4835; doi:10.3390/ijerph10104812

Lerer, L. (1999). How to do (or not to do). . . Health impact assessment. *Health Policy and Planning* 14(2), 198–203.

Luh, J. and J. Bartram (accepted, 2015) Progress towards universal access to improved water and sanitation and its correlation to country characteristics. *Bulletin of the World Health Organization.*

Luh J., R. Baum and J. Bartram. (2013). Equity in water and sanitation: developing an index to measure progressive realization of the human right. *International Journal of Hygiene and Environmental Health* 216, 662–671.

Meadows, D.H., D.L. Meadows, J. Randers and B. Behrens. (1972). *Limits to Growth.* New York: Universe Books.

Meier, B., G.L. Kayser, U.Q. Amjada and J. Bartram. (2013). Implementing an evolving human right through water and sanitation policy. *Water Policy* 15, 116–133.

Sims, J. (1994). *Women, Health and Environment: An Anthology.* Geneva: World Health Organization.

UN Human Rights Council. (2008). Resolution 7/22: Human Rights and Access to Safe Drinking Water and Sanitation. Geneva: United Nations.

UN Human Rights Council. (2010). Resolution on Human Rights and Access to Safe Drinking Water and Sanitation (A/HRC/15/L.14, 24 September 2010). New York: United Nations.

United Nations. (1948). Universal Declaration of Human Rights, 10 December 1948, 217 A (III). New York: United Nations. Available at: http://www.refworld.org/docid/3ae6b3712c.html [accessed 18 May 2015]

United Nations. (1970). UN General Assembly resolution A/RES/25/2626 (1970). 2626 (XXV). International Development Strategy for the Second United Nations Development Decade. New York: United Nations. http://www.un-documents.net/a25r2626.htm

United Nations. (1990).. Resolution on World Summit for Children. GA/Res/45/217 New York: United Nations.

United Nations. (2000). Millennium Development Goals. G.A. Res. 55/2, U.N. GAOR, 55th Session, 8th plenary meeting, UN Doc. A/RES/55/2. New York: United Nations.

United Nations. (2010). Resolution on Human Right to Water and Sanitation. UN General Assembly Resolution. A/64/292. New York: United Nations.

United Nations. (2013). *World Population Prospects: The 2012 Revision.* Population Division, Department of Economic and Social Affairs, New York: United Nations.

United Nations, Department of Economic and Social Affairs, Population Division. (2014) *World Urbanization Prospects: The 2014 Revision. Highlights* (ST/ESA/SER.A/352). New York: United Nations.

United Nations, General Assembly. (2014) Open Working Group Proposal for Sustainable Development Goals. A/68/970, available at http://undocs.org/A/68/970. New York: United Nations.

UN-Water. (2012). UN-Water Global Annual Assessment of Sanitation and Drinking-Water (GLAAS) 2012 Report: The Challenge of Extending and Sustaining Services. Geneva: World Health Organization.

Walsh, J.A. and K.S. Warren. (1979). Selective primary health care: an interim strategy for disease control in developing countries. *New England Journal of Medicine* 301(18), 967–974.

World Commission on Environment and Development. (1987). *Our Common Future.* Oxford: Oxford University Press.

WHO. (1948). *The Constitution of the World Health Organization.* Geneva: World Health Organization.

WHO. (2002). *The World Health Report – Reducing Risks, Promoting Healthy Life.* Geneva: World Health Organization.

WHO. (2003). *Guidelines for Safe Recreational Water Environments. Volume 1: Coastal and Fresh Waters.* Geneva: World Health Organization.

WHO. (2006a). *Guidelines for the Safe Use of Wastewater, Excreta, and Greywater in Agriculture and Aquaculture.* Geneva: World Health Organization.

WHO. (2006b). *Water, Sanitation and Hygiene Standards for Schools in Low-Cost Settings.* J. Adams, J. Bartram, Y. Chartier and J. Sims (eds). Geneva: World Health Organization.

WHO. (2006c). *Guidelines for Safe Recreational Water Environments. Volume 2: Swimming Pools and Similar Environments*. Geneva: World Health Organization.

WHO. (2008). *Essential Environmental Health Standards in Health Care*. J. Adams, J. Bartram and Y. Chartier (eds). Geneva: World Health Organization.

WHO. (2011). *Guidelines for Safe Drinking Water*. 4th edition. Geneva: World Health Organization.

WHO. (2012). Global Costs and Benefits of Drinking-water Supply and Interventions to Reach The MDG Target and Universal Coverage, WHO/HSEWSH/12.01. Geneva: World Health Organization.

WHO and UNICEF. (2014). *Progress on Drinking Water and Sanitation: 2014 Update*. Geneva and New York: World Health Organization and United Nations Children's Fund.

44

DRINKING WATER QUALITY REGULATIONS*

Katrina Charles

DEPARTMENTAL LECTURER, COURSE DIRECTOR ON MSc IN WATER
SCIENCE, POLICY & MANAGEMENT, UNIVERSITY OF OXFORD, UK

Katherine Pond

RESEARCH FELLOW, ROBENS CENTRE FOR PUBLIC AND
ENVIRONMENTAL HEALTH, DEPARTMENT OF CIVIL AND
ENVIRONMENTAL ENGINEERING, UNIVERSITY OF SURREY, UK

Learning objectives

1 Understand the basics of compliance-based and risk-based approaches to drinking water quality regulations.
2 Understand drinking water quality enforcement approaches.

Introduction

There is good evidence that the increases in access to improved drinking water facilities in the previous decades with the Millennium Development Goals have not always achieved the intended provision of safe drinking water (Bain et al., 2012). Current estimates suggest the numbers of people using unsafe water may be as high as 1.8 billion (28 percent of the global population), more than double the 783 million (11 percent) who lack access to an improved drinking water facility (Onda et al., 2012). Regulation is one tool that, with other management, technical and behavioral interventions, can help achieve safe drinking water.

This chapter will start with an introduction to drinking water regulation and provide a review of two approaches to regulating drinking water supplies: compliance-based monitoring and risk-based approaches. It will then consider how to ensure that regulations are followed

* Recommended citation: Charles, K. and Pond, K. 2015. 'Drinking water quality regulations', in Bartram, J., with Baum, R., Coclanis, P.A., Gute, D.M., Kay, D., McFadyen, S., Pond, K., Robertson, W. and Rouse, M.J. (eds) *Routledge Handbook of Water and Health*. London and New York: Routledge.

through enforcement approaches. As suitable regulation and enforcement approaches vary with context, it will conclude with a short discussion of regulation of small water supplies.

What is drinking water quality regulation?

Regulation has been described as "a system that allows government to formalize and institutionalize its commitments to protect consumers and investors" (Tenenbaum, 1995) or "the process of interpreting and implementing laws, policies and regulations, to achieve what was intended in their formation" (Rouse, 2013). However, regulations are not just a system of control (the "stick"), but can be used to provide incentives to facilitate change (the "carrot").

With regards to water, the regulatory framework will address issues of drinking water quality and distribution, setting of tariffs, bathing water quality, abstractions and discharges. The focus of this chapter is on the issue of drinking water quality.

The primary aim of drinking water quality regulation is to protect public health by providing safe drinking water (WHO, 2011). In order to achieve this aim, the water must "not represent any significant risk to health over the lifetime of consumption" (WHO, 2011). This presents challenges in determining what levels of chemical and microbiological parameters are safe and how we can detect them, as well as in managing the water supply to *protect* it from contamination, *treat* contamination and *distribute* the water safely to the customers. To achieve the aim of protecting public health, more than the safety of the water must be considered:

- To reduce the use of unsafe sources, consumers need to have ready access to sufficient quantities of drinking water for domestic purposes; and
- To reduce recontamination of safe supplies, users need to trust the suppliers that the water source can be used safely without additional treatment and will be available on demand such that additional household treatment or storage is not used.

In drinking water quality regulation, if legal measures are applied it indicates that there has been provision of unsafe water, which can have significant impacts on human health and consumer trust. Hence, there is a need for the regulatory approach to promote good management of water supplies to prevent the provision of unsafe water, rather than just focusing on punishing breaches.

Drinking water quality regulations may include aspects of compliance monitoring values for a range of parameters: quality assurance and analytical quality control, to ensure the validity of compliance and other monitoring operations; risk assessment methodologies; and management responses to emergencies. Two approaches to regulating drinking water supplies are considered below: compliance-based monitoring and risk-based approaches, with key points summarized in Table 44.1.

Compliance monitoring

Compliance monitoring involves the setting of standards for drinking water quality, with final water quality assessed at a specific point (e.g. after treatment or at the consumer tap) against these standards. The standards for compliance monitoring are set to protect public health and ensure that water quality is acceptable to consumers. As well as parameters directly related to health outcomes, compliance monitoring can include process variables,

Table 44.1 Comparison of compliance monitoring and risk-based approaches for drinking water quality regulations

Regulatory approach	Compliance-based	Risk-based
Approach	Measuring final water quality against standard criteria	Targeting resources at and monitoring the greater risks to health
Benefits	Scientifically developed global standards reduce reliance on local skills	Proactive: address highest risks; more cost-effective in long term
Limitations	Reactive: results often available only after water is consumed	Skills/resources needed to establish health targets, and initiate risk assessments

such as chlorine residuals and turbidity, which indicate that the treatment processes are operating within or outside appropriate ranges, and which may have consequences for health or acceptability.

There is good agreement globally on the science behind the setting of health-based standards for drinking water and this expert evidence is documented by the World Health Organization in the *Guidelines for Drinking-Water Quality* (WHO, 2011). In a recent report these guidelines were identified as playing a significant role in the derivation of guideline values in over half of national drinking water quality guidance documents from the 100 countries and territories surveyed (Drury, 2013).

Standards for drinking water quality typically include a range of chemical, radiological and microbiological parameters with criteria based on health; however, they will also include organoleptic properties, that is aspects of drinking water that are sensed and, while they don't have direct health effects, they are likely to affect people's acceptance and use of the water for drinking purposes. These include color, taste and smell.

There are a number of limitations of relying solely on compliance monitoring. Compliance monitoring relies on tests being available to detect, within a reasonable time and cost, the relevant water quality parameters. In the case of microbiological hazards the industry has historically relied on faecal indicator bacteria. There is no doubt that water supplies which consistently meet the standards and regulations set for faecal indicator bacteria provide a reduced public health risk. However, as discussed in Chapter 59, faecal indicator bacteria are not as robust as other organisms such as viruses and protozoa, and are therefore not representative of the types of microorganisms most likely to survive water treatment processes. In addition, results of microbiological testing are retrospective, meaning that once a problem has been detected and confirmed by a laboratory test, consumers will have been exposed to the contaminated water. In addition, samples are representative of only a small volume of water and are generally of low frequency. At best, sampling can only provide a "snapshot" of the quality of the water.

The focus on a prescribed list of parameters can also lead to a focus on achieving the standard for those targets rather than being based on a holistic understanding of how to improve the water supply and the actual health risk occurring due to exposure. This is particularly problematic when we consider the risks from emerging contaminants and contaminants that cannot currently be adequately monitored due to the availability or affordability of the technology.

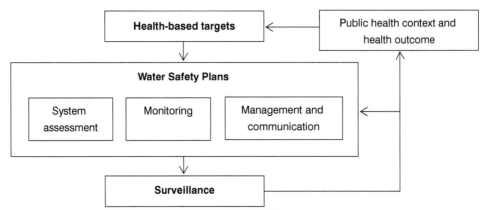

Figure 44.1 Framework for Safe Drinking Water (adapted from WHO, 2011)

Risk-based approaches

In response to these limitations there has been an increasing shift towards less emphasis on the reporting and monitoring regimes that dominate the water industry. This has been driven by the adoption of a risk-based approach to drinking water quality management in the second edition of the WHO *Guidelines for Drinking-Water Quality* in 2004, as well as by the inclusion of risk-based approaches in the second key Principle of the Bonn Charter for Safe Drinking Water.[1]

In a recent review of national drinking water quality guidance documents, Drury (2013) reported that many countries included statements that suggested a risk-based approach (i.e. that drinking water should not contain concentrations of compounds which could, either alone or in combination with other compounds, be harmful to human health; and not contain any microorganisms and parasites in such a number that could be harmful to human health); however, the risk-based approach was not explicitly stated.

As part of this risk-based approached, WHO recommend setting context-specific health-based targets (Figure 44.1). Health-based targets include measureable health, water quality or performance objectives based on assessment of the risk, including health outcome targets (e.g. tolerable burdens of disease); water quality targets (e.g. guideline values for chemical hazards); performance targets (e.g. log reductions of specific pathogens); and specified technology targets (e.g. application of defined treatment processes) (WHO, 2011). This approach is designed to make water supply systems more resilient to changes in water quality and emerging pathogens and chemicals of concerns. It should also ensure that the water quality is managed appropriately to meet the needs of sensitive sub-populations, such as babies, young children, the elderly and the immunocompromised.

However, despite a number of benefits resulting from the use of health targets, their development requires considerable human and financial resources in order to obtain good quality and reliable data. Health outcome targets and water quality targets both require significant scientific and technical support in order to obtain reliable data. This is a major constraint in some countries. Data are often unavailable and of poor quality. Many countries do not have the resources to establish independent data quality assurance schemes or systems for analyzing, sharing and disseminating data. The availability of accurate, timely and consistent data at the national and sub-national levels is crucial for countries to be able to effectively set health targets and subsequently manage their health systems, allocate resources according to need, and ensure accountability for delivering on health commitments.

Even where resources are available to collect sufficient data, other issues arise when trying to set health-based targets. In particular, there may be difficulty in separating drinking water quality influences from other factors. For example, for diseases where infection is universal, such as viral gastroenteritis, it is very difficult to attribute the cause to one factor, for example drinking water. Determinants of ill health such as variations in exposure throughout the population has an impact on the risk of infection; pre-existing health conditions, social conditions and acquired immunity all impact on the ability of an individual to deal with infectious agents. All these factors may therefore impact on identifying an acceptable risk for a population and subsequently a health-based target.

Water Safety Plans, a risk-based approach to water supply management discussed in Chapter 23, are now acknowledged widely as being "the most effective means of consistently ensuring the safety of a drinking-water supply" (WHO, 2011).

This approach does not remove the need for compliance monitoring but reduces the reliance on monitoring at the end of the process and enables reporting of treatment and management processes that ensure continuous supply of good quality water. This ensures that the regulations support a proactive approach that is more likely to identify a problem before a public health incident occurs; it ensures there is a holistic focus to management allowing better mitigation of risks that may not be addressed by compliance monitoring, including emerging contaminants; and it allows prioritization of resources on the issues that pose the biggest risk to public health.

Enforcement of drinking water regulations

Drinking water quality regulations are only effective if there is effective monitoring and they are enforced. Monitoring may be of the drinking water quality, as well as the processes, systems and risk assessments developed to protect water quality. How regulations are enforced vary between countries, regions and states. This section will discuss regulation tools and the importance of considering the local context for enforcement in design of the regulations. More information on monitoring approaches and Water Safety Plans are provided in other chapters.

Enforcement does not have to be limited to punishment, but can include a wider range of tools that can be used to promote good management of water supplies to prevent breaches. These are particularly important in drinking water regulation as if a breach has occurred it can have significant impacts on human health and consumer trust. Ayres and Braithwaite (1992) considered that an enforcement strategy is more effective when there are multiple tools available, and that most efforts to obtain compliance should be based on persuasion. Examples of where this has been adapted in drinking water quality regulation include in England and Wales (Figure 44.2), where there is an emphasis on "soft regulation", such as developing best practice guidance, with "hard regulation" such as sanctions through to prosecution reserved for serious breaches. A similar approach is used by the Washington State Department of Health in the USA which develops bilateral compliance agreements with the water providers to identify commitments and deadlines for improving water supplies and meeting the regulations, as well as having stages of compliance. This is followed by four steps to enforcement:

1 Notifying the water system of a violation and offering technical assistance.
2 Coming to an informal compliance agreement.
3 Issuing a departmental order.
4 Issuing fines (Washington State Department of Health, 2008).

Furthermore, in order for effective enforcement, the standards and regulations need to be appropriate for the context, including for the level of resources that are available, and facilitate constant improvement rather than enforcing unobtainable standards. This can be particularly challenging in resource-poor settings in developing countries, and also in rural areas of developed countries. In such cases, there may be a need for interim standards to provide a medium-term goal as a step towards the achievement of guideline values in the longer term.

Interim standards allow resources to be prioritized towards those communities with the greatest problems. They provide incentives to upgrade which is particularly important in countries subject to severe economic constraints. In some countries, health authorities have adopted interim standards for long-persistent natural contaminants. For example, while the WHO drinking water quality guideline for arsenic was reduced from 50 μg/L to 10 μg/L in 1993, Bangladesh has maintained an interim guideline of 50 μg/L to enable it to prioritize water sources with the highest concentration of arsenic (Yamamura, 2003). It should be emphasized that priority must be placed on the control of microbiological contamination with long-term chemical quality deferred until affordable.

Interim processes are also used to support development. In South Africa, interim audits are used to support capacity development of the provincial government water suppliers, with the intention that they are phased out as a water provider demonstrates sustained improvements in compliance (Hodgson and Manus, 2006).

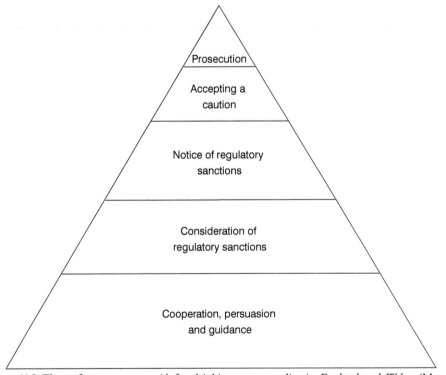

Figure 44.2 The enforcement pyramid for drinking water quality in England and Wales (May & Colbourne, 2009)

Drinking water regulation for small water supplies

Small water supplies, ranging from a single household to a community supply, present very specific problems due to the small resource base available for treatment, management and regulation, as well as the decentralized nature of the supplies. Management of these small supplies is a concern worldwide in both developed and less-developed countries. While the definition of small supplies may vary between countries, there are clear operating and management challenges that make them distinct from larger supplies, and because of this some countries choose to regulate small water supplies differently from that for municipal supplies to make them appropriate for the resources available.

In developed countries, small water supplies are often differentiated from larger supplies in regulations. In England and Wales, where there are an estimated 48,000 small water supplies, there are separate regulations (Private Water Supply Regulations, 2009) which apply to any water supply not provided by a water company, with monitoring frequency and responsibilities designed to reflect the size of the supply (and therefore assumed to reflect resource availability). The USA uses a similar approach whereby public water supplies are regulated at a national level under the US Environmental Protection Agency's (EPA) Safe Drinking Water Act. These include water supplies, regardless of ownership, that serve at least 25 people for at least 60 days of the year, and/or have at least 15 service connections. Supplies that serve fewer people or with fewer connections are considered private water systems, and are not covered by these regulations such that "owners of individual water systems are responsible for ensuring that their water is safe from contaminants" (CDC, 2014).

In developing countries where often not all the population will have access to safe drinking water, there is often less differentiation of regulation of small and community water supplies from larger supplies. For example, in South Africa, the municipalities take responsibility for infrastructure regardless of the size of the water supply; however, they may be managed by the communities or privately. However, in Kenya, where there are Service Provision Agreements (SPA) between the provider and the Water Services Board, these SPAs will differ between communities and medium to large water supplies.

Summary

Regulation of drinking water quality is an important tool for the management of safe, accessible and affordable drinking water supplies that contribute to improved public health. Increasingly, regulations are adopting a risk-based approach to encourage improvement in water supplies as a whole, with a reduced emphasis on the monitoring of final water quality and legal sanctions. Effective drinking water quality regulations need to reflect the local contexts in which they are enforced, and require transparency to militate against corrupt practices and engender consumer trust.

Key recommended readings (open access)

1 WHO (2011). *Guidelines for Drinking-Water Quality*. Fourth edition. World Health Organization, Geneva. Guidelines for drinking water quality developed by an international network of scientists that are widely used as the basis for establishing national water quality regulations.
2 DWI (2009). Drinking water safety. Guidance to health and water professionals. Drinking Water Inspectorate, London. http://dwi.defra.gov.uk/stakeholders/

information-letters/2009/09_2009Annex.pdf. Provides a review of how drinking water is regulated in England and Wales, including how different agencies operate, how consumer complaints are managed and learning points from incidents.

Note

1 Principle 2 "Systems to ensure drinking water quality should not be based solely on end-of-pipe verification (testing against predetermined standards). Rather, management control systems should be implemented to assess risks at all points throughout water supply systems and to manage such risks" (IWA, 2004).

References

Ayres, I. and Braithwaite, J. (1992). *Responsive Regulation: Transcending the Deregulation Debate*. Oxford: Oxford University Press.

Bain, R. E., Gundry, S. W., Wright, J. A., Yang, H., Pedley, S. and Bartram, J. K. (2012). Accounting for water quality in monitoring access to safe drinking-water as part of the Millennium Development Goals: lessons from five countries. *Bulletin of the World Health Organization*, 90(3), 228–235.

CDC (2014). Private Water Systems. http://www.cdc.gov/healthywater/drinking/private/index.html. Accessed 26 June 2014.

Drury, D. (2013). Draft report on regulations and standards for drinking water quality. Geneva: World Health Organization.

Hodgson, K. and Manus, L. (2006). A drinking water quality framework for South Africa. *Water SA*, 32(5), 673–678.

IWA (2004). *The Bonn Charter for Safe Drinking Water*. London: International Water Association.

May, A. and Colbourne, J. (2009). Regulatory risk assessment roll-out for England and Wales. *Water Utility Management International*, 4, 12–14.

Onda, K., LoBuglio, J. and Bartram, J. (2012). Global access to safe water: accounting for water quality and the resulting impact on MDG progress. *International Journal of Environmental Research and Public Health*, 9(3), 880–894.

Rouse, M. (2013). *Institutional Governance and Regulation of Water Services. The essential elements*. Second edition. London: International Water Association.

Tenenbaum, B. (1995). The real world of power sector regulation. Viewpoint Note No 50. Washington, DC: The World Bank.

Washington State Department of Health (2008). Fact sheet. How enforcement affects operating permits. November. DOH 331-339. http://www.doh.wa.gov/Portals/1/Documents/Pubs/331-339.pdf. Accessed 26 June 2014.

WHO (2004). *Guidelines for Drinking-Water Quality*. Second edition. Geneva: World Health Organization.

WHO (2011). *Guidelines for Drinking-Water Quality*. Fourth edition. Geneva: World Health Organization.

Yamamura, S. (2003). Drinking water guidelines and standards. In H. Hashizume and S. Yamamura (eds) *Arsenic, Water, and Health: The State of the Art*. Geneva: World Health Organization. http://www.who.int/water_sanitation_health/dwq/arsenicun5.pdf. Accessed 24 June 2014.

45

RECREATIONAL OUTDOOR WATER REGULATIONS*

Julie Kinzelman

PhD, MT (ASCP), Laboratory Director and Research Scientist,
City of Racine, Department of Public Health, Racine, WI, USA

Learning objectives

1 Describe adverse health outcomes related to outdoor recreational water exposure.
2 Recognize the role fecal indicator organisms play in determining outdoor recreational water exposure risk despite their inherent drawbacks.
3 Gain an appreciation of similarities and differences in outdoor recreational water quality standards, guidelines and classification schemes, and the importance of implementation, through the global examples provided.

Introduction

Human well-being is dependent upon the management of Earth's ecosystems to ensure their conservation and sustainable use; without this preservation ecosystem services, including those related to water recreation, will be compromised (WHO 2005). Agricultural practices, animal husbandry, increasing urbanization of coastal areas, the return of sewage over large distances as a result of urban sprawl, inadequately treated sewage and the delivery of water-washed pollutants via storm water runoff have put a burden on aquatic systems in general as well as outdoor recreational water resources, creating routes by which we are exposed to a variety of physical, chemical and microbiological hazards (Chapters 2–11) as well as straining those interventions designed to keep us safe (Karn and Harada 2001, Aitken 2003, Burak et al. 2004, Dwight et al. 2005) (see Chapters 17–32). For example, sewage releases from overburdened or inadequate treatment facilities have been proven to contribute to an increase in bathing water quality failures, in some cases depriving people of recreational use and in others increasing the potential risk to human health due to exposure (Griffin

---

* Recommended citation: Kinzelman, J. 2015. 'Recreational outdoor water regulations', in Bartram, J., with Baum, R., Coclanis, P.A., Gute, D.M., Kay, D., McFadyen, S., Pond, K., Robertson, W. and Rouse, M.J. (eds) *Routledge Handbook of Water and Health*. London and New York: Routledge.

et al. 2001). The loss of recreational use of our bodies of water can have a negative, direct economic impact as well as create public dissatisfaction through loss of utility (Koteen et al. 2002, Houston 2013). In order for the quality of our surface waters to improve or be maintained, public decision makers at municipal and national levels must intervene through the development of policies, regulations and enforcement schemes combining both political and technical elements which are flexible and adaptive in order to meet changes both in the ecosystem and in the field of science as it evolves (Chapters 42–47). This chapter will provide an overview of the regulation of outdoor recreational water in the context of regulatory compliance monitoring and implementation. Outdoor recreational waters are inclusive of fresh (Great Lakes coastal waters, inland lakes and tributary systems), marine (ocean, sea and in a few instances inland lakes) and estuarine locations (interface between fresh and marine waters).

Recreational water and health

Swimming and other recreational activities are a shared experience; whatever illnesses a person nringsto the lake or ocean will result in a common exposure to those around them and, therefore, hygienic practices of adults (carrying on average 0.14 g of feces on their bottom at any given time) and children (especially those in diapers) can contribute to the overall risk of disease transmission to co-bathers (Gerba 2001; CDC, http://www.cdc.gov/healthyswimming/, last accessed 24 November 2014). Other bathers, however, are not the sole source of fecal contamination resulting in recreational water illness (RWI) (diarrheal, eye, skin, ear and respiratory illnesses). Sources such as sewage, domestic or wild animal waste, storm water, agricultural or urban runoff, and free-living organisms may also contribute to the risk of adverse health outcomes in the context of a bathing water encounter (Table 45.1) (WHO 2003). While direct pollution sources such as sewage may be obvious, if more costly to repair, indirect sources of pollution are harder to identify and yet are by far the most common source of contamination (Scott et al. 2002, US EPA 2014).

The single most frequent adverse health outcome from swimming in contaminated water is gastrointestinal illness (van Asperen et al. 1998, WHO 2003, Sanborn and Takaro 2013) which can result from either bacterial (Sorvillo et al. 1988, Lauber et al 2003) or viral (Griffin et al. 2003) agents. However, a survey of adverse health outcomes associated with bathing waters demonstrated variations in actual disease endpoints (respiratory, gastrointestinal (GI), ear, eye, nose and skin) between countries in temperate regions, with respiratory illnesses being most often reported in England (24 per cent) while areas of the US reported greater GI illness rates (Santa Monica Bay, California, 37 per cent; Key West, Florida, 33 per cent) (Rose 2004). In sub-tropical and tropical areas parasitic infections of zoonotic (animal) origin, for example schistosomiasis, may also be prevalent (Kinzelman and McPhail 2012). Due to the morbidity and mortality associated with disease outbreaks associated with outdoor recreational waters, regulatory approaches, based on compliance monitoring programs, derived from epidemiological studies (Chapter 55), and driven by legislature, have been developed for the protection of public health (Table 45.2).

Marine and freshwater fecal indicator organisms

While epidemiological studies worldwide indicate a disease association with swimming in contaminated surface waters, the approach to regulating these waters varies dramatically geographically (Georgiou and Langford 2002). Worldwide regulatory schemes for

Table 45.1 Pathogens and swimming-associated illnesses listed by causative agent and resulting disease

Pathogenic agents of disease

Bacterial		Viral		Protozoan	
Organism	*Disease*	*Organism*	*Disease*	*Organism*	*Disease*
E. coli 0157:H7	Gastroenteritis (HUS)	Rotavirus	Gastroenteritis	*Cryptosporidium* spp.	Gastroenteritis
Salmonella typhi	Typhoid fever	Norovirus	Gastroenteritis	*Giardia lamblia*	Diarrhea
Salmonella spp.	Gastroenteritis	Poliovirus	Poliomyelitis	*Entamoeba histolytica*	Amoebic dysentery
Shigella dysenteriae	Dysentery	Adenovirus	Respiratory, gastrointestinal infections	*Isospora belli*	Gastrointestinal infections
Vibrio cholera	Cholera	Hepatitis A and E	Infectious hepatitis	*Balantidium coli*	Dysentery, ulcers
Helicobacter pylori	Peptic ulcers, gastritis	Enterovirus (67 types, e.g. polio, echo, Coxsackie)	Gastroenteritis	*Naegleria fowleri*	Amebic meningo-encephalitis
Pseudomonas spp.	Urinary tract, respiratory and ear infections	Rotavirus	Gastroenteritis	Non-human schistosomes	Swimmer's itch (cercarial dermatitis)

©Julie Kinzelman 2005.

Note: HUS = hemolytic uremic syndrome

Sources: van Asperen et al. (1995), US EPA (2001), Farahnak and Essalat (2003), CDC (2004).

Table 45.2 Reported waterborne disease outbreaks associated with recreational waters, US, 1997–2002

US outbreaks of waterborne disease – recreational water			
Reporting period	No. of outbreaks	No. individuals affected	No. of deaths
1997–1998	32	1,000	0
1999–2000	59	2,093	4★
2001–2002	65	2,536	8★

©Julie Kinzelman 2005.

★ All deaths were due to primary amoebic meningoencephalitis (PAM) caused by *Naegleria fowleri.*

Sources: Barwick et al. (2000), CDC (2000), CDC (2004).

microbiological quality of recreational water, although differing in some aspects, are primarily or exclusively based on percentage compliance with fecal indicator organism (FIO) levels, typically a genus or species of bacteria (Chapter 59). While the use of a FIO paradigm to assign health risk has been called into question in recent years, especially when nonpoint source pollution is the suspected culprit, there have been no global standards put forward for a single or suite of pathogens that would adequately characterize health risk across all aquatic environments and exposure scenarios (Colford et al. 2007). Although direct measurements of pathogens in outdoor recreational water samples would be a more relevant measure of human exposure risk, it is not yet feasible to test for every type of pathogen that may be present within a water environment. There are a great number and the detection methods are often difficult, costly and require a high level of technical expertise (WHO 1999, Health Canada 2012). Therefore, FIO are used as surrogates to indicate when the presence of pathogens is more likely to occur, despite these constraints and based on the strength of evidence derived from global epidemiological studies (Prüss 1998, Wade et al. 2006, Kay et al. 2009a, 2009b) (Chapter 55).

In addition to their inability to fully characterize pathogen presence, regulation of recreational waters using FIO also suffers due to the inherent time lag resulting from the requisite culture-based assays, typically 24 hours or more. This analytical delay, coupled with the fact that many recreational waters are monitored infrequently, for example once weekly throughout the bathing season, results in retrospective management actions. For example, results from samples collected on Monday would not be generated until Tuesday, leaving those who swam on Monday exposed. To provide better protection of public heath, alternatives to culture-based enumeration of FIO for the regulation of designated bathing waters exist. These alternatives include more rapid analytical assays and pre-emptive or predictive models, which estimate FIO values or regulatory threshold value exceedances based on predictable relationships between FIO density in the aquatic environment and a suite of readily measured independent variables, for example turbidity, wave height and antecedent precipitation (Lavender and Kinzelman 2009, McPhail and Stidson 2009).

Measurements of FIO also do not account for other hazards that may be found in the context of recreational exposure, nor do national/international regulatory compliance schemes always address them outright. Toxic algae, cyanobacterial blooms, strong tidal/ rip currents, exposure to chemicals, dangerous/venomous organisms and infections (sometimes fatal) resulting from amoeba, parasites (at various stages within their life cycle) and protozoa are responsible for morbidity and mortality arising from exposure to recreation waters. Therefore, local, state and national agencies have developed guidance to be used in conjunction with compliance monitoring schemes as an additional level of public health protection (Australian National Health and Medical Research Council 2008).

Bathing water standards, guidelines and classification schemes

Select compliance monitoring schemes

FIOs, chiefly enterococci and *E. coli*, have been almost exclusively chosen as surrogates for pathogens for monitoring the microbiological quality of fresh and marine bathing waters, flowing and still. Secondary indicators, such as *Clostridium perfringens* for tropical environments such as Hawaii, or *Bacteriodales* as an indication of human sewage, may also be utilized (Fujioka 1997, Walters et al. 2007). Although the use of these FIOs is almost universal, the way in which they are applied within worldwide compliance monitoring schemes may differ. In some instances levels of these organisms form the basis of regulatory standards (US: US EPA 2012; EU: European Union 2006) where in other instances microbial indicator concentration is used for guidance purposes only (Australia: WHO 2003).

In either of these two scenarios, that is, guidelines or standards, microbial assessment may or may not be combined with a beach classification category (excellent, good or satisfactory; where officially designated beaches are present) based on sanitary inspections (WHO 1999). Sanitary inspections/beach profiles (EU), sanitary surveys (US) or environmental health and safety surveys (Canada) are guided data collection tools which allow for an estimation of health risk based on site-specific pollution sources. The most recent revisions to the European Bathing Water Directive, US Environmental Protection Agency's (EPA) Recreational Water Quality Criteria and Guidelines for Canadian Recreational Water Quality all require or recommend this type of multi-tiered approach in conjunction with regulatory monitoring of outdoor recreational waters (European Union 2006, Health Canada 2012, US EPA 2012). Whether or not beach classification is required as part of the regulatory framework, environmental assessment tools have value in the determination of pollution sources, leading up to mitigation (the only finite source of public health protection), as well as being useful for the development of predictive models (Kinzelman et al. 2012).

The revised World Health Organization (WHO 2003) guidelines form the framework on which many federal water quality regulations have been modeled, as well as providing a framework for local decision making (WHO 2003, Kay et al. 2004). Developed using the results of epidemiological studies which delineated an association between exposure to contaminated bathing water and an increase in swimmer-associated illnesses, the primary aim of the WHO (2003) Guidelines for Safe Recreational Water Environments is the protection of public health when individuals are employed in the full body immersion activities in coastal (marine or freshwater), inland (freshwater) and estuarine recreational water outdoor environments. The purpose of these guidelines, like all other guidelines and standards, is not to deter the public from recreational water use but to ensure that these recreational water areas are operated as safely as possible in order for the largest segment of the population to get the maximum benefit.

To ensure the safety of the public in the context of bathing water encounters, the WHO suggests that local authorities employ a combination of sanitary inspections and laboratory analysis of surface water using microbial indicators. According to the WHO, this approach will provide information on the possible sources of contamination as well as the actual level of fecal pollution (enumerated through the use of approved FIOs). Combining data obtained from comprehensive sanitary inspections/surveys, which aid in the identification of all potential direct and indirect pollution sources, with FIO levels allows for a graded beach classification system (Figure 45.1) rather than a purely binary testing scheme (open or safe for swimming versus swimming advised against/prohibited) such as is currently employed in the United States. Using this dual type of testing scheme not only provides a measure of enhanced regulatory compliance but allows for effective remediation strategies (via the identification of pollution sources via the sanitary survey process) and informed decision making.

Informed decision making, as a result of a robust regulatory monitoring scheme, can extend to both the decision makers and the recreating public. Decision makers can use historic compliance data to adjust future monitoring frequency; decreasing monitoring at those locations where water quality is consistently good or poor (until such time as the mitigation occurs) as a way to stretch available funding while still protecting public health (WHO 1999). Managers can also use sanitary survey data to develop predictive models, an inexpensive and real time approach to regulating outdoor recreational waters once validated (McPhail and Stidson 2009, Nevers et al. 2009). Modeling may also be attractive for lower and middle income settings as a means of providing protection while reducing the analytical and staff costs associated with traditional FIO-based compliance monitoring.

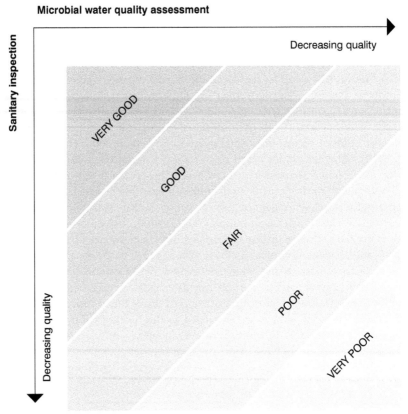

Figure 45.1 Beach classification matrix based on microbial indicator assessment performed in conjunction with sanitary surveys
©WHO 2003, http://www.who.int/water_sanitation_health/bathing/srwe1/en/index.html, last accessed 20 November 2013

For marine outdoor recreational water models, the WHO recommends the use of enterococci with levels based on a calculation of risk at the 95th percentile for bathing encounters involving healthy adults having at least three head immersions within ten minutes (derived from the results of epidemiological studies). The 95th percentile level of enterococci is to be calculated for each beach based on historical data (preferably 100 or more data points from the previous three to five years). It is permissible to adjust these values based on local conditions noted in the sanitary survey and/or to take action on a single exceedance if imminent risk to health is present (WHO 2003, 2009 Addendum). At present the WHO feels that insufficient data exists for the use of *E. coli* as a regulatory tool for fresh outdoor recreational waters although it has been adopted into the regulatory frameworks nonetheless. Some countries follow an approach the same as, or very similar to, the WHO guidelines, for example the European Union (EU), Australia, New Zealand and Hong Kong. Recreational water quality criteria apply to all surface waters where contact occurs, inclusive of coastal, inland, river and estuarine bathing waters.

More than 22,000 bathing waters were monitored in the EU in 2012; of those two thirds were coastal (marine) and the rest inland lakes and rivers (European Environment Agency 2012). In 2012 the majority of EU bathing waters conformed to the criteria set forth under the revised EC Bathing Water Directive (rBWD) (European Union 2006) (Table 45.3). All EU member countries monitor and report intestinal enterococci and *E. coli* values over a four-year period to determine if the bathing water in question meets standards (European Union 2006).

The United States also uses enterococci (freshwater or marine), with the addition of *E. coli* (freshwater), as indicators of outdoor recreational water quality. The US EPA

Table 45.3 Outdoor recreational water quality criteria under the 2006 revised EC Bathing Water Directive (rBWD) for monitoring fresh and marine bathing waters

Inland (fresh) waters

Parameter	Excellent quality	Good quality	Satisfactory/ acceptable quality	Reference method
Intestinal enterococci (IE/dL)	200★	400★	360★★	ISO 7899-1 ISO 7899-2
E. coli (EC/dL)	500★	1,000★	900★★	ISO 9308-3 ISO 9308-1

Coastal (marine) and transitional waters

Parameter	Excellent quality	Good quality	Satisfactory/ acceptable quality	Reference method
Intestinal enterococci (IE/dL)	100★	200★	185★★	ISO 7899-1 ISO 7899-2
E. coli (EC/dL)	250★	500★	500★★	ISO 9308-3 ISO 9308-1

Notes:

★ Based upon a 95th percentile evaluation.

★★Based upon a 90th percentile evaluation.

Source: European Union (2006).

Table 45.4 Recreational water quality criteria, 2012

Criteria elements	Estimated illness rate (NGI): 36 per 1,000 primary contact recreators[a]		Estimated illness rate (NGI): 32 per 1,000 primary contact recreators	
	Magnitude		*Magnitude*	
Indicator via culture-based methods	GM (cfu/100 mL)	STV (cfu/100 mL)	GM (cfu/100 mL)	STV (cfu/100 mL)
Enterococci – marine and fresh [US EPA Method 1600] OR	35	130	30	110
E. coli – fresh [US EPA Method 1603]	126	410	100	320
Indicator via qPCR	GM (cce/100 mL)	STV (cce/100 mL)	GM (cce/100 mL)	STV (cce/100 mL)
Enterococci – marine and fresh [US EPA Method 1611]	470	2,000	300	1,280

Duration and frequency: The waterbody GM should not be greater than the selected GM magnitude in any 30-day interval. There should no more than a 10% excursion frequency of the selected STV magnitude in the same 30-day interval.

[a – corresponds to 1986 illness rates, cfu = colony forming unit, cce = calibrator cell equivalent]

©US EPA 2012, http://water.epa.gov/scitech/swguidance/standards/criteria/health/recreation/upload/RWQC2012.pdf, last accessed 9 December 2014.

recreational water quality criteria, revised in 2012, express regulatory breakpoints as the number/rate of anticipated illnesses per 1000 bathers using the magnitude of these FIOs, as determined by approved analytical methods (Chapter 59) and with frequency (duration) of 30 days (US EPA 2012). The revised criteria recommend the use of both a geometric mean (GM, corresponding to the 50th percentile) and a statistical value threshold (STV, corresponding to the 75th percentile) believing that together they indicate whether water quality meets levels required for primary contact recreation (full body immersion) while not being more restrictive than is the intent of the criteria (US EPA 2012) (Table 45.4). These breakpoints roughly fit within the confidence intervals delineated by the WHO (2003). In addition to approved culture-based criteria, the US EPA has also developed criteria values based using quantitative, real-time, polymerase chain reaction (qPCR) (Table 45.4). All recreational waters fall under these criteria, including riverine and estuarine waters. Criteria for incidental contact, such as encountered during boating and fishing, when one is not anticipated to fully immerse themselves has been removed with the most recent revision.

In Canada the regulation of recreational water quality is left to the jurisdiction of the individual provinces and territories using nationally published guidelines which were first established in 1983. Revised recommendations at the federal level were recently published (Health Canada 2012). As with the United States and Australia/New Zealand, the Canadian federal guidelines call for 1) the use of enterococci (marine) and *E. coli* (freshwater) and 2) recommend multiple points of sampling for large recreational beaches in order to determine representative water quality throughout the entire bather exposure area. These points should be selected to include those areas which could adversely impact surface water quality such as streams, sewer outfalls and storm water outlets.

Table 45.5 Summary of selected worldwide compliance monitoring approaches employed in regulating outdoor recreational waters

Select global compliance monitoring approaches			
	Indicator employed	*Sampling frequency*	*Classification scheme?*
EU	*E. coli* (FW) Enterococci (M)	1 pre-season sample, then monthly.	Yes
US EPA	*E. coli* (FW) Enterococci (M)	Based on priority (high, moderate, low). No less than 5 samples within a 30-day period. One sample for every 500 meters in length. Applies to all outdoor recreational waters.	No
WHO	Enterococci (M) No standard (FW)	5 samples per season for very low, low or very high quality waters. Others waters require 4 sample replicates collected 5 times per season.	Yes
Australia/NZ	Enterococci (FW, M)	Minimum of 5 samples per season for low risk and 20 samples per season for high risk with ideal of 100 samples over 5 years.	Yes
Canada	*E. coli* (FW, M) Enterococci (M)	Once per week. Recommend multiple point sampling for large beaches.	No
Hong Kong	*E. coli* (FW, M)	Once weekly unless non-compliant.	Yes

Note: FW = freshwater, M = marine, NZ = New Zealand

As previously stated, hazards exist within the outdoor recreational water environment that are inadequately addressed through traditional, FIO-based compliance monitoring schemes (Table 45.1). One example of a pathogenic organism poorly characterized by FIO assessments is the parasitic trematode, *Schistosoma* spp. Larvae of this worm are present in tropical aquatic environments, with an obligatory passage through aquatic or amphibious snails. Schistosomiasis, the disease state in humans, occurs when the larvae released from the intermediate host (snails) penetrate the skin of persons washing, bathing, or participating in recreational activities (wading, swimming or rafting) in contaminated water (Schwartz et al. 2005). In this case, education is the best preventative measure.

A summary of select global compliance monitoring approaches for the regulation of outdoor recreational waters is presented in Table 45.5.

Regulatory action decisions based on compliance monitoring

Regulatory action decisions are only as strong as the results of the compliance monitoring programs on which they are built. Regardless of the approach to regulating outdoor recreational waters, the likelihood of incorrectly assigning risk is real. In compliance monitoring a Type I error (false positive) occurs when swimming is prohibited or recommended against but the level of fecal indicator organisms does not exceed recommended standards or guidelines. Conversely, a Type II error (false negative) occurs when public exposure is allowed in the presence of elevated bacterial indicator levels, that is, a heightened health risk. Current monitoring schemes are likely to commit both of these errors with some degree of frequency due to the inherent lag in the receipt of culture-based analytical results, financial constraints of the testing authorities and minimal implementation of real-time test methods and reliable predictive models. Sampling fortnightly, weekly or even daily at single or multiple sampling points does not account for the within day variation or day-to-day variation in water quality, nor do currently approved test methods requiring a minimum of 18 to 24 hours before results become available provide results in a manner in which to be immediately protective of public health.

Other considerations with respect to the regulation of outdoor recreational waters and protection of public health are related to notification, education and compliance. Information (Chapter 50) about water quality can be relayed in a variety of ways: signs posted at the point of contact, flag systems, radio announcements, press releases, television and web-based applications. Whatever media is used it must be formulated in a manner that is understandable and accessible to the intended audience. Education about hazards and exposure routes associated with outdoor recreational water contact must be also conveyed in a way that is compelling to ensure compliance with regulations. Public health education is one method of exposure intervention (Chapters 17–25) which, if properly crafted and delivered, can reach all segments of the intended population as part of a comprehensive implementation strategy (Chapters 26–32).

Education regarding water quality is important when one considers the extensive coastlines (China – 15,273 km, Greece – 13,600 km, Italy – 7,600 and the United Kingdom – 12,500 km) and likelihood of public interaction. The United States alone has nearly 37,000 km of ocean shoreline (excluding Alaska) and over 8,000 km of Great Lakes shoreline. Shorelines provide direct access to oceans, freshwater lakes and rivers. For example, outdoor recreational waters are visited by over one third of all Americans resulting in tens of billions of dollars in revenue (goods and services) and tens of millions of jobs being generated. The loss of recreational use can have a negative, direct economic impact as well as create public dissatisfaction through loss of utility, even in the absence of adverse health outcomes (Koteen et al. 2002, Houston 2013). In order for the quality of global recreational surface waters to improve or be maintained, public decision makers at municipal and national levels must intervene through the development regulations and enforcement schemes, combining both political and technical elements which are flexible and adaptive in order to meet changes both in the ecosystem and in the field of science as it evolves.

Key recommended readings

1 World Health Organization. 2003. Guidelines for Safe Recreational Water Environments: Volume 1 Coastal and Fresh Waters. http://www.who.int/water_sanitation_health/bathing/srwe1/en/index.html. Presents framework upon which many global outdoor recreational water quality regulations are based. Also provides pertinent information related to other chapters in this text.

2 US EPA. 2012. Recreational Water Quality Criteria (EPA 820-F-12-058). http://water.epa.gov/scitech/swguidance/standards/criteria/health/recreation/upload/RWQC2012.pdf. Presents a good statistical framework for the revision and derivation of statistical rational for regulatory monitoring breakpoints from the perspective of health risk. Includes information on the derivation of regulatory standards based on new molecular analytical methods; the first of its kind worldwide.

3 Health Canada. 2012. *Guidelines for Canadian Recreational Water Quality* – 3rd edition. http://www.hc-sc.gc.ca/ewh-semt/pubs/water-eau/guide_water-2012-guide_eau/index-eng.php. Presents regulation of outdoor recreational water quality from the perspective of a shared responsibility between federal, provincial and local governments. Due to its

background, the Canadian government gives a view of regulation combining elements of both the US EPA and WHO approaches.

4   Australian Government. Guidelines for Managing Risks in Recreational Water. 2008. http://www.nhmrc.gov.au/_files_nhmrc/publications/attachments/eh38.pdf.   Puts regulatory compliance monitoring of outdoor recreational waters via fecal indicator organisms in context to other potential hazards found in the larger marine aquatic environment. Comprehensive in nature with tie backs to the WHO (2003) document.

References

Aitken, MN. 2003. Impact of agricultural practices and river catchment characteristics on river and bathing water quality. *Wat. Sci. Technol.* 48(10): 217–224.

Australian National Health and Medical Research Council. 2008. *Guidelines for Managing Risk in Recreational Water.* Canberra: Australian Government.

Barwick, RS, Levy, DA, Craun, GF, Beach, MJ, and Calderon, RL. 2000. Surveillance of waterborne-disease outbreaks 1997–1998. *MMWR Surveill. Summ.* May 26, 49(SS04): 1–35.

Burak, S, Dogan, E, and Gazioglu, C. 2004. Impact of urbanization and tourism on coastal environment. *Ocean Coastal Manage.* 47(9–10): 515–527.

Centers for Disease Control and Prevention. 2000. Surveillance of waterborne-disease outbreaks – United States, 1997–1998. *MMWR.* 49(SS04): 1–35.

Centers for Disease Control and Prevention. 2004. Surveillance for waterborne-disease outbreaks associated with recreational water – United States, 2001–2002. *MMWR.* 53(SS-8): 1–22.

Colford, JM, Wade, TJ, Schiff, KC, Wright, CC, Griffith, JF, Sandhu, SK, Burns, S, Sobsey, M, Lovelace, G, and Weisberg, SB. 2007. Water quality indicators and the risk of illness at beaches with nonpoint sources of fecal contamination. *Epidemiol.* 18(1): 27–35.

Dwight, RH, Fernandez, LM, Baker, DB, Semenza, JC, and Olson, BH. 2005. Estimating the economic burden from illnesses associated with recreational coastal water pollution—a case study in Orange County, California. *J. Environ Manage.* 76(2): 95–103.

European Environment Agency. 2012. *European Bathing Water Quality in 2012.* Copenhagen: European Environment Agency.

European Union. 2006. Directive 2006/7/EC of the European Parliament and of the Council. *Official J. European Union* 4.3.2006, L (64) 37 – 51. http://eur-lex.europa.eu/LexUriServ/LexUriServ.do?uri=OJ:L:2006:064:0037:0051:EN:PDF.

Farahnak, A. and Essalat, M. 2003. A study on cercarial dermatitis in Khuzestan province, southwestern Iran. *BMC Public Health* 3: 35.

Fujioka, R. 1997. Indicators of recreational water quality. In: CJ Hurst, GR Knudsen, MJ McInerney, LD Stetzenbach, MV Walter eds. *Manual of Environmental Microbiology.* Washington, DC: American Society for Microbiology. p. 180.

Georgiou, S, and Langford, IH. 2002. Coastal bathing water quality and human health risks: a review of legislation, policy, and epidemiology, with an assessment of current UK water quality, proposed standards, and disease burden in England and Wales. (Working Paper ECM 02-06). Norwich: CSERGE.

Gerba, CP. 2001. Assessment of enteric pathogen shedding by bathers during recreational activity and its impact on water quality. *Quant Microbiol.* 2: 55–68.

Griffin, DW, Donaldson, KA, Paul, JH, and Rose, JB. 2003. Pathogenic human viruses in coastal waters. *Clin. Microbiol.* 16(1): 129–143.

Griffin, DW, Lipp, EK, McLaughline, MR, and Rose, JB. 2001. Marine recreation and public health microbiology: quest for the ideal indicator. *Bioscience* 51(10): 817–825.

Health Canada. 2012. *Guidelines for Canadian Recreational Water Quality.* Ottawa: Health Canada, Ottawa.

Houston, JR. 2013. The economic value of beaches – a 2013 update. *Shore and Beach* 81(1): 3–11.

Karn, SK, and Harada, H. 2001. Surface water pollution in three urban territories of Nepal, India, and Bangladesh. *Environ. Manage.* 28(4): 483–496.

Kay, D, Bartram, J, Prüss, A, Ashbolt, N, Wyer, MD, Fleisher, JM, Fewtrell, L, Rogers, A, and Rees, G. 2004. Derivation of numerical values for the World Health Organization guidelines for recreational waters. *Water Research* 38(5): 1296–1304.

Kay, D, Wyer, MD, Au-Yeung, C, Bartram, J, Figueras, M, Hunter, P, Kadar, M, Vargha, M, Salmon, R, and Thomas, D. (2009a) Epibathe Period 2 and Final Activity Report Submitted to the European Community DG-RTD, Contract 022618, under Framework Programme 6. http://www.epibathe. eu/.

Kay, D, Wyer, MD, Au-Yeung, C, Bartram, J, Figueras, M, Hunter, P, Kadar, M, Vargha, M, Salmon, R, and Thomas, D. (2009b) Epibathe 'Accessible' Public Report. Submitted to the European Community DGRTD, Contract 022618, under Framework Programme 6. http://www.epibathe. eu/.

Kinzelman, JL. 2005. Investigating Bathing Water Quality Failures and Initiating Remediation for the Protection of Public Health. PhD Dissertation. Guildford, UK: University of Surrey.

Kinzelman, JL, and McPhail, CD. 2012. Exposure interventions. In: A Dufour, J Bartram, R Bos and V Gannon eds. *Animal Waste, Water Quality and Human Health*. London: IWA Publishing.

Kinzelman, JL, Field, KG, Green, HC, Harwood, VJ, and McPhail, C. 2012. Indicators, sanitary surveys and source attribution techniques. In: A Dufour, J Bartram, R Bos and V Gannon eds. *Animal Waste, Water Quality and Human Health*. London: IWA Publishing.

Koteen, J, Alexander, SJ, Loomis, JB. 2002. Evaluating benefits and costs of changes in water quality. Gen. Tech. Rep. PNW-GTR-548. Portland, OR: U.S. Department of Agriculture, Forest Service, Pacific Northwest Research Station.

Lauber, C, Glatzer, L, and Sinsabaugh, R. 2003. Prevalence of pathogenic *Escherichia coli* in recreational waters. *J. Great Lakes Res.* 29(2): 301–306.

Lavender, JS, and Kinzelman, JL. 2009. A cross comparison of QPCR to agar-based or defined substrate test methods for the determination of *Escherichia coli* and enterococci in municipal water quality monitoring programs. *Wat. Res.* 43(19): 4967–4979.

McPhail, C, and Stidson, R. 2009. Bathing water signage and predictive water quality models in Scotland. *Aquatic Ecosyst. Health Manag.* 12(2): 183–186.

Nevers, M, Shively, D, Kleinheinz, G, McDermott, C, Schuster, W, Chomeau, V, and Whitman, R. 2009. Geographic relatedness and predictability of *Escherichia coli* along a peninsular beach complex of Lake Michigan. *J. Environ. Qual.* 38: 2357–2364.

Prüss, A. 1998. Review of epidemiological studies on health effects from exposure to recreational water. *Int J Epidemiol.* 27(1): 1–9.

Rose, J. 2004. Waterborne pathogens and indicators: a pathway forward. Proceedings of the 2004 National Beaches Conference. http://www.epa.gov/ost/beaches/meetings/2004/.

Sanborn, M, and Takaro, T. 2013. Recreational water-related illness: office management and prevention. *Canadian Family Physician* 59(5): 491–495.

Schwartz, E, Kozarsky, P, Wilson, M, and Cetron, M. 2005. Schistosome infection among river rafters on Omo River, Ethiopia. *J. Travel Med.* 12(1): 3–8.

Scott, T, Rose, J, Jenkins, T, Farrah, S, and Lukasik, J. 2002. Microbial source tracking: current methodology and future directions. *Appl. Environ. Microbiol.* 68(12): 5796–5803.

Sorvillo, FJ, Waterman, SH, Vogt, JK, and England, B. 1988. Shigellosis associated with recreational water contact in Los Angeles County. *Am J Trop Med Hyg.* 38(3): 613–617.

United States Environmental Protection Agency (US EPA). 2001. Guidelines establishing test procedures for the analysis of pollutants in ambient water; proposed rule. United States Federal Register – 40 CFR Part 136, Vol. 66, No. 169, August 30.

US EPA. 2012. Recreational water quality criteria. Office of Water Regulation and Standards, Criteria and Standards Division, United States Environmental Protection Agency, Washington, DC (EPA 820-F-12-058).

US EPA. 2014. Nonpoint source pollution: the nation's largest water quality problem. Nonpoint Source Control Branch, United States Environmental Protection Agency, Washington, DC (EPA841-F-96-004A). http://water.epa.gov/polwaste/nps/outreach/point1.cfm.

van Asperen, IA, Medema, G, Borgdorff, MW, Sprenger, MJ, and Havelaar, AH. 1998. Risk of gastroenteritis among triathletes in relation to faecal pollution of fresh waters. *International J. Epidemiol.* 27: 309–315.

van Asperen, IA, Rover, CM, Schijven, JF, Oetomo, SB, Schellekens, JFP, van Leeuwen, NJ, Collé, C, Havelaar, AH, Kromhout, D, and Sprenger, MWJ. 1995. Risk of otitis externa after swimming in recreational freshwater lakes containing *Pseudomonas aeruginosa*. *British Medical Journal* 311: 1407–1410.

Wade, TJ, Calderon, RL, Sams, E, Beach, M, Brenner, KP, Williams, AH, and Dufour, AP. 2006. Rapidly measured indicators of recreational water quality are predictive of swimming-associated gastrointestinal illness. *Environ Health Perspect*. 114(1): 24–28.

Walters, S, Gannon, V, and Field, K. 2007. Detection of *Bacteroidales* faecal indicators and the zoonotic pathogens *E. coli* O157:H7, *Salmonella*, and *Campylobacter* in river water. *Environ. Sci. Technol*. 41(6): 1856–1862.

WHO. 1999. Health based monitoring of recreational waters: the feasibility of a new approach (the Annapolis Protocol). Geneva: World Health Organization.

WHO. 2003. *Guidelines for Safe Recreational Water Environments – Volume 1 Coastal and Fresh Waters*. Geneva: WHO.

WHO. 2005. Ecosystems and human well-being: health synthesis: a report of the Millennium Ecosystem Assessment. http://www.who.int/globalchange/ecosystems/ecosys.pdf.

WHO. 2009. Addendum to the guidelines for safe recreational water environments - Volume 1 coastal and fresh waters – List of agreed amendments. WHO/HSE/WSH/10.04. Geneva: WHO.

46

REGULATION OF
SWIMMING POOLS[*]

Katherine Pond and Lowell Lewis

University of Surrey, Guildford, UK

Learning objectives

1 Understand the challenges in maintaining appropriate water quality of swimming pools.
2 Describe the global trends in regulations related to swimming pool water quality.
3 Understand the process of regulation development through a case study of the UK.

The need for regulation

Whilst swimming is advocated as one of the best forms of exercise, there are also negative health effects. Firstly, drowning is a leading cause of child mortality globally. In 2008, the United Nations Children's Funds (UNICEF) reported that there was a global incidence of 7.2 per 100,000 children drowning. Of all fatal injuries in children, 22 per cent were due to road traffic injuries and 17 per cent due to drowning. In addition, respiratory tract infections, gastrointestinal illnesses, ear, nose and throat infections and concerns about the carcinogenic effects of disinfection by-products such as trihalomethanes have been documented (Zwiener et al., 2007). Aside from the risks from drowning, infant swimming pool attendance is a contentious subject. Many pools offer swimming classes to children from three months of age but studies have shown an increased incidence of recurrent respiratory tract infections and otitis media in children using swimming pools from an early age and later on potentially to the development of allergic diseases. Children also show a high sensitivity to swimming-associated gastrointestinal illnesses (Bernard et al., 2007; Schoefer et al., 2008).

There are over 10 million swimming/leisure pools in use worldwide, and approximately 314 million visits to recreational water venues, including treated venues (e.g., pools), each year in the USA alone (US Census Bureau, 2007). An increase in participation in swimming-related activities is therefore not without issue. In the USA 12 per cent of public pools recently inspected were closed for serious violations, such as improper chlorine

---

[*] Recommended citation: Pond, K. and Lewis, L. 2015. 'Regulation of swimming pools', in Bartram, J., with Baum, R., Coclanis, P.A., Gute, D.M., Kay, D., McFadyen, S., Pond, K., Robertson, W. and Rouse, M.J. (eds) *Routledge Handbook of Water and Health*. London and New York: Routledge.

468

levels for example (CDC, 2010). Poor management as well as little or no awareness by users of potential health implications warrant the requirement for such establishments to be regulated.

There are a number of types of pools as described by the World Health Organization (2006). For example, pools may be private (domestic), semi-public (e.g., hotel, school, health club, condominium, cruise ship) or public (municipal or governmental). Pools may be located indoors, outdoors or both. In terms of structure, the conventional pool is often referred to as the "main" pool or public or municipal pool. It is by tradition rectangular, with no extra water "features," and it is used by people of all ages and abilities. However, there are many "specialist" pools for a particular user type – for example, paddling pools, diving pools, pools with special features such as water slides, and spas. There are also pools containing thermal or medicinal waters, such as physical therapy pools. Each of these types of pools poses particular hazards.

The concerns associated with the operation of pools are numerous, for example the production of disinfectant by-products, and water contamination issues, but also energy and water use. Large bather loads have the potential to exacerbate these if they are not managed effectively. In addition, different pool types present specific challenges with regard to safety issues. For example, hotels and clubs may have intermittent use which may place a burden on the water treatment plant; in addition, they may be managed by non-technically trained staff. Likewise school pools may be subject to similar issues. Outdoor pools may be especially easy to access and therefore may require particular attention towards security features. Cruise ships may utilise marine water which may present special challenges with regard to treatment. Hydrotherapy pools also require specific considerations in terms of controls and monitoring due to their nature – small size, warm water, easy access. Home pools face particular issues as they are easily accessed by children, posing a particular drowning risk, and are often maintained by owners who have little or no knowledge of the potential risks. Further details on how these specialised pools should be operated can be found in the Pool Water Treatment Advisory Group guidelines (PWTAG, 2009).

Despite the clear need for controls in the pool environment, globally, the regulation of swimming pools is poorly covered. In the United States, for example, there is no federal regulatory authority responsible for disinfected swimming pools, water parks, and so on; all pool codes are developed, reviewed and approved by state and/or local public health officials. As a result, there are no uniform, national standards governing the design, construction, operation and maintenance of swimming pools and other treated aquatic facilities. Thus, the requirements for preventing and responding to recreational water illnesses can vary significantly amongst local and state agencies (CDC, 2013). Similarly, in the UK, there are no specific regulations governing the design, construction and management of swimming pool facilities; however, designers and operators are liable for their actions under the provisions of the Health and Safety at Work etc. Act 1974 (HMSO, 1974). There is an increasing requirement for designers and operators to consider health implications associated with pool use and implement procedures for mitigating the potential risks inherent in the design and use of swimming pools.

The potential legal consequences of incidents associated with swimming pool use has driven the publication of a number of guidance documents that aim to assist pool designers and operators in ensuring that the risks are minimised. A wide range of organisations have published guidance over the past two decades focusing on different aspects of traditional pool operation (see for example PWTAG, 2009). At the global level this includes the World Health Organization who, in 2006, developed guidelines which include both specific guideline values and good practice. The guidelines address a range of hazards such as those leading to drowning and injury, water quality, contamination of associated facilities and air quality. It should be noted that these guidelines are not mandatory but are designed to be used as the basis for the development of approaches (including legislation) to controlling hazards that may be found in swimming pools (WHO, 2006).

In the USA, the CDC and the National Swimming Pool Foundation is developing a Model Aquatic Health Code (MAHC; http://www.cdc.gov/healthywater/swimming/pools/mahc/structure-content/index.html#open). It is intended that the MAHC will serve as a model and guide for local and state agencies needing to update or implement swimming pool and spa code, rules, regulations, guidance, law or standards governing the design, construction, operation and maintenance of swimming pools, spas, hot tubs and other treated or disinfected aquatic facilities. The code covers a range of subjects such as: recirculation systems and filtration; contamination burden; disinfection and water quality; facility design and construction; facility maintenance and operation; hygiene facilities; lifeguarding and bather supervision; monitoring and testing; operator training; regulatory programme administration; risk management/safety; ventilation and air quality; faecal/blood/vomit contamination response.

The Royal Life Saving Society Australia has published Guidelines for Safe Pool Operation. This is an example of a venue-based risk management tool provided as a voluntary guide for operators which helps an operator to meet their legislative duties and provide a high standard of care for visitors. Guidelines also exist for specialist areas such as pools in caravan parks or hotels. Further details can be found at http://www.royallifesaving.com.au/aquatic-centres/managers/guidelines-for-safe-aquatic-venues/hotels,-motels,-camping-and-caravan-grounds.

European regulations

Whilst coastal bathing waters are regulated by a dedicated directive (Bathing Water Directive), swimming pools across the European Community are not. In fact there are no enforceable European Union-wide swimming pool safety laws, and this is the case globally. The European Parliament has passed advisory laws governing pool safety products and procedures relating to both public and private swimming pools. Currently, these have not yet become enforceable directives throughout the whole of Europe. Of relevance to swimming pool operation is the EU Biocidal Products Directive which includes disinfectants. This provides a list of approved chemicals complying with the directive.

France enforces nationwide pool safety laws and imposes heavy fines for non-compliance for private pools and communal pools. Other European countries have specific regulations but these are not universal across Europe. In Spain, for example, pools are subject to a specific regulation, composed of national, regional (or of the autonomous communities) and local rules. These cover aspects such as pool fencing around communal pools and requirements for lifeguards. In addition, the Spanish Government is committed to adopt the European Child Safety Action Plan's Water Safety Guidelines, so it is likely that similar laws will soon be put in place with regard to private pools.

Mavridou et al. (2013) undertook a survey of pool and spa regulations in 20 Mediterranean countries – all very popular tourist destinations. The study highlighted the difference in emphasis on swimming pool regulations throughout this region. In summary, 17 of the countries surveyed had national regulations and 2 apply the WHO guidelines. Variations exist in terms of regulations, for example only 11 countries surveyed had any regulations for spa pools; 14 countries require the presence of a lifeguard and have provisions for the prevention of accidents; 16 countries require certain standards for microbiological and chemical quality of the water; 13 countries have specific regulations on filtration.

Outside the developed world there is very little information regarding controls relating to swimming pools other than specific regulations imposed by hotels where they have pools. However, the WHO guidelines are intended to be generally applicable to public pools and spas, semi-public pools and spas (as encountered in clubs, hotels and schools, for example) and private (domestic) pools and spas throughout the world. The preferred approaches adopted by national or local authorities towards implementation of guideline values and conditions vary between these types of environment and across different countries (WHO, 2006). There is currently no documented data relating to the level of use of the guidelines in developed or developing countries.

Singapore

In Singapore, swimming pools (or public pools) are regulated by the National Environment Agency. The Environmental Public Health (Swimming Pools) Regulations, 2010, define the treatment strategy to ensure the quality of the swimming pool water. The regulations state that "No person shall establish, manage, operate or run a swimming pool without first obtaining a licence from the Director-General." The licensees must comply with a set of water quality parameters such as pH, chlorine, turbidity, and so on, and the bacteriological quality of the water is defined as:

1   no sample of water shall contain any Escherichia coliform organism in 100 millilitres of water;
2   not more than one out of 5 consecutive samples of water shall contain any coliform organism in 100 millilitres of water, and in any case none of the 5 consecutive samples shall contain more than 10 coliform organisms in 100 millilitres of water; and
3   no sample of water shall contain more than 200 bacteria per millilitre as determined by the 24-hour plate count at 37° Celsius or by the membrane filter method.

The Director-General has the power to close a swimming pool if it fails to comply with the regulations or there is an outbreak of an infectious disease, which may endanger the health of any person using the swimming pool. In addition, the licensee may be fined or imprisoned for contravening the regulations.

Case study – review of UK guidance

The UK guidance for pools has been included here as a case study. In the UK, there are a wide range of organisations that produce guidance for different aspects of pool design and operation, but even in a country that is well regulated for coastal bathing waters, as discussed above, the design and operation of swimming pool facilities is unregulated in the most part. However, some aspects are covered by broader health and safety legislation and, in the case of competition pools, the sport's governing bodies. The Health and Safety Executive (HSE) have consolidated the relevant legal requirements for pool designers and operators into a single guidance document that aims to assist facilities to be constructed and managed in accordance with current UK regulations (HSE, 2007). The HSE first issued a guidance document specifically targeted towards swimming pools in 1988. The information within this guidance has been updated and amended over the years to incorporate the various developments in health and safety law, pool design, technologies and use. The latest edition of the guidance was published in 2007 and contains general guidance on the management of health and safety at swimming pool facilities.

The Management of Health and Safety in Swimming Pools Guidance (HSG179) provides a summary of the more generic legal requirements relevant to designers and operators of pool facilities and includes appropriate references to particular Health and Safety Regulations enforced by the Health and Safety at Work etc. Act 1974 (HMSO, 1974). The following regulations have been identified in the guidance document as being of significant relevance to the aquatic leisure industry: Electricity at Work Regulations 1989 (HMSO 1989); Workplace (Health, Safety and Welfare) Regulations 1992 (HMSO 1992); Reporting of Injuries, Diseases and Dangerous Occurrences Regulations 1995 (HMSO 1995); Health and Safety (Safety Signs and Signals) Regulations 1996 (HMSO 1996); Confined Spaces Regulations 1997 (HMSO 1997a); Diving at Work Regulations 1997 (HMSO 1997b); Provision and Use of Work Equipment Regulations 1998 (HMSO 1998); Management of Health and Safety at Work Regulations 1999 (HMSO 1999); Control of Substances Hazardous to Health Regulations 2002 (HMSO 2002); Construction (Design and Management) Regulations 2007 (HMSO 2007).

Additional guidance on specific microbial health hazards have been published in conjunction with other health agencies (HSE, 2007). Some aspects of competition pool design are also regulated to ensure that it does not impede or enhance athlete performance (Sport England, 2013). Repeated incidents of entrapment and other injuries led to the development of European Standards for pool equipment (BSI, 2001). These standards focused solely on the prevention of injury to users and operators of the pool equipment. The incorporation of new technologies in the pool water treatment process led to the development of new European Standards for pool design and operation that supersede the former code of practice (BSI, 2008a). These modified standards, however, do not consider the effects of these new technologies on the fundamental behaviour of the pool water. Legislation is slow to evolve due to the complex approval process and is therefore often found to be inconsistent with latest guidance or unable to cover new technologies.

The British Standards Institution is the UK's national standard body and is responsible for developing and certifying both national and international standards. These standards range from manufacturing specifications through to corporate management systems. Some of the standards that have been developed for the UK have been adopted globally and have become international standards, such as the quality and environmental management systems ISO 9001 and ISO 14001 (ISO, 2004; ISO, 2008).

Over the last decade a number of codes of practice and British Standards have been developed specifically for swimming pools. Some of these have come directly as a result of European implementation of standards (BSI, 2008a; BSI, 2008b) whilst others have been developed through the consolidation of best practice in the UK (BSI, 2003; BSI, 2004b). The British Standard, BS EN 15288: *Swimming Pools* (BSI, 2008a; BSI, 2008b), for example, is the UK implementation of the European standard for the safety requirements for the design and operation of swimming pool facilities.

This British standard, BS EN 13451: *Swimming Pool Equipment* (BSI, 2001; BSI, 2004a), is the UK's implementation of the corresponding European standard that was first published in 2001 and covers safety requirements and testing procedures for specific pool equipment. The standard is segregated into 11 parts, each focusing on a different type of equipment. The majority of the standards are focused around the physical hazards associated with pool components such as steps and treatment equipment. However, some of the standards include requirements that directly impact the water circulation of the pool.

Outside the health and safety related aspects of pool design and operation there is very limited guidance available for aquatic facilities. The Carbon Trust has developed guidance for the sport and leisure industry that aims to assist facilities to reduce their energy and water use and in turn become a more sustainable entity (Carbon Trust, 2006). The content of the guidance is limited to the technological amendments facilities can make to reduce their operational costs. There is currently no guidance that integrates the impacts of new technologies on the operational practices required to maintain safe bathing conditions with the sustainability benefits achievable through the use of these technologies and designs.

Future

The potential health impacts associated with microbial and chemical substances in pool water has been the driver for the development of guidance for pool managers and pool designers. However, the absence of standard regulations for swimming pools is very much at odds with the situation for surface recreational waters for which regulations and standards are quite well defined (see Chapter 45).

The increase in demand for aquatic activities, coupled with the increasing cost of construction and operation of swimming pools, has resulted in pools being designed to incorporate multiple activities. The installation of internal structures results in pool designs that are far more complex than the rectangular tank on which all the traditional guidance has been based. Similarly the adoption of new treatment technologies has made significant alterations to the operation of swimming pools. It is therefore important to continually review the current status of pool guidance and its ability to be effectively applied to the management of these new-generation facilities.

The majority of this guidance has been developed following the investigation and successful litigation of frequent incidents that have occurred in these facilities. Although this has resulted in some detailed specifications with respect to the physical form of the pools and its associated equipment, less emphasis has been placed on producing standards for aspects that do not have a direct impact on the safety of the users.

There are currently no standards for critical operational aspects such as water circulation within the pool tank and there is also little reference to energy or health aspects regarding the use of moveable floors and booms or ultraviolet (UV) disinfection systems. It is also apparent that many of the best practice recommendations surrounding water quality and water treatment have been developed based on experience or lab-scale experimental work rather than pool-based experiments. There is therefore a need for new research based on pool experiments in order to develop appropriate legislation.

Conclusions

Swimming pool regulations lag behind those of drinking water. Research in this area of public health is limited primarily due to the lack of funding. However, due to the popularity of this sport / leisure activity there is a need for consistency in standards across countries and regions. The WHO guidelines, in particular, provide applicable information to feed into the development of national legislation; however, there is a clear need for more evidence in relation to the health implications and risks associated with swimming pool use, including the implications of the more complicated designs being developed for leisure pools.

Key recommended reading (open access)

WHO (2006). Guidelines for safe recreational water environments. Volume 2. Swimming pools and similar environments. World Health Organization.

This publication provides a referenced review and assessment of health hazards associated with swimming pools and spas.

References

Bernard, A., Carbonnelle, S., Dumont, X. and Nickmilder, M. (2007). Infant swimming practice, pulmonary epithelium integrity, and the risk of allergic and respiratory diseases later in childhood. *Pediatrics*, 119(6): 1095–1103.

BSI (2001). BS EN 13451-1 – *Swimming Pool Equipment, Part 1: General Safety Requirements and Test Methods*. London: British Standards Institution.

BSI (2003). PAS 39 – *Management of Public Swimming Pools, Water Treatment Systems, Water Treatment Plant and Heating and Ventilation Systems Code of Practice*. London: British Standards Institution.

BSI (2004a). BS EN 13451-11 – *Swimming Pool Equipment, Part 11: Additional Specific Requirements and Test Methods for Moveable Pool Floors and Moveable Bulkheads*. British Standards Institution, London.

BSI (2004b). PAS 65 – *Management of Public Swimming Pools, General Management Code of Practice*. London: British Standards Institution.

BSI (2008a). BS EN 15288-1 – *Swimming Pools, Part 1: Safety Requirements for Design*. London: British Standards Institution.

BSI (2008b). BS EN 15288-2 – *Swimming Pools, Part 2: Safety Requirements for Operation*. London: British Standards Institution.

Carbon Trust (2006). CTV006 – Sports and Leisure, Introducing Energy Saving Opportunities for Business, London: Carbon Trust.

CDC (2010). Violations identified from routine swimming pool inspections — selected states and counties, United States, 2008. *MMWR*, May 21, 59(19): 582–587.

CDC (2013). Recreational water illnesses. http://www.cdc.gov/healthywater/swimming/rwi/

Environmental Public Health (Swimming Pools) Regulations (2010) Rg 10 G.N. No. S 308/1992 Revised Edition 2000 (31st January 2000).

HMSO (1974). Health and Safety at Work etc. Act. 1974 Chapter 37. London: HMSO.

HMSO (1989). Electricity at Work Regulations 1989. SI 1989/635. London: HMSO.

HMSO (1992). Workplace (Health, Safety and Welfare) Regulations 1992. SI 1992/3004. London: HMSO.

HMSO (1995). Reporting of Injuries, Diseases and Dangerous Occurrences Regulations 1995. SI 1995/3163. London: HMSO.

HMSO (1996). Health and Safety (Safety Signs and Signals) Regulations 1996. SI 1996/341. London: HMSO.

HMSO (1997a). Confined Spaces Regulations 1997. SI 1997/1713. London: HMSO.

HMSO (1997b). Diving at Work Regulations 1997. SI 1997/2776. London: HMSO.

HMSO (1998). Provision and Use of Work Equipment Regulations 1998. SI 1998/2306. London: HMSO.

HMSO (1999). Management of Health and Safety at Work Regulations 1999. SI 1999/3242. London: HMSO.

HMSO (2002). Control of Substances Hazardous to Health Regulations 2002. SI 2002/2677. London: HMSO.

HMSO (2007). Construction (Design and Management) Regulations 2007. SI 2007/320. London: HMSO.

HSE (2007). *HSG 179 – Managing Health and Safety in Swimming Pools*, 3rd Edition. London: Health and Safety Executive.

ISO (2004). *ISO 14001:2004 Environmental Management Systems – Requirements with Guidance for Use*. Geneva: ISO.

ISO (2008). *ISO 9001:2008 Quality Management Systems – Requirements*. Geneva: ISO.

Mavridou, A., Pappa, O., Papatzitze, O. and Drossos, P (2013). An overview of poll and spa regulations in Mediterranean countries. Proceedings of the fifth international conference swimming pools and spas. Rome, Italy.

PWTAG Ltd (2009). *Swimming Pool Water. Treatment and Quality Standards for Pools and Spas*. Diss: PWTAG.

Schoefer, Y., Zutavern, A., Brockow, I., Schafer, T., Kramer, U., Schaaf, B., Herbrath, O., von Berg, A., Wichmann, H.-E. and Heinrich, J. (2008). Health risks of early swimming attendance. *International Journal of Hygiene and Environmental Health*, 211 (3–4): 367–373.

US Census Bureau. (2007). Recreation and leisure activities: participation in selected sports activities 2007. http://www.census.gov/compendia/statab/2010/tables/10s1212.pdf

WHO (2006). Guidelines for safe recreational water environments. Volume 2. Swimming pools and similar environments. Geneva: World Health Organization.

Zwiener, C., Richardson, S.D., DeMarini, D.M., Grummt, T., Glauner, T. and Frimmel, F.H. (2007). Drowning in disinfection byproducts? Assessing swimming pool water. *Environ Sci Technol*. 41(2): 363–372.

47

WASTEWATER REGULATIONS*

Laura Sima

Johns Hopkins Bloomberg School of Public Health, Baltimore, MD, USA

Learning objectives

1 Describe the major categories of pollutants in wastewater.
2 Understand key challenges to enforcing wastewater regulation in developing nations.
3 Distinguish between water reuse and discharge into the environment.

Introduction

Wastewater[1] can contain a range of pollutants and its untreated release into natural water sources has the potential to significantly increase the incidence of water related diseases, expose ecosystems to toxic chemicals and contaminate precious aquifers (International Finance Corporation 2007a). Clear regulations, efficient management and effective technologies for treating wastewater are fundamental to appropriately balance the benefits and risks of wastewater generation and reuse.

Wastewater regulations vary significantly amongst countries, and often between regions of countries, and urban and rural regions within countries (Cunha Margues and Simoes 2010). In many developed countries, wastewater regulation drives advancements in wastewater treatment and collection processes while protecting public health and aquatic ecosystems (see Chapter 20 on Wastewater Treatment), but regulations of developing countries, even when such regulations are appropriate, may not be monitored or enforced, rendering regulations inconsequential. Since water and wastewater treatment are capital-intensive investments (Rouse 2014), clear and consistently enforced regulations are critical to development of the sector (Global Water Intelligence 2012). This chapter examines wastewater regulations around the world, but, as you approach these critical issues, bear in mind that the strength of law and regulation enforcement in any country will be a primary driver in the efficacy of any regulation (Cunha Margues and Simoes 2010).

---

* Recommended citation: Sima, L. 2015. 'Wastewater regulations', in Bartram, J., with Baum, R., Coclanis, P.A., Gute, D.M., Kay, D., McFadyen, S., Pond, K., Robertson, W. and Rouse, M.J. (eds) *Routledge Handbook of Water and Health*. London and New York: Routledge.

Regulations differ for industrial and domestic sources of wastewater, since they are each likely to contain different contaminants and pollutants. For example, wastewater from a mining site may need to be treated for heavy metals, while domestic wastewater may be higher in biochemical oxygen demand and pathogens. This chapter briefly discusses industrial regulations, but given that there are different regulations for the diverse classes of industrial pollutants and that these differ by country, we direct the reader elsewhere for a more detailed analysis of industrial pollution.

Regulations rely on transparent, credible and rigorous enforcement to incentivize compliance. In many developing countries enforcement can be weak, corrupt or irregular, undermining the utility of regulations for wastewater (Johnstone 2014), and further complicated by overlapping government jurisdiction for regulation and enforcement (UNEP 2010). A survey of South and Southeast Asian cities reveals that nearly all cities failed to meet nearly all applicable wastewater quality standards (Asian Development Bank 2009). Recent consideration of the outcomes of the privatization of utility monopolies reinforces the central prerequisite of strengthening governance and regulation enforcement to achieve environment and public health goals, especially in instances where monopoly private or state-owned entities control water and wastewater utilities (Beecher 2013; Scott and De Gouvello 2014). Exacerbating the problem of governance, many countries have elected to use developed country standards, which pose an unreasonable financial burden, rather than setting interim standards that can be attained at reasonable costs (von Sperling and Fattal 2001). In some cases standards have been unclear or even misleading, leading to unclear implementation (Johnstone 2003).

Wastewater treated to appropriate levels can be discharged into natural or artificial watercourses or water bodies, treatment ponds or wetlands, directly used for crop irrigation or further treated in advanced water treatment facilities for reuse as low quality or reused as drinking water source water, depending on regulations and treatment process. Treatment standards are designed to ensure that wastewater is safe enough for the most sensitive end use, taking into account the capacity of a receiving environment to withstand nutrient, pathogen and chemical loads without posing significant risks to people and ecosystems. As such, standards will differ based on how wastewater is reused. This section is segmented by options for wastewater discharge and reuse.

Pollutants

Wastewater contains several pollutant classes that can pose a risk to receiving waters.

Oxygen-demanding substances

Many of the pollutants in wastewater demand a certain supply of oxygen, referred to as biochemical oxygen demand, or BOD. Since dissolved oxygen is needed to support life, water with a high BOD can place a significant strain on aquatic ecosystems, making BOD a key determinant of wastewater treatment plant effluent quality.

Total suspended solids

Suspended particles can create a cloudy appearance for untreated, unfiltered water, which prevents aquatic species from getting light, and may indicate a certain level of contamination.

The term total suspended solids (TSS) is a measure of the total particulates within a given volume of water that are of a small enough particle size to remain suspended in water.

Nutrients

Large nutrient releases from sewage to receiving waters, especially phosphorus and nitrogen, can lead to excessive algae growth, which in turn blocks sunlight and oxygen from other aquatic organisms. At worst, nutrient release can lead to eutrophication of receiving water bodies.

Pathogens

Pathogens can be released into surface and groundwater sources from untreated, or inadequately treated wastewater of human, agricultural or industrial origin. Drinking is not the only way to come into contact with pathogens; often swimming or fishing in wastewater contaminated surface waters can aid in disease transmission. Wastewater treatment can help reduce the likelihood that diseases are transmitted within a population.

Inorganic and synthetic organic chemicals

Many substances of domestic use or that industries use for production can be harmful to aquatic species, and humans, in turn. Some chemicals include household detergents, pharmaceuticals, heavy metals, pesticides and manufacturing wastes. Chemical oxygen demand (COD) is a test that indirectly measures the amount of organic compounds in water. COD is a useful measure for water quality that is able to approximate the concentration of organic pollutants found in wastewater or surface water. COD measures the mass of oxygen consumed per gram of solution. Inorganic chemicals are often measured and characterized independently.

Discharge to the environment

Surface water can become unsafe for human drinking, fishing or swimming, and ecosystems vibrancy can be reduced where wastewater is released. Treatment standards can be either technology standards, specifying the treatment process that must be used, or effluent standards, specifying physical, biological and chemical criteria for water to be acceptably treated. These generally differ in each country or region where they exist. In the US, the Environmental Protection Agency uses both plant design standards to prescribe components characteristics for a variety of plant designs (US Environmental Protection Agency 1986) and additional requirements as part of the National Pollutant Discharge Elimination System, as authorized under the Clean Water Act.

Recent agreements under the Equator Principles have simplified standards within and among many developing countries (Leader and Ong 2011). Now, wastewater treatment and discharge standards of the International Finance Corporation's Environmental Health and Safety (IFC EHS) Guidelines apply to projects financed by the World Bank Group, the Organisation for Economic Co-operation and Development (OECD), the regional development banks, the United States Agency for International Development (USAID), the Millennium Challenge Corporation (MCC), the US Export-Import Bank and other international development organizations. As a result, IFC EHS Guidelines have been adopted as national standards in many of the world's developing nations (Leader and Ong 2011).

Table 47.1 Indicative values for treated sanitary sewer discharges from non-municipal sources

Pollutant	Units	Guideline values
pH	pH	6–9
BOD	mg/l	30
COD	mg/l	125
Total nitrogen	mg/l	10
Total phosphorus	mg/l	2
Oil and grease	mg/l	10
Total suspended solids	mg/l	50
Total coliform bacteria	MPN*/100ml	400

Note: * Most probable number.

Source: International Finance Corporation (2007).

IFC EHS Guidelines use both technology and effluent standards in different sections. As shown in Table 47.1, IFC EHS Guidelines specify indicative values for sanitary sewer discharges from decentralized, non-municipal sources, including industrial and independent household waste, which include limits for allowable concentrations of biological oxygen demand (BOD), COD, TSS, nitrogen, phosphorus and bacterial counts (International Finance Corporation 2007b).

For municipal wastewater treatment plants, the IFC EHS Guidelines reference relevant best practice technology standards in the European Union, US, Brazil, India, China and Mexico, without recommending preference for one set of standards over another (International Finance Corporation 2007b). In some countries, effluent standards may only be applied to large municipal plants, if at all.

Industrial wastewater[2] can have its own standards. Between 5 and 20 percent of all water goes to industrial use, and nearly 70 percent of the resulting wastewater is untreated in developing countries (WWAP 2009). Depending on the industrial sector, wastewater can include nutrient rich organics, synthetic organic chemicals and heavy metals, potentially threatening human health and the environment if improperly treated. A variety of technologies are available to treat contaminated industrial wastewater, and sector-specific regulations exist for wastewater emissions for sectors as diverse as mining, landfills, power generation, food processing and chemicals manufacturing (International Finance Corporation 2007a), but lax or ineffective enforcement of these standards reduces the incentive for industry to purchase expensive treatment and comply with the standards.

Wastewater reuse

Water scarcity, growing populations, urbanization, economic growth and climate change are increasing pressure on existing fresh water sources (Zimmerman et al. 2008). If managed properly wastewater reuse has the potential to address many of these challenges by recycling valued fresh water resources.

Reuse in agriculture

Wastewater reuse for irrigation or aquaculture can be an excellent means to recycle nutrients while benefiting from them to produce timber, fiber, fodder and food. If well managed, water reuse for agriculture can be an economically advantageous solution for treated wastewater disposal (Winpenny et al. 2010). In 2006 the World Health Organization released its third iteration of wastewater reuse for agriculture guidelines (World Health Organization 2006). Unlike previous iterations of the guidelines, which were very rigid and inapplicable to the socio-economic environment in developing countries, the current standards were designed to be flexible enough to benefit realistically in a wide variety of local conditions. This is done by using a risk-management approach to inform health-based outcomes using strict metrics, which allows governments to determine compliance requirements by first determining what a tolerable level of disease risk is for each agricultural application.

Indirect potable reuse

Planned indirect potable reuse is the release of treated wastewater effluent into an environmental buffer that is part of a municipality's water supply, through injection into groundwater aquifers to release into surface water reservoirs with the intent to increase supply. Although the effect is similar to direct potable reuse, it is more socially acceptable, and, as a result, it is becoming increasingly common (see Table 47.2). Still, municipalities are finding that they need to invest significant resources in public education campaigns to gain public acceptance for these projects.

In the United States, direct potable reuse has had to face considerable hurdles at the federal level, and, although state regulations have existed, projects like Montebello Forebay in Los Angeles County, Water Factory 21 in Orange County and the Scottsdale Water Campus have helped to achieve growing public acceptance of the concept. The projects not only enhanced public confidence in water reuse but also improved the knowledge base that we have to implement such projects. In 2012 the EPA released *Guidelines for Water Reuse* consolidating information about state policies for reuse in the US (US Environmental Protection Agency 2012).

Direct potable reuse

Direct potable water reuse is the planned introduction of recycled water either directly into a public water system or into a raw water supply upstream of a water treatment plant. This is already taking place in areas like Southern California, Singapore and Israel. Technology enables us to reclaim wastewater for potable consumption (see water treatment section), but public acceptance of the concept and lack of confidence have slowed its progress in development. Recently, increasing water scarcity and energy costs, combined with recent technological advances, are showing that these systems offer an undeniable potential (Shannon et al. 2007).

There are no US regulations for direct potable reuse in the United States for multiple reasons, including continued hesitancy and perception of public risk by some, and because it may call into question some of the existing laws under the Safe Drinking Water Act. But some direct potable reuse is nonetheless happening (US Environmental Protection Agency 2012). Internationally, in areas with extreme water resource constraints, direct potable reuse has been embraced more rapidly. In Windhoek, Namibia, however, water is valuable and scarce enough that direct reuse is a reality. Thus, Goreangab Reclamation Plant was designed to be the first direct potable water reuse plant in the world (Du Pisani 2005). In Singapore unprecedented efforts at public information sharing, including a state-of-the-art visitors' center, were used to sell the public on consuming NEWater, the reclaimed wastewater

Table 47.2 Non-exhaustive overview of some selected planned indirect and direct potable reuse installations worldwide

Country	City	Project capacity (mgd)	Description of advanced system for potable reuse
Belgium	Wulpen	1.9	Reclaimed water is returned to the aquifer before being reused as a potable water source
India	Bangalore (planned)	36	Reclaimed water will be blended in the reservoir, which is a major drinking water source
Namibia	Windhoek	5.5	Reclaimed water is blended with conventionally-treated surface water for potable reuse
United States	Big Spring, Texas	3	Reclaimed water is blended with raw surface water for potable reuse
United States	Upper Occoquan, Virginia	54	Reclaimed water is blended in the reservoir, which is a major drinking water source
United States	Orange County, California	40	Reclaimed water is returned to the aquifer before being reused as a potable water source
United Kingdom	Langford	10.5	Reclaimed water is returned upstream to a river, which is the potable water source
Singapore	Singapore	122	Reclaimed water is blended in the reservoir, which is a major drinking water source
South Africa	Malahleni	4.2	Reclaimed water from a mine is supplied as drinking water to the municipality

Source: US Environmental Protection Agency (2012).

product water (PUB 2012). Even with these success stories, on an international level the application of these direct reuse systems has been so rare that international quality standards have yet to be developed.

Key recommended readings (open access)

1 International Finance Corporation. (2007). "Sanitation" in *Environmental, Health and Safety Guidelines for Water and Sanitation*. Retrieved June 11, 2014, from IFC Environmental, Health and Safety Guidelines: http://www.ifc.org/ehsguidelines. IFC EHS Guidelines for water and sanitation include information about the treatment of collected sewage at centralized facilities.

2 Winpenny, J., Heinz, I., and Koo-Oshima, S. (2010). Executive Summary of *The Wealth of Waste: The Economics of Wastewater Use in Agriculture*. FAO Water Report 35, www.fao.org/nr/water. The United Nations Food and Agriculture Organization (FAO) report highlights the economic benefits of water reuse for agriculture. The executive summary discusses how various levels of treatment may make sense for a range of applications. The paper focuses on developing countries.

3 PUB. (2012). *Every Drop More than Once: Determination Drives Innovation*. Retrieved June 6, 2014, from From the First Drop: http://www.pub.gov.sg/annualreport2013/

article04_01.html. This is a brief explanation of the NEWater system and motivations in Singapore for the public. It shows pictures of the system and speaks to its history. For more fun, an online video explaining NEWater process: http://www.pub.gov.sg/ water/newater/newatertech/PublishingImages/newater_process.swf

Notes

1 Wastewater is an umbrella term for water degraded by human use, with many distinct interpretations. For this chapter, we use the term broadly to describe domestic effluent, including black-water (excreta, urine and fecal sludge) and grey-water (kitchen and bathing water), industrial effluent, including from hospitals, commercial institutions and manufacturing plants, and agricultural and aquaculture effluent in either dissolved or suspended form (UNEP 2010).

2 Given that there are different regulations for the diverse classes of industrial pollutants and that these differ by country, we direct the reader to Annex 1.3.1 of the General IFC Environmental and Safety Guidelines for Wastewater for a concise table of wastewater contaminant types and corresponding treatment options (International Finance Corporation 2007b). For an international comparison of regulations, treatment performance and regulations compliance costs we direct the reader to other research papers (Wheeler and Pargal 1999; Salmoaa and Watkins 2011; Global Water Intelligence 2012).

References

Asian Development Bank. (2009). *Asian Sanitation Data Book 2008: Achieving Sanitation for All.* Manila, Philippines: ADB.

Beecher, J. A. (2013). "What matters to performance? Structural and institutional dimensions of water utility governence." *International Review of Applied Economics* 27(2): 150–173.

Cunha Margues, R., and Simoes, P. (2010). *Regulation of Water and Wastewater Services: An International Comparison.* London: IWA Publishing.

Du Pisani, P. L. (2005). "Direct reclamation of potable water at Windhoek's Goreangab Reclamation Plant." In *Integrated Concepts of Water Recycling*, by S. J. Khan, M. H. Muston and A. I. Schafer. Wollongong: University of Wollongong Printing Services.

Global Water Intelligence. (2012). *Global Water and Wastewater Quality Regulations 2012.* Oxford: GWI.

International Finance Corporation. (2007a). *Environmental, Health and Safety Guidelines for Water and Sanitation.* http://www.ifc.org/ehsguidelines (accessed June 11, 2014).

International Finance Corporation. (2007b). *General Guidelines: 1.3 Wastewater and Ambient Water Quality.* 2007. http://www.ifc.org/ehsguidelines (accessed June 11, 2014).

Johnstone, D. W. (2003). "Effluent discharge standards." In D. Mara and N. Horan (eds) *Handbook of Water and Wastewater Microbiology.* London: Academic Press.

Johnstone, D. W. M. (2014). "Regulation and reality: some reflections on 50 years of international experience in water and wastewater." *International Journal of Water Resources Development* 30(2): 345–354.

Leader, S. L., and Ong, D. M. (2011). "Case study of a transnational, non-state actor agreement: the EP." In *Global Project Finance, Human Rights and Sustainable Development.* Cambridge University Press, 85–97.

PUB. (2012). *Every Drop More than Once: Determination Drives Innovation.* http://www.pub.gov.sg/ annualreport2013/article04_01.html (accessed June 6, 2014).

Rouse, M. (2014). "The worldwide urban water and wastewater infrastructure challange." *International Journal of Water Resources Development* 30(1): 20–27.

Salmoaa, E., and Watkins, G. (2011). "Environmental performance and compliance costs for industrial wastewater treatment – an international comparison." *Sustainable Development* 19(5): 325–336.

Scott, C., and De Gouvello, B. (2014). *The Future of Public Water Governance – Has Water Privatization Peaked?* London: Routledge.

Shannon, M. A., Bohn, P. W., Elimelech, M., Georgiadis, J. G., Marinas, B. J., and Mayes, A. M. (2007). "Science and technology for water purification in the coming decades." *Nature* 452: 301–310.

UNEP. (2010). *Sick Water? The Central Role of Wastewater Management in Sustainable Development.* Oslo: Birkeland Trykkeri AS.

US Environmental Protection Agency. (1986). *Design Manual: Municipal Wastewater Disinfection.* Cincinnati, OH: National Service Center for Environmental Publications.

US Environmental Protection Agency. (2012). *Guidelines for Water Reuse.* Washington, DC: National.

Von Sperling, M., and Fattal, B. (2001). "Implementation of guidelines: Some practical aspects." In *Water Quality: Guidelines, Standards and Health*, by L. Fewtrell and J. Bartram (eds). London: IWA Publishing, Chapter 16.

Wheeler, D., and Pargal, S. (1999). "Informal regulation of industrial pollution in developing countries: evidence from Indonesia." Policy Research Working Papers, http://dx.doi.org/10.1596/1813-9450-1416.

Winpenny, J., Heinz, I., and Koo-Oshima, S. (2010). "The wealth of waste: the economics of wastewater use in agriculture." www.fao.org/nr/water, (accessed November 14, 2013). FAO Water Report 35.

World Health Organization. (2006). *Volume 1: Policy and Regulatory Aspects.* http://www.who.int/water_sanitation_health/wastewater/usinghumanwaste/en/ (accessed 2014).

WWAP. (2009). *The United Nations World Water Development Report 3: Water in a Changing World.* UNESCO, London: Earthscan.

Zimmerman, J. B., MIhelcic, J. R., and Smith, J. A. (2008). "Global stressors on water quality and quantity." *Environmental Science and Technology* 42(12): 4247–4254.

48

WATER CHARGES
AND SUBSIDIES*

Richard Franceys

SENIOR LECTURER WATER AND SANITATION MANAGEMENT,
CRANFIELD UNIVERSITY, BEDFORDSHIRE, UK

Learning objectives

1 Understand the scale and pattern of the costs of water.
2 Understand the different approaches to charging for water.
3 Recognize the varying ways in which subsidies are and may be used.

The costs of water

It is a truth globally acknowledged that water is a gift of nature and God (or the gods) and no one should have to pay for it. Apart, that is, from the cost of the labour involved in carrying water from streams and rivers or in digging a well (this for the forebears of most readers here, but still today a reality for many people in the world) and carrying water back to the household. Those non-monetary (economic) costs can too easily be ignored (particularly, it appears, as it is women who continue to take on so much of the cost of carrying water).

By the time water service delivery involves handpumps, let alone sophisticated treatment and a network of pipes to distribute that water, it is clear that there are financial costs involved as well; initial costs for investment in plant and equipment (in addition to the costs of establishing and training village water committees in some parts of the world; communicating with customers about new developments or price increases in other parts of the world); and recurrent costs for operating pumping and treatment processes and minor maintenance. All of these costs have to be borne by society if water supply for public health is to be, at least, effective and sustainable (if not necessarily efficient or equitable in the first instance).

In addition to the needs for capital expenditure and operating expenditure, one of the inconvenient truths of water supply (as for all infrastructural investments) is that the assets delivered by the initial investment in capital expenditure become 'used up' or 'time-expired' and require renewal and rehabilitation – capital maintenance expenditure in

---

* Recommended citation: Franceys, R. 2015. 'Water charges and subsidies', in Bartram, J., with Baum, R., Coclanis, P.A., Gute, D. M., Kay, D., McFadyen, S., Pond, K., Robertson, W. and Rouse, M.J. (eds) *Routledge Handbook of Water and Health*. London and New York: Routledge.

modern regulatory accounting terminology, usually accounted for initially under the term 'depreciation'.

The capital intensive nature of water supply, that is an apparently 'zero' cost resource requiring expensive infrastructure to deliver it to where we want to use it, also incurs a cost of accessing the financial capital to fund those investments and re-investments – interest payments and perhaps dividends to the providers of capital, together referred to as the 'weighted average cost of capital', weighted according to the balance between debt (lenders) and equity (owners) capital. In accounting terms these costs of using capital are referred to as 'profit', something of great concern to some campaigners working to improve services to all – though the costs are real and have to be paid for somehow. Where there is effective economic regulation of monopoly water providers (rarely true when governments take on that role; not always true when 'semi-autonomous' economic regulators are responsible), then these costs of accessing the capital necessary to fund the fixed assets in water supply are considered a critical part of the overall costs of supply.

These cost components can be aggregated into a total cost of supply (operating expenditure + depreciation / capital maintenance expenditure + cost of capital) which has to be paid for if people are to continue to receive improved and, ideally, potable water. This 'cost plus' approach is different from the pricing of normal consumer goods, recognizing the monopolistic nature of water supply where it is never economic to have competing pipes supplying water to homes (though it has been attempted in times past). This is the reason for having economic regulation or detailed government oversight to ensure that it is only the efficient costs which are being paid for, not excessive costs due to over-employment or excessive 'economic' profits above the cost of capital.

So, whether it is a broken handpump in a low-income country village requiring maintenance to function again, or a diesel pump on a multi-village scheme in South Asia which requires continuing purchase of diesel, or a sophisticated water distribution network anywhere in the urban world which requires funds for the personnel to manage it as well as the chemicals to treat the water and the power to pump it … the delivery of potable water incurs financial costs, irrespective of the generosity of the initial gift.

Considering the scale of these costs, in assessments through the four country WASHCost project (Burkina Faso, Ghana, India and Mozambique), Burr and Fonseca (2013) report that for lower-income countries in rural areas the capital cost for preparing and installing a borehole and handpump (at 2011 prices) range from US$20 to just over US$60 per person. For small piped schemes in villages and small towns, including powered pumping systems from boreholes, costs range from US$30 to just over US$130 per person. For intermediate and larger schemes, investment costs vary widely from US$20 to US$152 per person. However, the WASHCost research showed that investment costs appeared to be the 'easy' part – household surveys revealed the extent of service failure and lack of support, from consumers or from government, to meet the recurrent costs. Burr (2015) reports that

> insufficient operational and capital maintenance expenditure means service delivery is stagnating. Across [five researched (WASHCost four plus Sierra Leone)] country contexts the failure to maintain or rehabilitate water supply systems results in extended periods of asset breakdown. Comparison between the amounts currently being spent and the estimated capital maintenance requirements of these systems suggests a very significant financial shortfall. Amongst the African countries these shortfalls are most profound, with maintenance requirements estimated at between five and six times existing expenditure for piped systems and over forty

times existing expenditure on community point sources. In Andhra Pradesh, India, expected required maintenance expenditure is estimated at between 1.3 and 3.0 times current levels.

It is not possible to ensure the health benefits of improved water supply without ensuring the continuation of that service beyond initial project implementation. One of the challenges of using low-cost, 'intermediate' technology, such as handpumps, is that the technology has a significantly shorter asset life, therefore requiring renewal sooner than often expected. Economic growth will not have been sufficient in that time to deliver community capability to renew assets which have initially been funded by donors and governments with 90–95 per cent grants (under the 'demand responsive approach'[1]) due to lack of affordability.

For larger-scale urban networked water supply systems in low- and middle-income countries, where initial investment has never been required to be funded directly by consumers in the first place, analysis of information from the IBNET database[2] (www.ib-net.org) showed that the median water fixed assets per person was US$66, with an average US$106 per person. This is a relatively small amount, partly influenced by the failure of utility accounting systems to revalue their past asset investments to reflect current value. IBNET (2014) reports directly that there is 'a wide variance between the O&M [operating and maintenance] cost per cubic meter of water sold by income categories. These cost differences are largely due to variations in service levels provided.' The challenge to water supply and resulting health impacts in urban areas is the likelihood that water will not have been treated adequately and will have been distributed without adequate final disinfection. With the subsequent challenge that the pipe network will distribute that water on an intermittent basis, commonly for two or three hours twice a day but not uncommonly for two or three hours once every two or three days (with outliers being once a week, once a fortnight and, at the extreme, once every six weeks). Intermittent supply is usually a direct result of insufficient financial resources being available to manage the system properly – but means that unpressurized pipes allow the ingress of contaminated groundwater back into the clean water pipes which is then delivered direct to household taps – a clear public health risk. Not to mention the impact on the conditions of water mains and the increased water leakage. IBNET reports median operations and maintenance expenditure in urban low-income country utilities of US$15 per person per year, US$26 in middle-income countries and US$93 in high-income countries. However, the underlying costs (as opposed to the actual expenditure) do not vary that significantly between countries (lower labour costs in low-income countries often being counterbalanced by significantly higher staffing levels). The limited expenditure reveals the likelihood of the service being equally limited both in quality terms and with regard to the population able to access that service. These both reflect a challenge to likely health outcomes.

In contrast, analysis of information from Ofwat, the economic regulator for England and Wales (and by way of a challenging benchmark, demonstrating apparently representative costs of fully potable water in continuous supply in high-income countries), indicates investment levels based on a 'modern equivalent asset value'[3] of US$4,472 per person for water assets[4] with US$132 per person recurrent annual costs of water supply[5] of which approximately US$38 per person annually represents the critical capital maintenance expenditure. If these are realistic full service costs which deliver (nearly) all the desired health benefits (whilst recognizing there are additional economic 'externality' costs which we could also consider (over-abstraction of groundwater, pollution of receiving waters; see Rogers et al., 2002), many to do with sanitation and wastewater, not addressed in this chapter), the question comes as to who should pay for these cost? And how should the costs be paid?

Charging for water

The default assumption is usually that the consumer of the water should pay directly – through user charges based on a system of tariffs differentiated between consumer segments, with some protection for the poorest. But the reality is that many consumers access water supply which has directly or indirectly been subsidized through government's use of taxpayer funds (or in low-income countries, donors as well), contributing to the initial investment with only the recurrent costs to be paid for directly through user charges. And then the evidence from the IBNET (2014) analysis suggests that in many locations recurrent capital maintenance expenditure is also not being recovered fully, the operating and maintenance (O&M) expenditure reported above reflecting almost no capital maintenance spending, another form of indirect subsidy.

This is true in rural areas where charges for handpump access have not worked particularly well (though a prepaid meter for handpumps has been developed – perhaps an overly expensive technical solution for a social challenge) and the occasional 'lumpy' costs of repair through 'fix on failure' are usually beyond the capacity of households which are too poor and too cash limited to have savings. Paying regularly for fuel for motorized boreholes appears to function satisfactorily, being a cost that consumers understand (but not usually sufficient to pay for occasional pump replacement). Paying for fixed costs through user charges can deliver if a village water committee has been sufficiently trained and supported (adding to the necessary ongoing support costs, with the charges sometimes costing more to collect than they are worth). Therefore paying for water supply in rural areas of lower-income countries through user charges may not be an appropriate solution, particularly where there are cost free solutions (streams, open wells, rainwater catchment) and apparently (recurrent) cost free solutions (handpumps and gravity flow water) available.

In urban areas where there is 'no alternative', then user charges should be easier to manage as a means of covering the costs of water supply. However, the revenues that are achieved by urban utilities, even the extremely limited US$15 cost per person per year low-income country urban expenditure reported earlier, are only supported by median revenues of US$12 per person per year (IBNET, 2014). For middle-income countries the figure is US$28 (relative to the US$26 median O&M expenditure) and US$132 for high-income countries (relative to median US$93 O&M expenditure). Only higher-income countries therefore demonstrate a sufficient margin of revenues over operating costs to allow for capital maintenance expenditure and the cost of capital.

The Organisation for Economic Co-operation and Development (OECD, 2009), following on from the Camdessus Report (Winpenny, 2003), brought clarity to this issue of paying and charging for water with their explanation of the 'Three Ts' – tariffs, taxes and transfers – as the only ultimate sources of paying for the costs of water. Water delivery can be paid for through user charges, that is, direct tariffs based on pipe size or number of people in the household or the amount of water consumed, through national taxation (a variety of types) but which requires the ongoing commitment of government to continue paying the costs if water is to be safely delivered – and transfers from international partners, mostly derived from taxpayers in higher-income countries (or 'other' philanthropic giving). Or any and every combination of those three 'T's.

The underlying principle is that water delivery costs; and someone, somewhere has to pay for it. There may be good reasons for harnessing a 'progressive' taxation system to ensure potable water for all, with particular social and health benefits for the poorest. However, experience has shown that governments can rarely be trusted to allocate sufficient taxes to pay all the costs – particularly capital maintenance. There is a fourth 'T', that of timing (or 'inter-generational transfer'), to refer to the common practice of delaying renewal of the expensive fixed assets so that a subsequent generation of consumers then has to pay more to undertake the by-then even more necessary work. The water distribution network in London, capital city of the high-income United Kingdom, was found by its new private owners to have one third of its previously public managed pipe network older than 150 years, half being older than 100 years – representing a significant backlog of renewal investment. Similarly the American Society of Civil Engineers[6] reports that water 'pipes and mains are frequently more than 100 years old and in need of replacement' at an estimated cost of 'more than $1 trillion' (though interestingly they also note that 'outbreaks of disease attributable to drinking water are rare'). Such inter-generational transfer should be (but never is!) used with care. Too often postponing renewal works leads to supply failure, perhaps more noticeable in convenience rather than public health terms, and the opportunity costs of accessing alternative, much lower quality sources, a greater cost to society than the initial deferred investment.

Even though the need for capital maintenance on pipe networks is much delayed, compared to the short asset lives of pumps for example (handpumps or motorized pumps), there is a charging principle of allowing for 'depreciation' so that consumers now pay their share of the cost of using up the value of fixed assets in delivering their water now, the depreciation 'fund' then being allowed to build up over time so that resources are available when eventually needed. In low-income countries ignoring some depreciation charges may well be an appropriate solution to ensure clean water now for the poorest, in the expectation that the next generation will have benefited from sufficient economic growth in order then to be able to pay for the needed renewal costs. Funding for renewal of shorter-life mechanical and electrical equipment still has to be provided for in present paying mechanisms – and necessarily depends on adequate governance structures which allow for the accumulation of these present savings and restricts the temptation to spend those funds on apparently more urgent needs.

A key question remains. Are consumers willing to pay the full cost of an improved water supply? What has been called 'the paradox of demand' reflects the idea that consumers in many societies like having clean water delivered to their homes, or being accessible at a relatively near standpost or handpump. But they do not want to pay the cost of that service … unless there is no alternative. If the alternative is to accept poorer quality water and more leaks from pipes, or a longer distance to walk … then society often prefers to use scarce financial resources on other things (households in small towns in Ghana paying three times as much for mobile phones as for water; Braimah, 2010). If there is no alternative then a community will find funds to pay for the repair of a failed handpump, to purchase fuel for the shared motorized borehole, or to pay for very small amounts of drinking water through plastic sachets and bottles. And there is the particular case of informal housing settlement or slum dwellers in many low-income countries (70 per cent of urban dwellers in Africa) having to pay a 'ten times tariff' in a slum for a 'jerrycan' or two of water – not representing 'willingness to pay', to misrepresent the economic terminology, but desperation to pay where there is no alternative and where, in many locations, the additional costs of intermediary vendors and water deliverers have to be paid directly at the point of consumption.

But it seems there has to be no alternative if consumers are to pay. In general, consumers in lower-income countries (and as reflected in resulting societal and governance approaches) do not value the change from a sub-standard (in quality and health outcome terms) service to an improved service sufficiently to be prepared to pay for the full costs now. And there are indications that they only overcome that unwillingness to pay when the average wealth of a country is high enough for the majority not to notice their water (and by then their wastewater) bill too much such that paying for the full costs of water becomes acceptable. A trend line of improving service quality against average economic wealth in a secondary city in a now high-income country (Franceys and Gerlach, 2008) can be recognized in the development of the water and sanitation sector in many other countries. Paying for water becoming acceptable that is until the tariff is raised and consultations reveal the only grudging acceptance to pay in a higher-income country for a very high quality service very efficiently delivered. The same consultations showing even less enthusiasm to support lower-income consumers through cross-subsidy social tariffs as demonstrated by the profusion of consumer surveys in the most recent five-yearly price review in England and Wales.

In the meantime, and even for a part of society in higher-income countries, to ensure that particularly the poorest can access the benefits of clean water it is necessary to consider how to share out the costs of water such that those who need it most can get it. Sharing out the costs, even with very low levels of affordability but to encourage all to pay something, enables people to understand their responsibility for the wise use of a 'scarce' and valuable resource.

Subsidizing water

Sharing out the costs of water amongst those who are best able to pay is an appropriate response to ensure a service which delivers significant health benefits to society as a whole. Economists speak of 'progressive' charging or taxation where the rich pay a greater share as opposed to regressive charging where all income groups pay the same. Paying different rates according to income, or need, leads to the concept of one group of users subsidizing another group.

Very significant subsidies are already present in most low-income country water systems, based on international transfers (usually only for capital expenditure) where richer taxpayers in richer countries pay, in addition to payments from national general taxation which are normally equally 'progressive' in collection (if not always in use). This ideal, however, is based on how any particular country collects taxes and the use of default subsidies described earlier through inter-generational transfers (postponing necessary capital maintenance to the next generation of consumers), all relative to who receives the benefits of potable water. A further example of default subsidies can be seen in normal accounting mechanisms of water utilities 'writing-off' bad debt from consumers who cannot be traced and 'enabled' to pay. This subsidy, progressive if it is only the 'can't pays' whose debts are being written off after six years of trying (as opposed to the 'won't pays'), is currently estimated to be just over US$10 per person per year in England.

Subsidies may therefore be external to the sector (through taxes and transfers) or internal where they are then referred to as 'cross-subsidies'.

Targeted subsidies may be through social welfare payments to very low-income consumers which include payment for water bills. Chile has a sophisticated system of assessing the needs of low-income households and, with respect to water, paying a proportion of the water bill directly to the service provider (Franceys and Gerlach, 2008), saying that it is the task of water utilities to supply water rather than manage social welfare. Similar approaches are to be found in any high-income country welfare approach, some of which allow for water payments to be deducted from social welfare payments ahead of receipt, to protect both the water supplier and the consumer (from an increase in their debts).

Similarly a targeted progressive subsidy for water is South Africa's 'Free Basic Water' approach whereby every household receives 6m^3 water per month at no cost to the consumer.

The cost to the service provider is recovered through support from government (taxes) and higher-income and industrial commercial consumers (tariff cross-subsidies).

An alternative approach to cross-subsidies, illustrating the variation in how water is charged for and subsidized, is shown in England and Wales where the fixed costs of water were recovered from households with a fixed bi-annual charge based on property values used in local taxation. This was a relatively progressive approach whereby the rich in larger, more expensive houses paid more than the poor, perhaps with larger families and greater water use but in less expensive properties. To illustrate the complexity of charging for water, the complementary desire for environmental best use of water has now led to metering being increased from 3 per cent at privatization to about 50 per cent now, at a cost considerably greater than the cost of the water saved, whilst simultaneously dismantling the earlier progressive charging approach, the rich in larger houses being the first to take advantage of 'free' metering in order to reduce their water bills. Those in new homes who have been compulsorily metered and who have particular health conditions (or three or more children) and are on social welfare benefits have their metered bills capped at the average metered bill. In addition some water companies established 'charitable trust funds' (non-governmental organizations (NGOs) in lower-income country terminology) with funding from their shareholders to support those who can't pay, with sharing mechanisms to write off debt as customers begin to pay again. And, apparently arbitrarily, the government has used national taxes to subsidize all consumers in one region with respect to their higher than average sewerage bills (well separated habitations on an extended hilly coastline leading to high costs in water pumping and sewage treatment). In the next step, the government has recently issued guidance based on earlier legislation allowing water companies to introduce 'social tariffs' to restore some level of cross-subsidy within the charging system, having earlier banned disconnections of domestic water consumers to ensure access for all. However, water companies have to consult their entire customer base on their preparedness to cross-subsidize poor households – the results of which show very limited willingness to pay for others, most customers believing that it is the task of government (through taxes) to manage welfare.

This illustrates the complexity of charging for water (and sewerage) when attempting to recognize all of the competing interests. For that proportion of the costs which can realistically be recovered directly from consumers (a mixture of affordability, willingness to pay and politics),

> the goals of user fees and cost recovery are:
> - to ensure sufficient revenue to deliver services over the long term;
> - to ensure sufficient revenue to support improved quality of services;
> - to ensure sufficient revenue to support extending service coverage, particularly to serve low-income consumers;
> - to ensure better use of scarce water resources and management of waste water disposal;
> - to conserve the natural environment by signalling to consumers the cost to the economy of the resources used by the services.
>
> *(Fonseca et al., 2010)*

And in the particular context of this publication, to emphasize that the goal should be to ensure adequate ongoing access to water for health benefits.

Balancing the competing interests is a challenge. A higher volumetric charge, for example, encourages environmental conservation but may limit consumption by the poorest, thereby threatening their health interests. In addition volumetric charging requires metering which adds to the overall cost paid by consumers (where costs are >80 per cent fixed, that is, overall costs are not so much a function of the amount of water consumed), further challenging affordability and use for health purposes, a particular challenge to low-income large families or multi-household compound housing with a single connection who are then unable to access the lifeline tariff.

This challenge can lead to 'multi-part tariffs' which are a mixture of fixed charges, so that everyone pays part of the overwhelmingly fixed costs of being connected to the supply system (approximately 90% of the costs of water supply being 'fixed', that is they do not vary relative to the amount of water being delivered, a notable characteristic of a capital intensive industry) and volumetric charges to influence consumption.

Metered (volumetric) water charges may be priced at a uniform rate per cubic metre consumed, irrespective of consumption, or through a variety of variable rates. Most common is the rising or increasing block tariff (IBT), with progressively increasing charges per cubic metre consumed per month (paid by consumption specified within each block or according to the highest rate block corresponding to total consumption). Alternative options are decreasing block tariffs (which better reflect the cost of supply of water: the more water supplied through a fixed network, the lower the average cost) and mixed block tariffs (MBTs) featuring a combination of the two.

The first block of consumption is often referred to as the 'lifeline' block, ideally representing the average consumption of a low-income household in a month, priced relative to likely affordability (and in South Africa priced at zero, being the 'free basic water' approach). In some countries the consumption allowed for in the initial lifeline block is so high that the majority of domestic consumers are able to purchase all their water within that subsidized level; in other countries the jump from the lifeline block to the normal tariff is so large that it encourages too many households to constrain their consumption to remain within that block, which leaves the utility with insufficient revenue to support ongoing operations and maintenance. Alternatively, very low-income consumers may well be accessing water through a shared tap or standpost where total monthly consumption easily surpasses the lifeline amount, thereby forcing the lowest-income households for whom the lifeline block was established to go without the subsidy and likely overly constrain their consumption.

Water usage charges may also be differentiated by type of user and also be adapted to reflect seasonal variations in supply and demand. Tariffs can be varied within a service area to reflect, for instance, administrative boundaries, pumping zones or historical precedent by including zoning differentials. Contingency charges may be included in response to droughts or other external events. Finally, conservation payments or credits may be provided in a tariff for customers demonstrating effective usage reduction. In many cases, tariffs will be a combination of these options. The ongoing challenge is the most appropriate allocation of costs between fixed standing charges (appropriate for a 'fixed cost' business) and measured volumetric charges (appropriate for a 'consumer pays' approach).

A move towards volumetric pricing is considered to be more efficient, as metering discourages the wasteful consumption patterns promoted by fixed charges. Metering has significant benefits, including providing customers with a sense of equity in that they pay according to their measured consumption. High fixed charges reduce customers' ability to influence the size of their water bills, while low fixed charges create revenue uncertainty for water companies.

The author's study for African Development Bank (2010) on user charges revealed the following picture:

> Large proportion of unmetered (or non-functioning metered) connections: The lack of incentives to conserve water leads to high costs for additional water supplies which are not usually matched with high-value uses.
>
> Large sizes of low-priced initial blocks in IBT structures distribute heavily subsidised water to the majority of households. Over-consumption leads to rationing, which in turn induces customers to invest in storage facilities and even the counter-productive use of suction pumps.
>
> Low average prices adversely affect quality and reliability of service and provide no financial incentives to utilities to expand into presently unconnected areas, leaving substantial minorities unserved.
>
> Shared connections in combination with IBTs, which were originally designed for exclusive use of a private connection by a single household, lead to affected poor households paying higher per unit costs than their middle and high income counterparts.

That study also suggested

> that where incremental block pricing cannot be avoided it is recommended that, in addition to specific connection size fixed charges: large users are charged at the long-run incremental cost of water (that is the cost of water supplied by the next large investment in reservoir and treatment capacity); average domestic (and institutional) consumers are charged at the average historical accounting cost of water and there is a limited lifeline block (5 or 6m^3 per connection per month maximum allowance) for the poorest only with, ideally, some provision to reflect household size and/or numbers of households per connection.

These recommendations are rarely implemented systematically.

The level of the lifeline block is judged not only by affordability but also by the amount of surplus revenue raised through the excess of large user incremental tariffs above the average cost of providing water. This is because of the imperative of adequate revenue collection to ensure capital maintenance for serviceability. The charge for the lifeline block should not be less than the operating and minor maintenance costs so as to limit the potential for distorting customers' understanding of the value of water. In addition, and potentially controversially, lifeline blocks should be targeted only at the poorest housing areas, that is, informal settlements including slums, shanties and tenements. Recognized formal housing areas should not be eligible for the lifeline block. The principles of simplicity and enforceability in tariffs, allied with the principles of integrated water resource management (IWRM) with consumers understanding something of the economic value of the resources being consumed, require a move towards a single volumetric tariff for the majority of users.

The different approaches are summarized in Table 48.1.

Supporting the slum dwellers – subsidizing standpost consumers and subsidizing connection charges

The billion slum dwellers in our rapidly urbanizing world are particularly at risk from poor quality water supply. Living in informal, often illegal, housing settlements, accessing water of even poorer quality than the conventional mains supply through legal and illegal vendors they pay a price between five and twenty times more per cubic metre than regularized connected consumers, managing by using very small quantities of water, likely sub-optimal for health. There are additional costs to the informal distribution system which have to be paid: the informal pipe extensions, the costs of the intermediary carriers and carters and the costs of the vendors. Traditionally these costs have been passed on directly to the poorest consumers, leading to the price differential. It can be argued that these costs, representing the failure of the public utility (whether publicly or privately managed) to extend its service to all urban consumers, the reason for having a public water supply in the first place, should be absorbed by that utility and added to the overall operating cost base.

This does not happen, representing a failure in public health policy terms, except in the now special case of prepaid electronic meters. These new devices, whereby the consumer has a prepaid token which slots into the standpost and discharges a fixed amount, have been very well received by consumers in Africa (Heymans et al., 2014) both for convenience and, more importantly, for the way in which they allow poor consumers to access the lifeline tariff. The recent World Bank study demonstrated the benefits but also the costs of attempting to maintain sophisticated electronic and mechanical equipment in difficult conditions. Whilst invaluable in some locations it should be that the approach of absorbing those vending costs, which is done by utilities in the case of 'robot' prepaid vending points, is extended to human vendors also.

There are other approaches to enhancing water services to the slum dwellers (Jacobs and Franceys, 2008) but in the longer term the most important development is to subsidize the initial connection cost (Franceys, 2005) such that the poorest can receive water directly to their homes, without risk of contamination through communal water points, water carrying and transfer between containers in the home. The National Water and Sewerage Corporation in Uganda has demonstrated a remarkable improvement in take-up of household connections in low-income areas as a result of absorbing the additional costs of new connections into their overall fixed asset base, rather than charging them directly to new connectees as has been usual practice.

Conclusions

The costs of delivering potable water for health are clear, that is, relatively low operating costs supported by high capital investment costs with resulting recurrent capital maintenance costs and costs of capital employed.

Whether through fixed charges per household or variable charges per volume consumed, society has generally shown an unwillingness to pay those costs of an improved supply until average wealth is high enough to marginalize the bill in consumers' perception.

The levels of default subsidies, through postponing capital maintenance and, more significantly, delivering a sub-standard service, at present mean that the full health outcomes of an improved water supply are not being realized by the poorest, particularly in lower-income countries. It appears that 'society' accepts or demands higher standards only as 'surplus' wealth becomes available. If global society wants to achieve most effective health

Table 48.1 Summary of charging options

Type	Description	Impact on revenue adequacy recognizing simplicity and enforceability			Impact on social fairness and health 'Developing poor', 'Coping poor', 'Very poor', 'Destitute'	Impact on conserving the water environment
		Stability and predictability of revenues	Ease of administration	Enforceability		
Household/property-based	Fixed per connection size	High	High	High	Medium	Low
	Variable and/or 'progressive' (pro-poor) By number of persons in household By property characteristics e.g. taps/bathrooms	High	Medium	Low over time	Medium	Low
	By property valuation By property size/frontage/built area Addition to council tax	High	High	High	High	Low
Volume limited	Entitlement to water is defined (absolutely, or qualified by actual availability) Fixed charge for limited volume per day Flow limiters Intermittent supply Time-based household tank filling Volumetric controllers	High	High	High	Medium (for large and/or multiple households on connection)	High
Volumetric metered	A fixed rate per unit water received, where the service charge is directly related to, and proportional to, the volume of water received	High	High	High	Medium	Medium
	Incremental block systems Lifeline blocks Average accounting cost blocks Average incremental cost blocks Large user discount blocks	Low	Low	Medium	Apparently high but not for multiple households per connection	High

Mixed	Some combination of fixed charge plus volumetric charge	Medium	Low	Low	Medium	Medium
Seasonal and interruptible	Higher charges in dry season/season of peak demand	Medium	Medium	Medium	Not applicable	High
	Large users accepting interruptions in peak demand season in exchange for lower tariffs					
Connection charges and infrastructure development charges		Medium	High	High	Low	Not applicable

Source: Fonseca, Franceys and Perry, African Development Bank, 2010

outcomes in advance of that socio-economic trend line in lower-income countries, it seems that international 'transfers' will have to be deployed to a greater extent than at present and targeted at recurrent operations (in rural areas) and capital maintenance expenditure (rural and urban areas) as well as at the more common initial capital expenditure stage. The other chapters in this book describe those health challenges and outcomes in detail and the reader will judge to what extent global society is willing to pay.

The recent experiences of Detroit (mass disconnections for not paying), Ireland (mass protests against the introduction of direct charging through tariffs) and England (very limited consumer acceptance to fund social tariffs) demonstrate the challenges in charging for and subsidizing water even in high-income countries. More promising is the experience of Chile where appropriate tariffs with well-designed taxation-based subsidies, supported by a drive to reduce costs of supply through reformed utility management linked with economic regulation, have delivered a transformed and viable water and sanitation service within a generation.

Key recommended readings (open access)

1 Fonseca, C., Franceys, R. and Perry, C. (2010) 'Guidelines for User Fees and Cost Recovery for Urban, Networked Water and Sanitation Delivery' and 'Guidelines for User Fees and Cost Recovery for Rural, Non-Networked Water and Sanitation Delivery', African Development Bank, October, Tunis. This report provides guidance for determining affordable tariffs for lower-income countries, within the context of appropriate external support, in the two different settings of networked, predominantly urban, water supply and non-networked, predominantly rural, supply.

2 Franceys, R. and Gerlach, E. (2008) *Regulating Water and Sanitation for the Poor – economic regulation for public and private partnerships*, Earthscan, London. This publication explores ten case studies across the world from both high- and low-income countries and explores how economic regulation (or pre-regulation) can ensure that any combination of tariffs and subsidies actually deliver adequate serves to the poor.

3 Shugart, C. and Alexander, I. (2009) 'Tariff Setting Guidelines – A Reduced Discretion Approach for Regulators of Water and Sanitation Services', PPIAF, World Bank, Washington DC. This article provides specific advice on methodologies and approaches to economic regulation as a means of facilitating tariff-setting at an appropriate level, ideally promoting transparency to consumers and giving additional information to politicians and policy-makers.

Notes

1 The rural water supply oriented 'demand responsive approach' aims to ensure local 'ownership' of systems in order to promote long-term sustainability of supply. It has been based around the idea of i) giving priority to communities that were seen to be seeking improvements, ii) it then enables communities to make informed decisions based on a range of approaches and iii) the community's willingness to pay the resulting costs (Sara and Katz, 1998). In practice, the demand responsive approach has often defaulted to communities being required to pay 10 per cent of the initial investment cost, with that often being allowed to be 'paid' through volunteer labour or the contractor eager to make progress on the scheme.

2 Analysis of the 267 middle- and low-income country utilities with service populations greater than one million giving 63 utilities reporting fixed asset values > US$20 per person out of 267 middle- and low-income country utilities with service populations greater than one million in the IBNET database.

3 A modern equivalent asset (MEA) value is the cost of replacing an old asset with a technically up-to-date asset with the same service capability – Ofwat 'Financial performance and expenditure of the water companies in England and Wales 2009–10'.
4 2010 most recently available figures.
5 2014 figures.
6 http://www.infrastructurereportcard.org/a/#p/drinking-water/overview

References

Braimah, C. (2010) 'Management of small towns water supply, Ghana', unpublished PhD Thesis, Cranfield University

Burr, P. (2015) 'The financial costs of delivering rural water and sanitation services in lower-income countries', unpublished PhD thesis, Cranfield University

Burr, P. and Fonseca, C. (2013) 'Applying a life-cycle costs approach to water costs and service levels in rural and small town areas in Andhra Pradesh (India), Burkina Faso, Ghana and Mozambique', IRC International Water and Sanitation Centre, January, Full Working Paper

Franceys, R. (2005) 'Charging to enter the water shop? The costs of urban water connections for the poor', *Water Science & Technology: Water Supply,* Vol 5, No 6, pp 209–216

Gerlach, E. and Franceys, R. (2010) 'Regulating water services in developing economies – regulating for universal access to water', *World Development,* Vol 38, No 9, pp 1229–1240

Heymans, C., Eales, K. and Franceys, R. (2014) 'The limits and possibilities of prepaid water in urban Africa: lessons from the field', World Bank Group WSP Report, Washington DC: World Bank

Jacobs, J. and Franceys, R. (2008) 'Better practice in supplying water to the poor in global PPPs', *Municipal Engineer,* Vol 161, No ME4, pp 247–254

OECD (2009) 'Managing Water for All – An OECD Perspective on Pricing and Financing', Paris: OECD

Rogers, P., de Silvab, R. and Raesh Bhatiac, R. (2002) 'Water is an economic good: How to use prices to promote equity, efficiency, and sustainability', *Water Policy,* Vol 4, No 1, pp 1–17

Sara, J. and Katz, T. (1998) 'Making Rural Water Supply Sustainable: Report on the Impact of Project Rules, Water and Sanitation Program', Washington DC: World Bank

Winpenny, J. (2003) 'Financing Water for All', Report of the World Panel on Financing Water Infrastructure, Marseille: World Water Council

49

WATER EXCHANGE SYSTEMS[1]

Srinivas Sridharan

ASSOCIATE PROFESSOR, MONASH UNIVERSITY, MELBOURNE, VICTORIA, AUSTRALIA

Dani J. Barrington

RESEARCH FELLOW, MONASH UNIVERSITY, MELBOURNE, VICTORIA, AUSTRALIA
AND INTERNATIONAL WATERCENTRE, BRISBANE, QUEENSLAND, AUSTRALIA

Stephen G. Saunders

SENIOR LECTURER, MONASH UNIVERSITY, MELBOURNE, VICTORIA, AUSTRALIA

Learning objectives

1 Understand the concept of marketing exchange systems in the context of the water sector.
2 Consider whether water supply systems are producing optimal outcomes for human and natural ecosystem needs.
3 Develop ideas toward new hybrid models for equitable and inclusive satisfaction of water demand.

Introduction

With the increasing pollution of freshwater resources and a corresponding rise in consumption of bottled water, there is an emerging view that the world is facing a water management crisis. This is compounded by aging and malfunctioning supply infrastructure and rapid urbanisation that is resulting in areas which existing supply systems do not reach. One global visible response to these inter-related threats has been to advocate for market-based

---

1 Recommended citation: Sridharan, S., Barrington, D.J. and Saunders, S.G. 2015. 'Water exchange systems', in Bartram, J., with Baum, R., Coclanis, P.A., Gute, D.M., Kay, D., McFadyen, S., Pond, K., Robertson, W. and Rouse, M.J. (eds) *Routledge Handbook of Water and Health*. London and New York: Routledge.

solutions that offer an increasing role for the private sector and enable better governance that is sensitive to local demand conditions. That is, market-based solutions that attempt to satisfy people's water needs and wants through a market pricing mechanism (Gulyani et al., 2005). However, the fundamental aim of the proposed post-2015 Sustainable Development Goal Six is to balance market demand with sustainable and equitable access to water and sanitation for all (United Nations Department of Economic and Social Affairs, 2014), and there is increasing uncertainty as to whether market-based solutions will be able to provide such universal access (Bakker, 2007; Ahlers et al., 2013a).

In this chapter we present an exchange systems approach to evaluate the potential of water markets to provide sustainable and equitable access for all. The approach is sourced from marketing theory, which in turn is constructed on a social exchange and general systems theory platform (Bagozzi, 1978; Layton, 2007). The exchange systems concept offers a systemic view of exchange that allows for the integration of broader economic and social considerations into the design of water markets. We illustrate different types of exchange systems to offer the reader a broader perspective on water markets. Such a perspective can enable new thinking in how water provision systems can contribute to improved human health. We begin with a general introduction to the marketing exchange concept and subsequently trace its evolution to exchange systems.

Marketing exchange concept

Marketing theorists have described marketing as the creation and resolution of exchange relationships (Bagozzi, 1978). What is exchanged in the relationship is "value", which can be social or economic in nature (e.g., time, energy, feelings versus goods, services, money) (Kotler, 1972). Whether marketing or the exchange process is seen as a predominantly economic or societal activity depends upon the prevalent mediating mechanism. In a market economy, for example, the market becomes a focal point of exchange relationships; the values exchanged tend to become more economic; and the exchange partners resemble economic institutions and consumers (i.e., willing buyers and sellers). By the same token, public-interest platforms can also be focal points of exchange relationships where the values exchanged are predominantly societal (e.g., voters and politicians, citizens and governments, communities and social organisations).

Hunt (1983, p13) emphasises that there are three inter-related areas of investigation that are of particular importance to marketing scholars when studying marketing exchange relationships:

1 The behaviours of exchange partners.
2 The institutional frameworks that facilitate the exchange.
3 The consequences of the exchange relationship on societal well-being.

The first area of investigation may be categorised as a micro-level topic, which involves developing a detailed understanding of human exchange behaviours (e.g., a study of perceptions and preferences that explain why and when an individual will seek to satisfy their needs and wants through exchange). The second area may be categorised as a meso-level topic, which raises questions about the larger institutional frameworks in society that broker exchanges (e.g., What kinds of institutions develop to facilitate exchange? Under what conditions will they develop? Once developed, what kinds of functions will they perform?). The third area may be categorised as a macro-level topic (Alderson, 1965), which allows for broader questions to be asked about the consequences of the exchange relationship on societal well-being (e.g., Is the exchange relationship producing optimal effects on a human or natural ecosystem? Is the relationship harmonious? Equitable? Inclusive?). Elevating the study of marketing exchange to meso- and macro-levels allows a transition to marketing exchange systems, the core subject matter of a specific school of thought, macromarketing (Layton, 2007).

Marketing exchange systems concept

Layton (2014) defines marketing exchange systems as "complex, adaptive social networks of individuals and groups linked through shared participation in the creation and delivery of economic value through exchange". One can view the (marketing) exchange system concept as the culmination of a logical ordering of analytical ideas from isolated exchange acts, to stable exchange relationships, to complex, adaptive social networks of exchange. Whilst the outcome of a single exchange act may be a sale or donation or installation of one product, a key outcome of an exchange system is the "assortment" of products that are generated (Layton, 2007). These could include tangible products or intangible services and experiences that are differentiated in terms of their form, price, location, promotion, availability, or a range of other exchange factors.

For example, a government's provision of piped water to household taps and its supply of water tankers could represent two distinct members of a "water assortment" made available to a particular community. Installing community standpipes would then further deepen and widen the water assortment that the community has access to. One can ask important micro-, meso- and macro-level questions when adopting such an exchange system frame of analysis. For example, why would a poor slum dweller sometimes prefer to procure water from a private seller, rather than from a community tap, even though the privately acquired water may cost five times as much as that from a community tap (micro-level question)? Under what conditions do informal water sellers proliferate (meso-level question)? Why does a community get access to one assortment of products rather than another, and why is this different (is there a disparity or inequity?) for another community (macro-level question)?

Therefore, the overall goal is to interrogate the marketing exchange system for the breadth and depth of assortments it can generate, which contributes positively to the quality of life of a community. This can be contrasted with "success" measures of an individual exchange act or relationship, which tends to be defined in terms of isolated benefit or profit.

Macromarketing theorists find it useful to classify four types of exchange systems: *market-based, command-based, culturally determined* and *non-market-based exchange systems*. These exchange systems are not mutually exclusive, and often co-exist in complementary ways on the ground. They may exist as several systems interlinked at their boundaries and working together, or as an amalgamated hybrid system. Regardless of its structure the analytical benefit of specifying and evaluating each type of system is that it highlights key features as to the behaviours of willing exchange partners (micro-level), the institutional frameworks that facilitate the exchange (meso-level), and consequences of the exchange on societal well-being (macro-level); all of which would not be evident through a traditional demand analysis. Examples of different types of water exchange systems which reside within individual archetypes are shown in Table 49.1.

Table 49.1 Examples of water exchange systems

		Level of aggregation		
		Large-scale	*Medium-scale*	*Small-scale*
Marketing exchange system	*Market-based*	Private multi-national bottled water seller	Private water tank provider	Private local water kiosk/seller
	Command-based	National government water utility provider	Provincial government water tank provider	Municipal water distributor
	Culturally determined	Community managed water systems	Community collective water sharing and bill splitting arrangement	Intra-household water sharing and distribution
	Non-market	Global non-governmental organisation (NGO) or charity providing water utilities and aid	Country-level infrastructural subsidies	Local NGO or charity distributing household water treatment filters

Market-based exchange systems

As previously mentioned, a market-based exchange system occurs when a willing buyer and seller enter into an exchange through a market pricing mechanism. From a buyer perspective a market-based exchange is a way to access goods or services, while from a seller perspective it usually provides an opportunity to pursue economic profit. The market-based exchange system has been criticised in the water and sanitation sector for not being universally inclusive and sustainable, for its relentless drive for profit at any cost and for failing to adequately address human rights (Davis, 2005). For example, the multi-national bottled water industry, which sells brand-name water in both developed and developing countries, has come under close scrutiny for its exploitative practices (Clarke, 2007), even though it may have achieved some broader social good (e.g., improving health outcomes by selling bottled water in developing countries in areas where the local water source is polluted). Despite much criticism, market-based exchange systems operate widely across the water sector – from small-scale water kiosks to large-scale, multi-national bottled water sellers.

Command-based exchange systems

A command-based exchange system occurs when an authority (often governments or regulatory authorities) provides goods and services through a regulatory institution that pursues a provision motive rather than a profit motive. The goal is often to ensure the health and well-being of a population by ensuring their right to water is upheld. However, in pursuing a non-profit motive, this system has been criticised for not providing people with an optimal range of choices (i.e., assortments), not fully engaging recipients in the decision-making process, and for being rigid and bureaucratic with low responsiveness to local economic and social dynamics (Mitlin, 2004). In the water sector, command-based exchange systems range from local government provision of water (e.g., community boreholes or wells) to large-scale water supply infrastructural projects. In addition, spurred by market-centric reforms, command-based exchange systems can often involve water authorities that regulate environmental water flow so as to provide ecosystem services (Commonwealth of Australia, 2007).

Culturally determined exchange systems

A culturally determined exchange system occurs when the provider and recipient enter into an exchange relationship primarily sanctioned by social traditions and norms rather than by economic institutions (Thapar, 1987). Given its culturally rooted motivations of reciprocity and locally equitable redistribution of resources (Layton, 2007), a culturally determined exchange system tends to produce some collectively beneficial outcome, rather than purely an individual gain (Levy and Zaltman, 1975). For example, in rural areas a community-scale water system may be maintained by an elected committee of representatives with the aim of providing all villagers with access to sufficient water resources. Alternatively, in informal settlements of some of the rapidly growing cities of the world, it is common to see households splitting their water bills with other households. This is potentially a win–win outcome in that the water on-seller shares fixed costs, while the recipient overcomes their difficulties to access or pay for individual water supply connections.

Non-market exchange systems

A non-market exchange system occurs when the supplier receives no explicit form of payment from the recipient when the good or service is provided (Kotler, 1972). Typically, non-market providers include charitable organisations, non-government organisations and other stakeholders that rely on donations or subsidies to fund their operations. For example, on a small-scale, charity organisations often provide water treatment technologies (e.g., filters) to individual households. Alternatively, global agencies such as the International Committee of the Red Cross can provide large-scale water aid projects or humanitarian missions. A common criticism in the water sector of the non-market exchange system is that the recipients often do not take ownership or feel invested in the eventual water supply solution provided (Marks and Davis, 2012).

Integration of multiple exchange systems

In reality, even in the simplest of cases, water exchange systems are likely to be complex hybrid systems. This section provides practical examples of water exchange systems and highlights how multiple forms of exchange can be involved in the provision of water from source to use.

Example 1: water marketing exchange system in England and Wales

Public water supply in England and Wales is provided to consumers (households, industrial and commercial customers) at a retail level by private companies (i.e., a market-based exchange between retailers and consumers); however, the price is regulated by the Office of Water Services, a government authority (i.e., a command-based exchange between regulator and retailers), resulting in a hybrid command–market exchange system (OFWAT, 2014) (see Figure 49.1). This hybrid system creates a "water assortment" that contains different constituent members – based on the various retail pricing and discount schemes, service features and guarantees offered by the various retailers. All of these share a common thread – the unchanging price component pertaining to the physical transmission of water to the local district, and the overall price cap imposed by the water regulators. The hybrid system can be analysed for its innovation in a number of exchange need areas (e.g., a better flow of information to end consumers, flexibility in contracts, instituting competitive forces).

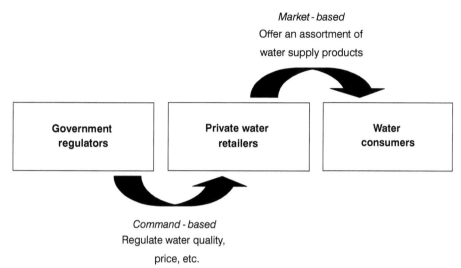

Figure 49.1 The relationships between water exchange partners in England and Wales

Example 2: complex source-to-use water exchange systems

In peri-urban Honiara, Solomon Islands, the authors have observed a complex system where all four marketing exchange systems are involved in providing households with water. The following explains how such systems can arise, particularly in low-income settings.

Often water utilities serve high-density urban centres through command-based exchange systems mandated by government regulations. However, while one part of an urban centre may receive water through pipes plumbed into houses, another part (e.g., an informal settlement) may only be provided with community-scale taps susceptible to uncertain water flow and breakdowns. In the latter context, the under-met need for water has stimulated local entrepreneurs in some cities to become informal water traders through running private trucker or tanker services or water kiosks (i.e., market-based exchange systems). For example, the size of the informal water market in Luanda, Angola, in sub-Saharan Africa, is estimated at US$250mn a year (Cain, 2014). The command and market-based systems produce their own assortments. The latter sometimes contain more numerous member products, as multiple providers can be involved (Ahlers et al., 2013b), which speaks to the breadth of its assortment; on the other hand, its pricing may be exorbitant and exploitative, denting its ability to be inclusive or enhance the quality of life of the community.

Such a situation has also arisen in peri-urban settlements where command-based water systems are unable to reach all consumers (Njiru, 2004). Here water utilities sometimes "unofficially" rely on small-scale informal providers (SSIP) as complementary "partners" to deliver water (Solo, 1999) – an unofficial hybrid exchange system. If the safety of water provided by the SSIPs becomes problematic, the utilities may choose to formally sub-contract them, thus moving to an official hybrid command–market exchange system where the utility can regulate the water quality. In such a case, the prices of SSIP services are regulated by the command-based portion of the system, but the assortment presented to consumer households remains in the market-based portion.

Further, it has been observed that households sometimes share the water they have purchased from small scale providers with neighbouring low-income households in interesting locally determined arrangements (Solo, 1999; Zuin et al., 2011). This has the

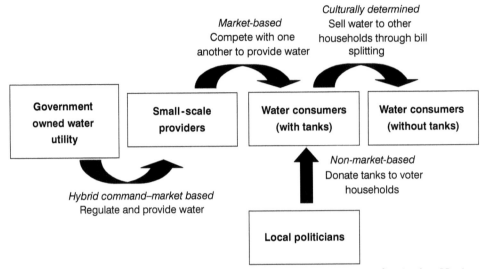

Figure 49.2 The relationships between water exchange partners in some parts of peri-urban Honiara, Solomon Islands

effect of inducing further complexity in how assortments are created and made available. It is an adaptive response to the worldwide reality that poor households cannot shoulder the upfront capital and hardware expenses for permissions, meters, pipelines and tanks (even though water sold volumetrically through such infrastructure may be more affordable per se). In such cases, those households with reliable water supply (such as a municipal connection) may demonstrate the initiative to satisfy the neighbourhood's water demand, and may either re-sell their water (a community-scale, market-based exchange system) or engage in a culturally determined exchange system such as splitting bills with their neighbours. To add complexity, in some instances households have a water connection or tank which they have acquired through a non-market exchange mechanism. For example, it may have been donated by a local politician.

Our example from the Solomon Islands highlights how all of these exchange systems can come together to provide a single household with water (Figure 49.2). In a particular informal, peri-urban community, some households split water bills with a neighbour as part of a culturally determined exchange system, where their neighbour received the tank as part of a non-market exchange system, where their water is provided by a small-scale provider working within a market-based exchange system but sub-contracted by a utility in a command-based exchange system.

Conclusion

The global debate around water supply systems has alternatingly supported private versus public water management, and more recently hybrid public–private partnership methods. However, such analysis, which is focused on governance and control issues, can obscure the nuanced nature of water demand, and not adequately consider the goal of achieving universal sustainable and equitable access to water services.

This chapter has presented readers with a broader perspective by which to move toward better understanding a demand-driven paradigm. The exchange systems lens highlights that becoming demand-driven is not merely about responding to what citizens desire or what they are willing to pay for in a consumerist sense, but rather about creating and delivering assortments that address both citizen demand and quality of life. By incorporating marketing exchange systems theory into our analysis of water systems, we can observe that it is the ability of the system to provide needs-fulfilling assortments that will determine its success or failure, rather than whether it is public or private. The subjects of analysis become the identities, behaviours, roles and relationships of exchange partners, the exchange flows they generate, the institutional frameworks within which they act, and the consequences such water provision systems bring about to the quality of living of communities.

We have also described and illustrated four water exchange systems types. It is unlikely that a single form will be a panacea to the global water management crisis; it is therefore unhelpful to engage in debate around which exchange system is universally the "best". Instead, it would be productive to analyse the functioning of existing systems using this analytical frame, and generate new insights into how the four system types can be constructively combined to provide water in a way that is responsive to people's needs and wants whilst moving towards universal equitable access.

Our premise is that water demand is a complex, multilayered concept. This demand may not even be about water per se, but rather about what the availability of clean and safe water enables people to achieve – such as avoiding disease, enjoying better health and well-being, and achieving economic progress. Hence it is imperative that water exchange systems are designed to generate innovative and adaptive water assortments that are economically, socially and environmentally enriching within the context of individual countries and communities.

Key recommended readings

1 Chenoweth, J. (2004), "Changing ownership structures in the water supply and sanitation sector," *Water International*, Vol. 29 No. 2, pp. 138–147.
   This reading investigates the changing role of the private sector in water supply in four countries, and how this has occurred within the country's enabling environment.
2 Layton, R.A. (2014), "Formation, growth, and adaptive change in marketing systems," *Journal of Macromarketing*, (in press) DOI: 10.1177/0276146714550314.
   This reading outlines a theoretical framework that explains how economic exchange causes marketing systems to form, grow and adaptively change.
3 Mankiw, N.G. (2014), *Principles of Microeconomics*. 7th Edition, Stamford, CT: Cengage Learning.
   This reading is a classic text on microeconomics, which is a branch of economics focusing on how the decisions and behaviours of individuals and institutions affect the supply and demand, and therefore prices, for goods and services.

References

Ahlers, R., Schwartz, K. and Guida, V.P. (2013b), "The myth of 'healthy' competition in the water sector: the case of small scale water providers", *Habitat International*, Vol. 38, pp. 175–182.

Ahlers, R., Guida, V.P., Rusca, M. and Schwartz, K. (2013a), "Unleashing entrepreneurs or controlling unruly providers? The formalisation of small-scale water providers in Greater Maputo, Mozambique", *Journal of Development Studies*, Vol. 49 No. 4, pp. 470–482.

Alderson, W. (1965), *Dynamic Marketing Behavior: A Functionalist Theory of Marketing*. Homewood, IL: R.D. Irwin.

Bagozzi, R.P. (1978), "Marketing as exchange: a theory of transactions in the marketplace", *American Behavioral Scientist*, Vol. 21 No. 4, pp. 535–556.

Bakker, K. (2007), "The 'commons' versus the 'commodity': alter-globalization, anti-privatization and the human right to water in the global south", *Antipode*, Vol. 39 No. 3, pp. 430–455.

Cain, A. (2014), "The Challenges of Community Water Management", Development Workshop, Angola. http://www.dw.angonet.org/content/water-sanitation (accessed October 15, 2014).

Clarke, T. (2007), "Inside the Bottle: An Exposé of the Bottled Water Industry", Canadian Center for Policy Alternatives, Ottawa, Canada.

Commonwealth of Australia (2007), Water Act. Basin Plan. Canberra, Australia.

Davis, J. (2005), "Private-sector participation in the water and sanitation sector", *Annual Review of Environment and Resources*, Vol. 30, pp. 145–183.

Gulyani, S., Talukdar, D. and Mukami Kariuki, R. (2005), "Universal (non)service? Water markets, household demand and the poor in Urban Kenya", *Urban Studies*, Vol. 42 No. 8, pp. 1247–1274.

Hunt, S.D. (1983), "General theories and the fundamental explananda of marketing", *The Journal of Marketing*, Vol. Fall No. 47, pp. 9–17.

Kotler, P. (1972), "A generic concept of marketing", *The Journal of Marketing*, Vol. 36 No. 2, pp. 46–54.

Layton, R.A. (2007), "Marketing systems—A core macromarketing concept", *Journal of Macromarketing*, Vol. 27 No. 3, pp. 227–242.

Layton, R.A. (2014), "Formation, growth, and adaptive change in marketing systems", *Journal of Macromarketing*, DOI: 10.1177/0276146714550314.

Levy, S.J. and Zaltman, G. (1975), *Marketing, Society, and Conflict,* Englewood Cliffs, NJ: Prentice Hall.

Marks, S.J. and Davis, J. (2012), "Does user participation lead to sense of ownership for rural water systems? Evidence from Kenya", *World Development*, Vol. 40 No. 8, pp. 1569–1576.

Mitlin, D. (2004), "Competition, regulation and the urban poor: a case study of water". In: Cook, P., Kirkpatrick, C., Minogue, M. and Parker, D. (eds) *Leading Issues in Competition, Regulation and Development.* Manchester: Centre on Regulation and Competition, Institute for Development Policy and Management, University of Manchester.

Njiru, C. (2004), "Utility-small water enterprise partnerships: serving informal urban settlements in Africa", *Water Policy*, Vol. 6 No. 5, pp. 443–452.

OFWAT. (2014), Regulating the Industry. Available: http://www.ofwat.gov.uk/regulating/. (accessed 20 May 2015).

Solo, T.M. (1999), "Small-scale entrepreneurs in the urban water and sanitation market", *Environment and Urbanization*, Vol. 11 No. 1, pp. 117–131.

Thapar, R. (1987), "Cultural transaction and early India: tradition and patronage", *Social Scientist*, Vol. 15 No. 2, pp. 3–31.

United Nations Department of Economic and Social Affairs. (2014), Outcome Document – Open Working Group on Sustainable Development Goals. United Nations Department of Economic and Social Affairs. Available: http://sustainabledevelopment.un.org/focussdgs.html (accessed September 3, 2014).

Zuin, V., Ortolano, L., Alvarinho, M., Russel, K., Thebo, A., Muximpua, O. and Davis, J. (2011), "Water supply services for Africa's urban poor: the role of resale", *Journal of Water and Health*, Vol. 9 No. 4, pp. 773–784.

50

INFORMATION IN WATER AND HEALTH[*]

Pamela Furniss

SENIOR LECTURER IN ENVIRONMENTAL SYSTEMS, THE OPEN UNIVERSITY, UK

Learning objectives

1 Recognise the role of information as a tool for change in water and health.
2 Understand how and why information is used in the water, sanitation and hygiene (WASH) sector.
3 Describe some of the ways in which WASH practitioners collect and provide information.

Introduction

In the context of water and health, information has multiple roles. It is a key link in the communication chain between communities, practitioners and decision makers; it is an essential component of education, advocacy and health promotion; and it underpins policy and practice by water and health professionals at all levels. In an overview of these different aspects of information, this chapter broadens the focus from water and health to include the wider sector of water, sanitation and hygiene (WASH).

Defining information

Information can be characterised as one stage in a hierarchy of content of the human mind. Russell Ackoff, a systems theorist, described the hierarchy as: 'Descending from wisdom there are understanding, knowledge, information and, at the bottom, data' (Ackoff, 1989). Information, therefore, is not an end in itself but a means to achieve the goals of greater knowledge, improved understanding and, hopefully, wisdom.

The hierarchy makes a distinction between information and data. Information is not the same as data but is derived from it. Data are raw facts. Information is data that has been processed or organised to give it meaning (Bellinger et al., 2004). For example, a survey

---

[*] Recommended citation: Furniss, P. 2015. 'Information in water and health', in Bartram, J., with Baum, R., Coclanis, P.A., Gute, D.M., Kay, D., McFadyen, S., Pond, K., Robertson, W. and Rouse, M.J. (eds) *Routledge Handbook of Water and Health*. London and New York: Routledge.

may produce data on the number of hand pumps in a district. It becomes information when linked with other items of data such as comparison with other districts, or correlated with the number of people using the pumps, or the depth of the boreholes, or some other variable.

Using information

An enormous amount of information is available these days, especially to anyone with access to the internet. A few key words entered in a search engine will take you to a vast range of sources from which to choose. Information on subjects related to water and health in the form of leaflets and guidance notes (online and print) from official sources is available in many countries. For example, in the UK, the Drinking Water Inspectorate (DWI) publishes leaflets for consumers on the quality of drinking water covering a wide range of topics including chlorine, fluoride, *Cryptosporidium*, pesticides, lead, nitrates and many more (DWI, various dates). The DWI also offers an enquiry service for anyone with additional questions not covered by the leaflets.

Having access to information brings many benefits but these days the complaint is probably more likely to be about too much (information overload) than too little. In order to make sense of it all, we need processes for filtering and selecting. How do you decide whether information is of value to you? Consciously or otherwise, you probably apply the following criteria:

- Accuracy: Is the information true? Can you trust the source?
- Relevance: Is it what you need/want to know?
- Availability: Can you access the information? Is it in a form that's useful to you?

Evaluating information in this way should help you to reach conclusions about its value, but using information is not only an individual exercise. Information exchange between two or more actors and wider sharing among multiple stakeholders are vital to ensure transparency, accountability and collective awareness of issues and problems. For example, in emergency response situations there may be many non-governmental organisations (NGOs) and other groups involved. To work effectively, it's important that they share information and communicate with each other about what they are each doing, and how and where they are doing it (Global WASH Cluster, 2008). Adopting information management processes that involve regular sharing should ensure that coordinated strategic decisions are taken, which in turn should result in complementary actions rather than overlapping or conflicting interventions.

In some contexts, an effective way of rapidly sharing information with many people is by use of Twitter, Facebook or other social media. This can provide the opportunity to communicate with a large audience in a transparent and accessible way. However, as users will be aware, social media offer potential to spread scare stories as well as accurate information so applying evaluative criteria, such as those suggested above, may be advisable.

Despite the general ubiquity of information, it is recognised in the WASH sector that there are many gaps in information required for effective governance and provision of water and sanitation services. Water utilities and governments often lack information on the needs of populations; funders and NGOs lack information on the continued functioning of projects; and communities lack information on how to make their case for improved facilities (Hutchings et al., 2012). Filling these gaps by collecting, delivering and sharing reliable and useful information is an integral part of projects and programmes for change.

Collecting information

According to Ackoff's hierarchy, the foundation of information is data. Before embarking on data collection it helps to have a clearly defined statement of the aim of your investigation: what do you want to find out? and why? This is essential to avoid wandering down irrelevant avenues that may be interesting but are not going to produce useful information.

Many investigations require data collection at several stages. Typically, baseline data, collected at the start of a project, is followed by regular monitoring and assessment of results. Ideally, these stages should be planned together. You don't want to reach the end of a project and discover you don't have the necessary baseline data to compare with the final outcomes.

You may be interested in quantitative or qualitative data, or both. Quantitative data, usually involving measuring or counting, answers questions such as how many? how much? and how often? Examples include demographic data (e.g. age, household size), climate (e.g. rainfall amount, seasonal pattern) and hydrology (e.g. water table depth). Qualitative data, useful for assessing people's attitudes and opinions, answers questions such as why? and how? You would use qualitative methods if you were interested in people's beliefs about disease transmission or their attitudes to home water treatment methods, for example.

Some information can be gathered informally by reading, observation or conversation but this is likely to be incomplete. More rigorous data collection processes such as surveys, questionnaires, interviews and focus groups may come into play. The use of information and communication technologies (ICTs) as a tool for data gathering in the WASH sector is increasing rapidly, though with varying degrees of success (Schaub-Jones, 2013). In particular, the rapid expansion of mobile phone network coverage has created many opportunities for innovative methods of data gathering and information exchange. For example, Ball et al. (2013) describe projects in Vietnam, Cambodia and Mozambique that use mobile phones for transmitting water quality data from users to water suppliers (see also Hutchings et al., 2012). Beyond information transfer, mobile networks can be used to transfer money and make payments for water supply services using systems that are available to anyone with a phone. This can bring several advantages including security, flexibility, ease of payment and availability to people without access to formal banking services (Hope et al., 2012).

Discussion of other key aspects of collecting information, such as questionnaire design, the need for representative sampling and statistical analysis of data, is beyond the scope of this chapter but one important point to keep in mind is that data gathering is in itself an intervention. If you are the researcher, you need to be conscious of the implications of your interaction with the people from whom you are gathering data, and consider any ethical issues that may arise. For example, research inquiries may delve into personally sensitive issues such as latrine use, hygiene practices and, more broadly, people's wider experience of poverty. In communities, the possible influence of social hierarchy, gender issues, economic status and religious beliefs are all important backgrounds to be aware of when planning any data collection activity. In academic contexts, most universities and other higher education institutions require research proposals to be assessed and approved by a human ethics committee or institutional review board if the research might involve sensitive issues of this nature.

Providing information

In some situations, participatory approaches to providing information that involve information exchange among many stakeholders may be appropriate and more effective than one-way processes (Akudago et al., 2013). However, whether one or more people are involved, part of the process will be conveying useful information to others. To be an effective communicator you should aim to understand the context in which you are operating and consider similar criteria to those mentioned above: accuracy (get your facts right), relevance (to your target audience) and availability (accessibility).

If you are planning to provide information, some key questions to consider are discussed below.

What is your aim? A clearly defined goal is a good starting point for any project. For water and health practitioners there are many possible reasons for providing information such as to raise awareness, to educate, to encourage behaviour change, to report on project progress or outcomes, to request funds, to influence policy, and many others. As well as a clear statement of the aim, you should also establish the scope of your project by deciding on issues such as the quantity of information to be provided and the timescale.

Who is your audience? This is a critical question. You need to understand your audience in order to plan and deliver the information in an appropriate way. For effective communication, it's important to avoid making assumptions about other people's lives, especially if your personal background is very different from that of your audience. You should try to put yourself in their shoes and see things from their perspective.

In more formal report-writing settings, the emphasis is on clearly and succinctly providing the information that is required in order to meet the needs of the person/people to whom you will submit your report. For community-based activities, language, levels of literacy, and awareness of the wider social and cultural context are important. What language(s) should you use to present information? Will people be able to read and understand written materials? Do you need to consider using several different approaches to ensure equal access to information for all members of the community?

You need to consider the characteristics and context of the people you are aiming to inform and pitch your information at the right level. For example, for a school discussion on hygiene, you would need to consider the age of the children concerned, their prior experiences and their understanding of the issues, and you'd also need to know about their local situation. There would be no point in enthusiastically explaining the importance of handwashing with soap after using the school latrines if the children have no soap or water next to the latrine block.

How will you deliver the information? Whatever the scenario, you need to select and organise the information to make sure you convey your message clearly, and then consider the best way to present it, remembering this is not necessarily a one-way process. Choosing a delivery medium that is appropriate for your target audience is obviously vital. There are many options including written documents, websites, blogs, posters, leaflets, maps, social media (e.g. Facebook, Twitter), conversation, discussion, group meetings, practical demonstrations, presentations, text messages, video, radio, drama, role play, participatory activities (e.g. community mapping, transect walks), and more.

In each case the information you wish to impart needs to be tailored to the medium and the audience you are aiming at. To elaborate on just one example from the list, well-designed posters are very effective devices for communicating simple information to raise awareness of an issue and, if displayed appropriately, can provide a regular reminder of good practice. Box 50.1 lists some general guidelines for preparing posters for community health education that were developed for use in Ethiopia.

Box 50.1 Guidelines for preparing a community health education poster

- Written messages should be co-ordinated with pictures or symbols.
- All words in a poster should be in the local language or two languages.
- The words should be few and simple to understand. A slogan might contain a maximum of seven words.
- Any symbols used should be understood by everyone, whatever their educational status.
- The colours should be eye-catching and the pictures meaningful to local people.
- Put only one or two ideas on a poster.
- The poster should encourage practice-action oriented messages.
- It should attract attention from at least 10 metres away.

(adapted from HEAT, 2010, Box 10.3)

What are your measures of success? How will you know if the information you are providing hits the spot? In some circumstances, measuring success can be straightforward. If your goal is to provide information as part of a funding bid then you know if this was successful when the bid is approved and you get the money. If you were submitting a report to a manager or project partner, the response you receive will tell you if you've met their requirements. In many situations, however, it is not so simple. If the purpose of providing information was to elicit a change in behaviour then you will need to consider how this can be assessed. For example, to measure the impact of a poster campaign advocating good hygiene practices, you might consider asking people if they had noticed the posters. You might also consider more challenging approaches such as conducting a survey to ask people if and how their behaviour had changed, or collect data on numbers and usage of household latrines, or how often mothers washed their hands after cleaning a baby's bottom, or monitor levels of diarrhoea in the community. In practice, a combination of methods is probably desirable and, with appropriate planning, should enable the measured impacts to be linked more accurately to the intervention(s) in question (Curtis et al., 2001). This returns to the need for baseline data mentioned above. Any of the options for evaluation require comparison with 'before' data in order for information on the 'after' situation to be meaningful. In accordance with Ackoff's hierarchy, the data you collect will provide information that increases your knowledge and understanding about the best ways to provide information effectively.

Conclusion

This chapter has attempted to cover a very broad subject. Inevitably it has been an overview, only able to touch on selected aspects of the role of information as a tool for change in WASH. It has briefly described some of the ways in which information is used within WASH projects and programmes. In conclusion, it is worth remembering that information alone is not enough to elicit change; what matters is what you do with it. Having read this general introduction, you might like to consider the question: how can you use information to make a difference?

Key recommended readings

1 WASH Cluster Information Management toolkit for use in WASH emergency response. http://washcluster.net/im-toolkitpage/. The WASH Information Management (IM) Toolkit is a resource to help Information Management Officers, Coordinators, Partners or other interested groups improve the flow of critical information during an emergency response. The Toolkit consists of an overall Guidance Note and a Support Pack with multiple sections that cover all the processes and requirements of IM during an emergency response. The Toolkit also includes a section of capacity building tools that are free to use for all agencies for training in preparedness or response.

2 Thomas, A. and Mohan, G. (eds) (2007) *Research skills for policy and development: How to find out fast*, London: Sage Publications/The Open University. Guide to planning, conducting and evaluating policy-related research. Designed for development managers in NGOs and public sector agencies globally, students of development studies and all those involved in policy making and research with development goals.

3 Hutchings, M.T., Dev, A., Palaniappan, M., Srinivasan, V., Ramanathan, N. and Taylor, J. (2012) mWASH: Mobile Phone Applications for the Water, Sanitation and Hygiene Sector, Pacific Institute/Nexleaf Analytics, http://www.pacinst.org/wp-content/uploads/2013/02/full_report36.pdf. Comprehensive review of the use of mobile phones for information exchange in the WASH sector (115 pp).

4 Mobile/Water for Development (mw4d), http://oxwater.co.uk/# mw4d is a research initiative based at the University of Oxford that seeks to design and test mobile communications technologies within the water sector.

5 Expanding our understanding of K★, United Nations University – Institute for Water, Environment and Health, http://inweh.unu.edu/wp-content/uploads/2013/05/KStar_ConceptPaper_FINAL_Oct29_WEB.pdf. A concept paper from the K★ conference held in Hamilton, Ontario, Canada, April 2012. K★= Knowledge Management, Transfer, Translation, Exchange, Brokering, Mobilisation, etc.

References

Ackoff, R.L. (1989) 'From data to wisdom: presidential address to ISGR', June 1988. *Journal of Applied Systems Analysis*, Volume 16, pp 3–9.

Akudago, J.A., Hutchings, M.T. and Nanedo, N.A. (2013) 'Participatory processes for designing and implementing information and communication technology tools for the WASH sector', IRC Symposium 2013, http://www.irc.nl/page/79357 (accessed 28 August 2013).

Ball, M., Rahman, Z., Champanis, M., Rivett, U. and Khush, R. (2013) 'Mobile data tools for improving information flow in WASH: Lessons from three field pilots', IRC Symposium 2013, http://www.irc.nl/page/79361 (accessed 28 August 2013).

Bellinger, G., Castro, D. and Mills, A. (2004). 'Data, Information, knowledge, and wisdom', http://www.systems-thinking.org/dikw/dikw.htm (accessed 28 August 2013).

Curtis, V., Kanki, B., Cousens, S., Diallo, I., Kpozehouen, A., Sangare, M. and Nikiema, M. (2001) 'Evidence of behaviour change following a hygiene promotion programme in Burkina Faso', *Bulletin of the World Health Organization*, vol 79, no 6, pp 518–527, http://www.who.int/bulletin/archives/79(6)518.pdf (accessed 19 September 2014).

DWI (various dates) Drinking Water Inspectorate, information leaflets, http://dwi.defra.gov.uk/consumers/advice-leaflets/ (accessed 19 September 2014).

Global WASH Cluster (2008) 'Global WASH Cluster Information Management Toolkit', http://washcluster.net/im-toolkitpage/ (accessed 3 September 2013).

Health Education and Training (HEAT) (2010) 'Health Education, Advocacy and Community Mobilization Module', http://www.open.ac.uk/africa/heat/heat-resources (accessed 6 August 2013).

Hope, R.A., Foster, T., Money, A. and Rouse, M. (2012) 'Harnessing mobile communications innovations for water security', *Global Policy*, vol 3, no 4, pp 433–442.

Hutchings, M.T., Dev, A., Palaniappan, M., Srinivasan, V., Ramanathan, N. and Taylor, J. (2012) 'mWASH: mobile phone applications for the water, sanitation and hygiene sector', Pacific Institute/Nexleaf Analytics, http://www.pacinst.org/wp-content/uploads/2013/02/full_report36.pdf (accessed 3 September 2013).

Schaub-Jones, D. (2013) 'Considerations for the successful design and implementation of ICT systems in the WASH sector', IRC Symposium 2013, http://www.irc.nl/page/79357 (accessed 28 August 2013).

51

THE HUMAN RIGHTS FRAMEWORK FOR WATER SERVICES*

Inga T. Winkler

SCHOLAR IN RESIDENCE, CENTER FOR HUMAN RIGHTS
& GLOBAL JUSTICE, NEW YORK UNIVERSITY

Virginia Roaf

INDEPENDENT CONSULTANT, BERLIN, GERMANY

Learning objectives

1 Understand the importance of the right to water in achieving universal access to water.
2 Understand what the right to water entails and what obligations states have to realize the right to water.
3 Understand the main contributions of using the human rights framework, in particular the focus on the most marginalized and disadvantaged individuals and groups.

The human rights framework provides important guidance for the water sector, and can make a significant contribution to improved policy-making and implementation towards ensuring access to water for personal and domestic uses for all. Human rights are enshrined in human rights instruments agreed by states. While the reference to a 'human rights-based approach' has become quite common, human rights provide, in fact, not an approach that can be chosen, but entail legally binding obligations on states. States have agreed at the United Nations (UN) General Assembly that water is a human right through the adoption of the resolution on the 'Human Right to Water and Sanitation' (UN General Assembly 2010). The UN Human Rights Council as the main UN body dealing with human rights affirmed by consensus, 'that the human right to safe drinking water and sanitation is derived from the right to an adequate standard of living and inextricably related to the right to the highest

* Recommended citation: Winkler, I.T. and Roaf, V. 2015. 'The human rights framework for water services', in Bartram, J., with Baum, R., Coclanis, P.A., Gute, D.M., Kay, D., McFadyen, S., Pond, K., Robertson, W. and Rouse, M.J. (eds) *Routledge Handbook of Water and Health*. London and New York: Routledge.

attainable standard of physical and mental health, as well as the right to life and human dignity' (UN Human Rights Council 2010).

The human right to water has a solid legal basis in international human rights law, and is derived from provisions in binding human rights treaties agreed by states, such as the International Covenant on Economic, Social and Cultural Rights, in particular as a component of the human right to an adequate standard of living. In 2002, the Committee on Economic, Social and Cultural Rights, the treaty body responsible for monitoring state compliance with the covenant, adopted General Comment No. 15 on the right to water as an authoritative interpretation of the covenant, which clarifies the content of the right to water.

The human right to water requires that services are available, safe, acceptable, accessible and affordable. These criteria define the normative content of the human right to water, and have evolved as a common typology used more generally by human rights bodies to define the normative content of human rights. In addition to these criteria, the right to water must be guaranteed without discrimination on the basis of equality, while ensuring participatory processes and accountable institutions.

Normative criteria to define the right to water

Availability

Water must be available in a sufficient quantity to meet requirements of drinking and personal hygiene, including menstrual hygiene for women and girls (see Winkler and Roaf 2014 and Sommer and Caruso, Chapter 52 in this volume), as well as further personal and domestic uses, such as cooking and food preparation, dish and laundry washing as well as cleaning (CESCR 2003, para. 12(a)). The exact quantity required cannot be determined in the abstract, since individual requirements for water consumption vary, for instance due to climatic conditions, level of physical activity as well as personal health conditions. While amounts should be adapted to specific circumstances, estimates and international recommendations can provide broad guidance for assessing whether the availability criterion is being met. The World Health Organization (WHO) estimates that all domestic needs can be met with about 100 liters per capita per day (Howard and Bartram 2003, p. 7). An absolute minimum amount in the context of disaster response is set at 15 liters per capita per day (The Sphere Project 2011, p. 98). However, such an amount raises health and dignity concerns, as it is insufficient to meet hygiene requirements. Similarly, the WHO identifies 20 liters per capita per day as basic access, but notes that this does not guarantee crucial hygiene requirements (Howard and Bartram 2003, p. 23). Fifteen to 20 liters per capita per day thus must not be understood to correspond to the full realization of the right to water.

Quality / safety

Water must be of such a quality that does not pose a threat to human health (CESCR 2003, para. 12). The transmission of water-borne diseases via contaminated water must be avoided. The WHO *Guidelines for Drinking-water Quality* define safe drinking water as water that 'does not represent any significant risk to health over a lifetime of consumption, including different sensitivities that may occur between life stages' (WHO 2011, p. 1), and they contain limits for a wide range of potentially harmful substances. To ensure water quality, facilities must be managed safely, which can be guaranteed through the adoption of water safety plans (WHO, 2011).

Acceptability

Apart from safety, which directly links to health requirements, water should also be of an acceptable color, odor and taste (CESCR 2003, para. 12 (b)). In many cases, water may be perfectly safe, but without attention to these features of acceptability, users may refuse to use it, or prefer to use an alternative unsafe source.

Accessibility

Water has to be accessible. A tap in the home is the best way to ensure both sufficient quality and quantity, but this may not be realistic to achieve in many settings in the short term. At the minimum, the human right to water requires that a water source is in reach of every household, bearing in mind the special needs of certain individuals and groups such as persons with disabilities. When it takes longer (up to 30 minutes for a round trip including waiting times), only basic access, with significant health risks, can be assumed (Howard and Bartram 2003, p. 18).

The route to the water source itself should be secure and convenient for all users, including children, older persons, persons with disabilities, women, including pregnant women, and chronically ill people. The risk of attack from animals or people, particularly for women and children, must be minimized.

Moreover, supply needs to be accessible on a reliable and continuous basis, to allow for the collection of a sufficient quantity of water to satisfy all needs, day and night, without compromising the quality of water.

Beyond the household, water must also be supplied in public places and institutions wherever people spend a significant amount of time, including at work, school, in health centers, places of detention and in public places, to be able to satisfy needs throughout the day.

Affordability

Use of water facilities and services must be available at a price that is affordable to all people. This must include all costs associated with it, ranging from regular tariffs in the case of networked provision, to connection fees that contribute to capital costs, to costs of on-site solutions such as wells or boreholes (capital costs as well as operation, maintenance and repair) and prices paid at nearby water kiosks or standpipes.

Paying for these services must not limit people's capacity to acquire other basic goods and services guaranteed by human rights, such as food, housing, sanitation, health services and education, which are interdependent with the right to water. Affordability does not necessarily require services to be provided free of charge. However, when people are unable, for reasons beyond their control, to access water through their own means, the state is obliged to find solutions for ensuring their access to water. This does not imply that providers may not recoup the costs of providing services – the human rights framework acknowledges the need for financial sustainability to ensure sustainable services. However, affordability has implications for how services are being paid for, for instance through tariffs that allow for cross-subsidization or through direct subsidies where people are unable to pay.

Obligations to realize the human right to water

The characteristic feature of the human rights framework, from which it derives its strength, is that it endows individuals with inalienable human rights and imposes corresponding obligations on states to realize these human rights. States' obligations can be understood as obligations to respect, protect and fulfill human rights. For the right to water, the obligation to respect implies that states must not prevent people already enjoying the right from continuing to enjoy access to water, for example by disconnecting water services without an adequate alternative being provided. The obligation to protect the right to water suggests that states must prevent non-state actors such as companies from interfering with the right, for example by polluting a water source. States must also regulate private companies and other actors involved in service provision. At the same time, non-state actors have responsibilities to respect the human right to water and to exercise due diligence to avoid any action which would result in human rights abuses.

The obligation to fulfill the right to water requires states to ensure that the conditions are in place for everyone to realize their rights. Governments are obliged to ensure that everybody gains access to these services over an acceptable time frame, through adopting appropriate legislation, policies and programs, and ensuring that these are adequately resourced and monitored. This does not imply that individuals are not responsible for ensuring their own access to water. Individuals are expected to contribute to the realization of human rights up to their means, for example by paying water tariffs, or by connecting to a mains service, and constructing and maintaining a tap in their own home. However, the state must ensure that everyone can access services and may need to put into place subsidies or other mechanisms. In some circumstances, where people cannot access their rights otherwise, the state may be required to provide water directly.

The obligation to fulfill does not imply that the state has to provide services itself; but rather, that it must make provision for the services to be delivered, either via a public company, municipal services, or another service provider. Crucially, however, even in the case of provision by non-state actors, the state retains the primary obligation for the realization of human rights and must exercise regulatory, monitoring and oversight functions over other actors as well as put in place the overall framework for the realization of the human right to water.

The human rights framework requires states to realize the right to water progressively, that is, to move towards the goal of full realization as expeditiously and effectively as possible, by taking deliberate, concrete, and targeted steps, using the maximum of their available resources (CESCR 1990, paras. 2 and 9). The obligation to utilize the 'maximum available' resources entails the duty of the state to raise adequate revenues through tariffs, taxes and other mechanisms and to seek international assistance where necessary (CESCR 1990, para. 13). At the same time, the principle of progressive realization recognizes that the full realization of human rights is a long-term process that is frequently beset by technical, economic and political constraints (CESCR 1990, para. 9). Progressive realization is not intended to provide states with an excuse not to act; rather, it acknowledges the fact that full realization is normally achieved incrementally, and that improved conditions are always possible. As such, the human rights framework does not demand the impossible, but rather provides a flexible framework that guides state action. It recognizes that resources and capacity may be limited but provides importance guidance on how to determine priorities. The human rights framework requires setting priorities in a way to meet the most basic needs first. Hence actions by states have to start and focus first on the un- and underserved. This means that basic access for all must be achieved before moving to higher levels of service.

Progressive realization relates to moving towards universal basic access for all, and it also relates to improving the level of services. The human rights framework requires progressive improvements, starting with minimum standards in terms of quantity, quality and accessibility, for instance referring to water in the proximity of the dwelling. However, human rights do not settle for minimum standards. They ultimately require achieving an adequate standard of living, which could, for instance, imply water supplied within the home. States that have already achieved basic access have to move beyond this in order to ensure the full realization of the human right to water.

Cross-cutting human rights principles

The human rights framework not only states the goal of universal access in line with human rights criteria, but also outlines the process of how to achieve this goal.

Non-discrimination and equality

Determining why particular individuals and groups do not have access to water will uncover discrimination and inequalities, often built into the fabric of society. Human rights challenge the existing power relations, by stating that inequalities in access to water are not only morally unacceptable, but also prohibited by international law.

The principles of non-discrimination and equality recognize that people have different needs as a result of inherent characteristics or past discrimination and therefore require different support. All human beings are entitled to their human rights without discrimination of any kind, based on race, color, sex or gender, ethnicity, age, language, religion, political or other opinion, national or social origin, disability, property, birth or other status. Discrimination can either be *de jure*, meaning that it is enshrined in law, or *de facto*, which includes indirect discrimination resulting from laws, policies or practices that are neutral on paper but have a discriminatory impact in practice. Both of these forms of discrimination are prohibited; however, the second type can be harder to identify and address.

States are required to ensure that individuals and groups enjoy substantive rather than just formal equality. States must actively diminish or eliminate conditions that cause or perpetuate discrimination. They must take active, affirmative and targeted measures aimed at ensuring all people enjoy the right to equality and the right to water. This means that states must prioritize the needs of people who are targets of discrimination or are marginalized or at risk of human rights violations. This will include analyzing whether existing water policies and practices are discriminatory, whether explicit in law, or stemming more from historical discrimination or social and cultural practices. The right to equality requires states to ensure that legislation, policies, programs and other measures are reformed to address and remedy discrimination, marginalization and exclusion.

Participation, transparency and access to information

Participation and access to information have long been key aspects of good development practice, helping to ensure acceptability, affordability and sustainability of water services. The human rights framework obliges states to ensure participation that is active, free and meaningful and provides people with an opportunity to influence decision-making.

Access to a participatory process must be facilitated for all individuals and groups concerned by a decision, including those who are marginalized, or who are stigmatized in the community or who may require particular accommodations to fully participate. The element of state obligation is important, as this ensures that participation is a continuing process, rather than simply suggesting that participation is a good idea.

Transparency and access to information are essential for participation to be meaningful. This requires providing information in accessible formats, languages and various channels and media. People must be able to understand the information presented. Only when this is the case can they meaningfully participate in decision-making.

Participation must be ensured at all relevant levels. While participation at community level regarding particular projects such as the siting of a water point is often seen as relatively straightforward, challenges arise even in this context. In particular, it must be ensured that the process is not captured by the majority group – the powerful, the male, the predominant ethnic group – but that it is truly inclusive. Women, persons with disabilities, older persons, those with lower social status, people belonging to ethnic minorities, and others who tend to have lesser opportunities to voice their views and needs, must be heard and their opinions taken into account. Women's increased engagement, particularly in programs designed to improve access to water, which tends to be very much women's responsibility, is recognized as having had a positive effect on the sustainability and appropriateness of water services.

However, the right to participation does not stop at the local level, but extends to the sub-national and national levels, including in planning and budgeting processes. Processes must be devised that allow for the representation of all those concerned in the process. While not every individual might have the opportunity to participate directly and express his or her opinions, democratic processes must ensure that inputs are gathered through various administrative structures and feed into the national level decision-making. Again, from a human rights perspective, it is crucial to ensure the inclusiveness of this process, inter alia through avoiding that it is captured by a few well-established civil society organizations.

Accountability

Accountability is at the heart of the human rights framework. States must comply with legal norms and standards enshrined in human rights instruments, and rights-holders are entitled to appropriate redress where states fail to comply with human rights. Human rights define the relationship between the state as duty-bearer and individuals as rights-holders. The human rights framework emphasizes that access to water is not dependent on the charitable benevolence of certain actors, but is in fact a legal entitlement.

Accountability provides the framework for monitoring services and providing redress for human rights violations. Accountability mechanisms can take various forms, including complaints mechanisms at different levels such as service providers, regulators or administrative bodies, national human rights institutions and international human rights treaty bodies. States have an obligation to put into place accountability mechanisms that are accessible, affordable, timely and effective. Accountability can also be established at the political level through parliamentary review committees, petition committees or similar structures. Apart from these more formal mechanisms, achieving accountability can be complemented through social mobilization and activism, media reporting, campaigning and lobbying. Civil society organizations can play an important role in holding states accountable to their obligation to progressively realize the right to water.

Ultimately, however, the human rights framework allows for alleged violations of the right to water to be taken to the courts, and enables individuals to claim their rights. While litigation is a last resort, the possibility of it strengthens the position of individuals. Moreover, access to an independent and functioning judicial system is crucial in case other forms of accountability fail to respond effectively to the violations in question. Access to justice must be ensured not only on paper but also in practical terms. States must put into place mechanisms to overcome physical, economic, linguistic, administrative and other barriers for accessing the courts, including through legal aid schemes.

Conclusion

Using the human rights framework to advance access to water for all does not require an entirely new approach. Human rights do not exist in isolation and independent from good practice in the sector – they draw on standards developed in the sector for specifying the content of the right to water. However, using the human rights framework brings significant benefits. Access to water must not be perceived as a matter of charity, but rather a legal entitlement. Human rights give legitimacy and voice to those who are usually not heard, and they provide a legally binding framework as the basis for accountability. They require shifting the focus to the most disadvantaged, marginalized, vulnerable and un- and underserved, making sure that no one is left behind. All too often those who lack access to adequate water are also disadvantaged in other ways, highlighting the need to address the interdependence of the human rights to water, sanitation, food, health, education and others. By seeking to make sure that contextualized, local solutions are found that respond to people's needs and concerns, human rights help to achieve more sustainable interventions.

Realizing the right to water will almost invariably require challenging the existing power structures, to ensure that people who do not enjoy their right to water are given the opportunity to claim this right. This happens not only through protest or through the courts, but also by means of policy, legislation and regulation, understanding and respecting the key principles of human rights and prioritizing the needs of marginalized and vulnerable individuals and groups.

Human rights principles and standards provide a framework that states and other actors can use to assess current access to water and sanitation services, and to design approaches to improve access for those who lack it. The human right to water requires as much attention in countries where only a few people do not have access to water as they do in countries where significant numbers of people do not have access.

Key recommended readings

1 Committee on Economic, Social and Cultural Rights (2003), General Comment No. 15, The right to water (arts. 11 and 12 of the International Covenant on Economic, Social and Cultural Rights), UN Doc. E/C.12/2002/11, 20 January 2003, available at www2.ohchr.org/english/bodies/cescr/comments.htm (accessed 12 August 2013). General Comment No. 15 provides an authoritative interpretation of the International Covenant on Economic, Social and Cultural Rights (a binding human rights treaty) to include an implicit right to water and specifying its normative content and state obligations corresponding to the right to water.

2 de Albuquerque, C. (2014), *Realising the Human Rights to Water and Sanitation: A Handbook*, available at www.ohchr.org/EN/Issues/WaterAndSanitation/SRWater/Pages/SRWaterIndex.aspx. This handbook explains the meaning and legal obligations that arise from the human rights to water and sanitation, translating the often complex technical and legal language into accessible information. It provides concrete guidance

and checklists for states to follow on how to implement these human rights, with a strong focus on how states can be held to account for delivering on their obligations.

3 de Albuquerque, C. and Roaf, V. (2012), *On the Right Track, Good Practices in Realising the Rights to Water and Sanitation*, available at www.ohchr.org/Documents/Issues/Water/BookonGoodPractices_en.pdf (accessed 12 August 2013). A compilation of good practices in realizing the rights to water and sanitation, offering solutions, ideas and pragmatic examples of legislation, policies, programs, advocacy approaches and accountability mechanisms to demonstrate how human rights become reality.

4 Centre on Housing Rights and Evictions et al. (2007), *Manual on the Right to Water and Sanitation,* available at www.unhabitat.org/pmss/getElectronicVersion. aspx?nr=2536&alt=1 (accessed 12 August 2013). A guidance tool offering insights into using human rights standards and principles for addressing practical difficulties in water and sanitation provision, describing a range of policy measures that could be adopted to make the water and sanitation sector operate in a more pro-poor, inclusive and accountable manner.

References

de Albuquerque, C. (2012), Special Rapporteur on the human right to safe drinking water and sanitation, Catarina de Albuquerque. Stigma and the realization of the human rights to water and sanitation, UN Doc. A/HRC/21/42

de Albuquerque, C. and Winkler, I. (2011), 'Neither friend nor foe – Why the commercialization of water and sanitation services is not the main issue in the realization of human rights', *Brown Journal of World Affairs*, 17(1), pp. 167–179

Freshwater Action Network (2010), *Rights to Water and Sanitation: A Handbook for Activists*, London: Freshwater Action Network

Howard, G. and Bartram, J. (2003). *Domestic Water Quantity, Service Level and Health*, Geneva: World Health Organization

Langford, M. and Russell, A. (eds.) (2015), *The Right to Water: Theory, Practice and Prospects*, Cambridge: Cambridge University Press

The Sphere Project (2011), *Sphere Handbook – Humanitarian Charter and Minimum Standards in Humanitarian Response*, 3rd edn, Bourton on Dunsmore: Practical Action Publishing

UN Committee on Economic, Social and Cultural Rights (CESCR) (1990), General Comment No. 3: The Nature of States Parties' Obligations (Art. 2, Para. 1 of the Covenant), UN Doc. E/1991/23

UN Committee on Economic, Social and Cultural Rights (CESCR) (2003), General Comment No. 15, The right to water (arts. 11 and 12 of the International Covenant on Economic, Social and Cultural Rights), UN Doc. E/C.12/2002/11

UN General Assembly (2010), The Right to Water and Sanitation, Resolution, UN Doc. A/RES/64/292

UN Human Rights Council (2010), Human Rights and Access to Safe Drinking Water and Sanitation, Resolution, UN Doc. A/RES/HRC/15/9

United Nations, Office of the High Commissioner for Human Rights (OHCHR) (2010), The Right to Water, Fact sheet no. 35, Geneva

WASH United, Freshwater Action Network and WaterLex (2012), *The Human Right to Safe Drinking Water and Sanitation in Law and Policy – A Sourcebook*, Berlin: WASH United, Freshwater Action Network and WaterLex

WHO (2011), *Guidelines for Drinking-water Quality*. 4th edn, Geneva: World Health Organization

Winkler, I. (2012), *The Human Right to Water: Significance, Legal Status and Implications for Water Allocation*, Oxford: Hart Publishing

Winkler, I. and Roaf, V. (2014), 'Taking the bloody linen out of the closet – menstrual hygiene as a priority for achieving gender equality', *Cardozo Journal of Law and Gender*, 21(1), pp. 1–37

52

MENSTRUAL HYGIENE MANAGEMENT AND WASH*

Marni Sommer

DrPH, MSN, RN Associate Professor of Sociomedical Sciences,
Mailman School of Public Health, Columbia University, USA

Bethany A. Caruso

MPH, Center for Global Safe Water, Department of Behavioral Sciences and
Health Education, Rollins School of Public Health, Emory University, USA

Learning objectives

1 Describe the menstrual hygiene management (MHM) challenges facing girls in school and other contexts.
2 Explain why MHM is an important gendered topic within water, sanitation and hygiene (WASH) in need of attention.
3 Identify the key components of an MHM in WASH intervention.

Introduction

In recent years, the specific challenges related to menstrual hygiene management (MHM) for adolescent girls and women in low-resource contexts have been increasingly documented and researched. A recent systematic review of health and social effects of MHM found that reproductive tract infections related to MHM are plausible, but that intensity and type of infection are not clear. Additionally, the routes of transmission are not well understood, including what the role of inadequate water, sanitation and hygiene (WASH) may be (Sumpter and Torondel, 2013). Nonetheless, WASH facilities are critical for women and girls to manage menses with dignity, and there is growing attention to the importance of integrating MHM into WASH guidelines. In 2012, the Joint Monitoring Programme (JMP)

---

* Recommended citation: Sommer, M. and Caruso, B.A. 2015. 'Menstrual hygiene management and WASH', in Bartram, J., with Baum, R., Coclanis, P.A., Gute, D.M., Kay, D., McFadyen, S., Pond, K., Robertson, W. and Rouse, M.J. (eds) *Routledge Handbook of Water and Health*. London and New York: Routledge.

for Water Supply and Sanitation of the World Health Organization and the United Nations International Children's Fund (UNICEF) developed a working definition for MHM as part of an effort to define goals, targets and indicators for post-2015 global monitoring of WASH:

> Women and adolescent girls are using a clean menstrual management material to absorb or collect menstrual blood, that can be changed in privacy as often as necessary for the duration of a menstrual period, using soap and water for washing the body as required, and having access to facilities to dispose of used menstrual management materials.
>
> *(JMP, 2012)*

A recently published 2014 handbook outlining the human rights to water and sanitation also gives specific attention to menstruation (Albuquerque, 2014). Specifically, it is noted that women and girls require toilets that enable MHM, including receptacles for the disposal of materials and appropriate hygiene facilities, and that they have the right to be free of cultural practices and taboos that compromise dignity.

Both the MHM definition and the menstruation-specific aspects outlined in the human rights document (see the preceding chapter on "The human rights framework for water services") can serve as a useful guide for WASH practitioners, policy makers, and researchers focused on addressing adolescent girls' and women's MHM-related needs in WASH service delivery.

In this chapter we provide a brief overview of the gendered needs of girls and women utilizing WASH facilities in low-resource settings as a backdrop in which to discuss MHM in those locations. We then specifically address 1) the MHM challenges that arise for girls and women in many low-resource contexts, 2) the particular MHM in WASH needs of schools, humanitarian contexts and for special populations, and 3) the specific MHM interventions that can be integrated into existing WASH service delivery design and implementation.

Brief overview: gendered WASH needs of women and girls in low-resource settings

WASH projects are increasingly being implemented in low-resource contexts globally. A critical aspect of implementation that needs to be addressed is the gendered nature of WASH facility needs and uses, specifically those of women and girls. The gendered experience of water and sanitation usage in low-resource contexts—specifically how women and girls interact with water and sanitation because of their culturally defined roles and expected responsibilities—is an important contextual backdrop for understanding MHM challenges facing girls and women.

In many low-resource contexts (defined as low-income countries or select contexts within countries that have limited WASH-related facilities and products) the daily utilization of, and access to, water and sanitation can incur time, energy and opportunity costs for the users, and may pose risks to an individual's or community's safety and health. Globally, women and girls inequitably experience such hardships due to the gendered division of responsibility that puts WASH-related work in their hands, yet also keeps them from engaging in decision-making about water and sanitation. In many low-income contexts, girls and women are the most common water fetchers, particularly in countries where half or less than half of the households in a country lack access to an improved water source (Sorenson et al., 2011). In such contexts, the longer the time required to fetch water, the less likely men are to fetch it

(Crow et al., 2012). The documented time costs associated with water fetching for women are immense. Data from 25 sub-Saharan countries estimated that women spend a collective total of 16 million hours each day fetching water, including the time needed to get to the source, stand in line, access the water and bring it home (WHO/UNICEF, 2012).

Water insecurity—which is defined as "insufficient and uncertain access to adequate water for an active and healthy lifestyle"—can make collection and use more difficult for girls and women (Hadley and Wutich, 2009). A study in Ethiopia found water insecurity may have profound negative impacts on individual and family health, including contributing to a water fetcher's stress; fear of assault or rape en route to a water source; decisions to collect insufficient water to meet basic needs or to collect water from known dirty sources; and economization of water needs that may compromise personal hygiene and contribute further to feelings of shame or stress (Stevenson et al., 2012b).

The burden of carrying water can also be physically damaging to an individual's health (Evans et al., 2013). The amount of water required to meet basic needs—including for drinking, cooking and cleaning—is estimated to be between 20 to 50 liters (4.4–11 gallons) of water per day, which is approximately 20 to 50 kilograms (44–110 pounds) (Gleick, 1996; UNWATER, 2013). If women and girls are solely responsible, this burden is multiplied by the number of people that require water in a given household. Carrying such loads poses dangers to girls' and women's backs and necks, and can result in significant energy loss from carrying heavy loads long distances over uneven terrain for extended periods of time (Mehretu and Mutambirwa, 1992; Fisher, 2006; Robson et al., 2013). Because women have specific gendered responsibilities—culturally defined responsibilities that compel them to perform duties as a mother or a wife—they have been reported to have sacrificed their own water needs, namely their own consumption of water for drinking and personal hygiene, in order to have enough water to provide for the needs of husbands and children (Coelho et al., 2004; Remigios, 2011; Stevenson et al., 2012a). Women are further constrained in their ability to provide for their families, and for themselves, when they are denied access to water sources specifically because they are women. Research in rural Northern Kenya revealed that women fetching water for domestic purposes faced conflict at water sources when men came to use the same source for watering their livestock. Women were forced to wait and had little voice in water management institutions that were dominated by men (Yerian et al., 2014).

Access to sanitation also poses challenges for women and girls in particular. Social norms and taboos may hinder girls' and women's abilities to practice sanitation-related behaviors, with their use restricted to certain times of the day (such as early morning or after dark) due to shame, embarrassment or simply a basic need for privacy. Female slum-dwellers in India have reported that shared sanitation facilities are not always available or are located a 30-minute distance away, with long lines and filthy conditions. As a result, women reported defecating at night for privacy on railroad tracks, along highways or in municipal dumps (Bapat and Agarwal, 2003). When facilities are available, they are not always constructed with women's needs in mind. In Uganda, women had to pay more than men to access shared pay-for-service facilities in an urban location. Their costs were higher because they needed to pay for their children's access and for their own urination, defecation and menstruation needs. Men, however, only needed to pay when going to defecate as they could urinate freely in the open (Massey, 2011). While there is a need for further rigorous study, limited access to sanitation may result in reduced food and water intake to minimize the need to defecate or urinate, urinary tract infections and chronic constipation, all of which may be compounded if women are sick, pregnant or post-partum (Bapat and Agarwal, 2003). Assault is also a concern; rape of young girls in India has been reported as "caused by lack of toilets" though the full extent of this phenomenon is unknown (Tewary, 2013).

Despite their prescribed WASH responsibilities and the challenges they face, women are underrepresented on community-level water and sanitation management committees (Ray, 2007) and also lack power to negotiate changes at the household level (Ivens, 2008). Girls' and women's MHM-related needs are managed in the context of these broader gender-specific WASH challenges and are important to consider when deliberating research, policy and programs related to MHM. Similarly, there is a need for local and national advocacy around the gendered needs of girls and women in relation to MHM, as shifting government investments in water and sanitation, and community norms around the importance of providing safe, private and clean facilities for girls and women despite lingering menstrual taboos, takes locally relevant and sustained advocacy efforts.

MHM: evidence to date

In the last decade, there has been growing evidence of the specific challenges girls and women in low-resource contexts face in managing menses comfortably, safely and hygienically. The vast majority of the research to date has focused on the MHM needs of girls in school, which has included some mention of the needs of female teachers as well. A much smaller body of literature has begun to explore the unique MHM needs of girls and woman in humanitarian contexts. We also discuss the limited availability of research among specific populations of women and girls. Specifically, some grey literature has highlighted the unique needs girls and women with disabilities, fistula or female genital cutting have faced while managing menses in low-resource contexts. The existing evidence on each of these will be discussed below, and we encourage further research among populations whose needs are under-researched and not well understood.

MHM and WASH in schools

There is a growing body of literature documenting the numerous challenges facing schoolgirls and female teachers in managing menses successfully in educational environments in low-resource settings (Sommer and Sahin, 2013). In 2001, a series of case studies on sexual maturation and schools was conducted in Uganda, Zimbabwe, Kenya and Ghana, and indicated the specific MHM-related challenges facing girls in school (Stewart, 2004). More recent research has supplemented these findings, including studies conducted on menstruation and education with adolescent girls and teachers in Tanzania (Sommer, 2009, 2010b), Ghana (Montgomery et al., 2012; Sommer and Ackatia-Armah, 2012), Nepal (Oster and Thornton, 2010), Kenya (McMahon et al., 2011; Alexander et al., 2014), Ethiopia (Fehr, 2010), Bolivia (Long et al., 2013), Philippines (Haver et al., 2013), Sierra Leone (Caruso et al., 2013) and elsewhere (Mahon and Fernandes, 2010; Sommer et al., 2013).

The majority of the existing research utilized qualitative and participatory methodologies, and focused on documenting the specific challenges facing schoolgirls in managing menses successfully (or unsuccessfully) in school. The studies all identified similar challenges facing girls in schools in a range of social and cultural contexts, including a lack of information about menstruation prior to menarche; insufficient practical knowledge for managing menses; feelings of confusion, fear, shame and embarrassment related to revealing menstrual status or having leaks and stains; and inadequate WASH facilities for managing menstruation en route to or in school (Grant et al., 2013).

Such evidence provides important insights into the essential MHM components needed as part of WASH in schools service delivery (see Freeman in this text (Chapter 28) for more information about WASH in schools generally). These MHM components include: 1) adequate numbers of safe, private and clean latrines or toilets that have locks on the doors; 2) accessible water and soap located inside the toilet stall in a bucket or tap; 3) mechanisms for disposal of used sanitary materials such as a dustbin, a nearby incinerator or pit for burning; 4) pragmatic menstrual management guidance provided to girls in a confidential and comfortable manner; and 5) adequate materials for managing menses such as quality cloth, pads and underwear (Sommer, 2010a). Although additional research is needed to identify the most effective and efficient aspects of MHM interventions to include in WASH service delivery in school settings, there is growing consensus that these five aspects represent the minimum requirements for schoolgirls and female teachers (Sommer et al., 2013), along with the importance of addressing the maintenance of sanitation systems themselves (Kjellen et al., 2012). The proposed MHM interventions are recommended to be culturally adapted given numerous cultural and social taboos around menstruation and, in particular, the disposal of menstrual waste (Sommer et al., 2013). The evaluation of proposed interventions can be appropriately designed to effectively measure the impact of various WASH-focused MHM interventions on girls' school participation, comfort, self-efficacy around menstrual management, and related measures, which will in turn provide valuable evidence enabling the most appropriate targeting of limited resources.

MHM in humanitarian and other contexts

The WASH-related MHM evidence for girls out of school and for women in the household, in work environments in resource-poor settings, and in humanitarian contexts, is more limited and in need of further research. The UN Special Rapporteur on Human Rights has also indicated that the lack of access to water and sanitation in the workplace for managing menses can negatively impact the right of women to work.

In humanitarian contexts, there are a number of mandates and organizations that attempt to respond to the MHM needs of girls and women (Sommer, 2012). Girls and women in such contexts frequently exist in conditions with even less privacy than those in stable situations, and may face increased concerns around safety given the existing evidence on sexual violence in many humanitarian contexts. The minimal evidence that exists also suggests that humanitarian organizations may not adequately consider the MHM needs of girls and women, nor sufficiently consult with girls and women about their preferences for sanitary materials, and the location and design of water and sanitation facilities. There is also often inadequate coordination amongst organizations, and insufficient inter-sectoral responses. While WASH practitioners are often charged with providing the necessary water and sanitation for girls and women in such contexts, they may not have the requisite expertise to address the gendered and health aspects of MHM. There is a great need to better document current efforts about how MHM has been incorporated into humanitarian contexts for learning purposes, and to more rigorously evaluate whether or not these efforts have been effective from the perspective of the populations served.

MHM and special populations

There exists minimal to no evidence on the unique needs of special populations of girls and women, such as those with disabilities or who have had experienced fistula or female genital cutting, that describe specific challenges they may face when managing menses with insufficient water and sanitation facilities. The grey literature, such as the recently published "Menstrual Hygiene Matters" manual for practitioners, provides important suggestions for addressing the needs of these populations; however, qualitative research is needed to specifically explore the unique challenges that are faced, and the most effective approaches for improving access to adequate water and sanitation during menses among these populations.

Integrating MHM into a WASH project

At the center of all future effective MHM-related programming within WASH is the incorporation of the target population's gendered experiences, perceptions and recommendations on proposed interventions. In each new cultural and social context— inclusive of humanitarian settings—there may be unique gendered aspects to girls' and women's WASH-related experiences, along with specific menstrual-related taboos that need to be addressed (Sommer, 2012). The recommendation is to explore local insights and norms prior to implementation in each new context, as girls and women may have unique experiences to incorporate into proposed interventions. The social and cultural aspects of MHM in relation to WASH facility use are important not to overlook, including those pertaining to class, religion and cultural practices, as they are critically important to both assuring the uptake of WASH interventions, and the maintenance and sustained use of implemented interventions in respective communities into the future. A booklet of tools used to explore MHM and WASH in schools in Bolivia, the Philippines and Sierra Leone can be adapted to other populations and contexts by researchers and practitioners to understand specific needs and priorities of local contexts (Caruso, 2014).

Of equal importance is promoting the *integration* of MHM into new and existing WASH service delivery. The attention to girls' and women's MHM-related WASH needs is not a stand-alone program needing specific attention and resources. Rather, the recommendation for proposed essential MHM-related interventions is to implement them as enhanced aspects of WASH-related service delivery, so they are not gender discriminatory. For example, if latrines are being built in a school or health facility, it is recommended that an adequate number of safe, private, easily accessible latrines with water and soap nearby are provided, along with sufficient disposal mechanisms inside to meet the needs of menstruating girls and women. If water and sanitation facilities are built on the exterior areas of a refugee camp where users are at risk of conflict and violence, then attention may be needed to assure the safe placement of WASH facilities for this gendered reality. If governments draft water and sanitation-related policies, these policies would be strengthened if attention is given to women and girls' needs during menstruation.

Meeting basic WASH-related requirements requires more than just assuring structures are well-designed and put into place. WASH practitioners are encouraged to consider what software components are needed to accompany the hardware, both for the successful use of facilities by all and for the longevity of the facilities themselves through effective operation and maintenance. For example, in schools, practitioners may need to engage teachers, parents or other adults to make sure that girls have the adequate MHM guidance (see girls' puberty books under listed publications at: www.growandknow.org) needed on how to use the available facilities. Specifically, girls may need instruction on how to appropriately dispose of menstrual management materials (e.g. not dropping them into the latrine or toilet facilities). In communities, health workers may be important for conveying messages and answering questions related to menstruation that are linked to WASH, such as how often to change and where to dispose of used menstrual materials safely, how to hygienically clean re-usable sanitary materials, and that it is important and safe to bathe or wash during menstruation (contrary to what may be believed in some contexts). However, simply having instruction about the importance and acceptability of regular hygiene during menstruation may not facilitate women's ability to practice these behaviors. In rural Orissa, India, women describe having limited privacy to bathe, an inability to bathe when they need to, and conflict from fellow community members if they are known to be menstruating and are using— and therefore considered polluting—a public water source (Caruso, unpublished data). Successful WASH service delivery may require collaboration with practitioners outside the WASH sphere, including educators and gender and health specialists, to formulate effective, well-integrated and targeted software components.

Lastly, monitoring and evaluating the implementation of all MHM aspects of WASH service delivery is essential, along with continuing to grow the evidence base on the most effective and efficient MHM-specific components of WASH service delivery and policy for low-resource and humanitarian emergency (Sommer, 2012) contexts in particular. Such monitoring, evaluation and basic research can continue to improve the WASH environments for women and girls so they can manage menstruation safely, with dignity and without compromise to their participation in school, work or other aspect of life.

Key recommended readings

1 House, S., Mahon, T. & Cavill, S. 2012. *Menstrual Hygiene Matters: A Resource for Improving Menstrual Hygiene around the World.* http://www.wateraid.org/what-we-do/ our-approach/research-and-publications/view-publication?id=02309d73-8e41- 4d04-b2ef-6641f6616a4f. House et al. have compiled programmatic materials, policy documents, and research related to MHM in lower- and middle-income countries. This open resource features nine modules and associated toolkits for improving programs related to MHM, including chapters explaining MHM basics and MHM in schools, the workplace and emergency settings.

2 Sommer, M. 2010. Putting menstrual hygiene management on to the school water and sanitation agenda. *Waterlines.* 29(4), 268–277. Sommer highlighted the long overlooked nature of MHM into WASH interventions in schools. The article additionally emphasized the various research methodologies to be utilized in studying MHM, including participatory and quantitative approaches.

3 Sommer, M. 2012. Menstrual hygiene management in humanitarian emergencies: gaps and recommendations. *Waterlines,* 31(1–2), 1–2. Sommer conducted a review of the policy and practice literature on addressing MHM in humanitarian contexts, supplemented by key informant interviews conducted with humanitarian response experts around the globe. Gaps in attention to addressing MHM in emergencies were identified with recommendations for improving MHM responses.

References

Albuquerque, C. D. 2014. *Realising the Human Rights to Water and Sanitation: A Handbook.* http://www. righttowater.info/handbook/

Alexander, K. T., Oduor, C., Nyothach, E., Laserson, K. F., Amek, N., Eleveld, A., Mason, L., Rheingans, R., Beynon, C. & Mohammed, A. 2014. Water, sanitation and hygiene conditions in Kenyan rural schools: are schools meeting the needs of menstruating girls? *Water,* 6, 1453–1466.

Bapat, M. & Agarwal, I. 2003. Our needs, our priorities; women and men from the slums in Mumbai and Pune talk about their needs for water and sanitation. *Environment and Urbanization*, 15, 71–86.

Caruso, B. A. 2014. *WASH in Schools Empowers Girls' Education: Tools for Assessing Menstrual Hygiene Management in Schools*. New York: United Nations Children's Fund.

Caruso, B. A., Fehr, A., Inden, K., Ellis, A., Andes, K. L. & Freeman, M. C. 2013. *WASH in Schools Empowers Girls' Education in Freetown, Sierra Leone: An Assessment of Menstrual Hygiene Management in Schools*. New York: UNICEF.

Coelho, A. E., Adair, J. G. & Mocellin, J. S. P. 2004. Psychological responses to drought in northeastern Brazil. *Revista interamericana de psicologia (Interamerican Journal of Psychology)*, 38, 95–103.

Crow, B., Swallow, B. & Asamba, I. 2012. Community organized household water increases not only rural incomes, but also men's work. *World Development*, 40, 528–541.

Evans, B., Bartram, J., Hunter, P., Rhoderick Williams, A., Geere, J., Majuru, B., Bates, L., Fisher, M., Overbo, A. & Schmidt, W.-P. 2013. *Public Health and Social Benefits of At-House Water Supplies*. Leeds: University of Leeds.

Fehr, A. 2010. *Stress, Menstruation, and School Attendance: Effects of Water Access Among Adolescent Girls in South Gondar, Ethiopia*. CARE Ethiopia.

Fisher, J. 2006. For her it's the big issue: putting women at the centre of water supply, sanitation and hygiene. https://dspace.lboro.ac.uk/dspace-jspui/handle/2134/9970

Gleick, P. H. 1996. Basic water requirements for human activities: meeting basic needs. *Water International*, 21, 83–92.

Grant, M., Lloyd, C. & Mensch, B. 2013. Menstruation and school absenteeism: evidence from Rural Malawi. *Comparative Education Review*, 57, 260–284.

Hadley, C. & Wutich, A. 2009. Experience-based measures of food and water security: biocultural approaches to grounded measures of insecurity. *Human Organization*, 68, 451–460.

Haver, J., Caruso, B. A., Ellis, A., Sahin, M., Villasenor, J. M., Andes, K. L. & Freeman, M. C. 2013. *WASH in Schools Empowers Girls' Education in Masbate Province and Metro Manila, Philippines: An Assessment of Menstrual Hygiene Management in Schools*. New York: United Nations Children's Fund.

House, S., Mahon, T. & Cavill, S. 2012. *Menstrual Hygiene Matters: A Resource for Improving Menstrual Hygiene around the World*. http://www.wateraid.org/what-we-do/our-approach/research-and-publications/view-publication?id=02309d73-8e41-4d04-b2ef-6641f6616a4f

Ivens, S. 2008. Does increased water access empower women? *Development*, 51, 63–67.

JMP 2012. Consultation on Draft Long List of Goal, Target and Indicator Options for Future Global Monitoring of Water, Sanitation and Hygiene. New York: WHO/UNICEF Joint Monitoring Programme.

Kjellen, M., Pensulo, C., Nordquist, P. & Fogde, M. 2012. *Global Review of Sanitation System Trends and Interactions with Menstrual Management Practices*. Stockholm: Stockholm Environment Institute.

Long, J., Caruso, B. A., Lopez, D., Vancraeynest, K., Sahin, M., Andes, K. L. & Freeman, M. C. 2013. *WASH in Schools Empowers Girls' Education in Rural Cochabamba, Bolivia: An Assessment of Menstrual Hygiene Management in Schools*. New York: United Nations Children's Fund.

Mahon, T. & Fernandes, M. 2010. Menstrual hygiene in South Asia: a neglected issue for WASH (water, sanitation and hygiene) programmes. *Gender & Development*, 18, 99–113.

Massey, K. 2011. *Insecurity and Shame: Exploration of the Impact of the Lack of Sanitation on Women in the Slums of Kampala, Uganda*. London: SHARE and WaterAid.

McMahon, S. A., Winch, P. J., Caruso, B. A., Obure, A. F., Ogutu, E. A., Ochari, I. A. & Rheingans, R. D. 2011. 'The girl with her period is the one to hang her head': reflections on menstrual management among schoolgirls in rural Kenya. *BMC Int Health Hum Rights*, 11, 7.

Mehretu, A. & Mutambirwa, C. 1992. Gender differences in time and energy costs of distance for regular domestic chores in rural Zimbabwe: a case study in the Chiduku Communal Area. *World Development*, 20, 1675–1683.

Montgomery, P., Ryus, C. R., Dolan, C. S., Dopson, S. & Scott, L. M. 2012. Sanitary pad interventions for girls' education in Ghana: a pilot study. *PLoS One*, 7, e48274.

Oster, E. F. & Thornton, R. L. 2010. Menstruation, Sanitary Products and School Attendance: Evidence from a Randomized Evaluation. NBER Working Paper Series. http://ssrn.com/abstract=1376156

Ray, I. 2007. Women, water, and development. *Annu. Rev. Environ. Resour.*, 32, 421–449.

Remigios, M. V. 2011. Women–water–sanitation: the case of Rimuka high-density suburb in Kadoma, Zimbabwe. *Agenda*, 25, 113–121.

Robson, E., Porter, G., Hampshire, K. & Munthali, A. 2013. Heavy loads: children's burdens of water carrying in Malawi. *Waterlines*, 32, 23–35.

Sommer, M. 2009. Ideologies of sexuality, menstruation and risk: girls' experiences of puberty and schooling in northern Tanzania. *Cult Health Sex*, 11, 383–398.

Sommer, M. 2010a. Putting menstrual hygiene management on to the school water and sanitation agenda. *Waterlines*, 29, 268–277.

Sommer, M. 2010b. Where the education system and women's bodies collide: the social and health impact of girls' experiences of menstruation and schooling in Tanzania. *J Adolesc.*, 33, 521–529.

Sommer, M. 2012. Menstrual hygiene management in humanitarian emergencies: gaps and recommendations. *Waterlines*, 31, 1–2.

Sommer, M. & Ackatia-Armah, N. M. 2012. The gendered nature of schooling in Ghana: hurdles to girls, menstrual management in school. *JENdA: A Journal of Culture and African Women Studies*. http://www.africaknowledgeproject.org/index.php/jenda/article/view/1578

Sommer, M. & Sahin, M. 2013. Overcoming the taboo: advancing the global agenda for menstrual hygiene management for schoolgirls. *American Journal of Public Health*, 103, 1556–1559.

Sommer, M., Vasquez, E., Worthington, N. & Sahin, M. 2013. WASH in Schools Empowers Girls' Education. Proceedings of the Menstrual Hygiene Management in Schools Virtual Conference 2012. New York: United Nations Children's Fund and Columbia University.

Sorenson, S. B., Morssink, C. & Campos, P. A. 2011. Safe access to safe water in low income countries: water fetching in current times. *Social Science & Medicine,* 72, 1522–1526.

Stevenson, E. G., Greene, L. E., Maes, K. C., Ambelu, A., Tesfaye, Y. A., Rheingans, R. & Hadley, C. 2012a. Water insecurity in 3 dimensions: an anthropological perspective on water and women's psychosocial distress in Ethiopia. *Social Science & Medicine*, 75, 392–400.

Stevenson, E. G., Greene, L. E., Maes, K. C., Ambelu, A., Tesfaye, Y. A., Rheingans, R. & Hadley, C. 2012b. Water insecurity in 3 dimensions: an anthropological perspective on water and women, psychosocial distress in Ethiopia. *Social Science & Medicine*, 75(2), 392–400.

Stewart, J. E. 2004. *Life Skills, Sexual Maturation, and Sanitation: What's (not) Happening in Our Schools?: An Exploratory Study from Zimbabwe*. Harare: Weaver Press.

Sumpter, C. & Torondel, B. 2013. A systematic review of the health and social effects of menstrual hygiene management. *PLoS One*, 8, e62004.

Tewary, A. 2013. India Bihar rapes 'caused by lack of toilets'. BBC News. Available: http://www.bbc.co.uk/news/world-asia-india-22460871 (accessed 22 May 2013).

UNWATER. 2013. Graphs and Maps: Drinking Water, Sanitation & Hygiene. UN Water. Available: http://www.unwater.org/statistics_san.html (accessed 22 May 2013).

WHO/UNICEF. 2012. *Progress on Drinking Water and Sanitation: 2012 Update*. Geneva: World Health Organization.

Yerian, S., Hennink, M., Greene, L., Kiptugen, D., Buri, J. & Freeman, M. 2014. The role of women in water management and conflict resolution in Marsabit, Kenya. *Environmental Management*, 54(6), 1320–1330.

53

HEALTH IMPACT
ASSESSMENT*

Lorna Fewtrell
CREH, University of Wales, Aberystwyth, UK

Learning objectives

1 Understand the two broad definitions of health and how they impact on Health Impact Assessment (HIA).
2 Understand the key elements of an HIA.
3 Understand why implementing HIA in advance of water-related projects could be beneficial.

What is Health Impact Assessment and why is it useful?

Health Impact Assessment (HIA) has been defined in a number of ways but, essentially, it is a way of making sure that health is on the agenda as it aims to predict public health issues 'upstream' of the introduction or start of any policy, programme or project and thus influence the decision-making process. It provides a framework or formal process for thinking through the possible health consequences *before* implementing something and, as will be clear from the section 'HIA process' below, the HIA process will draw on many of the 'investigative tools' outlined in later chapters.

In the past, development was generally undertaken without any sort of assessment of possible health impacts, largely because 'health' was considered the exclusive responsibility of the health sector (a view that still prevails in some countries). The upshot of this stance was a range of adverse health effects, largely experienced by vulnerable populations, which could have been avoided. On top of this, most development also missed opportunities for health promotion (such as the development of numerous irrigation schemes in developing countries without the inclusion of components to improve access to safe drinking water and sanitation). Examples of water-related unintended negative health effects include the spread and intensification of vector-borne disease transmission, particularly schistosomiasis and malaria associated with hydropower projects and irrigation systems, intensification of

---

* Recommended citation: Fewtrell, L. 2015. 'Health Impact Assessment', in Bartram, J., with Baum, R., Coclanis, P.A., Gute, D.M., Kay, D., McFadyen, S., Pond, K., Robertson, W. and Rouse, M.J. (eds) *Routledge Handbook of Water and Health*. London and New York: Routledge.

531

dengue transmission due to the introduction of storage jars for drinking water in Viet Nam, arsenic poisoning as a result of using contaminated boreholes for supplying drinking water in Bangladesh, and the psycho-social disorders resulting from forced resettlement out of reservoir areas or irrigation schemes.

Although, generally, not a legal requirement, HIA offers a structured approach to determine how a policy, plan or project will affect health and, as such offers, benefits to public health, communities and decision makers.

HIA process

There are numerous different ways in which to conduct a HIA; it has been noted that there is 'no single "blueprint" for HIA that will be appropriate in all circumstances' (Douglas et al., 2001) and different methods are often chosen depending upon the disciplines of those doing the work, an approach which is recognised in the definition of HIA as 'a combination of procedures, methods and tools that systematically judges the potential and sometimes unintended effects of a policy, plan, programme or project on the health of a population and the distribution of those effects within the population. HIA identifies appropriate actions to manage those effects' (Quigley et al., 2006). Despite the scope for variety there are some constants in the process. Two key features of an HIA have been identified, namely:

- it is concerned with the prediction of different options and how these contemplated decisions would affect the health of a population; and
- it is intended to influence decision making and to assist decision makers.

Therefore, by definition and as indicated above, an HIA needs to be undertaken prior to the implementation of a policy, project or programme.

Health and determinants of health

An important pre-requisite of an HIA is the definition of health; there are two broad paradigms, namely the reductionist 'biomedical' approach and the 'social or public health model' as outlined below:

- Reductionist approach: where health is considered within a series of disease categories (such as communicable disease, non-communicable disease, nutrition, injury, mental disorder and so on) and the health sector structure and operations deal with these through the delivery of health services.
- Public health model: where health is considered to be 'a state of complete physical, mental and social well-being and not merely the absence of disease or infirmity' as defined by the World Health Organization in their constitution.

A range of complex factors (determinants) are acknowledged to play a role in community health and it is these determinants that are affected by development decisions. Health determinants refer to a range of personal, social, economic and environmental factors (as outlined in Table 53.1) that determine the health of individuals and populations.

It can be seen from this table that virtually every area of human activity influences health.

Table 53.1 Example determinants of health

Determinant	Examples
Fixed	Ageing
	Gender
	Genes
Social and economic	Community structure and infrastructure
	Employment
	Poverty
	Social exclusion
Lifestyle and behaviour	Alcohol
	Coping skills
	Diet
	Drugs
	Physical activity
	Sexual behaviour
	Smoking
Access to services	Drinking water (see Chapters 18 and 22)
	Education
	Health services
	Leisure
	Sanitation
	Social services
	Transport
Environment	Air quality
	Climate
	Disease vectors
	Housing
	Noise
	Risk of injury
	Social environment
	Sun exposure
	Water quality (see Chapters 13, 18 and 19)

Stages of HIA

A full HIA generally consists of a number of steps encompassing the following (or a slight variation thereof).

Screening

This step is used to determine whether an HIA is required and, if so, at what level. Useful questions to be considered in the screening process are shown in Figure 53.1.

Scoping

This step sets the boundaries for the HIA. It outlines the likely negative (see Chapters 2 to 11 in 'Water-related hazards' for a good starting point) and positive health effects, the population affected (in particular, the likely vulnerable population groups), the timescale

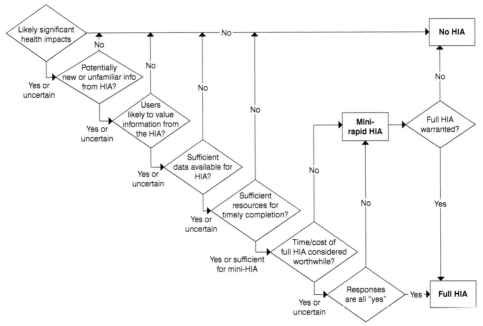

Figure 53.1 Screening process to guide the decision on whether or not to conduct an HIA

Source: Reprinted from Cole et al. (2005). Copyright (2005) with permission from Elsevier.

to be considered (e.g. up to five years after completion/implementation), the geographical scope (which, in many cases, will extend beyond the immediate project locality), identifies stakeholders and describes the baseline situation (i.e. the current state of health and the determinants of health in the affected population). A key issue for HIAs conducted in developing countries can be the lack of baseline health data.

Risk assessment

This is the investigation and appraisal (both qualitative and quantitative) of the impacts. It would be expected to include:

- the nature of the impacts (e.g. infection, changes to well-being);
- the direction of the change (i.e. increased or decreased);
- the magnitude of the change (in terms of both numbers affected and the level/severity of the impact); and
- the distribution (who is affected).

It is likely that, in a large-scale project this would be split into different project phases, such as construction, operation and decommissioning, as health impacts would be expected to be different during these different phases.

In a UK-based HIA on the operation of sustainable drainage options based around a housing estate, the risk assessment considered hazards associated with drowning and near-drowning, infection (from microbial contamination of the environment from wildlife and domestic animals attracted to the ponds and wetlands) and anxiety (from, for example, parents worrying about their child drowning and/or concerns over an increase in the numbers of mosquitoes). The health benefits resulting from perceived amenity value, the presence of green space and the potential for increased activity were also considered.

Risk assessment is the area where the greatest differences between HIAs tend to be seen in terms of the different evidence examined, the methods used (e.g. stakeholder participation) and the degree of quantification of impacts.

Reporting including recommendations/health action plan

The report would be expected to lay out the health impacts, the severity, the people affected and so on and make recommendations on removing/minimising the negative impacts and maximising the benefits.

Decision making

This is not part of the HIA process but, as the HIA is intended to influence decision making and be a decision support tool, shows where this fits into the overall HIA framework.

HIA levels

HIAs vary in depth and complexity as shown in Table 53.2. The deciding factors, for the level at which an HIA is implemented, are often the availability of financial resources and the time frame in which the assessment is required.

Although the depth and complexity of HIAs have been neatly categorised into levels in Table 53.2, in reality each individual HIA lies somewhere on a continuum, with aspects and methodologies combined to provide the most appropriate assessment given time and financial constraints.

Challenges

While the number of HIAs conducted has increased dramatically over the last few years, undertaking an HIA is still far from being routine, and there are a number of challenges associated with conducting an HIA.

Table 53.2 Levels of HIA

Level of HIA	Characteristics
Desk-based (or mini)	• Provides a broad overview of possible health impacts. • Collection and analysis of existing and readily accessible data.
Rapid (or standard)	• Provides more detailed information of possible health impacts. • Involves collecting and analysing existing data and some new qualitative data from stakeholders and key informants.
Full/comprehensive (or maxi)	• Provides a comprehensive assessment of potential health impacts. • Robust definition of impacts. • Involves collecting and analysing data using multiple methods and sources (quantitative and qualitative, including participatory approaches involving stakeholders and key informants). • Lasts six months or more.

Timing

A major challenge for HIA, especially with regard to influencing the decision-making process, is the limited time that is often available for the assessment, coupled with the notorious lack of high-quality baseline health and socioeconomic data, particularly in developing country settings. A good example of this is provided by the Nam Theun 2 hydroelectric project (see below) where only roughly six weeks were available for the conduct and reporting of the HIA (Krieger et al., 2008).

Stakeholder participation

The inclusion of stakeholder participation (and the level of that participation) will, largely, depend upon the HIA method chosen and the level of HIA undertaken but, in large-scale projects, it is considered pivotal by the World Bank and the International Petroleum Industry Environmental Conservation Association. The likely degree of community participation in HIA is contextual and it depends on a number of factors including:

- the prevailing governance system;
- the constitutional framework;
- local cultural practice;
- determinants of the capacity of communities to participate (such as literacy rates); and
- the capacity among professionals to apply participatory methods.

The International Finance Corporation's introduction to HIA includes a section on stakeholder participation in developing country settings (IFC, 2009 – see 'Key recommended readings').

Data needs, data availability and quantification

One of the greatest challenges relates to the data needs for HIA and the availability of the data; this is particularly the case where quantification is an objective and when HIAs relate to a developing country setting. HIA has been described by one practitioner as being 'at the mercy of the evidence available' (Harris et al., 2014).

As a result of the wide range of effects that an HIA can expect to consider, a broad and varied evidence base is often called upon. Thus, in an HIA looking at a proposed irrigation project in Zimbabwe (Konradsen et al., 2008), evidence was gathered on hazards associated with water-washed disease (see Chapter 4), malaria (Chapter 6), schistosomiasis (Chapter 5), agrochemical poisoning and sexually transmitted disease.

The use of a diverse evidence base may require some lateral thinking in terms of locating data and published reviews, and may require information obtained from fields outside the usual expertise of, even, a multi-disciplinary team. A guide to reviewing published evidence for use in HIA has been produced which is useful for both HIA practitioners and people charged with appraising the quality of the information used in an HIA (Mindell et al., 2006 – see 'Key recommended readings').

In desk-based or rapid HIAs there will be little or no opportunity to collect new data, thus the best must be made of what is readily available. This may mean using routine surveillance data (although this may not exist, or be of extremely dubious quality in some countries). Where health surveillance data and population statistics do exist they often may not be available in a form that allows their disaggregation to explore the possible impacts on vulnerable groups.

Quantification is still relatively rare in HIA and while quantification can clearly provide a measure of the magnitude of an effect and, thus, help decision makers, it also requires greater levels of input information and can lead to controversies over the validity of the methods used and/or the choice of assumptions employed. Where quantification is used it is important that assumptions are clearly outlined and that limitations of the estimation methods are discussed. There is also the risk that quantified impacts will be given more weight in the overall assessment, essentially, just because they can be measured. As Einstein pointed out, 'not everything that counts can be counted, and not everything that can be counted counts'.

Vulnerable groups

Vulnerable groups require special consideration in HIAs, as they often suffer a disproportionate impact. Vulnerable groups may include the very young, the elderly, pregnant women, people with poor health and pre-existing disease (especially the immunocompromised), certain occupational groups and some religious groups. These groups may be vulnerable due to inherent characteristics (e.g. age, gender), differential exposure (e.g. farmers using untreated wastewater to irrigate crops), behaviour (e.g. lack of belief in health services) or lack of resources (e.g. poor access to safe drinking water and/or adequate sanitation).

Possible biases

It can be difficult to ensure that HIAs are free from bias in either procedure and/or adopted methods. HIAs need to be objective, comprehensive and based on available evidence. Problems can creep in where there is a conflict of interest between the client and the consultant, where the timing of the HIA is inappropriate or insufficient, where there aren't sufficient funds available, where there is a lack of expertise, experience or skill, and where there may be a lack of access to information.

Western/developed world focus

Despite the great benefits that could derive from performing HIA in developing countries, they are still relatively rare with (according to Erlanger et al.'s 2008 study of published HIAs) only 6 per cent of identified work having an explicit developing country focus. This may, at least in part, be related to the methods, procedures and tools for performing HIA, which have all developed with a strong eurocentric or, more broadly, 'western' focus. This focus is clear in the core values of HIA (democracy, equity, sustainability and ethical use of evidence) which were outlined in the Gothenburg consensus document (for an early attempt to outline the main concepts and suggested approach for HIA, see WHO ECHP, 1999). While these core values are laudable, they are difficult to apply in many developing country settings, where democracy may be a nice idea, equity a relative term, many projects (such as mining) are inherently unsustainable and evidence may, at best, be limited (Krieger et al., 2012).

HIA and a water-related example

There are few published HIAs looking at specific water-related issues, particularly in the sense of major development projects in developing countries; a notable exception being the Nam Theun 2 hydroelectric project in Lao PDR in South East Asia (Krieger et al., 2008). Issues associated with large hydroelectric projects, including adverse impacts on local populations and ecosystems as a result of profound demographic, ecologic and socioeconomic transformations, are well known and it was against this background that an HIA was conducted at a late stage in the project development.

The Nam Theun HIA was divided into discrete 'potential impact areas of concern' (which included villages within the watershed, the plateau resettlement area and construction camps) and the impacts determined according to the stage of the project (construction, reservoir filling and operation).

The HIA team identified and assessed potential health impacts in a number of ways:

- in-depth review of available regional, national, provincial and district health data;
- analysis and synthesis of project study area-specific baseline economic, health and nutritional surveys;
- comparison of study area data to national, provincial and district health data;
- field survey visit by the core HIA study team; and
- consultation with relevant stakeholders, particularly Ministry of Health officials and regional representatives of the World Health Organization and the World Bank.

Overall, the HIA used a modified environmental health area approach, defining 11 different areas to capture the identified health impacts. The results for a selected area are shown in Table 53.3.

It can be seen that high negative impacts are expected for all age groups across most of the environmental health areas, especially during the construction process and, to some degree, during reservoir filling. A rather different risk profile is seen with mitigation strategies in place (Table 53.4).

The development went ahead with many of the mitigation strategies in place and has been subject to some post-implementation monitoring which showed that many of the strategies were performing well, although it was noted that there was scope for further improvement.

Conclusion

HIA can provide an estimation of potential health impacts, whether these are:

- positive or negative;
- intended or unintended;
- direct or indirect;
- single, multiple or cumulative.

Based on the health impacts identified, it can be used to define health safeguards, mitigation measures and health promotional activities for the design, construction and operational project phases. HIA, therefore, supports minimization of predicted health risks and makes it possible to take optimal advantage of health opportunities.

Table 53.3 Risk profiling of the Nakai plateau resettlement area

	Project timing														
	Construction					Reservoir filling					Operation				
Environmental health area	A	B	C	D	E	A	B	C	D	E	A	B	C	D	E
Respiratory diseases															
Vector-related diseases															
Sexually-transmitted infections															
Food, water and soil-borne															
Accidents and injuries															
Exposure to hazardous materials															
Nutrition, food source															
Psychosocial															
Cultural health practices															
Health infrastructure and capacity															
Programme management system delivery															

Risk profile legend

Population affected		Impact level	
A	Children/infants < 5 years	High	■
B	Children aged 5–13 years	Medium	■
C	Women of reproductive age	Low	■
D	Men aged 14–60 years	None	
E	Elderly > 60 years	Enhanced	+

Source: Reproduced from Krieger et al. (2008) with permission from the copyright holders, IWA Publishing.

Table 53.4 Post-mitigation risk profiling Nakai plateau resettlement area

	Project timing														
	Construction					Reservoir filling					Operation				
Environmental health area	A	B	C	D	E	A	B	C	D	E	A	B	C	D	E
Respiratory diseases												+	+	+	+	+
Vector-related diseases																
Sexually-transmitted infections																
Food, water and soil-borne												+	+	+	+	+
Accidents and injuries																
Exposure to hazardous materials																
Nutrition, food source																
Psychosocial																
Cultural health practices																
Health infrastructure and capacity												+	+	+	+	+
Programme management system delivery																

Source: Reproduced from Krieger et al. (2008) with permission from the copyright holders, IWA Publishing.

Key recommended readings

1 Birley, M. (2011) *Health Impact Assessment: Principles and Practice*, London: Routledge. Text book with a broad scope, a chapter on water resources development and a more developing country focus.
2 BMA (1998) *Health and Environmental Impact Assessment. An Integrated Approach*, British Medical Association, Earthscan Publications Ltd., London. An insight into fairly early developments of HIA and a medical perspective.
3 enHealth (2001) 'Health Impact Assessment Guidelines', Commonwealth of Australia. http://www.health.gov.au/internet/main/publishing.nsf/Content/9BA012184863E206 CA257BF0001C1B0E/$File/env_impact.pdf (accessed 20 May 2014). This provides good guidance on how to manage and how to conduct an HIA in a wealthy country setting, including information on suitable stakeholder groups and public engagement.
4 Fewtrell, L. and Kay, D. (2008) *Health Impact Assessment for Sustainable Water Management*, IWA Publishing, London. The source for much of the material in this chapter and a number of water-based HIA examples, including a chapter on the Nam Theun project.
5 HIA Gateway (part of Public Health England). http://www.apho.org.uk/default. aspx?RID=40141 (accessed 20 May 2014). A UK-based website, with links to numerous published HIAs and a useful HIA bibliography which is periodically updated.
6 IFC (2009) 'Introduction to health impact assessment', International Finance Corporation, Washington DC http://www.ifc.org/wps/wcm/connect/a0f112004885 5a5a85dcd76a6515bb18/HealthImpact.pdf?MOD=AJPERES&CACHEID=a0f1120 048855a5a85dcd76a6515bb18 (accessed 20 May 2014). This provides good practice guidance on conducting an HIA in less-developed settings and specifically considers expansion of existing facilities or projects and the development of new projects or new locations (i.e. it does not consider policies).
7 Kemm, J. (2013) *Health Impact Assessment. Past Achievement, Current Understanding and Future Progress*, Oxford University Press, Oxford. Update on Kemm et al. (2004) with a slightly wider scope.
8 Kemm, J., Parry, J. and Palmer, S. (2004) *Health Impact Assessment: Concepts, Theory, Techniques, and Applications*, Oxford University Press, Oxford. Some useful background and some developed country HIA examples.
9 Mindell, J., Biddulph, J.P., Boaz, A., Boltong, A., Curtis, S., Joffe, M., Lock, K. and Taylor, L. (2006) 'A guide to reviewing evidence for use in Health Impact Assessment', London Health Observatory. http://www.lho.org.uk/viewResource.aspx?id=10846 (accessed 22 May 2014). A straightforward guide to reviewing the evidence used in HIAs.
10 WHO website Health Impact Assessment: http://www.who.int/hia/en/ (accessed 20 May 2014). Includes information on World Health Organization activities in HIA, examples of HIAs and use of evidence.

References

Cole, B.L., Shimkhada, R., Fielding, J.E., Kominski, G. and Morgenstern, H. (2005) 'Methodologies for realizing the potential of health impact assessment', *American Journal of Preventative Medicine*, vol 28, no 4, 382–389.
Erlanger, T.E., Krieger, G.R., Singer, B.H., and Utzinger, J. (2008) 'The 6/94 gap in health impact assessment. *Environmental Impact Assessment Review*' vol 28, 349–358.
Douglas, M., Conway, L., Gorman, D., Gavin, S. and Hanlon, P. (2001) 'Developing principles for health impact assessment', *Journal of Public Health Medicine*, vol 23, 148–154.

Harris, P., Sainsbury, P. and Kemp, L. (2014) 'The fit between health impact assessment and public policy: practice meets theory', *Social Science & Medicine*, vol 108, 46–53.

Konradsen, F., Chimbari, M. and Furu, P. (2008) 'Mupfure irrigation project, Zimbabwe. HIA of a water resources development', in L. Fewtrell and D. Kay (eds) *Health Impact Assessment for Sustainable Water Management*, London: IWA Publishing.

Krieger, G., Singer, B., Winkler, M., Divall, M., Tanner, M. and Utzinger, J. (2012) 'Health impact assessment in developing countries', in J. Kemm (ed.) *Health Impact Assessment: Past Achievement, Current Understanding and Future Progress*, Oxford: Oxford University Press.

Krieger, G.R., Balge, M.Z., Chanthapone, S., Tanner, M., Singer, B.H., Fewtrell, L., Kaul, S., Sananikhom, P., Odermatt, P. and Utzinger, J. (2008) 'Nam Theun 2 hydroelectric project, Lao PDR', in L. Fewtrell and D. Kay (eds) *Health Impact Assessment for Sustainable Water Management*, London: IWA Publishing.

Quigley, R., den Broeder, L., Furu, P., Bond, A., Cave, B. and Bos, R. (2006) 'Health Impact Assessment best practice principles', International Association for Impact Assessment. http://www.iaia.org/publicdocuments/special-publications/SP5.pdf (accessed 20 May 2014).

WHO ECHP (1999) 'Health Impact Assessment: Main concepts and suggested approach (Gothenburg consensus paper)', World Health Organization European Centre for Health Policy, Brussels.

PART VII

Investigative tools

54

INTRODUCTION: INVESTIGATIVE TOOLS*

David Kay

CREH, University of Wales, Aberystwyth, UK

Learning objectives

1 Better understand the interactions of water and health through analytical methods.
2 Understand and use investigative tools for managing water and health.

The present context

Maintaining the health of water consumers and users is a central aim of the global water management community. This can be seen in the goals set for United Nations (UN) agencies and member countries expressed first in the Millennium Development Goals (MDGs) and more recently in the Sustainable Development Goals (SDGs) published in 2014 (see Bartram and Baum, 2015; in the introduction to this volume; and Anon (2014)). Water and sanitation management actions designed to deliver public health improvements at the global scale are outlined in Goal 6 which is designed to 'Ensure availability and sustainable management of water and sanitation for all'. Goal 6 is itself split into six key actions, namely:

6.1 By 2030, achieve universal and equitable access to safe and affordable drinking water for all.

6.2 By 2030, achieve access to adequate and equitable sanitation and hygiene for all and end open defecation, paying special attention to the needs of women and girls and those in vulnerable situations.

6.3 By 2030, improve water quality by reducing pollution, eliminating dumping and minimizing release of hazardous chemicals and materials, halving the proportion of untreated wastewater and increasing recycling and safe reuse by [x] per cent globally.

---

* Recommended citation: Kay, D. 2015. 'Introduction: investigative tools', in Bartram, J., with Baum, R., Coclanis, P.A., Gute, D.M., Kay, D., McFadyen, S., Pond, K., Robertson, W. and Rouse, M.J. (eds) *Routledge Handbook of Water and Health*. London and New York: Routledge.

6.4 By 2030, substantially increase water-use efficiency across all sectors and ensure sustainable withdrawals and supply of freshwater to address water scarcity and substantially reduce the number of people suffering from water scarcity.

6.5 By 2030, implement integrated water resources management at all levels, including through trans-boundary cooperation as appropriate.

6.6 By 2020, protect and restore water-related ecosystems, including mountains, forests, wetlands, rivers, aquifers and lakes.

(Anon, 2014)

These ambitious goals are designed to produce the complementary and interlinked objectives of human health protection and improvement through the reduction of microbial and toxic hazards from water and the parallel protection of aquatic ecosystems. This is also essential to the delivery of sustainable ecosystem services which are vital for societies worldwide requiring: (i) assimilation and purification of pollutants; (ii) flood protection; (iii) maintenance of biodiversity; and (iv) sufficient quantity of clean water for industrial processes including cooling, irrigation of crops and hydro-power generation.

Water quality is clearly at the heart of SDG 6 and is specifically addressed in 6.3 above. Here, the mechanisms suggested for achieving good water quality include reducing untreated wastewater, which would certainly have a major impact worldwide. But, importantly, this is only a surrogate measure for water quality and, as such, its real effect will be very different in different regions or locations. For example, if the untreated wastewater input represents only a tiny proportion of a river's discharge then the effect of the original discharge (or indeed its reduction by 50 per cent – see 6.3 above) on real water quality parameters, such as dissolved oxygen in the river, would be small compared to a site where the untreated effluent comprises a considerable proportion of the flow of the river in question. In the latter case, the oxygen demand of the untreated effluent could well cause ecosystem damage such as fish kills, and the probability of encountering infectious doses of human microbial pathogens from the sewage input would be much higher and present a more serious health risk. Hence, direct measurement of water quality provides the best index of health risk where this is feasible.

The SDGs often imply the need for good water quality which is an essential requirement, or a significant outcome, for most of the Goal 6 targets, that is,

- provision of 'safe' drinking water (6.1) implies good water quality;
- provision of hygiene for all (6.2) requires a sufficiency of clean water for washing;
- water use efficiency and sustainable withdrawals (6.4) maintains healthy ecosystems able to operate as sustainable receiving environments for human waste streams;
- integrated water resource management (6.5) includes water quality maintenance; and
- ecosystem protection and restoration (6.6) is founded on adequate water quality to maintain ecosystem health.

It is therefore of great significance that no truly global water quality database, or monitoring systems, exists at the present time (i.e. 2015). This has important implications for the ability of UN agencies and member countries to chart their success in achieving the targets in SDG 6. Past efforts to rectify this deficiency have been made and, here, the principal work has been delivered by a United Nations Environment Programme (UNEP) initiative 'GEMS Water' through construction of the GEMStat database which was launched on World Water

Day in March 2005 to help kick-off the 'Decade for Action: Water for Life'. However, the existing GEMStat data resource cannot deliver the water quality data needed at the global scale.

For example, at a GEMS Water workshop on global water resource modelling in Saskatoon, Canada, Silva et al. (2010) noted: 'Unfortunately, there is a complete disconnect between the reporting conducted and the level of data needed for accurate assessments.' This patchy coverage of available data is seen in the GEMStat pathogens database, available for download, which covers only US water utilities in its present form (2014). GEMS Water has noted this weakness in the empirical data coverage in many reports and the issue has been addressed by its technical advisory groups. GEMStat has sought to rectify this deficiency by more proactive data acquisition; the 2005 Environmental Sustainability Index (ESI) report noted the effects of this approach and stated: 'As a result of these changes, participation in GEMS Water has grown from less than 40 countries when the ESI first started using the data to over 100 countries today, although data coverage is still low' (see UNEP GEMS Water Technical Advisory paper No 3 Sept 2006, Page 7).

However, the present GEMS website notes the continued data heterogeneity and urges caution where these data are used for water quality index calculations: 'The WQI calculations are based on water quality data currently housed in GEMStat only. For some participating countries, data may have a limited geographic representation and / or temporal coverage. Therefore, results must be carefully assessed' (http://www.gemstat.org/waterqualityindexguide.aspx).

This pattern of patchy water quality information, in the only global investigative tool, with gaps evident in the key developing nations where water quality has the highest potential impacts on human health, is of great concern and, probably, surprising to managers and scientists with experience, principally, of the developed world where harmonised monitoring, driven by established regulatory systems, is ubiquitous. The lack of capacity and available funding in developing nations in this area is a significant reason for this current pattern and could be a significant brake on progress in the delivery of human health improvements.

What tools do we have and how do they contribute?

It is the dearth of existing empirical data at the global scale in water and health that makes investigative tools so essential to take forward health improvements through both water quality assessment and remediation. Papers presented in this part first consider how water quality links to, and predicts, health risk. Jacqueline MacDonald Gibson sets the scene for the key metric in assessing the impacts of drinking water on health which is known as a 'burden of disease assessment'. This type of analysis has become central to the international policy community in designing the priorities for action to improve public health. Huw Taylor then provides detailed guidance on water quality measurement and its role in building a credible management evidence base leading to public health protection. Karin Yeatts sets out the principles of modern epidemiology in the water sector. Epidemiology (the study of illness patterns in space and time and their causes) can provide the best evidence of cause and effect together with dose–response relationships linking specific water quality exposures to resultant disease patterns. Importantly, the statistical tools in this area also facilitate determination and quantification of non-water-related exposures and characteristics such as age, gender, diet, historical and existing health status, domestic arrangements and social class of exposed and non-exposed individuals. These are often termed confounding factors. This may imply a 'nuisance' status for such confounders but, in reality, their quantification can furnish useful information to the policy community: to direct efforts in reducing risks from the full spectrum of water and non-water-related exposures but also to provide a risk metric for exposures that the public traditionally accept in their normal daily lives. This concept can make a useful contribution to standards design and application as seen in the World Health Organization (WHO) *Guidelines for safe recreational water quality* (WHO, 2003; Kay et al., 2004). A range of protocols, or research designs, have been used in epidemiological studies in the water and health area. Generally, these include comparison of exposed and non-exposed groups and they require data on: (i) the level of exposure (in this case, water quality) for the exposed group; and (ii) the incidence of illness in both groups in the pre- and post-exposure periods. These studies, at best, use specifically designed and carefully controlled studies of cases (exposed) and controls (non-exposed) who may be randomised into bather and non-bather groups (Kay et al., 1994; Widenmann et al., 2006) or self-selecting. Alternatively, they can adopt a longitudinal (before and after) design for the retrospective examination of outbreaks (Logan, 2013).

In many locations, we do not have targeted epidemiological studies. In such situations, the health risk associated with adverse water quality can be quantified through the, now well established, tool of quantitative microbial risk assessment (QMRA), illustrated in Medema's chapter. This examines the risk from a range of potential pathogens and uses data on pathogen presence to derive a calculated risk to water users for a specific water environment and exposure type. There are a series of assumptions required for the application of this approach, such as the likely ingestion volume of water associated with a specific activity, and empirical data to underpin such assumptions is growing in the science literature (e.g. Dorevitch et al., 2011).

The pathogens used in QMRA are generally difficult and expensive to analyse in routine water quality testing. For this reason, indicator microbes have been used for many years as more useful measures of water quality. They indicate a potential connectivity between a pollution source, often a faecal contamination, and the point of use of a water resource which could be in the water body itself or after collection and transport to provide a household water supply. Thus, the indicators of microbial quality discussed by Brown and Grammer indicate a potential risk of present or future contribution by pathogen carriers to the faecal loading observed at a site. For much of the time, particularly if the population contributing faecal contamination is small and healthy, there may be few or no pathogens present. However, the manager can use the microbial indicator presence to warn of potential future risk, if and when the contributing population begins shedding pathogens at times of illness in the population. Thus the faecal indicator bacteria, in particular, are the main tools in water regulation to protect health from drinking and recreational water ingestion.

Integrated water resource management is seen as an essential to the SDGs if we are to achieve the health and environmental quality objectives set in Goal 6 above. This principle has driven a catchment paradigm in modern water resource management and regulation, as exemplified in the US Clean Water Act and the EU Water Framework Directive. Risk distribution (i.e. both spatially and temporally) in such systems is determined by the timing and magnitude of the accumulation, transport and storage of microbial and toxin fluxes in catchment systems. Kay addresses the emerging science of catchment microbial dynamics which is at the centre of efforts to manage health risk in water catchment systems. The importance of this topic is seen in US Environmental Protection Agency (USEPA) data published in real time to quantify the number of MDL water quality impairments in the USA. Consistently, microbial pollutants top the list of CWA 'impairments' needing management action through 'total maximum daily load' (TMDL) studies illustrating the pre-eminence of the heath-related parameters in modern catchment management (see 'Useful websites' below).

Integrated management of local and global water quality is increasingly urgent as seen in the UN initiatives of MDGs and SDGs. The most promising tool addressing global water resource management is undoubtedly emerging geospatial technologies, which are increasingly being applied to water quality problems and associated disease patterns (Mayorga et al., 2010; Lehner et al., 2012). This is addressed by Wright's chapter in this part who presents a novel overview of the issues involved in reporting disease patterns, risk mapping and related water safety planning, in a geospatial context, providing analytical insights which can be derived from the intelligent combination of a range of geospatial data resources.

In two related chapters, Jeuland presents introductions to, and analysis of, both cost–benefit analysis and demand assessment and valuation, the latter related to water, sanitation and hygiene interventions and policy approaches developed by WHO in addressing the challenges of the MDGs and the emergent water, sanitation and hygiene (WaSH) agenda. In this critical evaluation, the place of economic assessment in providing empirical data for the policy community is outlined and covers both market approaches and non-market tools (revealed preference and stated preference) for demand valuation.

Useful websites

EU Water Framework Directive: http://ec.europa.eu/environment/pubs/pdf/factsheets/water-framework-directive.pdf

GEMS Water: http://www.unep.org/gemswater/

GEMStat pathogens database: http://www.unep.org/gemswater/FreshwaterAssessments/PathogensProject/tabid/78261/Default.aspx

GEMStat 3rd Technical Advisory Group on water quality September 2006: http://www.unep.org/gemswater/Portals/24154/publications/pdfs/technical_advisory_paper3.pdf

UN Water and SDGs: http://www.unwater.org/topics/water-in-the-post-2015-development-agenda/en/

USEPA Clean Water Act Impairments and total maximum daily loads (TMDLs): http://iaspub.epa.gov/waters10/attains_nation_cy.control?p_report_type=T

WHO (2003) Recreational Water Quality Guidelines: http://www.who.int/water_sanitation_health/bathing/en/

WHO Water Sanitation and Hygiene/Health WaSH: http://www.who.int/water_sanitation_health/en/; http://www.who.int/water_sanitation_health/hygiene/en/

References

Anon (2014) Open Working Group of the UN General Assembly on Development Goals. Document A/68/970. Downloadable from http://undocs.org/A/68/970. 24pages.

Dorevitch, S., Panthi, S., Huang, Y., Li, H., Michalek, A.M., Pratap, P., Wroblewski, M., Lui, L., Scheff, P.A. and Li, A. (2011) Water ingestion during water recreation. *Water Research* 42, 2020–2028.

Kay, D., Fleisher, J.M., Salmon, R.L., Jones, F., Wyer, M.D., Godfree, A.F., Zelenauchjacquotte, Z., and Shore, R. (1994) Predicting the likelihood of gastroenteritis from sea bathing – Results from randomized exposure. *Lancet* 344, 905–909.

Kay, D., Bartram, J., Prüss, A., Ashbolt, N., Wyer, M.D., Fleisher, J.M., Fewtrell, L., Rogers, A. and Rees, G. (2004) Derivation of numerical values for the World Health Organization guidelines for recreational waters. *Water Research* 38, 1296–1304. doi:10.1016/j.watres.2003.11.032.

Lehner, B., Verdin, K. and Jarvis, A. (2012) New global hydrography derived from spaceborne elevation data. *Eos Transactions, AGU*, 89(10), 93–94, 2008.

Logan, J. (2013) The Strathclyde Loch Norovirus Outbreak. In *Public Health 2012/13. The annual report of the Director of Public Health, Lanarkshire, Scotland*, pp 21–22. 68p. Available from: http://www. nhslanarkshire.org.uk/publications/Documents/PublicHealth-2012-13.pdf

Mayorga, E., Seitzinger, S.P., Harrison, J.A., Dumont, E., Beusen, A.H.W., Bouwman, A.F., Fekete, B.M., Kroeze, C. and Van Drecht, G. (2010) Global nutrient export from watersheds 2 (NEWS 2): Model development and implementation. *Environmental Modelling & Software* 25, 837–853. doi:10.1016/j.envsoft.2010.01.007

Silva, G., Dube, M. and Robarts, R. (2010) Workshop on Global Water Quality Modelling, October 13th and 14th, The National Hydrology Research Centre (NHRC), Saskatoon, SK, Canada. http:// www.unep.org/gemswater/Portals/24154/common/pdfs/modelling_workshop-final_report.pdf

WHO (2003) *Guidelines for safe recreational water environments Volume 1: Coastal and freshwaters.* Geneva: World Health Organization.

Wiedenmann, A., Krüger, P., Dietz, K., López-Pila, J., Szewzyk, R., and Botzenhart, K. (2005) A randomized controlled trial assessing infectious disease risks from bathing in fresh recreational waters in relation to the concentration of *Escherichia coli*, intestinal enterococci, *Clostridium perfringens* and somatic coliphages. *Environmental Health Perspectives,* 8115, 1–41.

55

EPIDEMIOLOGY*

Karin B. Yeatts

RESEARCH ASSISTANT PROFESSOR, DEPARTMENT OF EPIDEMIOLOGY,
GILLINGS SCHOOL OF GLOBAL PUBLIC HEALTH UNIVERSITY OF
NORTH CAROLINA AT CHAPEL HILL, CHAPEL HILL, NC, USA

Learning objectives

1 Calculate and interpret common epidemiologic measures of disease occurrence: prevalence, risk, rate, odds; calculate person-time.
2 List characteristics of various types of epidemiologic study designs.
3 Define and calculate the different types of measures of ratios.

Epidemiology is the study of the distribution and determinants of disease in the *population*. In addition to studying the patterns of disease in the population, epidemiology can also be used to investigate other health outcomes, such as injuries, events, or health behaviors. In the field of global water and health, epidemiology can be used to describe the patterns of waterborne diseases in the population. In the historic development of epidemiology as a field, John Snow's famous 1849 work was seminal – he determined the source of the waterborne outbreak (cholera) by calculating and comparing rates of deaths in different neighborhoods of London.

Epidemiology can also be used to evaluate community-based interventions to improve water quality and reduce diarrheal diseases from unclean water supplies, or investigate topics such as cancer and adverse reproductive outcomes from exposures to waterborne pesticides, recreational water contaminants, vectorborne disease, or water-related diseases caused by insect vectors, such as mosquitoes carrying malaria. Basic epidemiologic tools for investigative research include measures of disease occurrence, measures of association, and study designs. Measures of disease occurrence are often used to describe the patterns of disease or health outcomes by "person, place, and time."

Measures of disease occurrence

In epidemiology, specific measures are used to describe the occurrence or frequency of disease in a population. This helps researchers quantify the magnitude of the potential problem or

* Recommended citation: Yeatts, K.B. 2015. 'Epidemiology', in Bartram, J., with Baum, R., Coclanis, P.A., Gute, D. M., Kay, D.,McFadyen, S., Pond, K., Robertson, W. and Rouse, M.J. (eds) *Routledge Handbook of Water and Health*. London and New York: Routledge.

Table 55.1 Summary table of measures of disease frequency formulas

Prevalence	Risk	Rate	Odds
# of prevalent cases	# of incident cases	# of incident cases	probability
total study population at specific time period	total # of at-risk individuals	total person-time at risk	1 − probability or cases/non-cases

investigate potential risk factors for a disease. Measures we will define and describe here include prevalence, risk, rate, and odds (Table 55.1). For these terms, first it is important to understand the distinction between prevalent and incident cases. Prevalent cases include both existing and new cases of the disease. Incident cases are only *new* cases of disease.

Our example will include malaria, which has a devastating global impact, and is an important water-related disease spread by mosquitoes. Prevalent cases of malaria would include both existing and new cases of malaria among a group of 1,081 pregnant women attending antenatal clinics for the first time during their current pregnancy in two hospitals in semi-urban areas of Lagos, Nigeria (Agomo and Oyibo, 2013). Of the 1,081 women who consented and enrolled in this cross-sectional study between March 2007 and February 2008 (Agomo and Oyibo, 2013), 83 had a positive laboratory test for *Plasmodium falciparum* and/or *P. malariae*. If, instead, a group of 1,081 pregnant women in Lagos who had never had malaria were followed over the course of their pregnancy and 300 developed *new* cases of malaria during their pregnancy, these women would be considered incident cases of malaria.

Prevalence

Prevalence is defined as the proportion of the population living with a disease at a specific point or period in time. The numerator is prevalent cases of the disease and the denominator is the total study population. Prevalence is often scaled to a percentage. Prevalence is a useful measure as it quantifies the magnitude of existing disease in the study population. In our example of pregnant women attending antenatal clinics in Lagos, we calculate the prevalence of malaria infection as follows (assuming a total study population of 1,081 pregnant women, and 83 prevalent cases of malaria in a one-year period):

> Prevalence = 83 prevalent cases/1,081 total study population = 0.0768 in a one-year period

This can then be converted to a percentage, by multiplying by 100, to 7.68 percent. Thus, the prevalence of malaria in pregnant women attending antenatal clinics in Lagos, Nigeria, over a one-year period between 2007 and 2008 was 7.68 percent.

Risk

Risk is defined as the proportion of the population who developed the disease over the specified amount of follow-up time; it is also the average probability of disease occurrence. The numerator consists of incident or new cases of disease, and the denominator is the total at-risk population at the start of the study. The denominator is fixed by the size of the study population at the beginning of the observation period; this is considered a closed population.

Also, study subjects must be disease-free at the start of the study. Units are in cases per person over a specified time period. Often risks are converted to cases per 100, 1,000, or 100,000 persons so that cases are whole numbers rather than fractions. Synonyms for risk in the literature include "cumulative incidence" or "incidence proportion."

Rate

A rate is defined as the proportion of the population who developed the disease per unit of time at risk over the specified amount of follow-up time. A rate is considered the average rate of new disease occurrence and indicates how rapidly new events are occurring in the population. As with "risk," the numerator consists of incident cases of disease or the health outcome, and the denominator only includes people at risk of getting the disease or health outcome. However, in contrast with "risk," the denominator of a rate is "person-time," which is calculated by the length of time each study subject contributes to the study. Using the units of "person-time" allows the denominator to be constantly adjusted for losses and additions of subjects that occur in a dynamic cohort. Common time units are person-years and person-days. Rates can be scaled and reported as per 1,000 or 100,000 person-years. Synonyms for rate include "incidence rate" or "incidence density."

For example, in Teschke et al.'s study (2010), the authors reported the rates of physician visits and hospitalizations for water-related endemic intestinal infectious disease. The authors calculated crude incidence rates of 1,353 physician visits and 33.8 hospitalizations for intestinal infections per 100,000 person-years. These incidence rates are for the time period from 1995 through 2003. In a study of arsenic exposure in drinking water among pregnant women in Bangladesh, Rahman et al. (2011) calculated the incidence rates of lower respiratory tract infection (LRTI) and severe LRTI in the infants during their first year of life; these rates were 2.96 (95 percent confidence interval (CI), 2.78–3.16) and 2.35 (95 percent CI, 2.18–2.52) episodes per person-year, respectively.

Odds

Odds are often used in epidemiology as a component of the "odds ratio," a measure of association described below. However, they are rarely used on their own to describe disease occurrence. They are a familiar concept in describing chances in gambling or games of luck. For example, a coin which has a 50:50 chance of coming up heads is a description of the odds. Odds are defined as the probability of an event occurring divided by the probability of it not occurring (i.e. P/1-P, where P = probability). Odds are often presented as the ratio of cases to non-cases, or the ratio of exposed to non-exposed, and are used when meaningful prevalence, risk, or rate data are not available.

Study designs

To investigate research questions, epidemiologists have a variety of study designs from which to choose. The choice of study design depends upon the research question. Common studies include randomized control trials/interventions, cohort, case-control, and cross-sectional.

Experimental study designs: interventions and randomized control trials

Intervention studies can be conducted with individuals or communities. First the investigator defines and assembles the study population of interest. The investigator assigns the intervention, randomly or not. One group of individuals or a community receives the intervention and the control group (individuals or community) does not receive the intervention. For example, one could design an intervention study to evaluate the effects of fluoridation in drinking water on dental caries in children, with one community's water supply fluoridated and the control community's drinking water not fluoridated. Both the intervention and control community would be followed over time and the outcome of interest (incident or new dental caries in this example) measured.

Randomized control trials (RCTs) are often designed to evaluate the effect of medication or treatment. A key feature of RCTs is that the intervention of interest is randomized. The randomization reduces the influence of other determinants of exposure and outcome (i.e., confounding factors). RCTs can provide strong evidence of cause and effect. However, a disadvantage of RCT design is that many exposures of interest in populations cannot be ethically randomized, for example exposures to known toxic chemicals such as arsenic in drinking water or water contaminated with cholera or *Shigella*. Using an intervention design, Colwell et al. (2003) showed that simple sari cloth filtration protected villagers from cholera in Matlab, Bangladesh. Twenty-seven villages (with 2,750 households and 14,709 individuals) received the sari cloth intervention; thirteen villages (with 2,750 households and 15,662 individuals) did not. Participants who received the sari cloth filtration intervention showed a 48 percent reduction in cholera rate compared with those in the thirteen control villages.

Observational study designs

Observational study designs are those in which the investigator does not assign exposures to populations but "observes" exposures and health effects as they occur in the population – many exposures cannot ethically be randomized. These designs include cohort, case-control, and cross-sectional studies and they are commonly used by epidemiologists to assess the distribution and determinants of disease in the population. Other designs not discussed here include ecologic studies, time series, and case-crossover.

Cohort

A cohort is typically defined as a group of persons sharing a common characteristic. Epidemiologists may define their cohort geographically or may choose individuals based on occupation, gender, race, ethnicity, or other factors. Cohort studies are also called longitudinal or follow-up studies. Individuals in the cohort are free of the disease of interest at the start of the study. Cohorts with and without an *exposure* of interest are identified at baseline, and *followed over time* for *incident* (new) occurrences of the outcome of interest (disease, death, change in health status or behavior). As exposure precedes disease in this type of study design, cohort studies are the best observational designs to contribute evidence of causality. An additional key advantage of cohort studies is that they allow direct estimation of risks or rates.

One example of a cohort study is the study by Teschke et al. (2010), which evaluated different rates of endemic gastrointestinal infections from different water supplies and sewage disposal systems; these authors assembled a cohort using population-based Canadian health insurance data from 1995 to 2003.

Case-control

Case-control studies are an efficient and common epidemiologic method to study causal factors for specific cancers or other rare diseases (with prevalence $< = 10$ percent). In contrast to the cohort design, the investigator begins by identifying diseased and non-diseased subjects (cases and controls) and then "looks back" to measure and compare the exposure of interest in each group. In the first step in conducting a case-control study, the investigators identify a group that has the disease outcome of interest. They then select some or all of the cases from this group. Cases are often selected from a hospital, clinic, or disease registry. Next, a group of controls must be identified. The controls need to be representative of the same source population as the cases, that is, if the controls were to develop the disease they would be identified as a case. Sources of controls can be selected from the hospital or from the general population. Once cases and controls have been identified and enrolled in the study, researchers then measure the odds of the exposure in both cases and controls. The investigators then compare the exposure odds among cases and controls using either an odds ratio or odds difference. The case-control design is best used in the following situations: exposure data are expensive or difficult to obtain; the disease has a long induction and latent period; the disease is rare; or little is known about the disease. For example, in the Global Enteric Multicenter Study (GEMS), Kotloff et al. (2012, 2013) used a case-control design to evaluate the etiology and burden of moderate-to-severe pediatric diarrheal disease across four sites in Africa and three in Asia. Severe-to-moderate cases are only a small fraction of all pediatric diarrhea cases among children in developing countries, so this outcome meets the rare outcome requirement for case-control studies.

Cross-sectional

In a cross-sectional study, the exposure and the outcome are assessed at the same time point. This study design can be described as a "snapshot" in time. Cross-sectional studies begin by defining and selecting a study population, sampling, then obtaining data from study participants, classifying study participants as having the health outcome or not, and lastly comparing exposure prevalence in these two categories.

One disadvantage of this type of study is that one cannot be sure that exposure preceded disease, since both are ascertained at same time (analogous to "which came first, the chicken or the egg?" question, also known as "antecedent–consequent" bias). Advantages of cross-sectional studies include that they are generally inexpensive and quickly conducted compared with other types of studies. Cross-sectional studies are also useful for generating hypotheses or describing disease, potential exposures, or risk-factor frequencies in populations in which little is known. They can also be used for planning preventive or health care services, as well as surveillance programs and conducting surveys and polls.

Measures of association

Measures of association such as "risk ratios," "rate differences," or "odds ratios" are used to compare disease occurrence in exposed and unexposed groups. These comparisons can be made by division (*ratio* effect measures) or subtraction (*difference* effect measures). In cohort studies, the risk or rate in the exposed group is compared with the risk or rate in the unexposed group. Risk ratios are interpreted as follows: Those who were exposed were [risk ratio] times as likely to develop the outcome compared with those who were unexposed over the follow-up timeframe of the study. A rate difference would be interpreted as: Among those who were exposed, the rate of the disease was [rate difference] higher/lower than among those who were unexposed. In a case-control study, we calculate the exposure odds ratio (OR). The OR approximates the rate ratio or risk ratio under the rare disease assumption, in which the prevalence is less than 10 percent. ORs are interpreted as: Those who were exposed had [OR times] the odds of the disease compared with those who were unexposed. For example, pregnant women of young maternal age (20 years of age or younger) were 2.61 times as likely to be develop a malaria infection compared with pregnant women of old ages (21 years of age or older) (Agomo and Oyibo 2013).

Key recommended readings

1 Colwell, R.R., Huq, A., Islam, M.S., Aziz, K.M., Yunus, M., Khan, N.H., Mahmud, A., Sack, R.B., Nair, G.B., Chakraborty, J., Sack, D.A., and Russek-Cohen, E. 2003. Reduction of cholera in Bangladeshi villages by simple filtration. *Proc Natl Acad Sci USA* 100(3):1051–5. The authors used an intervention design. Colwell et al. showed that simple sari cloth filtration protected villagers from cholera in Matlab, Bangladesh. (Free PMC access.)
2 Kotloff, K.L., Blackwelder, W.C., Nasrin, D., Nataro, J.P., Farag, T.H., Van Eijk, A., Adegbola, R.A., Alonso, P.L., Breiman, R.F., Faruque, A.S., Saha, D., Sow, S.O., Sur, D., Zaidi, A.K., Biswas, K., Panchalingam, S., Clemens, J.D., Cohen, D., Glass, R.I., Mintz, E.D., Sommerfelt, H., and Levine, M.M. 2012. The Global Enteric Multicenter Study (GEMS) of diarrheal disease in infants and young children in developing countries: epidemiologic and clinical methods of the case/control study. *Clin Infect Dis.* Suppl 4:S232–45. doi: 10.1093/cid/cis753. Erratum in: *Clin Infect Dis.* 57(1):165. This study is an example of a case control study design. The study design is applied to answering the question, "What are the top causes of diarrheal diseases in infants and young children in developing countries?"
3 Teschke, K., Bellack, N., Shen, H., Atwater, J., Chu, R., Koehoorn, M., MacNab, Y.C., Schreier, H., and Isaac-Renton, J.L. 2010. Water and sewage systems, socio-demographics, and duration of residence associated with endemic intestinal infectious diseases: a cohort study. *BMC Public Health.* 10:767. doi: 10.1186/1471-2458-10-767. One example of a cohort study is the study by Teschke et al., who calculate different rates of endemic gastrointestinal infections from different water supplies and sewage disposal systems; these authors assembled a cohort using population-based Canadian health insurance data from 1995 to 2003. (Open access.)

Epidemiologic textbooks for reference

Aschengrau, A. 2013. *Essentials of Epidemiology in Public Health*, 3rd edition. Burlington, MA: Jones and Bartlett.
Gordis, L. 2013. *Epidemiology.* 5th edition. Philadelphia, PA: WB Sanders.

References

Agomo, C.O., and Oyibo, W.A. 2013. Factors associated with risk of malaria infection among pregnant women in Lagos, Nigeria. *Infect Dis Poverty* 2(1):19. (Open access.)

Colwell, R.R., Huq, A., Islam, M.S., Aziz, K.M., Yunus, M., Khan, N.H., Mahmud, A., Sack, R.B., Nair, G.B., Chakraborty, J., Sack, D.A., and Russek-Cohen, E. 2003. Reduction of cholera in Bangladeshi villages by simple filtration. *Proc Natl Acad Sci USA.* 100(3):1,051–5. (Free PMC access.)

Kotloff, K.L., Blackwelder, W.C., Nasrin, D., Nataro, J.P., Farag, T.H., Van Eijk, A., Adegbola, R.A., Alonso, P.L., Breiman, R.F., Faruque, A.S., Saha, D., Sow, S.O., Sur, D., Zaidi, A.K., Biswas, K., Panchalingam, S., Clemens, J.D., Cohen, D., Glass, R.I., Mintz, E.D., Sommerfelt, H., and Levine, M.M. 2012. The Global Enteric Multicenter Study (GEMS) of diarrheal disease in infants and young children in developing countries: epidemiologic and clinical methods of the case/control study. *Clin Infect Dis.* 55 Suppl 4:S232–45. doi: 10.1093/cid/cis753. Erratum in: *Clin Infect Dis.* 57(1):165. (Free PMC article.)

Kotloff, K.L., Nataro, J.P., Blackwelder, W.C., Nasrin, D., Farag, T.H., Panchalingam, S., Wu, Y., Sow, S.O., Sur, D., Breiman, R.F., Faruque, A.S., Zaidi, A.K., Saha, D., Alonso, P.L., Tamboura, B., Sanogo, D., Onwuchekwa, U., Manna, B., Ramamurthy, T., Kanungo, S., Ochieng, J.B., Omore, R., Oundo, J.O., Hossain, A., Das, S.K., Ahmed, S., Qureshi, S., Quadri, F., Adegbola, R.A., Antonio, M., Hossain, M.J., Akinsola, A., Mandomando, I., Nhampossa, T., Acácio, S., Biswas, K., O'Reilly, C.E., Mintz, E.D., Berkeley, L.Y., Muhsen, K., Sommerfelt, H., Robins-Browne, R.M., and Levine, M.M. 2013. Burden and aetiology of diarrhoeal disease in infants and young children in developing countries (the Global Enteric Multicenter Study, GEMS): a prospective, case-control study. *Lancet* 382(9888):209–22. doi: 10.1016/S0140-6736(13)60844-2.

Rahman, A., Vahter, M., Ekström, E.C., and Persson, L.Å. 2011. Arsenic exposure in pregnancy increases the risk of lower respiratory tract infection and diarrhea during infancy in Bangladesh. *Environ Health Perspect.* 119(5):719–24. doi: 10.1289/ehp.1002265.

Snow, J. 1849. *On the Mode of Communication of Cholera.* London: Churchill.

Teschke, K., Bellack, N., Shen, H., Atwater, J., Chu, R., Koehoorn, M., MacNab, Y.C., Schreier, H., and Isaac-Renton, J.L. 2010. Water and sewage systems, socio-demographics, and duration of residence associated with endemic intestinal infectious diseases: a cohort study. *BMC Public Health* 10:767. doi: 10.1186/1471-2458-10-767. (Open access.)

56

QUANTITATIVE MICROBIAL RISK ASSESSMENT[*]

Gertjan Medema

Water Quality and Health, KWR Watercycle Research
Institute, Nieuwegein, Sanitary Engineering, Delft
University of Technology, Delft, the Netherlands

Learning objectives

1 Understand the purposes and benefits of quantitative microbial risk assessment (QMRA).
2 Describe how to carry out a QMRA.
3 Recognize the value of QMRA.

Introduction

Waterborne pathogens are associated with a high burden of disease. Water safety management is instigating effective and proportional control strategies for these waterborne pathogens. This requires information about sources, fate and transport of pathogens in water systems, about exposure of humans to these water systems and the resulting health risks. Risk assessment aims to aid decision makers by collating and evaluating this type of information: the likelihood of exposure of humans to pathogens via water and the consequence thereof, the probable magnitude of health effects. For risks of waterborne pathogenic microbes, the use of risk assessment was first proposed in the early 1990s (Regli et al., 1991). The World Health Organization has been instrumental in the introduction of microbial risk assessment (MRA) as a basis for safety management of the water we use for drinking, recreation and food crop irrigation (WHO, 2003, 2006, 2011).

---

[*] Recommended citation: Medema, G. 2015. 'Quantitative microbial risk assessment', in Bartram, J., with Baum, R., Coclanis, P.A., Gute, D.M., Kay, D., McFadyen, S., Pond, K., Robertson, W. and Rouse, M.J. (eds) *Routledge Handbook of Water and Health*. London and New York: Routledge.

Microbial risk assessment

QMRA paradigms have been described in several water and food-related international guidelines, using different headings for the steps of QMRA (CAC, 1997, 1999; Haas et al., 1999, 2014). However, the basic elements of each of the paradigms are very similar. Petterson et al. (2014) combined descriptions in the World Health Organization's (WHO's) guidelines on drinking water, bathing water and wastewater reuse in agriculture into a consistent framework for water-related QMRA (Figure 56.1).

Problem formulation

Before embarking on a MRA, it is crucial to determine the context, scope and purpose of the MRA in the context of management of the health risks. The intended use of the results of the MRA will target the MRA process and help to determine the most efficient MRA approach. In the problem formulation phase, the scope is translated into a conceptual model and analysis plan. It is important for the risk manager to be involved to understand technical limitations, such as limited availability of data, and the need for assumptions and for process simplifications will arise in this phase.

In the problem formulation, the pathogens (hazards) are identified. It is a qualitative process and serves to select the pathogen (or pathogens) of concern, document important pathogen characteristics, exposure pathways, including its (potential) presence in water and

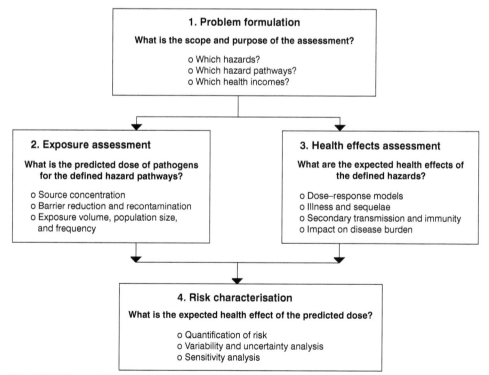

Figure 56.1 Harmonized framework for water-related QMRA

Source: Petterson et al., 2015, based on QMRA outlined in WHO's water-related guidelines and the Stockholm Safe Water Framework (Fewtrell and Bartram, 2001).

produce, and health effects associated with the pathogen (mode of action, symptoms, disease, etc.). Recent books give an overview of available information (Fan et al., 2009; Drechsel et al., 2009). Databases that collate the information about pathogens, exposure routes, disease are emerging (see for instance http://www.foodrisk.org and http://qmrawiki.msu.edu).

Exposure assessment

In the exposure assessment, the likelihood of the actual or anticipated intake, inhalation and/or dermal exposure of pathogens through water (or related media, such as irrigated fresh produce) is assessed. The likelihood of exposure is usually determined by a chain of events: for example in the case of irrigation of fresh produce the presence of pathogens in the water used for irrigation, irrigation practice, transmission of pathogens to the fresh produce, survival (and potentially growth of some) of the pathogens on/in the produce through the food chain, food preparation and consumption practice and so on. Exposure assessment requires (preferably quantitative) information about all the steps in this chain. Questions that need quantitative answers during the exposure assessment are, on the one hand, the pathogen concentrations and fate and transport in water and in barriers (e.g. water treatment processes, inactivation during withholding periods etc.). On the other hand, data are needed on the exposure of humans, for example 1) size of the exposed population, 2) nature of the exposed population (e.g. vulnerable groups), 3) how often they are exposed (e.g. daily, etc.). Relevant data on exposure assessment can be found for instance in the US Environmental Protection Agency's (1997) Exposure factors handbook; NHMRC (2006); Drechsel et al. (2010) and http://www.foodrisk.org.

Health effects assessment

This step provides a qualitative or (preferably) quantitative description of the nature of adverse health effects associated with the pathogens. This includes the type of health effects (including sequellae), severity, duration that may occur after ingestion of the pathogen and available information on the relationship between ingested dose and the probability that health effects (infection, illness) occur (dose–response relation). The dose–response model is the critical link between pathogen exposure and estimated health outcomes. The most commonly applied models within QMRA are based on the single-hit theory: where every ingested pathogen particle is assumed to act independently, and has an individual probability of causing infection (Haas et al., 2014). A model needs to be selected from the published literature, and appreciation of the details of the dose–response studies will help to judge which dose–response data and model are most appropriate for the particular MRA. A resource for dose–response models for pathogens that are relevant for water is http://wiki.camra.msu.edu/.

Risk characterization

This step combines the information from the previous steps into an assessment of the probability of occurrence and severity of adverse health effects in the exposed population. The timescale in which the risk is expressed may differ, from single exposure events or a series of exposure events or for a year. The risk characterization is either deterministic (single values) or probabilistic (statistical distributions). In a deterministic MRA, the best estimate of each of the MRA model parameters in the exposure and effect assessment is selected and combined to compute the resulting health risk. In a probabilistic MRA, statistical distributions are used to describe the model parameters. The type of distribution selected to describe the model parameters is a combination of knowledge of (pathogens in) water systems and of statistics. The health risk is computed by combining the statistical distributions, using Monte Carlo methods (Haas et al., 2014). Burmaster and Anderson (1994) published principles of good practice for the use of Monte Carlo simulation in health risk assessments. Statistical software that can be used for the stochastic models are Microsoft Excel add-ins such as Modelrisk and @Risk, and packages like R, Analytica and so on. Dedicated QMRA models and software tools are increasingly available to aid this risk characterization step (ILSI, 2012).

The risk may be quantified in different end-points including: the probability of infection, probability of illness, expected number of illness cases, and measures for burden of disease, such as disability adjusted life years (DALYs). The DALY is the metric used in WHO guidelines for overall community health burden (WHO, 2011). DALYs has the advantage that it allows weighting to be given to illnesses that lead to more serious health outcomes and comparison of different types of health risks. When health-based targets are set, QMRA can also be performed in reverse, to derive health-based operational targets. The WHO *Guidelines for Drinking Water Quality* (WHO, 2011) show an example of using QMRA to translate the tolerable disease burden of 10^{-6} DALY per person per year to the required log-removal of water treatment.

It is important to include the variability (natural dispersion in a system, such as pathogen concentrations in a river) and uncertainty (lack of understanding and/or inability to measure) in all steps of the risk characterization. Sources of uncertainty in QMRA include: extrapolation from dose–response data (though, unlike with toxic chemicals, many dose–response data are from human exposure), limitations of pathogen detection methods, estimates of exposure. To determine how the variability and uncertainty in the information at individual steps of the risk assessment affects the overall risk estimate, sensitivity analysis is used. Sensitivity analysis can be conducted in several ways. The nominal range sensitivity analysis is relatively robust; here, the value of each parameter in the risk assessment is varied, one at a time, along the confidence range of that parameter (e.g. average and maximum concentration of a pathogen in water) to determine the effect on the final risk estimate. This procedure generates 1) the range of possible values of the final risk estimate and 2) which of the parameters contribute most to the uncertainty of the final risk estimate.

Tiered approach

Depending on the scope of the MRA, as defined at the start of the MRA process, different grades of risk characterization are possible: qualitative, deterministic, stochastic (ILSI, 2012). In practice, there is no sharp division between the different grades of risk characterization. They do not differ in their nature, only in the level of data quantification and mathematical sophistication. The intended use of the MRA should determine the required level of sophistication. A tiered approach with increasing levels of quantification/sophistication is good practice. The idea is that you start with the most simple assessment and then increase complexity as required. The basic tiers help to determine whether more advanced tiers are needed and where the MRA is still weak because good/specific data are lacking.

Good MRA practice

Several international bodies have produced guidance on good microbial risk assessment practice (WHO, 1999; EFSA, 2009; US EPA and USDA FSIS, 2012). General principles are:

- The scope and objectives of the risk assessment should be clearly defined and stated at the onset, in collaboration with the risk manager who is going to apply the results.
- Risk assessment should be clearly separated from risk management.
- Risk assessment should be soundly based on science.
- Risk assessment should be transparent: clear, understandable and reproducible. It should follow a harmonized procedure based on the accepted standards of best practice.
- The data used are evaluated to determine their quality and relevance to the assessment (taking into account their overall weight in the risk and uncertainty). If data are judged irrelevant or of too low quality, this conclusion should be justified. All data that are used are referenced.
- The variability in data should be documented and taken into account in the risk assessment.
- All assumptions are documented and explained. Where alternative assumptions could have been made, they can be evaluated together with other uncertainties (preferable in a sensitivity analysis).
- The risk assessment should include a description of the uncertainties encountered in the risk assessment process. Their relative influence on the risk assessment outcome should be described, preferably in a quantitative (probabilistic) manner. Where point estimates are used for uncertain (or variable) quantities, the selected values should be justified and their influence on the assessment included in the sensitivity analysis.
- Conclusions should reflect the objectives and scope of the risk assessment, and include uncertainties and data gaps to aid risk managers.

Case study: enteric pathogen MRA for wastewater reuse in crop irrigation

Problem formulation

The use of wastewater for irrigation of food crops that are eaten raw is a common practice in many arid and semi-arid regions, such as in Catalonia (Spain). This case study illustrates a MRA that addressed the question of the risk manager about the need for enhanced treatment of secondary wastewater effluent for application for irrigation of fresh produce. Norovirus was selected as reference pathogen, since it was the most common cause of acute gastroenteritis outbreaks in the study area. Sales-Ortells et al. (2015) constructed a model to evaluate the norovirus health risk associated with the consumption of wastewater irrigated lettuce crops.

Exposure assessment

For exposure assessment, quantitative data were needed on the following model parameters:

- Occurrence of Norovirus in secondary (after activated sludge) and tertiary (after filtration) effluent.

- Removal of virus by ultraviolet (UV) and chlorination.
- Fraction of viruses that attach to lettuce during irrigation (and internalize).
- Inactivation of norovirus on lettuce before and after harvesting.
- Removal of norovirus from lettuce during washing.
- Consumption of lettuce by Catalan population.

Data on several of these parameters were collected at the study site: norovirus concentration was measured with quantative polymerase chain reaction (qPCR) in secondary and tertiary effluent, and the removal of viruses by the UV and chlorination was determined using data on F-specific phages before and after the UV and/or chlorination (see Table 56.1). Data on consumption of lettuce by the Catalan people were available from food consumption surveys (as daily averages). For other parameters (fraction of virus that attaches to the lettuce during irrigation, the inactivation in the field and after harvesting, and the effect of washing the lettuce during the preparation of the meal) no site-specific data were available. Data on these parameters were collected from scientific literature. All these data showed variability and a probabilistic QMRA was conducted to incorporate this variability into the risk assessment.

The daily dose of virus on lettuce surface (*ds*) ingested by the consumers of the market where the reclaimed water-irrigated lettuce are sold was calculated as shown by equation 56.1:

$$d_s = 10^{\left(log10\left(C_{eff} x V_{surf} x 0.001\right) - LR - R_s - R_T - R_W\right)} xl \qquad (56.1)$$

The impact of additional wastewater treatment with UV and/or chlorination was modelled as alternative scenarios.

Table 56.1 Exposure assessment inputs, units, distribution and parameter values, and references

Model inputs	Units	Distribution (parameter values)[a]	Source of data
Ceff: concentration of norovirus in secondary effluent	Genome copies/L	Gamma (0.3, 1.2×10⁻⁶)	Norovirus data collected on site
LR: log reduction due to tertiary treatment (alternative scenario)	Log10 units	PERT (0.57, 0.94, 1.30)	Data of F-specific phage inactivation collected on site (Montemayor et al., 2008)
Vsurf: water that clings to lettuce surface through sprinkler irrigation	mL/g	Lognormal3 (–4.57, 0.50, 0.006)	Data on volume of water on lettuce after spray irrigation (Mok and Hamilton, 2014)
Rs: in-field reduction of surface viruses	Log10 units	Uniform (1, 2)	Inactivation data of MS2 on lettuce (Carratalà et al., 2013)
Rt: reduction of viruses during transport and storage	Log10 units	Uniform (0, 1)	Inactivation data of MS2 on lettuce (Carratalà et al., 2013)
Rw: reduction of surface viruses due to washing	Log10 units	PERT (0.1, 1, 2)	Data on virus removal from lettuce during washing (Mok and Hamilton, 2014)
I: daily consumption of lettuce	g person⁻¹, day⁻¹	Lognormal (20.72, 26.35, inf=0, sup=120)	Catalan food consumption survey (AESAN, 2011)

Health effects assessment

The dose–response model for norovirus as described by Teunis et al., (2008) was used (equation 56.2):

$$Pd = 1 - {}_1F_1\left(\alpha, \alpha + \beta, -d\right) \tag{56.2}$$

where ${}_1F_1$ is the Kummer confluent hypergeometric function, α and β are shape parameters with values 0.04 and 0.055, respectively, and d the dose. The illness given infection risk (Pd,ill) is calculated by multiplying the infection risk by an illness given infection ratio, which for norovirus is 0.67 (Atmar et al., 2013).

Risk characterization

Farmers irrigate the crops with tertiary effluent during warm months, and not in winter. Therefore, frequency of ingestion of reclaimed water irrigated lettuce (f) happens from April to October, that is, 214 days per year, assuming all the lettuce consumed comes from the street market. The annual probability is estimated using equation 56.3 (Haas et al., 1999).

$$Py = 1 - (1 - Pd)^f \tag{56.3}$$

Annual disease burden was calculated using DALYs. DALYs account for the years lived with disability (YLD) (equation 56.4) plus the years of life lost (YLL) (equation 56.5) due to the hazard, as compared to the average expected age of death in a community.

$$YLD = Py, ill \times Dw \times Dt \tag{56.4}$$

$$YLL = Py, ill \times Nd \times L \tag{56.5}$$

where Dw is the disability weight, Dt the duration of illness, Nd the number of deaths per illness and L the average years lost per fatality. The Dw for acute gastroenteritis is 0.0007 for cases who do not visit the general practitioner (GP), which constitute 83.1 per cent of norovirus disease cases in Catalonia. The Dw is 0.0062 for patients who do, which are 16.9 per cent in Catalonia (Kemmeren et al., 2006; Domínguez et al., 2008). The Dt was 3 to 6 days, with average 3.8 days, in people who did not visit the GP, and from 5.73 to 7.23 in those who did (Kemmeren et al., 2006). This variability was modelled by defining a lognormal distribution (mean=log10(3.8), sd=0.1) and uniform (5.73–7.23) days, for non-visiting and visiting GP respectively.

No information on norovirus mortality in Spain was available, so data on norovirus from the Netherlands was used: mortality: 0.009 per cent; years lost due to premature death: 20.7 (Havelaar et al., 2012). Risks were calculated using Monte Carlo simulations with random sampling of 10,000 values from each distribution input. Monte Carlo simulations, and uncertainty and sensitivity analysis, were performed using R version 3.0.1 (Team, 2013).

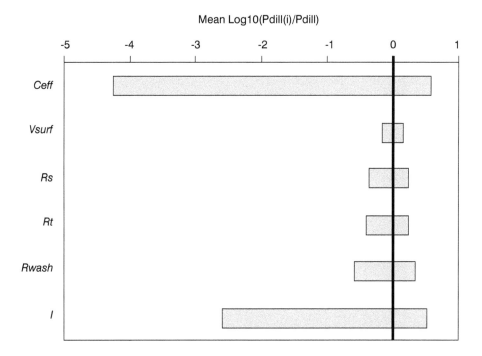

Mean Log10(Pdill(i)/Pdill)

Figure 56.2 Sensitivity of *Pd,ill* to setting a parameter to their 95 per cent values

Sensitivity analysis

Sensitivity analysis was conducted to know how sensitive the model outputs are to the inputs. The effect of the value of a model parameter on the probability of illness was calculated by varying a model parameter to the 95 per cent confidence limits of its variability, while keeping the variability of the other parameters (Table 56.1). The sensitivity analysis showed that the concentration of virus in the tertiary effluent and the consumption of lettuce were major factors influencing the variability of the risks (Figure 56.2). Washing the lettuce, in-field inactivation, inactivation during transport and storage, and the volume of water clinging to the lettuce surface had little effect on the variability of the probability of disease.

Appraisal for risk management

Without the UV/chlorination, the annual disease burden of consuming lettuce irrigated with reclaimed water exceeds the recommended threshold of 10^{-6} DALYs by almost a factor of 1,000 (Figure 56.3). Although the tertiary treatment with UV/chlorination was efficient enough to reduce the concentration of *Escherichia coli* to below the regulatory threshold for reclaimed water uses for crops irrigation, additional removal is needed in the system in order to meet the 10^{-6} DALY guideline value. To reduce the virus load in reclaimed water, improvement of the UV system should be considered, for example adding a pre-treatment step that increases the transmittance of the water, as done in other wastewater treatment plant (WWTP) (Montemayor et al., 2008). Measures to prevent contamination include using other sources of water during the warmest period (when the transmittance of the water decreases and, hence, disinfection

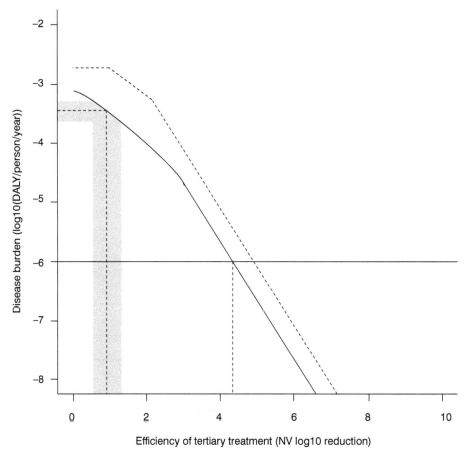

Figure 56.3 Mean (solid) and 95 per cent (dashed) of the annual disease burden of norovirus via consumption of lettuce irrigated with treated wastewater as function of the efficiency of the tertiary wastewater treatment. Horizontal line is the recommended tolerable risk of 10^{-6} DALY/person/year (WHO, 2011), with dotted vertical lines showing the required treatment efficiency to achieve this tolerable risk. The dashed black vertical line corresponds to the known efficiency of the treatment plant and the corresponding disease burden, the grey area representing the 95% confidence interval

efficiency is lower) or using an irrigation method that results in lower surface deposition (such as subsurface drip irrigation). To inactivate the viruses after contamination, farmers could be advised to irrigate with a different source water on the last irrigation event, or increase the time between the last irrigation and the harvest, increasing the inactivation of viruses already deposited on the surface.

Value of QMRA for risk management

Assessing the microbial risks of water systems is a relatively young field of science. It has the capacity to further professionalize safety management in water by providing science-based, objective, credible and proportionate information to help risk managers make informed decisions, as shown in the case study. Adequate and efficient water safety management requires a scientifically sound understanding of the potential health risks associated with specific water systems, and the efficacy of alternative measures to control these risks. Implementing a QMRA approach will provide a quantitative, evidence-based system understanding of numerous important factors relevant to the management of waterborne risks. There are many examples from wastewater reuse, recreational water, and drinking water and grey water in the academic literature that demonstrate how the systematic approach of implementing QMRA informed the understanding of the system including: the identification of the most hazardous risk pathways for a particular scheme (Westrell et al., 2004; Oesterholt et al., 2007; Diallo et al., 2008); the identification of the factors that drive risk (Signor and Asbolt, 2006; Hunter et al., 2009); identification of the most important barriers (Medema et al., 2003; Hamilton et al., 2006; Åstrom et al., 2007); identification of the most sensitive components within the risk model (Petterson et al., 2001; Hamilton et al., 2006; Jaidi et al., 2009); and for determining the relative significance of major, well-controlled and minor, less well-controlled routes of exposure and the moments of reduced treatment performance (Gale 2002; Smeets et al. 2010). This information is of high value for evaluating water system safety and robustness, to inform the formulation of risk mitigation options or subsequent data collection. Most importantly, QMRA provides a systemic approach for managers to get to know their water systems from a pathogen risk perspective (Ashbolt et al., 2010.)

In conclusion, adequate risk assessment is essential to determine the appropriate level of risk management and the resources needed to address water-associated health risks proportionally.

Key recommended readings

1   Haas CN, Rose JB and Gerba CP (2014). *Quantitative Microbial Risk Assessment*, Wiley, New York, 440p. Textbook on quantitative microbial risk assessment with examples from water-, food- and fomite-borne disease
2   McBride G (2005). *Using Statistical Methods for Water Quality Management: Issues, Problems and Solutions*. Wiley, New York, 344p. Textbook for using statistical methods in water resource management, hypothesis test procedures and practical solutions on environmental topics: formulating water quality standards; determining compliance with standards; MPNs and microbiology; water-related, human health risk modeling; trends, impacts, concordance, and detection limits
3   U.S. EPA and USDA/FSIS (2012). Microbial Risk Assessment Guideline: Pathogenic Microorganisms with Focus on Food and Water. Prepared by the Interagency Microbiological Risk Assessment Guideline Workgroup. EPA/100/J-12/001; USDA/FSIS/2012-001. U.S. Environmental Protection Agency (EPA) and U.S. Department of Agriculture/Food Safety and Inspection Service (USDA/FSIS), Washington DC. Guidelines for performing quantitative microbial risk assessment in food and water, with best practices on scoping, performing and using QMRA including sensitivity analysis

Suggested websites

http://climate-adapt.eea.europa.eu/ecdc-tool
http://wiki.camra.msu.edu/
http://www.foodrisk.org

Bibliography

AESAN (2011). *Encuesta Nacional de Ingesta Dietética Española* (ENIDE). Madrid: AESAN.

Asano, T, Leong, LYC, Rigby, MG and Sakaji, RH (1992). Evaluation of the California wastewater reclamation criteria using enteric viruses monitoring data. *Water, Science and Technology* 26(7–8): 1513–1524.

Ashbolt, NJ, Schoen, ME, Soller, JA and Roser, DJ (2010). Predicting pathogen risks to aid beach management: the real value of quantitative microbial risk assessment (QMRA). *Water Res* 44: 4692–4703.

Åström, J, Petterson, S, Bergstedt, O, Pettersson, TJ and Stenstrom, TA (2007). Evaluation of the microbial risk reduction due to selective closure of the raw water intake before drinking water treatment. *Journal of Water and Health* 5(Suppl 1): 81–97.

Atmar, RL, Opekun, AR, Gilger, MA, Estes, MK, Crawford, SE, Neill, FH, Ramani, S, Hill, H, Ferreira, J and Graham, DY (2014). Determination of the 50% human infectious dose for Norwalk virus. *Journal of Infectious Diseases* 209(7): 1016–1022.

Burmaster, DE and Anderson, PD (1994). Principles of good practice for the use of Monte Carlo techniques in human health and ecological risk assessments. *Risk Anal* 14(4): 477–481.

Carratalà, A, Rodriguez-Manzano, J, Hundesa, A, Rusiñol, M, Fresno, S, Cook, N and Girones, R (2013). Effect of temperature and sunlight on the stability of human adenoviruses and MS2 as fecal contaminants on fresh produce surfaces. *International Journal of Food Microbiology* 164(2–3): 128–134.

Codex Alimentarius Commission (CAC) (1997). Hazard Analysis and Critical Control Point (HACCP) System and Guidelines for Its Application. Annex to the Recommended International Code of Practice – General Principles of Food Hygiene, CAC/RCP 1-1969, Rev. 3 –1997. Secretariat of the Joint FAO/WHO Food Standards Programme, FAO, Rome.

Codex Alimentarius Commission (CAC) (1999). Principles and Guidelines for the Conduct of Microbiological Risk Assessment, CAC/GL 30-1999. Secretariat of the Joint FAO/WHO Food Standards Programme, FAO, Rome.

Diallo, MB, Anceno, AJ, Tawatsupa, B, Houpt, ER, Wangsuphachart, V and Shipin, OV (2008). Infection risk assessment of diarrhea-related pathogens in a tropical canal network. *Science of the Total Environment* 407(1): 223–232.

Domínguez, A, Torner, N, Ruíz, L, Martínez, A, Barrabeig, I, Camps, N, Godoy, P, Minguell, S, Parrón, I, Pumarés, A, Sala, MR, Bartolomé, R, Pérez, U, de Simón, M, Montava, R and Buesa, J (2008). Aetiology and epidemiology of viral gastroenteritis outbreaks in Catalonia (Spain) in 2004–2005. *Journal of Clinical Virology* 43(1): 126–131.

Drechsel, P, Scott, CA, Raschid-Sally, L, Redwood, M and Bahriet, A (2009). *Wastewater Irrigation and Health*. London: Earthscan.

EFSA (2009). Transparency in risk assessment – scientific aspects. Scientific opinion. *EFSA Journal* 1051: 1–22.

Fan, X, Niemira, BA, Doona, CJ, Feeherry, FE and Gravani, RB (2009). *Microbial Safety of Fresh Produce*. Chichester: Wiley/Blackwell.

Fewtrell, L and Bartram, J (2001). *Guidelines, Standards and Health: Assessment of Risk and Risk Management for Water-Related Infectious Disease*. London: IWA Publishing.

Gale, P (2002). Using risk assessment to identify future research requirements. *Journal of American Water Works Association* 94(9): 30–38.

Haas, CN, Rose, JB and Gerba, CP (1999, 2014). *Quantitative Microbial Risk Assessment*. New York: John Wiley and Sons Inc.

Hamilton, AJ, Stagnitti, F, Premier, R, Boland, AM and Hale, G (2006). Quantitative microbial risk assessment models for consumption of raw vegetables irrigated with reclaimed water. *Applied and Environmental Microbiology* 72(5): 3284–3290.

Havelaar, AH, Haagsma, JA, Mangen, MJ, Kemmeren, JM, Verhoef, LP, Vijgen, SM, Wilson, M, Friesema, IH, Kortbeek, LM, van Duynhoven, YT and van Pelt, W (2012). Disease burden of foodborne pathogens in the Netherlands, 2009. *Int. J. Food Microbiol.*156(3): 231–238.

Hunter, PR, Zmirou-Navier, D and Hartemann, P (2009). Estimating the impact on health of poor reliability of drinking water interventions in developing countries. *Science of the Total Environment* 407(8): 2621–2624.

ILSI (2012). Tool for microbiological risk assessment. Brussels: ILSI Europe.

Kemmeren, JM, Mangen, M-JJ, Duynhoven, YV and Havelaar, AH (2006). *Priority Setting of Foodborne Pathogens: Disease Burden and Costs of Selected Enteric Pathogens.* Biltoven: RIVM.

Medema, GJ, Hoogenboezem, W, van der Veer, AJ, Ketelaars, HAM, Hijnen, WAM and Nobel, PJ (2003). Quantitative risk assessment of *Cryptosporidium* in surface water treatment systems. *Water Science and Technology* 47(3): 241–247.

Mok, HF and Hamilton, AJ (2014). Exposure factors for wastewater-irrigated Asian vegetables and a probabilistic rotavirus disease burden model for their consumption. *Risk Anal* 34: 602–613.

Montemayor, M, Costan, A, Lucena, F, Jofre, J, Munoz, J, Dalmau, E, Mujeriego, R and Sala, L (2008). The combined performance of UV light and chlorine during reclaimed water disinfection. *Water Science & Technology* 57(6): 935–940.

Oesterholt, F, Martijnse, G, Medema, G and van der Kooij, D (2007). Health risk assessment of non-potable domestic water supplies in the Netherlands. *Journal of Water Supply: Research and Technology – AQUA* 56(3): 171–179.

Petterson, SR, Ashbolt, NJ and Sharma, A (2001). Microbial risks from wastewater irrigation of salad crops: a screening-level risk assessment. *Water Environ Res* 73(6): 667–672. Errata: *Water Environ Res* 674(664): 411.

Petterson, SR et al. (2015). *Use of Quantitative Microbial Risk Assessment for Water Safety Management.* Geneva: WHO.

Regli, S, Rose, JB, Haas, CN and Gerba, CP (1991). Modeling the risk from *Giardia* and viruses in drinking water. *Journal American Water Works Association* 83: 76–84.

Sales-Ortells, H, Fernandez-Cassi, X, Timoneda, N, Dürig, W, Girones, R and Medema, G (2015). Health risks derived from consumption of lettuces irrigated with tertiary effluent containing norovirus. *Food Research International* 68: 70–77.

Signor, RS and Ashbolt, NJ (2006). Pathogen monitoring offers questionable protection against drinking-water risks: a QMRA (quantitative microbial risk analysis) approach to assess management strategies. *Water Science and Technology* 54(3): 261–268.

Smeets, PW, Rietveld, LC, van Dijk, JC and Medema, GJ (2010). Practical applications of quantitative microbial risk assessment (QMRA) for water safety plans. *Water Science and Technology* 61(6): 1561–1568.

Team (2013). The R Project for Statistical Computing. http://www.r-project.org

Teunis, PF, Moe, CL, Liu, PE, Miller, S, Lindesmith, L, Baric, RS, Le Pendu, J and Calderon, RL (2008). Norwalk virus: How infectious is it? *Journal of Medical Virology* 80(8): 1468–1476.

US Environmental Protection Agency (1997). *Exposure Factors Handbook, vol. II. Ingestion factors.* EPA/600/P-95/002Fb. National Center for Environmental Assessment, U.S. Environmental Protection Agency, Washington, DC.

US EPA and USDA/FSIS (2012). *Microbial Risk Assessment Guideline: Pathogenic Microorganisms with Focus on Food and Water.* Prepared by the Interagency Microbiological Risk Assessment Guideline Workgroup. EPA/100/J-12/001; USDA/FSIS/2012-001. U.S. Environmental Protection Agency (EPA) and U.S. Department of Agriculture/Food Safety and Inspection Service (USDA/FSIS), Washington DC.

Westrell, T, Schönning, C, Stenström, TA and Ashbolt, NJ (2004). QMRA (quantitative microbial risk assessment) and HACCP (Hazard Analysis and Critical Control Points) for management of pathogens in wastewater and sewage sludge treatment and reuse. *Water Science and Technology* 50(2): 23–30.

WHO (1999). *Principles and Guidelines for the Conduct of Microbiological Risk Assessment.* CAC/GL-30. Geneva: World Health Organization.

WHO (2003). *Guidelines for Safe Recreational Water Environments. Volume 1: Coastal and Fresh Waters.* Geneva: World Health Organization.

WHO (2006). *Guidelines for the Safe Use of Wastewater, Excreta and Grey Water.* Geneva: World Health Organization.

WHO (2011). *Guidelines for Drinking Water Quality*, third edition. Geneva: World Health Organization.

57

BURDEN OF DISEASE
ASSESSMENT*

Jacqueline MacDonald Gibson

GILLINGS SCHOOL OF GLOBAL PUBLIC HEALTH, UNIVERSITY
OF NORTH CAROLINA AT CHAPEL HILL, USA

Learning objectives

1 Understand the history and uses of the burden of disease assessment approach in the field of water, sanitation, and hygiene (WASH).
2 Understand the data required to conduct a burden of disease assessment and sources for such data.
3 Compute estimates of the burden of disease attributable to contaminants in drinking water and/or to deficiencies in WASH services, including computation of uncertainties in the burden of disease estimates.

Overview

The overarching purposes of a burden of disease assessment are to quantify the disease rates in a population by age and to attribute observed diseases to underlying risk factors. This chapter focuses on the latter purpose—that is, to identify the contributions of different risk factors to observed disease rates. For example, a decision-maker in the water, sanitation, and hygiene (WASH) sector may wish to know what fraction of acute gastrointestinal illnesses documented in a population arose from pathogens in drinking water, as compared to other causes, such as malnutrition or food contamination. Similarly, an environmental agency may be interested in what fraction of bladder cancers arose from disinfection byproducts in drinking water, as compared to other causes such as smoking, occupational exposures, or chronic urinary infections. A burden of disease assessment can answer such questions.

While burden of disease assessment can serve similar purposes as risk assessment, there are important conceptual and methodological differences between the two approaches. Risk assessment seeks to estimate the probability of a future illness arising due to exposure to a particular risk factor, for example a contaminant in water. To predict risks, the risk assessment

---

* Recommended citation: MacDonald Gibson, J. 2015. 'Burden of disease assessment', in Bartram, J., with Baum, R., Coclanis, P.A., Gute, D.M., Kay, D., McFadyen, S., Pond, K., Robertson, W. and Rouse, M.J. (eds) *Routledge Handbook of Water and Health*. London and New York: Routledge.

framework requires information on population or individual exposure to the contaminant, and information from previous studies to estimate the probability of illness for a given exposure dose. No information on the observed incidence of the health outcome of concern is needed. It is possible, therefore, that a risk assessment could predict a greater number of adverse health outcomes (for example, bladder cancers from disinfection byproducts in drinking water) than are actually observed in the population. Methodologically, the burden of disease assessment approach prevents this possibility by linking exposure estimates to observed health outcomes, so that the estimated number of health outcomes never exceeds the observed number. Risk assessment generally is more appropriate when evaluating the potential impacts of a specific intervention, such as, for example, installation of a specific new kind of water treatment system designed to decrease the concentration of a contaminant by a percentage that can be estimated from empirical data. Generally, burden of disease assessment is the best method when seeking to understand the underlying causes of illness at a broad, population scale. Nonetheless, the approaches overlap and can be used for similar purposes.

The burden of disease assessment approach arose from World Bank efforts to develop guidance for international donor organizations and national governments on prioritizing investments designed to improve population health, beginning with the 1993 report *Disease Control Priorities in Developing Countries*.[1] This effort led donor organizations and policymakers to recognize that to prioritize investments, they needed to know not only the types of diseases claiming the most lives and causing the greatest burden but also the relative importance of underlying risk factors for disease that could be modified through financial investments and other interventions. As a result, the World Health Organization (WHO) led an effort to quantify contributions to the global disease burden by risk factor, the results of which were published in 1996 using 1990 as a base year for comparison (Table 57.1).[2] As part of this effort, the WHO recommended a new summary measure, the disability-adjusted life year (DALY), to compare disparate health outcomes (such as chronic diabetes and premature deaths).[3] The DALY is essentially an accounting technique for tallying the years of life lost or lived in a less-than-healthy state due to a particular disease. Since the 1996 estimate of disease burden by cause, the WHO and others have issued several updated estimates. To encourage local and national governments to adopt burden of disease assessment techniques as part of their policy planning, the WHO also developed a series of guidance documents on methods for quantifying the burden of disease due to specific risk factors.[4]

Table 57.1 The global burden of disease attributable to selected risk factors in 1990[5]

	Deaths ($\times 10^3$)	% of deaths	DALYs ($\times 10^6$)	% of DALYs
Malnutrition	5,881	11.7	219.6	15.9
Poor water supply, sanitation, and hygiene	2,668	5.3	93.4	6.8
Unsafe sex	1,095	2.2	48.7	3.5
Tobacco use	3,038	6.0	36.2	2.6
Alcohol use	774	1.5	47.7	3.5
Occupation	1,129	2.2	37.9	2.7
Hypertension	2,918	5.8	19.1	1.4
Physical inactivity	1,991	3.9	13.7	1.0
Illicit drug use	100	0.2	8.5	0.6
Air pollution	568	1.1	7.3	0.5

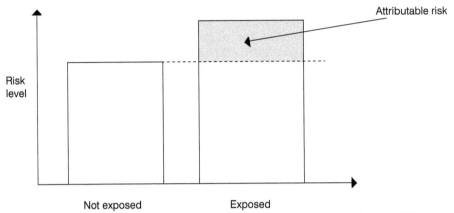

Figure 57.1 Representation of attributable risk, or attributable fraction, as the fraction of illness in an exposed population that could be prevented if the exposure were eliminated

Source: Rothman et al., 2008.[6]

Burden of disease assessment methods

Burden of disease estimates are based on calculating a measure known as the *excess fraction*, also known as an *attributable fraction (AF)*. The *AF*, a basic concept in epidemiology, is the excess caseload of a particular disease occurring in a population exposed to a risk factor, as compared to that occurring in an unexposed population (Figure 57.1).[6] Equivalently, *AF* represents the fraction of an observed health outcome that could be prevented if exposure to the risk factor were completely eliminated, all else being equal.

Calculating the *AF* requires two kinds of information: the distribution of risk factor exposure within the population and an estimate of the probability of illness for those exposed to the risk factor as compared to the illness probability for those unexposed. The latter factor is known as the *relative risk (RR)* or *risk ratio* (Chapter 55) and is defined as

$$RR_i = \frac{P(ill \mid exposed_i)}{P(ill \mid unexposed)} \qquad (57.1)$$

where $P(ill \mid exposed_i)$ represents the probability of illness among those exposed to a risk factor at some level *i*, and $P(ill \mid unexposed)$ represents the illness probability for an equivalent but unexposed population. The exposure level *i* could represent, for example, the dose of a contaminant, such as a virus or a chemical ingested via drinking water. In other cases, *i* could represent an exposure category, such as access or lack of access to improved water and sanitation systems.

Distribution of risk factor exposure in a population

The distribution of risk factor exposure in a population represents the varying levels of exposure to the risk factor within the population. When continuous exposure data (for example, contaminant concentrations in a water supply) are available, exposure can be represented as a probability distribution function (PDF). For example, Figure 57.2 shows an example estimation of the concentration of arsenic in private wells in Clay County, North Carolina (USA). The fraction of the total area under the curve contained between

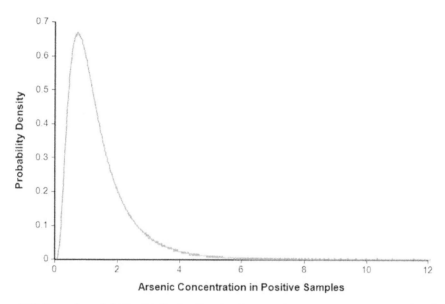

Figure 57.2 Example probability distribution function for a continuous exposure concentration (in this case, μg/l of arsenic in private wells testing positive for arsenic in a North Carolina county)

any two concentration values on the x-axis shows the probability of detecting arsenic at a concentration between these two values. To develop this PDF, a lognormal distribution was fitted to observed concentrations from 1,188 private well water samples sent to the North Carolina Division of Public Health for analysis. Many nations collect such water quality data.

Where potable water quality data are unavailable (as is the case for many developing nations), estimates of exposure to waterborne contaminants can be based on surrogate exposure measures that can serve as an indicator of contamination potential. For example, studies of the burden of acute gastrointestinal illness (AGI) attributable to microbial contaminants often are based on information about levels of access to disinfected water supplies.[7] In some cases, as well, burden of disease studies of cancer risks from disinfection byproducts use the proportion of the population drinking chlorinated water as a surrogate exposure indicator.[8]

Relative risk

Most burden of disease analyses use *RR*s from previous epidemiologic studies. However, *RR*s also can be estimated using toxicologic data. In addition, if sufficient exposure and health outcome data are available for a population, *RR*s can be calculated by fitting regression models to these data sets. Data sufficient to support the latter approach generally are difficult to obtain and therefore have not been routinely used in burden of disease assessments, but DeFelice et al. recently employed such data to estimate the fraction of AGI cases attributable to microbial contaminants in drinking water in North Carolina, United States.[9]

RR estimates from previous epidemiologic studies

Most burden of disease studies use *RR* estimates from previous epidemiologic studies. Since epidemiologic study results may vary depending on the population studied, *RR*s should be

Table 57.2 RR of lung cancer for different arsenic exposure levels as estimated by previous epidemiologic studies in Taiwan and Chile (means and standard deviations[10]

Exposure range (µg/l)	Taiwan		Chile	
	Males	*Females*	*Males*	*Females*
Greater than 0 but less than 4.5	4 (0.076)	5.4 (0.15)	20 (2.8)	14 (1.8)
4.5–7.5	6.8 (0.13)	8.9 (0.23)	33 (4.6)	24 (2.8)
7.5–15	14 (0.26)	18 (0.51)	67 (8.9)	48 (4.3)
>15	27 (0.51)	36 (1.02)	130 (19)	95 (12)

obtained from studies of populations with similar demographic and exposure characteristics as the population for which the burden of disease is being estimated and/or from meta-analyses including multiple studies. As an example of *RR* estimates that could be used in a burden of disease assessment focused on arsenic in drinking water, Table 57.2 shows *RR*s of lung cancer from exposure to arsenic in drinking water at various concentrations using data from epidemiologic studies in two different populations, one in Taiwan and one in Chile. Based on meta-analyses of studies of these populations, the U.S. National Academy of Sciences developed *RR*s recommended for use in assessing U.S. risks from arsenic in drinking water.

RR estimates from toxicologic studies

For most chemicals of potential concern in drinking water in industrialized countries, epidemiologic studies of exposure–response relationships have not been carried out, so *RR*s must be estimated from toxicologic studies in animals. Rather than reporting *RR*s, toxicologic studies usually report a slope factor (*SF*), which describes the slope of a line predicting the probability of illness as a function of dose. Typically, *SF*s are reported in units of kg-day/mg, where the mass in the denominator represents the chemical mass and that in the numerator represents body mass. Agencies in the United States and elsewhere have developed databases of such slope factors for a variety of different chemicals of potential concern (for example, the Integrated Risk Information System, known as IRIS). Slope factors can be converted to relative risk estimates using the following equation:

$$RR_i = \frac{SF \times LADD_i + I_u}{I_u} \tag{57.2}$$

where $LADD_i$ is the lifetime average daily dose of the chemical in drinking water (in units of mg/kg-day) and I_u is the cancer rate in an unexposed population. $LADD_i$ can be estimated by multiplying the chemical concentration i by the daily volume of water consumed and then dividing by body weight. I_u can be estimated from national health statistics, assuming exposure is rare. A recent analysis of the burden of disease from chemicals in North Carolina community water systems provides an example use of this method.[11]

Population-specific RR estimates

Typically, burden of disease studies rely on publicly available data from environmental and health agencies. For patient confidentiality reasons, publicly available health data normally can be obtained only at a spatial resolution too coarse to match the health outcomes to specific environmental exposures. In rare cases, however, spatially resolved data may be available to researchers upon special request, assuming confidentiality provisions are in place. In such cases, it is possible to estimate exposure–response relationships specific to the population for which the burden of disease analysis is being conducted. Such an approach is known as a "population intervention model."[12,13] A recent study in North Carolina provides an example of use of this approach to estimate AGI risks from microbial contaminants in community water supplies.[14]

Attributable fraction and burden of disease calculation

Once information on the population distribution of exposure and *RR*s for each exposure level is compiled, the *AF* can be calculated as follows:[4]

$$AF = \frac{\sum_{i=1}^{n} P_i RR_i - \sum_{i=1}^{n} P'_i RR_i}{\sum_{i=1}^{n} P_i RR_i} \tag{57.3}$$

where P_i represents the proportion of the population exposed at level i (for example, the population proportion lacking access to any improved water or sanitation or the proportion exposed to a contaminant at a given concentration range) under current conditions, and P'_i represents the exposure distribution under a desired counterfactual (reduced exposure) scenario. If the reduced exposure scenario involves eliminating all exposure to the contamination (or, for example, providing the maximum improved water and sanitation access), then the right-hand term in the numerator of equation 57.3 reduces to 1.

Equation 57.3 represents the exposure distribution as categorical with n possible exposure levels. When continuous exposure and *RR* functions are available, the equivalent expression is

$$AF = \frac{\int_0^\infty RR(x) f(x) dx - \int_0^\infty RR(x) f'(x) dx}{\int_0^\infty RR(x) f(x) dx} \tag{57.4}$$

where $f(x)$ and $f'(x)$ represent the PDFs of exposure in the population under current conditions and a counterfactual condition. For example, chemical contaminant concentrations in water often can be approximated as lognormal distributions. In such cases, $f(x)$ and $f'(x)$ are the PDFs for lognormally distributed variables with parameters representing the current and counterfactual exposure conditions (see worked example 3 below). As in equation 57.3, if the counterfactual condition would eliminate all exposure, then the right-hand term in numerator of equation 57.4 simplifies to 1.

Once the AF is calculated, the total excess cases attributable to the adverse exposure is estimated as

$$Y = AF \times I_0 \tag{57.5}$$

where Y is the attributable number of cases and I_0 is the observed case rate (for example, deaths per year due to bladder cancer or AGI cases per year). Note that the case definition must match the definition used in the studies from which *RR* estimates were derived.

Estimating uncertainty

Estimates of the population distribution of exposure and *RR* are subject to many sources of uncertainty, due to limitations of available measurement methods, an inability to fully capture spatial and temporal variability, and epidemiologic study design limitations. Uncertainty in these key input variables for burden of disease calculations in turn creates uncertainty in the estimated disease burden—uncertainty that is important to communicate to decision-makers. For example, a decision-maker may prefer to invest more in preventing a risk with a relatively small mean value if there is a relatively high potential for a catastrophic event, such as a major disease outbreak affecting thousands of people, than if there is no such catastrophic potential. The purpose of uncertainty analysis is to calculate the potential for such "worst-case" scenarios.

The textbook *Uncertainty: A Guide to Dealing with Uncertainty in Quantitative Risk and Policy Analysis* provides information on methods for carrying out uncertainty analysis.[15] The book *Environmental Burden of Disease Assessment: A Case Study in the United Arab Emirates*[16] provides detailed information on quantifying uncertainty in the context of a national burden of disease assessment. Commonly, uncertainty is estimated using Monte Carlo simulation methods, in which the burden of disease calculation is repeated thousands of times, each time selecting at random a value of each input variable from the possible distribution of values. The mean and 95 percent confidence intervals of the estimated burden of disease can then be estimated from the output of the simulations. Example uncertainty analyses are provided in the worked examples.

Comparing disparate health outcomes

Public health decision-makers typically are interested in obtaining information on the number of deaths and nonfatal illnesses attributable to particular risk factors. Nonfatal illnesses associated with drinking water contamination can range from mild gastrointestinal distress to debilitating chronic illnesses, such as Guillain-Barré syndrome. As previously mentioned, the WHO developed the DALY metric to enable comparisons of disparate health outcomes. In brief, DALYs provide a measure of the difference between the current state of health in the population and an ideal state in which everyone lives disease-free to the standard life expectancy, which is assumed to be 80 years for men and 82.5 years for women. Total DALYs are the sum of the years of life lost (YLL) due to premature death and the years of life spent in less than perfect health, denoted as years of life lived with disability (YLD). For any single premature death, YLL is the product of the age at death and the years of life that would remain if the individual lived to full life expectancy. YLD is the product of the duration of the illness and a "disability weight," which is intended to represent social preferences for various states of health, with 0 representing perfect health and 1 representing death. The WHO Environmental Burden of Disease series publications provide details on calculation of YLD for various health outcomes.[4]

Example burden of disease calculations

The data available to support burden of disease estimates vary substantially from one country to the next. Ideally, such estimates will be based on local contaminant concentration and health data, but often complete, reliable data are unavailable, and approximation methods are needed. Where contaminant concentration data are unavailable or unreliable, WASH-related risks can be estimated using categorical exposure indicators. The first example below shows an example calculation for a case in which direct, continuous exposure measures were unavailable. The second and third examples show examples of burden of disease analyses in locations where detailed exposure concentration information was available. Example 2 shows a burden of disease estimate for a location with detailed local microbiological water quality and infectious disease health data. Example 3 shows a calculation for a situation where chemical concentration data were available.

Example 1: disease burden attributable to WASH conditions in the United Arab Emirates

The Secretary General of the Environment Agency-Abu Dhabi commissioned a team of public health specialists to quantify the environmental burden of disease in the United Arab Emirates (UAE) attributable to multiple contaminants in multiple environmental media (drinking water, coastal water, outdoor air, indoor air, seafood, fruits and vegetables, and soil). Measurements of contaminants in the UAE's potable water supply were not readily available to the Secretary General, since the UAE's drinking water system is managed by a separate agency. However, according to publicly available data, 98 percent of the UAE population is served by a regulated water supply and has access to improved sanitation. (In the UAE, many communities are served by septic systems, and waste from these systems is trucked to a sewage treatment plant prior to disposal but, due to overloads at these plants, is sometimes dumped illegally.) Following WHO guidance,[4,7] the public health team used the level of water and sanitation service plus previous studies linking these access levels to AGI risk, in order to characterize the number of annual AGI cases that potentially could be attributed to contaminants in drinking water.

To support burden of disease analyses for situations in which water quality data are not available, the WHO conducted analyses of previous studies to estimate the *RR* of diarrheal illness corresponding to different levels of water supply and sanitation service (for details, see [7,17]). The WHO grouped these different service levels into six categories. In the UAE, approximately 98 percent of the population has service at what the WHO calls Level II, defined as "regulated water supply and full sanitation coverage, with partial treatment for sewage, corresponding to a situation typically occurring in developed countries."[7] The WHO estimated that the relative risks of AGI among those with Level II water and sanitation service is 2.5, compared to a perfect situation in which there is zero risk of disposal of untreated fecal matter into the environment. The remaining 2 percent of the population is assumed to be in water and sanitation Level Vb, defined as "improved water supply and no basic sanitation." This small population fraction is assumed to live primarily in rural areas, where water service is unavailable and instead households obtain their water from wells. The WHO estimated that the relative risk of AGI for such populations is 8.7, compared to the perfect situation in which there is a complete absence of disease transmission through water and insufficient sanitation and hygiene.

Therefore, using equation 57.3, the fraction of cases potentially attributable to microbiological contamination in drinking water is

$$AF = \frac{0.98 \times 2.5 + 0.02 \times 8.7 - 1}{0.98 \times 2.5 + 0.02 \times 8.7} = 0.62 \tag{57.6}$$

A local public health agency in the UAE that processes insurance claims provided data on the number of AGI cases potentially associated with waterborne pathogens (those with

International Classification of Disease 9 codes 008-9 and 558.9). In total in 208, there were 81,110 health-care facility visits in which these codes were indicated as the primary diagnosis. Therefore, using equation 57.5, the total AGI disease burden attributed to deficiencies in water and sanitation services is

$$Y = 0.62 \times 81,110 = 50,000 \tag{57.7}$$

The computations in equations 57.6–57.7 should be regarded as approximations subject to variability and uncertainty that are not captured in the deterministic estimate. Ideally, these estimates would consider variability and uncertainty—the former arising from geographic differences in water and sanitation systems, climate, exposure pathways, and other sources and the latter arising from the limitations of studies assessing the relationships between AGI risk and water and sanitation access. WHO guidance documents do not provide information on this variability and uncertainty. Nonetheless, the UAE analysis team conducted a crude variability and uncertainty analysis by representing RR as a uniformly distributed random variable (that is, a random variable for which any value in a particular interval is equally likely as any other). RR was represented as uniformly distributed on the interval [1, 4] or [7.2, 10.2] for those with Level II and Level Vb water and sanitation access, respectively. In this case, equations 57.6–57.7 become

$$Y = \frac{0.98 \times RR_{II} + 0.02 \times RR_{Vb} - 1}{0.98 \times RR_{II} + 0.02 \times RR_{Vb}} \times 81,110 \tag{57.8}$$

where RR_{II} and RR_{Vb} are uniformly distributed random variables with the previously described parameters. To estimate the uncertainty in Y, equation 57.8 is computed repeatedly (such as 1,000 times), each time selecting values at random from the probability distributions of those variables. The result is a probability distribution of possible values of Y, shown in Figure 57.3 as a cumulative distribution function. A confidence interval on the estimate of Y can be

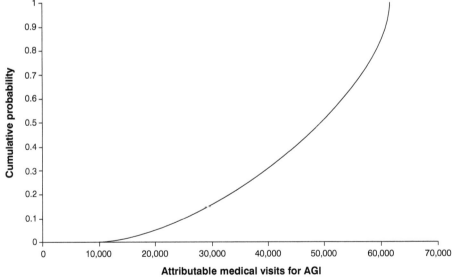

Figure 57.3 Probability distribution of medical visits in the year 2008 potentially attributable to deficiencies in water, sanitation, and hygiene in the United Arab Emirates

read from this figure. For example, a 95 percent confidence interval is given by reading the values of attributable cases for y-axis values of 0.025 and 0.975, yielding a confidence range of 15,000–61,000. The mean estimate from this simulation is 47,000, slightly lower than the deterministic estimate computed by setting the *RR* values at their mean values.

Example 2: disease burden attributable to microbial contamination in North Carolina drinking water

In North Carolina, in the United States, public health and environmental officials collected geographically referenced data on microbiological and chemical contaminants in community water supply systems and private water wells (the latter used for households lacking connections to community water supplies, as is the case for about 26 percent of the state's population). In addition, since 2009 health officials have gathered data on primary diagnostic codes recorded for each visitor to each hospital emergency department in the state (122 emergency departments in all). Concerned about the potential risks associated with inadequate monitoring and maintenance of private wells, health officials and researchers from the University of North Carolina (UNC) collaborated to estimate the number of emergency department (ED) visits for AGI that could be attributed to microbial contaminants in private wells. Since in this case geographically referenced contaminant concentration and public health data were available, the study team could develop an *AF* estimate that better reflects local conditions than one developed from *RRs* estimated in other locations.

In this case, the research team used the available data to fit a regression model to estimate the number of AGI ED visits in any one of North Carolina's 100 counties during any given month based on multiple explanatory factors, including drinking water quality in community water systems and private wells in each county. This regression model then formed the basis for calculating the *AF.* The regression model was of the following form:

$$V_{i,j} = \alpha + \beta_1 C_{CWS_{i,j}} + \beta_2 E_{CWS_{i,j}} + \beta_3 C_{DWS_i} + \beta_4 N_i + \beta_5 Pov_i + \left(\sum_{m=6}^{16} \beta_m I_{j.m} \right) + \varepsilon_{i,j} \quad (57.9)$$

where $V_{i,j}$ is the number of observed AGI ED visits in county i during month j, $C_{CWS,i,j}$ is the potential number of community water system (CWS) customers in county i exposed to a monthly maximum contaminant level (MCL) violation during month j, $E_{CWS_{i,j}}$ is the number of CWS customers exposed to an acute MCL violation in month j, C_{DWS_i} is the number of private well (DWS) users exposed to total coliform bacteria in county i, N_i is the total county population, Pov_i is the population in poverty, $I_{j,m}$ is an indicator variable for month, and ε_{ij} is the serially correlated error term. For this case, the *AF* for any county and month can be computed as follows:

$$AF_{i,j} = \frac{V_{i,j} - V_{i,j-counterfactual}}{V_{i,j}} \quad (57.10)$$

where $V_{i,j-counterfactual}$ represents the case of zero exposure to contamination: $C_{CWS,i,j} = E_{CWS i,j} = C_{DWSi} = 0$. Substituting these zero values in equation 57.9 and them combining Equations 57.9 and 5.10 gives

$$AF_{i,j} = \frac{\beta_1 C_{CWS_{i,j}} + \beta_2 E_{CWS_{i,j}} + \beta_3 C_{DWS_i}}{V_{i,j}} \quad (57.11)$$

Equivalently, the attributable disease burden in any given county and month can be computed from equations 57.5 and 57.11 by recognizing that $I_0 = V_{i,j}$, giving

$$Y_{i,j} = AF_{i,j} \times V_{i,j} = \frac{\beta_1 C_{CWS_{i,j}} + \beta_2 E_{CWS_{i,j}} + \beta_3 C_{DWS_i}}{V_{i,j}} \times V_{i,j} = \beta_1 C_{CWS_{i,j}} + \beta_2 E_{CWS_{i,j}} + \beta_3 C_{DWS_i}$$

(57.12)

In North Carolina, the average attributable fraction across all counties in the state as computed using this approach was estimated as 11.7 percent, with fractions varying by county from 0.9–30 percent, reflecting wide variations in the types of water and sanitation service across the state (ranging from reliance on untreated private well water and failing septic systems to access to state-of-the-art community water supply and sanitation systems). Of the attributable cases, 99 percent arose from exposure to microbial contamination in private wells. Details of this analysis are provided elsewhere.[18]

Example 3: disease burden attributable to arsenic in North Carolina private wells

Parts of North Carolina contain natural arsenic deposits, posing potential risks to water supplies. Arsenic deposits are especially high in Clay County. North Carolina public health officials have collected water quality samples from 1,188 private wells, and one-third of the samples tested positive for arsenic. Among the positive samples, the arsenic concentration fit a lognormal distribution (shown in Figure 57.2) with mean and standard deviation equal to 1.4 and 1.04 $\mu g/l$, respectively. According to the U.S. National Academy of Sciences, chronic, low-level exposure to arsenic may increase the risk of lung cancer (Table 57.2). In Clay County, the lung cancer mortality rates among males and females are approximately 92 and 56 cases per 100,000 people, respectively. Approximately 8,725 people (51 percent female and 49 percent male) are served by private wells.

A burden of disease approach can be used to estimate the cases of lung and bladder cancer attributable to arsenic in private well water in Clay County. Since the exposure concentration is continuous, equation 57.4 must be used. Furthermore, since the RR estimates are uncertain, RR also must be represented as a random variable. Within each exposure concentration range in Table 57.2, the RR can be represented as normally distributed with the given mean and standard deviation. As a result of combining these two different probability distributions (one for the exposure concentration and one for RR), equation 57.4 becomes quite complex and cannot be solved analytically. Instead, AF can be estimated for each gender by simulation, as follows.

1 Choose a value of the arsenic exposure concentration at random from a probability distribution represented by $X = A \times B$, where A is a Bernoulli random variable with parameter one-third (representing the probability of detecting arsenic in a randomly selected well) and B is a lognormally distributed random variable with mean and standard deviation of 1.4 and 1.04 $\mu g/liter$, respectively (representing the arsenic concentration in wells testing positive for this chemical).

2 If $X = 0$, set $RR = 1$. Otherwise, choose RR at random from one of the two probability distributions in Table 57.2. For example, if $X = 5$ $\mu g/l$, then, using the Taiwan study data, the RR for males is selected at random from a normal probability distribution with mean and standard deviation 6.8 and 0.13, respectively.

3 Using the *RR* from step 2, compute

$$AF = \frac{RR - 1}{RR} \qquad (57.13)$$

4 Compute the attributable cancer burden by multiplying *AF* by the gender-specific population and lung cancer rate. For example, if *AF* has been computed for males, then

$$Y_{males} = AF \times 0.49 \times 8,725 \times \frac{92}{100,000} \qquad (57.14)$$

Repeat steps 1–4 many times (such as 1,000 times) for each gender and using each study shown in Table 57.2. The resulting outputs provide a distribution of possible values of the attributable disease burden.

Many software packages, or even a basic spreadsheet, can be used to conduct such a simulation. One option is Analytica (Lumina Decision Systems, Los Gatos, California). Analytica allows the construction of user-friendly simulation models using a pictorial interface. Figure 57.4 shows a picture of the interface for a model constructed for this example. Table 57.3 shows the attributable annual lung cancer deaths estimated using the procedure outlined here. As shown, the *RR* estimates from Chile and Taiwan produce significantly different burdens of disease estimates, demonstrating the importance of analyzing the sensitivity of burden of disease estimates to alternative plausible *RR* estimates, especially when local estimates are unavailable.

Table 57.3 Estimated annual lung cancer deaths attributable to arsenic exposure in example 3 (mean and 95% confidence interval)

	Male	Female
Taiwan	0.32 (0.31–0.33)	0.23 (0.23–0.24)
Chile	0.54 (0.52–0.56)	0.32 (0.30–0.33)

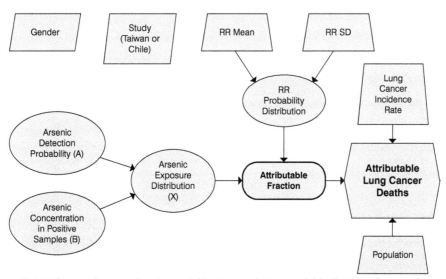

Figure 57.4 Influence diagram showing variables in simulation model built in Analytica software for example 3

Conclusions

Burden of disease assessment can be used as a tool to identify the relative contributions of different WASH risk factors to the observed disease burden in a population, in order to target interventions and help to guide investment priorities. In addition, the method can be used to identify high-risk populations, both by geography and, within a given geographic location, by age and gender. An example use for these purposes was a burden of disease assessment project commissions by the Environment Agency-Abu Dhabi for the United Arab Emirates.[8,16,19] The results of this project helped the agency to prioritize its investments intended to reduce environmental risks to health. For example, key priorities that emerged included the need for better monitoring of and stronger controls on air pollution and coastal water pollution. If carried out regularly and systematically, burden of disease estimates can be used to track changes in population health risk over time. Indeed, the WHO and academic partner institutions periodically update the global disease burden estimates, with the most recent update published in 2012.[20–22]

Additional resources and readings

Prüss-Üstün A, Mathers CD, Corvalan C, Woodward A (2003) *Introduction and Methods: Assessing the Environmental Burden of Disease at National and Local Levels.* Geneva: World Health Organization. Available at: http://www.who.int/quantifying_ehimpacts/publications/9241546204/en/ Accessed June 2, 2015

MacDonald Gibson J, Thomsen J, Launay F, Harder E, DeFelice N (2013) Deaths and medical visits attributable to environmental pollution in the United Arab Emirates. *PLoS One.* 8(3):e57536. doi:10.1371/journal.pone.0057536. Available at: http://www.plosone.org/article/info%3Adoi%2F10.1371%2Fjournal.pone.0057536

Prüss A, Kay D, Fewtrell L, Bartram J (2002) Estimating the burden of disease from water, sanitation, and hygiene at a global level. *Environ Health Perspect.* 110(5):537–542. Available at: http://www.pubmedcentral.nih.gov/articlerender.fcgi?artid=1240845&tool=pmcentrez&rendertype=abstract Accessed June 2, 2015

References

1. Lopez, AD, Mathers, CD, Ezzati, M, Jamison, DT, and Murray, CJL (2006) *Global Burden of Disease and Risk Factors.* New York: Oxford University Press. Available at: http://books.google.com/books?hl=en&lr=&id=F8Abr-ofOwIC&pgis=1. Accessed August 14, 2014.
2. Murray, C, and Lopez, A (1996) *The Global Burden of Disease and Injury Series, Volume I: A Comprehensive Assessment of Mortality and Disability from Diseases, Injuries, and Risk Factors in 1990 and Projected to 2020.* Geneva: World Health Organization. Available at: http://scholar.google.com/scholar?start=10&q=murray+lopez+global+burden+of+disease&hl=en&as_sdt=1,34&as_ylo=1996&as_yhi=1996#3. Accessed August 14, 2014.
3. Lopez, AD (2005) The evolution of the Global Burden of Disease framework for disease, injury and risk factor quantification: developing the evidence base for national, regional and global public health action. *Global Health.* 1(1):5. doi:10.1186/1744-8603-1-5.
4. Prüss-Üstün, A, Mathers, CD, Corvalan, C, and Woodward, A (2003) *Introduction and Methods: Assessing the Environmental Burden of Disease at National and Local Levels.* Geneva: World Health Organization. Available at: http://www.who.int/quantifying_ehimpacts/publications/9241546204/en/. Accessed June 2, 2015.
5. Murray, CJL, and Lopez, AD (1996) Evidence-based health policy: lessons from the Global Burden of Disease Study. *Science.* 274:740–743.

6. Rothman, KJ, Greenland, S, and Lash, T (2008) *Modern Epidemiology*. Lippincott Williams & Wilkins; 758. Available at: http://books.google.com/books?hl=en&lr=&id=Z3vjT9ALxHUC&pgis=1. Accessed August 5, 2014.

7. Prüss, A, Kay, D, Fewtrell, L, and Bartram, J (2002) Estimating the burden of disease from water, sanitation, and hygiene at a global level. *Environ Health Perspect*. 110(5):537–542. Available at: http://www.pubmedcentral.nih.gov/articlerender.fcgi?artid=1240845&tool=pmcentrez&rendertype=abstract. Accessed June 2, 2015.

8. MacDonald Gibson, J, Thomsen, J, Launay, F, Harder, E, and DeFelice, N (2013) Deaths and medical visits attributable to environmental pollution in the United Arab Emirates. *PLoS One*. 8(3):e57536. doi:10.1371/journal.pone.0057536.

9. DeFelice, NB, Johnston, JE, and MacDonald Gibson, J. (in revision) Burden of acute gastrointestinal illness from microbial contaminants in North Carolina community water systems. *Environ Sci Technol*.

10. National Research Council (2001) *Arsenic in drinking water : 2001 update*. Washington, D.C.: National Academy Press. Available at: http://search.lib.unc.edu/search?R=UNCb4104000. Accessed May 24, 2013.

11. DeFelice, NB, Leker, H, and MacDonald Gibson, J. (in revision) Burden of cancer from chemicals in North Carolina drinking water. *Risk Anal*.

12. Hubbard, AE, and Van Der Laan, MJ (2008) Population intervention models in causal inference. *Biometrika*. 95(1):35–47. Available at: /pmc/articles/PMC2464276/?report=abstract. Accessed February 11, 2014.

13. Ahernm J, Hubbardm A, and Galea, S (2009) Estimating the effects of potential public health interventions on population disease burden: a step-by-step illustration of causal inference methods. *Am J Epidemiol*. 169(9):1140–1147. doi:10.1093/aje/kwp015.

14. DeFelice, NB (2014) Drinking Water Risks to Health 40 Years After Passage of the Safe Drinking Water Act: A County-by-County Analysis in North Carolina, Ph.D. Thesis. Department of Environmental Sciences and Engineering. University of North Carolina at Chapel Hill.

15. Morgan, MG, Millett, G, and Henrion, M (1990) *Uncertainty: A Guide to Dealing with Uncertainty in Quantitative Risk and Policy Analysis*. Cambridge; New York: Cambridge University Press.

16. MacDonald Gibson, J, Brammer, A, Davidson, C, Folley, TJ, Launay, F, and Thomsen, J (2013) *Environmental Burden of Disease Assessment: A Case Study in the United Arab Emirates*. Dordrecht, The Netherlands: Springer.

17. Fewtrell, LA, Prüss-Üstün, A, Bartram, J, and Bos, R (2007) Water, sanitation and hygiene: quantifying the health impact at national and local levels in countries with incomplete water supply and sanitation coverage. *Public Health*. 215(15):1–15. Available at: http://bases.bireme.br/cgi-bin/wxislind.exe/iah/online/?IsisScript=iah/iah.xis&src=google&base=REPIDISCA&lang=p&nextAction=lnk&exprSearch=174394&indexSearch=ID. Accessed May 29, 2012.

18. DeFelice, NB (2014) Drinking Water Risks to Health 40 Years After Passage of the Safe Drinking Water Act: A County-by-County Analysis in North Carolina, Ph.D. Thesis. Department of Environmental Sciences and Engineering. University of North Carolina at Chapel Hill.

19. MacDonald Gibson, J, and Farah, ZS (2012) Environmental risks to public health in the United Arab Emirates: a quantitative assessment and strategic plan. *Environ Health Perspect*. 120(5):681–686. doi:10.1289/ehp.1104064.

20. Rodgers, A, Ezzati, M, Vander Hoorn, S, Lopez, AD, Lin, R-B, and Murray, CJL (2004) Distribution of major health risks: findings from the Global Burden of Disease study. *PLoS Med*. 1(1):e27. doi:10.1371/journal.pmed.0010027.

21. Lopez, AD, Mathers, CD, Ezzati, M, Jamison, DT, and Murray, CJL (2006) Global and regional burden of disease and risk factors, 2001: systematic analysis of population health data. *Lancet*. 367(9524):1747–1757. doi:10.1016/S0140-6736(06)68770-9.

22. Lim, SS, Vos, T, Flaxman, AD et al. (2013) A comparative risk assessment of burden of disease and injury attributable to 67 risk factors and risk factor clusters in 21 regions, 19902010: a systematic analysis for the Global Burden of Disease Study 2010. *Lancet*. 380:2224–2260. doi:10.1016/S0140-6736(12)61766-8.

58

WATER MONITORING AND TESTING[*]

Huw Taylor

PROFESSOR OF MICROBIAL ECOLOGY, UNIVERSITY OF BRIGHTON, UK

Learning objectives

1 Demonstrate how a growing understanding of waterborne pathogen ecology and the environmental behavior of enteric organisms has provided the tools currently used to monitor the hygienic quality of water.
2 Describe how recently developed strategies, systems and tools use information gained from water monitoring and testing to *prevent* waterborne disease transmission.
3 Consider the future contribution of water monitoring and testing within the multi-disciplinary challenge of providing safe drinking water for all.

Introduction

This chapter considers how water monitoring and testing are currently used to protect human health, with a focus on the provision of potable drinking water (and, to a lesser extent, the protection of recreational waters).

'Pure and wholesome'

Before the widespread acceptance of the 'germ theory' and the understanding that many common infectious diseases are waterborne, the concept of drinking water quality was fairly nebulous. Although the potential for drinking water to cause harm was probably widely recognized in many cultures, water quality was often defined in terms of 'purity' and 'wholesomeness'. The precise meaning of these ill-defined words was seldom tested in law ,and water quality was interpreted through the unaided human senses of sight, taste, smell and touch. However, from a modern perspective, the term 'purity' can probably be considered to be related to chemical contamination and 'wholesomeness' to the tendency of the water to transmit infectious diseases to consumers. The enumeration of specific indicators

[*] Recommended citation: Taylor, H. 2015. 'Water monitoring and testing', in Bartram, J., with Baum, R., Coclanis, P.A., Gute, D.M., Kay, D., McFadyen, S., Pond, K., Robertson, W. and Rouse, M.J. (eds) *Routledge Handbook of Water and Health*. London and New York: Routledge.

of potential harm to human health had to await the discoveries of nineteenth century science and, although the transmission of both infectious and non-infectious waterborne diseases is now much better understood, the search for effective, rapid and low-cost ways to assess the potential health impact of drinking water continues.

In the nineteenth century, John Snow's extensive epidemiological investigations, which included a study that demonstrated an outbreak of cholera in London's Soho district to be waterborne, took considerable time to gain widespread acceptance, chiefly because they contradicted the widely believed 'miasma theory' of disease transmission. However, the development of bacteriology as a laboratory science gradually provided practical and relatively simple tools that could be used to estimate the degree of fecal contamination of raw and processed drinking water supplies. These embryonic laboratory tools provided valuable information to support early waterborne disease control measures.

Laboratory science in support of water monitoring and testing

Since its earliest days, the science underlying the hygienic quality assessment of drinking and recreational waters has been based on the 'indicator paradigm' advocated by Frankland and Frankland (1894), which uses the enumeration of specific organisms, or groups of organisms, of (assumed) enteric origin to demonstrate the presence of fecal pollution and, indirectly, to suggest the potential presence of enteric pathogens (disease-causing organisms). As a result, the destruction and removal of fecal microorganisms in both the natural aquatic environment and engineered water and wastewater treatment processes may be routinely monitored as a key component of an integrated health protection strategy that can potentially encompass the entire water cycle. The approach continues to be most commonly used to monitor the quality of water intended for direct human consumption (drinking water) or water that may be consumed accidently (recreational waters). In many parts of the world, the hygienic quality of treated wastewater effluents is not routinely tested, other than when the effluent is directly reused for human activities, such as in agriculture and aquaculture.

The theoretical 'ideal' indicator organism, or indicator group, would meet a number of essential criteria (Bonde, 1963), namely: that it is itself non-pathogenic, but it is consistently identified in pathogen-contaminated water; that it does not multiply outside the mammalian gut, particularly in aquatic environments; that it may be reliably detected at low concentrations; and that it is present in greater numbers than, and have similar survival times to, waterborne pathogens. No single bacterial or viral group meets all these requirements in all contexts and, in practice, advances in protocols for water microbiology testing have been decidedly slow during the past one hundred and fifty years. The hygienic quality of water therefore continues to be assessed with reference to a heterotrophic plate count plus the enumeration of a relatively small number of organisms of assumed enteric origin, the ecological behavior of which has not been fully elucidated.

By the 1890s, the fecal indicator organism approach to water quality testing had led to the 'Wurtz method', in which the enteric bacterium '*Bacillus coli*' was identified by its growth on litmus lactose agar. Lactose fermentation and the production of acid (and later the production of gas) were considered to be a characteristic of a broader group of bacteria considered to be of largely enteric origin and by 1901 the concept of 'coliform bacteria' was included in UK guidelines. Further studies demonstrated that the coliform group contained bacteria of environmental origin, but that those bacteria that met the definition of *B. coli* were more likely to be of enteric origin. The methods of identifying fecal indicator bacteria continued for many years to be based on their phenotypic growth characteristics (such as carbon source utilization, gas production and optimal growth temperature), though definitions became gradually more precise; the term 'total coliforms' being used to signify bacteria that can ferment lactose with the production of gas at 37 °C, and the more exacting term 'thermotolerant (or fecal) coliforms' referring to a subgroup of coliforms that are able to multiply at 44 °C. Members of this second subgroup of coliform bacteria are far more likely to be of enteric origin, but the group is still defined by a limited number of phenotypic tests and is taxonomically heterogeneous. However, *Escherichia coli* is normally the predominant species to be identified among the thermotolerant coliforms and enzymatic tests now enable this species to be enumerated using fairly basic laboratory resources.

Increasingly it has been recognized that there is no direct correlation between numbers of any indicator and enteric pathogens. As a consequence of the possible ambiguity of the term 'microbial indicator', Ashbolt et al. (2001) note that the following terms are now recognized: 1. general (process) microbial indicators, 2. fecal indicators and 3. index and model organisms. Meanwhile, recent advances in molecular microbiology have supported the standardization and automation of methods to enumerate genes copies from specific waterborne pathogens (Girones et al., 2010) and tools are emerging that may further elucidate the ecological behavior of enteric organisms beyond the gut environment. Although concerns continue to be expressed about the uncertain relationship between numbers of infectious organisms and concentrations of nucleic acid targets in environmental samples, quantitative real-time polymerase chain reaction (qPCR) is now extensively used to support the burgeoning field of quantitative microbial risk assessment (QMRA). In addition, in the evolving field of quantitative microbial source tracking (QMST) the identification and enumeration of microbial species and groups of specifically human origin (or other specified animal groups) have begun to inform decision-making in the fields of watershed and recreational water management.

Efforts continue within the academic field of health-related water microbiology, partly to provide new microbial detection methods that may be used to protect human health in a wide range of situations. Molecular microbiology is gradually providing a means of enumerating pathogens and indicator organisms directly from a wide range of aquatic matrices but, in parallel, efforts continue to provide relatively simple, low-cost and rapid techniques that can readily identify the presence of fecal contamination, particularly in low-resource settings.

The legislative framework within which the hygienic quality of water is assessed and managed is inherently conservative. Both national and international agencies charged with establishing reliable water quality testing protocols understandably remain wary of the claims of researchers who champion novel methods and targets. This is because, in many cases, important questions about the survival of novel microbial targets in natural and engineered environments, and about their geographic ubiquity and source specificity, have not yet been adequately answered.

The changing rationale for water monitoring and testing

The chronic and continuing impact of waterborne disease on human health in low-income countries and more occasional but well-publicized outbreaks of waterborne disease in industrialized countries have highlighted a need for a risk-based preventive approach to reducing the burden of waterborne disease on human society. Within such a risk-based framework, a number of approaches, described below and then summarized in Table 58.1, haas been developed that may be combined to generate a toolbox approach to watershed management and the provision of sustainable drinking water supplies.

Table 58.1 A summary of recently developed systems and tools to support waterborne disease control

Process	Application	Benefits	Limitations
Water safety planning	Extensively used in Europe and elsewhere as a key component of legislation to ensure the sustainable provision of safe drinking water.	The preventive nature of the risk-based approach has been demonstrated to support water safety at a lower cost to previous management approaches.	There have been problems associated with adapting the principles of the approach to low-resource settings, notably in rural sub-Saharan Africa.
Quantitative microbial risk assessment	Increasingly used in Europe and elsewhere to support decision-making in the drinking water supply sector.	Provides a transparent methodology to compare the potential health impact of alternative water treatment and supply options.	Knowledge gaps remain, including dose–response relationships for all but a limited number of index waterborne pathogens.
Quantitative microbial source tracking	Has been used to support total mean daily load (TMDL) studies in the USA and has been proposed as a tool to support 'bathing water profiling' in the European Union.	Potentially provides information to focus resources on tackling sources of fecal contamination of recreational waters and drinking water sources.	Few methods have been sufficiently standardized and validated in a wide range of settings. Further, the use of novel fecal source markers may appear to 'contradict' information from established FIB enumeration.
Quantitative microbial source apportionment	The tool has been proposed as a relatively simple way to explain FIB dynamics in European watershed studies.	Method is based on readily available data, e.g., river discharge and FIB counts.	The approach is a way to manage FIB compliance rather than a way to investigate risk of human waterborne disease

The water safety planning framework

Drinking water quality testing aims to protect human health from the deleterious impact of contamination by toxic chemicals and disease-causing organisms. Increasingly, testing is undertaken within a water safety planning (WSP) framework that aims to ensure the consistent safety and acceptability of a drinking water supply through the use of comprehensive risk assessment and risk management approaches (Bartram et al., 2009). This preventative WSP framework is beginning to influence water monitoring and testing strategies and is leading to more cost-effective watershed management practices that focus on 'critical control-points' of water contamination (Nnane et al., 2011). These use a variety of emerging tools to explain and predict the spatial and temporal dynamics of water quality within a watershed. Some of the more promising approaches are described below.

Quantitative microbial risk assessment (QMRA)

QMRA evolved from risk assessment protocols used in chemical industries, and its application to quantify accurately risks to human health from living, waterborne, infectious agents of disease presents considerable challenges. Knowledge gaps remain in areas essential to the construction of a useful QMRA process, but it potentially provides a more transparent and rational framework to support the construction and communication of risk management strategies. It also provides a way to rank alternative water supply schemes in terms of public health risk. Therefore, QMRA may provide useful quantitative inputs to water safety plans (Smeets et al., 2010). Details of the component parts of the QMRA process are described in Chapter 56. Water resource management decisions based entirely on the findings of the process may, however, be criticized for being too 'reductionist', focusing on a limited number of specific known hazards, but QMRA is now used widely to aid the design of new drinking water treatment systems and to improve the performance of existing systems. As long as the inherent limitations of the approach are fully appreciated, QMRA will undoubtedly in the future provide valuable evidence to support more effective water safety plans.

Quantitative microbial source tracking (QMST)

Fecal indicator bacteria (FIB) were used throughout the twentieth century to monitor the hygienic quality of water. In more recent years, the target species (or group) has been modified on occasion, either to seek organisms that are more consistently of enteric origin (e.g., the enumeration of *E. coli* rather than thermotolerant coliforms) or to seek organisms that survive in water for similar periods to common disease-causing pathogens (e.g., the enumeration of intestinal enterococci in coastal bathing waters). However, as with their predecessors, these FIB remain of wide enteric provenance so their detection does not assign the source of the fecal contamination either to human or to specific non-human sources. Further, regrowth of these bacteria in the natural environment has been suggested in some settings. Although co-evolution has resulted in a degree of pathogen-host specificity, our growing understanding of the gut biome reveals a permissive ecology in which zoonotic pathogens, as well as a variety of commensal organisms, survive transit through the gut of several animal species, contributing to a surprisingly widespread transfer of genetic material within the gut biome. Notwithstanding this ecological complexity and the continuing impact of waterborne zoonotic diseases on human health (Cotruvo et al., 2004), the search continues for host-specific microbial targets that may support improved protection of recreational waters and drinking water abstraction points.

The focus of microbial source tracking (MST) has evolved since its inception from library-based approaches to methods based on the detection and enumeration of microbial species and genetic sequences that are associated with specific host animal groups (QMST) (Hagedorn et al., 2011). In the search for 'specificity' and 'sensitivity', the contribution of Field and her co-workers is particularly noteworthy (Field et al., 2003). This approach, which focuses on the identification and enumeration of genetic sequences of the anaerobic Bacteriodales family, continues to provide markers that can be used to identify the fecal contribution to water bodies of humans, and importantly species of reared and wild animals, including birds. The expansion and successes of the field of microbial source tracking was demonstrated in a recent international comparative study of forty-one methods (Boehm et al., 2013), which showed that numerous library-independent methods developed in recent years, including those for detecting human, cow, gull, ruminant, pig and horse feces, are both sensitive and specific and that field validation of the methods is warranted. Meanwhile, efforts to develop human-specific markers of fecal pollution at lower cost (based on the detection of bacteriophage lysis of a host-specific strain of *Bacteroides*) are beginning to show promise (Ebdon et al., 2011; McMinn et al., 2013), though low phage titers encountered in some matrices continue to limit the method's wider application.

Quantitative microbial source apportionment (QMSA)

In recent years, the sanitary quality of water has continued to rely on the enumeration of FIB as a means of indicating fecal contamination and the potential for the transmission of enteric disease. Where microbial source tracking has been used, it has often been in response to specific pollution concerns, and more specifically as a response to non-compliance with standards and guidelines for FIB concentrations. The application of the two approaches (enumeration of traditional fecal indicator bacteria and enumeration of source-specific microbial markers) may potentially lead to confusion and it is important to decide whether it is the source of fecal pollution, the source of human pathogens or the source of the FIB that needs to be defined. As the science stands, and with considerable uncertainty about the relationship between traditional FIB, novel MST markers and the prevalence of specific human waterborne diseases, it is perhaps understandable that attempts to communicate risk to public health can cause confusion. In the absence of routine monitoring of common waterborne infectious pathogens and with uncertainty as to the relationship between FIB, MST markers and pathogens in different matrices, Kay et al. (2010) suggested a more pragmatic approach in which FIB are apportioned to specific sources within a watershed, based on an understanding of watershed hydrology under dry and wet conditions and the resulting contribution of diffuse overland flow and point sources of wastewaters of either human or non-human origin.

Simplified water monitoring and testing

The emergence of molecular methods to identify a wide variety of microbial entities in the aquatic environment has inspired microbiologists to develop novel methods that can be used to quantify numbers of waterborne pathogens (or at least their nucleic acids) with greater specificity and sensitivity. These exciting advances in microbial technology offer opportunities to develop a comprehensive and quantitative risk-based strategy for managing human waterborne disease. However, a growing gulf is emerging between what can now be routinely achieved in the laboratories of industrialized countries and what is possible in the low-income countries in which waterborne disease has its greatest impact on human well-being. It is therefore important that effective but affordable water monitoring testing protocols are made readily available in those countries in which the impact of human waterborne disease is greatest.

From the perspective of quantifying the risk to human health of specific waterborne diseases, it is easy to list the limitations of simply enumerating levels of fecal indicator bacteria in the target water body. However, these organisms continue to provide useful information, at relatively low cost, and this information can be used to support decisions about both the initial choice of water source and subsequent protection of the source from fecal contamination. If we are to achieve the greatest possible reduction in human waterborne disease, efforts to detect and enumerate specific disease-causing organisms should at least be matched by greater efforts to understand better the fecal transmission mechanisms within the most impoverished communities of the world. The idea of 'Rapid Assessment of Drinking Water Quality' (RADWQ) was therefore developed in order to investigate to what extent the quality of drinking water from 'improved' sources (according to the Millennium Goal 7 Target C, namely to halve, by 2015, the proportion of the population without sustainable access to safe drinking water and basic sanitation) met international guideline standards for water quality (WHO/UNICEF, 2012).

To some extent, the resurgent focus on practical tools to help the poor is being met by the development and application of simplified techniques to enumerate FIB using simplified enzymatic methods (Bain et al., 2012) and by the development of much simpler methods, such as the hydrogen sulfide test (McMahon et al., 2012; Wright et al., 2012), that indicate the possible presence of bacteria of enteric origin. However, in practice, direct water quality monitoring will continue to be limited (at best) to the earliest stages of providing drinking water (i.e., immediately following water-point construction), particularly in rural communities and it has been suggested that future monitoring strategies should combine measures of water quality with more easily monitored indicators of sanitary inspection (Bain et al., 2014). In order to support a more sustainable risk-based water quality management framework that is understood and managed by the community consuming the water, an expert knowledge of the ecology of waterborne pathogens that may be prevalent within a specific environment needs to be combined with local understanding of pollution sources, hydrology and local societal needs and behavior. A vehicle for this approach may be provided by the concept of WSP, though obstacles remain to the wider application of its principles beyond the sphere of piped urban water supplies.

Future prospects

Despite the successes of the Millennium Development Goals in targeting the provision of improved drinking water for the world's poorest communities, the ultimate target of safe drinking water for all requires a stronger evidence base on which to make difficult decisions. Water quality monitoring and testing using newly emerging techniques will inform our understanding of waterborne disease transmission, which will in turn hopefully inform the design and operation of more sustainable drinking water supplies. However, it is increasingly evident that a much greater emphasis needs to be placed on the development and implementation of simplified community-based approaches to water quality monitoring that can be readily used to control the risk of human waterborne disease as a core component of practical community-led water safety plans.

Key recommended readings (open access)

1 Bartram, J., Corrales, L., Davison, A., Deere, D., Drury, D., Gordon, B., Howard, G., Rinehold, A., and Stevens, M. (2009). *Water Safety Plan Manual: Step-by-Step Risk Management for Drinking-water Suppliers*. World Health Organization: Geneva. http://www.who.int/water_sanitation_health/publication_9789241562638/en/. This manual provides the rationale for water safety planning and offers those involved with the process the protocols they need for their successful design and operation. The publication has been an essential foundation for more recent research and advice on the application of WSP to a wide range of situations in various regions of the world.

2 World Health Organization (WHO) (2001). *Water Quality: Guidelines, Standards and Health*. Edited by Lorna Fewtrell and Jamie Bartram. IWA Publishing: London. ISBN: 1 900222 28 0. http://www.who.int/water_sanitation_health/dwq/whoiwa/en/. The World Health Organization has developed a series of normative guidelines that present an authoritative assessment of the health risks associated with exposure to health

hazards through water and of the effectiveness of approaches to their control. The first of these, 'Guidelines for drinking water quality', is particularly useful to all those concerned with issues related to water quality and health, including environmental and public health scientists, water scientists, policy-makers and those responsible for developing standards and regulations.

3 Chapman, D. (ed.) (1996). *Water Quality Assessments. A Guide to the Use of Biota, Sediments and Water in Environmental Monitoring.* Second edition. E & FN Spon, an imprint of Chapman & Hall: London. http://www.who.int/water_sanitation_health/resourcesquality/ watqualassess.pdf. This book takes into account developments in strategies, technologies and methods and is particularly useful for water resources managers charged with the monitoring, assessment and control of water quality for a variety of purposes. It contributes to the capacity-building initiatives launched in a number of countries in the aftermath of the Rio de Janeiro conference by supporting the scientifically sound assessment of water resources. The guide complements its companion handbook, *Water Quality Monitoring: A Practical Guide to the Design and Implementation of Freshwater Quality Studies and Monitoring Programmes*, by giving an overall strategy for assessments of the quality of the main types of water body. Together the two books cover all the major aspects of water quality, its measurement and its evaluation.

4 Bartram, J., and Balance, R. (1996). *Water Quality Monitoring: A Practical Guide to the Design and Implementation of Freshwater Quality Studies and Environmental Monitoring.* E & FN Spon. An imprint of Chapman & Hall: London. http://whqlibdoc.who.int/ Publications/1996/0419217304_eng.pdf. The approaches and methods for water quality monitoring described in this handbook are based on experience gained, over two decades, with the design and establishment of the global freshwater quality monitoring network, GEMS/WATER, a goal of which was 'to strengthen national water quality monitoring networks in developing countries, including the improvement of analytical capabilities and data quality assurance'. This handbook supported this goal by providing a practical tool for use in water quality management by national and local agencies and departments dealing with water quality issues.

References

Ashbolt, N. J., Grabow, W. O. K., and Snozzi, M. (2001) Indicators of microbial water quality. In: *Water Quality: Guidelines, Standards and Health.* World Health Organization (WHO). Edited by Lorna Fewtrell and Jamie Bartram. London: IWA Publishing.

Bain, R., Cronk, R., Wright, J., Yang, H., Slaymaker, T., and Bartram, J. (2014). Fecal contamination of drinking-water in low- and middle-income countries: A systematic review and meta-analysis. *PLOS Medicine*, 11, 5, e1001644.

Bain, R., Bartram, J., Elliott, M., Matthews, R., McMahon, L., Tung, R., Chuang, P., and Gundry, S. (2012). A summary catalogue of microbial drinking water tests for low and medium resource settings. *International Journal of Environmental Research and Public Health*, 9(5), 1609–1625.

Bartram, J., Corrales, L., Davison, A., Deere, D., Drury, D., Gordon, B., Howard, G., Rinehold, A., and Stevens, M. (2009). *Water Safety Plan Manual: Step-by-Step Risk Management for Drinking-water Suppliers.* Geneva: World Health Organization.

Boehm, A. B., Van de Werfhorst, L. C., Griffith, J. F., Holden, P. A., Jay, J. A., Shanks, O. C., Wang, D., and Weisberg, S. B. (2013). Performance of forty-one microbial source tracking methods: A twenty-seven lab evaluation study. *Water Research*, 47, 6812–6828.

Bonde, G. J. (1963). *Bacterial Indicators of Water Pollution.* Copenhagen: Teknisk Forlag.

Cotruvo, J. A., Dufour, A., Rees, G., Bartram, J., Carr, R., Cliver, D. O., Craun, G. F., Fayer, R., and Gannon, V. P. J. (2004). *Waterborne Zoonoses: Identification, Causes and Control.* London: IWA Publishing for the World Health Organization.

Ebdon, J. E., Sellwood, J., Shore, J., and Taylor H. D. (2011). Phages of *Bacteroides* (GB-124): A novel tool for viral waterborne disease control? *Environmental Science & Technology*, 46(2), 1163–1169.

Field, K. G., Chern, E. C., Dick, L. K., Fuhrman, J. A., Griffith, J. F., Holden, P. A., LaMontagne, M. G., Le, J., Olson, B. H., and Simonich, M. T. (2003). A comparative study of culture-independent, library-independent genotypic methods of fecal source tracking. *Journal of Water and Health*, 1(4): 181–194.

Frankland, P., and Frankland, P. (1894). *Micro-organisms in Water; Their Significance, Identification and Removal*. London: Longmans, Green and Co.

Girones, R., Ferrús, M. A., Alonso, J. L., Rodriguez-Manzano, J., Calgua, B., Corrêa Ade, A., Hundesa, A., Carratala, A., and Bofill-Mas, S. (2010). Molecular detection of pathogens in water – the pros and cons of molecular techniques. *Water Research*, 44(15), 4325–4339.

Hagedorn, C., Blanch, A. R., and Harwood, V. J. (eds.) (2011). *Microbial Source Tracking: Methods, Applications and Case Studies*. New York: Springer.

Kay, D., Anthony, S., Crowther, J., Chambers, B. J., Nicholson, F. A., Chadwick, D., Stapleton, C. M., and Wyer, M. D. (2010). Microbial water pollution: A screening tool for initial catchment-scale assessment and source apportionment. *Science of the Total Environment*, 408, 5649–5656.

McMahon, L., Gunden, A. M., Devine, A. A., and Sobsey, M. D. (2012). Evaluation of a quantitative H2S MPN test for fecal microbes analysis of water using biochemical and molecular identification. *Water Research*, 46(6), 1693–1704.

McMinn, B. R., Korajkic, A., and Ashbolt, N. J. (2013). Evaluation of *Bacteroides fragilis* GB-124 bacteriophages as novel human-associated faecal indicators in the United States. *Letters in Applied Microbiology*, 59(1), 115–121.

Nnane, D. E., Ebdon, J. E., and Taylor, H. D. (2011). Integrated analysis of water quality parameters for cost-effective faecal pollution management in river catchments. *Water Research*, 45, 2235–2246.

Smeets, P. W. M. H., Rietweld, L. C., van Dijk, J. C., and Medema, G J. (2010). Practical applications of microbial risk assessment (QMRA) for water safety plans. *Water Science and Technology*, 61(6), 1561–1568.

World Health Organization/UNICEF. (2012). *Rapid Assessment of Drinking-water Quality. A Handbook for Implementation*. Geneva: World Health Organization/UNICEF.

Wright, J. A., Yang, H., Walker, K., Pedley, S., Elliott, J., and Gundry, S. W. (2012). The H2S test versus standard indicator bacteria tests for faecal contamination of water: Systematic review and meta-analysis. *Tropical Medicine and International Health*, 17(1), 10.

59

INDICATORS OF
MICROBIAL QUALITY*

Joe Brown

SCHOOL OF CIVIL AND ENVIRONMENTAL ENGINEERING, GEORGIA
INSTITUTE OF TECHNOLOGY, ATLANTA, GEORGIA, USA

Phillip Grammer

DEPARTMENT OF CIVIL, CONSTRUCTION, AND ENVIRONMENTAL ENGINEERING,
UNIVERSITY OF ALABAMA, TUSCALOOSA, ALABAMA, USA

Learning objectives

1 Define *microbial indicator* and describe ideal indicator characteristics.
2 Describe limitations of microbial indicators in assessing water safety.
3 Identify common and emerging fecal indicators.

Introduction

In 1677, Antonie van Leeuwenhoek reported the first microbial count in a water sample: 2,730,000 "animalcules" in a volume he estimated to be the size of a pea (Egerton 2006). From that time we have known that water is teeming with microbial life. Some microbes present in water may be readily detected using basic culture methods. Others require advanced and relatively recently developed molecular techniques. Our knowledge about the microbial communities that inhabit water is still basic and mostly utilitarian: we pay especially careful attention to those microbes we know can make us sick. When we refer to "microbial quality" of water, we mean its safety for drinking, recreation, irrigation, and other uses that may result in human contact. Therefore we mean the presence or potential presence of pathogenic microbes.

Waterborne infectious diseases are caused primarily by pathogenic bacteria, viruses, and protozoa and are associated with the presence of human or animal fecal material, since these microbes are generally transmitted via the "fecal–oral route" (Chapter 2). Of the several

---

* Recommended citation: Brown, J. and Grammer, P. 2015. 'Indicators of microbial quality', in Bartram, J., with Baum, R., Coclanis, P.A., Gute, D.M., Kay, D., McFadyen, S., Pond, K., Robertson, W. and Rouse, M.J. (eds) *Routledge Handbook of Water and Health*. London and New York: Routledge.

dozen known and suspected pathogens transmitted in water, recent evidence suggests that the relative contribution of specific microbes to the burden of disease varies widely by setting and population. Rotavirus, *Cryptosporidium*, enterotoxigenic *Escherichia coli* (ETEC), *Shigella*, *Vibrio cholera*, and *Salmonella* are among the most important globally (Kotloff et al. 2013). Although rapid, low-cost, simultaneous methods for detection of multiple pathogens in water is the focus of intense research, current methodological constraints limit our ability to rapidly and accurately identify all pathogens that may be present and determine whether they exist in quantities that pose risks to public health. Because some pathogens have low infectious doses, even as few as a single microbe, even small numbers of pathogens in water may contribute to increased risk of disease. Detecting them directly may therefore require concentration of large volumes of water. So, to assess water safety, we look for the common element that is associated with most of these microbes: human or animal feces, markers of which are more easily detected than specific pathogens.

Although no known indicator meets all of these criteria, an ideal microbial indicator of fecal contamination in water should:

1 Be present whenever enteric pathogens are present, ideally in greater quantities for ease of detection and to be conservative in estimates of risk.
2 Be absent when enteric pathogens are absent or at levels that pose no increased risk.
3 Survive as long or longer than the most environmentally persistent enteric pathogens.
4 Not proliferate independently in the environment.
5 Be detectable in all types of water that may lead to human exposure.
6 Be shed in the feces of species who share fecal–oral pathogens with humans.
7 Be reliably, rapidly, and unambiguously detectible at low cost.
8 Be randomly distributed in a given sample (i.e., not "clumpy").[1]

The following sections and Table 59.1 briefly describe some of the currently used and emerging indicators for microbial quality of water. It is important to note two qualifications that are frequently overlooked in discussions about indicators of fecal contamination. First, although the presence of these indicators may be associated with feces, positive identification of indicators may not correlate with the presence of pathogens. This is because the presence and quantity of pathogens in feces is determined by the (seasonally variable) infection prevalence in those persons contributing fecal material to the water; dilution, survival, and growth (among some bacteria, such as *E. coli*) after shedding; the fact that some indicators may thrive in environmental conditions; and other factors that are highly context specific. The second is that indicators may or may not be associated with higher risk of disease in humans who ingest (or, less frequently, inhale) the water, partly for the same reason but also because these pathogens are transmitted via several other important pathways in any given setting (Wagner and Lanoix 1958). Indicator counts in water may be highly variable due to these and other reasons. It is common to detect pathogens in waters that test negative for indicators or for counts from different indicators to be inconsistent. Indicator count data should be interpreted with caution in light of this uncertainty.

Members of the coliform group: total coliform, fecal coliform, *E. coli*

Coliform bacteria are a group of microorganisms from the genera *Escherichia*, *Enterobacter*, *Citrobacter*, and *Klebsiella*. Some members of the coliform group are native to the environment and therefore are found in uncontaminated water. Because of this, total coliform is an unreliable indicator of fecal contamination but is commonly used as a process indicator.[2] Fecal (or thermotolerant) coliform are coliforms capable of surviving at 44.5° C and are thought to be more specific to feces, though they may not be reliable in predicting health risks (Gruber et al. 2014).

E. coli, one thermotolerant member of the coliform group, is present in the gut of all warm-blooded animals, making up about 1 percent of fecal biomass. *E. coli* has been observed to survive for moderately long periods in the environment and, under certain conditions, to proliferate (Fujioka and Unutoa 2006) in tropical and sub-tropical waters. Some *E. coli* are now known to be adapted to living apart from a host (Oh et al. 2012). *E. coli* remains the most widely used indicator of water safety risk globally, despite these caveats. The World Health Organization (WHO) *Guidelines for Drinking-water Quality* considers water with no more than 10 *E. coli*/100 ml to be of relatively low risk and of low priority for action if also rated to be of low risk based on sanitary inspection (WHO 2011). *E. coli* counts in drinking water may be associated with increased risk of diarrheal disease, though the association is not strong or consistent across studies (Moe et al. 1991; Gundry et al. 2004; Brown et al. 2008; Gruber et al. 2014).

Fecal *Streptococcus* and *Enterococcus*

Fecal *streptococci* (FS) are found in the gut of warm-blooded mammals, and were formerly measured alongside fecal coliforms (FC) to compute the FC/FS ratio, thought to be the basis (now known to be unreliable) for determining whether fecal contamination was of human or animal origin. Enterococci are a subgroup of FS that are thought to be both more closely associated with human fecal contamination as well as more persistent in salt water environments, which are two reasons enterococci are now recommended by the United States Environmental Protection Agency (USEPA) for beach and other recreational water quality monitoring in the USA. There is emerging evidence that enterococci may be more reliable than *E. coli* or members of the coliform group for drinking water quality as well as recreational water quality (Noble et al. 2003; Byappanahalli et al. 2012).

F+RNA coliphages

Bacteriophages (or, phages) are viruses that infect bacteria. They are closely associated with their bacterial hosts, and so may be found in the same settings. Unlike their bacterial hosts, there is some evidence that they are often present in greater quantities, are more environmentally persistent, and may be better models of virus fate and transport in the environment than bacterial indicators. *Bacteriodes fragilis* phages, *Salmonella* phages, and *E. coli* phages (coliphages) have all been proposed as indicators of fecal contamination. Of the phages that have been studied, F-pilus-specific (F+) coliphages with RNA genomes (sometimes known as "male-specific" coliphages) have received the most attention (Sobsey et al. 1995). Less is known about F+DNA coliphages and somatic coliphages as fecal and viral indicators, though somatic coliphages may be more common in feces as well as more stable in the environment (Jofre 2003). F+RNA coliphages infect *E. coli* and potentially other members of the coliform group. They have been used in microbial source tracking applications (Love and Sobsey 2007; Lee et al. 2011), although their associations to health outcomes in humans have not been well characterized.

H2S-producing bacteria

Manja et al. (1982) first reported the development of a simple, reliable, and low-cost field microbiological test to detect bacteria that produce H_2S (including *Citrobacter*, *Salmonella*, *Clostridium perfringens*, *Staphylococcus*, *Klebsiella*, and others, though not *E. coli*). These microorganisms reduce organic sulfur to H_2S gas, which then readily reacts with iron to form a black iron precipitate. The H_2S test is unable to differentiate between naturally occurring H_2S-producing bacteria and those associated with fecal contamination of water. The H_2S test has been shown to be largely consistent with other indicators (Khush et al. 2013), though it is often used as a presence/absence or most probable number (MPN) method that may not be directly comparable to membrane filtration-based assays. H_2S-producing bacteria may be useful as indicators of diarrheal disease risk (Levy et al. 2012), though additional studies are needed.

Bacteroides

Bacteroides spp. (order Bacteriodales) are gram-negative, rod shaped bacteria that are the most abundant bacteria in human feces, outnumbering *E. coli* and *Enterococcus* (Madigan et al. 2003). *Bacteroides* are facultative anaerobic, meaning they can tolerate oxygen but prefer and will only thrive in anaerobic environments. Consequently, re-growth of *Bacteroides* in the environment is limited (Kreader,1998). Although detection of anaerobic bacteria to assess fecal contamination has traditionally been avoided due to difficulty associated with cultivating anaerobic bacteria, modern molecular methods including the polymerase chain reaction (PCR/qPCR) now make detection of these organisms practical in most laboratories with molecular biology capabilities (Bernhard and Field 2000). One study of swimming-associated illness showed a positive association with Bacteriodales but not with *Bacteroides* (Wade et al. 2010). The primary advantage of using *Bacteroides* is its usefulness in microbial source tracking: *Bacteroides* HF183 has been shown to be highly specific to humans (Boehm et al. 2013).

Clostridium perfringens

Clostridium perfringens is a gram-positive, rod-shaped, spore-forming bacterium that occurs naturally in the environment and in the gut of all warm-blooded animals; it is a member of the sulfite-reducing clostridia (SRC) group. *C. perfringens* are obligate anaerobes, meaning they do not tolerate oxygen, and will therefore not grow and divide in the environment. However, spores of *C. perfringens* resist disinfection and persist in the environment, making them more suitable as indicators of disinfection performance than of fecal pollution in environmental waters.

Emerging indicators

Non-human viruses

The pepper mild mottle virus (PMMoV), a virus that infects and can cause significant damage to pepper plants, has been demonstrated to be a promising emerging indicator (Rosario et al. 2009). When consumed by humans, the virus is not infective and is passed through the stool. PMMoV is frequently detected in human sewage with little seasonal variation, and is more stable in the environment than many human enteric viruses (Hamza et al. 2011). More work is needed to characterize the survival and persistence of plant viruses as indicators of recent fecal pollution in water.

Human DNA

Host cells are shed in high numbers in fecal material, so detection of human mitochondrial DNA has been proposed for use as a source-specific indicator of fecal contamination (Martellini et al. 2005; Kapoor et al. 2013). DNA (sometimes referred to as eDNA in the environment) is relatively stable and can be an unambiguous indicator of human fecal contamination, though more work is needed on fate, transport, and correlation with other indicators, pathogens, and health outcomes.

Non-microbial markers of fecal contamination

The ability of modern instrumentation to detect trace amounts of compounds in water has led to the development of new non-microbial indicators of human fecal contamination. Many compounds, such as caffeine and pharmaceuticals, are consumed exclusively by humans, and

Table 59.1 Comparative summary of common and emerging fecal indicators

Fecal indicator	Association with pathogens in water	Association with health outcomes in humans	Persistence in the environment	Potential for regrowth in the environment	Relative cost per sample	Suitability for source tracking
Turbidity	+	+	N/A	-	+	-
Heterotrophic plate count (HPC)	+	+	+ +	+ + +	+	-
Total coliform	+	+	+ +	+ + +	+	-
Fecal coliform	+ +	+	+ +	+ +	+	-
E. coli	+ +	+ +	+ +	+ +	+	+
Bifidobacteria	+	Unknown	-	-	+ +	+
Fecal streptococcus	+	+ +	+ + +	+	+	+
Enterococci	+ +	+ + +	+ + +	+	+	+
H_2S producers	+	+	+	+ + +	+	-
C. perfringens / SRC	+	+	+ + +	-	+ +	-
B. fragilis phages	+ +	Unknown	+ + +	-	+ +	+ + +
Somatic coliphages	+	Unknown	+ +	-	+ +	-
F+DNA coliphages	+	Unknown	+ +	-	+ +	-
F+RNA coliphages	+	+	+ +	-	+ +	+
PMMoV	Unknown	Unknown	Unknown	-	+ + +	+ + +
Bacteroides	Unknown	Unknown	+	-	+ + +	+ + +
Human DNA	Unknown	Unknown	Unknown	-	+ + +	+ + +
Chemicals	Unknown	Unknown	Varies	-	+ + +	+ + +

Symbol key: - (minimal), + (low), + + (moderate), + + + (high)

Sources: Kueh et al. (1995), Leclerc et al. (2000), Scott et al. (2002), Lemarchand and Lebaron (2003), Hörman et al. (2004), Grabow (2004), Harwood et al. (2005), Savichtcheva and Okabe (2006), Field and Samadpour (2007), McMahan et al. (2011) and Wright et al. (2012).

a portion of what is consumed is excreted unmetabolized from the body. Human-specific chemicals present several advantages to biological indicators: sample analysis is generally easy and quick, they are source specific, and some compounds are stable in the environment with no regrowth. Disadvantages to the use of chemicals are that they require expensive specialized equipment to detect and occur at low concentrations that can be lost if diluted. Some chemicals that demonstrate potential as chemical indicators are caffeine, fluorescent whitening agents (from detergents), coprostanol, and two pharmaceuticals: carbamazepine and diphenhydramine (Glassmeyer et al. 2005). These markers are likely to be very context specific.

Key recommended readings

1 NRC (National Research Council). 2004. *Indicators for Waterborne Pathogens*. Washington DC: The National Academies Press. This book presents a straightforward summary of the use of indicators, including specifics on relative advantages, disadvantages, and applications.
2 Sen, K. and Ashbolt, N. (eds). 2011. *Environmental Microbiology: Current Technology and Water Applications*. Norfolk, UK: Caister Academic Press. This book includes up-to-date information on emerging methods and approaches in environmental microbiology, including the use of indicators for microbial source tracking.
3 Bitton, G. 2005. Microbial Indicators of Fecal Contamination, in *Wastewater Microbiology*, Third Edition, Hoboken, NJ: John Wiley & Sons, Inc. doi: 10.1002/0471717967.ch5 Chapter 5 in this book presents a very readable summary of currently used indicators and their application in water and wastewater.

Notes

1 List modified from NRC (2004).
2 Two kinds of microbial indicators are commonly used: fecal indictors (the subject of this chapter) and process indicators. *Fecal indicators* are used to indicate the potential presence of fecal pathogens, and are primarily used to identify health risks associated with water. *Process indicators* may be used as surrogate microbes to evaluate the effectiveness of treatment processes intended to reduce or eliminate fecal pathogens (which may or may not be present). Confusingly, some of the same microbes are used for both purposes and the two uses are often conflated.

References

Bernhard, A. E. and Field, K. G. 2000. A PCR assay to discriminate human and ruminant feces on the basis of host differences in *Bacteroides-Prevotella* genes encoding 16S rRNA. *Appl Environ Microbiol.* 66(10), 4571–4.

Boehm, A. B., Van De Werfhorst, L. C., Griffith, J. F., Holden, P. A., Jay, J. A., Shanks, O. C., Wang, D. and Weisberg, S. B. 2013. Performance of forty-one microbial source tracking methods: a twenty-seven lab evaluation study. *Water Res.* 47(18), 6812–28. doi: 10.1016/j.watres.2012.12.046.

Brown, J., Proum, S. and Sobsey, M. 2008. *E. coli* in household drinking water and diarrheal disease risk: evidence from Cambodia. *Water Science and Technology* 58(4): 757–63.

Byappanahalli, M. N., Nevers, M. B., Korajkic, A., Staley, Z. R. and Harwood, V. J. 2012. Enterococci in the environment. *Microbiology and Molecular Biology Reviews* 76(4), 685–706.

Colford Jr, J. M. et al. 2012. Using rapid indicators for *Enterococcus* to assess the risk of illness after exposure to urban runoff contaminated marine water. *Water Research* 46(7), 2176–86.

Egerton, F. N. 2006. A history of the ecological sciences, part 19: Leeuwenhoek's microscopic natural history. *Bulletin of the Ecological Society of America* 87, 47.

Field, K. G. and Samadpour, M. 2007. Fecal source tracking, the indicator paradigm, and managing water quality. *Water Research* 41(16), 3517–38.

Fujioka, R. S. and Unutoa, T. M. 2006. Comparative stability and growth requirements of *S. aureus* and faecal indicator bacteria in seawater. *Water Sci Technol.* 54(3), 169–75.

Glassmeyer, S. T., Furlong, E. T., Kolpin, D. W., Cahill, J. D., Zaugg, S. D., Werner, S. L., Meyer, M. T. and Kryak, D. D. 2005. Transport of chemical and microbial compounds from known wastewater discharges: potential for use as indicators of human fecal contamination. *Environ Sci Technol.* 39(14), 5157–69.

Grabow, W. O. K. 2004. Bacteriophages: update on application as models for viruses in water. *Water SA* 27(2), 251–68.

Gruber, J. S., Ercumen, A. and Colford, J. 2014. Coliform bacteria as indicators of diarrheal risk in household drinking water: systematic review and meta-analysis. *PLoS ONE* 9(9), e107429. doi:10.1371/journal.pone.0107429

Gundry, S., Wright, J. and Conroy, R. 2004. A systematic review of the health outcomes related to household water quality in developing countries. *Journal of Water and Health* 2(1), 1–13.

Hamza, I. A., Jurzik, L., Uberla, K. and Wilhelm, M. 2011. Evaluation of pepper mild mottle virus, human picobirnavirus and Torque teno virus as indicators of fecal contamination in river water. *Water Res.* 45(3), 1358–68. doi: 10.1016/j.watres.2010.10.021. Epub 2010 Oct 23.

Harwood, V. J., Levine, A. D., Scott, T. M., Chivukula, V., Lukasik, J., Farrah, S. R. and Rose, J. B. 2005. Validity of the indicator organism paradigm for pathogen reduction in reclaimed water and public health protection. *Applied and Environmental Microbiology* 71(6), 3163–70.

Hörman, A., Rimhanen-Finne, R., Maunula, L., von Bonsdorff, C. H., Torvela, N., Heikinheimo, A. and Hänninen, M. L. 2004. *Campylobacter* spp., *Giardia* spp., *Cryptosporidium* spp., noroviruses, and indicator organisms in surface water in southwestern Finland, 2000–2001. *Applied and Environmental Microbiology* 70(1), 87–95.

Jofre, J. 2003. Bacteriophage as indicators, in Bitton. G (ed.) *Encyclopedia of Environmental Microbiology*. New York: John Wiley & Sons, Inc.

Kapoor, V., Smith, C., Santo Domingo, J. W., Lu, T. and Wendell, D. 2013. Correlative assessment of fecal indicators using human mitochondrial DNA as a direct marker. *Environ Sci Technol.* 47(18), 10485–93. doi: 10.1021/es4020458. Epub 2013 Aug 26.

Khush, R. S., Arnold, B. F., Srikanth, P., Sudharsanam, S., Ramaswamy, P., Durairaj, N., London, A. G., Ramaprabha, P., Rajkumar, P., Balakrishnan, K. and Colford, J. M. Jr. 2013. H2S as an indicator of water supply vulnerability and health risk in low-resource settings: a prospective cohort study. *Am J Trop Med Hyg.* 89(2), 251–9. doi: 10.4269/ajtmh.13-0067. Epub 2013 May 28.

Kotloff, K. L. et al. 2013. Burden and aetiology of diarrhoeal disease in infants and young children in developing countries (the Global Enteric Multicenter Study, GEMS): a prospective, case-control study. *Lancet* 382(9888), 209–22. doi: 10.1016/S0140-6736(13)60844-2. Epub 2013 May 14.

Kreader, C. A. 1998. Persistence of PCR-detectable *Bacteroides distasonis* from human feces in river water. *Appl Environ Microbiol.* 64(10), 4103–5.

Kueh, C. S. W. et al. 1995. Epidemiological study of swimming-associated illnesses relating to bathing-beach water quality. *Water Science and Technology* 31(5), 1–4.

Leclerc, H., Edberg, S., Pierzo, V. and Delattre, J. M. 2000. Bacteriophages as indicators of enteric viruses and public health risk in groundwaters. *Journal of Applied Microbiology* 88(1), 5–21.

Lee, J. E., Lee, H., Cho, Y. H., Hur, H. G., Ko, G. 2011. F+RNA coliphage-based microbial source tracking in water resources of South Korea. *Sci Total Environ.* 412–13, 127–31. doi: 10.1016/j.scitotenv.2011.09.061. Epub 2011 Oct 27.

Lemarchand, K. and Lebaron, P. 2003. Occurrence of *Salmonella* spp. and *Cryptosporidium* spp. in a French coastal watershed: relationship with fecal indicators. *FEMS Microbiology Letters* 218(1), 203–9.

Levy, K., Nelson, K. L., Hubbard, A. and Eisenberg, J. N. S. 2012. Rethinking indicators of microbial drinking water quality for health studies in tropical developing countries: case study in Northern Coastal Ecuador. *American Journal of Tropical Medicine and Hygiene* 86(3), 499–507.

Love, D. C. and Sobsey, M. D. 2007. Simple and rapid F+ coliphage culture, latex agglutination, and typing assay to detect and source track fecal contamination. *Appl Environ Microbiol.* 73(13), 4110–18. Epub 2007 May 4.

Madigan, M. M., Martinko, J. M. and Parker, J. 2003. *Brock Biology of Microorganisms*, 10th ed. Upper Saddle River, NJ: Prentice Hall.

Manja, K. S., Maurya, M. S. and Rao, K. M. (1982). A simple field test for the detection of faecal pollution in drinking water. *Bulletin of the World Health Organization.* 60(5), 797–801.

Martellini, A., Payment, P. and Villemur, R. 2005. Use of eukaryotic mitochondrial DNA to differentiate human, bovine, porcine and ovine sources in fecally contaminated surface water. *Water Res.* 39(4), 541–8. Epub 2004 Dec 24.

McMahan, L., Devine, A. A., Grunden, A. M., Sobsey, M. D. 2011. Validation of the H$_2$S method to detect bacteria of fecal origin by cultured and molecular methods. *Appl Microbiol Biotechnol.* 92(6), 1287–95. doi: 10.1007/s00253-011-3520-z. Epub 2011 Oct 26.

Moe, C. L., Sobsey, M. D., Samsa, G. P. and Mesolo, V. 1991. Bacterial indicators of risk of diarrheal disease from drinking-water in the Philippines. *Bulletin of the World Health Organization* 69(3), 305–17.

Noble, R. T., Moore, D. F., Leecaster, M. K., McGee, C. D. and Weisberg, S. B. 2003. Comparison of total coliform, fecal coliform, and *Enterococcus* bacterial indicator response for ocean recreational water quality testing. *Water Research* 37(7), 1637–43.

NRC (National Research Council). 2004. *Indicators for Waterborne Pathogens.* Washington DC: The National Academies Press.

Oh, S., Buddenborg, S., Yoder-Himes, D. R., Tiedje, J. M. and Konstantinidis, K. T. 2012. Genomic diversity of *Escherichia* isolates from diverse habitats. *PLoS One* 7(10), e47005. doi: 10.1371/journal.pone.0047005. Epub 2012 Oct 8.

Rosario, K., Symonds, E. M., Sinigalliano, C., Stewart, J. and Breitbart, M. 2009. Pepper mild mottle virus as an indicator of fecal pollution. *Applied and Environmental Microbiology* 75(22), 7261–7.

Savichtcheva, O. & Okabe, S. 2006. Alternative indicators of fecal pollution: relations with pathogens and conventional indicators, current methodologies for direct pathogen monitoring and future application perspectives. *Water Research* 40(13), 2463–76.

Scott, T. M., Rose, J. B., Jenkins, T. M., Farrah, S. R. and Lukasik, J. 2002. Microbial source tracking: current methodology and future directions. *Applied and Environmental Microbiology* 68(12), 5796–803.

Sobsey, M. D., Battigelli, D. A., Handzel, T. R. and Schwab, K. J. 1995. *Male-specific Coliphages as Indicators of Viral Contamination in Drinking Water.* Denver, CO: AWWA Research Foundation.

Wade, T. J., Sams, E., Brenner, K. P., Haugland, R., Chern, E., Beach, M., Wymer, L., Rankin, C. C., Love, D., Li, Q., Noble, R. and Dufour, A. P. 2010. Rapidly measured indicators of recreational water quality and swimming-associated illness at marine beaches: a prospective cohort study. *Environ Health* 9, 66. doi: 10.1186/1476-069X-9-66.

Wagner, E. G. and Lanoix, J. N. 1958. *Excreta Disposal for Rural Areas and Small Communities.* WHO Monograph series No 39 Geneva: WHO.

WHO. 2011. *Guidelines for Drinking-water Quality,* 3rd edn. Geneva: World Health Organization.

Wright, J. A., Yang, H., Walker, K., Pedley, S., Elliott, J. and Gundry, S. W. 2012. The H(2)S test versus standard indicator bacteria tests for faecal contamination of water: systematic review and meta-analysis. *Trop Med Int Health* 17(1), 94–105. doi: 10.1111/j.1365-3156.2011.02887.x. Epub 2011 Sep 22.

60

POLLUTANT TRANSPORT MODELLING*

David Kay

CENTRE FOR RESEARCH INTO ENVIRONMENT AND
HEALTH, ABERYSTWYTH UNIVERSITY, UK

Learning objectives

1 Distinguish between black-box and process-based pollutant transport models.
2 Design a data capture protocol to drive calibration of a black-box model with the assistance of statistically trained collaborators.
3 Evaluate the outcomes of the modelling strategy in terms of explained variance (R^2) and the degree of observed correct classification produced by the model.

Introduction

Modelling pollutant transport is an established tool for the design of remediation strategies intended to reduce impairment of both fresh and marine waters by anthropogenic inputs of infectious agents and/or toxic chemicals. Examples can be seen in the design and calibration of near-shore hydrodynamic and water quality models designed to limit microbial and chemical pollution of bathing and shellfish harvesting waters. Parallel models of terrestrial drainage basins are deployed to provide input data for the near-shore systems and these often address both the riverine systems and their contributing catchment areas and the artificial sewerage network, removing foul effluent and surface waters away from urban areas and delivering this flow to waste water treatment plants. These river and network models are generally treated as discrete systems but the network models are often used to provide modelled flows of treated effluent discharged to the river and/or local coastal waters. This demand for pollution transport models has produced an industry of consultancies building generic model platforms which are available to the designers of specific improvement schemes as well as utilities and regulators responsible for delivery

---

* Recommended citation: Kay, D. 2015. 'Pollutant transport modelling', in Bartram, J., with Baum, R., Coclanis, P.A., Gute, D.M., Kay, D., McFadyen, S., Pond, K., Robertson, W. and Rouse, M.J. (eds) *Routledge Handbook of Water and Health*. London and New York: Routledge.

of legally required environmental standards. Catchment and near-shore models are often integrated as modules in a family of modelling tools as is seen in the Danish Hydraulics Institute's Mike 11 (river catchment[1]), Mike Urban (network modelling) and Mike 21 (near-shore) commercial modelling products.

The example of near-shore water quality modelling

This infrastructure design function has been the main reason for the construction of pollutant transport models to date. However, the regulatory landscape is changing radically. The principal driver for this change is the novel application of pollutant transport models as a regulatory tool which was recommended in the Guidelines for Safe Recreational Water Environments published by the World Health Organization (WHO, 2003).[2] These incorporated the recommendations of an expert advisory group assembled by WHO in Annapolis, USA (WHO, 1999).[3] This group suggested the incorporation of pollutant transport model predictions into the regulatory process. This was justified, in the management of recreational waters, because the water quality parameters measured to determine both health risk and associated compliance of the bathing waters were, and are, enteric indicator bacteria, that is, coliforms and enterococci. Both groups suggest faecal contamination by humans and/or warm blooded animals and have been used since the late 19th century as measures of drinking water quality. Their measurement is based on growing the bacteria on a filter plate or in a nutrient broth, commonly termed 'culture methods'. The presence of faecal indicator bacteria does not prove the actual presence of a health risk but, if the human or animal source becomes infected and begins to shed pathogens in their faeces, then there will be a health risk to any human population exposed to the environment containing these faecal indicators by drinking water, ingesting water whilst bathing or, potentially, eating shellfish from such environments. In effect, therefore, faecal indicator bacteria prove the presence of connectivity (i.e. contamination) to a faecal source which might, at some point, present a health risk of pathogen contamination. There is, however, credible epidemiological evidence that those exposed to elevated concentrations of faecal indicator organisms during common exposures, such as bathing in fresh and marine waters, are more likely to get ill than control groups who are not so exposed (Kay et al., 1994; Pruss, 1998; Wiedenmann et al., 2006; Wade et al., 2006, 2008, 2010). Water quality regulators are generally national agencies with responsibility for a single nation but they may apply internationally agreed standards across broad regions such as the European Union (EU) where 27 nations have agreed to apply EU standards, defined in relevant directives which constitute legal requirements for member states, to protect public health at drinking, bathing and shellfish harvesting waters.

The EU Bathing Water Directive (EU, 2006)[4] was revised following the publication in 2003 of the WHO guidelines for recreational water quality (WHO, 2003) and the EU sought to incorporate the earlier WHO recommendations into the new directive. This applied to the acceptable levels of faecal indicators required to maintain public health (i.e. both organisations suggest that a 95 percentile value of 200[5] intestinal enterococci per 100ml of recreational water should be achieved for an acceptable bathing water) and, importantly, the EU was the first international authority to build predictive pollutant transport modelling into the legislative framework for bathing waters throughout the community. This will first apply in 2015 to almost 22,000 EU bathing waters.

The main public health rationale for the use of modelling and prediction, as part of a regulatory package, is to facilitate near-real-time prediction of the faecal indicator organisms and, by implication, health risk. This can be used to inform potential bathers, in real time,

Figure 60.1 Electronic signage used by Scottish Environmental Protection Agency (SEPA) at Scottish beaches to provide public information on bathing water quality

of the risk present at the bathing water, thus, 'informed choice' can be exercised by potential bathers. Traditional sampling, followed by analysis using culture methods, cannot deliver this real-time informed choice because the methods used to measure the faecal indicator bacteria take a minimum of 24, and commonly 48, hours to provide a definitive result. Thus, it has only been possible, to date, to inform a member of the public the day after the exposure took place that they had been exposed to unacceptable levels of pollution. The historical regulatory sampling has, therefore, been useful in defining the general quality of a bathing water at the end of a bathing season. This might guide future pollution management and remediation strategies but it has little relevance to the protection of public health within a bathing season, which tends to vary very rapidly in response to the drivers of microbial transport such as rainfall, river flow and sunlight irradiance.

The use of prediction, with associated real-time communication to the public using advisory signs on beaches and/or internet communication tools, has been deployed in the UK and is used as an active regulatory tool in Scotland (McPhail and Stidson, 2009; Stidson et al., 2012) where electronic signs are used to inform the public that adverse water quality may be expected (Figure 60.1). Parallel developments in the USA have resulted in software systems developed by the US Geological Survey (USGS) and US Environmental Protection Agency (USEPA[6]) for use by beach management agencies and, in a valuable recent contribution, Thoe et al. (2014) compared five different pollutant modelling approaches to prediction of bathing water quality at Santa Monica beach, California.

Whilst real-time pollution prediction has real concrete health benefits through communication of real-time risk levels supporting a novel 'predict and protect' approach to public health, it is also highly cost-effective when applied at a national scale. In the EU

(2006) application of this approach, member countries are allowed to 'discount' up to 15 per cent of sample results taken over a four year sampling period from the data sequence where: (i) they predicted and communicated adverse water quality conditions to the public for a specific bathing water through advisory signage, often combined with internet and smartphone alerts; and (ii) they have taken a replacement sample (because the scheduled sample is not used in the formal compliance assessment) for the scheduled sample predicted as adversely impaired by poor water quality. The samples so discounted do not have their water quality result counted within the percentile calculations required by the European Commission to determine the classification of the bathing water. This provision will ensure that the UK can retain its present number of Blue Flag awards.[7] Without this provision, the new microbial standards derived from the WHO (2003) guidelines as incorporated into the EU (2006) standards would mean that the UK would lose approximately 50 percent of its present blue flag awards, with unknown but significant economic impacts at coastal resorts using this award as part of their marketing. A parallel regulatory impact assessment,[8] completed for the UK Government to quantify the economic costs and benefits derived from the application of prediction and discounting within the 2006 Bathing Water Directive (EFTEC, 2002), suggested that the benefits of pollutant transport modelling designed to deliver prediction and discounting ranged from £1.5 to £5.4 billion.[9]

Models used for this type of real-time prediction to date have been relatively simple, using statistical association between potential drivers of microbial transport (termed the predictor or independent variables) and pollutant levels at some point used for compliance measurement (termed the dependent variable). These are often termed 'black-box models' because they do not require process relationships between predictors and dependents to be specified in the model calibration process. Often, multiple predictors are used which seek to characterise: (i) antecedent rainfall; (ii) flow in local rivers; (iii) tidal state at the time of sampling; (iv) tidal range on the sampling day or antecedent period; (v) irradiance received in an antecedent period; (vi) cloud cover; (vii) operation of sewerage infrastructure in an antecedent period (e.g. combined sewage overflow or storm tank overflows); (viii) water temperature; (ix) air temperature; (x) wind speed; (xi) wind direction; (xii) site specific hydrographic conditions such as upwelling; and (xiii) other water quality parameters measured at the site. This can result in many hundreds of potential predictors for a given site as different combinations are calculated for different antecedent conditions. The analyst then controls the selection of variables for inclusion into the prediction model or equation using the significance of the variable's correlation with the dependent variable and generally seeks to maximise the additional contribution or explanation provided by the new variable added to the model at each step in the analysis. This type of 'stepwise' inclusion is common for multivariate regression models where models are generally sought to limit the inter-correlation between predictors, termed multicollinearity. The model's predictive power is best described using the 'explained variance' (i.e. the variance in the dependent variable, generally faecal indicator concentration at the compliance site, which is explained by the combination of predictor variable entered into the equation). This statistic is the square of the multiple correlation coefficient between all the predictors and the dependent variable. It takes a value of +1.0 (perfect positive correlation) to −1.0 (perfect negative correlation) and is often expressed as a percentage where +1.0 is expressed as 100 per cent, indicating the percentage of variance in the dependent variable explained by the independent variables used in the analysis. This is a statistical measure of model quality but the most useful measure of model utility to the operational management community is given by the number of days in a bathing season or compliance period when the model correctly predicts the quality of the bathing water. Of particular management relevance is the number of misclassified days, that is, when the model predicts: (a) poor water quality when it is good; or (b) good water quality when it is poor. The former (a) is still protective, if inconvenient in discouraging use of the bathing water, whist the latter (b) is a more serious model error because use of the bathing water is not advised against when the water quality is actually poor.

In any area with public health significance, the predictive power of a management model is crucial and many models of this type reported in the literature have relatively low explained variance, possibly below 50 per cent, which would be associated with significant misclassification (Crowther et al., 2001). Such models are generally statistically significant if assessed at the accepted 0.05 probability that the model is not produced by chance factors. This suggests that a higher statistical threshold is appropriate for the acceptance of models where health protection is a principal management objective. Thoe et al. (2014) present a useful model evaluation framework applied to a range of black-box modelling tools.

It should be stated that there are significant limitations in the application of a black-box modelling approach to any environmental prediction problem. For many models using parametric statistical techniques, such as least squares multiple regression analysis, factor analysis or discriminant function analysis, these relate to the parametricity of the data used and its appropriateness for the modelling tools employed. This is beyond the scope of this introduction but should be addressed in any new model building effort. As significant is the veracity of the assumptions used in building the model and assessing its predictive capacity. The most significant is the utility of the dependent variable, in this case the water quality measured on a sampling day. An historical data sequence of 'compliance' data is the starting point for such modelling. The analyst often acquires data on the predictor variables, listed (i) to (xiii) above and constructs a data matrix where every row in the spreadsheet represents one sampling day's water quality and contains data for a set of predictors describing antecedent conditions prior to the sampling event. Often, 10 years of compliance data are used; thus, assuming 20 samples per bathing season, this would produce 200 data rows with many 10s of predictors.

At the core of this approach is the assumption that the compliance water quality measurement characterises the water quality on the bathing day and, by definition, that the water quality on any day is constant. There is growing evidence that this may not be the case (Boehm et al., 2002) and recent data from European beaches suggests that the range of faecal indicator concentration on most days averages 1000 fold on both wet and dry days and where bathing waters are very remote from pollution sources (Davies et al., 2008; Wyer et al., 2013). Associated modelling suggests that this variability is essentially diurnal and driven by bactericidal irradiance in the absence of rainfall or tidal drivers. The time in the bathing day when the compliance samples was taken therefore becomes an important consideration in compliance outcomes. This observation has great significance for any modelling effort. Faced with an actual 1000 fold variability within each bathing day, which would appear to the analyst as random variation, any attempt to model an assumed constant daily quality is likely to result in poor predictive power and hence misclassification. There is not sufficient detailed temporal within-day data available from multiple sites worldwide to assess whether this observation at EU beaches is replicated elsewhere but this observation is worthy of further study.

Many of these disadvantages, associated with simple black-box models, could be overcome by the application of more complex and process-based models such as the Mike 21 family linked to the Mike 11 family used to predict the input pollutant flux to the near-shore area of interest. The big advantage of this approach would be that real-time prediction could be directly linked to infrastructure design which would represent a significant advance over the black-box family of models outlined by Thoe et al. (2014). However, to date this has not been possible for two reasons. First, the data needed to quantify key parameters in the process-based models is insufficient properly to represent key model drivers such as microbial decay, often expressed as a T_{90} value or the time for 90 per cent decay of the faecal indicator bacteria. If the diurnal variation observed at EU bathing waters is a generic observation, it is likely that the assumption of constant daytime and night-time T_{90} values is far too simplistic for accurate model prediction. Other model parameters describing sedimentary cycling of microbial flux from and to the water column in response to riverine and tidal entrainment is also lacking empirical parameterisation at this time. Second, it is very difficult and time-consuming first to model the multitude of anthropogenic and livestock inputs to a given bathing water, thence to model the near-shore transport process for specific weather conditions prior to making a real-time prediction of the water quality at the compliance point. Recent UK investigations suggest that the inputs requiring river catchment and sewer network modelling can commonly exceed 50 inputs for a single compliance monitoring point, all with their own input data requirements. Thus, model run times often stretch to 10s of hours making real time prediction impractical at this point in time for most process-based modelling strategies.

Thus, most current attention in this area resides with a black-box approach. Recent work suggests that this can be adapted to predict the complex within-day pattern of faecal indicator organism concentration, facilitating much finer prediction at hourly intervals, which can drive real-time signage and internet information (Wyer et al., 2013). However, this does require intensive bespoke model build data rather than historical compliance data. In the case of Wyer et al. (2013) this model build data involved half-hourly sampling over 60 bathing season days with triplicate filtration of samples to enhance enumeration precision of the dependent variable (i.e. faecal indicator organism concentration). The resultant model with real-time bathing water management will result in this bathing water staying open. Previous historical compliance data may have been less precise as the model samples now comply with the standards in EU (2006), with the associated risk of de-designation.

Notes

1 http://www.mikebydhi.com/download/mike-by-dhi-2014
2 http://www.who.int/water_sanitation_health/bathing/srwe1/en/
3 http://whqlibdoc.who.int/hq/2010/WHO_HSE_WSH_10.04_eng.pdf
4 http://eur-lex.europa.eu/legal-content/EN/TXT/?uri=CELEX:32006L0007
5 See Table 4.7, Page 70 of WHO (2003) and Annex 1 of EU (2006) Page L64/46.
6 http://www2.epa.gov/exposure-assessment-models/virtual-beach-30-download-page
7 http://www.blueflag.org/
8 http://archive.defra.gov.uk/environment/quality/water/waterquality/bathing/documents/partial-ria.pdf
9 £1 billion is £1000 million.

References

Boehm, A.B., Grant, S.B., Kim, J.H., Mowbray, S.L., McGee, C.D., Clark, C.D., Foley, D.M., and Wellman, D.E., 2002. Decadal and shorter period variability of surf zone water quality at Huntington Beach, California. *Environ. Sci. Technol.* 36 (18), 3885–3892.
Crowther, J., Kay, D., and Wyer, M.D., 2001. Relationships between microbial water quality and environmental conditions in coastal recreational waters: the Fylde Coast, UK. *Water Res.* 35 (17), 4029–4038.

Davies, C., Kay, D., Kay, C., McDonald, A., Moore, H., Stapleton, C., Watkins, J., and Wyer, M., 2008. Microbial tracer study of selected inputs into the Guernsey coastal zone. A report to the States of Guernsey, Public services Department. CREH. 26p. Available online at: http://www.gov.gg/ccm/cms-service/download/asset/?asset_id=11226081

EFTEC, 2002. *Valuation of Benefits to England and Wales of a Revised Bathing Water Quality Directive and Other Beach Characteristics Using the Choice Experiment Methodology*. London: DEFRA.

EU (European Union), 2006. Directive 2006/7/EC of the European Parliament, concerning the management of bathing water quality and repealing Directive 76/160EEC. *Official J. Eur. Union* 64, 37–51.

Kay, D., Fleisher, J.M., Salmon, R.L., Jones, F., Wyer, M.D., Godfree, A.F., Zelenauch-Jacquotte, Z., and Shore, R. 1994. Predicting likelihood of gastroenteritis from sea bathing: results from randomised exposure. *Lancet* 344(8927), 905–909.

McPhail, C.D., and Stidson, R.T., 2009. Bathing water signage and predictive water quality models in Scotland. *Aquat. Ecosyst. Health & Manag.* 12 (2), 183–186.

Pruss, A., 1998. Review of epidemiological studies on health effects from exposure to recreational water. *Int. J. Epidemiol.* 27 (1), 1–9.

Stidson, R.T., Gray, C.A., and McPhail, C.D., 2012. Development and use of modelling techniques for real-time bathing water quality predictions. *Water Environ. J.* 26 (1), 7–18.

Thoe, W., Gold, M., Grisbach, A., Grimmer, M, Taggart, M.L. and Boehm, A.B., 2014. Predicting water quality at Santa Monica Beach: evaluation of five different models for public notification of unsafe swimming conditions. *Water Research* 67, 105–117. doi.org/10.1016/j.watres.2014.09.001

Wade, T., Calderon, R., Sams, E., Beach, M., Brenner, K., Williams, A., and Dufour, A., 2006. Rapidly measured indicators of recreational water quality are predictive of swimming-associated gastrointestinal illness. *Environ. Health Persp.* 114 (1), 24–28.

Wade, T., Calderon, R., Brenner, K., Sams, E., Beach, M., Haugland, R., Wymer, L., and Dufour, A., 2008. High sensitivity of children to swimming-associated gastrointestinal illness results using a rapid assay of recreational water quality. *Epidemiology* 19, 375–383.

Wade, T., Sams, E., Brenner, K., Haugland, R., Chern, E., Beach, M., Wymer, L., Rankin, C., Love, D., Li, Q., Noble, R., and Dufour, A., 2010. Rapidly measured indicators of recreational water quality and swimming-associated illness at marine beaches: a prospective cohort study. *Environ. Health* 9, 66. (http://www.ehjournal.net/content/9/1/66)

WHO, 2003. *Guidelines for Safe Recreational Water Environments. Coastal and Freshwaters, vol. 1.* Geneva: World Health Organization.

Wiedenmann, A., Krüger, P., Dietz, K., López-Pila, J.M., Szewzyk, R., and Botzenhart, K., 2006. A randomized controlled trial assessing infectious disease risks from bathing in fresh recreational waters in relation to the concentration of *Escherichia coli*, intestinal enterococci, *Clostridium perfringens*, and somatic coliphages. *Environ. Health Persp.* 114, 228–236.

Wyer, M., Kay, D., Morgan, H., Naylor, S., Govier, P., Clark, S., Watkins, J., Davies, C., Osborn, H., and Bennett, S., 2013. Statistical modelling of faecal indicator organisms at a marine bathing water site: results of an intensive study at Swansea Bay, UK. *Smart Coasts = Sustainable Communities*. August. https://www.aber.ac.uk/en/news/archive/2010/11/title-92900-en.html

61

GEOGRAPHIC INFORMATION SYSTEMS AND SPATIAL ANALYSIS*

Jim Wright

Senior Lecturer, University of Southampton, Southampton, Hampshire, UK

Learning objectives

1 Understand the functions of a Geographic Information System (GIS) that are most relevant to water and global health.
2 Describe the potential contribution of GIS to three areas of water and global health, namely water safety planning at local level, international monitoring of safe water access and the spatial analysis of water-related disease.
3 Recognise some of the emerging technical developments in GIS that could support the planning and provision of safe drinking-water.

Introduction

In public health, the importance of spatial patterns has long been recognised. The earliest and most well-known example of geographic patterns being analysed for public health purposes is John Snow's investigation of a cholera outbreak in nineteenth century London (McLeod, 2000). Snow's study mapped the locations of cholera deaths in Soho, a central London neighbourhood, providing evidence pointing towards a contaminated well in Broad Street as the outbreak's source. Geographical Information Systems (GIS) have a long history as a means of storing and analysing spatial data and today provide insights into water and public health in several ways (Cromley and McLafferty, 2011). One contribution is through the visualisation of geographic patterns through mapping of disease cases or hazard locations. A second contribution is through spatial analysis, for example in understanding whether cases of diarrhoeal disease are clustered spatially. Thirdly, GIS can combine data held in different map layers through map overlay, thereby enabling the analyst to examine the interrelationship between variables such as climate change, water resources, and the distribution of water-related disease, for example.

---

* Recommended citation: Wright, J. 2015. 'GIS and spatial analysis', in Bartram, J., with Baum, R., Coclanis, P.A., Gute, D.M., Kay, D., McFadyen, S., Pond, K., Robertson, W. and Rouse, M.J. (eds) *Routledge Handbook of Water and Health*. London and New York: Routledge.

A GIS is a digital data management system for the capture, storage, retrieval at will, analysis and display of spatial information (Schmandt, 2013). GIS is often used to bring together data about different aspects of a resource within a single management system. In the context of drinking-water, relevant map layers include water supply infrastructure, the geographic distribution of consumers, source catchments and contamination hazards (Kistemann et al., 2001). Despite their ability to integrate such data, digital spatial data can be costly to capture and need to be operated by specialist technical staff. For these reasons, in practice the widespread use of GIS in water management and safety planning is restricted to high income countries. In low income countries, GIS use is often restricted to donor-funded and research projects.

Water management and water safety planning at local level

Water safety planning is a water supply management methodology that helps to identify and control hazards and risks within a drinking-water supply system. Water safety plans cover all stages of water supply from the catchment, through treatment and distribution to the consumer (see Chapter 23 by Charles for further details) and both large and small systems. With its ability to integrate map layers depicting all of these supply components, GIS can be used to support water safety planning as outlined below.

Utility network analysis

One of the most fundamental management uses of a GIS is in recording the locations of water supply assets. For piped systems, the locations of treatment and storage facilities and piped infrastructure can be stored as a network of interconnected points and lines representing the various component parts. Within a GIS, a topological data structure is one that organises spatial data based on the principles of feature adjacency and connectivity. Within a topological structure, connectivity between sections of pipe is explicitly recorded, so that the manager can understand how the parts are interconnected. Analytically, such a data structure has a number of advantages (see Schmandt, 2013, Chapter 5 on network analysis). In particular, it enables network tracing to be undertaken, such that interconnected sections of a reticulated network can be identified. For example, if a contaminant is detected at a monitoring point within a network, it is possible to use topology to trace those parts of a piped distribution system that are upstream of the relevant monitoring point, and thereby provide guidance on potential contamination sources (e.g. Trepanier et al., 2006). However, network analysis techniques rely on the existence of an up-to-date, accurate spatial database depicting the location of the pipes, valves, treatment units, tanks and other elements of a distribution system. In many settings, this information may not exist and so the potential benefits of network analyses may remain unrealised.

Water point mapping

For the small-scale community-managed supplies common in rural areas of low and middle income countries, GIS also has a potential role to play. WaterAid have pioneered a technique known as water point mapping (Jimenez and Perez-Foguet, 2008), in which the locations of boreholes, protected wells and other point sources can be systematically mapped. Waterpoint mapping provides a basis for mapping contamination, identifying the spatial distribution of non-functional supplies, assessing patterns of household drinking-water access (e.g. by calculating the distance to the nearest groundwater source), and in targeting investments to upgrade existing water points or develop new water points. The technique does, however, require investment to underpin the costs of both field surveys and database design and maintenance.

Mapping hazards

Beyond asset inventory, GIS provides a mechanism for identifying and quantifying hazards. Water safety plans involve risk management from catchment right through to consumer, and the identification of catchment hazards is a key part of such a plan. Catchment land use, livestock densities, and patterns of human access within a catchment are all mappable within a GIS (Foster and MacDonald, 2000). In this context, the use of GIS also relates to Integrated Water Resource Management, the idea that water resource management for domestic purposes should be balanced against other uses in a sustainable and equitable manner. GIS can also be used for hazard identification in relation to small-scale community supplies. For example,

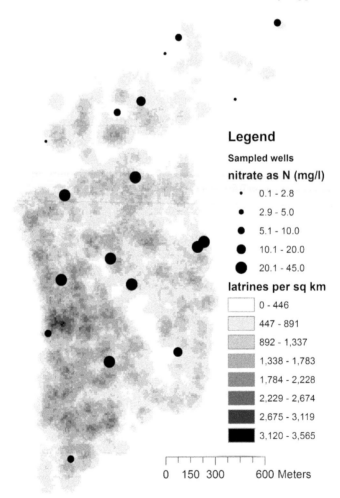

Figure 61.1 Interpolated nitrate concentrations in groundwater and distribution of pit latrines in a suburb of Kisumu, Kenya

Source: Derived from Wright et al. (2012).

Figure 61.1 shows nitrate levels in shallow wells in two suburbs of Kisumu, Kenya, in relation to a potential hazard, namely the local density of pit latrines within the two neighbourhoods.

GIS can also be used to design and plan monitoring networks. Figure 61.2, for example, shows the spatial distribution of surveillance samples required from urban piped supplies in central Colombia. Not only can monitoring sites be randomly selected or identified based on stratification of a given area, but spatially balanced monitoring designs can be produced. Such designs seek to maximise the spatial independence of sample locations, such that each monitoring site produces more useful information, making for a more efficient monitoring network overall.

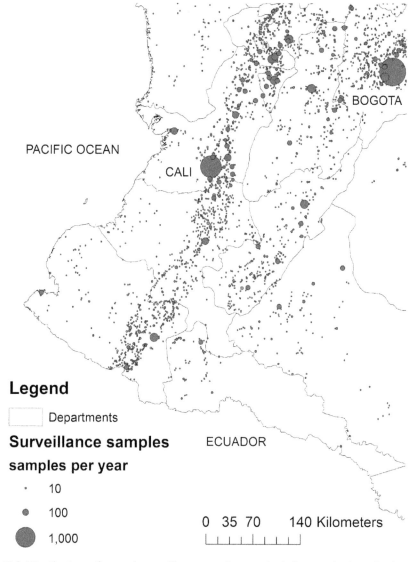

Figure 61.2 Distribution of annual surveillance samples required for monitoring piped water in southwestern Colombia

Source: Adapted from Wright et al. (2014).

International monitoring of safe water access

Beyond these more local level uses of GIS, the technology also has a role to play at the more strategic, national and international level where it can inform policy. One of the key functions of many GISs is the ability to interpolate data from known measurement points and estimate the values of a parameter at an unknown location. Interpolation relies on the property of spatial autocorrelation, commonly encountered with many environmental and socio-economic phenomena, by which neighbouring locations have more similar characteristics than locations that are distant from one another. Amini et al. (2008), for example, used interpolation techniques to estimate the global distribution of the geogenic contaminants arsenic and fluoride, thereby helping to quantify potential population exposure to these hazards via groundwater.

Examination of inequalities in household access to adequate sanitation and safe water suggests that these often strongly relate to location, particularly the gap between rural and urban areas (Molinas et al., 2010). Measurement of inequalities is gaining prominence in international monitoring, with a renewed emphasis on targeting service delivery at the poorest households following the expiry of the Millennium Development Goals post-2015. GIS has long been used to measure interregional variations in material deprivation (e.g. lack of access to services, employment and housing), and such techniques can also be applied to the measurement of inequalities in access to safe water and sanitation. One such measure is the dissimilarity index (see Molinas et al., 2010: pp. 41–47), which measures the percentage of population who would need to change location in order for there to be a completely even distribution of safe water access. GIS-based analyses of safe water access necessarily have to focus on those aspects of water safety where spatial data are available. However, in doing so, potentially important aspects of water access may be overlooked: for example, nationally representative spatial data sets on drinking-water quality do not exist in many countries. Furthermore, classifications of water source type can vary between data sources, making international comparisons challenging.

Spatial analysis of water-related disease

As well as being used to examine spatial patterns in water supply characteristics and catchment hazards, GIS also forms a tool for the analysis of water-related disease. Drawing on John Snow's pioneering study of cholera in London, GIS is one of several tools that can be deployed as part of an outbreak investigation. In particular, spatial analysis can be used to identify whether the number of reported cases of a water-related disease in a given part of a study area is unexpectedly high and forms a so-called 'disease cluster'. One of the most widely used spatial analysis tools, the spatial scan statistic, systematically examines subareas within a study location, assessing the number of disease cases in each subarea relative to the population at risk. If the number of cases exceeds that which would be expected under a prior statistical distribution, such as the Poisson distribution, then the subarea can be flagged as a potential disease cluster. Sarkar et al. (2007) used the spatial scan statistic alongside local mapping of risk factors such as open defecation to understand the causes and nature of an outbreak of acute diarrhoea in a southern Indian village. Elsewhere, Carrel et al. (2009) used the spatial scan statistic to examine variation in spatio-temporal patterns in cholera cases in Bangladesh in response to flood control.

The application of GIS to water-related disease analysis varies depending on the nature of the disease. For dracunculiasis (guineaworm), the World Health Organization is coordinating international efforts to eradicate the disease, which is now concentrated in West Africa. In supporting disease eradication, GIS can be used to target surveillance efforts and active case detection on the ground. Clarke et al. (1991) used GIS and remote sensing to identify remote villages in areas where dracunculiasis was endemic, so control measures like water filtration could be targeted towards populations that might otherwise have been missed.

Cholera is perhaps the water-borne disease that has been the greatest focus of GIS-based work. Following recognition of the symbiotic relationship between cholera vibrios and copepod species of zooplankton, several studies have examined the relationship between reported cholera cases and remotely sensed variables such as sea surface temperature and marine vegetation indices, the latter being indicative of conditions favourable to zooplankton (de Magny et al., 2008). Such studies not only provide a potential early warning system for cholera outbreaks, but also provide one means of modelling the possible impact of climate change on the distribution of water-related disease. For example, GIS and remote sensing provide a means of linking the ecological niche occupied by the cholera vibrio to predicted sea surface temperatures under various emissions scenarios.

New developments in geospatial technology

The emergence of new geospatial data sets and technologies provides the water and health community with a potential new set of tools for tackling water-related problems. These technologies include remote sensing, online collaborative tools, ground-based sensors, web mining and cell phone technology.

Exploratory research projects are also investigating the ability of groups of users to generate geospatial data through online collaboration, with OpenStreetMap perhaps the most well-known online collaborative mapping project. In Kenya, the high density neighbourhood of Kibera was among the first where water and sanitation data were mapped through an online collaboration of residents (http://www.mapkibera.org/) via the Map Kibera Trust. Similar collaborative mapping efforts include the MajiData initiative, which has collated community-generated spatial data sets depicting, for example, perceived drinking-water quality (http://www.majidata.go.ke/). However, such approaches require investment to promote their use and systems for collaborative data capture. The data produced may also be variable in quality and patchy in coverage.

Alongside this, the proliferation of cell phone coverage and use across many parts of Africa provides a means of updating spatial databases in areas where internet access and desktop computer use are limited. Rivett et al. (2013) used a cell phone-based information system to transmit water quality test results in four rural South African municipalities, adapting the information system design in response to local conditions. Thomson et al. (2012) piloted the use of a cell phone network to transmit data from devices attached to rural handpumps, so as to monitor groundwater use.

There are other emerging, novel techniques for capturing geospatial data relevant to water and health. Griffith et al. (2006) used postings from an online public health forum known as ProMed to assess the spatio-temporal distribution of cholera in Africa, with others exploring ways of using web mining to capture information from such sources.

Summary

GIS is an integrative technology that provides a means of bringing together disparate data in a single, integrated geospatial database. As such, it can support a wide range of public health applications in relation to water and global health, such as identifying hazards within catchments, managing water supply systems, measuring inequality in safe water access, and assessing population exposure to contaminants in domestic water. It also provides tools for outbreak investigation, early warning and water-related disease control efforts. Some of the barriers to uptake of these tools in low and middle income countries are a comparative lack of both geospatial data and technical capacity. However, some recent developments – such as cell phone technology and online collaborative database development – may provide a means of overcoming these obstacles and realising the potential benefits of geospatial technology for water and public health.

Acknowledgements

Joseph Okotto-Okotto (Victoria Institute for Research on Environment and Development International, Kenya), Aidan Cronin and Steve Pedley (University of Surrey, UK), Stephen Gundry (University of Bristol, UK), and Hong Yang (University of Southampton, UK) contributed to the study underpinning Figure 61.1. Jenny Liu (University of Southampton, UK), Rob Bain, Jamie Bartram and Jonny Crocker (University of North Carolina) and Andrea Perez (Universidad de Boyacá, Colombia) contributed to the research underpinning Figure 61.2.

Key recommended readings (open access)

1 Schmandt, M. (2013) *GIS Commons: an Introductory Textbook on Geographic Information Systems*. Available at http://giscommons.org/, accessed 10 Sept 2014. This resource is a freely available, online, open source textbook that explains the principles of GIS to those new to the technology.

2 Cromley, E. and McLafferty, S. (2011) *GIS and Public Health Supplemental Exercises*. Available at http://www.guilford.com/cgi-bin/cartscript.cgi?page=add/cromley/index. html, accessed 10 Sept 2014. If you wish to learn to use the popular ArcGIS software, this companion website to the helpful Cromley and McLafferty textbook provides a series of health-related GIS practical instructions. Exercises 1 to 6 provide a particularly useful introduction to the software.

3 WaterAid (2013) *Water Point Mapper*. Available at http://www.waterpointmapper.org/, accessed 10 Sept 2014. This site provides software and resources, including a manual for mapping water points and sanitation, and provides an easily accessible resource for those wishing to develop geospatial WatSan databases in rural areas in particular.

References

Amini, M., Abbaspour, K., Berg, M., Winkel, L., Huq, S., Hoehn, E., Yang, H., and Johnson, C. (2008) 'Statistical modeling of global geogenic arsenic contamination in groundwater', *Environmental Science and Technology* Vol. 42, No. 10, pp. 3669–3675.

Carrel, M., Emch, M., Streatfield, P., and Yunus, M. (2009) 'Spatio-temporal clustering of cholera: the impact of flood control in Matlab, Bangladesh, 1983–2003', *Health and Place* Vol. 15, No. 3, pp. 771–782.

Clarke, K.C., Osleeb, J.P., Sherry, J.M., Meert, J.P., and Larsson, R.W. (1991) 'The use of remote sensing and geographic information systems in UNICEF's dracunculiasis (guineaworm) eradication effort', *Preventive Veterinary Medicine* Vol. 11, No. 3–4, pp. 229–235.

Cromley, E. and McLafferty, S. (2011) *GIS and Public Health*, Guilford Press, New York, 2nd edition.

De Magny, G., Murtugudde, R., Sapiano, M., Nizam, A., Brown, C., Busalacchi, A., Yunus, M., Nair, G., Gil, A., Lanata, C., Calkins, J., Manna, B., Rajendran, K., Bhattacharya, M., Huq, A., Sack, R., and Colwell, R. (2008) 'Environmental signatures associated with cholera epidemics', *Proceedings of the National Academy of Sciences of the USA* Vol. 105, No. 46, pp. 17676–17681.

Foster, J. and MacDonald, A. (2000) 'Assessing pollution risks to water supply intakes using geographical information systems (GIS)', *Environmental Modelling and Software* Vol. 15, pp. 225–234.

Griffith, D., Kelly-Hope, L., and Miller, M. (2006) 'Review of reported cholera outbreaks worldwide, 1995-2005', *American Journal of Tropical Medicine and Hygiene.* Vol. 75, No. 5, pp. 973–977.

Jimenez, A. and Perez-Foguet, A. (2008) 'Improving water access indicators in developing countries: a proposal using water point mapping methodology', *Water Science and Technology – Water Supply* Vol. 8, No. 3, pp. 279–287.

Kistemann, T., Herbst, S., Dangendorf, F., and Exner, M. (2001) 'GIS-based analysis of drinking-water supply structures: a module for microbial risk assessment', *International Journal of Hygiene and Environmental Health* Vol. 203, pp. 301–310.

McLeod, L. (2000) 'Our sense of Snow: the myth of John Snow in medical geography', *Social Science and Medicine* Vol. 50, No. 7–8, pp. 923–935.

Molinas, J., Paes de Barros, R., Saavedra, J., and Giugale, M. (2010) Do our children have a chance? The 2010 human opportunity report for Latin America and the Caribbean. Washington, DC: The World Bank.

Rivett, U., Champanis, M., and Wilson-Jones, T. (2013) 'Monitoring drinking water quality in South Africa: designing information systems for local needs', *Water SA* Vol. 39, No. 3, pp. 409–414.

Sarkar, R., Prabhakar, A.T., Manickam, S., Selvapandian, D., Raghava, M.V., Kang, G., and Balraj, V. (2007) 'Epidemiological investigation of an outbreak of acute diarrhoeal disease using geographic information systems', *Transactions of the Royal Society of Tropical Medicine and Hygiene* Vol. 101, No. 6, pp. 587–593

Schmandt, M. (2013) *GIS Commons: An Introductory Textbook on Geographic Information Systems.* Available at http://giscommons.org/, accessed 10 Sept 2014.

Thomson, P., Hope, R., and Foster, T. (2012) 'GSM-enabled remote monitoring of rural handpumps: a proof-of-concept study', *Journal of Hydroinformatics* Vol. 14, No. 4, pp. 829–839.

Trepanier, M., Gauthier, V., Besner, M.-C., and Prevost, M. (2006) 'A GIS-based tool for distribution system data integration and analysis', *Journal of Hydroinformatics* Vol. 7, pp. 13–24.

Wright, J.A., Cronin, A., Okotto-Okotto, J., Yang, H., Pedley, S., and Gundry, S.W. (2012) 'A spatial analysis of pit latrine density and groundwater source contamination', *Environmental Monitoring and Assessment* Vol. 185, No. 5, pp. 4261–4272.

Wright, J.A., Liu, J., Bain, R., Perez, A., Crocker, J., Bartram, J., and Gundry, S. (2014) 'Water quality laboratories in Colombia: a GIS-based study of urban and rural accessibility', *Science of The Total Environment* Vol. 485–486, pp. 643–652.

62

DEMAND ASSESSMENT AND VALUATION[*]

Marc Jeuland

ASSISTANT PROFESSOR, SANFORD SCHOOL OF PUBLIC POLICY AND DUKE
GLOBAL HEALTH INSTITUTE, DUKE UNIVERSITY, DURHAM, NC, USA

Learning objectives

By the end of this chapter, readers will be able to:
1 Explain the concept of demand, and how it relates to private and social marginal benefit.
2 Describe the challenges associated with application of the standard approach to demand assessment of water, sanitation and hygiene (WASH) goods and services.
3 Assess the relative strengths and weaknesses of alternative revealed and stated preference approaches for measuring demand of WASH goods and services.

Introduction

In economics, an individual's *demand* is generally deemed to be equivalent to his/her willingness (and ability) to pay (WTP) for a specific unit of a good or service. The concept of demand is theoretically simple; in practice, however, measuring demand poses a number of challenges. These challenges can be broadly grouped into three categories: 1) insufficient information on how WTP varies over a policy-relevant range of quantities; 2) an inability to observe demand in competitive markets for the good or service of interest, because markets are missing or highly distorted; and 3) the existence of external effects that cause deviations between private and social WTP.

This chapter explains the economic concept of demand, and then describes briefly a variety of approaches – specifically market-based (including those simulated in randomized controlled trials (RCTs)) and nonmarket valuation techniques – that economists use to measure WTP. As will be discussed, all three of the problems identified above are common for goods related to water, sanitation and hygiene (WASH). Those seeking to measure and understand demand for WASH goods and services should therefore exercise caution in applying these alternative methods, and should carefully consider their relative strengths and weaknesses.

[*] Recommended citation: Jeuland, M. 2015. 'Demand assessment and valuation', in Bartram, J., with Baum, R., Coclanis, P.A., Gute, D.M., Kay, D., McFadyen, S., Pond, K., Robertson, W. and Rouse, M.J. (eds) *Routledge Handbook of Water and Health*. London and New York: Routledge.

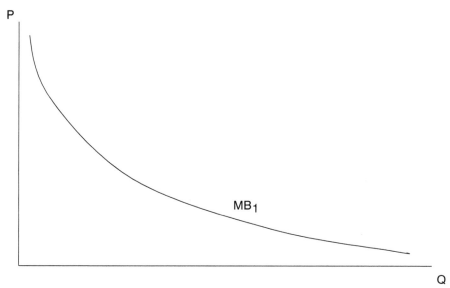

Figure 62.1 A typical downward-sloping demand, or marginal benefit (MB$_1$), curve. Note: The y-axis is the price of a good, and the x-axis is the quantity demanded

Methods for conventional demand assessment and applications to WASH

The economic concept of demand

In economics, an individual's *demand* is equivalent to their willingness (and ability) to pay for a specific quantity of a good or service.[1] In addition, when considering a good or service that is available at a specific price of p, an individual i's (or market) demand is the quantity q_i (or $Q = \sum_i q_i$ for all individuals in that market) that is purchased at that price. Due to the concept of diminishing marginal utility whereby each additional unit of a good consumed contributes reduced utility compared to previously consumed units, a typical *demand curve* is downward sloping with respect to price. In other words, higher prices lead to lower purchases of the good or service by an individual (and market) (Figure 62.1). For example, a consumer of water places greater value on the first units of water supply they use, which might be used for drinking, cooking, and basic hygiene purposes, than on later ones allocated to a flower garden. A demand curve is frequently referred to as a *marginal benefit curve*, where marginal benefits are the benefits derived from the consumption of the last unit of the good that is purchased at price p. For a normal good, all other units purchased prior to this final unit would have been purchased at a higher price than p and therefore provide benefits greater than the marginal benefits derived from this final unit.[2]

A market demand curve is obtained by adding individual demand curves horizontally. If the good or service is a pure private good whose consumption only affects the purchasing consumer (in terms of benefits), this market demand curve will also be equal to the *social benefit curve*, that is, the totality of the societal benefits (including effects on non-purchasers) derived from the consumption of the last unit of the good that is purchased in that market. When this is not the case, as in the case of a good whose consumption generates positive or negative externalities that spill over to non-consumers, the measurement of the social benefit curve (or true social demand for a good or service) will deviate from the market demand curve.

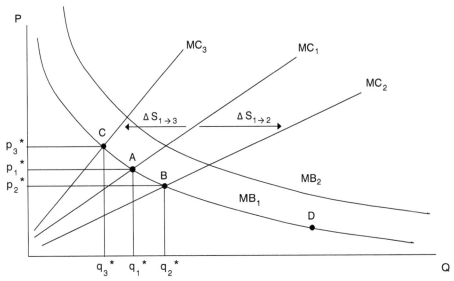

Figure 62.2 Shifting supply or marginal cost (MC) curves allows one to observe movements along a demand curve MB$_1$, as shown by the effects of the shift from point A to points B and C, all of which lie along the original demand curve MB$_1$. MB$_2$ represents an outward shift in demand

Returning to our water supply example, this could occur if one individual's water consumption for hand washing decreases the risk of disease transmission to others in the same community.

Measuring demand in conventional (competitive) markets without externalities

In conventional competitive markets without externalities, the demand curve is typically measured by economists through construction of a demand schedule. Put simply, a demand schedule relates different price levels p_j to different quantities Q_j purchased at that price:

$$Q_j = f(p_j). \tag{62.1}$$

Constructing a demand curve requires measurement of a sufficient number of such price–quantity combinations to have confidence that the true nature of the relationship has been detected, which can be difficult in practice. In a competitive market, the price–quantity pairs that are observed are actually co-determined by an equilibrium between supply and demand. Thus in order to observe movements along the demand curve, one must rely on a level shift in the supply curve as shown in Figure 62.2. The fact that demand curves may also shift up or down means that it is not always easy to determine whether an observed change in market prices was due to movements of the supply curve (desired for tracing out demand and shown by movement from A to B or C) or the demand curve (undesirable since what is being measured in such cases are points on multiple demand curves, for example as shown by movement to MB$_2$). In addition, in the case of many goods and services, this equilibrium does not typically move very far from the initial market price, so it may not be easy to observe large portions of the demand curve (for example point D).

Nonetheless, these difficulties do not necessarily present problems for the measurement of marginal benefits, which do correspond to the observed price in a competitive market.

This is important given that the relevant concept for analyzing the benefits associated with an additional unit of consumption of the good or service is that of *marginal benefits*. Thus, unless a policy or supply-based intervention results in a very large movement away from the existing equilibrium, this measure of marginal benefits will be the correct one for valuing changes in consumption.

Applications of the standard approach to WASH

There are very few examples of the standard approach to developing a demand schedule for WASH goods and services, for several reasons. First, the economics of water are such that many WASH services are not provided in competitive markets, because of a variety of well-known distortions or market failures (Hanemann 2006). For example, many infrastructure-intensive aspects of water supply are subject to large economies of scale and therefore evolve towards natural monopoly, for example in utility or other community-based water supplies (Whittington et al. 2009). This situation leads to deviations from the typical competitive equilibrium conditions that equate price and marginal benefit. There is also a strong resistance among consumers and policy-makers to thinking of WASH services as commodities, because water is deemed to be a unique and extraordinary good that is "essential for life" (Savenije 2002). Such views are often reflected in the way WASH service provision is regulated and/or delivered to consumers (e.g., large subsidies or distribution of such services free of charge are not uncommon). Finally, WASH goods and services typically have strong external effects, for example when they provide or compromise significant non-use benefits (e.g., those arising from precious ecosystems), or generate spillovers (e.g., via reduction of community-level pollution or provision of environmental health benefits) (Carson and Mitchell 1993, Pattanayak and Pfaff 2009, Whittington et al. 2012).

Second, there are critical challenges in establishing market demand curves for many WASH goods and services that are relevant for planning purposes. Even where one can observe several pairs of price–quantity combinations for a particular good, these may not cover a range that allows extrapolation over a range of non-marginal changes that are often implied by large-scale improvements (in Figure 62.2, a shift from point A to D, whereby assuming $MB = p_2{}^*$ for all additional units would clearly overestimate total benefits).[3] In addition, many WASH goods and services, especially for infrastructure-heavy investments, deliver benefits over a long and dynamic time horizon, which requires projection and anticipation of shifting future demand, an exercise that is fraught with uncertainty (Jeuland and Whittington 2013).

Moreover, demand curves that are measured by observing market transactions may not correspond to those for the WASH goods and services in question. For example, one might attempt to evaluate the demand for a new filter technology at a particular time and place by looking to purchases of other water treatment products, but the new filter may provide different net benefits from those other technologies, for example changing the taste of water in some important way, or providing more effective protection against particular organisms (or chemical contaminants) that are of concern to health (Lantagne et al. 2008, Jeuland et al. 2015). Due to these various issues, the concept of marginal benefits obtained in the conventional way may mischaracterize demand for the specific WASH good or service being valued.

Despite these challenges, the recent trend towards the use of more RCTs in economics has allowed researchers to measure the demand for some (but not all) types of WASH goods and services directly using an experimental approach. The number of such studies is still small and the generalizability of their findings is unknown (Null et al. 2012), but the approach appears promising for better understanding demand for WASH technologies and interventions, especially because of the strong internal validity of the estimates that are obtained. In addition, RCTs allow testing of the extent to which this demand can be stimulated (equivalent to outward shifts in the demand curve) by marketing, peer pressure, or simple information provision (Jalan and Somanathan 2008, Brown et al. 2013). Important challenges of course remain with measuring demand for WASH goods and services (e.g., urban piped water supply) for which RCTs (and even quasi-experimental research designs) are more difficult to implement. There may also be ethical concerns with using RCTs to measure demand when higher prices are deemed to decrease adoption of a beneficial good.

Methods for nonmarket valuation of benefits from WASH interventions

Given the challenges associated with obtaining data on the demand for WASH goods and services, economists must often rely on other techniques for estimating benefits. Some of the methods are based on revealed behavior while others are based on surveys. These are commonly called revealed and stated preference methods.

Revealed preference methods

The basic insight behind revealed preference (RP) approaches to demand estimation rests in the idea that the value of a nonmarket good or service may be reflected indirectly in the market for a related good. Through analysis of this related market, an analyst can estimate the value of the nonmarket good.

The simplest form of RP estimation of demand considers nonmarket goods – for example latrines whose construction is highly subsidized by the government – that may have very close substitutes in the private market – such as similar latrines constructed by the private market. The key difficulty in using private values in such cases is to make sure that the individuals benefitting from the good in question are generally similar to those they are being compared with. In general one would expect that these individuals' WTP would be higher than the subsidy price (otherwise they would not construct the subsidized latrines) but also somewhat lower than the price of the private sector latrines (which would make the subsidies unnecessary). Similar considerations might apply to analysis of data on vending of drinking water in developing countries (Whittington et al. 1991).

Another RP approach is to look at real tradeoffs that are relevant to estimation of demand for a particular good. In the context of WASH interventions that provide mainly health benefits (e.g., a promotion of a new filter for water treatment that is directly attached to a water tap), the tradeoff methodology might look to the averting expenses that a household makes to avoid or cope with the negative health effects of diseases that are directly related to poor drinking water quality (Pattanayak et al. 2005). Such averting expenses might include the cost of prevention behavior such as boiling drinking water or cleaning household storage containers that hold boiled water. In general, the averting expenditures methodology will underestimate benefits, however. This is because interventions that reduce averting expenses, such as the filter intervention, apply a very narrow definition of benefits (Whittington et al. 2009), and represent an outward shift in the supply of a good such as safe water, which effectively lowers its price and thus increases the quantity consumed (Figure 62.3). The approach does not properly account for the net benefits associated with this extra consumption (Boardman et al. 2011).

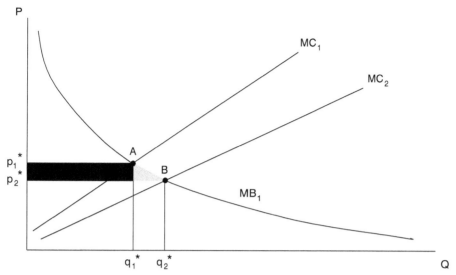

Figure 62.3 A new technology that increases the quantity of household health represents an outward shift in supply, which reduces the amount spent on averting expenditures to produce health up to $q_1{}^\star$ by the area shown in black. This also increases averting expenses and quantity of health up to $q_2{}^\star$; these extra net benefits (which in a naïve analysis look like net costs) are shown by the triangle

Two additional RP methods deserve mention, although detailed discussion of them is beyond the scope of this chapter. The first is the travel cost method, an approach that considers a complementary market – actual revealed expenses on travel to a location where specific benefits can be obtained – to make inferences about how much people really want those benefits. This method has been most widely applied to consider the value of recreational water use in developed countries since consumers are often willing to travel from far away to enjoy such environmental assets. It has also been used for valuation of improved water quality and health (Choe et al. 1996, Jeuland et al. 2010). The second is the hedonic pricing method, which uses multivariate regression to relate the price of some capitalized asset (e.g., property or wages) to the attribute of interest (e.g., reduced mortality risk from safer jobs, or better water/air quality in a particular town). This approach has also been applied for valuing environmental assets in developed countries, and for assessing the net benefits of better water supply and sanitation services in developing countries (Arimah 1996, Yusuf and Koundouri 2005).

Stated preference methods

The main alternatives to RP methods for assessing the demand for nonmarket goods and services are stated preference (SP) approaches: contingent valuation and stated choice. SP methods rely on surveys to ask individuals to make *hypothetical* tradeoffs between money and a particular improvement. Similarly to the more sophisticated RP methods, well-designed SP studies are challenging to implement. Yet they have been applied to a wide variety of valuation problems and have been the subject of an extensive literature focused on methodological aspects. Despite this extensive literature, many economists today remain skeptical of SP methods, however. Their critiques most frequently revolve around concerns over hypothetical bias, or the idea that respondents to survey questions will not answer truthfully given that no real money is at stake (Diamond and Hausman 1994).

In many situations, however, SP methods offer the only way to fully or even partially value a particular good or service (Boardman et al. 2011). Furthermore, comparisons of estimates obtained from well-designed SP and RP studies have not been found to diverge as much as one might expect (Carson et al. 1996). The most commonly applied SP approach, and the one most often applied in the context of demand for WASH improvements, is the *contingent valuation method* (CVM) (Whittington et al. 1990). When implementing this approach, researchers most often use the dichotomous choice method to minimize hypothetical bias and maintain incentive compatibility. In this approach, researchers randomly assign different price levels for a well-described improvement to different respondents in a sample. If the survey sample is sufficiently large, this allows the measurement of quantity–price pairs at different points along the demand curve without compromising the design by inducing strategic responses. A variety of other important issues for researchers conducting CVM studies to consider include selection of a realistic payment vehicle (particularly in the case of public goods which require a referendum vote design – see Carson et al. (1998)), minimization of non-response bias or rejection of the hypothetical scenario, and a whole set of potential biases arising from aspects related to survey design.

The main SP alternative to CVM that is used today is the stated choice or *discrete choice experiment* (DCE) (Louviere et al. 2000). This approach has long been used to assess consumer preferences for goods and services across multiple dimensions, and is increasingly applied in environmental valuation. In the context of studying demand for nonmarket goods such as WASH services for which well-developed markets do not exist, a particular advantage of DCE preference elicitation is to allow consumers to explicitly consider tradeoffs between hypothetical alternatives with varying levels of different attributes, such as quality, quantity, convenience, reliability, taste, and so on. Analysis of the responses then allows researchers to back out the WTP for changes in these specific attributes. In the WASH sector, DCEs have most often been applied to analyze aspects related to domestic or drinking water supply (Yang et al. 2007, Jeuland et al. 2015).

Brief reflections on alternative approaches to demand assessment

There are obviously advantages and disadvantages to each of the approaches to demand assessment discussed in this chapter. An exhaustive comparison of these strengths and weaknesses of the different methodologies is beyond the scope of this chapter, but Table 62.1 presents a brief summary of the main issues that have been discussed.

Though it is arguably most accurate for truly understanding demand, the direct market measurement approach has the least applicability to elicitation of WTP for WASH goods and services, primarily because few if any of the goods are actually available in competitive markets. Simulated markets using RCTs offer advantages that researchers can use on a small scale to assess demand. Then, RP methods for nonmarket valuation offer a middle road between direct measurement and SP approaches; these seek to balance the strengths of observing actual decisions made by individuals with the inability to observe competitive markets for such goods. Finally, SP methods, while potentially most flexible and relevant for measuring the WTP for WASH goods and services, may suffer from a variety of problems ranging from hypothetical bias to a general lack of realism that affects responses by those answering a survey.

Table 62.1 Comparison of methodologies for demand assessment

| Feature | Direct measurement | | Nonmarket valuation | |
	Real markets	Using RCTs	RP methods	SP methods
Basic approach	Measure price–quantity point(s); extrapolate	Test a range of prices and observe quantities that result	Observe demand through proxy markets	Construct relevant market through scenario
Policy relevance	Relevant if match between market good and policy good is close	Relevance depends on external validity of RCT result	Relevant if match between proxied good and policy good is close	Potentially most relevant; can be tailored to desired situation
Observed range of demand	Usually limited	Potentially extensive	Varies by approach	Potentially extensive
Types of values included	Generally only use value	Generally only use value	Use value only	Use and non-use value
Hypothetical or real	Real	Real	Real	Hypothetical
Capacity for dealing with market failure	Requires adjustments for price distortions and externalities	Requires adjustments for externalities	Requires adjustments for externalities	Can capture social demand (via referendum design)
Applicability to WASH goods and services	Limited applicability	Moderate applicability	Moderate applicability	Greatest applicability

Summary

This chapter presented the economic concept of demand, including the distinction between individual and social marginal benefits. Following this description, we discussed a variety of approaches for measuring demand, beginning with direct measurement and estimation of the relationship between price and quantity in conventional markets, or using randomized controlled trials. We then discussed a variety of nonmarket methodologies ranging from revealed preference (markets for close substitutes, averting expenditures, travel cost, and hedonics) and stated preference (contingent valuation and stated choice) approaches. We concluded with brief reflections and a partial summary of the strengths and weaknesses of these alternatives. In the context of WASH, all of the approaches reviewed have important shortcomings. Analysts working to understand the economic benefits of such goods and services therefore need to be mindful of such limitations when considering their implications for use in economic analyses of policies designed to promote WASH interventions.

Key recommended readings

1 Hanemann, W. M. (2006). "The economic conception of water". *Water Crisis: Myth or Reality*, 61–91. Available at: http://escholarship.org/uc/item/08n4410n.pdf
2 Null, C., M. Kremer, et al. (2012). Willingness to pay for cleaner water in developing countries: systematic review of the experimental evidence, 3ie. Available at: http://www.indiawaterportal.org/sites/indiawaterportal.org/files/sr006.pdf
3 Whittington, D., W. M. Hanemann, C. Sadoff and M. Jeuland (2009). *The Challenge of Improving Water and Sanitation Services in Less Developed Countries*, Now Publishers Inc. Available at: http://www.researchgate.net/publication/228385588_The_challenge_of_improving_water_and_sanitation_services_in_less_developed_countries/file/9fcfd509d4085c9e18.pdf

Notes

1 An alternative concept for demand is an individual's willingness to accept (WTA) giving up the good or service. We do not discuss differences between WTP and WTA in this chapter.
2 And total benefits for a good that is sold at price p correspond to the area under the demand curve up to price p.
3 In practice, demand analysts often assume a specific functional form for the demand curve in order to extrapolate over a larger range of price–quantity combinations. The most common forms assumed are the linear ($q = \alpha_0 + \beta_0 \cdot p$) or log-linear ($q = \alpha_1 \cdot p^{\beta_1}$) demand curves. The slopes may be estimated from a fit to one or more points along the demand curve, or may be specified based on the concept of demand elasticity η_D, which measures the rate of change of the ratio of price to quantity at a particular point along the curve. Specifically, for linear demand, $\eta_D = \beta_0 \cdot (p/q)$, and for log-linear demand, $\eta_D = \beta_1$. The greater the extrapolation required, the greater the sensitivity of estimates will be to the assumed functional form of demand.

References

Arimah, B. C. (1996). "Willingness to pay for improved environmental sanitation in a Nigerian City." *Journal of Environmental Management* 48(2): 127–138.
Boardman, A. E., D. H. Greenberg, A. R. Vining and D. Weimer (2011). *Cost-benefit Analysis: Concepts and Practice*. Upper Saddle River, NJ, Prentice Hall.
Brown, J., M. Jeuland, A. Hamoudi and G. Turrini (2013). Heterogeneous effects of information on household behaviors to improve water quality. Duke University Working Paper. Durham, NC.

Carson, R. T. and R. C. Mitchell (1993). "The value of clean water: The public's willingness to pay for boatable, fishable, and swimmable quality water." *Water Resources Research* 29(7): 2445–2454.

Carson, R. T., N. E. Flores, K. M. Martin and J. L. Wright (1996). "Contingent valuation and revealed preference methodologies: comparing the estimates for quasi-public goods." *Land Economics.* 72: 80–99.

Carson, R. T., W. M. Hanemann, R. J. Kopp, J. A. Krosnick, R. C. Mitchell, S. Presser, P. A. Ruud, V. K. Smith, M. Conaway and K. Martin (1998). "Referendum design and contingent valuation: the NOAA panel's no-vote recommendation." *Review of Economics and Statistics* 80(2): 335–338.

Choe, K., D. Whittington and D. T. Lauria (1996). "The economic benefits of surface water quality improvements in developing countries: a case study of Davao, Philippines." *Land Economics* 72(4): 519–537.

Diamond, P. A. and J. A. Hausman (1994). "Contingent valuation: is some number better than no number?" *The Journal of Economic Perspectives* 8(4): 45–64.

Hanemann, W. M. (2006). "The economic conception of water." In Peter P. Rogers, M. Ramon Llamas, and Luis Martinez-Cortina (eds) *Water Crisis: Myth or Reality*. London: Taylor and Francis.

Jalan, J. and E. Somanathan (2008). "The importance of being informed: experimental evidence on demand for environmental quality." *Journal of Development Economics* 87(1): 14–28.

Jeuland, M. and D. Whittington (2014). "Water resources planning under climate change: assessing the robustness of real options for the Blue Nile". *Water Resources Research* 50(3): 2086–2107.

Jeuland, M., M. Lucas, J. Clemens and D. Whittington (2010). "Estimating the private benefits of vaccination against cholera in Beira, Mozambique: a travel cost approach." *Journal of Development Economics* 91(2): 310–322.

Jeuland, M., J. Orgill, A. Shaheed, G. Revell and J. Brown (2015). "A matter of good taste: investigating preferences for in-house water treatment in peri-urban Cambodia." (Under Review)

Lantagne, D., R. Meierhofer, G. Allgood, K. McGuigan and R. Quick (2008). "Comment on 'Point of use household drinking water filtration: a practical, effective solution for providing sustained access to safe drinking water in the developing world'." *Environmental Science & Technology* 43(3): 968–969.

Louviere, J. J., D. A. Hensher and J. D. Swait (2000). *Stated Choice Methods: Analysis and Applications.* Cambridge: Cambridge University Press.

Null, C., M. Kremer, E. Miguel, J. G. Hombrados, R. Meeks and A. P. Zwane (2012). Willingness to pay for cleaner water in developing countries: systematic review of the experimental evidence. Available at: http://www.indiawaterportal.org/sites/indiawaterportal.org/files/sr006.pdf

Pattanayak, S. K. and A. Pfaff (2009). "Behavior, environment, and health in developing countries: evaluation and valuation." *Annual Review of Resource Economics* 1(1): 183–217.

Pattanayak, S. K., J.-C. Yang, D. Whittington and K. B. Kumar (2005). "Coping with unreliable public water supplies: averting expenditures by households in Kathmandu, Nepal." *Water Resources Research* 41(2): W02012.

Savenije, H. H. (2002). "Why water is not an ordinary economic good, or why the girl is special." *Physics and Chemistry of the Earth, Parts A/B/C* 27(11): 741–744.

Whittington, D., D. T. Lauria and X. Mu (1991). "A study of water vending and willingness to pay for water in Onitsha, Nigeria." *World Development* 19(2): 179–198.

Whittington, D., J. Briscoe, X. Mu and W. Barron (1990). "Estimating the willingness to pay for water services in developing countries: a case study of the use of contingent valuation surveys in southern Haiti." *Economic Development and Cultural Change* 38(2): 293–311.

Whittington, D., W. M. Hanemann, C. Sadoff and M. Jeuland (2009). *The Challenge of Improving Water and Sanitation Services in Less Developed Countries*. Delft: Now Publishers. Available at: http://www. researchgate.net/publication/228385588_The_challenge_of_improving_water_and_sanitation_ services_in_less_developed_countries/file/9fcfd509d4085c9e18.pdf

Whittington, D., M. Jeuland, K. Barker and Y. Yuen (2012). "Setting priorities, targeting subsidies among water, sanitation, and preventive health interventions in developing countries." *World Development* 40(8): 1546–1568.

Yang, J. C., S. K. Pattanayak, F. R. Johnson, C. Mansfield, C. van den Berg, J. C. Yang, and K. Jones (2006). "Un-packaging Demand for Water Service Quality: Evidence from Conjoint Surveys in Sri Lanka." World Bank Policy Research Working Paper No. 3817. Washington, DC: World Bank.

Yusuf, A. A. and P. Koundouri (2005). "Willingness to pay for water and location bias in hedonic price analysis: evidence from the Indonesian housing market." *Environment and Development Economics* 10(6): 821–836.

63

COST–BENEFIT ANALYSIS AND COST-EFFECTIVENESS ANALYSIS*

Marc Jeuland

Assistant Professor, Sanford School of Public Policy and Duke
Global Health Institute, Duke University, Durham, NC, USA

Learning objectives

1 To explain the steps of cost–benefit analysis (CBA) and cost-effectiveness analysis (CEA) methodologies, and to define the outcome measures they produce.
2 To discuss some of the key challenges associated with application of these approaches.
3 To provide comparisons of CEA and CBA, with reference to practical issues related to water, sanitation and hygiene (WASH) interventions.

Introduction

Cost–benefit analysis (CBA) and *Cost-effectiveness analysis* (CEA) are two specific methodologies that aim to provide guidance on how to best allocate scarce resources across potential interventions or investments (Petitti 1999, Boardman et al. 2011). The underlying motivation for such methods is to make *social* decision-making more rational, that is, to offer recommendations to policy-makers on how to more efficiently achieve beneficial outcomes for society as a whole. Here the efficiency concept is an economic one, which simply defined means that nothing more can be achieved for society given the resources available. Thus, these tools do not consider interests from a private perspective, but rather more appropriately emphasize the social (economic) efficiency of programs or policies.

Given that public resources are scarce and could theoretically be allocated to a range of competing programs, the CBA and CEA methodologies provide one useful – but not the only – perspective to decision-makers wanting to set priorities. Many of the specific steps of the CBA and CEA methodologies are similar, and both attempt a relative weighting of costs

---

* Recommended citation: Jeuland, M. 2015. 'Cost–benefit analysis and cost-effectiveness analysis', in Bartram, J., with Baum, R., Coclanis, P.A., Gute, D.M., Kay, D., McFadyen, S., Pond, K., Robertson, W. and Rouse, M.J. (eds) *Routledge Handbook of Water and Health*. London and New York: Routledge.

and benefits. Yet they also differ in very important ways, and nuanced policy analysis requires understanding and critical assessment of the implications of these differences.

Though variants of CBA or CEA have long been applied in the form of prioritization or other decision-making heuristics, the earliest applications of social CBA were to large water infrastructure projects in the United States, during the early 20th century era of development of large-scale investments to provide multi-purpose benefits (e.g., hydropower, irrigation, and flood control) (Hanemann 1992). Since 1981 (and President Reagan's Executive Order (EO) 12291), the US government has required CBAs to be done for every major regulatory initiative, a commitment that was confirmed and further specified during the Clinton administration (EO 12866). Many other countries and multilateral development banks – which seek to ensure positive social returns from their lending activities – also have similar requirements (Boardman et al. 2011). CEA, on the other hand, has been widely applied to different topics, most specifically in the context of health and defense policy (Boardman et al. 2011). For example, the National Health Service in the United Kingdom uses CEA to set a threshold that must be satisfied by treatment regimens (Appleby et al. 2009). Perhaps the most well-known example of comparative CEA has been the Disease Control Priorities Project, which grew out of a World Bank effort started in the late 1980s (and later promoted by the World Health Organization) to rank health interventions (Murray et al. 2000, Jamison et al. 2006, Edwards 2011).

Due to the overlap across the policy areas of water and health, we should not be surprised to see these two different competing methodologies being applied to assess the value of interventions in water, sanitation and hygiene (WASH). In fact, the literature offers several examples of CEA and CBA being applied (sometimes by the same researchers) to the same basic problems (see for example an application to assessment of the rationale for water treatment: Clasen et al. 2007, Hutton et al. 2007), without extensive critical discussion and interpretation of the fact that their outcome metrics may not be consistent (Jeuland et al. 2009).

In this chapter, we consider in more detail the application of the CBA and CEA methodologies in WASH. We first examine briefly the main steps that comprise these methodologies, and discuss some of their critical features. We then consider their appropriateness for making decisions about investments in water, sanitation and hygiene specifically. We conclude with general thoughts about the value of such methodologies for making *ex ante* investment decisions about what WASH interventions to pursue, thoughts which must inevitably address the issue of external validity and generalizability of CEA and CBA findings.

Cost–benefit analysis

CBA is a methodology for policy assessment or project appraisal that seeks to quantify in monetary terms the value of all consequences of that policy or project to all parties affected by it (Boardman et al. 2011). The total value of the project is indicated by its *net social benefits* NSB, which are equal to its benefits B (the value of all of the positive consequences of the project) minus its costs C (the value of its negative consequences):

$$NSB = B - C. \tag{63.1}$$

For example, a CBA of an investment in improved community water supply would need to consider various benefits – improved health, time savings, aesthetic benefits from

Table 63.1 Steps of a cost–benefit analyis (CBA)

1. Specify the set of alternative projects or policies (usually as changes relative to a consistent baseline)
2. Determine who has standing (i.e., whose costs and benefits should be included in the CBA)
3. Identify the types of costs and benefits (impacts) and determine how these should be measured
4. Predict / quantify those impacts over the life of the project
5. Monetize the quantified costs and benefits, using valuation techniques
6. Aggregate costs and benefits to obtain the project net present value (NPV), incorporating the time value of money
7. Perform sensitivity analysis
8. Make recommendations

Source: Based on Boardman et al. (2011), and modified slightly by the author.

increased water use, community empowerment – as well as costs – capital, operation and maintenance, community training and management costs, in time and money.

When using the NSB metric, it is most common (though not essential) to define the status quo against which a policy is being compared, and to measure B and C relative to that status quo. In this case, B is most appropriately assessed by aggregating the maximum willingness to pay (WTP) of all those gaining from a project for the specific change from the status quo, whereas C is measured by aggregating the minimum willingness to accept (WTA) for those losing from it. Alternatively, one can compare the NSB of the status quo policy directly to the NSB of a new project.[1] The NSB metric provides a straightforward way of assessing whether a project or policy will be beneficial. Policies or projects that provide net social value have NSB > 0, and the greater the NSB, the better the project is for society. Some prefer to use the benefit–cost ratio (B/C) because it allows for normalization across projects of very different scale. For a variety of technical reasons not discussed here, however, it is preferable to normalize NSB (not B) by C, when one is concerned about the relative resource intensiveness of different policies (Boardman et al. 2011).

The steps of a CBA

Undertaking a CBA is a comprehensive and technical exercise that comprises eight steps (Table 63.1). There are notable practical and conceptual challenges to each of these steps, and comprehensive discussion of those challenges is beyond the scope of this short chapter. Thus, this section briefly highlights several important features of this tool, particularly as it applies to assessment of WASH interventions.

Steps 1 and 2 in Table 63.1 largely pertain to the scope of the CBA. In theory, a CBA should assess all potential alternative policies or projects and measure their effects on all parties that are affected by them. In practice, given the important cognitive constraints facing analysts and decision-makers, it is usually necessary to limit the number of alternatives considered, and to define the boundaries over which their effects are considered. For example, WASH interventions may target different pathways of contamination (e.g., hand-washing versus simple sanitation versus phase-in of advanced water treatment like membrane filtration), may have slightly different technological designs (construction of ventilated pit latrines versus eco-sanitation versus sewerage) or implementation modalities (market-based versus free distribution versus utility-based service). They may be promoted at village, regional, national, or even global scales, with very different implications for who has standing. It will often be impossible to consider all possible alternatives and affected parties. At best, the boundaries and scope of the CBA will be pragmatic and allow assessment of the most critical policy tradeoffs in the decision-making problem. At worst, the institutions requesting a CBA may limit its scope in strategic ways, only focusing on particular alternatives or biasing the analysis to consider only effects on specific interests.

Steps 3 through 5 have to do with the identification, quantification, and monetization of the impacts of a project or policy, on those who have been assigned standing (Whittington et al. 2012 provide a useful example and explanatory tables for WASH and other health interventions). Theoretically, a CBA should be able to apply these steps to all conceivable impacts of a project. In practice, however, this is often challenging, and some impacts may be ignored (e.g., aesthetic benefits from hand-washing, or peace of mind from membrane removal of chemicals of unknown toxicity) if a) quantification is uncertain and/or causal attribution is hard, b) the methods to measure or value certain types of impacts are undeveloped, inaccurate or controversial (issues we highlight further below as they apply to WASH interventions), or c) if the project time horizon extends far into the future such that predictions become problematic. Step 6, which deals with aggregation of impacts, seems relatively straightforward, but can raise objections from those who disagree over the way future impacts should be discounted, or those who object to presentation of a single metric that masks important asymmetries in the distribution of impacts (e.g., benefits concentrated on those who are already better off).

Finally, as the literature makes clear and as we discuss further below (Jeuland et al. 2009, Whittington et al. 2012), CBA analyses often rely on a large number of assumptions, such that making recommendations (step 8) on the basis of a simple deterministic analysis of alternatives is often unwarranted. The degree to which these results are robust to other assumptions can potentially be tested using sensitivity analysis (step 7), but the extent of sensitivity testing that can be deemed informative is often limited by capacities for generating, interpreting, and understanding such results.

Cost-effectiveness analysis

CEA is a second methodology for policy assessment or project appraisal that compares alternatives in terms of the ratio of costs C and a single quantified measure of benefits, or effectiveness E. In mathematical terms, the cost-effectiveness ratio CER is expressed as:

$$CER = C/E. \qquad (63.2)$$

When using the CER, it is important to carefully define the status quo against which a policy is being compared, and to measure C and E as changes relative to that status quo. Unlike the NSB metric, the CER does not provide a straightforward way of assessing whether a project or policy will be beneficial for society or not (i.e., it does not weigh benefits directly against costs). Thus it does not necessarily reveal the most economically efficient policy. Instead, the CER is most useful when comparing similar policies or projects designed to achieve a particular type of benefit, as expressed in the effectiveness measure selected by the analyst. In such comparisons, the policy with the lowest CER is the one which costs the least per unit of effectiveness, or is most desirable. Alternatively, some apply CEA as a satisficing

Marc Jeuland

criterion whereby some threshold λ must not be exceeded by the CER if a particular policy is to be deemed attractive.

Returning to the water supply example from the previous section, a CEA needs to determine the most useful indicator of E, perhaps some measure of avoided disease burden or time saved by a household, and normalize E by the total intervention costs. The selection of the metric for E will largely depend on what the intervention is being compared to (e.g., disease burden measures would be most useful if making a comparison with other health interventions, whereas time savings would allow comparisons with other productivity enhancements).

The steps of a CEA

CEA requires many of the same steps as CBA, though the scope is narrower due to its focus on a single non-monetary measure of effectiveness (Table 63.2). Here we consider several of the most important ways in which these steps differ in a CEA as compared with the CBA process described above.

First, only one specific type of benefit (e.g., disability-adjusted life years (DALYs) avoided, number of households obtaining water supply) appears in the denominator of a CER. This greatly simplifies calculation of the metric since one need not do an exhaustive enumeration of project benefits. Second, only those who receive the particular type of benefit included in the CEA effectiveness measure and/or who incur the costs required to produce it have standing in a CEA, and the CEA can only compare policies that produce benefits that can be measured by the effectiveness metric that is selected. On the other hand, the handling of costs in a CEA is often inconsistent and can be very confusing. In many cases, analysts take a purely public sector perspective because the main concern is spending limited public money most efficiently, such that only costs incurred to the public sector appear in the numerator (Walker et al. 2010). A more holistic view acknowledges that other costs are important as well, however, and aims to include them. From an economic perspective, however, even this holistic view of CEA is theoretically incomplete given that benefits are limited to those included in the chosen effectiveness measure.

Table 63.2 Steps of a cost-effectiveness analysis (CEA)

1. Specify the set of alternative projects or policies (changes relative to a consistent baseline)

2. Determine who has standing (i.e., whose costs and benefits should be included in the CEA)

3. Identify the types of costs to include in the numerator of the ratio, and the type of benefit to include in the denominator, and determine how these should be measured

4. Predict / quantify those impacts over the life of the project

5. Monetize the costs, using valuation techniques if necessary

6. Discount any future costs, or benefit indicators, and aggregate them to obtain the project cost-effectiveness ratio

7. Perform sensitivity analysis

8. Make recommendations

Source: Author's analysis.

Finally, monetization (step 5) only applies to the costs in a CEA, which is convenient given that costs are typically more easily expressed in monetary units. This in turn affects the discounting in step 6, although many CEAs discount benefits such as future avoided DALYs or YLLs (years of life lost). Unlike CBA where discounting is always practiced, there continues to be debate among public health analysts about whether effectiveness measures should be discounted in this way (Murray and Acharya 1997, Edejer 2003).

Comparing CBA and CEA for the analysis of WASH interventions

Table 63.3 highlights some of the differences in key features of CEA- and CBA-based decision-making.

As presented above, the basic rationale and decision criteria (features 1 and 2 in the table below) of these two methodologies are fundamentally different. On the one hand, the focus in CBA is on achieving improvements in (or optimal) overall *social welfare*, as mandated by the requirement that NSB > 0 (or Max[NSB]). On the other hand, the focus in CEA is on *resource rationing*, that is, incurring a cost per unit of effectiveness in producing

Table 63.3 General comparison of cost–benefit and cost-effectiveness analyses (CBA and CEA)

Feature	CBA	CEA	Comments
1. Basic rationale	Social welfare-based	Resource rationing	Social welfare is based on NSB measure; resource rationing focuses on cost per unit of some specific effectiveness measure
2. Decision criterion	Choose alternative with NSB > 0, or Max[NSB])	Choose alternative with CER < λ, or Min[CER]	No economic rationale for CEA criterion (not grounded in a comparison of economic gains and losses)
3. Measurement of decision criterion	More difficult	Less difficult	CERs are less difficult to measure because effectiveness measure limits scope
4. Approach to valuation	Aims to value all impacts	No explicit valuation of benefits	Economic valuation may not always be possible; some may consider some valuation (e.g., mortality reductions) to be unethical
5. Alternatives that can be compared	Theoretically any alternatives	Only alternatives producing similar benefits	Practical considerations also limit the scope for comparison
6. Efficiency vs. equity	Often argued to be more efficient	Often argued to be more equitable	Such arguments are not necessarily true
7. Implications for distributional analysis of impacts	Meaning of NSB disaggregated to each affected party is retained	Meaning of CER disaggregated to each affected party is unclear	Disaggregated analysis is not advisable in the case of CER except when a policy is strictly targeted at a particular party

Notes: CER = cost-effectiveness ratio; NSB = Net social benefit.

Source: Author's analysis.

some type of benefit that is lower than a pre-determined threshold λ or that is minimized across alternative policies. Another way to see this crucial difference is to consider that the effectiveness measure in a CEA is selected by an expert who decides what weight should be given to the specific changes that parties affected by a change will experience. CBA metrics, however, depend on the idea that the most appropriate valuation of the changes faced by affected parties comes from those persons' WTP or WTA for the change.

Though some policies may satisfy both or neither of these two criteria, they are in practice rarely equivalent. Perhaps most important in the context of WASH interventions is the fact that the types of benefits included in CBA and CEA will not always correspond. For example, while it is true that many WASH interventions are motivated by a desire to reduce diarrheal disease, one of the key difficulties with applying CEA for the evaluation of water supply improvements is the inability to combine multiple benefits (e.g., disease reductions + increases in access, or aesthetic benefits; see Whittington et al. (2012) for a complete typology). In addition, as Jeuland et al. (2009) point out, there is no economic rationale for minimizing the cost per unit of effectiveness if additional benefits can still be generated at very low, but not minimum, incremental cost (for example, if five deaths would be avoided by Program A for $5, but an alternative Program B avoided six deaths for $10, the former would be more cost-effective, but Program B would clearly be better).

Still, carrying out a complete CBA is often much more resource intensive than CEA and may be impractical for many analysts, in part due to the challenge of expressing impacts in monetary terms when they are uncertain, hard to quantify, and/or not easily valued (features 3 and 4). The challenge of CBA is further increased by the nature of the social welfare improvement criterion: the burden for determining optimal policies can extend across a variety of very different and competing sectors (feature 5). Policy-makers working in WASH may for example resist an exercise that also compares their best-performing interventions to a wider set of nutritional, environmental, or infectious disease-reducing policies.

Finally, it is worth considering the equity and ethical implications of the CBA and CEA approaches carefully (feature 6). Many in the health sector emphasize that CEA is fundamentally more equitable than CBA, because health outcomes such as a DALY avoided are treated as equivalent, regardless of who benefits. In addition, CEA does not necessitate ethically challenging valuation of benefits such as mortality reductions. In contrast, CBA may be considered more efficient, given that it compares WTP for benefits directly against the costs incurred. When considering health benefits in CBA, for example, deaths avoided in a rich country will typically be worth more than deaths avoided in a poor country, simply because WTP for mortality risk reductions increases as a function of income (Hammitt and Robinson 2011). Nonetheless, because CBA is often unable to fully include all benefits, costs and alternative policies, it may not be more efficient than CEA.[2] Conversely, because CEA implicitly places value on effectiveness measures through the selection of outcomes by expert analysts, it may not be more equitable than CBA, if for example the improved outcomes are not equally welfare-improving for all affected parties. An additional problem with CEA with regards to equity relates to the fact that it is incorrect to make judgments about the distributional impacts of a policy based on disaggregation of the CER across parties affects by a policy (feature 7). This is because minimizing the cost per unit of effectiveness is problematic, as pointed out above. CBA on the other hand readily accommodates distributional analysis since NSB can be calculated and assessed for each affected party.

Brief additional reflections on the role of CBA and CEA

As several researchers have pointed out, it is understandable to want to make general (rather than location and time-specific) comparisons of the economic net benefits, or alternatively the cost-effectiveness, of different types of investments, yet this desire is also typically impractical (Whittington et al. 2012, Jeuland and Pattanayak 2012). As discussed above, a large number of categories of costs and benefits result from such interventions, and a systematic modeling approach to aggregating them typically requires a series of assumptions, judgment calls, and omissions. Once an analyst progresses beyond such issues, it is often also difficult to make sense of the potential variation in outcomes that emerges from careful sensitivity analysis. Alternatively, site-specific studies that measure and compare economic outcomes more precisely and comprehensively in particular places and at a specific point in time may have limited external validity, and therefore relevance, to aid decision-making across a wider array of settings.

Summary

This chapter briefly described the CBA and CEA methodologies, and considered their potential value in aiding decision-making about WASH interventions. The underlying motivation for such methods is to make *social* decision-making more rational, that is, to offer recommendations to policy-makers on how to more efficiently – in economic terms – achieve beneficial outcomes for society as a whole. Such a perspective is useful for avoiding policies that generate benefits that are not commensurate with the costs incurred to produce them. As discussed, these approaches have many similarities, yet also differ in fundamental and important ways, and their application requires careful consideration of scope, identification and measurement of impacts, and interpretation of outcome metrics. A nuanced understanding of these methods and their relative strengths and weaknesses is therefore advisable prior to their application.

Acknowledgments

Thanks are due to many research collaborators who have contributed to the arguments presented in this chapter – many of which can be found in the accompanying reference list. The author also thanks the students in his semester long course on economic analysis for health and the environment, whose incisive questions have led to refinement of these arguments over the past several years. Three anonymous reviewers also provided comments that were critical to improving the chapter.

Key recommended readings

1   Clasen, T., Cairncross, S., Haller, L., Bartram, J. & Walker, D. 2007. Cost-effectiveness of water quality interventions for preventing diarrhoeal disease in developing countries. *Journal of Water and Health*, 5, 599–608.
2   Whittington, D., Jeuland, M., Barker, K. & Yuen, Y. 2012. Setting priorities, targeting subsidies among water, sanitation, and preventive health interventions in developing countries. *World Development*, 40, 1546–1568.

Notes

1 In this alternative approach, the valuation task is concerned with "total values" in a particular state of the world; such values are usually considerably more difficult to measure than those associated with the incremental changes that result from policy changes, because they require definition of WTP relative to a state of the world that does not exist.

2 For example, it may be hard to quantify the value of aesthetic benefits from a piped water connection. Or there may be hard to value learning costs associated with use of point-of-use treatment technologies like chlorination, or start-up costs for filter ripening, or finally removal of chemicals of unknown toxicity by membrane filtration.

References

Appleby, J., Devlin, N., Parkin, D., Buxton, M. & Chalkidou, K. 2009. Searching for cost effectiveness thresholds in the NHS. *Health Policy*, 91, 239–245.

Boardman, A. E., Greenberg, D. H., Vining, A. R. & Weimer, D. 2011. *Cost–benefit Analysis: Concepts and Practice*. Upper Saddle River, NJ: Prentice Hall.

Clasen, T., Cairncross, S., Haller, L., Bartram, J. & Walker, D. 2007. Cost-effectiveness of water quality interventions for preventing diarrhoeal disease in developing countries. *Journal of Water and Health*, 5, 599–608.

Edejer, T. T.-T. 2003. Making choices in health: WHO guide to cost effectiveness analysis. Geneva: World Health Organization.

Edwards, C. 2011. Cost-effectiveness analysis in practice. In: J. Cameron, P. Hunter, P. Jagals, and K. Pond (eds) *Valuing Water, Valuing Livelihoods*. London: IWA Publishing.

Hammitt, J. K. & Robinson, L. A. 2011. The income elasticity of the value per statistical life: transferring estimates between high and low income populations. *Journal of Benefit-Cost Analysis*, 2, 2152–2812.

Hanemann, W. M. 1992. Preface. In: Navrud, S. (ed.) *Pricing the European Environment*. Oslo: Scandinavian University Press.

Hutton, G., Haller, L. & Bartram, J. 2007. Global cost–benefit analysis of water supply and sanitation interventions. *Journal of Water and Health*, 5, 481–502.

Jamison, D. T., Breman, J. G., Measham, A. R., Alleyne, G., Claeson, M., Evans, D. B., Jha, P., Mills, A. & Musgrove, P. 2006. *Disease Control Priorities in Developing Countries*. Washington, DC: World Bank Publications.

Jeuland, M., Lucas, M., Clemens, J. & Whittington, D. 2009. A cost–benefit analysis of cholera vaccination programs in Beira, Mozambique. *The World Bank Economic Review*, 23, 235–267.

Jeuland, M. A. & Pattanayak, S. K. 2012. Benefits and costs of improved cookstoves: assessing the implications of variability in health, forest and climate impacts. *PloS One*, 7, e30338.

Murray, C., Evans, D. B., Acharya, A. & Baltussen, R. 2000. Development of WHO guidelines on generalized cost-effectiveness analysis. *Health Economics*, 9, 235–251.

Murray, C. J. L. & Acharya, A. K. 1997. Understanding DALYs. *Journal of health Economics*, 16, 703–730.

Petitti, D. B. 1999. *Meta-analysis, Decision Analysis, and Cost-Effectiveness Analysis: Methods for Quantitative Synthesis in Medicine*. Oxford: Oxford University Press.

Walker, D. G., Hutubessy, R. & Beutels, P. 2010. WHO Guide for standardisation of economic evaluations of immunization programmes. *Vaccine*, 28, 2356–2359.

Whittington, D., Jeuland, M., Barker, K. & Yuen, Y. 2012. Setting priorities, targeting subsidies among water, sanitation, and preventive health interventions in developing countries. *World Development*, 40, 1546–1568.

PART VIII

Learning from history

64

INTRODUCTION

Learning from history[*]

Peter A. Coclanis

ALBERT R. NEWSOME DISTINGUISHED PROFESSOR OF HISTORY, AND DIRECTOR, GLOBAL
RESEARCH INSTITUTE, UNIVERSITY OF NORTH CAROLINA AT CHAPEL HILL, USA

Learning objectives

1 Explain the case-study approach and discuss why it can be useful both to scholars and
   to practitioners.
2 Lay out some key features of historical methodology and demonstrate how "thinking
   in time" can help lead to better decision making.
3 Provide some historical context for the seven "cases" included in this section.

Henry V. Sir Francis Drake. Father Marquette. Akbar the Great. Roger Sherman. Carl von
Clausewitz. G.W.F. Hegel. Nicolas Carnot. James K. Polk. Adam Mickiewicz. Prince Albert.
Leland Stanford, Jr. Gerard Manley Hopkins. Pyotr Tchaikovsky. Wilbur Wright. At first
blush, determining what these highly diverse eminences have in common would seem to
be difficult even for people experienced in cluster analysis and in possession of powerful
algorithms. Given the subject of this book, however, many readers probably would hazard
guesses that what the list of names above had in common had something to do with water
and health. And these readers would be right, for the fifteen luminaries above all died from a
water-borne bacterial infection of some type, in most cases, cholera, but some from typhoid
fever or dysentery. In so doing, they were joined over the centuries by hundreds of millions
of less well-known people all over the world.

   Although such diseases and allied water-borne bacterial infections continue to pose
huge problems in less-developed countries today, morbidity and mortality levels from these
diseases have fallen drastically in developed countries, largely as a result of the so-called
sanitary revolution that began in England in the 1840s, thence spreading over the course of
the next seventy-five years or so to other parts of Europe and to the US and ultimately to

[*] Recommended citation: Coclanis, P.A. 2015. 'Introduction: learning from history', in Bartram, J.,
   with Baum, R., Coclanis, P.A., Gute, D.M., Kay, D., McFadyen, S., Pond, K., Robertson, W. and
   Rouse, M.J. (eds) *Routledge Handbook of Water and Health*. London and New York: Routledge.

other parts of the world. In 2007, readers of the *BMJ* selected this revolution, which can be boiled down (as it were) to the provision of clean water and effective sewage disposal, as the single greatest medical advance since 1840 (*BMJ* 2007). They were right to do so. That the revolution is still incomplete is no reason to downplay the hugely positive impact it has had on human populations across the globe.

This part consists of seven historical cases spanning the period from the 1840s to the present and areas far removed. To be sure, the historical details of each of these chapters are both fascinating and important (not to mention dramatic at times), but the chapters were included as much for methodological reasons as for their content, however rich. For familiarity with the case-study approach and the historical mode of inquiry—the ability to think in time, as Richard E. Neustadt and Ernest R. May put it in 1986—adds significantly to both the scholar's and the practitioner's methodological skill set (Neustadt and May 1986). Moreover, the case study and history, when combined, are particularly useful for teaching and training, as they can provide an interesting and well-nigh inexhaustible supply of pedagogical materials.

In other words, one doesn't need a research specialization on cholera in late nineteenth-century Germany or arsenic poisoning in twenty-first century Bangladesh to profit from reading the pieces on these topics in this part. Rather, the editors believe that the "thick description" associated with the case-study approach and the richness of the temporal and contextual framing associated with historical inquiry can add value for researchers and practitioners regardless of geographical interest or field specialization.

The basic idea behind the case study, of course, is that detailed study of one single individual, group, event, period, process, and so on, can yield certain types of insights that are usually not possible to attain from more general approaches, particularly those proceeding deductively or operating from 30,000 feet. Obviously, cases must be well chosen and there are substantive problems that need to be kept in mind, *to wit*: Selection/confirmation biases, outlier cases, justifying case interpretations and the question of extrapolating therefrom, and so on. Moreover, the opportunity cost of mounting and completing case studies are often high, particularly in terms of time. That said, few other methods can provide the kind of in-depth qualitative and quantitative evidence that is possible in a well-chosen, well-documented case. Furthermore, as cases accumulate, evidence begins to build inductively that enables researchers to think and speak more confidently about the behavior of individuals or groups under certain conditions, about certain kinds of events or processes, and about both scientific and social scientific possibilities and limitations.

Each of the cases included in this part provides a close analysis of a specific public-health/sanitation problem in a particular place during a defined, generally relatively narrowly circumscribed, unit of time. Regarding temporality: The time frames for the cases vary from weeks to months to years. For example, the case on the contamination of the drinking-water system in Walkerton, Ontario, in May 2000—an episode that led to a disease outbreak that killed seven people and left more than 2,300 sick—covers about three weeks, while the cases on the cholera epidemic in Hamburg in 1892 and the *Cryptosporidium* outbreak in Milwaukee in 1993 take several months. In contrast, the case on the protracted arsenic crisis in Bangladesh, which crisis first came to world attention in the early 1990s, remains unresolved a quarter century later, and can be considered in a sense an interim report or update.

The cases included in the part typically proceed in standard case form, which is to say, the author of the case introduces the problem to be analyzed and situates it in time and space; the principals involved, whether individuals or institutions, are identified, and close, detailed accounts of the actions taken by the same are provided; the results and implications of such actions are laid out and assessed; the case closes with a discussion of lessons learned or at least lessons that *might be* learned.

Invoking the concept of "learned lessons" is tricky business, trickier than it appears on the surface, for it implies that history teaches or can be made to teach. Virtually everyone is familiar with the expression "history repeats itself," and many with philosopher George Santayana's famous remark that "[t]hose who cannot remember the past are condemned to repeat it" (Santayana 1905, p. 284). Do these seemingly authoritative assertions not suggest that lessons perforce flow freely from well-crafted case studies and that such lessons are readily transferable and applicable whenever situations/conditions that seem similar crop up? The short answer is no. As a historian, I can't go along with Henry Ford's position that history "is more or less bunk," but I don't buy into the idea that what philosophers of history call *historical recurrence* is the rule (Trompf 1979). More to my taste is the more modest stance often associated with Mark Twain: "History doesn't repeat itself, but it does rhyme." Let me now discuss why this less absolute position is at once more realistic and perhaps ultimately empowering to scholars and professionals in public health.

To cut to the chase: If history doesn't repeat itself, why bother trying to learn about it? Obviously, one can point to the aesthetic appeal of history and to the fact that history and the humanities take as their object human culture, the study of which can guide us toward more intellectually and emotionally rich and meaningful lives. But one can also make the case for history on more instrumental grounds. Just because history doesn't repeat itself hardly means that studying it is irrelevant. For as Marx famously pointed out in *The Eighteenth Brumaire of Louis Napoleon* (1964): "Men make their own history, but they do not make it just as they please; they do not make it under circumstances chosen by themselves, but under circumstances directly encountered, given and transmitted from the past" (Marx 1964, p. 15).

To use modern social scientific parlance, "history matters," sometimes a lot. Indeed, the "paths" we take today are often influenced or inflected by, if not dependent upon, decisions, actions, circumstances, and so on, made at earlier points in time. Viewed in this way, the quest for historical knowledge—at least historical knowledge of certain types—cannot be construed as the mindless hoarding of non-nutritious intellectual fodder suitable only for a game of Trivial Pursuit, but as the mindful accumulation of essential data that can help us to explain how and why we got—or did not get—from Point A to Point B or how and why we ended up instead at Point F, R, or Z. One needn't memorize every last detail regarding the contributions to sanitation made by the Romans or the Victorians to find value in knowing a little history, even just the basics about the Romans' *Cloaca Maxima* and John Snow's observations about what happened in 1854 to residents of Soho in London whose drinking water came from the Thames—just downstream of a sewage outfall pipe (Johnson 2006).

History is useful or can be made useful in other ways as well, as Harvard professors Richard E. Neustadt and Ernest R. May, alluded to earlier, made clear in their (once) well-known book *Thinking in Time: The Uses of History for Decision Makers*. In this book Neustadt, a political scientist, and May, a diplomatic historian—both now deceased—drew upon a large number of historical case studies, or, more accurately, case illustrations, to demonstrate how "thinking historically" can aid professionals in the decision-making process. In so doing, moreover, they developed and laid out a number of quick and easy methods, tools, techniques, and rubrics—drawn from their own careful study of history— that can help reduce costly (and, in the case of public health, alas, potentially lethal) errors and omissions (Neustadt and May 1986, pp. 273–275). In some ways, in fact, *Thinking in*

Time is thus analogous to and serves some of the same functions as Atul Gawande's 2009 book, *The Checklist Manifesto: How to Get Things Right*, but with applicability to an even broader range of professions (Gawande 2009).

What are these methods, tools, and so on? Not exactly back-of-the-envelope stuff, but nothing too arcane or elaborate either. Moreover, like Gawande's checklists, they are ordered, structured, and based on common sense. Crucial to Neustadt and May's methodology, first of all is the proper framing of the issue to be studied. Once an issue or concern is identified, they advise not to jump immediately to the conclusion that the answer—or even the problem—is self-evident. Miasmas as the cause of cholera in mid-nineteenth-century London, for example.

According to N & M, researchers should first ask a series of questions before moving forward very far. What is known about a particular issue? What is unknown? What is presumed? Known, unknown, and presumed, or KUP for short. (Just for the record: They are, thankfully, not asking us to speculate à la Donald Rumsfeld about "unknown unknowns".)

Such concerns are particularly important in retrospective analyses of disease outbreaks because the infection agents and routes of transmission were often unknown or erroneously identified at the time. Unless we can identify such agents and transmission routes and so on, we have not found the true keys to a given outbreak.

If correct answers to such questions are found—indeed, even if they aren't—it can also be useful to study crisis *management* during past disease outbreaks because the dynamics of management responses can help to guide decision makers both today and tomorrow (Dayton et al. 2013). Stephen Gradus's detailed piece in this part on the administrative response to the *Cryptosporidium* outbreak in Milwaukee in 1993 provides an excellent case in point. Even a little familiarity with what went wrong and what went right in the past can go a long way in improving administrative responses to similar situations.

I refer to "similar situations" with some trepidation, for surface similarities often disappear under close scrutiny. For this reason, Neustadt and May wisely suggest that researchers think very carefully before drawing historical analogies. Moreover, even when such a situation or situations comes/come to mind—and humans do like to think analogically—don't jump to the conclusion that the analogy is perfect. How are seemingly analogous situations similar? How are they different (Neustadt and May 1986, pp. 273–275)?

For example, the USA has sometimes gotten into trouble in diplomatic negotiations over the years because policymakers, fearing "another" Munich-like agreement, were overly aggressive and confrontational. Moreover, in disaster relief we've often set ourselves up for failure by making false analogies based on the Marshall Plan, which has resulted in policies and initiatives inappropriate for desperately poor less developed countries (LDCs) far different than Europe after World War II. That is to say, places such as Haiti after the devastating 2010 earthquake. In the realm of public health, influenza outbreaks provide another case in point. Policy makers are sometimes too quick to make plans based on the worst-case scenario—most notably, the "Spanish" flu pandemic of 1918–1919—with less than optimal results. In this regard, think of the overcharged policy response to the swine flu "epidemic" of 1976. A little more due diligence—and more rigorous examination of the "Spanish flu" analogy—would have gone a long way in that particular case (Neustadt and May 1986, pp. 48–57, 219–225).

Just as a little history and a little historical thinking can help us avoid bad analogies, history/historical thinking can help us avoid careless extrapolation, particularly careless linear extrapolation. This is so whether we are speaking of economic growth rates—poor countries often achieve high rates of growth because they start from low bases, but slow down as their bases enlarge—or of commodity prices, birth rates, death rates, progress toward meeting the Millennium Development Goals (MDGs), and so on. Lots of factors can intervene in—and, in so doing, interrupt, redirect, or even reverse—trends over the intermediate to long term, and exceptional care should be taken before projecting too far forward. To use statistical jargon, many "functions" are non-smooth, and non-smooth functions are difficult targets for accurate extrapolation. Especially "functions" in history!

With these considerations in mind, a researcher is in better position to figure out the "story" behind an issue or problem. And even here one has to be careful because story-telling is subjective stuff—think *Rashomon*. But even simple tools such as time lines can provide some help. Neustadt and May recommend them, to which recommendation the authors of the essays in this part wisely adhere. And in getting the story right—and thus figuring out what the issue or problem really is—N & M are bullish on the classic m.o. of journalists, that is to say, asking not only "when" (time lines) but also what, where, who, and how (Neustadt and May 1986, pp. 273–275).

If and when these questions are answered satisfactorily, one begins to feel a bit better about the story in which one is interested. But, before one gets too confident, N & M suggest a few further questions, the answers to which may still bring a researcher up short, to wit: What odds would you give that your story is correct? How much of your own money would you wager that it is? And what new facts, if they could be found, would get you to change the story (Neustadt and May 1986, p. 274)? If the odds are good, the wager high, and the need for new facts limited, a researcher can feel reasonably good about their explanation— with the emphasis on the word *reasonably*. For if history teaches anything at all, it is humility. Explanations of the past are perforce always contingent: History is not studied under lab conditions, information is either incomplete or ambiguous or messy (and often all three), and humans fallible. Learning to live with ambiguity is part and parcel of thinking historically.

Although I've spent a good bit of time above on Neustadt and May's approach to "thinking in time," theirs is but one of many. Indeed, there is a lively (and well-attended) biennial conference devoted to policy history; there are respected professional journals such as *History and Policy* in the U.K. and the *Journal of Policy History* in the USA; and more and more scholars are hard at work tilling the policy–history field (Berridge 2008; Seaman and Smith 2012; Coclanis 2014). An increasing number of historians are finding employment in government and in the corporate sector. Why? Because of the growing belief that knowing whence you came can better enable an individual or organization to move forward. With these considerations in mind, it is not surprising that knowing some history—often via well-chosen cases—and something about thinking historically (*à la* N & M, for example) is increasingly seen as a real asset, if not becoming de rigueur.

But there is more. One of the stronger virtues of history grows out of the *discipline* it provides researchers. History is often, perhaps *too* often, sobering, demonstrating time and again that human will is not boundless but circumscribed and that, as Kant famously put it, "out of the crooked timber of humanity, no straight thing was ever made" (Kant 1991, p. 46).[1] This austere, even severe view of the human condition need not be immobilizing, though. History matters and sets limits, but such limits are hardly absolute, which gives us more than a margin of hope. This is true in the realm of public health generally and water, sanitation and hygiene (WaSH) specifically.

The case studies included in this part are certainly sobering, illustrating time and again both the biological challenges that humans persistently face and the often imperfect policy responses humans have made to the same. Moreover, as Rosalind Stanwell-Smith makes clear in her chapter on John Snow and the Broad Street pump, if good explanatory models are not available—in this case, the germ theory—even accurate demonstration of the epidemiological method is insufficient to get people *quickly* to change their minds about the causes and routes of transmission of disease.

But at the end of the day we have endured, and in some cases done considerably better than that. In his new book *How We Got to Now: Six Innovations that Made the Modern World*, science/technology writer Steven Johnson devotes considerable attention to the history of sanitation in the Western world (Johnson 2014). Johnson—author as well of the celebrated 2006 book *The Ghost Map* (on Snow's discovery of the aetiology of cholera in London in 1854)—considers innovations in sanitation to be among the most important shapers of modernity, with pride of place in the sanitation realm going to the provision of clean water and the removal of sewage from European and North American cities beginning in the second half of the nineteenth century (Johnson 2006; Johnson 2014, pp. 127–160).

Such momentous innovations, one hastens to add, obviously came about as a result of the work of many others besides Snow. In the case of the U.K., Snow's predecessor, Edwin Chadwick (1800–1890), who was largely responsible for the Public Health Act of 1848, was especially important. Other sanitation "pioneers"—both individuals and institutional—proved instrumental wherever the sanitation revolution spread. Indeed, the need for such "champions," and the formidable costs—economic and political—often associated with long-term investment in water infrastructure pose challenges sufficiently rigorous to dampen the spirits of public-health practitioners in many LDCs today (Fisher and Cotton 2005).

Before despairing though, read on. For experts such as Steven Johnson, who know their history and are rightfully respectful of its often heavy hand, remain hopeful nonetheless. Why? Simply put, because just as new developments in ICT (information and communications technology) in recent years have enabled people in LDCs in many parts of the world to bypass landlines and move directly to cell phones, exciting new developments in the world of WaSH—developments fostered and supported by organizations ranging from the Gates Foundation to Caltech and from MIT's Poverty Action Lab to the World Health Organization (WHO)—may be on the verge of "reinventing" sanitation, thereby relaxing history's hold and bringing effective low-cost WaSH solutions to many of the world's poorest people (Johnson 2014, pp. 152–156). If they do, it will doubtless be cause for a rich case history such as the ones included here. Stay tuned.

Note

1 Note that the most famous part of this quote, found in "Proposition Six" in Kant's *Idea for a General History with a Cosmopolitan Purpose*, is rendered in various ways in English, but generally as either "crooked timber" or "warped wood." I am using "crooked timber," following Isaiah Berlin's usage.

References

Berridge, Virginia. 2008. "History matters? History's role in health policy making." *Medical History* 52: 311–326, http://www.ncbi.nlm.nih.gov/pmc/articles/PMC2448976/, accessed 6 January 2015.

BMJ. 2007. "BMJ readers choose the 'sanitary revolution' as greatest medical advance since 1840." 334: 111 (18 January 2007), http://www.bmj.com/content/334/7585/111.2, accessed 6 January 2015.

Coclanis, Peter A. 2014. "THEMAS is WWRN*: Why STEM students need an h." *Perspectives on History* 52 (December): 28–29, http://www.historians.org/publications-and-directories/perspectives-on-history/december-2014/themas-is-wwrn, accessed 6 January 2015.

Daton, Bruce W., Xuefeng Li, and Wendy Leasure Wicker. 2013. "Crisis and disaster management in the U.S.: research, methods, analysis, and findings." http://cdm.syr.edu/wpcontent/uploads/2014/10/Crisis_and_Disaster_Management_in_the_US-.pdf, accessed 6 January 2015.

Fisher, Julie and Andrew Cotton. 2005. Learning from the Past: Delivery of Water and Sanitation Services to the Poor in Nineteenth-Century London. WELL Briefing Note 10. Water Engineering and Development Centre (WEDC), Loughborough University. http://wedc.lboro.ac.uk/resources/well/WELL_BN10_Learning_from_the_past.pdf, accessed 6 January 2015.

Gawande, Atul. 2009. *The Checklist Manifesto: How to Get Things Right*. New York: Metropolitan Books.

Johnson, Steven. 2006. *The Ghost Map: The Story of London's Most Terrifying Epidemic—and How It Changed Science, Cities, and the Modern World*. New York: Riverhead Books.

Johnson, Steven. 2014. *How We Got to Now: Six Innovations That Made the Modern World*. New York: Riverhead Books.

Kant, Immanuel. 1991. "Idea for a universal history with a cosmopolitan purpose." In Kant, *Political Writings*, 2d ed. Ed. Hans Reiss, trans. H.B. Nisbet. Cambridge and New York: Cambridge University Press. Originally published in German in 1784.

Marx, Karl. 1964. *The Eighteenth Brumaire of Louis Napoleon*. New York: International Publishers. Originally published in German in 1852.

Neustadt, Richard E. and Ernest R. May. 1986. *Thinking in Time: The Uses of History for Decision Makers*. New York: The Free Press.

Santayana, George. 1905. *Reason in Common Sense*. New York: Charles Scribner's Sons.

Seaman, John T., Jr. and George David Smith. 2012. "Your company's history as a leadership tool." *Harvard Business Review* 90 (December): 44–52.

Trompf, G.W. 1979. *The Idea of Historical Recurrence in Western Thought: From Antiquity to the Reformation*. Berkeley, CA: University of California Press.

65

CHOLERA EPIDEMIC IN HAMBURG, GERMANY 1892*

Martin Exner

PROFESSOR OF HYGIENE AND PUBLIC HEALTH, UNIVERSITY OF BONN, BONN, GERMANY

Introduction

The cholera outbreak in Hamburg in 1892 was the biggest waterborne outbreak ever in Germany. More than 16,850 people were infected with cholera and 8,526 patients died. Robert Koch's contemporaneous investigation of the circumstances, causes, and management of the cholera outbreak remains even today one of the best examples of extremely efficient outbreak management, integrating clinical investigation, field inspection, epidemiology, and clear implications for sustainable control management and prevention strategies.[1, 4] The consequences of this investigation for water hygiene were tremendous.

The outbreak

The first two cases of cholera were identified on August 16, 1892. The next day, two patients died. Eight days later, on August 24, Robert Koch went to Hamburg, and based upon his experiences from his cholera expedition to Egypt and India in 1883 and 1884, he hypothesized in a conference with the senator of the free Hansa City of Hamburg, Mr. Hartmann, that drinking water was the transmission vehicle of the cholera pathogens. The next day, Robert Koch visited the barracks of a small ship that transported emigrants from Russia to the United States via the port of Hamburg. The barracks were rebuilt by the HAPAG Line (Hamburg American Line). Robert Koch also inspected the disposal of the sewage of this emigration camp. He was critical of the policy that allowed all of the faeces of the emigrants to be introduced into the Elbe River without any treatment and the fact that all laundry washers were not functioning correctly. Once he saw the bales of straw which were thrown into the Elbe River he was convinced that *Vibrio cholera* was flushed upstream with the high

---

* Recommended citation: Exner, M. 2015. 'Cholera epidemic in Hamburg, Germany 1892', in Bartram, J., with Baum, R., Coclanis, P.A., Gute, D.M., Kay, D., McFadyen, S., Pond, K., Robertson, W. and Rouse, M.J. (eds) *Routledge Handbook of Water and Health*. London and New York: Routledge.

tide, and introduced in the drinking-water supply because the raw water intake was situated a short distance upstream from the emigrant camp. At this time, up to 1,000 emigrants were in the camp, where they also washed their laundry. After this inspection visit, Robert Koch proposed that a boil-water advisory for drinking water should be published. This boiling advisory was published on August 26, 1892.[4]

The police department ordered that all water for drinking-water purposes and for washing must be boiled. Carriages were driven through the city to distribute clean water to the population of Hamburg. Fruits and vegetables were not allowed to be eaten. More than 40 groups of employees had the task of disinfecting houses and plumbing systems.

After publishing these proposals for controlling the epidemic, cholera incidence began to decrease immediately.

Consequences of the Hamburg cholera outbreak

Shortly after the acute control measures, Robert Koch proposed building a drinking-water filtration system because Hamburg's drinking water used water taken directly from the Elbe River without any filtration or treatment.

Through these control measures for water hygiene, cholera could be controlled, and an immediate decrease in child mortality of more than 50 per cent for children between the ages of one and five years was achieved (Figure 65.1).[2]

In addition, the so-called cholera commission was installed in 1894 which published the principles for cleaning surface water by central filtration.

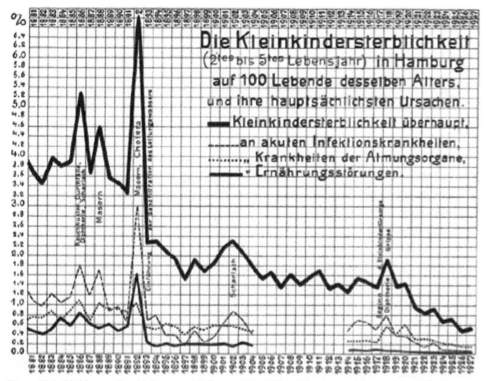

Figure 65.1 Child mortality (1–5 years old) from 1881–1897 in Hamburg with a peak in 1892 (the year of the cholera outbreak) and sharp decrease after implementation of drinking-water filtration system

Koch published an article in *Hygiene and Infectious Diseases* entitled: "Water filtration and Cholera". The most important parts of this article are mentioned below.[1]

Robert Koch wrote:

> To my surprise the cholera situation was completely different at the barrier between Hamburg and Altona. Altona was a city downstream of Hamburg but both cities took their water from the river Elbe. Interestingly, the cholera cases stopped at the barrier between the area of Hamburg and only few cases of cholera were seen in Altona. This was the barrier also from the public water line. (Figure 65.2)[3]

Robert Koch wrote further on:

> This is an experience which has been done with more than 100,000 people, but fulfilled even with these tremendous dimensions all prerequisites one would demand for an exact and evident laboratory experiment. In 2 groups of people all factors were equal but one factor is different, namely the public water supply system. The group which was supplied with unfiltrated raw water from the river Elbe were heavily infected by Cholera on the one side; the group which was supplied with filtered water was only affected by a few cases of cholera. This difference is important, because the water for the public water supply of Hamburg was taken from a point where the river Elbe was not so heavily contaminated. On the other side Altona had had to use water from the river Elbe which was contaminated by the sewage including faeces from more than 800,000 people. Under such circumstances there is – from the point of evidence – no other explanation that this difference must be explained by the difference of the public water supply system and that Altona was protected by the filtration of the water of the river Elbe against Cholera.
>
> Altona was supplied by water, of which the raw water was of much more worsened quality than that of Hamburg. But by cautious filtration this water was cleared nearly completely from the bacteria of Cholera. This explanation is consistent with our present knowledge on infectious diseases.
>
> The Cholera of Hamburg and Altona has given us the apodictic evidence that filtration of water by sand gives a sufficient protection against Cholera infection.

Figure 65.2 Cholera cases (every dot is one cholera case) at the frontier between the free state of Hamburg and the Prussian city of Altona from August until November 1892. The line marks the frontier between the two cities and is also the frontier between the two drinking-water systems

In Hamburg, for more than 18 years before the cholera outbreak, there had been a broad political discussion concerning the need for public water to be filtrated. In 1872 the medical commission in Hamburg declared that the public water system of Hamburg in its present condition was abject and that a central filtration system could supply water of good quality.

In a conference of the German Society of Gas and Water, the director of the Water Society of Altona pointed out:

> From a technical point of view the public water supply in Hamburg is in a worse condition and if technicians would be responsible these technical conditions would be immediately ameliorated. But in such questions of water quality not only hygienists and technicians are responsible but the politicians of the communities and cities which have a significant role. When in Hamburg the filtration of public drinking water is not introduced it is only in the responsibility of the politicians.

In Hamburg in 1890 a regulation was introduced concerning the installation of central filtration for the public water systems. For the implementation of the central-filtration system, it was estimated that building time would take up to four years. When the implementation of the central-filtration system began in August of 1892, cholera broke out. But the work was delayed by the cholera outbreak because there was a loss of working people caused by the cholera outbreak. In this situation, troops from the German army were detailed to fulfil the building of the central-filtration system. More than 1,000 men temporarily occupied these positions. In May 1893, after closing the old catchment area from the Elbe River for public drinking water, only filtrated drinking water was pumped into the public water pipeline.

Robert Koch described the circumstances in his well-known publication: "Water filtration and Cholera", in which he pointed out that Altona, which was protected from cholera, used only filtrated water, and there the heterotrophic plate counts / ml were never more than 100 CFU/ml. Therefore Koch proposed the guideline value of 100 CFU/ml at 20 °C for the verification of good drinking-water quality for filtrated water. Still today, these basic principles for the control of filtrated water have not lost their value.[1]

Koch avidly promoted the regular surveillance of the public water system by microbiological methods. He was convinced that only the state should have responsibility for monitoring water quality. And he pointed out that in a case of an incident or outbreak associated with the public water supply, not only individuals were affected, but the health and life of thousands of people were threatened.

The introduction of central filtration in Hamburg led to mortality rate decrease from 4 per cent to 2 per cent.[2] With the introduction of state guidelines for drinking-water filtration in Germany, cholera was brought under control. Since 1892, no cholera outbreaks have occurred in Germany. There is no better example of the significance of good water quality for the prevention of diseases like cholera and typhus fever.

References

1. Schwalbe, J. 1912. *Gesammelte Werke von Robert Koch*. Leipzig: Verlag von Georg Thieme.
2. Meyer-Delius, H. 1928. Die Sterblichkeit der Säuglinge und Kleinkinder in der Stadt Hamburg in: *Hygiene und Soziale Hygiene in Hamburg*. Hamburg: Paul Hartung Verlag
3. Gärtner, A. 1905. *Leitfaden der Hygiene*. Berlin: Verlag S. Karger.
4. Evans, R. 1987. *Death in Hamburg: Society and Politics in the Cholera Years 1830–1910*. Oxford: Oxford University Press.

66

THE DISCOVERY OF THE AETIOLOGY OF CHOLERA BY ROBERT KOCH 1883*

Martin Exner

PROFESSOR OF HYGIENE AND PUBLIC HEALTH, UNIVERSITY OF BONN, BONN, GERMANY

Introduction

No disease has influenced water hygiene and the prevention and control of waterborne diseases as much as cholera. The detection by microbiological methods of water as a vehicle for *Vibrio cholerae* was the basis for implementation of evidence-based measures for sanitation and the surveillance and verification of the quality of water hygiene.

Through the same water-hygiene-based strategies, cholera and a multitude of waterborne diseases were brought under control. Even up to today, these measures have remained basic hygiene measures and have not lost their significance. Following these hygiene rules will be key to preventing waterborne disease outbreaks in the future.

Through the implementation of hygiene-related strategies for prevention and control of waterborne diseases, child mortality was dramatically reduced. Cholera is seen as a disease associated with the lack of a high degree of civilisation and the lack of strong governance in implementing sanitation infrastructure, which can improve public health protection.

The detection of the cholera aetiology by Robert Koch is among one of the most important discoveries in the history of medicine, hygiene and public health.[1,2]

The cholera outbreak of 1883

On June 24, 1883, *Wolff's Typographic Circle* published the following telegram:

> The government has received by a physician in Damictte (Egypt) the message, that a malignant fever has occurred in the last days in Damiette; out of 20 cases, 6

---

* Recommended citation: Exner, M. 2015. 'The discovery of the aetiology of cholera by Robert Koch 1883', in Bartram, J., with Baum, R., Coclanis, P.A., Gute, D.M., Kay, D., McFadyen, S., Pond, K., Robertson, W. and Rouse, M.J. (eds) *Routledge Handbook of Water and Health*. London and New York: Routledge.

were fatal. The Egyptian sanitary commission therefore resorted to Damiette. The epidemic in Damiette occurred during the fair; meanwhile up to 19 persons died, 11 are under suspicion of cholera.[2]

European governments feared that Europe was at high risk of being affected by cholera, which was introduced from Egypt in 1865. Since 1817, cholera had been responsible for four cholera pandemics coming from the delta of the Ganges in India, which occurred in the years 1817–1823, 1826–1837, 1845–1862 and 1864–1875 in Asia, Africa, Europe and America with different speeds of spreading. In the last cholera pandemic, the disease, discovered in Egypt in May 1865, was introduced in a few weeks into different countries of Europe, including Italy, France, Spain, Turkey, Romania and Russia. England was afflicted in 1866 when cholera was introduced, coming from Rotterdam, the Netherlands. In Germany, cholera escalated so seriously that more than 100,000 people died from the disease, including 6,000 people in Berlin alone. Bismarck finished the war against Austria in 1866 after the battle of Königgrätz in the Prussia-Austrian war because cholera had affected Prussian troops.

The European governments therefore intended not to wait until cholera was introduced into their countries, but to try to find out adequate measures for control and prevention based on scientific methods.

The French government was the first government to deploy a scientific expedition to Egypt on behalf of a proposal by Louis Pasteur. After the approval of this plan by the Comité Consultative d'Hygiène, the French government granted a credit of 50,000 francs for the scientific expedition. The French commission (Mission Pasteur) arrived in Alexandria (Egypt) on August 15, 1883, to begin its scientific work in the "Hôpital Européen". The French commission consisted of the scientists Roux, Thuillier (assistants of Louis Pasteur), Strauß (Professor of Medical Faculty of Paris) and Nocart (Professor of the Veterinary Academia of Alfort, near Paris).

The German government also had an interest in sending out an expedition. Robert Koch, a member of the imperial department of public health, was appointed as leader of the commission. Dr. Gaffky, Dr. Fischer and Dr. Treskow were also members of the commission.

On August 16, 1883, the German expedition, under the leadership of Robert Koch, left Berlin. On August 23, 1883, the members arrived by steamship in Port Said (Egypt). On August 24, 1883, they arrived in Alexandria, where a laboratory was set up with the full support of the Egyptian government.

Cholera in Egypt in 1883 developed in a characteristic way. The first cases occurred on June 22, 1883, in Damiette, on June 27 in Port Said, on July 15 in Cairo and on July 16 in Alexandria. Based upon official statistics, the overall mortality from cholera in this epidemic in Egypt was 28,722 deaths. Robert Koch believed that this was a huge underestimation.

On September 18, 1883, the young and hopeful French scientist Louis Thuillier (1856–1883), a member of the French Mission Pasteur, died of cholera.

Robert Koch and his team not only performed pathological and microbiological investigations, but also investigated the living conditions of the people in Egypt. Koch also used excellent epidemiological statistical material on cholera mortality from 1865 to 1883.

From October 16–30, 1883, Koch and his team visited Cairo where the cholera epidemic had a dramatic impact compared to Alexandria. Koch described in detail the sanitary conditions in Cairo. In some of the suburbs, there was no public supply of filtered water. The commission also visited the water utility where members could inspect filtrated water, which showed contaminants and high turbidity levels. Koch reported that sometimes even 'little fishes' were found in filtrated water.

In addition, Koch also suggested that Ramadan (from July 6–August 14, 1883) had had severe consequences for the cholera outbreak. Later on, during the cholera conference in 1885, Robert Koch described the public water supply of Cairo in detail. He wrote:

> In the year 1865 severe cholera epidemics occurred in Alexandria and Cairo. After this time, both cities got public water supplies. The next epidemic in 1883 was very mild [in Alexandia]with low rates of cholera; in Cairo the cholera outbreak was as severe as in the year 1865. The epidemic curve of both years was the same in Cairo. Has the public water supply had any public health impact? Indeed the public water supply of Cairo would have a public health impact if the water supply would have been better constructed. The point of water-intake was situated in the Ismailia channel directly in the neighbourhood of the bridge connecting Cairo and the suburb of Boulacq. When I visited this point I saw an aspect which reminded me of India. At the riverside of the channel directly to the neighbourhood of the suction tube for the raw water of the public drinking water system, people were washing their dirty shirts, other people were swimming in the channel and there have been a lot of marks of faeces. Even when the water utility was filtering the water, it was made in such an insufficient way that people found in the drinking water tap little fishes. Such a public water tube is not able to prevent cholera infections.[2]

Koch and his co-workers could not verify in detail the microbiologic cause of cholera in Egypt. In all bigger cities of Egypt, the cholera epidemic stopped. Therefore, Robert Koch wrote a letter to the Secretary of Internal Affairs in Berlin to demand an allowance to continue the expedition on to India. He received a positive message that he had permission to continue his expedition. Robert Koch selected Calcutta as the site for the continuance of the expedition. He did so because British public health officers had informed him that cholera was endemic in Calcutta.

The commission arrived on December 11, 1884, in Calcutta. On December 12, 1883, Koch and his co-workers were welcomed by the surgeon general of the government of India, D. J. M. Cunningham, who assured them of his full support. Then he accommodated them in the medical hospital in Calcutta.

On December 13, 1883, the laboratory was set up in the hospital. On December 14, 1883, the commission began with the autopsy of bodies that had succumbed to cholera.

Up until this time, the number of cholera cases was increasing in Calcutta. Therefore Robert Koch thought that Calcutta would be an ideal place to continue his research on cholera. In a letter to the German Secretary of Internal Affairs, Herr von Boetticher, in Berlin on December 16, 1883, the topics of research work of the commission were formulated. These topics were:

1 Microscopic Research of Post-Mortem Autopsies concerning the occurrence of Bacteria in the mucous membranes of the intestinal tract from corpses of cholera patients, especially research on the specific characteristics of this bacteria by microscope to have a form different from other bacteria.
2 Research on the occurrence of cholera in animals including trials of infection to infect different animal species including direct injection into the intestinal tract.
3 Recovery of pure culture of the bacteria which were isolated from the intestinal tract of the corpses of cholera patients and use of these pure cultures for infections trials on animals.

4 Destination of the biological characteristics of this bacteria, especially spore formation, tenacity, properties in different media and a different temperature.
5 Trials to disinfect bacteria.
6 Research on soil, water, air in relation to the cholera infection agent, especially concerning questions whether this agent could exist in endemic regions of cholera independently from use and if there is an influence of destruction procedures in the soil. The special research on the cholera conditions in India included:
  a Associations of the occurrence of cholera in endemic regions with special characteristics in relationship to the population and its environment.
  b Investigations into cholera outbreaks in prisons and in troops on ships.
  c Characteristics in endemic areas of cholera and the most affected places on the one side and in the areas untroubled by cholera.
  d Type and degree of the carry-over of cholera over areas in which cholera is endemic, and the transmission areas in which the carry-over in India and other regions is occurring.
  e To look for strategies for the control of cholera in jails and in army troops which were successful, and the conditions under which some cities like Madras, Pondichery, Guntur, Calcutta showed a the remarkable decrease in cholera mortality.[2]

On January 7, 1884, Koch reported in a letter from Calcutta to the Secretary of Internal Affairs, Herr v. Boetticher, that six autopsies were done and material from six cholera patients had been investigated. The microscopic investigations confirmed the occurrence of the same bacteria that was found in cholera corpses in Egypt.

Now the isolation of these bacteria and cultivation in pure cultures were successful. In addition, the investigators succeeded in seeing characteristics concerning form and microbial growth in gelatine for culture. To validate that these were the causative bacteria, the plan was to investigate if these bacteria were occurring only in the intestinal tract of cholera patients or also in healthy patients. Further research demonstrated that these bacteria could only be isolated in the intestinal tract of cholera patients. These facts were an important indication for the causal relationship, even though the research with animals was not successful.

Later, Koch reported that the epidemiological research based on the British surveillance data showed that cholera in Calcutta decreased in 1870 dramatically. Before 1870, cholera mortality averaged 10.1 per 1,000 inhabitants. After 1870, cholera mortality decreased to 3 per 1,000 inhabitants. Koch postulated that this was related to the introduction of the central public water pipe, which opened in 1869. According to the opinion of the physicians in Calcutta, the decrease in cholera was associated causally with the introduction of the public water pipe.

Based on these insights the commission visited the public water utilities and also the public sewage system in Calcutta. In addition, the members investigated a lot of water samples of the river water in Pultah before and after filtration in the water utility. Koch confirmed, based on the bacteriological results, that the quality of the drinking water was excellent.

The public water supply of Calcutta was instituted on November 1, 1869. Since then the central city area of Calcutta has been supplied with a central public water pipe. The Pultah water supply enterprise was now situated 16 miles outside Calcutta on the left bank of the Hoogly River. Raw water was taken from the Hoogly River and after sedimentation was filtrated in 12 filter basins. The filtration zone consisted of – from the top to the bottom – stones like hen's eggs up to grain of gunpowder shot, a layer of yellow sand and a layer of river sand. The filtrated water was collected in big tanks and then introduced via a natural fall in iron tubes into the city of Calcutta.

In 1869 all the main streets and lanes of the city of Calcutta were supplied with the public water. By 1877, more than 10,471 houses were supplied with public drinking water.

In 1883, Robert Koch had developed the methods for the quantitative enumeration of bacteria in soil, water and air. Using this method, Koch and his co-workers investigated the quality of the drinking water and raw water. In the Hoogly water they found up to 220,000 CFU/ml. In the water of the sedimentation tank they found 20,000 CFU/ml and in the drinking water they found 50 CFU/ml. Robert Koch concluded that the water quality of the drinking water of Calcutta was the same as the drinking water in Berlin, Germany. These investigations were the first investigation of drinking water by modern bacteriological methods in India.

The commission also visited the public sewage channels. In addition it reviewed the extensive epidemiological material on mortality from cholera in Calcutta.

Calcutta was regarded as an endemic area for cholera at this time. Taking into account the statistical material of the morbidity of cholera in Bangladesh, the commission discussed the question of why cholera had been known to be endemic in Bangladesh since 1817.

Based on statistical surveillance data on the mortality of cholera there was a sharp decrease in cholera morbidity in conjunction with the opening of the public water supply (Figure 66.1).

Koch extensively discussed the validity of the results so that they would not be susceptible to misinterpretation. In later reports Koch also argued that with the introduction of a public water supply of high quality (documented by the bacteriological examination of the commission), there was a sustained decrease of cholera mortality in Calcutta to one third the death rates of former times. The commission also extensively considered whether the decrease in cholera mortality was associated with the introduction of the public sewage channels. Because the sewage channels were introduced considerably earlier than the public water supply without any decrease of cholera after introduction of the sewage channels, Robert Koch argued that the decrease of cholera mortality was mainly caused by the introduction of the public water supply.

In this report Koch also discussed the special characteristics of the different cholera mortality in Fort William in which – as in the European quarters of Calcutta – only a low cholera morbidity and mortality could be determined. In the quarters of Calcutta where filtrated drinking water was used, low or no cholera mortality was found. In the quarters, especially of the poor population, where people used drinking water from water tanks which were also used for laundry, cleaning, sewage and disposal of faeces, and for drinking water supply, cholera mortality remained very high (Figure 66.2).

In his letter dated February 2, 1884, to the German Secretary of Internal Affairs, Herr v. Boetticher, Robert Koch declared for the first time that the question of whether the bacteria found in the intestinal tract would be exclusively associated with cholera was now solved. He described the cholera bacteria in detail as bacteria with very active movement. On the agar with gelatine there could be seen colourless colonies like little fragments of glass. From 22 corpses of cholera patients, in 17 of them Koch isolated these – whereas in other non-cholera corpses the bacteria could not be found.

Koch wrote in his letter: "From these results it is possible to conclude that the Comma like bacteria are specific for cholera.[2]

This is the first official report from Robert Koch in which he definitely claimed that he had found the causal pathogen of cholera. The time of the detection of the pathogen of cholera must be dated to the year 1884.

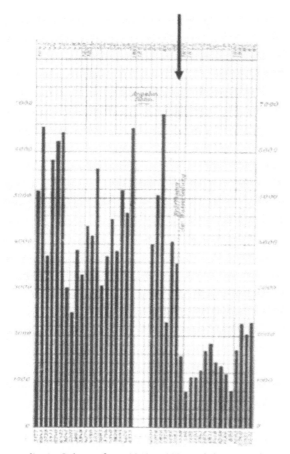

Figure 66.1 Cholera mortality in Calcutta from 1841 to 1884 and the time of the opening of the public water supply in 1869 (arrow)

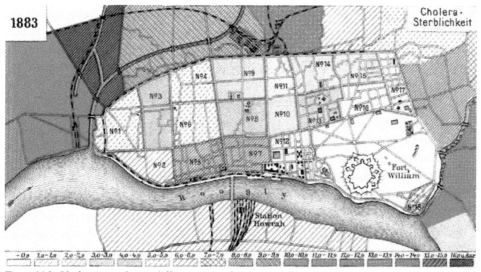

Figure 66.2 Cholera mortality in different parts of Calcutta and Fort William (dark grey: high mortality; white: low mortality) in 1883

He discussed very critically and extensively the reasons for causality. And he concluded that the causality of the *Vibrio cholerae* was now well-founded by additional facts. In detail he described the ecological characteristics of the *Vibrio cholerae*.

In the last point of his letter, he also described the visit to Fort William in Calcutta and the central prison in Alipore, Calcutta.

The epidemiological and hygienic situation of Fort William was of extreme interest to Robert Koch. In Fort William there had been stationed British and Indian troops. There had been a remarkable decrease of cholera some years before the decrease of cholera in Calcutta. Up to 3,300 soldiers were stationed in Fort William.

Fort William got their drinking water from the public drinking water system which was used as drinking water and as water for the preparation of food. Water was taken from two guarded tanks (Figure 66.3) with no access by unauthorized persons allowed. The water in these tanks was filtered before introduction to Fort William (Figure 66.4).

In former times before introduction of this kind of water supply, there had been high cholera mortality – which until 1848 was up to 7 per cent of the English troops in Fort William. After the introduction of the drinking water supply (Figure 66.4) in 1865 there was a drastic decrease of cholera morbidity. No deaths from cholera occurred after this time. Also these epidemiological facts provided evidence-based arguments for Koch that the dissemination of cholera by water was excluded in Fort William and that the amelioration of the water supply in 1865 was the reason that the cholera disappeared almost completely.

In the experiments, the cholera commission members also demonstrated that *Vibrio cholerae* increased in milk and that milk could also be an ideal transmission medium for *Vibrio cholerae*.

In his last letter to the German Secretary of Internal Affairs, dated March 2, 1884, from Calcutta, Robert Koch gave an account of surprising epidemiological characteristics and the first bacteriological-based evidence on the causality of water tanks for locally occurring cholera epidemics. Koch mentioned reports documenting the fact that in endemic areas small epidemics of cholera had occurred in neighbourhoods with water tanks. He described tanks as little ponds with huts in which people were bathing, washing laundry, and from which they also took their drinking water. In addition there were latrines introducing their sewage into these tanks. The British government had ordered that these tanks should be filled in. There was a consequent decrease of the number of such tanks, but in Calcutta alone there had been up to 800 tanks in 1883. Work to fill in these tanks was going on only slowly and the public health officer, Dr. McLeod, wrote in his report: "The work of filling up the tanks and wells has only, as a matter of fact, been commenced, and this must progress until the inhabitants of Calcutta are deprived of this means of *committing sanitary suicide.*"

Because of the association between cholera outbreaks and water tanks, Robert Koch begged the sanitary commissioner of the government to inform him whenever a new local cholera outbreak occurred.

Such a situation occurred in February 1884. From Sahab Baghan in Belliaghatta, one of the suburbs of Calcutta, an unusual increase of cholera cases was reported. The cases of cholera were occurring exclusively around one tank in this region where up to 100 persons lived. Out of these, 17 persons died of cholera, while further from the tank cholera cases were not reported.

Koch and his co-workers made detailed investigations on the origins and cause of the epidemic. They found out that the first patient who had died from cholera had washed his laundry in the tank.

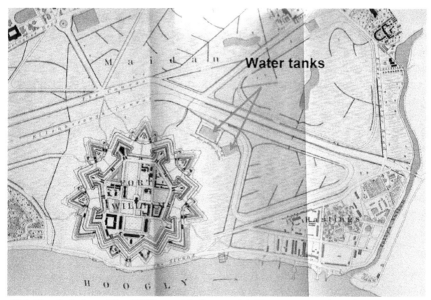

Figure 66.3 Detailed plan of Fort William with the two drinking water supply tanks

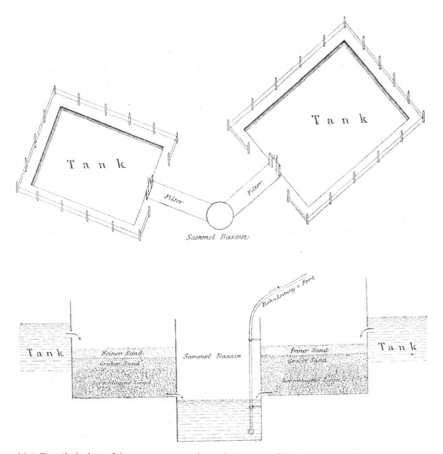

Figure 66.4 Detailed plan of the two water tanks and the water filtration system for Fort William

The commission took a lot of water samples from different areas of the tank and at different times (Figure 66.5). Cholera bacteria could be found in some of the water samples in high concentrations. In water samples at the end of the epidemic only in one of the samples could a few cholera bacteria be isolated. The investigations with the isolation of *Vibrio cholerae* in water tanks were done on February 8 (first visit) and February 11 (second visit). From the last visit, on February 21, after the end of the cholera epidemic, only one colony of *Vibrio cholerae* could be found in the water samples.

Koch described the discovery of the *Vibrio cholerae* in the water samples from the tanks of Sahab Baghan:

> In Calcutta by investigations of different samples from patients of cholera or corpses from cholera, *Vibrio-cholerae* could never be found. Therefore it was very important to answer the question, in how far it would be possible to isolate *Vibrio-cholerae* in the water of tank. This would mean that water would be the most probable transmitter of the infection to the inhabitants.[2]

So in February 1884 cholera bacteria could be found outside the human intestinal tract. It was the first time that *Vibrio cholerae* was found in water associated with a cholera epidemic.

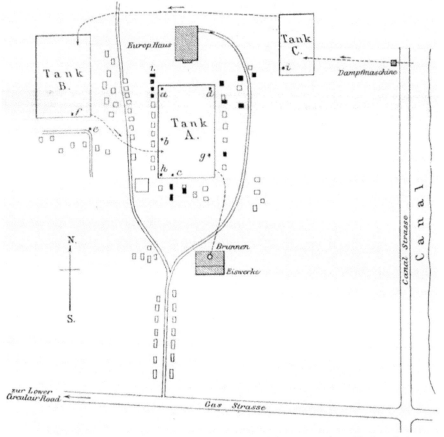

Figure 66.5 The tank of Saheb Bagan and its environment and the points where Koch took the samples (a – i). In samples b, c, d and h, *Vibrio cholerae* could be isolated

Through isolating *Vibrio cholerae*, the causal role of water was not only proven by epidemiological methods, in the same way as John Snow had shown in 1854, but also by a combination of epidemiological, geo-medicinal and hygiene-microbiological investigations of patients and water. This was the crucial supplement to the thesis formulated by John Snow in 1854 from his epidemiological investigation: that cholera cases in London were associated with the water pump in Broad Street.

References

1. Exner M. 2009. Die Entdeckung der Cholera-Ätiologie durch Robert Koch 1883/84. *Hyg & Med* 34(4):54–63.
2. Koch R, and Gaffky G. 1887. *Bericht über die Thätigkeit der zur Erforschung der Cholera im Jahre 1883 nach Egypten und Indien entsandten Kommission*. Berlin: Julius Springer.

67

DR. JOHN SNOW
AND THE BROAD
STREET PUMP[*]

Rosalind Stanwell-Smith

HONORARY SENIOR LECTURER, LONDON SCHOOL OF HYGIENE
AND TROPICAL MEDICINE, KEPPEL STREET, LONDON, UK

Introduction

History is, strictly speaking, the study of questions…
WH Auden (1907–1973), The Dyer's Hand, 1962

The story of Dr. John Snow's investigation of the 1854 London cholera outbreak has been taught to generations of epidemiologists, water scientists and others in public health. His demonstration that the epidemic was caused by contaminated pump water gained influence when later discoveries, particularly establishment of the germ theory, supported his conclusions. This contribution to epidemiology is now better known than his pioneering work as an early anaesthetist. Recent interpretations of the Broad Street pump story (Lax 2005, Hempel 2006, Johnson 2006, Morris 2007) include a novel intended for young readers (Hopkinson 2013). An academic biography encompassing the range of his work is the most comprehensive account (Vinten-Johansen et al. 2003). The John Snow Society, which I founded with other enthusiasts in the early 1990s, celebrates his memory with a growing international membership. A poll in a British medical magazine voted him as 'the greatest doctor of all time' in 2003, with Hippocrates, no slouch in water matters, coming second. Snow's iconic status has spread from anaesthesia and epidemiology to chemistry, water science, statistics and cartography. Public health is a relatively new specialty, with a corresponding need for heroic stories, yet the growth of 'Snow the Icon' has not gone uncriticised, not least as to why he should be singled out as such a hero (Vandenbroucke et al. 1991, McLeod 2000). Others were working on water as a cause of infections such as typhoid fever and on anaesthetics; and he was certainly not the first to use a map to study an outbreak. In response, it may be argued that this is how history works: the key people in place at the

---

[*] Recommended citation: Stanwell-Smith, R. 2015. 'Dr. John Snow and the Broad Street pump', in Bartram, J., with Baum, R., Coclanis, P.A., Gute, D.M., Kay, D., McFadyen, S., Pond, K., Robertson, W. and Rouse, M.J. (eds) *Routledge Handbook of Water and Health*. London and New York: Routledge.

time of decisive or well-documented events tend to be the ones remembered. Dr. John Snow was the ideal man to be in that place, with his experience of the first arrival of cholera in England and his meticulous analytical approach to medicine and related scientific questions of the day. Also, leading on from Auden's view of history, Snow was asking questions that others had not thought to pose, and attempting to answer them from empirical studies or research in the field.

This chapter aims to put the Broad Street cholera outbreak into a water and public health perspective as well as to review its iconic status. It starts with the background to the understanding of cholera in the 1840s and 1850s as an introduction to the events of the summer of 1854 in Soho, and concludes with thoughts on Snow's contribution and the implications of the Broad Street outbreak for water today.

Cholera in London

The arrival of 'King Cholera' in Europe

The intense cholera outbreak of 1854 in Soho was not the first experienced in London. Before it reached Europe in the 1830s, 'cholera' had been a term for the diarrhoeal disorders common in tropical regions, usually attributed to climate effects. Its name also related to the ancient theory of four humours as the base of disease, so that those with an irascible 'choleric' humour were thought more likely to contract it. The first recognised outbreak occurred in India, East Africa and South East Asia in 1817. Reports of this more serious form of cholera, renamed 'cholera morbus' to distinguish it from a passing summer symptom, caused great anxiety when it arrived in North East England in October 1831, eventually causing an estimated 30,000 deaths (Morris 1976). John Snow (1813–1858) witnessed early cases in mining villages near Newcastle while working as an apprentice general practitioner. By 1832 cholera was all over Europe and had reached the Americas. The rapid death, leaving victims with sunken eyes and shrivelled, discoloured skin, was terrifying in itself, but worse was the mystery surrounding its cause. The disease appeared to be *sui generis* (a law unto itself) (Alibrandi 2013) but theories were quickly presented. The miasma theory of disease causation predominated; this blamed the toxic atmosphere generated by decomposing waste or unventilated, dirty conditions. Contagionists, who argued for spread between people as well as sources other than 'miasma', were not taken seriously, since in many cases those attending the sick did not succumb to the disease. We know now that person-to-person transmission of cholera is very rare, but for people in the early and mid-nineteenth century, the miasma model fitted well enough. Cases were more common in the cramped conditions of the poor, in crowded military garrisons and in dirty, smelly areas. Inadequate diet, overwork and immoderate temper were cited amongst additional reasons for the victims being susceptible while others were spared (Davey-Smith 2002).

London, the largest city in the early nineteenth century world and the first to reach a population of a million people, experienced its first outbreak of cholera in 1832. It was an ideal environment for the spread of 'King Cholera'. The infrastructure was ill equipped for the rapid rise in population, with sewers based mainly on underground rivers and drinking water abstracted from an increasingly contaminated River Thames. Most citizens disposed of human waste via cesspits that were emptied infrequently in poor quarters, where people could not afford to pay disposal charges or their landlords could choose to ignore the problem. The liquid and solid waste ran into the streets, joining animal excreta and other waste, eventually draining to the Thames along with the sewer outfalls. Piped water was a luxury, although common in the houses of central London, but usually the supply was limited to the ground floor and intermittent, including no supply on Sundays. The river water was untreated, as was that provided by street pumps. Most streets and squares had these shallow pump-wells, benefitting from the abundant water table below. The water varied in quality, with the Broad Street pump ironically valued for its clear, good tasting water. Mains water in Soho mostly came from the man-made New River to the north, much cleaner than the central Thames and, until 1854, Soho had been spared any major outbreak.

Other parts of London were less fortunate, giving Dr. Snow the opportunity in the 1840s to study mains water supply from different companies in relation to cholera deaths and to conclude that water from the most contaminated section of the Thames posed the highest risk. His fieldwork included tasting as well as chemically analysing the salt content of the household supplies to check which part of the river it came from, a brave manoeuvre for Snow who drank only boiled water. An active member of medical societies in London, he also published his findings about water in the first edition of his book on cholera in 1849. His thesis was largely ignored. There was a deluge of publications about cholera, and more eminent doctors and scientists, as well as politicians and highly placed administrators, were banking on the miasma theory. This led to the first Public Health Act in the UK in 1848 (Parliament UK), where cleaning up street 'nuisances' and other sources of bad odours were ordered, although these were not specifically required for many areas, nor enforced. A related act required cesspits and indoor toilets to be connected to sewers, but did not otherwise recommend changes in water or sewage management. The rising popularity of flushing lavatories only increased the burden of sewage entering the Thames and its tributaries. Meanwhile, Dr. Snow developed his successful practice as a pioneer anaesthetist at his house in Frith Street, Soho, and later in Sackville Street in Piccadilly, both only a short distance from the explosive outbreak of cholera that was to provide compelling evidence for his theory.

The Broad Street and Golden Square cholera outbreak, 1854

When the first recorded death in the Soho outbreak occurred, cholera had already raged in South London since the previous year. Dr. Snow had been continuing his door-to-door investigation of links to different water supplies, particularly those receiving the most sewage laden Thames water. Knowing the Soho water to be relatively good and also of instances elsewhere in which pump water could have been responsible, he focused on the rapidly escalating cases around Golden Square and Broad Street (Figure 67.1). Within a short time there were over 500 deaths and it remains the single largest and most sudden outbreak experienced in London. In addition to the key events of the incident (Table 67.1), John Snow recognised the importance of the far fewer deaths at the Broad Street brewery and the nearby workhouse: both had their own wells, not polluted by cesspits. Snow also demonstrated the value of outlying cases, such as the 'widow of Hampstead' (Mrs. Eley) who had received regular deliveries of the Broad Street pump water; hers was the only case of cholera in the Hampstead area. There were others whose sole consumption had been a beverage diluted with the pump water, for example at local coffee shops. The famous map, documenting the 616 deaths traced by Snow, was not produced until he was revising his cholera book (Snow 1849) and when he was required to report to the Cholera Inquiry Committee following this incident. William Whitehead, a young clergyman, set out to disprove Snow's theory but his house-to-house survey, using a questionnaire developed by Snow and other committee members, gradually convinced him that Snow was right.

Table 67.1 Main events of the Soho outbreak in 1854

Date	Event	Related features
28 August 1854	Frances Lewis, infant at 40 Broad St. falls sick with diarrhoea	Water from rinsing the baby's nappies (diapers) is poured into the house's cesspool, located close to the Broad St. pump.
2 September	Baby Frances dies; also Susannah Eley in West End, village next to Hampstead	Mrs. Eley was the widow of a factory owner in Broad St. Many other cases occurring; Florence Nightingale, working at a hospital nearby, reported patients coming in 'every half hour' from the Broad St. area.
3 and 4 September	Dr. Snow takes water from pumps in the area	Notices white particles in otherwise clear water from Broad St. pump – takes sample to be examined by microscopist Dr. Arthur Hill Hassall, who reports oval 'animalcules' and organic matter.
5 September	Dr. Snow obtains list of cholera deaths	73 of 83 registered deaths occurred in those living closest to the Broad St. pump; others were known to have drunk the water.
7 September	Dr. Snow attends the vestry meeting of the local board of governors	Persuades the governors to have the pump handle removed.
8 September	Pump handle removed	Unpopular with locals – pump water considered cleaner than the mains supply, even when available.
27 November	Broad St. well inspected	Superficial check: pronounced to be sound.
4 December	Dr. Snow presents his cholera spot map to the Epidemiological Society of London	Shows deaths as black bars on the map. Final map shows points of equal walking distance to the pump. Further evidence presented over next few months.
April 1855	Pump and well re-inspected	Bricks in wall of well are found to be loose and backed up sewage could easily enter from cesspool and drain.

Of 896 residents in Broad Street, 90 had died and Whitehead was able to trace some 500 of the survivors, many of whom had fled the area when the outbreak struck. Of those who developed cholera, 80 per cent had drunk the water and 20 per cent had not, while only 17 per cent of those who reported drinking it were free of the disease (Vinten-Johansen et al. 2003). When properly excavated some nine months after the outbreak, a clear route for sewage to percolate into the well water was demonstrated. The General Board of Health, founded in response to the 1848 Public Health Act, dismissed both Snow's and Whitehead's evidence, and in September 1855 the local parish authority reopened the pump. Until mains water could be made available continuously to most homes, there was no other reliable source of water. Gradually, the dangers of superficial wells and sewage ingress were recognised and the wells were closed, mostly with little trace of their existence. We are still not sure what the Broad Street pump looked like; a replica of its probable structure was erected in the street in the 1990s (Figure 67.2).

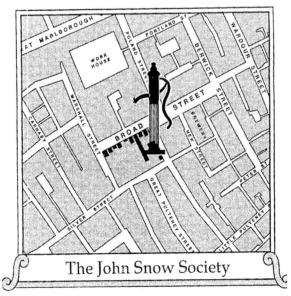

Figure 67.1 The area of the outbreak. Black bars indicate some of the cases in Broad Street. Golden Square was located south of Silver Street, with a corner of it showing in this drawing

Source: From artwork for the early John Snow Society mugs based on the original maps. Note that cases other than in one section of Broad Street were omitted in this design. (From author's collection of drawings made for the society.)

Figure 67.2 Replica of the Broad Street pump in Soho, London. The street is now 'Broadwick Street'. The public house on the site of 40 Broad Street, renamed 'The John Snow' in 1954, is the building in the background; the original pump was located near the second lamp on the pub. Its position is marked by a pink granite paving stone, not visible in this picture

Source: Photograph by author.

The importance of the Broad St. outbreak

Lessons from this outbreak

> It may be noticed that the deaths are most numerous near to the pump where the water could be more readily obtained.
>
> *(Snow, 1855)*

Most cholera epidemics in the early nineteenth century showed a continuous source pattern, as would be expected from frequent sewage contamination of the drinking water. A point source (single source) outbreak is a modern concept, but this unusual eruption of cholera could not have been better placed to grab the attention of influential people and newspapers. Timing was ideal for information gathering: standardised and centralised registration of deaths by cause had been established in 1837 and it was well developed in London, alongside the traditional records held by parish churches. The timing was also pertinent for the debate of contagion versus miasma, urgently discussed in London's medical societies as cholera continued to spread (Ryan 2013). Yet, with miasma, the 'non-contagionists' had an apparently logical cause that fitted with their beliefs and understanding of the environment. The existence of miniature, live agents that could cause disease seemed bizarre; Snow's arguments about routes of transmission were equally alien to them and deemed irrelevant, since the toxic effects of miasma were believed to reach anywhere in the body. It would be nearly 30 years before Robert Koch demonstrated the existence of cholera bacteria (*Vibrio cholerae*) and, even by the 1880s, the faecal–oral route was incompletely understood, nor acted upon in measures such as provision for hand washing. The latter was an optional extra in public toilets, for those who could afford to pay, even in the 1900s. The principal sanitary measure in the Soho outbreak was to douse the gutters and roads with chloride of lime to suppress the miasmic 'effluvia'. Sanitary studies, such as Edwin Chadwick's survey of the appalling conditions for many poor urban dwellers (Chadwick 1842), all seemed to point at the efficacy of cleaning up cities, as well as promoting the moral lifestyles that were thought to prevent the inhaled effluvia taking hold. Without a determined researcher such as Dr. Snow, the association with water could easily have been missed or simply added to the many proposed causes of cholera, particularly when the Broad Street well was initially declared to be in good order and the importance of Hassall's microscopy went unrecognised. Others had proposed polluted water as a cause of cholera, although none with the singular focus on 'mode of communication' shown in Snow's writings. Snow's contribution was not so much the map, produced well after the outbreak, as in his deductive reasoning about the importance of a sudden outbreak in an area with previously fairly good water, and on focusing on the likely mode of transmission. He had seen many cases of cholera in the north, noting the unhygienic conditions in coalmines where neither toilets nor hand washing facilities existed. His expertise in respiratory physiology made him doubt that the disease was due to inhalation of miasmic gases; surely, he reasoned, it was more likely to have been ingested orally, as borne out by the predominantly gastrointestinal symptoms.

His painstaking documentation of outlying cases is now used as an example of how to solve an epidemiological mystery; at the time, these were rejected or explained away. The removal of the pump handle when the main outbreak was subsiding has also been suggested as a more symbolic than effective action. Whitehead's detailed enquiries showed that soiled bed linen, from the cholera stricken father of the first case, had been rinsed into the cesspool at 40 Broad St. on 9 September. This could have caused further cases if the handle had not been just removed. So, while the repeated teaching of this outbreak has raised it far above its importance in the mid-nineteenth century, eclipsing the many others who contributed to the cholera and water story, it was a significant incident with an effective control measure. Within 50 years it was cited as one of the most instructive cases of the transmission of disease by polluted water by William Sedgwick, an influential public health expert in the United States, and Wade Hampton Frost then championed it in the 1930s (Vinten-Johansen et al. 2003). It was in American schools of public health that the Broad Street pump story was first applauded as an exemplary case study for epidemiology, with later adoption of the iconic tale, complete with beer barrel earrings to commemorate the brewery clue, for those training at the Centers for Disease Control and Prevention in Atlanta. Ironically, the names of infection heroes decorating the London School of Hygiene and Tropical Medicine include not Snow, but Max von Pettenkofer, who opposed the germ theory. The 23 names were carved in the late 1920s, showing that Snow's reputation was not prominent in the UK at this time; in their defence, decisions by the name committee appear to have been made mainly on the basis of teaching on hygiene, to which von Pettenkofer made a major contribution.

Contemporary relevance of the 1854 Soho outbreak

Unlike some mythic heroes, once the myth is removed, good reasons for remembering Snow remain, especially for epidemiologists…Once mode of communication is established, preventive measures nearly always follow.

(Paneth 2004)

The fact that the significance of Snow's research on cholera and water was not appreciated for so long after his lifetime does not diminish the quality and unique nature of his work; a major shift in thinking about disease causation was needed before Snow's ideas could be accepted.

Reasons for its epidemiological and water science importance, now recognised in training courses around the world, include:

1 Demonstration of the value of the epidemiological method – population studies and field work to document case (and control) characteristics.
2 Comprehensive analysis of the modes of transmission of disease.
3 Deduction, based on evidence, that causes and routes of transmission are specific and, most importantly, that they can be interrupted (e.g. removing the pump handle).
4 For water science, it provided a lucid demonstration of the need to separate sewage from water supplies and later for water treatment; its example has both inspired and justified continuing attempts to give everyone clean water and safe sanitation.

Few examples in public health are as clear as 'removing the pump handle'. This and the 'spot map' have helped to enshrine the story as visually attractive, memorable and a rare instance of readily understood interruption of a disease pathway.

After the outbreak, and the mystery of Snow's London home

Dr. John Snow died four years after the 1854 epidemic, and while he continued to collect evidence to prove his theory, he would have been astonished by the long-reaching influence of the Broad Street episode. Rather, he might have preferred his detailed study of water supplies and sewerage to be his main recognised contribution, along with his achievements in pioneering the use of chloroform and related respiratory research. His contemporaries did not see him as a hero and his immediate legacy was in anaesthesia, for which he became celebrated in the UK long before the pump handle story was disseminated. The emphasis on the pump incident – and simplification of the story for teaching – has prevented proper recognition that his anaesthetic knowledge informed his theories about the transmission of cholera, as did his extensive clinical experience and fieldwork. He also made important contributions to chemistry, such as in dose–response studies, commemorated in 2008 by a Royal Society of Chemistry plaque on the wall of the John Snow public house in Broadwick Street.

Snow was never aggressive or impolite to the misguided sanitarian 'miasma slayers', but he was no diplomat. His determination to refute the non-contagionists undoubtedly antagonised his medical colleagues, including his refusal to support legislation against the odorous emissions from bone boiling factories (Lilienfeld 2000, Paneth and Fine 2013). He correctly questioned the absence of evidence that such fumes caused disease; much later epidemiological research on air pollution showed he was wrong to be so dogmatic, despite making a valid point about evidence bases.

We know very little of Snow's personal life. He appears to have had few interests outside medicine, apart from his adherence to vegetarianism and teetotalism, which baffled most of his contemporaries. Even the exact address for his main home in London remained a mystery until recently (Stanwell-Smith and Zuck 2013). Due to haphazard and changing house numbering in the early nineteenth century, it was suggested that a plaque commemorating his home at 54 Frith Street was at the wrong address. Maps showing street numbers were not available for the years of Snow's occupancy (1838 to 1852) and the numbering question was researched by detailed investigation of census records for Frith Street residents, events such as fires and court cases concerning local robberies. These were then checked against fire insurance records and gazette notices, which showed that house numbers had changed before Snow had moved into the street. Number 54 is in the same location as when Snow lived there, although the building has been replaced. It is appropriate for Dr. Snow's contribution to public health science that this intriguing problem was resolved not by maps, but by studying person, time and place, the building blocks of epidemiology.

Key recommended readings (open access)

1 UCLA Department of Epidemiology and Public Health website on John Snow, edited by Frerichs RR. http://www.ph.ucla.edu/epi/snow.html. Wealth of information about Snow's life and work, including bibliography and links to papers.

2 John Snow Society website: http://www.johnsnowsociety.org. General information, past editions of the Broadsheet newsletter and transcripts, audio recordings/slides of the Pumphandle lectures held at the London School of Hygiene and Tropical Medicine.

3 Faruque SM, Albert MJ, Mekalanos JJ. 1998. Epidemiology, genetics and ecology of toxigenic *Vibrio cholerae*. *Microbiology and Molecular Biology Reviews* http://www.ncbi.nlm.nih.gov/pmc/articles/PMC98947/. Describes the aetiology and different pandemic strains of cholera, including those in the early nineteenth century.

4 A tour of Soho looking at the work and legacy of Dr. John Snow. YouTube audio recording: http://www.youtube.com/watch?v=8vBHbGnIsow. A recorded walk by the author in Soho made by London School of Hygiene and Tropical Medicine to celebrate the bicentenary of Snow's birth in 2013, with slides showing the plaque and the pink paving stone location of the pump.

5   Cholera and the Thames: http://www.choleraandthethames.co.uk/. Website dedicated to a project about London's battle against cholera, including a game that takes you through the events of the 1854 outbreak

References

Papers indicated with an asterisk below are available freely online, via the UCLA website where not otherwise specified.

Alibrandi, R. 2013. Epidemic and public institutions in early 19th century Naples. *Rivista di Storia della Medicina*. http://www.academia.edu/6650917/epidemic_and_public_institutions_in_the_early_19th_century_naples_at_the_time_of_cholera_1836-1837_

Chadwick, E. 1842. Report to her Majesty's Principal Secretary of State for the Home Department from the Poor Law Commissioners on an Inquiry into the Sanitary Condition of the Labouring Population of Great Britain London: HMSO. Online extracts: http://www.sochealth.co.uk/resources/public-health-and-wellbeing/report-of-the-poor-law-commissioners/report-on-the-sanitary-condition-of-the-labouring-population-of-great-britain/

Davey-Smith, G. 2002. Commentary: Behind the Broad Street pump: aetiology, epidemiology and prevention of cholera in the mid-19th century. *International J Epidemiology* 31(5): 920–932. http://ije.oxfordjournals.org/content/31/5/920.full

Hempel, S. 2006. *The Medical Detective: John Snow; Cholera and the Mystery of the Broad St. Pump*. London: Granta Books

Hopkinson, D. 2013. *The Great Trouble: A Mystery of London, the Blue Death and a Boy called Eel*. New York: Alfred A Knopf/ Random House Books

Johnson, S. 2006. *The Ghost Map*. London: Allen Lane

Lax, A. 2005. *Toxin: The Cunning of Bacterial Poisons*. Oxford: Oxford University Press

Lilienfeld, DE. 2000. John Snow: the first hired gun? *American Journal of Epidemiology* 152(1): 4–9

McLeod, KS. 2000. Our sense of Snow: the myth of John Snow in medical geography. *Social Science & Medicine* 50: 923–935

Morris, RD. 2007. *The Blue Death. Disease, Disaster and the Water We Drink*. Oxford: Oneworld Publications

Morris, RJ. 1976. *Cholera 1832 – the Social Response to an Epidemic*. London: Croom Helm

Paneth, N. 2004. Assessing the contributions of John Snow to epidemiology: 150 years after removal of the Broad Street pump handle. *Epidemiology* 15(5): 640–644

Paneth, N, and Fine, P. 2013. The singular science of John Snow. *The Lancet* 381: 1267–1268

Parliament UK. 1848. The 1848 Public Health Act. http://www.parliament.uk/about/living-heritage/transformingsociety/towncountry/towns/tyne-and-wear-case-study/about-the-group/public-administration/the-1848-public-health-act/ (includes commentary)

Ryan, ET. 2013. Eyes on the prize: lessons from the cholera wars for modern scientists, physicians and public health officials. *American J Tropical Medicine and Hygiene* 89(4): 610–614. http://www.ajtmh.org/content/89/4/610.full

Snow, J. 1849. *On the Mode of Communication of Cholera*. London: Churchill (revised to include the Broad Street outbreak and further evidence, 1855)

Stanwell-Smith, R, and Zuck, D. 2013. The John Snow plaque: Where exactly was 54 Frith Street? *Proceedings of the History of Anaesthesia Society*. http://www.histansoc.org.uk/uploads/9/5/5/2/9552670/anaesthesia_vol_46.pdf (pages 23–31)

Vandenbroucke, JP, Eeklman Rooda, HM, and Beukers, H. 1991. Who made John Snow a hero? *American Journal of Epidemiology* 133(10): 967–973

Vinten-Johansen, P, Brody, H, Paneth, N, Rachman, S and Rip, M. 2003. *Cholera, Chloroform and the Science of Medicine: A Life of John Snow*. Oxford: Oxford University Press

68

THE ARSENIC CRISIS
IN BANGLADESH*

Christine Marie George

ASSISTANT PROFESSOR, JOHNS HOPKINS BLOOMBERG SCHOOL OF PUBLIC
HEALTH, DEPARTMENT OF INTERNATIONAL HEALTH, GLOBAL DISEASE
EPIDEMIOLOGY AND CONTROL PROGRAM, BALTIMORE, MD, USA

Learning objectives

1 Understand the history of the arsenic problem in Bangladesh.
2 Identify health implications of chronic arsenic exposure from drinking water.
3 Understand the complexities of developing sustainable arsenic mitigation strategies.

History of the arsenic problem

During the 1970s, the government of Bangladesh in collaboration with the United Nations Children's Fund (UNICEF) encouraged a shift from using microbially contaminated surface water to wells that tapped groundwater (Kinniburgh, 2001). Wells were believed to represent a safe drinking water option relative to surface water that was easy to install, relatively cheap, and required low maintenance. However, by the early 1990s it was apparent that many of these wells tapped shallow aquifers with elevated levels of naturally occurring arsenic (Dhar et al., 1997). The shift to wells was estimated to expose a population of 28 to 35 million to water arsenic concentrations that exceeded the Bangladesh arsenic standard of 50 μg/L (Figure 68.1), and an even higher number if the World Health Organization (WHO) arsenic guideline of 10 μg/L is used (Ahmed et al., 2006).

Elevated levels of arsenic in the shallow groundwater aquifer have been found across Asia (Fendorf et al., 2010). An estimated 100 million people in India, Bangladesh, Vietnam, Nepal, and Cambodia are exposed to water arsenic concentrations exceeding the WHO arsenic guideline (Ahmed et al., 2006). The elevated naturally occurring arsenic in the region is hypothesized to be the result of arsenic-rich iron oxides in sediments being dissolved and released into the groundwater aquifer (Zheng et al., 2004; Fendorf et al., 2010).

---

* Recommended citation: George, C.M. 2015. 'The arsenic crisis in Bangladesh', in Bartram, J., with Baum, R., Coclanis, P.A., Gute, D.M., Kay, D., McFadyen, S., Pond, K., Robertson, W. and Rouse, M.J. (eds) *Routledge Handbook of Water and Health*. London and New York: Routledge.

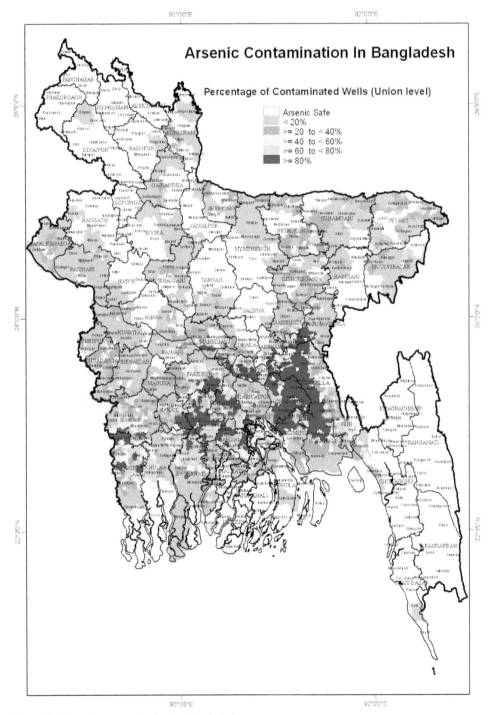

Figure 68.1 Arsenic contamination in Bangladesh

Health implications of arsenic exposure

Exposure to elevated levels of inorganic arsenic is associated with an increased risk for cancers of the lung, bladder, and skin (Chen and Ahsan, 2004), cardiovascular disease (Chen et al., 2011), neurological effects (Wright et al., 2006), skin lesions (Haque et al., 2003), respiratory effects (Parvez et al., 2008), and all-cause mortality (Argos et al., 2010). Chronic arsenic exposure is also associated with deficits in childhood cognitive and motor function (Wasserman et al., 2007).

Arsenicosis is defined as chronic clinical toxicity due to high levels of arsenic being present in the body. Clinical symptoms of arsenicosis can include: melanosis, spotted and diffuse keratosis on the palms and soles, and dorsal keratosis (Saha, 2003). These arsenic-induced skin lesions are the most common visible clinical sign of toxicity from chronic arsenic exposure (Yu et al., 2006), though overall morbidity and mortality are more strongly derived from internal organ damage which often occurs in individuals without skin lesions. Arsenic-induced skin lesions were observed in West Bengal, India, as early as the 1980s (Chakraborty and Saha, 1987). In contrast to other diseases associated with arsenic, arsenicosis has been reported after a short duration of exposure (Rahman et al., 2001).

In the years to come Bangladesh will face a rapid increase in arsenic-related diseases due to the latency period of many of the associated illnesses. Many of the new generations born after the switch to wells will likely face a lifetime of chronic arsenic exposure. It has been estimated that the arsenic crisis in Bangladesh will result in more than a doubling of the future excess cancer risk for this exposed population (229.6 versus 103.5 per 100,000 population) (Chen and Ahsan, 2004). Alarmingly, a recent study estimated as many as 42,000 deaths in Bangladesh annually are associated with chronic arsenic exposure above 10 μg/L. Over the next 20 years this arsenic-related mortality could lead to estimated economic losses of 13 billion USD if the current arsenic exposure in the population remains the same (Flanagan et al., 2012).

Policy and arsenic mitigation

In the early 1990s, after the scale of the arsenic problem was recognized, there was a national survey of water arsenic concentrations in groundwater sources conducted by the Bangladesh Department of Public Health Engineering (DPHE) and the British Geological Survey (Kinniburgh, 2001). Groundwater arsenic concentrations in Bangladesh were found to be the lowest in the northern part of the country and the highest in the southern part. In 1999, DPHE with support from the World Bank and UNICEF undertook a nationwide water arsenic testing campaign under the Bangladesh Arsenic Mitigation and Water Supply Program (BAMWSP) (DPHE, 2010a). Through this program they were able to test close to 5 million wells using field arsenic test kits throughout the country from 1999 to 2005 (DPHE). During the national testing campaign, wells were painted red if they exceeded the Bangladesh arsenic standard, and green if they were below this standard. Approximately 30 percent of these 5 million wells were found to have arsenic concentrations that exceeded the Bangladesh arsenic standard of 50 μg/L (Johnston and Sarker, 2007). This campaign was one-off in scope and did not disseminate messages on the health implications of arsenic, or provide arsenic testing over time as new wells were installed.

In 2004, the Bangladeshi government issued a National Policy for Arsenic Mitigation. This policy promoted the use of the following options for arsenic mitigation: improved dug wells, pond sand filters, large scale surface water treatment, deep wells, rainwater harvesting, arsenic removal technologies (ARTs), and piped water supply systems (National Arsenic Mitigation Information Center, 2004). The Bangladeshi government also set up the Bangladesh Environmental Technology Verification-Support to Arsenic Mitigation (BETV-SAM) program under the Bangladesh Council of Scientific and Industrial Research. BETV-SAM was responsible for evaluating the effectiveness of potential ARTs to be used in the country. Evaluations were based on the ability of the ART to consistently provide water below 50 μg/L arsenic, to produce manufacturer specified quantities of arsenic safe water, and to adhere to national water quality standards. To date, five household level ARTs and one community level ART have been approved under this program.

Of the population of 28–35 million initially exposed to arsenic above the Bangladesh standard, 57 percent are estimated to remain exposed. The most common arsenic mitigation option used by the affected population is well switching at 29 percent (Ahmed et al., 2006). This involves arsenic unsafe well users switching to arsenic safe wells relative to the Bangladesh arsenic standard (<50 μg/L As) located in their communities. This is possible due to the spatial heterogeneity in the distribution of arsenic in the shallow groundwater aquifer (Van Geen et al., 2002). Intervention studies in Bangladesh which encouraged well switching through health promotion and water arsenic testing have observed significant reductions in urinary arsenic concentrations, a biomarker of arsenic exposure, in individuals who reported switching to arsenic safe drinking water sources (Chen et al., 2007; George et al., 2012a). Well switching has been found to be most effective in communities with low to moderate arsenic contamination (0–60 percent of arsenic unsafe) relative to the Bangladesh arsenic standard (George et al., 2012a). This is estimated to include 77 percent of the population in Bangladesh. Well switching, however, is not an option in highly arsenic-contaminated areas of the country where more than 80 percent of shallow wells exceed the local arsenic standard. One potential concern for well switching is the temporal variability of the arsenic concentrations in a well over time. Previous studies, however, have found the temporal variability of arsenic concentrations in wells over time to be low (Dhar et al., 2008; Cheng et al., 2005; Steinmaus et al., 2005; Thundiyil et al., 2007; Fendorf et al., 2010). Rare events, such as the entry of arsenic contaminated groundwater through cracks in pipes, however, can lead to the contamination of wells originally labeled as arsenic safe relative to the Bangladesh arsenic standard (van Geen et al., 2007).

The second most common arsenic mitigation option is the use of deep wells, now used by an estimated 12 percent of the affected population (Ahmed et al., 2006). Studies have shown that deep groundwater aquifers (>150 m) generally have arsenic concentration below 10 μg/L (Radloff et al., 2011). Therefore the installation of deep wells is another viable option for areas that have arsenic in shallow groundwater. One barrier to the widespread use of deep wells is their high cost, which can be up to 1500 USD depending on the depth of the well.

Arsenic mitigation options such as rainwater collectors, dugwells, arsenic filters, and pond sand filter are estimated to be used by less than 2 percent of the arsenic affected population in Bangladesh (Ahmed et al., 2006). Barriers to the widespread use of arsenic removal devices have been mostly related to inadequate maintenance, frequent clogging of filters, and lack of user friendliness of the technologies used (Hossain et al., 2005). In addition, the use of dugwells and pond sand filters have been associated with microbial contamination (Howard et al., 2006).

A national drinking water quality survey conducted in 2009 found that 20 million and 45 million people were still exposed to arsenic that exceeded the Bangladesh arsenic standard of 50 μg/l and the WHO arsenic guideline of 10 μg/l, respectively (Pathey, 2009). A major barrier to arsenic mitigation in Bangladesh is the lack of access to water arsenic testing services. In many regions of Bangladesh it has been nearly 10 years since the previous nationwide arsenic testing program. However, the number of new wells being installed continues to grow rapidly (George et al., 2012b). The arsenic status of 44 percent of wells in arsenic affected sub-districts of Bangladesh is unknown (DPHE, 2010b). Furthermore, the paint used to label wells red or green relative to the Bangladesh arsenic standard in the previous nationwide arsenic testing campaign has long ago faded. Without information on the arsenic status of household drinking water, the population will continue to be exposed to elevated arsenic concentrations in drinking water.

There has been an emergence of rapid field tests for measuring arsenic in drinking water. These field kits are relatively low in cost (0.17–0.60 USD/test) and have a reaction time of 10 to 40 minutes (George et al., 2012c). Previous studies have demonstrated that village workers with relatively little training are capable of effectively using these water arsenic testing kits (George et al., 2012b). The Ministry of Local Government and Rural Development Cooperatives of Bangladesh, in collaboration with UNICEF and several other developmental agencies, recently piloted a fee-based well water testing program for arsenic. This program trained arsenic testers in eight sub-districts of Bangladesh to go door to door to offer fee-based water arsenic testing services. The revenue from the arsenic testing went to purchasing additional arsenic testing kits. An evaluation in Shibalaya sub-district found that the vast majority of households (93 percent) purchased an arsenic test for their well when fee-based arsenic testing was combined with an arsenic educational program (George et al., 2013). An advantage of fee-based arsenic testing is that it provides a financial incentive for the tester to seek out untested wells (George et al., 2013; van Geen, 2013).

Bangladesh currently lacks arsenic safe drinking water options to serve the entire arsenic affected populations. Without arsenic testing, the number of untested wells will rise as new wells are installed and millions will unknowingly be exposed to elevated arsenic in drinking water. As we approach the 40 year mark since the arsenic crisis first arose for this country, sustainable arsenic mitigation strategies are urgently needed. The most sustainable arsenic mitigation strategies will likely be those that build local capacity through fee-based arsenic testing implemented by villager workers. Another massive blanket water arsenic testing campaign that is free of charge would likely also reduce arsenic exposure, but would probably delay the viability of commercial arsenic testing for several years. Furthermore, this type of top-down approach leaves no infrastructure for future testing programs. In highly arsenic contaminated areas of the country, targeted installation of deep wells will likely be the most sustainable approach over the long term. In addition, government policies are needed that mandate the installation of deep wells in areas that are easily accessible by the majority of the affected populations. Focusing efforts on arsenic removal technologies will require a higher long term infrastructure cost in comparison to deep wells because of the maintenance, personnel, and replacement cost associated with these systems.

In 2009, DPHE worked in close collaboration with the Japan International Cooperation Agency to conduct a situational analysis of arsenic mitigation for the country. Through this program a database was established to collect information on the location of arsenic safe water options in each arsenic affected sub-district. This has allowed the country to identify and target high priority areas where the percent of arsenic safe water options is low and the proportion of the arsenic contaminated drinking water sources is high. The situational analysis found that 154,236 deep wells and 10,250 deep Tara pumps were functional in arsenic affected areas of the country, providing arsenic safe water to an estimated 23 percent of the affected population. This report did not evaluate the impact of well switching; however, it was estimated that approximately 70 percent of the arsenic safe water sources used by the affected population were shallow wells (DPHE, 2010b).

Key recommended readings

1 Chen, Y. & Ahsan, H. 2004. Cancer burden from arsenic in drinking water in Bangladesh. *Am J Public Health*, 94(5):741–4. These results indicate that the arsenic crisis in Bangladesh will result in more than a doubling of the future excess cancer risk for this exposed population (229.6 versus 103.5 per 100,000 population).

2 Ahmed, M. F., Ahuja, S., Alauddin, M., Hug, S. J., Lloyd, J. R., Pfaff, A., Pichler, T., Saltikov, C., Stute, M. & Van Geen, A. 2006. Epidemiology. Ensuring safe drinking water in Bangladesh. *Science*, 314(5806):1687–8. This article gives an overview of arsenic mitigation options available to arsenic exposed populations in rural Bangladesh.

3 Flanagan, S. V., Johnston, R. B. & Zheng, Y. 2012. Arsenic in tube well water in Bangladesh: health and economic impacts and implications for arsenic mitigation. *Bulletin of the World Health Organization*, 90(11):839–46. This article estimates arsenic-related mortality rates and economic losses associated with the arsenic crisis in Bangladesh.

References

Ahmed, M. F., Ahuja, S., Alauddin, M., Hug, S. J., Lloyd, J. R., Pfaff, A., Pichler, T., Saltikov, C., Stute, M. & Van Geen, A. 2006. Epidemiology. Ensuring safe drinking water in Bangladesh. *Science*, 314, 1687–8.

Argos, M., Kalra, T., Rathouz, P. J., Chen, Y., Pierce, B., Parvez, F., Islam, T., Ahmed, A., Rakibuz-Zaman, M., Hasan, R., Sarwar, G., Slavkovich, V., Van Geen, A., Graziano, J. & Ahsan, H. 2010. Arsenic exposure from drinking water, and all-cause and chronic-disease mortalities in Bangladesh (HEALS): a prospective cohort study. *Lancet*, 376, 252–8.

Chakraborty, A. K. & Saha, K. C. 1987. Arsenical dermatosis from tubewell water in West Bengal. *Indian J Med Res*, 85, 326–34.

Chen, Y. & Ahsan, H. 2004. Cancer burden from arsenic in drinking water in Bangladesh. *Am J Public Health*, 94, 741–4.

Chen, Y., Van Geen, A., Graziano, J. H., Pfaff, A., Madajewicz, M., Parvez, F., Hussain, A. Z. M. I., Slavkovich, V., Islam, T. & Ahsan, H. 2007. Reduction in urinary arsenic levels in response to arsenic mitigation efforts in Araihazar, Bangladesh. *Environmental Health Perspectives*, 115, 917–23.

Chen, Y., Graziano, J. H., Parvez, F., Liu, M., Slavkovich, V., Kalra, T., Argos, M., Islam, T., Ahmed, A., Rakibuz-Zaman, M., Hasan, R., Sarwar, G., Levy, D., Van Geen, A. & Ahsan, H. 2011. Arsenic exposure from drinking water and mortality from cardiovascular disease in Bangladesh: prospective cohort study. *BMJ*, 342, d2431.

Cheng, Z., van Geen, A., Seddique, A. A. & Ahmed, K. M. 2005. Limited temporal variability of arsenic concentrations in 20 wells monitored for 3 years in Araihazar, Bangladesh. *Environ. Sci. Technol*, 39, 4759–66.

Dhar, R., Zheng, Y., Stute, M., Vangeen, A., Cheng, Z., Shanewaz, M., Shamsudduha, M., Hoque, M., Rahman, M. & Ahmed, K. 2008. Temporal variability of groundwater chemistry in shallow and deep aquifers of Araihazar, Bangladesh. *Journal of Contaminant Hydrology*, 99, 97–111.

Dhar, R. K., Biswas, B. K., Samanta, G., Mandal, B. K., Chakraborti, D., Roy, S., Jafar, A., Islam, A., Ara, G., Kabir, S., Khan, A. W., Ahmed, S. A. & Hadi, S. A. 1997. Groundwater arsenic calamity in Bangladesh. *Current Science*, 73, 48–59.

DPHE. 2010a. Bangladesh Arsenic Mitigation Water Sample Project (BAMWSP). Available: www.bamwsp.org.

DPHE. 2010b. Situation Analysis of Arsenic Mitigation 2009. Department of Public Health Engineering Bangladesh and Japan International Cooperation Agency.

Fendorf, S., Michael, H. A. & Van Geen, A. 2010. Spatial and temporal variations of groundwater arsenic in south and Southeast Asia. *Science*, 328, 1123–7.

Flanagan, S. V., Johnston, R. B. & Zheng, Y. 2012. Arsenic in tube well water in Bangladesh: health and economic impacts and implications for arsenic mitigation. *Bulletin of the World Health Organization*, 90, 839–46.

George, C. M., Geen, A. V., Slavkovich, V. N., Singha, A., Levy, D., Islam, T., Ahmed, K., Moon-Howard, J., Tarozzi, A. & Liu, X. 2012a. A cluster-based randomized controlled trial promoting community participation in arsenic mitigation efforts in Singair, Bangladesh. *Environmental Health*, 11, 41.

George, C. M., Graziano, J. H., Mey, J. L. & Van Geen, A. 2012b. Impact on arsenic exposure of a growing proportion of untested wells in Bangladesh. *Environmental Health*, 11, 7.

George, C. M., Zheng, Y., Graziano, J., Hossain, Z., Rasul, S. B., Mey, J. & Van Geen, A. 2012c. Evaluation of an arsenic test kit for rapid well screening in Bangladesh. *Environmental Science & Technology* 46(20), 11213–11219.

George, C. M., Inauen, J., Rahman, S. M. & Zheng, Y. 2013. The effectiveness of educational interventions to enhance the adoption of fee-based arsenic testing in Bangladesh: a cluster randomized controlled trial. *The American Journal of Tropical Medicine and Hygiene* 89(1), 138–144.

Haque, R., Mazumder, D., Samanta, S., Ghosh, N., Kalman, D., Smith, M. M., Mitra, S., Santra, A., Lahiri, S. & Das, S. 2003. Arsenic in drinking water and skin lesions: dose-response data from West Bengal, India. *Epidemiology*, 14, 174.

Hossain, M. A., Sengupta, M. K., Ahamed, S., Rahman, M. M., Mondal, D., Lodh, D., Das, B., Nayak, B., Roy, B. K. & Mukherjee, A. 2005. Ineffectiveness and poor reliability of arsenic removal plants in West Bengal, India. *Environmental Science & Technology*, 39, 4300–6.

Howard, G., Ahmed, M. F., Shamsuddin, A. J., Mahmud, S. G. & Deere, D. 2006. Risk assessment of arsenic mitigation options in Bangladesh. *Journal of Health, Population, and Nutrition*, 24, 346.

Johnston, R. B. & Sarker, M. H. 2007. Arsenic mitigation in Bangladesh: national screening data and case studies in three upazilas. *Journal of Environmental Science and Health Part A*, 42, 1889–96.

Kinniburgh, D. 2001. *Arsenic Contamination of Groundwater in Bangladesh* (Final Report: BGS Technical Report WC/00/19). Nottingham: British Geological Survey.

National Arsenic Mitigation Information Center 2004. Implementation Plan for Arsenic Mitigation in Bangladesh. Dhaka: National Arsenic Mitigation Information Center.

Parvez, F., Chen, Y., Brandt-Rauf, P. W., Bernard, A., Dumont, X., Slavkovich, V., Argos, M., D'armiento, J., Foronjy, R., Hasan, M. R., Eunus, H. E., Graziano, J. H. & Ahsan, H. 2008. Nonmalignant respiratory effects of chronic arsenic exposure from drinking water among never-smokers in Bangladesh. *Environmental Health Perspectives*, 116, 190–5.

Pathey, P. 2009. *Monitoring the Situation of Children and Women: Multiple Indicator Cluster Survey 2009. Volume 1: Technical Report*. Dhaka: Bangladesh Bureau of Statistics and United Nations Children's Fund (UNICEF).

Radloff, K. A., Zheng, Y., Michael, H. A., Stute, M., Bostick, B. C., Mihajlov, I., Bounds, M., Huq, M. R., Choudhury, I., Rahman, M. W., Schlosser, P., Ahmed, K. M. & Van Geen, A. 2011. Arsenic migration to deep groundwater in Bangladesh influenced by adsorption and water demand. *Nature Geosci*, 4, 793–8.

Rahman, M. M., Chowdhury, U. K., Mukherjee, S. C., Mondal, B. K., Paul, K., Lodh, D., Biswas, B. K., Chanda, C. R., Basu, G. K., Saha, K. C., Roy, S., Das, R., Palit, S. K., Quamruzzaman, Q. & Chakraborti, D. 2001. Chronic arsenic toxicity in Bangladesh and West Bengal, India–a review and commentary. *J Toxicol Clin Toxicol*, 39, 683–700.

Saha, K. C. 2003. Diagnosis of arsenicosis. *J Environ Sci Health A Tox Hazard Subst Environ Eng*, 38, 255–72.

Steinmaus, C. M., Yuan, Y. & Smith, A. H. 2005. The temporal stability of arsenic concentrations in well water in western Nevada. *Environ Res*, 99, 164–8.

Thundiyil, J. G., Yuan, Y., Smith, A. H. & Steinmaus, C. 2007. Seasonal variation of arsenic concentration in wells in Nevada. *Environ Res*, 104, 367–73.

Van Geen, A., Cheng, Z., Jia, Q., Seddique, A. A., Rahman, M. W., Rahman, M. M. & Ahmed, K. M. 2007. Monitoring 51 community wells in Araihazar, Bangladesh, for up to 5 years: implications for arsenic mitigation. *J Environ Sci Health A Tox Hazard Subst Environ Eng*, 42, 1729–40.

Van Geen, A., Ahsan, H., Horneman, A. H., Dhar, R. K., Zheng, Y., Hussain, I., Ahmed, K. M., Gelman, A., Stute, M., Simpson, H. J., Wallace, S., Small, C., Parvez, F., Slavkovich, V., Loiacono, N. J., Becker, M., Cheng, Z., Momotaj, H., Shahnewaz, M., Seddique, A. A. & Graziano, J. H. 2002. Promotion of well-switching to mitigate the current arsenic crisis in Bangladesh. *Bulletin of the World Health Organization*, 80, 732–7.

Van Geen, A. S. C. 2013. Piloting a novel delivery mechanism of a critical public health service in India: arsenic testing of tubewell in the field for a fee. Policy Note 13/0238. April. Available: http://www.theigc.org/sites/default/files/13_0238_POLICYNOTE_V4_LR.pdf.

Wasserman, G. A., Liu, X., Parvez, F., Ahsan, H., Factor-Litvak, P., Kline, J., Van Geen, A., Slavkovich, V., Loiacono, N. J., Levy, D., Cheng, Z. & Graziano, J. H. 2007. Water arsenic exposure and intellectual function in 6-year-old children in Araihazar, Bangladesh. *Environmental Health Perspectives*, 115, 285–9.

Wright, R. O., Amarasiriwardena, C., Woolf, A. D., Jim, R. & Bellinger, D. C. 2006. Neuropsychological correlates of hair arsenic, manganese, and cadmium levels in school-age children residing near a hazardous waste site. *Neurotoxicology*, 27, 210–16.

Yu, H.-S., Liao, W.-T. & Chai, C.-Y. 2006. Arsenic carcinogenesis in the skin. *Journal of Biomedical Science*, 13, 657–66.

Zheng, Y., Stute, M., Van Geen, A., Gavrieli, I., Dhar, R., Simpson, H., Schlosser, P. & Ahmed, K. 2004. Redox control of arsenic mobilization in Bangladesh groundwater. *Applied Geochemistry*, 19, 201–14.

69

WALKERTON

Systemic flaws allow a fatal outbreak[*]

Steve E. Hrudey

PROFESSOR EMERITUS, UNIVERSITY OF ALBERTA, EDMONTON, ALBERTA, CANADA

Elizabeth J. Hrudey

RESEARCHER, STEVE E. HRUDEY & ASSOCIATES LTD. CANMORE, ALBERTA, CANADA

Learning objectives

1 Appreciate the relative importance of dealing effectively with microbial pathogens for ensuring safe drinking water.
2 Recognize that vulnerable conditions can exist for a long time in a drinking water system before disaster strikes.
3 Understand that drinking water safety is not primarily achieved by mere stringency of numerical water quality criteria (i.e. simply pursuing lower numbers for contaminants in regulations or guidelines).

Introduction

Walkerton is a community of about 5,000 population located in rural southwestern Ontario, Canada, 175 km northwest of Toronto. In May of 2000, this community experienced a disastrous drinking water outbreak that killed seven people and made over 2,300 ill with gastroenteritis as a result of allowing groundwater subject only to chlorine disinfection to become contaminated with *Escherichia coli* O157:H7 and *Campylobacter* spp. in manure from a nearby farm. This disastrous event captured national media attention across Canada, resulted in a \$9 million public inquiry which, over a period of 10 months, heard testimony under oath from 114 witnesses, including two former ministers of environment (responsible for the drinking water regulator) and the premier of Ontario. The Walkerton Inquiry reported

---

[*] Recommended citation: Hrudey, S.E. and Hrudey, E.J. 2015. 'Walkerton – systemic flaws allow a fatal outbreak', in Bartram, J., with Baum, R., Coclanis, P.A., Gute, D.M., Kay, D., McFadyen, S., Pond, K., Robertson, W. and Rouse, M.J. (eds) *Routledge Handbook of Water and Health*. London and New York: Routledge.

in two parts (O'Connor 2002a, b) in January and May 2002. Slightly over three months after testifying at the inquiry at the end of June 2001, the premier announced he was stepping down for personal reasons and his party lost the next provincial election in October 2003.

The exhaustive evidence obtained by the inquiry revealed a history of weak regulatory oversight by the Ontario Ministry of Environment for more than 20 years, a period with governments led by all major political parties in Ontario combined with widespread complacency among water utilities, regulators and the public about the challenges and requirements necessary to ensure safe drinking water. The second inquiry report (O'Connor 2002b) addressed what Ontario needed to do in seeking to avoid having a re-occurrence of another drinking water disaster like Walkerton. While the new Ontario government embraced this report fully and officially adopted all of the recommendations, a major structural flaw regarding competence to produce safe drinking water is not generally resolved in Canada (Hrudey 2011). This problem, which is most prevalent across North America (explained below), is the assigning of responsibility for producing drinking water to municipalities which, in the case of Canada, means numerous very small towns and villages, most much smaller than Walkerton, that are expected to meet all of the challenges of supplying safe drinking water to residents (Hrudey 2008). This downloading of responsibility for providing safe drinking water to very small communities contrasts with much of Britain, Europe and Australia where large regional water corporations or authorities manage this responsibility.

This chapter will review some of the history leading up to the failure events in May 2000, summarize how those events unfolded, review the short and long term consequences, explain the underlying causes of failure and conclude by discussing what lessons must be learned to prevent similar failures in the future. The primary facts for this case were taken from evidence introduced at the inquiry (testimony and reports submitted as exhibits) and the first inquiry report (O'Connor 2002a). An account of this tragedy that includes a substantially more detailed description of the personal perspectives of those who failed to ensure that Walkerton's drinking water was safe has recently been published (Hrudey and Hrudey 2014). This book also provides 20 other case study accounts of both microbial pathogen and chemical contamination, all written, as far as accessible information allowed, from the perspective of frontline personnel.

History prior to May 2000

The beginning – well #5 constructed in 1978

Walkerton's water supply was operated by the Walkerton Public Utilities Commission (PUC) which was also responsible for the electrical power utility. Well #5 was a shallow well located on the edge of town that was drilled in 1978 to a depth of 15 m, with about 2.5 m of overburden, protective casing provided to 5 m depth and production from a shallow unconfined aquifer at depths from 5.8 to 7.4 m. This well could provide 1.8 ML/d, more than half of the town's peak water demand in 2000. Two deeper wells #6 (72 m) and #7 (76 m) were drilled in 1982 and 1987 respectively. Water from well #5 was preferred in the community both residentially and with some commercial outlets because it provided soft water compared with that from the other wells. The original wells for Walkerton, wells #1, #2 and #3, had produced extremely hard water, making well #5 a welcome addition for the community in 1978.

The ministry of environment (MOE) issued an approval for well #5 in 1978 with no conditions (treatment, monitoring, notification or maintenance of a chlorine residual) as would later become standard practice for new water supplies. The MOE personnel at the time expressed concerns about the location of well #5 and its vulnerability to surface water contamination, but these concerns were not translated into regulatory requirements that the Walkerton PUC was required to satisfy to reduce the risk. The PUC had applied for authorization to construct well #5 at the end of September 1978 indicating it would take two months to complete. In fact, this well had been constructed in June 1978, but it had not yet been connected to the PUC distribution system. This was an early indication that the PUC staff and management held little respect for the drinking water regulatory processes.

A hydrogeologist had performed an evaluation of well #5 in a report at the end of July 1978 that documented the vulnerability of this well. Pump testing of the well produced fecal coliform contamination at 24 hours. Furthermore, the vulnerability of well #5 to contamination was described in terms of noting the nearby agricultural activity, the thin overburden and shallow aquifer and, perhaps most damning, that when the pump at well #5 was turned on, two nearby springs would stop flowing. Later it would be noted that the water level in a shallow surface water pond near well #5 would decline when the pump was turned on, clearly demonstrating what is now termed groundwater under the direct influence (GUDI) of surface water. The hydrogeologist recommended that the PUC acquire a land buffer zone around this well because the nearby farms posed risks of contaminating this shallow groundwater supply (Figure 69.1). Likewise, he noted that this water supply should definitely be chlorinated. This report was considered in the approval process by the MOE district office, and the approvals staff wrote a memorandum expressing concern for this vulnerability and openly questioned whether well #5 should be treated as if it was a surface water supply and be required to chlorinate with provision for continuous chlorine residual monitoring.

A meeting was held near the end of November 1978 to deal with the PUC having constructed well #5 with no approval from the MOE. As a result of this meeting, an

Figure 69.1 View from well #5 in Walkerton of nearby farm and fenced cattle enclosure

Source: Photo by S.E. Hrudey.

"understanding" was reached that the PUC would chlorinate well #5, that it would maintain a combined chlorine residual (the majority to be free chlorine) greater than 0.5 mg/L for at least 15 minutes contact time before the first consumer and that it would measure the chlorine residual daily and record these matters on an operating log that would report the daily results for each month. The minutes for this November 1978 meeting stated: "[t]he importance of maintaining a chlorine residual at all times was emphasized in light of the presence of bacteria in the well water" (O'Connor 2002a).

Unfortunately, the MOE did not follow through on demonstrating the importance of these requirements because there had been discussions with the town's engineering consultant about how to ensure the 15 minute contact time that was necessary to achieve adequate disinfection. The consultant had proposed to the MOE a means to achieve this by constructing an over-sized water main parallel to the existing main to cause the water supply to circulate for at least 15 minutes before reaching the first consumer. The approval for well #5 was issued near the end of January 1979 with no requirement for this contact time feature, so it was not built, meaning that the 15 minute contact time specified could not be ensured for the earliest consumers. These conditions were condoned for another seven months until Walkerton proposed another means for ensuring the 15 minutes contact time, which was implemented in August 1979. The apparent disinterest by MOE in ensuring adequate disinfection surely sent a message to the Walkerton PUC that these requirements must not be very important. Likewise, although the hydrogeologist's recommendation about acquiring adjacent farmland to provide a well catchment protection zone was discussed at these meetings, no action was ever taken.

Operations 1978 until 2000

Over the next 21.5 years after being commissioned, water from well #5 was undoubtedly contaminated from time to time and it may have even been causing low levels of illness in the community, but below a level detectable by passive public health surveillance. There were many red flags for the MOE over this period that the PUC was not performing in accordance with reasonable expectations, particularly as the MOE requirements were being upgraded and made more rigorous in the 1990s. On a number of occasions, it became apparent to MOE inspectors that the PUC was not performing its specified microbiological water monitoring. The PUC was told it must do so and would agree, but then the PUC would fail to follow through on its commitment. The continued pattern of tolerance by the MOE of the repeated poor performance by the PUC reinforced an implied message to the PUC that these regulatory requirements must not be very important.

These circumstances might never have happened if the PUC had employed properly trained operators who understood the reasons for the various regulatory requirements in order to ensure safe water for Walkerton consumers. Most important, properly trained operators would have understood the vulnerability of well #5 and the vital need to ensure adequate disinfection.

The PUC general manager had been hired as a laborer at the age of 19 in 1972, having completed grade 11. He became a power lineman from 1976 to 1980, in 1981 he was appointed as PUC foreman, and finally, in 1988, he was promoted to general manager. Similarly, his younger brother had started with the PUC in 1975 at the age of 17 and he was promoted to foreman in 1988 when the incumbent general manager had retired. The general manager, a long term resident of Walkerton, was widely regarded as quiet, hard working and conscientious. His brother was more brash and outspoken, but he had developed a problem with alcohol and proved difficult for the older brother to motivate to do things that were required from time to time by the MOE. Both men were competent in keeping the mechanical aspects of the water supply functioning and they regarded the water sources in Walkerton, particularly from well #5 with its soft water, to be high quality and inherently safe. They did not believe well #5 required chlorination (except as an arbitrary regulatory requirement). They routinely drank raw well water taken upstream of the chlorination, preferring its "natural" taste.

Both men learned their obligations on the job, unfortunately adopting many improper or inadequate practices under the leadership of the previous general manager. When the MOE brought in voluntary certification of drinking water operators in 1988, both men were certified under a grandfathering provision that required no training, nor any competence testing. The general manager was upgraded to a level 3 operator in 1996, again with no training or competence testing.

The MOE conducted inspections of the Walkerton PUC water systems in 1991, 1995 and 1998. The 1991 inspection revealed that the PUC was not following the monitoring required by the Ontario Drinking Water Objectives and the MOE "recommended" that the PUC should upgrade its microbiological monitoring program. The 1995 MOE inspection revealed that the required monitoring program had not been implemented. The general manager stated that the PUC would comply, but as of 1997, they were still on a list of municipalities that were not complying and were supposed to receive MOE Director's Orders to do so. The general manager indicated to the MOE that the PUC would comply and it was taken off the compliance list, but the MOE took no steps to verify that compliance was being achieved. When inspected again in 1998, the PUC was still not meeting the requirements of a minimum sampling program. The general manager told the MOE inspector that the PUC would comply immediately, but in a letter five months after the 1998 inspection, the PUC indicated that it would comply two months hence. No further inspection of the PUC was conducted by the MOE prior to the events of May 2000.

The three inspections also revealed that the PUC was not maintaining the required levels of chlorination. The 1991 inspection found chlorine residuals of 0.30 and 0.35 mg/L, but the MOE inspector mistakenly had advised the PUC that these low values were acceptable. The 1995 inspection showed that four of five samples taken were below 0.30 mg/L, another was 0.40 mg/L and one was only 0.12 mg/L. The inquiry revealed that PUC operators were not actually measuring chlorine residual, but would make fictitious entries on the daily operating log, with all entries being either 0.50 or 0.75 mg/L, a level of measurement consistency that would be impossible to achieve and was fictitious on the face of it for anyone knowing such measurements.

The deficiencies in chlorine residual were compounded by the more frequent detection of *E. coli* in treated water, with three cases of *E. coli* in distribution system samples mentioned in the 1995 inspection report and the 1998 inspection report referring to eight cases of treated water failing the microbiological monitoring criterion for adverse water quality, with several containing *E. coli* both in distribution system samples and treated water samples at well #5. Because *E. coli* is exquisitely sensitive to chlorine disinfection, these adverse results indicated both fecal contamination and inadequate chlorination. The MOE inspectors apparently failed to appreciate the serious implications of so many repeated failures occurring. By then, those inspectors were no longer drinking water specialists, but were generic environmental compliance inspectors, an unfortunate trend with many drinking water regulators in Canada. The failure of the MOE to take meaningful compliance action with the PUC simply reinforced to the PUC operators that the MOE requirements must not be very important. They lacked the knowledge and training to conclude on their own that these requirements were essential for ensuring safe drinking water.

The issue of customer complaints about chlorine taste and odor in the town water was also a recurring challenge for the PUC operators. Such complaints represent a common challenge facing drinking water providers that will be exacerbated if they fail to maintain a consistent chlorine residual because fluctuating chlorine levels will be more readily noticed by consumers. Most consumers will adapt to the chlorine residual being consistent. Consumer pressure to reduce or eliminate chlorine in drinking water is an unfortunate irony, particularly in this case where inadequate chlorination was a key factor in allowing this outbreak to occur.

The events of May 2000

How events unfolded

During the week of May 8 to 12, 2000, Walkerton experienced unusually heavy rainfall, totaling about 134 mm (5.3 in) with about 70 mm (2.8 in) falling on Friday, May 12. The general manager was away from Walkerton this whole week to attend an Ontario-wide annual drinking water technical convention. On the prior Wednesday, May 3, the general manager finally succeeded in getting his brother the foreman and other operational staff to begin replacing a faulty chlorinator on well #7. They succeeded in removing the chlorinator but, unknown to the general manager, they did not install the new chlorinator, so well #7 had no chlorination when it was brought into service. Well #7 was not seriously vulnerable to contamination like well #5, but operation of any of the PUC wells without chlorination was strictly contrary to the current MOE requirements. During the period of heavy rain, well #5 was the primary source of water to Walkerton and it remained so until Monday, May 15 when the general manager returned to work and switched the Walkerton supply over to well #7 without knowing that it had no chlorinator installed. He did not find out about this deficiency until a few hours later but, regardless, he left well #7 in service until Saturday, May 20. Well #5 was shut down in the early afternoon of Monday, May 15.

A new water main was being installed at this time (the Highway 9 project). On May 15, the contractor for this project asked the general manager if he could submit to the commercial lab serving the PUC the microbiological samples, checking for adequacy of the disinfection of the new water main. The general manager agreed and submitted those water main construction samples along with the routine PUC raw and treated water samples collected from well #7 and two distribution system samples.

On the morning of Wednesday, May 17, the lab phoned the general manager to advise him that all of the construction project samples and the distribution system samples were positive for total coliforms and *E. coli* by a presence / absence test. Only the well #7 treated sample was analyzed by the membrane filtration procedure allowing determination of microbial counts. Both total coliforms and *E. coli* produced overgrown plates meaning numbers greater than 200 CFU (colony forming unit) per 100 mL. The inquiry concluded that, based on several lines of evidence, this sample was mis-labeled and was actually taken at well #5. Because the PUC was now using a new commercial lab (its previous lab had stopped doing microbial analysis for them by the end of April, 2000), the new lab did not follow the voluntary convention of the previous lab of notifying the MOE and local medical officer of health about any adverse microbiological samples for drinking water systems. The commercial lab did fax these results to the PUC office. The first fax on the morning of Wednesday, May 17, reported the water main construction samples and a second fax on that afternoon reported the distribution and well #7 results. The general manager was busy all of that day, not returning to the office until the following morning, Thursday, May 18, when he would have found these faxes among other papers left on his desk.

The general manager remained busy all day Thursday, May 18, being preoccupied with the water main construction project and the fact that it had failed its disinfection test among several other concerns. Furthermore, he had to prepare a presentation for and attend the regular meeting of the PUC board that evening. He later testified that he had been told on the phone the previous day about the failed distribution system microbiological monitoring results, but he later claimed not to have paid attention to the faxed results because of other job-related pressures. Meanwhile, illness began to emerge in the community. Two children, a seven and a nine year old, both suffering from bloody diarrhea were admitted to hospital in Owen Sound, 65 km from Walkerton. Stool samples obtained from these two young patients eventually provided the first evidence of the severity of the emerging outbreak by revealing infection caused by *E. coli* O157:H7 (Chapter 4).

On Friday, May 19, evidence of the outbreak was emerging on several fronts. A large number of children were absent from Walkerton schools and others were sent home suffering from stomach pain, diarrhea and nausea. A local retirement home and a long term care facility reported residents were suffering from diarrhea, two with bloody diarrhea. The Owen Sound physician who had hospitalized the Walkerton children she had seen the previous day contacted the local public health unit, also located in Owen Sound, to advise their staff that gastrointestinal disease issues were emerging in Walkerton and that she was being told by Walkerton residents about widespread illness in the community. Meanwhile, a school receptionist also contacted the public health unit to advise it of an emerging view among parents in Walkerton that something was "going on" in Walkerton. A receptionist at one of the schools also conveyed to the public health unit the opinion of one parent that there was something wrong with Walkerton's water supply. An administrator from the same school advised the Walkerton office of the public health unit that 25 students were absent because of illness on Friday, May 19.

The public health response

The local public health inspector called the PUC general manager on the afternoon of Friday, May 19, to inquire whether there were any problems with Walkerton's drinking water supply because of the emerging reports of illness. The general manager told him that he "thought the water was okay" even though he had been told about adverse microbiological results two days earlier and had received the faxed results the day before this conversation. The public health inspector assumed that this must be a foodborne outbreak, so he did not press the PUC general manager; he did what his own manager had requested of him, that is, to make the call. The PUC general manager mentioned that he was concerned about the water main replacement project and suggested he would undertake some flushing of water mains. The public health inspector agreed that flushing would be a worthwhile precautionary measure and also recommended measuring chlorine residuals and keeping notes. However, it was apparent that drinking water was not a primary issue for the local public health inspector who generally assumed that the MOE was the responsible regulator. The chlorinator had finally been installed on well #7 the previous afternoon. This remained the current source of the Walkerton water supply, so the general manager went to well #7 and increased the chlorination level.

By Saturday, May 20, on a holiday long weekend, the outbreak had severely overloaded the Walkerton hospital which had fielded more than 120 calls from concerned residents, with more than half reporting bloody diarrhea. Meanwhile, the Owen Sound hospital had found a presumptive positive for the serious human pathogen, *E. coli* O157:H7, in one of the children admitted there on Thursday, May 18. The emerging outbreak kept the public health staff fully engaged. The general manager received two more calls from the public health unit to inquire about the chlorine residuals he was measuring. Public health officials were reassured that measureable chlorine residuals were appearing in the distribution system assuming that must indicate that the water was safe. The medical officer had been out of town to spend the long weekend with his wife who had been recently diagnosed with cancer, but the health unit administrator contacted him and requested he return to help the local health unit deal with the emerging crisis.

By Sunday, May 21, the public health unit had concluded that notwithstanding the reassurances about the water safety provided by the PUC general manager, the widespread distribution and large number of cases ruled out a foodborne outbreak. As a result, they decided to call a boil water advisory that was provided to the local radio station. A later survey found that only 44 percent of residents were aware of this boil water advisory and only 34 percent heard the announcement on the radio. The health unit also collected its own water samples from the Walkerton distribution system.

By Monday, May 22, the public health unit had received reports of about a hundred cases of infection. The MOE, despite being the primary drinking water regulatory, did not become involved until the medical officer of health advised the MOE district official that this situation had become urgent, with many cases of severe illness. The MOE sent an officer from Owen Sound to visit the PUC general manager and obtain monitoring records from him. The MOE official was provided with a copy of the fax reporting the adverse microbiological samples taken on Monday, May 15, but remarkably, although he admitted having seen this document, he did not report it to his supervisor based on his belief that the boil water advisory removed any urgency about these adverse microbial results. The public health unit began to plot the epidemic curve that suggested the peak of disease onset was Wednesday, May 17, which, for the estimated incubation time, suggested contamination occurred between May 12 and 14. That Monday, the first victim, a 66-year-old woman, died.

On Tuesday, May 23 the health unit received results for distribution system samples it collected on Sunday, May 21. Two samples from dead-end locations in the distribution system showed substantial microbial contamination. When the health unit administrator contacted the general manager to advise him of these results and ask for an explanation, the general manager broke down after admitting he had been advised of the adverse monitoring results for samples taken more than a week earlier on Monday, May 15. This stunning revelation resulted in a series of tense interactions between the public health unit and the municipal government. Meanwhile, a second victim died – a two-year-old infant who was not even a Walkerton resident. She had come to Walkerton for a Mothers' Day meal on May 14 when she consumed only one glass of water.

Analysis of this outbreak

This severe drinking water outbreak was caused by multiple factors. Various physical circumstances provided the source of the pathogens and treatment barriers were unable to inactivate those pathogens before they were ingested by the consumers of Walkerton's drinking water supply. But equally, if not more, important for the purposes of generalizing this experience to other circumstances, where the physical factors will differ in detail and substance, are the numerous systematic failures that allowed the outbreak to occur.

The physical causes of this outbreak

The critical physical causes of the Walkerton outbreak, as derived from the evidence produced for the inquiry (O'Connor 2002a), have been summarized and analyzed in detail (Hrudey and Walker 2005, Hrudey and Hrudey 2004, Hrudey et al. 2003).

Well #5 at Walkerton was located close to two farms posing a water contamination risk, only one of which was still active in May 2000 (Figure 69.1). The inquiry concluded the active farm was the source of the microbial outbreak, even though the small scale cattle-rearing operation run by a local veterinarian followed model environmental farm management practices. The inquiry commissioner made a point of noting that this local farmer was not responsible for the failures that occurred. The geological conditions at well #5 involved a shallow production zone of highly weathered bedrock with closely spaced horizontal and vertical fractures. The inquiry expert hydrogeologist concluded that localized breaching of the thin soil overburden by surface water contaminated by manure followed by rapid horizontal transport in the fractured bedrock was the most plausible explanation for the massive microbial contamination that occurred. Substantial microbial contamination of that shallow, unconfined aquifer was confirmed by investigation after the outbreak. The contamination most likely entered well #5 on May 12 when 70 mm of rainfall fell overnight.

The Walkerton operators were expected to provide a total chlorine residual (the majority to be free chlorine) of 0.5 mg/L after 15 minutes. However, the PUC chlorine dosage practice at well #5 was insufficient to achieve a 0.5 mg/L residual, even in the absence of any chlorine demand. Maintaining a 0.5 mg/L free chlorine residual after 15 minutes would have satisfied the concentration–contact time (CT) requirement for 99 percent kill of *E. coli* O157 at 5 °C, pH 7.0 (Rice 1999) by 20 fold and the CT for 99.99 percent kill by 7.5 fold. Although it is likely that the chlorination equipment at well #5 could not have provided a high enough chlorine dose to deal with the chlorine demand at the peak of the slug of contamination, the failure of operators to measure chlorine residual was a critical opportunity missed for minimizing the consequences of this disaster. Such monitoring could have allowed early recognition of serious trouble to invoke a rapid boil water advisory and a general community warning.

There were also several scenarios by which the distribution system at Walkerton could have been contaminated including:

- installation of new water mains at three locations,
- a fire event with potential for associated depressurization,
- main breaks and repairs at four locations in March 2000,
- potential for contamination of two treated water storage standpipes,
- cross connections found at eight private wells and many private cisterns,
- potential for cross contamination of water mains by sanitary sewers, and
- surface flooding in the town during the heavy rainfall of May 12, 2000.

Although one or more of these elements could have caused contamination of the Walkerton distribution system, none alone or in combination provided an adequate explanation for the widespread distribution of illness found in this outbreak. Subsequent application of molecular methods to identify strains of the pathogens that infected Walkerton residents were also used to confirm strain-type matches with isolates obtained from manure samples collected at the nearby farm, but not found at other farms in the area.

The systemic causes of this outbreak

As outlined in the lead-in explanation about what happened in Walkerton, there was a long history of poor practice and neglect by the Walkerton PUC, its management and employees. However, there was an even more troubling lack of oversight and commitment by the primary drinking water regulator, the MOE and, to some degree, by the local public health inspector who largely presumed that the MOE was responsible for safe drinking water. The MOE had allowed the commissioning of a highly vulnerable well 22 years earlier and then ignored repeated signals of poor operation and water quality from well #5. The MOE provided certification to operators primarily according to their ability to run a water system without providing the training needed for operators to appreciate the critical responsibilities that water operators carry. In particular, there was a disconnect between the reprehensible behavior of the PUC general manager (withholding adverse microbiological results and condoning falsification of monitoring results) that can only be logically reconciled with his favorable, long-standing reputation in the community by concluding that he truly had no idea what harm he could cause his friends and neighbors.

The sad irony of this tragedy is that most of the severe health consequences in the community could have been avoided if the PUC had measured the chlorine residuals within 24 hours of the contamination occurring as they were supposed to do. Such monitoring would have allowed them to know that because they were dosing chlorine, and the finding that they were unable to measure any chlorine residual, provided a meaningful real time signal that there was a massive chlorine demand being exerted in the system, a strong signal of massive contamination. Provided with adequate training about what such a circumstance must mean, they would have been equipped with the grounds to take immediate action leading to a boil water advisory and serious warning to the community, if not a complete system shut down. The failure to recognize that there was a serious problem meant that Walkerton residents consumed contaminated water for nine days before the boil water advisory was finally issued on Sunday, May 21. In the meantime, residents with diarrhea seeking medical attention were advised to drink lots of fluids to avoid dehydration. In at least one case, a mother seeking medical assistance for her severely ill 2-year-old son was told to force water, which may still have been contaminated, into her son using a syringe.

Ironically, the fact that the general manager was running well #7 without a chlorinator from Monday, May 15, until Thursday, May 18, something he knew was a clear violation of MOE requirements, led him to suspect that contamination must have happened at well #7 and that he would be held responsible for having allowed that to happen. Likely this situation may have caused the general manager to cover up, trying to buy time until the chlorination being provided by well #7, after May 18, had time to work. Of course, not only would he have been wrong to believe that well #7 was the problem, but with illness becoming evident in the community by May 18, it was already too late to "solve" the problem. Only the chlorine residual monitoring on Saturday, May 13, had the potential to provide enough warning of water quality trouble to substantially reduce or avoid this major outbreak. That monitoring was supposed to be done by the foreman because the general manager was out of town, and the foreman performed his usual sham of not measuring the chlorine residual at well #5, but recording a fictitious value of 0.50 mg/L to satisfy the MOE.

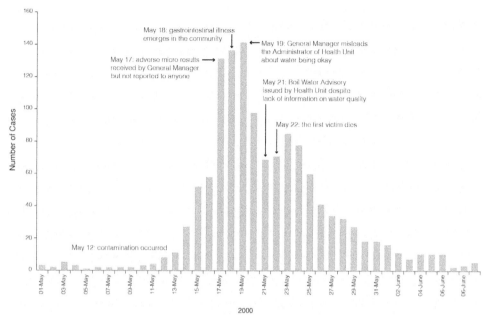

Figure 69.2 Epidemiologic curve for the Walkerton outbreak (after Hrudey and Hrudey 2014)

Source: BGOSHU, 2000, reproduced with permission of Grey-Bruce Health Unit.

The consequences of this outbreak

Ultimately, this outbreak caused a total of seven deaths, 27 cases (with a median age of four) of hemolytic uremic syndrome (HUS), a life-threatening kidney condition that may subsequently require kidney transplantation, and an estimated 2,300 cases of gastrointestinal illness. Stool cultures from some ill patients confirmed exposure to *E. coli* O157:H7, *C. jejuni* and other enteric pathogens. The epidemic curve for this outbreak is provided in Figure 69.2.

The distribution of cases across the community showed how extensively the system had been allowed to be contaminated (Figure 69.3). This widespread distribution suggests the obvious potential for having been able to reduce the impact of this outbreak if the contamination had been recognized as early as it could have been (Hrudey and Walker 2005) by measuring chlorine residual at Well #5, as the operators were supposed to have done.

Several long term health studies have been performed on a cohort from Walkerton to provide evidence about whether acute infection as happened in this outbreak may be responsible for chronic outcomes. Clark et al. (2010) have concluded that *E. coli* O157:H7 infection among Walkerton residents has caused an increased risk of hypertension, renal impairment and self-reported cardiovascular disease. Likewise, Thabane et al. (2010) has found strong evidence of increased incidence of irritable bowel syndrome among children who were under the age of 16 at the time of the outbreak.

Lessons from this outbreak

Sadly, there are countless lessons that should be taken from this public health failure. The more generalizable lessons are noted below:

○ Campylobacter Positive

▢ E. Coli 0157:H7 positive

△ Other (including not cultured or negative)

Figure 69.3 Distribution of cases of illness (after Hrudey and Hrudey 2014)

Source: BGOSHU, 2000, reproduced with permission of Grey-Bruce Health Unit.

1 Severe latent flaws can exist in a water system for years (about 22 years in the case of Walkerton) before disaster strikes.
2 Complacency can exist at many levels among many individuals who were responsible for ensuring a safe community drinking water.
3 If operational personnel are told they must do various things by a drinking water regulator, but they are provided with no understanding of why these requirements are necessary or with no comprehension of the severe health consequences of a drinking water failure, they may fail totally to discharge their responsibilities.
4 A water system may have multiple vulnerabilities. But, addition of only one critical circumstance is necessary to combine with pre-existing vulnerabilities to cause a disaster. In the case of Walkerton, the heavy rainfall that led to the manure contamination of the shallow aquiver caused this outbreak. This contamination escaped detection because the operators lacked the knowledge that measuring chlorine residual was vitally important for ensuring safety of the drinking water supply.
5 Because of the complexity of factors that can align to cause disaster, misleading signals about what may be a problem can and often do occur, with the effect of distracting responses from the true problem.

Conclusions

The Walkerton drinking water outbreak was caused by various physical factors that allowed microbial pathogens to enter a vulnerable shallow well water supply. This system had been approved and allowed to operate in a vulnerable state for about 22 years before disaster occurred. Ultimately the severe health consequences for the community were caused by a combination of systemic factors involving inadequate and ineffective regulatory oversight that permitted this system to operate in a manner that was hopelessly inadequate to ensure the safety of the community. Perhaps most important among these systemic factors were the failures to ensure:

* full comprehension by the operators of their public health responsibilities,
* the means available to operators to discharge those responsibilities, and
* effective support systems that operators can and will access when problems exceed their capacity to understand what is happening in problem circumstances.

In many ways, these failures are a result of society not placing enough value on the critical roles and responsibilities of drinking water operators in ensuring safe community water supplies.

Key recommended readings (open access)

1 O'Connor, D.R. 2002a. Report of the Walkerton Inquiry. Part 1. The Events of May 2000 and Related Issues. Toronto, The Walkerton Inquiry: 504 pp. http://www. attorneygeneral.jus.gov.on.ca/english/about/pubs/walkerton/part1/. This report provides a very readable summary of the evidence and findings of the inquiry about what contributed to the outbreak in May 2000. In addition to providing a detailed account about the various physical causal factors, it reviews the roles of the Public Utilities Commission operators and commissioners, the municipalities and mayors, the public health authorities and the ministry of environment.

2 O'Connor, D.R. 2002b. Report of the Walkerton Inquiry. Part 2. A Strategy for Safe Water. Toronto, The Walkerton Inquiry: 582 pp. http://www.attorneygeneral.jus.gov.on.ca/english/about/pubs/walkerton/part2/. This report provides a very readable summary of the findings of the inquiry about what measures needed to be introduced in Ontario to prevent a disaster like Walkerton from happening again considering issues like: the multiple barrier approach, treatment technologies, laboratories, monitoring and measurement, standards, certification and training of operators, and risk management.

3 Hrudey, S.E. 2011. Safe Drinking Water for Canada – Turning Hindsight into Foresight. C.D. Howe Institute. Commentary. The Water Series. Number 323. February 2011. http://www.cdhowe.org/pdf/Commentary_323.pdf. This report outlines that the problem is not that numerical water safety criteria are inadequately stringent; the documented failures in safe drinking water have been caused by an inability to operate water systems effectively, pointing to inadequate competence – pointing to the need to adopt a know your own system approach based on drinking water safety plans.

References

BGOSHU 2000. *The Investigative Report of the Walkerton Outbreak of Waterborne Gastroenteritis May–June 2000.* Owen Sound, Ontario: Bruce-Grey-Owen Sound Health Unit.

Clark, W.F., J.M. Sontrop, J.J. Macnab, M. Salvadori, L. Moist, R. Suri and A. Garg. 2010. Long term risk for hypertension, renal impairment, and cardiovascular disease after gastroenteritis from drinking water contaminated with *Escherichia coli* O157:H7: a prospective cohort study. *BMJ* 341 (c6020): 9 pages.

Hrudey, S.E. 2011. Safe Drinking Water for Canada – Turning Hindsight into Foresight. *C.D. Howe Institute Commentary: The Water Series.* Number 323. February. http://www.cdhowe.org/pdf/Commentary_323.pdf

Hrudey, S.E. 2008. Safe water? Depends on where you live. *Canadian Medical Association Journal* 178(8): 975.

Hrudey, S.E. and E.J. Hrudey. 2014. *Ensuring Safe Drinking Water – Learning from Frontline Experience with Contamination.* Denver, CO: American Water Works Association.

Hrudey, S.E. and E.J. Hrudey. 2004. *Safe Drinking Water – Lessons from Recent Outbreaks in Affluent Nations.* London: IWA Publishing.

Hrudey, S.E. and R. Walker. 2005. Walkerton – 5 years later. Tragedy could have been prevented. *Opflow.* 31(6): 1, 4–7.

Hrudey, S.E., P.M. Huck, P. Payment, R.W. Gillham and E.J. Hrudey. 2003. A fatal waterborne disease outbreak in Walkerton, Ontario: comparison with other waterborne outbreaks in the developed world. *Water Sci. Technol.* 47(3): 7–14.

O'Connor, D.R. 2002a. *Report of the Walkerton Inquiry. Part 1. The Events of May 2000 and Related Issues* (The Walkerton Inquiry). http://www.attorneygeneral.jus.gov.on.ca/english/about/pubs/walkerton/part1/

O'Connor, D.R. 2002b. *Report of the Walkerton Inquiry. Part 2. A Strategy for Safe Water.* (The Walkerton Inquiry) http://www.attorneygeneral.jus.gov.on.ca/english/about/pubs/walkerton/part2/

Rice, E.W. 1999. *Escherichia coli.* In: M. Marshall, M. Abbaszadegan, E. Geldreich and D. Fredericksen (eds) *Waterborne Pathogens.* Denver, CO: American Water Works Association.

Thabane, M., M. Simunovic, N. Akhtar-Danesh, A.X. Garg, W.F. Clark, S.M. Collins, M. Salvadori and J.K. Marshall. 2010. An outbreak of acute bacterial gastroenteritis is associated with an increased incidence of irritable bowel syndrome in children. *Am. J. Gastroenterol.* 105: 933–939.

70
MILWAUKEE AND THE *CRYPTOSPORIDIUM* OUTBREAK 1993[*]

M. Stephen Gradus

Ph.D., (D)ABMM, Laboratory Director, City of Milwaukee Public Health Laboratory; Adjunct Faculty, University of Wisconsin-Milwaukee, Zilber School of Public Health and the College of Health Sciences Biomedical Sciences Department, Milwaukee, WI, USA

Learning objectives

1 Describe the implications and necessity of local and state governments understanding water in the context of a public health issue rather than or in addition to a public works issue.
2 Identify lessons learned from this outbreak.
3 Discuss the importance of community-wide laboratory communications in the understanding and timely response to this outbreak.

Introduction

The largest documented waterborne outbreak in United States history occurred in Milwaukee, Wisconsin, during March and April of 1993 (Edwards, 1993; MacKenzie et al., 1994; Gradus et al., 1994; Batchelor, 2004; Davis, 2010). As the largest municipal public health laboratory in Milwaukee, with a history of public water quality analysis dating back to 1874, the City of Milwaukee Public Health Laboratory played a pivotal role in the outbreak investigation. My first-hand account of this incident, as a clinical microbiologist and the laboratory director for the Milwaukee Health Department, is included in this report. To provide context for this outbreak, metro Milwaukee is located 90 miles north of Chicago on Lake Michigan's western shore. Milwaukee, population 594,833, is the nation's 28th largest city. The four-county metro area has more than 1.5 million people, making it the nation's 39th largest metro area. Milwaukee in the early 1990s, like many larger Midwestern cities,

[*] Recommended citation: Gradus, M.S. 2015. 'Milwaukee and the *Cryptosporidium* outbreak 1993', in Bartram, J., with Baum, R., Coclanis, P.A., Gute, D.M., Kay, D., McFadyen, S., Pond, K., Robertson, W. and Rouse, M.J. (eds) *Routledge Handbook of Water and Health*. London and New York: Routledge.

689

was suffering from a decline in industrialization, as part of the 'Rust Belt' with movement to a more service-oriented and technologically driven economy. Milwaukee is the most racially and ethnically diverse city in Wisconsin – substantially more diverse than Milwaukee County, the metropolitan area, and the State as a whole – and also one of the most segregated. The south side of Milwaukee, where the outbreak originated, originally had strong Polish roots but had transitioned into a more multicultural community, including a growing Hispanic population, at the time of the outbreak, whereas the African American population was and continues to be more concentrated on the city's north side. The outbreak ultimately affected the entire city.

Event timeline

Day 1: Monday, April 5

On Monday morning, April 5, 1993, the City of Milwaukee Public Health Laboratory's (MHDL) chief virologist and the commissioner of health received calls from citizens and local media inquiring about the nature of apparent gastrointestinal (GI) illness reports in the city. The director of nursing had anecdotal information that some pharmacies were selling out of anti-diarrheal medications. Unknown to the health department at that time, the Milwaukee Water Works (MWW) had received some complaints regarding the aesthetic quality of tap water. At that time our MHDL community-wide virology surveillance program revealed no obvious viral etiology to the suspect increase in diarrheal disease in the community. We then proceeded to call local hospital microbiology laboratories and emergency rooms (ER) and determined there was extreme GI illness throughout the city, notably on the south side of Milwaukee, based on dramatically increased testing for enteric disease and much higher ER patient numbers than normal. Local laboratories reported they had not identified a causative agent at this point. Other MHD staff were also seeing similar indicators throughout the city, such as increased absenteeism in schools and businesses. I then contacted MWW who indicated that chlorine and filtration systems were 'OK.' MHD staff also conferred with the mayor's office and the Wisconsin State Division of Public Health. Given these negative findings, we coordinated with the city epidemiologist to arrange for the urgent collection of fresh 'outbreak' stool specimens from around the city for evaluation of a full battery of tests for infectious agents.

Day 2: Tuesday, April 6

Several meetings ensued internally to address the outbreak. Most MHD sections were besieged by phone calls regarding the widespread illness. I called both the north and south water treatment plants (WTP) to obtain daily water quality data going back to March 1, 1993, which indicated the only changing trend was the increasing turbidity readings from the south water treatment plant, which still remained within federal limits. This spike in turbidity subsequently was determined to be the largest in 10 years and was similar to our knowledge of a large waterborne outbreak of *Cryptosporidium* infections in 1987 in Carrollton, Georgia, among customers of a local municipal water supply (Hayes et al., 1989; Gradus, 1989). Outbreak stool specimens began to arrive in our laboratory late on day 2. Ongoing discussions and consultation with the state's medical officer and his team of epidemiologists 90 miles away in Madison, on day 1 and day 2, resulted in their team arriving to provide assistance in Milwaukee on day 3.

Day 3: Wednesday, April 7

Our laboratory started a rapid fax-back survey to city clinical microbiology labs that confirmed a dramatic increase in testing for GI pathogens citywide, yet identified no agents that would reflect such widespread illness. Ten faxed surveys were sent between 12:50 and 1:45 p.m. and six responses with data were received between 2:03 and 4:32 p.m. that same day. Testing of the outbreak stools proceeded at MHDL to detect *Cryptosporidium* in three stool specimens, one the wife of a laboratory employee. During that time, a 2 p.m. call from a colleague and infectious disease physician, who had interned in our laboratory, reported a case of *Cryptosporidium* from an otherwise healthy individual that was detected by an observant clinical microbiologist even though a test for *Cryptosporidium* had not been requested. We documented through our fax-back survey that virtually no clinical microbiology laboratories tested for the presence of *Cryptosporidium* without a physician's request. These findings prompted an initial alert to three major hospital laboratories to start testing all diarrheal stools for *Cryptosporidium*. By the end of day 3 we had documented eight laboratory confirmed cases of *Cryptosporidium* compared to only 12 positive of 600 tested over the past six years in our laboratory.

By 8:00 p.m., 60 hours after the realization of widespread community illness, the mayor issued a boil water advisory for the 880,000 customers serviced by the MWW. The decision was based on: 1) widespread community illness, 2) a review of complaints of taste and color of the water from the south treatment plant in late March, 3) an increase in turbidity of finished water during that time, and 4) laboratory confirmation of unsuspected *Cryptosporidium*, a known waterborne pathogen, in the stool specimens of eight ill persons. Subsequently, we alerted other local clinical microbiology laboratories to test all diarrheal stools for the presence of *Cryptosporidium*, and a massive search for the presence of *Cryptosporidium* in both clinical and environmental samples was underway.

Days 4–10: Thursday, April 8–Wednesday, April 14

The south WTP would remain open for water samples to be collected for testing for *Cryptosporidium* on day 4 (Thursday, April 8) and then was shut down on day 5 (Friday, April 9) through June. Those samples, both raw and effluent water, would be positive for *Cryptosporidium*. The city would be served solely by the northern WTP for the duration of the closure. By day 6 (Saturday, April 10), 62 laboratory-confirmed cases of *Cryptosporidium* had been identified. The boil water advisory was lifted on Wednesday, April 14, one week after it had been issued. With the remarkable cooperation of 14 local clinical microbiology laboratories, MHDL was able to document over 700 laboratory-confirmed cases of *Cryptosporidium* through July.

Eventually, epidemiologists determined that approximately 403,000, 25 percent of the 1.6 million people in the five-county area, had become ill after exposure, from March 11 through April 9, to *Cryptosporidium* in the drinking water. Subsequent studies suggested the initial estimate of disease was an underestimation (McDonald et al., 2001).

Aftermath and investigations

Public outreach and response

The public had many questions regarding the outbreak and the news media proved to be a critical source of information, with daily headlines and articles providing a steady flow of information for many weeks. Initially a phone bank was set up at a local TV station to field questions from the public as well. The boil water advisory would affect food production and recalls, certain industry operations, medical and pet care, and the food establishment industry to name but a few impacts on daily life for Milwaukeeans and surrounding communities. Many national media outlets, print and electronic, reported the ongoing development of the outbreak.

Regularly scheduled press conferences by the Health Department and other officials also worked well in providing information to the public. Although all water quality indicators were within federal guidelines, there was quite a bit of anger in the community regarding the outbreak, as well as loss of trust in the use of potable water.

Investigations, causes and impact on Milwaukee

A number of key epidemiological studies (Davis et al., 2009) identified the scope, timing, environmental and other aspects of this outbreak. The state health department would coordinate nine teams to investigate these components, which included databases and surveillance of, for example: laboratory-confirmed cases, emergency room logs, nursing homes, random digit dialing surveys, single-day, short duration of exposure to Milwaukee potable water, immunocompromised populations, childcare settings, MWW plant data, river and estuary data, efficacy of point of use filters, meteorological data analysis, and laboratory testing of stored large blocks of ice. As to the exact cause of the outbreak, no single factor

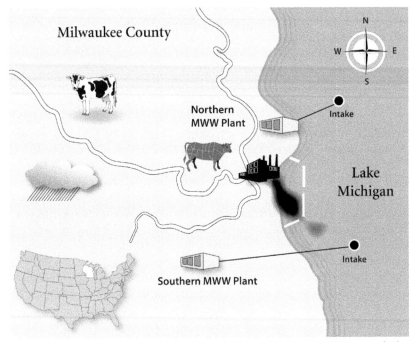

Figure 70.1 Location of the three rivers flowing through Milwaukee County, Wisconsin, the breakfront protecting the city of Milwaukee harbor, the MWW WTPs and their intake grids, and the Milwaukee Metropolitan Sewerage District (MMSD). The creation and southerly flow of a plume toward the southern plant and through a gap in the breakfront near the plant's intake grid in Lake Michigan is shown

Source: Reprinted with permission from *Global Issues in Water, Sanitation, and Health: Workshop Summary, 2009* by the National Academy of Sciences, Courtesy of the National Academies Press, Washington, D.C.

was identified; however, a number of conditions came together that likely contributed to the outbreak (Davis, 2010) (Figure 70.1).

These included: the location of the southern WTP intake near the outflow plume of three rivers' confluence in Milwaukee harbor (Figures 70.2 and 70.3), unusual weather conditions of record rainfall, snow melt, run-off and wind conditions, cross connection between an abattoir kill floor sanitary sewer and the storm sewer, change in use of a coagulant at the south WTP and associated difficulty in correct dosing (Fox and Lytle 1996), recycled backwash at the WTPs, human amplification of *Cryptosporidium* oocysts of thousands ill into overflowing storm sewers discharging into the harbor, limited testing of patient specimens

Figure 70.2 Skyline photo of Milwaukee showing the confluence where three rivers merge into Milwaukee harbor. Note just south of the convergence and west of the overpass is the Milwaukee Metropolitan Sewerage District (MMSD) plant

Source: City of Milwaukee Health Department.

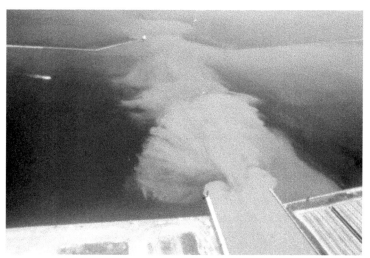

Figure 70.3 The formation of a plume and its southerly movement is shown, as the Milwaukee River empties into the Lake Michigan harbor after a period of high flow

Source: City of Milwaukee Health Department.

prior to outbreak recognition, and inability of the waste water treatment plan at the harbor to handle the volume of influent into the plant.

The outbreak resulted in an estimated 44,000 persons seen as outpatients, 4,400 hospitalized, and an official outbreak-related attributable mortality of 69 deaths, of which 93 percent occurred in persons with AIDS. An estimated 725,000 days of lost productivity were attributed to the outbreak. The total cost of outbreak-associated illness was $96.2 million – $31.7 million in medical costs and $64.6 million in lost productivity (Corso et al., 2003).

Impact and response on water quality policy and practice: local, national and worldwide

Local

Since 1993, MWW, with the endorsement of the mayor and common council, in ongoing investment has committed $417 million in its infrastructure to ensure high-quality water (as reported to the Public Service Commission of Wisconsin). The capital budget is based on long-term planning to replace or upgrade existing infrastructure, and to install new infrastructure as needed. The Capital Improvements Program prioritizes projects based on results of water-related research, new technology and condition assessments of existing systems.

The immediate response was a renovation of facilities from 1993 to 1998 to strengthen the barriers related to source water protection, disinfection and filtration. The detailed improvements that MWW has put forth are available at www.milwaukee.gov/water. These improvements have led Milwaukee to be a leader in water quality and water testing.

A key effort that also came out of the experience has been a collaboration of MWW with the Milwaukee Health Department that we call our Interagency Clean Water Advisory Council (IACWAC). IACWAC tracks and can respond to public health issues that may be related to water. Groundbreaking in the 1990s, the ongoing partnership is now recognized nationally for its effectiveness in protecting public health. The utility relays critical information about emerging contaminants, water treatment and water quality monitoring via communications with news media and customer service representatives, and on their website.

As a result of the outbreak, the MHDL, funded by the city, set up its own testing laboratory, and was one of the original labs to participate in the validation study of the United States Environmental Protection Agency (US EPA) Method 1622 for the detection and identification of *Cryptosporidium* and *Giardia* in water. MHDL has been testing for these parasites as well as culturable viruses, also an EPA method (Sedmak et al., 2003), since 1994. Scientists, locally and worldwide, have visited MHDL to observe and learn the EPA methods. More recently, MHDL is implementing the new EPA Method 1615 for virus detection by culture and quantitative molecular assays for enterovirus and norovirus genogroups GI and GII. This quantitative real-time polymerase chain reaction (qPCR) analysis may be completed within 24 to 48 hours, allowing MHDL to provide valuable information to our WTPs in a more timely manner while offering excess capacity to other utilities. Method 1615 is part of the third Unregulated Contaminant Monitoring Regulation which will occur during 2013–2015 to monitor 30 contaminants (28 chemicals and two viruses), and will provide a basis for future regulatory actions to protect public health.

As a result of improvements and lessons learned, Milwaukee's water is some of the highest quality in the nation. We have an effective, multiple-barrier process of source water protection, ozone disinfection, chlorine disinfection, biologically active filtration, and continuous water quality monitoring. Milwaukee's drinking water quality meets or exceeds all Wisconsin Department of Natural Resources (DNR) and EPA standards. The water utility's water quality monitoring program tests for many more illness-causing pathogens and contaminants than are required by the EPA. In fact, the utility now tests source and treated water for more than 500 contaminants. Our interagency collaboration also continues, and will continue, in order to promote collaboration in protecting public health.

National/international

At the national level to address public health and regulatory issues prompted by the concern of waterborne cryptosporidiosis, in particular the Milwaukee *Cryptosporidium* outbreak, the Centers for Disease Control and Prevention (CDC) and EPA called a meeting at CDC in September 1994 (Morbidity and Mortality Weekly Reports, 1995). Represented at this meeting were experts from water industry and utilities, the U.S. Department of Agriculture (USDA), U.S. Food and Drug Administration (FDA), local (including several from Milwaukee) state and federal public health agencies, laboratorians, and advocacy groups, totaling more than 300 individuals from 40 states. From the two-day meeting, the Working Group on Waterborne Cryptosporidiosis was created, which included 17 task forces to address specific topics related to waterborne cryptosporidiosis. CDC produced a document from these efforts, entitled *Cryptosporidium* and water: A public health handbook (Working Group on Waterborne Cryptosporidiosis, 1997). The handbook was developed to assist local health departments and water utilities in preparing for and responding to reports of *Cryptosporidium* oocysts in tap water or in a community's source of drinking water (river, lake, well).

The outbreak has been well-documented and written about both by news media and academic researchers, and we at the City of Milwaukee Health Department continue to receive calls from media and those in academia for the historical perspective.

The *Cryptosporidium* outbreak and lessons learned (Table 70.1) in Milwaukee in 1993 led to research and improvements worldwide (WHO and OECD, 2003) in water quality treatment processes, water quality monitoring and regulations that continue to contribute to the protection of public health.

Table 70.1 Lessons learned from the 1993 Milwaukee *Cryptosporidium* outbreak

	Lesson	Need
1	Consistent application of stringent water quality standards.	More stringent federal water quality standards, which had been under development for several years, were implemented shortly after this massive waterborne outbreak. The vastly improved attention to monitoring and to the quality of water filtration is a powerful impact of this investigation.
2	Application of technical advances to monitor water and minimize inadequately filtered water to the public.	The post-filtration turbidity (and particle counts, if possible) of treated water should be monitored continuously for each filter to detect changes in filtration status.
3	Testing of source and finished water for *Cryptosporidium*.	Testing and improved methods were needed to detect risk for an outbreak and to determine when the water was safe to drink afterward.

	Lesson	Need
4	Environmental studies.	A coordinated plan was needed to investigate the environment following a waterborne outbreak of *Cryptosporidium* infection.
5	Surveillance.	*Cryptosporidium* infection was not reportable at the time of the outbreak. A more refined clinical case definition was necessary to detect sporadic cases. Random-digit dialing surveys were very valuable in assessing the scope and progress of this large community outbreak. Nursing home surveillance was very effective. It would have been useful to have a surveillance system in place to analyze consumer complaints to the water authority.
6	Testing of human stool and serum.	Infrequent use of these tests likely contributed to the delay in outbreak recognition and improved assays were clearly needed. The striking data demonstrated the value of serologic assays to assess background occurrence and the magnitude and impact of the outbreak.
7	Routine assays for *Cryptosporidium*.	Physicians, other clinicians and public health officials clearly needed to broaden and sustain the index of suspicion for *Cryptosporidium* infection.
8	Communication.	Good interagency communication and working closely with communities of individuals at greatest risk proved very advantageous. For example, the AIDS service organization in Milwaukee had access to over 700 case patients, and we could monitor morbidity and mortality.
9	The media.	Electronic and print media were essential to communicating risk and delivering important public health messages; daily news conferences and televised updates occurred through the lifting of the boil water advisory and articles appeared daily in the news for weeks.
10	Other lessons.	Recurrence of diarrhea after a period of apparent recovery was documented frequently. HIV-positive persons found that the severity of illness, but not the attack rate, was significantly greater in persons with HIV infection. Young children attending daycare centers, asymptomatic or minimally symptomatic, may have contributed to several outbreaks associated with recreational water. Surveillance studies revealed the potential usefulness of monitoring sales of over-the-counter antidiarrheal drugs as an early indicator of community-wide outbreaks. Effectiveness of control measures and the absence of drinking water as a risk factor for the relatively low level of transmission during the post-outbreak period. Importance of testing more than one stool specimen. Usefulness of death certificate review for estimating outbreak-related mortality. Effectiveness of point-of-use water filters with pore diameters of less than 1 micron. Detailed cost analysis revealed the enormous economic impact of the outbreak. A near real-time reporting system of hospital labs to public health during the outbreak proved effective in identifying as well as ruling out probable disease agents.

Source: Adapted and abbreviated from Davis et al., 2009.

Key recommended readings (open access)

1 MacKenzie, W.R., N.J. Hoxie, M.E. Proctor, M.S. Gradus, K.A. Blair, D.E. Peterson, J.J. Kazmierczak, K.R. Fox, D.G. Addiss, J.B. Rose, and J.P. Davis. (1994) 'A massive outbreak in Milwaukee of *Cryptosporidium* infection transmitted through the public water supply', *New England Journal of Medicine*, vol 331, no 3, pp. 161–167. The definitive article on the Milwaukee *Cryptosporidium* outbreak describing the extent, magnitude and various studies characterizing multiple aspects of the outbreak.

2 Davis, J.P., W.R. MacKenzie, and D.G. Addiss. (2009) 'Lessons from the massive waterborne outbreak of *Cryptosporidium* infections, Milwaukee, 1993'. In: *Global Issues in Water, Sanitation, and Health*. Washington DC: National Academies Press, pp. 108–126. http://www.ncbi.nlm.nih.gov/books/NBK28462/pdf/TOC.pdf, accessed 14 January 2014. Written several years after the outbreak, this article by the primary investigators of the outbreak provides additional perspective and information learned in the immediate years following the outbreak.

3 Working Group on Waterborne Cryptosporidiosis. (1997) '*Cryptosporidium* and water: A public health handbook', Centers for Disease Control and Prevention, http://www.cdc.gov/ncidod/diseases/crypto/crypto.pdf, accessed 13 January 2014. This public health handbook, '*Cryptosporidium* and water', was developed by the Working Group on Waterborne Cryptosporidiosis (WGWC) – a multi-disciplinary group composed of representatives from the national CDC, EPA, FDA, USDA, state and local health departments, the drinking water industry, and organizations representing the concerns of immunocompromised persons. The handbook was developed to assist local health departments and water utilities in preparing for and responding to reports of *Cryptosporidium* oocysts in tap water or in a community's source of drinking water (river, lake, well).

References

Batchelor, C. (2004) 'An analysis of crisis management during the Cryptosporidium infection outbreak in the greater Milwaukee, Wisconsin area in the spring of 1993', MS thesis, Syracuse University.

Corso, P.S., M.H. Kramer, K.A. Blair, D.G. Addiss, J.P. Davis, and A.C. Haddix. (2003) 'Costs of illness in the 1993 waterborne *Cryptosporidium* outbreak, Milwaukee, Wisconsin', *Emerging Infectious Disease*, vol 9, no 4, pp. 426–431.

Davis, J.P. (2010) 'The Massive Waterborne Outbreak of *Cryptosporidium*, Milwaukee, Wisconsin, 1993'. In: MS Dworkin (ed), *Outbreak Investigations Around the World: case studies in field epidemiology*. Sudbury, MA: Jones and Bartlett Publishers.

Davis, J.P., W.R. MacKenzie, and D.G. Addiss. (2009) 'Lessons from the massive waterborne outbreak of *Cryptosporidium* infections, Milwaukee, 1993'. In: *Global Issues in Water, Sanitation, and Health: Workshop Summary*. Washington DC: National Academies Press. http://www.ncbi.nlm.nih.gov/books/NBK28462/pdf/TOC.pdf, accessed 14 January 2014.

Edwards, D. (1993) 'Troubled waters in Milwaukee', *ASM News*, vol 59, no 7, pp. 342–345.

Fox, K.R. and D.A. Lytle. (1996) 'Milwaukee's crypto outbreak: investigation and recommendations', *Journal American Water Works Association*, vol 88, no 9, pp. 87–94.

Gradus, M.S. (1989) 'Water quality and waterborne protozoa', *Clinical Microbiology Newsletter*, vol 11, no 16, pp. 121–125.

Gradus, M.S., A.S. Singh, and G.V. Sedmak. (1994) 'The Milwaukee *Cryptosporidium* outbreak: Its impact on drinking water standards, laboratory diagnosis, and public health surveillance', *Clinical Microbiology Newsletter*, vol 16, no 8, pp. 57–61.

Hayes, E.B., T.D. Matte, T.R. O'Brien, T.W. McKinley, G.S. Logsdon, J.B. Rose, B.L.P. Ungar, D.M. Word, M.A. Wilson, E.G. Long, E.S. Hurwitz, and D.D. Juranek. (1989) 'Large community outbreak of cryptosporidiosis due to contamination of a filtered public water supply', *New England Journal of Medicine*, vol 320, no 21, pp.1372–1376.

MacKenzie, W.R., N.J. Hoxie, M.E. Proctor, M.S. Gradus, K.A. Blair, D.E. Peterson, J.J. Kazmierczak, K.R. Fox, D.G. Addiss, J.B. Rose, and J.P. Davis. (1994) 'A massive outbreak in Milwaukee of *Cryptosporidium* infection transmitted through the public water supply', *New England Journal of Medicine*, vol 331, no 3, pp. 161–167.

McDonald, A.C., W.R. McKenzie, D.G. Addiss, M.S. Gradus, G. Linke, E. Zembrowski, M.R. Hurd, M.J. Arrowood, P.J. Lammie, and J.W. Priest. (2001) '*Cryptosporidium parvum*-specific antibody responses among children residing in Milwaukee during the 1993 waterborne outbreak', *Journal of Infectious Diseases*, vol 183, no 9, pp. 1373–1379.

Morbidity and Mortality Weekly Reports, Recommendations and Reports, (1995) 'Assessing the public health threat associated with waterborne cryptosporidiosis: Report of a workshop', Centers for Disease Control and Prevention, 44(RR-6); 1–19, http://www.cdc.gov/mmwr/preview/mmwrhtml/00037331.htm, accessed 13 January 2014.

Sedmak, G., D. Bina, J. MacDonald, and L. Couillard. (2003) 'Nine-Year study of the occurrence of culturable viruses in source water for two drinking water treatment plants and the influent and effluent of a wastewater treatment plant in Milwaukee, Wisconsin (August 1994 through July 2003)', *Applied and Environmental Microbiology*, vol 71, no 2, pp. 1042–1050.

WHO and OECD (2003) *Assessing Microbial Safety of Drinking Water. Improving approaches and methods.* London: IWA Publishing. http://www.who.int/water_sanitation_health/dwq/9241546301full.pdf, accessed 16 January 2014.

Working Group on Waterborne Cryptosporidiosis. (1997) '*Cryptosporidium* and water: A public health handbook', Centers for Disease Control and Prevention, http://www.cdc.gov/ncidod/diseases/crypto/crypto.pdf, accessed 13 January 2014.

71

EDWIN CHADWICK AND THE PUBLIC HEALTH ACT 1848

Principal architect of sanitary reform[*]

Martin Exner

PROFESSOR OF HYGIENE AND PUBLIC HEALTH, UNIVERSITY OF BONN, BONN, GERMANY

Learning objectives

1 Describe the work of Edwin Chadwick.
2 Describe the development of Chadwick's tremendous work leading to the Public Health Act and the sanitary revolution based on the inquiry into the sanitary conditions of the labouring population of Great Britain.
3 Describe the importance of the Public Health Act and its consequences in the following century and up to today.

Introduction

The basis for "public health" and "public wealth" in a society is to guarantee good sanitation in cities and homes and a safe public water supply. For safeguarding these principles, a strong political will and legal basis are needed.

The responsibility for building up this basis in the nineteenth century is largely due to one man – Edwin Chadwick (1800–1890).[1,2,3,4]

He was the driving force behind the so-called "Public Health Act" in 1848 for England and Wales, which marked the start of a commitment to proactive, rather than reactive, public health policy in which the state became guarantor of standards of health and environmental quality and provided the means for local governments to meet those standards.[1,2,3,4]

---

[*] Recommended citation: Exner, M. 2015. 'Edwin Chadwick and the Public Health Act 1848 – principal architect of sanitary reform', in Bartram, J., with Baum, R., Coclanis, P.A., Gute, D.M., Kay, D., McFadyen, S., Pond, K., Robertson, W. and Rouse, M.J. (eds) *Routledge Handbook of Water and Health*. London and New York: Routledge.

Curriculum vitae of Edwin Chadwick

Edwin Chadwick was born in Manchester on 24 January 1800. He studied law in London and made his money first by writing essays for publications such as the *Westminster Review*. Despite his training in law, he based his essays on scientific principles and how they could enforce political strategies in democratic governments. In November 1830 he was called to the bar which allowed him to become a licensed barrister. His essays were published in the *Westminster Review* in which he described the different methods of applying scientific knowledge to the practice of government. He became friends with two of the leading philosophers of the day, John Stuart Mill and Jeremy Bentham. Bentham (1748–1832), a British philosopher, jurist and social reformer, engaged him as his literary assistant. Chadwick's work on social reform and the influence of his philosopher friends led to his devoted efforts at sanitary reform.

In 1832 Chadwick was employed by the royal commission appointed to inquire into the operation of the poor laws and in 1833 he was made a full member of that body.

In 1842 he published his report on the sanitary conditions of the labouring population.

Chadwick's argument was economic, as he was convinced that if the health of the poor were improved, it would result in fewer people seeking poor relief; much poor relief was given to the families of men who had died from infectious diseases. Money spent on improving public health was therefore cost-effective, as it would save money in the long term.

He considered that the most important steps to improve the health of the public were:

- Improved drainage and provision of sewers.
- The removal of all refuse from houses, streets and roads.
- The provision of clean drinking water.
- The appointment of a medical officer for each town.

These national and local movements and proposals contributed to the passing of the Public Health Act of 1848 . [1, 2, 3, 4]

In his later life Chadwick was a commissioner of the Metropolitan Commission of Sewers in London from 1848 to 1849 and he was also commissioner of the General Board of Health from its establishment in 1848 to its abolition in 1855.

In January 1884 he was appointed as the first president of the Association of Public Sanitary Inspectors, which is now the Chartered Institute of Environmental Health in England and Wales. In recognition of his public service, Chadwick was knighted in 1889.

Edwin Chadwick died in Surrey on 16 July 1890.

Hamlin and Sheard wrote on the personality of Edwin Chadwick:[3]

> As a junior member of the 1832 royal commission on the poor law, Chadwick transformed policy analysis: He documented conditions far more comprehensively than had his predecessors and equally was creative in discovering acceptable solutions to long standing conflicts. He was the main architect of the new poor law of 1844 and, as its administrator, was set to be the most hated man in England. Issues of poor law administration led him into education and law enforcement as well as public health. Chadwick's personality was his success and his undoing: He was tenacious in pushing a reform by all available means until action was taken, but he was over bearing and unresponsive for the use of others. He did not negotiate or converse but lectured at people again and again, until they

acted. With no facultative accommodation differences of opinion, he failed as a practical politician, notwithstanding his ability as a political analyst. After his expulsion from the general board of health in 1854 he never again served in public administration.

The poor law

Today we are probably unable to appreciate the disorganization and lack of planning characteristic of the nineteenth century. Roughly 3 million people (slightly over 40 per cent) were urban in 1801 in England and Wales, compared with 28.5 million (almost 80 per cent) in 1901. Growth rates in some textile boom towns exceeded 60 per cent per decade; this despite the fact that towns were acting as sinks for human life. In Liverpool average life expectancy by class ranged from 15 years for unemployed persons or poor to 35 years for the well-to-do.[3]

Edwin Chadwick was the widely hated architect and enforcer of the new poor law of 1834. Its guiding principle was to make the conditions under which public relief could be given so unpleasant that most people would refuse to request it. Ever under pressure to cut costs, Chadwick began to focus on the causes of indigence: prevention was cheaper than relief. By 1838 he was looking mainly at one cause: acute infectious diseases that were fatal to male breadwinners, leaving families dependent on relief. These diseases, Chadwick insisted, had physical causes in poor urban drainage, which left towns covered in a residue of filth that contaminated the air in some ill defined way and caused disease.[3]

Starting in the late eighteenth century with the extension of the market economy, the introduction of steam power, the growth of transportation and the increasing dominance of the factory system of production, the Industrial Revolution demanded a constant supply of labourers to feed the growth of machine production. Workers had to be brought into the factories, located in industrial towns and cities.[2]

Mobilizing this industrial labour force required abolition of the older system of poor relief. Landowners, rationalizing agricultural production, had enlarged and enclosed their holdings and thus began to drive rural labourers off the land. At first, provision was made for the landless and unemployed persons by the parishes of their birth, following the Elizabethan poor law system. But as the ranks of the unemployed swelled and poor rates rose, the old poor law came to be viewed – at least by landowners, industrialists and taxpayers – as a constraint on the mobility of labour and an impediment to progress[2].

The organization and financing of poor relief was a central social policy problem of the early nineteenth century. A royal commission was appointed in 1832 to examine the operation and administration of the poor laws, and its report, largely written by Edwin Chadwick, appeared in 1834. The poor law amendment act of 1834, incorporating the principles of the report, decreed that no able pauper could be given assistance except in a workhouse. The conditions of labour in the workhouse were to be made more miserable than those of the worst situated labour outside the workhouse. The immediate intent and result of the act was to reduce the burden of the poor rates, but it also served to drive the poor out of the rural areas and into the new industrial towns. Within 20 years, the proportion of the population living in industrial cities doubled and the mushrooming of towns and cities, speculative building practices, ramshackle housing and congestion led to an explosion of disease rates. Builders rarely troubled themselves to supply sewers, water closets or privies, and little was done to supply fresh water, clean the streets or remove the garbage.[2]

The cholera epidemic of 1831 and 1832 had drawn attention to the deplorable lack of sanitation in the industrial cities. It was obvious that cholera was concentrated in the poorest districts, where sanitation was most neglected and slum housing most befouled by excremental filth and other dirt. The relationship between disease, dirt and destitution clarified the need for sanitary reform because, in the crowded and congested cities, disease could fairly readily spread from the homes of the poor to those of the wealthy.[2]

Chadwick became convinced that the health of the labouring population was largely determined by the state of the physical environment. In 1838 the poor law commission reported that it had employed three medical inspectors to look into the prevalence and cause of preventable disease in London, and that these physicians had reported that the expenditure needed to prevent disease would ultimately amount to less than the cost of the disease being created – the latter measured in lost productivity as well as the costs of hospital and burial care and the poor law support of widows and surviving dependents. Sanitary measures were justified on grounds of economy as well as of humanity. In 1840 the Select Committee on the Health of Towns declared that preventive measures were required for reasons of humanity and justice to the poor, but equally for the safety of property and the security of the rich.[2]

The Public Health Act of 1848

The Public Health Act of 1848 is one of the great milestones of public health history: "the beginning of a commitment to proactive rather than reactive public health. For the first time the state became the guarantor of standards of health and environmental quality and provided resources to local units of government to make the necessary changes to achieve those standards".[3]

The Public Health Act established a general board of health empowered to create local boards of health. The local boards had authority to deal with water supplies, sewage, control of offensive threats, quality of foods, removal of garbage and other sanitary matters. A local board could appoint a medical office of health, an inspector of nuisances, a surveyor, a treasurer and a clerk.[2]

In 1839 the government had instructed the poor law commission to examine the health of the working population in England and Wales. Over three years its members collected a large amount of information which provided the basis for Chadwick's report from the poor law commissioners on an "enquiry into the sanitary condition of the labouring population of Great Britain". The report provided a compelling argument that diseases among the working class were related to filthy environmental conditions caused by the lack of water supplies, drainage, and sewers and any effective means of moving dirt and refuse from house and streets. The problem of public health was declared to be largely environmental rather than a medical problem.

Chadwick wrote:

> The great preventives drainage, street and house cleaning by means of supplies of water and improved sewage and especially the introduction of cheaper and more efficient modes of removing all noxious refuse from the towns are operations for which aid must be sought from the science of the civil engineer, not from the physician who has done his work when he has pointed out the disease that results from the neglect of proper administrative measures and has alleviated the sufferings of the victims.[2]

Chadwick had the leading role in drafting the reports of 1844 and 1845 and he argued that property, crime, ill health and high mortality were all closely associated with the environmental conditions of the industrial cities. He proposed that the central government assume basic responsibility for public health with the creation of a new government department, and that, in each locality, a single administrative body be responsible for all water supplies, draining, paving, street cleaning and other necessary sanitary measures.

In 1848 a new wave of cholera was sweeping westwards, across Europe. By June an epidemic was raging in Moscow and by September it had reached Hamburg and Paris. Watching its spread with anxiety, the British government, after several failed attempts, passed the Public Health Act on the last days of August 1848, establishing a general board of health for a provisional five-year period. Fee and Brown wrote:

> Indeed, the Public Health Act of 1848 can be viewed as a powerful catalyst for the development of local government and for local government responsibility for public health. From this perspective the Public Health Act is less a key step in the growth of central state authority and more a marker in an ongoing struggle to sort out jurisdictional levels of government and to solve at all levels of government questions of rights and responsibilities in relation to the public health.[2]

The consequences of the Public Health Act and the sanitary reform up to today

On the legal basis of the Public Health Act there began a big investment in the sanitation of the cities. In England and Wales, but also in parts of central Europe such as France and Germany, the big cities began to build up adequate sanitation and public drinking water systems which led also to other environmental changes – green spaces, better ventilation and even better road surfaces.

The effect of these changes was reported by Robert Koch and G. Gaffky on the detection of cholera. They described the cholera mortality in Calcutta from 1840 to 1884. With the introduction of the public water system in Calcutta with filtered water, cholera mortality decreased immediately and dramatically (Figure 71.1).

The introduction of the public drinking water system in Calcutta, the capital of the viceroyalty of India, was also based on the Public Health Act of 1848.

In 2007 more than 11,300 readers of the *British Medical Journal* chose the introduction of clean water and sewage disposal – "the sanitary revolution" – as the most important milestone since 1840 when the *British Medical Journal* was first published.[6] Readers were given 10 days to vote on a shortlist of 15 milestones, and sanitation topped the poll, followed closely by the discovery of antibiotics and the development of anaesthesia. The work of the nineteenth century lawyer Edwin Chadwick, who pioneered the introduction of piped water to people's homes and sewers rinsed by water, attracted 15.8 per cent of the votes, while antibiotics took 15 per cent and anaesthesia took 14 per cent. The next two most popular choices were the introduction of vaccines, with 12 per cent, and the discovery of the structure of DNA. The general lesson which still holds is that passive protection against health hazards is often the best way to improve population health.

The original champions of the sanitary revolution were John Snow, who showed that cholera was spread by water, and Edwin Chadwick, who came up with the idea of sewage disposal and piping water into homes.

In Germany the mortality rate began to decrease continuously in the middle of the nineteenth century, long before antibiotics or vaccines were introduced as is demonstrated in Figure 71.2.

Since the systematic sanitation of German towns with public drinking-water supply and sewage drainage systems began at that time, there is an obvious correlation with the beginning of the decrease of the mortality.

Inadequate sanitation is still a major problem in the developing world. This can be seen in a report in the *New York Times* by Gardiner Harris, dated 3 December 2014, entitled, "'Superbugs' kill India's babies and pose an overseas threat". The article states: "A deadly epidemic that could have global implications is quietly sweeping India, and among its many victims are tens of thousands of newborns dying because once-miraculous cures no longer work." These infants are born with bacterial infections that are resistant to most known antibiotics, and more than 58,000 died last year as a result, a recent study found.

These babies are part of a disquieting outbreak. A growing chorus of researchers say the evidence is now overwhelming that a significant share of the bacteria present in India – in its water, sewage, animals, soil and even its mothers – are immune to nearly all antibiotics.

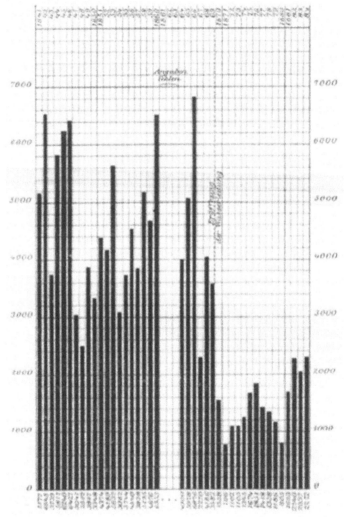

Figure 71.1 Cholera mortality in Calcutta, 1841–1884. The broken line marked the introduction of public water supply with filtered water[5]

"India's dreadful sanitation, uncontrolled use of antibiotics and overcrowding coupled with a complete lack of monitoring the problem has created a tsunami of antibiotic resistance that is reaching just about every country in the world," Dr. Timothy R. Walsh, a professor of microbiology at Cardiff University is cited as saying.

Bacteria spread easily in India, because half of the Indian population defecates outdoors, and much of the sewage generated by those who do use toilets is untreated. As a result, Indians have among the highest rates of bacterial infections in the world and collectively take more antibiotics, which are sold over the counter there, than any other nationality.

"In the absence of better sanitation and hygiene, we are forced to rely heavily on antibiotics to reduce infections," said Ramanan Laxminarayan, vice president for research and policy at the Public Health Foundation of India. "The result is that we are losing these drugs, and our newborns are already facing the consequences of untreatable sepsis, or blood infections".

Even today you find on the streets in Calcutta water pumps installed in the nineteenth century during the lifetime of Edwin Chadwick. Edwin Chadwick's idea of sanitary reform is still alive today.

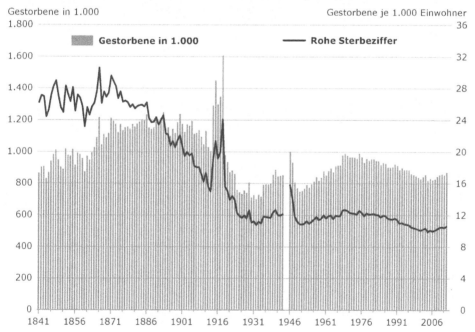

Gestorbene und rohe Sterbeziffer in Deutschland*, 1841 bis 2012**

Gestorbene in 1.000 — Gestorbene je 1.000 Einwohner

* bis 1943 Reichsgebiet, jeweiliger Gebietsstand, ab 1946 Früheres Bundesgebiet und DDR insgesamt
** Die Berechnungen der rohen Sterbeziffer für 2011 und 2012 beruhen noch auf den alten
Bevölkerungszahlenaus der Fortschreibung früherer Volkszählungen.

Datenquelle: Statistisches Bundesamt, Berechnungen: BiB   © BiB 2014

Figure 71.2 Mortality rate in Germany, 1841–2012 © BiB 2014

Source: http://www.bib-demografie.de/DE/ZahlenundFakten/08/sterblichkeit_node.html

References

1 Calman, K. 1998. The 1848 Public Health Act and its relevance to improving public health in England now. *BMJ* 317(7158):596–8.
2 Fee, E, and Brown, TM. 2005. The Public Health Act of 1848. *Bull World Health Organ* 83(11):866–7.
3 Hamlin, C, and Sheard, S. 1998. Revolutions in public health: 1848, and 1998? *BMJ* 317(7158):587–91.
4 Sram, I, and Ashton, J. 1998. Millennium report to Sir Edwin Chadwick. *BMJ* 317(7158):592–6.
5 Koch, R, and Gaffky, G. 1887. *Bericht über die Thätigkeit der zur Erforschung der Cholera im Jahre 1883 nach Egypten und Indien entsandten Kommission*. Berlin: Julius Springer.
6 Ferriman, A. 2007. *BMJ* readers choose sanitation as greatest medical advance since 1840. *BMJ* 334:111.

INDEX

INDEX

585–6, 587, 588, *589*, 590, 593–8, 602–3, 604, 605, 606; bacteriophages 595, *597 see also* bacteriophages; *Bacteroides* 588, 596, *597*; and beach classification matrix *460*; *Clostridium perfringens* 152, 459, 596, *597*; coliform group members 594–5 *see also* coliform bacteria; emerging indicators 596–8, *597*; enterococci *see* enterococci; faecal *streptococci* 595 *see also* enterococci; H2S-producing bacteria 596, *597*; human DNA 156, 597, *597*; non-human viruses 596–7, *597*; and QMST 587; recreational water contamination 458–60, *460*, 461, 462, *463*; summary of common and emerging indicators 597
faecally transmitted infection 4, 17, 38–48, 139, 146, 177, 593 *see also* manure; sewage; toilets; bacteria important for *24*, 46; and the carrier state 43; contributions of different faecal–oral pathogens to diarrheal disease and death 47–8; factors influencing faecal–oral microbe transport, survival and fate in the environment 44–5; faecal–oral disease transmission, 'F diagram' *23*, 38–9; faecal presence, pathogen shedding and other faecal–oral pathogen sources in the environment 44; and foodborne contaminated water 141; gastrointestinal/enteric infection categories 40; helminths important for 47–8, *47*; host factors 41–2; host immunity and response to infection 42; indicators of faecal presence in water *see* faecal contamination markers, non-microbial; faecal indicator organisms/bacteria (FIO/FIB); infecting enteric microorganism categories 40–1; overview of sites, processes and events of infection in the host 39; protozoa important for *25*, 46; risks of infection/disease and dose–response relationships 42–3; and safe containment of excreta 392–3, 394; through recreational water 456–63; transmission routes from infected to healthy individual *138*; virulence factors and properties of microbes 40; viruses important for *24–5*, 45; WASH interventions *see* water, sanitation and hygiene; waterborne 22–7; zoonotic pathogens 43–4
faecal *streptococci* (FS) 595, *597 see also* enterococci
Falkenmark Index 361
Fasciola spp. 62; *F. gigantica* 62; *F. hepatica 25*, 62
Fasciolopsis spp. 62
fertility rates 400
fetal damage 160
fetching water *see* water fetching/carrying
fever 40, 46, 59, 60, 61, 66, 92, 100, 103, 160, 161, 162; dengue *see* dengue; paratypoid

176; Pontiac *see* Pontiac fever; Rift Valley *see* Rift Valley fever (RVF); typhoid 23, 48, 176, 200, *391*, *457*; yellow 31, 71, 72, 76–7
filtration 173, 198, 207–8; and the 1890 cholera epidemic in Hamburg 644–7; backwashing 207; ceramic filter program, Cambodia 297–9; charcoal 131; membrane 113, 208; microfiltration (MF) 405; rapid gravity 207, 208; slow sand 207; ultrafiltration (UF) 405
Finland 125, 163, 176
fishing 318
fleas 71
flooding 30, 64, 76, 139, 142, 317, 373, 390, *391*, 419; and climate change 371
flukes (trematodes) *56*, *57*, 58–9, 61–2, *153*, 162; *Schistosoma* spp. see *Schistosoma* spp.
fluorescent whitening agents 598
fluoride 26, *108*, 110–11, 263–4
fluorine 276
fly-tipping 139
folliculitis 66
foodborne contamination, and water 159–67; with bacteria 160–1; biological contamination 160–4; chemical contamination 165–7; control and remediation 164, 166–7; exposure pathways and transmission routes 141, 159–67; with fungi 162, 163; with heavy metals 165, 166; with helminths 161–2, 163–4; with pesticides 165, 166; with protozoa 161; with radionuclides 165, 166; role of water in contamination of food supply 166; role of water in microbial contamination of foodstuffs 162–4; with viruses 161, 163
foodborne disease 159–60, 179, 323; and water contamination *see* foodborne contamination, and water
food hygiene 164
food poisoning, shellfish *54*, 68, 103
food production 194, 214, 364–5, 437, 692; and water scarcity 361–2, 364, 365
food security 166
Food Summit (1996) 438
forced migration *see* displacement
formative research 225, 226, 227, 228, 230, 231, 238
Fort William, Calcutta 652, *653*, 654, *655*
France 166, 329, 330, 399, 401, 470, 649, 703
Francisella tularensis 24
Frankland, P. and Frankland, P. 585
frank pathogens 41
Frontinus, Sextus Julius 270–4
frostbite 319
Frost, Wade Hampton 664
Fukushima disaster 130
fumonisin 162
fungi 328; contamination of food products 162, 163; mold 317–18

www.ingramcontent.com/pod-product-compliance
Ingram Content Group UK Ltd.
Pitfield, Milton Keynes, MK11 3LW, UK
UKHW051833180425
457613UK00022B/1230

9 781138 495302